# Major Patterns in
# Vertebrate Evolution

# NATO ADVANCED STUDY INSTITUTES SERIES

A series of edited volumes comprising multifaceted studies of contemporary scientific issues by some of the best scientific minds in the world, assembled in cooperation with NATO Scientific Affairs Division.

**Series A: Life Sciences**

*Recent Volumes in this Series*

The series is published by an international board of publishers in conjunction with NATO Scientific Affairs Division

| | | |
|---|---|---|
| **A** | **Life Sciences** | Plenum Publishing Corporation |
| **B** | **Physics** | New York and London |
| **C** | **Mathematical and Physical Sciences** | D. Reidel Publishing Company Dordrecht and Boston |
| **D** | **Behavioral and Social Sciences** | Sijthoff International Publishing Company Leiden |
| **E** | **Applied Sciences** | Noordhoff International Publishing Leiden |

# Major Patterns in Vertebrate Evolution

Edited by

## Max K. Hecht

*Queens College of the City University of New York*
*Flushing, New York*

## Peter C. Goody

*Royal Veterinary College of the University of London*
*London, England*

and

## Bessie M. Hecht

*Queens College of the City University of New York*
*Flushing, New York*

**PLENUM PRESS • NEW YORK AND LONDON**
Published in cooperation with NATO Scientific Affairs Division

Library of Congress Cataloging in Publication Data

Nato Advanced Study Institute on Major Patterns in Vertebrate
    Evolution, Royal Holloway College, 1976.
    Major patterns in vertebrate evolution.

    (Nato advanced study institutes series: Series A, Life sciences; v. 14)
    Includes index.
    1. Vertebrates–Evolution–Congresses. I. Hecht, Max K., 1925-          II. Goody,
Peter Charles. III. Hecht, Bessie M. IV. Title. V. Series.
QL607.5.N37    1976                              596'.03'8                    77-6440
ISBN 0-306-35614-7

Lectures presented at the NATO Advanced Study Institute on Major Patterns in
Vertebrate Evolution held at the Royal Holloway College (London University)
Surrey, England, July, 1976

© 1977 Plenum Press, New York
A Division of Plenum Publishing Corporation
227 West 17th Street, New York, N.Y. 10011

Printed in the United States of America

# PREFACE

This volume is the result of a NATO Advanced Study Institute held in England at Kingswood Hall of Residence, Royal Holloway College (London University), Surrey, during the last two weeks of July, 1976. The ASI was organized within the guide lines laid down by the Scientific Affairs Division of the North Atlantic Treaty Organization.

During the past two decades, significant advances have been made in our understanding of vertebrate evolution. The purpose of the Institute was to present the current status of our know-ledge of vertebrate evolution above the species level. Since the subject matter was obviously too broad to be covered adequately in the limited time available, selected topics, problems, and areas which are applicable to vertebrate zoology as a whole were reviewed.

The program was divided into three areas: (1) the theory and methodology of phyletic inference and approaches to the an-alysis of macroevolutionary trends as applied to vertebrates; (2) the application of these methodological principles and an-alytical processes to different groups and structures, particular-ly in anatomy and paleontology; (3) the application of these re-sults to classification. The basic principles considered in the first area were outlined in lectures covering the problems of character analysis, functional morphology, karyological evidence, biochemical evidence, morphogenesis, and biogeography. The second area dealt with more specific problems of vertebrate evol-ution in lectures ranging from adaptive radiation of mammals and teleosts, the origin of tetrapod, the effect of insularity, and the evolution of specific structures such as embryonic membranes. The third area was devoted to presentations of the three major philosophies of classification, pheneticism, cladism, and classi-cal or evolutionary taxonomy. The lectures presented were not necessarily in agreement and indeed often presented conflicting views, nevertheless they did present the problems as seen by practicing vertebrate systematists and biologists.

The proceedings present the major lectures in greater detail than was possible in the oral presentations.  It is also the case that various authors have incorporated some modifications and changes that resulted from the discussion following their presentation.  Much of the discussion cannot be presented, but where there was important disagreement or amplification of salient points, the comments are included, following the main article.

It is hoped that the articles included in this volume will be of general interest to a broad spectrum of biologists, as well as vertebrate phylogeneticists and systematists.


The Editors

# CONTENTS

PART 1

# Approaches to the Analysis of Macroevolutionary Trends

# THE METHODOLOGY OF PHYLOGENETIC INFERENCE ABOVE THE SPECIES LEVEL

Max K. HECHT
Dept. of Biology, Queens College of C.U.N.Y.
Flushing, New York   11367 USA
 -and-
James L. EDWARDS
Zoology Dept, Michigan State University
East Lansing, Michigan   48824

## INTRODUCTION

The determination of the relationships of organisms in time and space is the process called phylogenetic inference.  The process is termed an inference because one cannot directly observe evolution above the species level, and therefore, one must infer such relationships.  Furthermore, it is impossible to devise experiments to directly test such relationships, but that does not mean that they cannot be falsified by new data, such as the discovery of new fossils or recent organisms, or the re-analysis of the characters of previously known organisms.

Phylogenetic inference is often confused with classification, but they are not the same problem.  Classification may attempt to be phylogenetic in organization, but not all classifications must be so.  For those classifications which are natural or phylogenetic, the first question to be determined is that of relationship, using the methodology of phylogenetic inference to establish a scheme which best approximates the single real phylogeny.  There may be several schemes of relationships but there is only one true set of relationships.

An important consideration in phylogenetic inference is the nature of the units of comparison.  The basic unit in any system of evolutionary relationships must be the evolving unit which is generally agreed to be the population (or deme) or the biological species (a deme or group of demes reproductively isolated from its

3

nearest relatives).  These are the only units which actually maintain common gene pools and so can experience evolutionary changes
in gene frequencies through time or from generation to generation.
Within such units, evolutionary changes can, in principle, be
studied by direct observation or by experimentation.  However, such
observations are restricted to living forms and populations.  The
determination of relationships between different species or other
groups above the species level or of fossil organisms requires
techniques other than direct observation, and it is in these
situations that phylogenetic inference is most useful.


                           THE PHYLOGENY

     The goal of phylogenetic inference is to produce a scheme or
construct which depicts the genealogical relationships of the taxa
under consideration.  Such a diagram is called a phylogeny.
Implicit in the concept of the phylogeny is the idea that two (or
more) taxa which share an immediate common ancestor are more closely
related to each other than either is to any other group.  Hennig
(1966) has termed such taxa "sister groups," and thus the erection
of phylogenies has often been considered to revolve around the
"search for the sister group."  Other workers (especially Mayr,
1965) have stressed the genetic (or patristic) relationships of
organisms.  This kind of relationship is certainly important, for
the genotype of an organism is its only link to its ancestors.
However, genetic relationships cannot be measured directly, but
must be inferred from morphological data (in the widest sense, including biochemical, electrophoretic and karyological analyses).
It is assumed that a phylogeny will reflect the genetic similarities of the taxa in it, but such similarities are not the primary
concern in phylogeny construction.

     The groups of organisms whose relationships are under con-
sideration are called taxa, and they are defined by morphological
similarities.  All phylogeneticists agree that the taxa depicted
on a phylogeny should be monophyletic, but unfortunately there are
two major definitions for the term monophyly.  Simpson (1961:124)
defines monophyly as "the derivation of a taxon through one or
more lineages....from one immediately ancestral taxon of the same
or lower rank."  This concept intertwines phylogeny and classification, for one must somehow know the ranks of the involved taxa
before the definition can be applied.  The other, or Hennigian,
definition, as modified by Bonde (1975:293), states that "a
monophyletic groups includes (only) a species and all its descendents."  In actuality, it is not always possible to trace back a
group to a single species, but this remains a goal to be approached.
Monophyly is usually applied to taxa, but one can also trace the
evolution of characters apart from the taxa to which they belong,

and therefore speak of a monophyly of characters (see below).

Taxa are identified by examining the phenotypes of the organisms under consideration. Such data are intrinsic to the organisms themselves, since the phenotype is primarily determined by the interaction of the genotype with the external environment. Other attributes of an organism, such as its distribution in space and time, are extrinsic (Hecht, 1976) to the organism, and are not directly derived from its genotype. Schaeffer et al. (1972), Eldredge and Tattersall (1975) and Hecht (1976) have argued that only intrinsic data should be used in phylogenetic reconstruction. The characteristics of an organism's distributional pattern, for example, may in part be determined by its phenotype, but the pattern itself is most often due to historical events in the non-biological world, such as plate tectonics, glaciation, current patterns, etc., that have no relation to the organism's genetic history. Therefore, such extrinsic data can tell us very little about the genealogical relationships of the organism.

The most important kind of intrinsic data for indicating phylogenetic relationships is the shared derived character state (the synapomorphy of the cladist). Shared primitive character states (= symplesiomorphies) cannot be used to indicate monophyletic groups (Hennig, 1966) or lineages.

CHARACTER, CHARACTER STATE, AND HOMOLOGY

As implied above, there are two approaches to the problem in the determination of relationships; these are: analysis of taxa and analysis of characters (and their states). Although there is a bit of reciprocal illumination in these two processes, the basic process is analysis of characters.

All taxa are identified and described by descriptive terms which are called characters or attributes. These characters are part of the clues as to genealogical relationships. A character can be defined as a set of limited homologous features that are distributed among two or more taxa. The different expressions of the character among the taxa to be compared are called character states (Michener and Sokal, 1957; Kluge and Farris, 1969; Marx and Rabb, 1970). The whole suite of character states, which are assumed to be homologous, can be called a morphocline or morpho-logical transformation series. A series of character states must be assumed to be homologous and therefore share a common ancestral condition. In every morphoclinal series there is always a single ancestral or primitive condition, but there may be one or more derived states. The direction of change in a morphoclinal series is called polarity (Kluge and Farris, 1969), which may be a uni- or

multidirectional series (Marx and Rabb, 1972). Every student in
comparative anatomy can certainly recall examples of these basic
statements. For example, it is generally assumed that the uteri
of therian mammals are homologous structures and that the simplis-
tically categorized duplex, bicornuate, bipartite, and simplex
uteri can be considered character states in this transformation
series.

In order to make comparisons and to determine relationships,
one must be able to compare similar structures or structures which
are derived from a common ancestral condition. Two or more character states which are derived from a common ancestral character
state are said to be homologous and, therefore, homology is the
study of the monophyly of structures. In order to determine
homology there are four types of evidence which are commonly used;
they are: topographic, developmental, structural and biochemical.
Topographic evidence is most often used, and involves the relative
position of several elements to some common landmark, as demonstrated by the bones of the rhipidistian and primitive amphibian
skull (Jardine and Jardine, 1967). Developmental or embryological
evidence is usually restricted to living organisms and has been
used to determine homologies in such classic cases as the middle
ear bones of mammals and reptiles, and the endostyle of protochordates and thyroid gland of vertebrates. (In the latter
example, biochemical evidence supports the developmental evidence.)
Structural or morphological evidence is the second commonest type
used and is indicated by the steretyped relationships of several
structures, such as the elements comprising the basic plans of the
tetrapod manus. The problem of biochemical homology will be dealt
with elsewhere in this volume and presents such interesting problems
as paralogous and orthologous homologies (see Fitch, 1970).

Bock (1965, 1973) has pointed out the conditional nature of
homologies. The exact level of homology is related to the
hierarchical status of the organisms compared. For example, the
wings of bats, pterodactyls and birds are homologous as tetrapod
limbs, but since the common ancestor lacked wings (and the basic
plan of the structures for flight are different), these structures
are not homologous as wings. Therefore, Bock (1973) suggests the
use of a qualifier to specify the level of homology.

Homoplasy (Simpson, 1961) is the condition in which the compared structures appear to be similar in morphology and possibly
in development, but are not derived from the same immediate
lineage. In determining the reality of the homologous state, one
must always recognize the possibility that the conditions studied
are really homoplasic. In homoplasy there are two basic conditions,
parallelism and convergence, which have often been interchanged in
usage, or poorly distinguished. For reasons of clarity, they are

here defined in their extreme simplistic forms. Convergence is the
independent attainment of similar characters in at least two lineages here the most recent common ancestral form is assumed not to
have the structures being compared. The classic example of convergence is the structural similarity of the chordate and cephalopod
eye. Parallelism is more difficult to demonstrate, but as defined
here, the character is present in the ancestral form but a common
derived character state has been independently evolved in each
descendent form (Hecht and Edwards, 1976). For example, the avian
and mammalian atlas-axis complexes are very similar, yet in the
primitive archosaurian and primitive synapsid-mammalian lineages,
there is not a functional unit such as the atlas-axis complex.
However, the structural and embryological precursors were present.
The presence of the morphological and embryological anlagen in the
ancestors of these two groups makes this clearly an example of
parallelism and not convergence. By our definitions of parallelism
and homology, derived character states which have evolved in
parallel are also homologous.

The importance of recognizing structures that have evolved in
parallel should not be underestimated. Hecht and Edwards (1976)
and Edwards (1976) have demonstrated multiple parallelisms in the
evolution of paedogenetic salamanders. The important question is
whether parallel evolution of similar morphology, physiology, etc.
is a frequent process or an unusual one. As will be demonstrated
below, we believe it is not an infrequent process and is probably
more frequent than expected. The major cause for its apparent low
frequency in phylogenetic schemes is the failure to develop criteria
which will point out possible examples. In fact, most phylogeneticists and taxonomists use criteria such as maximum parsimony
and unweighted characters which purposely minimize the recognition
of parallelism. At the levels of genera and families and perhaps
even orders, parallelism should be frequent, since the forms being
compared share similar genetic backgrounds (as in the case of Homo
and Pan demonstrated by King and Wilson, 1975), similar developmental processes and the same general adaptations to a basic
environment. Therefore, parallelism is probably the most frequent
cause in the misinterpretation of phyletic relationships.


THE REALITY OF THE CHARACTER, OR, WHAT IS A CHARACTER?

In taxon analysis, the basic question is the reality of the
monophyletic taxon. In character analysis, the basic problem is to
determine the unit character. In a system of unweighted characters,
how does one treat characters which must obviously coevolve because of interrelated development and physiology? In a study of
proteid salamanders, Hecht and Edwards (1976) were faced with the
following problems. Other workers have suggested that the genera
Necturus and Proteus are related because they share several derived

characters, but Hecht and Edwards (1976) found nine of these to be
the result of paedogenesis.  Actually, these were not nine independent characters as used by previous authors, but, at least at this
level of discrimination, they were really only one character, the
state of paedogenesis.  The nine characters were merely the expression of a single growth process.  Similarly, Orton (1957), Starrett
(1973) and Sokol (1975) distinguished four types of frog tadpoles
which are found in various frog families.  Each type was characterized by at least five characters, each with different character
states.  In the determination of relationships of frog families,
the question remains as to whether frog tadpoles should be evaluated
as a single character (the entire larva) with four character states
(the four types) or the tadpole should be split into five different
characters each with four character states.  One possible interpretation is that a character may be considered a suite of correlated
attributes whose homologous states vary concordantly within the
taxa under consideration and are usually functionally and/or
developmentally related.  Using this reasoning we could conclude
that the different types of tadpoles are actually complex character
states rather than assemblages of independent characters.  This
conclusion does not preclude any variation at lower taxonomic levels
in each of the tadpole types, but it does state that, at this level
of hierarchical discrimination, we can consider each type to be a
complex character with a minimum of four character states.

Pheneticists (Sneath and Sokal, 1973) and various cladists
(McKenna, 1975) prefer to weight all characters equally for systematic and/or phylogenetic purposes.  It appears from the above
analysis that equal weighing is only an illusion.  Furthermore,
equal weight does not consider important factors which determine
the level of confidence that can be placed in a given set of relationships.

In determining relationships it is obviously preferable to use
more information than less, but the number of characters used is
dependent upon how they are partitioned, as described above, and
how much informational content is inherent in a character or
character transformational series.  On cursory examination, it
might appear that the most unbiased approach would be to give all
characters (or character states) equal weight.  However, it is
clear that this method biases the analysis in favor of those
characters which are improperly understood with regard to homology,
development, function and degree of complexity, since these data
are unknown for most characters.  Although the number of terms or
attributes used in characterizing groups may seem great when listed,
their information content indicating relationships may be relatively
small.  Therefore, those characters with the most information
should be weighted heavily.  In fact, examination of most cladograms or phylogenetic schema indicates only one or two derived

characters at each branching point or node.  This actual limited
amount of data emphasizes the importance of proper analysis of
morphoclines.

## THE MORPHOCLINE AND POLARITY

As stated above, the morphocline is an homologous series of
character states distributed among related taxa.  It is not sufficient to merely recognize the homologous series, but one must
also determine the evolutionary trends in the series, or its
polarity.  In order to determine the polarity the different character states of the morphocline must be clearly described and their
distribution among taxa be well known.

Polarity is the most important problem in the analysis of a
morphocline or transformation series.  For any series of comparisons, there can be only one primitive state, and the important
problem is to develop criteria to determine the primitive state.
Kluge and Farris (1969:5) have succinctly stated the criteria for
the determination of the primitive state.  "In order of reliability
these criteria are:

(1)  The primitive state of a character for a particular group
is likely to be present in many of the representatives of closely
related groups.

(2)  A primitive state is more likely to be widespread within a group than is any one advanced state.

(3)  The primitive state is likely to be associated with
states of other characters known from other evidence to be
primitive."

The most commonly used criterion is the distribution pattern
of character states among compared taxa, which includes the first
and second categories of Kluge and Farris (1969).  For this purpose one must use a minimum of three groups, the two taxa under
consideration and an outgroup (of several other taxa) related to
these two.  A character state that has the widest distribution
among the taxa is generally considered most primitive.  This has
been called the commonality principle of character state distribution (Schaeffer et al., 1972).

The monophyletic origin of the primitive character state is
the most parsimonious explanation for its widespread distribution
among taxa.  While this is the most reasonable interpretation, it
is important to remember that the principle of parsimony is part
of a method of analysis and not an evolutionary process.  In fact,

the more closely related two groups are, the more likely that they
may evolve the same character state in parallel from another more
primitive state.  The determination of the primitive state remains
an hypothesis to be continually reevaluated or tested.

In the determination of the primitive state, the distribution
of the various character states among taxa is of primary importance, but the developmental pattern and structural complexity of
the character state offer important secondary evidence.  Most often
the interpretation is heavily weighted in the selection of the
simplest state to represent the primitive state.  Yet, among
vertebrate types, particularly in the secondary adaptive radiations,
the simplification of structure by fusion, reduction or loss is a
common process and is often developed independently by different
taxa.  In this case, the widespread distribution of such reduced
states among taxa must be carefully considered since they present
the crucial problem in the analysis of any morphocline or in any
character analysis, the problem of parallelism.  This problem will
be treated below.

Although the distribution of the character states among taxa
is the most important observation in the determination of the
primitive state, there may be other factors involved.  In the
comparison of two closely related taxa with an outgroup, there remains the possibility that none of the observed states is primitive.  In this case, the primitive state is unknown and should be
reconstructed.  If all character states are equidistant from the
hypothetical primitive state, then the reconstructed morphotype
will be difficult to hypothesize.  In a morphoclinal series, the
derived states are usually recognized by their restricted distribution among taxa.  The more unique the distribution of the character state, the more recent is its probable time of origin.

Once the primitive and derived states have been deciphered,
then one can reconstruct a hypothetical polarity.  The polarity
may be simple, as in the step-wise, four character state morphocline a → b → c → d; or it may be the exact opposite, where three
character states are each independently derived from the primitive
state (Marx and Rabb, 1972).  In a simple dichotomously arranged
phylogeny based on four derived character states, there are minimally 15 possible cladograms.  If one allows schemes other than
dichotomies, the number of possible cladograms increases exponen-
tially.  All of these possible polarities should be examined.

The number of polarities to be examined can be reduced by the
following procedures:  1)  The direction and type of polarity may
be determined by (a) distribution among taxa and (b) the morphological and developmental patterns known for characters.  2)  Improbable polarities, such as the resurrection of complex structures

from loss or reduced states, can be given low priority of analysis, although they remain a possibility.

After the examination of a single morphoclinal series, other morphoclinal series should be examined to determine the degree of concordance and discordance. The greater the concordance between different morphoclines, the more likelihood that the original hypothesis was correct. This is essentially the third criterion of Kluge and Farris (1969).

Marx and Rabb (1970, 1972) and Heyer (1975) were also concerned with the problem of polarity. They utilized the distribution of character states among taxa as primary indicators of primitive and derived states. For further verification or determination where distribution patterns were not satisfactory, they utilized two criteria which are essentially extrinsic character values: ecological specialization and geographic restriction. For example, as categories of ecological restriction they used such broad groups as aquatic, fossorial and arboreal habits. In geographic restriction they used the distribution of character states by geographic regions or centers of diversity. We have argued above that extrinsic characters should not be used in phylogenetic inference. In this case, the use of these two criteria is also improper because of the assumption that the similarities in characters are the result of a single adaptive radiation.

In the analysis of a morphocline, Schaeffer et al. (1972) pointed out the dangers of uncritically considering the temporal distribution of the taxa involved. If a morphocline is studied among taxa with a fossil history with good diversity and temporal distribution, then a chronocline (a morphocline on a time axis) can be set up. Contrary to Harper (1976) and Simpson (1975), time by itself is no proof of primitiveness of a character state or ancestral position of a taxon (Throckmorton, 1968). On the other hand, if morphological analysis, comparing all taxa simultaneously, indicates that a given character state is primitive, then its occurrence among the earliest fossil taxa is merely corroboration for the hypothesis. Certainly, the distribution of characters and character states among the Equidae and other perissodactyls indicates that Hyracotherium (at the generic level) represents an ancestral or primitive type for the Equidae. Distribution in time, like morphological simplicity or complexity, cannot alone indicate a primitive or advanced state. An analysis of the morphology, function and development of a character is essential to determine which state is primitive.

Simpson (1975) correctly pointed out that the erection of a transformational or morphoclinal series may lead to a priori

orthogenetic interpretations.  There is certainly some danger of
such interpretation, but that does not mean that morphocline
erection is an incorrect methodology.

## CONSTRUCTING THE PHYLOGENETIC SCHEME

The series or set of apparently related taxa, originally
recognized by overall similarities, should number minimally four
groups.  Three of the groups comprise the taxa whose close relationships are under consideration, and the fourth group should be the
out-group sister to these taxa.  A minimum of two independent
morphoclines, each with a minimum of two character states, is required for a meaningful analysis of relationships.  One of the
morphoclines must indicate the proper out-group, and the other is
used to determine the relationships of the taxa under consideration.
The larger the number of reliable morphoclines used, the more reliable the results.

The patterns of polarity among the morphoclines should be
compared for concordance and discordance.  In the rare cases of
complete concordance of polarity, the interpretation is simple,
particularly should a single taxon bear all or the majority of
the primitive states.  In such a case where the taxa are all
modern or from a single time dimension, the process of erecting
the phylogeny is relatively simple.  One of the taxa may represent
an approximation of the ancestral morphotype, but there is no
problem as to an actual ancestral-descendent relationship.  If, on
the other hand, there is only some concordance, or actual discordance, in the distribution of character states, the data must be
carefully considered.  Each of the morphoclines must be considered
as to the source of the discordance, which may be the result of
parallel evolution, improper analysis or description of the characters, or improper arrangement of the taxa.  The most common cause
of discordance will be parallelism.

As stated above, the most important problem in analyzing
transformation series, particularly among closely related forms,
is the recognition of parallelism.  In attempting to show phylogenetic trends within species, it is almost impossible to eliminate
parallelisms.  Populations sharing a common gene pool (and therefore having the same mutational probabilities) and existing under
very similar selective forces, will respond in the same manner by
independently producing similar phenotypes.  The simpler the
genetic basis for the phenotype, the more likely it is that
parallelism will occur.  On the other hand, convergences and
parallelisms should be easier to detect at higher taxonomic levels.
It is at the generic to ordinal levels in vertebrates that parallelism obscures our development of a correct phylogeny.

In order to recognize parallelism in a single morphocline, one may examine the chronocline, and if there is discordance in distribution along the time axis, then parallel evolution may be suspected. The most common type of parallelism, within a single morphocline or character complex, is easy to detect because it will be contradicted by discordance with other morphoclines. Less commonly, but nevertheless occurring frequently enough, are whole sets of morphoclines which parallel each other and result in mosaics, making it difficult to differentiate true lineages from grades or levels of organization (Schaeffer and Hecht, 1965). In order to separate true lineages from grades it is necessary to recognize and search for shared and derived character states and distinguish them from primitive states and from other conflicting polarities.

Realizing that his methodology might allow hidden parallelisms, Heyer (1975) thought (as did Marx and Rabb, 1970 and Kluge and Farris, 1969) that the use of great numbers of characters would override the few mistaken paralellisms. They all failed to realize that, in the recognition of branching points in a phylogeny, only one reliable character is necessary to indicate lineage. The clouding of the pattern of relationships by unnecessary data prevents the recognition of the real lineage. Heyer (1975:2) stated ". . .a large sample size of characters is needed to swamp the effect of those possible rare instances where evolution has not operated in a logical manner." Evolution is a process which cannot be logical or illogical, but it is our interpretation of it which is logical or illogical. What Heyer implies here is that parallelism and reversals are illogical because they confuse our analysis of polarity. Heyer (1975:38) was amazed at the amount of parallelism still remaining in his analysis of the relationship of the genera of the frog family Leptodactylidae. We feel the cause was not the intransigence of nature but the interpretation of the data. From the viewpoint of lineage detection, it is more important to use a few well-analyzed morphoclines than many poorly or improperly analyzed ones. Heyer (1975) and others (Inger, 1967; Rabb and Marx, 1972) have felt that it is more important to utilize all available characters, even if polarities of some of the morphoclines are ambiguous, indeterminable or obscured by parallelism or reversal. We feel that it is better to use fewer well-interpreted morphoclines than to allow obfuscation by mere quantity, and that the apparently contradictory evidence of conflicting morphoclines can be analyzed by the use of character weighting.

The importance of character weighting has been recognized by many students, but only a few need be noted (Inger, 1967; Kluge and Farris, 1969; Bock, 1973). It is most evident to those students who use shared and derived characters to indicate dichotomies in lineages that some characters have more reliable informa-

tion within them than do other characters. Inger (1967), following Wilson's (1965) concept of looking for unique and unreversed
character states, suggested four criteria for recognizing characters with a large degree of reliability. These criteria are:
1) There is no obvious selective difference between the states
of the character, 2) The states occur widely in the taxa studied,
3) The character has low variability within taxa, 4) The unique
state has an unusual developmental pattern. Kluge and Farris
(1969) correctly pointed out that the first and last criteria are
non-operational because there is rarely sufficient information.
Recognizing the validity of the second and third criteria and the
need to recognize conservative shared and derived characters, they
proceeded to develop a weighting methodology that they believed
was quantitative and non-arbitrary.

Kluge and Farris (1969) actually weighted their characters in
two ways: First, they stated that there is a correlation between
the variability of a character on the species level and the ability
of the character to change, or its evolutionary rate. The greater
the variability of a character (or the number of character states),
the less reliable is the character as an indicator of lineage,
and therefore the less weight it should be given in phylogeny
reconstruction. The basic weakness in the proposal is the assumption that the specific variations or any other variations at any
given time level are an indication of the potentiality of the
particular morphocline or character suite at all time levels.
Whereas this may be true in some groups, there is no indication
that this holds true for all time intervals and all species.
Directional, stabilizing and disruptive selection are not fixed in
any lineage for all time. Certainly each biological species, as
an independent evolutionary unit, will develop its own variability
for different characters. There is no evidence that there is a
correlation between present intragroup variability and past variation and successive character transformation. Based on this
dubious biological hypothesis, Kluge and Farris (1969) disregarded
or downgraded various morphoclines. These procedures were done
although they had recognized the importance of parallelism, reduction and reversal.

The second way that Kluge and Farris weighted characters was
a result of the way they codified the character states. Because
of the binary nature of their coding system, no character could
possess more than two character states. In order to accommodate
complexity, Kluge and Farris divided more complex characters into
several smaller characters, each with two character states. Each
resulting character was then given equal weight with all other
characters, unless it showed variability. However, such a process
automatically gives greater weight to complex characters. As we
argue below, the various parts of an integrated functional complex

have to change together to maintain biological efficiency, so
separating these into smaller components is unjustified.  This
second method of weighting was not explicitly recognized by Kluge
and Farris.  Their analysis of the characters themselves was
excellent, but the codification was oversimplified.  As a result,
their phylogeny showed heavy weighting for a few characters, but
even so, the resulting phylogeny was not unreasonable.

Hecht and Edwards (1976) used a system of character weighting
in dealing with the relationships and evolution of proteid salamanders, in which they found a remarkable degree of parallelism.
In making this evaluation, Hecht and Edwards (1976) developed a
weighting scheme that may have general applicability, particularly
in the weighting of shared derived characters (see below).  The
system emphasizes the amount of new information that is actually
revealed by the analysis of characters or character states in relation to a given phyletic arrangement.  This weighting system is
not a statement of evolutionary change, but it provides us with a
measure of the level of confidence in the character analysis.

The weighting system (Hecht, 1976) contains five categories:

Characters and character states of the lowest value are
those involving loss of a structure.  Character weighting
group I has zero information (as to monophyly) because
there is no way of determining whether the state has been
derived by a single change or by two or more independent
processes.  Loss characters are here defined as characters which have no developmental information to indicate
the pathway by which the loss occurred.

The second category, character weighting group II, has
more informational content than the preceding, and includes simplification or reduction of complex characters.
These are indicated by comparative or developmental
anatomy.  Loss characters, such as eyelessness in cave
salamanders or fish, in which the developmental mechanism
leading to the loss is known, should be included in this
category.  Independent reduction of the same character
by two closely related taxa may show a different developmental process in some minor detail.

The third category, character weighting group III, includes those character states that are the result of
common growth processes for the taxa being compared.
These similarities are due to growth and developmental
processes dependent on size, age or hormonal and other
physiological relationships, such as allometry or
neoteny.

The fourth category of weighted characters includes
all those states which are part of a highly integrated
functional complex.  It is not surprising that many
characters that are closely integrated should change
together in order to maintain biological efficiency or
permit the organism to remain viable.  Therefore, the
separation of such a functional complex into more or
less arbitrary components may result in undue weighting.
The complexity of characters in this weighting state
makes them important indicators of polarity and useful
to distinguish parallelism, and their reliability is
greater than any of the preceding types.  For example,
the fact that the squamosal-dentary articulation was
evolved independently by several lineages at the
synapsid reptile-mammalian transition was determined
by examination of the functional complex including
the musculature.

The fifth and most informative type of weighting group
is that which is innovative and unique for the morphoclinal series and, therefore, most useful as a shared
and derived character state to distinguish a new
lineage.  The more complex the innovative character
state, the more reliable an indicator of lineage it
is.  At the higher hierarchical levels, this character
weighting group usually indicates new functional or
adaptive trends.  If the characters or character states
are complex enough, they can preclude parallelism.  Such
characters as the amniote egg, the artiodactyl astragalus,
or the gekkonid cochlea could be considered examples of
this.

The character weighting groups proposed above are admitted to
be artificially clearly defined.  In fact, there is overlap between the categories, but this is a useful beginning for analysis.
Whether there are three, five or ten categories depends on the
level of resolution of the analysis.


THE PROBLEMS OF REVERSAL, PARALLELISM, AND PARSIMONY

In the construction of a scheme of phylogenetic relationships,
the most parsimonious interpretation is that evolution is progressive and that, between any two branching points on a cladogram,
there have been no reversals.  The entire concept is based on a
strict adherence to the generalization called Dollo's Law.  Kurten
(1963) has demonstrated that reversals do occur, although usually
these are in simple characters and they arise merely by the break-
down of the genetic suppression of a character, thus allowing a

"lost" character to reappear.  These reversals can only be detected
by careful analysis of the several available morphoclines.  The
importance of chronocline analysis in the detection of reversals
in self-evident.  As will be demonstrated below, the reversal of a
character state in the vertebral column of the Gekkota resulted in
the redevelopment of an amphicoelous condition which is developmentally and morphologically indistinguishable from the primitive
state.  Only the discordances evident upon comparison of different
well-analyzed morphoclines will reveal such reversals.

     The problems of reversal and parallelism are the severest
tests of the principle of parsimony.  These problems are most
crucial in the interpretation of the evolution of protein structure,
where the operating principle of maximum parsimony does not allow
the detection of such phenomena.

     Parsimony is a general axiom accepted by most phylogeneticists
as an operating principle in the interpretation of data.  Parsimonious phylogenetic schemes can be constructed in two ways:
1)  By minimizing the character difference distance between
branching points (Kluge and Farris, 1969), or 2)  By minimizing
the total number of parallelisms and reversals (Camin and Sokal,
1965).  Parsimonious interpretation is a method of analysis, but
not an evolutionary process (Rogers, Fleming, and Estabrook, 1967;
Inger, 1967).  Therefore, minimizing character distance may not
detect true parallelisms and reversals which, as argued above, are
common occurrences.  Only if we accept that all selection is unidirectional can we deny the possibility of reversals.  On the
other hand, unidirectional selection, such as overriding requirements of a particular environment (e.g. biomechanical requirements),
would produce parallel changes in related lineages.

     A method is required to detect true parallelisms and reversals.
The weighting methodology outlined above, which indicates the
degree of confidence that one can place in various morphoclines or
transformation series, provides us with such a method.  Parallelism
and reversal should be relatively easy to detect in high weight
characters belonging to weighting groups IV and V because of the
complexity of such characters.  Characters belonging to these
categories should be the primary data used in phylogenetic reconstruction.  Parallelism and reversal in lower weight categories
are more difficult to detect because of the low informational content in such character states.  Therefore characters belonging to
weighting groups I-III should not be considered important.

     The recognition of parallelisms and reversals will increase
the complexity of a given phylogenetic scheme by increasing the
number of nodes.  Many phylogeneticists (including McKenna, 1975;

Heyer, 1975) apparently believe that a node represents a speciational event; this idea is based on the concept of punctuated
equilibria (Eldredge and Gould, 1972). This narrow interpretation
assumes that phyletic evolution does not occur and that morphological change only takes place in the genetic revolution associated
with speciation (Mayr, 1963). As demonstrated by Ayala (1975),
accumulations of differences can occur gradually in lineages and
no dramatic differences are associated with the attainment of the
species level.

## THE CONTRIBUTION OF PALEONTOLOGY

Paleontologists have generally assumed that the discovery of
a single specimen or population sample of a previously existing
organism would reveal special insight into the evolution of
lineages. Therefore, the discovery of a new fossil has been
thought to bring the observer or the interpreter nearer to the
origin or ancestry of a lineage. No doubt this is true in the
temporal sense, if generation time and/or evolutionary rate are
assumed to be fixed. Yet, these assumptions are only true in a
few cases. The most important contribution of the fossil record
is that it presents a picture of past diversities of organisms
and environments, an indication of the minimal longevities of
lineages and a test of phylogenetic inferences previously made.
For years the primary endeavor of many paleontologists has been
the search for ancestor-descendent relationships both with extinct
and living lineages. Harper (1976) and Cracraft (1974) present
opposite points of view on this subject. Cracraft (1974) states
that the search for ancestors is fruitless and hopeless, but this
conclusion is based on an a priori system of cladistic analysis in
which all groups compared are sister groups and are all diagnosed
by unweighted shared and derived characters. Furthermore, all
branching points represent theoretical morphotypes which can never
be real ancestors because all groups are assumed to be monophyletic
at the species level. This artifact of cladistic classification
was recognized by Hennig (1966), who was more optimistic about the
inferences that can be drawn from paleontology. Certainly, in the
search for ancestors many paleontologists, such as Gingerich (1976),
have been overzealous. Gingerich attempted to demonstrate species
succession in Hyopsodus by examining the dimensions of only a
single tooth from the whole tooth row. He believed he had demonstrated phyletic evolution, but did not consider other possible
explanations, such as sexual dimorphism or the existence of sibling
species. The ancestral-descendent relationships presented by
Gingerich are certainly over-interpreted. On the other hand, the
position of Cracraft (1974), that among the thousands of types of
fossil vertebrates described there are no ancestor-descendent
relationships, is ludicrous. For example, if a fossil form fulfills

all the requirements for a hypothetical morphotype, and its distribution in time and space are substantiated by chronocline analysis,
there can only be one parsimonious conclusion, that the fossil
form represents either the actual ancestral group or a form very
similar to the actual ancestor.  Cracraft or similar interpreters
of the fossil record would criticize this conclusion by stating
that in all cases one can always find some shared and derived
character to differentiate the supposed ancestral taxon from the
hypothetical morphotype.  If this character is of low weight or
subject to loss, the presence of the derived character should not
preclude the fossil taxon from being ancestral.  The lack of the
character in the derived taxon can easily be attained by loss,
reduction or fusion, usually characters of low weight.  Just as
one cannot assume that fossils which occur at the right time and
place are automatically ancestral, one cannot make the opposite
assumption.  Fossils which fulfill the morphotype predictions,
occurring at the appropriate place in the time axis, must be considered as possible true ancestors until new data falsify the
interpretation.

In contrast to Cracraft (1974), Harper (1976) has attempted
to outline procedures that incorporate paleontological data into
phylogenetic schemes.  His emphasis is on the utilization of the
principle of shared and derived character states, the recognition
of primitive and derived states, and the principle of commonality
of distribution and its utilization in the recognition of primitive states.

Harper's principles of morphological similarity, minimal
morphologic gaps, minimal geographic gaps, and minimal stratigraphic
gaps have fundamental problems built into them that we believe he
only partly recognizes.  Although cognizant of the problem of
parallelism and limited reversal (particularly of loss and reduction), he did not consider them as fully as he might.  His
principle of morphological similarity does not describe a procedure
to analyze such similarities.

Fundamentally, the problem of morphological similarity is the
problem of understanding the polarities in the morphocline.  The
comparison leading to polarity clarification must be done across
a homologous series or morphocline and not on a taxon by taxon
comparison.  It is only after at least three independent morphoclines have been analyzed that the meaning of character states
and what they indicate about relationships can be interpreted.

The principle of minimal morphological gaps suffers from the
same inadequacy as that of the preceding; it is not until the
direction of polarity is known, and the possibility of parallelism
is evaluated, that phylogenetic relationships can be hypothesized.

Morphological gaps may be the result of extinction or an incomplete
fossil record, but morphological similarity between closely related groups may not necessarily be due to monophyly.  The utilization of Occam's razor (Harper, 1976:188) will result in a simple
hypothesis, but not necessarily in a true phylogenetic relationship.

The principles of minimal geographic and stratigraphic gaps
are essentially the same principles as criticized by Schaeffer
et al. (1972).  As stated above, these patterns are negative
statements and extrinsic features of the organism.  The theory of
relationships developed by the first principle of Harper and the
methodology described in the present study must be compatible with,
but not derived from, distributional data.

Harper's version of Hennig's principle of veracity is simply
a statement that the greater the concordance between different
types of data, the more likely it is that we have arrived at a
close approximation of truth.  Seemingly, it would appear that he
favors weighing extrinsic and intrinsic features equally.  He does
not allow for the recognition of the level of resolution and discrimination allowed by the basic data.


CONSTRUCTING THE PHYLOGENY:  AN EXAMPLE

We will now use the method described above to construct a
phylogeny of salamanders (class Amphibia, Order Caudata) at the
family level.  This phylogeny exemplifies the dangers of relying
on strict parsimony when using unweighted data in phylogeny construction.

Most salamander taxonomists have not presented phylogenies to
accompany their classifications, nor have they explained in detail
why they ally various groups with each other.  However, we believe
that the phylogenies in Figures 1b, 2b and 3a correctly represent
the general ideas of the authors indicated.  The phylogenies in
Figures 1a, 2a and 3b are slightly modified from the diagrams
presented by the authors who proposed them.

All previous salamander taxonomists have agreed that the
family Hynobiidae most closely resembles the ancestral caudate.
This idea is based primarily on three primitive character states
(external fertilization, angular bone in lower jaw, and high
chromosome number with microchromosomes) found for certain only in
this family and the Cryptobranchidae.  Noble (1931) considered the
hynobiids to be the actual ancestors of all other groups (Fig. 1a).
The cryptobranchids were thought to retain most of the primitive
characters of the hynobiids but to be derived in being partially
neotenic and aquatic.  The Ambystomatidae were characterized by

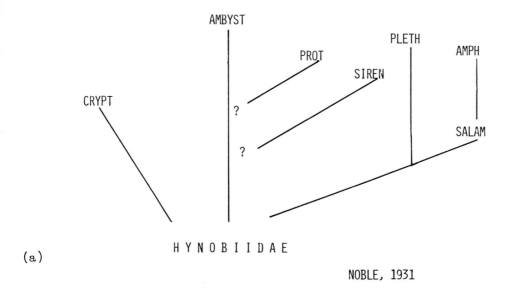

(a)

HYNOBIIDAE

NOBLE, 1931

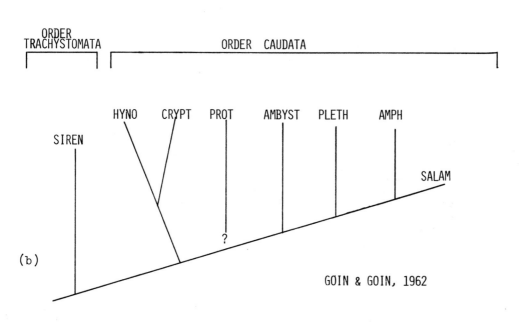

(b)

GOIN & GOIN, 1962

Figure 1.  Phylogenies of salamander families as proposed by (a) Noble, 1931, and (b) Goin and Goin, 1962.  Abbreviations for all figures and tables as follows:  Ambyst = Ambystomatidae; Amph = Amphiumidae; Crypt = Cryptobranchidae; Dicamp = Dicamptodontidae; Hyno = Hynobiidae; Nect = Necturidae; Pleth = Plethodontidae; Prot = Proteidae; Salam = Salamandridae; Siren = Sirenidae.

short vomerine tooth rows, while in the Plethodontidae, Salamandridae and Amphiumidae the vomerine tooth rows extend well behind
the internal nares.  The correct placement of the Proteidae and
Sirenidae was hampered by the paedogenetic nature of the adults of
these families.

The phylogeny (Fig. 1b) based on the classification of Goin
and Goin (1962) is very similar to that of Noble, the only major
difference being the removal of the family Sirenidae to a separate
order, the Trachystomata.  This was based on their vertebral structure; lack of cloacal glands, pelvic girdle and hind limbs; and a
horny sheath over the jaws.  Later workers have rejected this
classification and have retained the sirenids in the Order Caudata.

Herre (1935) felt that the two paedogenetic families, the
Proteidae and Sirenidae, and the partially metamorphosing Amphiumidae were derived from certain unknown genera of "advanced" salamandrids (Fig. 2a).  This concept was based on the presence of
posteriorly projecting paroccipital processes in these taxa, and
on a fossil form, Palaeoproteus klatti, which was considered intermediate between salamandrids and proteids.  In modern terminology,
Herre thought that the family Salamandridae was a paraphyletic
group (sensu Nelson, 1971).

Wake (1966) proposed a phylogeny (Fig. 2b) in which a pro-
hynobiid ancestral group gave rise to three suborders:  (1) Crypto-
branchoidea, which retained all the primitive character states;
(2) Ambystomatoidea, predominantly New World forms characterized
by the presence of the septomaxillary bone and of vomerine pre-
orbital processes which bear laterally oriented tooth rows; and
(3) Salamandroidea, predominantly Old World forms which lack both
the septomaxilla and the laterally oriented vomerine tooth rows.
Wake also adopted the suggestion of Hecht (1957) that the genera
Proteus and Necturus, previously united in the family Proteidae,
had actually evolved in parallel, and thus should be placed in
separate families, the Proteidae and Necturidae.

Regal (1966) based his phylogeny (Fig. 3a) on feeding specializations, including tongue morphology and vomerine tooth replacement patterns.

In the most recent study, Edwards (1976) used 13 characters
in constructing a modified cladistic classification of salamanders.
His phylogeny (Fig. 3b) differed significantly from all previously
proposed schemes, although, interestingly, his classification was
similar to that of Regal (1966).  (Edwards proposed a new family,
the Dicamptodontidae, based mainly on spinal nerve patterns; his
family Ambystomatidae equals the subfamily Ambystomatinae of previous workers.)

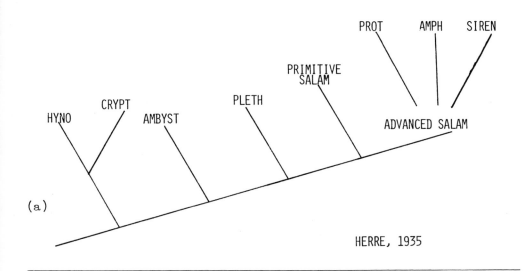

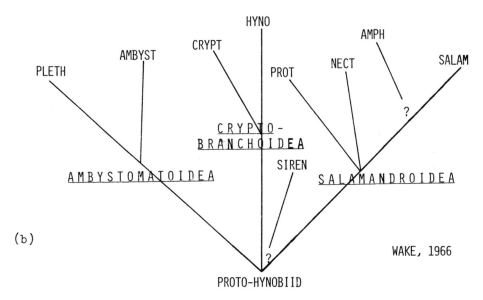

Figure 2. Phylogenies of salamander families as proposed by (a) Herre, 1935, and (b) Wake, 1966.

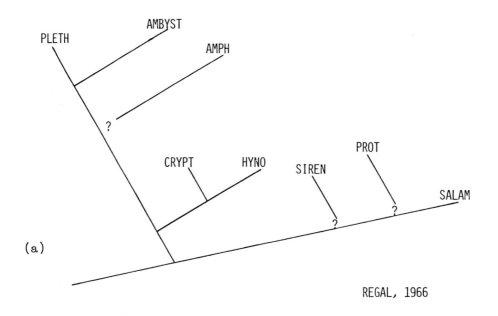

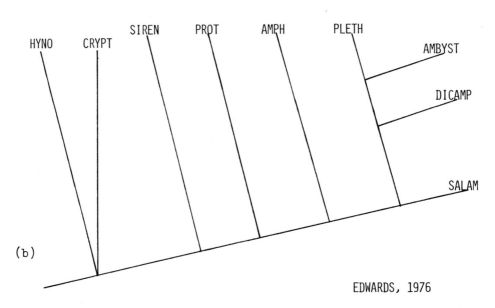

Figure 3.   Phylogenies of salamander families as proposed by
(a) Regal, 1966, and (b) Edwards, 1976.

Examination of Figures 1, 2 and 3 shows that, aside from the fact that all schemes retain hynobiids and cryptobranchids as the most primitive salamanders, there is hardly any agreement concerning the phylogeny of salamanders. For example, sirenids have been considered rather primitive offshoots from the basal stock (Figs. 1a, 2b), one of the most derived of all salamander families (Fig. 2a), or not even salamanders at all (Fig. 1b).

In order to test these phylogenies, we will use 18 characters which have previously been used in salamander classification (Table I). Characters 1-13 are the same as the similarly numbered characters in Edwards (1976). Characters 14-18 come from a larger suite of characters which has been used by various authors as evidence for the monophyletic origin of the genera Necturus and Proteus; these are discussed more fully in Hecht and Edwards (1976). The character definitions and morphoclines are as follows; morphocline polarities were determined by using the methodology in the section "The Morphocline and Polarity" of this paper, especially the principle of outgroup comparisons, and are discussed more fully in Edwards (1976), Hecht and Edwards (1976) and Hecht (1976).

1. Angular bone present (1+) or absent (1-) 1+ → 1-.

2. Fertilization external, cloacal glands absent (2-); fertilization external, one pair of cloacal glands (2±); fertilization internal by means of spermatophore, three pairs of cloacal glands (2+) 2- → 2± → 2+ (coded differently than in Edwards, 1976). Fertilization has not been observed in the Sirenidae. However, these animals lack the cloacal glands necessary for production of spermatophores by the male and retention of the sperm by the female. Therefore we assume that fertilization is external in this family (2-). Fertilization is also external in the families Hynobiidae and Cryptobranchidae, even though these animals have one of the three pairs of glands normally associated with spermatophore production and internal fertilization in other salamanders (2±). Bannikov (1958) reported that in one species of hynobiid (Ranodon sibiricus) the male actually produces a rudimentary spermatophore. Even in this species, however, fertilization remains external since the female attaches her egg sacs to the spermatophore rather than taking the sperm into her cloaca. This situation could be viewed as a preadaptive stage to true internal fertilization or as a reduction from that state as was initially suggested by Noble (1925, 1931). Regal (1966) took the position that without further data this issue was unresolvable, and therefore fertilization pattern could not be used as a major systematic character. However, we feel that the cryptobranchid-hynobiid condition is not the result of reduction for the following reasons: (1) since cryptobranchoids possess the primitive states for almost all other characters, the association criterion of Kluge and Farris (1969) suggests that external fertilization as seen in these families is

TABLE I

Distribution of 18 Characters among Families of Salamanders
(See text for explanation)

| | Hyno | Crypt | Siren | Prot Nect | Amph | Salam | Dicamp | Ambyst | Pleth |
|---|---|---|---|---|---|---|---|---|---|
| 1. Angular bone | + | + | − | − | − | − | − | − | − |
| 2. Internal fertilization | ± | ± | − | + | + | + | + | + | + |
| 3. Septomaxilla | + | − | − | + | − | − | + | + | + |
| 4. Vomerine tooth replacement | P,PL | larv | larv | larv | larv | M | P | P | PL |
| 5. Haploid chromosome number | HM | HM | H | 19 | 14 | 12,11 | 14,13 | 14 | 14,13 |
| 6. Spinal nerve pattern | I | I | III'' | I | III | III'' | III' | III'' | III'' |
| 7. Number of larval gill slits | 4 | 4 | 3,1 | 2 | 3 | 4 | 4 | 4 | 4,3 |
| 8. Second epibranchial | + | + | + | − | − | − | − | − | − |
| 9. Columella | a,b | b | b | b | b | c | a,b | a,c | d,e |
| 10. Lacrimal bone | + | − | − | − | − | − | + | − | − |
| 11. Ypsiloid cartilage | + | + | − | − | + | + | + | + | + |
| 12. Maxilla | + | + | ± | − | + | + | + | + | + |
| 13. Nasolabial groove | − | − | − | + | − | − | − | − | ± |
| 14. Perennibranchiate | − | − | + | + | − | − | ± | ± | ± |
| 15. Separate opisthotic* | − | − | (+) | + | − | − | (+) | (+) | − |
| 16. Fourth branchial arch | + | + | + | − | + | + | + | + | ± |
| 17. Columellar process | − | − | − | + | − | − | − | ± | − |
| 18. Jugularis nerve above ligament | − | − | − | + | − | − | − | − | ± |

* (+) = present in larva only

also relatively primitive; (2) since the primitive state in anurans
and caecilians is undoubtedly external fertilization, this mode of
reproduction must be primitive for Amphibia as a whole, and postulating the gain and subsequent loss of internal fertilization in
cryptobranchoids adds two extra evolutionary steps for which there
is no evidence; (3) once a group had achieved internal fertilization by means of a spermatophore, it does not seem likely to us
that this very specialized means of inserting the sperm into the
female would have been abandoned for a presumably much more
hazardous method.

   3.   Septomaxilla present (3+) or absent (3-) 3+ → 3-.

   4.   Adult vomerine tooth replacement from posterior side of
tooth row (4P); replacement from posterior and lateral sides (4PL);
replacement from medial side (4M); larval replacement pattern
(r larv).

$$4P \nearrow^{4PL}_{\searrow 4M} \quad ; \quad 4 \text{ larv}$$

   This character was coded incorrectly in Edwards (1976).
The adult vomerine tooth patterns of terrestrial salamanders result
from changes in the shape of the vomers which occur at metamorphosis
(see review in Regal, 1966).  In larval caudates, the palatal dentition parallels the marginal dentition.  Palatal teeth are attached
to the vomers and to the palatopterygoids.  Replacement denticles
form in the mucus membrane of the oral cavity medial to the palatal
row and then move peripherally to replace the palatal teeth as
they are resorbed (4 larv).  This larval type of tooth replacement
is retained in the paedogenetic families Sirenidae, Proteidae and
Necturidae, and in the differentially metamorphosing Cryptobranchidae
and Amphiumidae.  After metamorphosis teeth are found only on the
vomer.  The reshaping of the vomer occurs through the fusion of
denticles to each other and to the vomer.  In the families Ambystomatidae and Dicamptodontidae a few denticles are added to the
medial sides of the vomers, so that the adult vomerine teeth form
short transverse rows in the front of the mouth; replacement teeth
form posterior to the tooth rows (4P).  In plethodontids a large
number of denticles typically attaches to the medial side of each
vomer, thus producing two inverted J-shaped tooth rows with the
stem of the J formed by the fused denticles and the hook of the J
by the remnants of the larval vomer.  In some plethodontids
(Typhlotriton, Stereocheilus, Pseudotriton, Gyrinophilus) the
tooth rows remain intact, but in most genera resorption of denticles at the anterior end of the accretion leads to a condition in
which an anterior transverse row is separated from a longitudinal
tooth patch.  In all plethodontids replacement teeth form posterior
and lateral to the tooth rows (4PL).  In the family Hynobiidae one

finds an array of forms which bridge the gap between the
dicamptodontid-ambystomatid condition and the primitive plethodontid condition in which transverse and longitudinal rows remain
attached (4P, PL). Finally, in the family Salamandridae a row of
denticles attaches to the lateral ends of the vomers at metamorphosis, thus leading to a condition in which the replacement teeth
form medial to the tooth rows (4M). Replacement from the posterior
side (4P) is considered the primitive state for adults because of
the criteria of Kluge and Farris (1969) which state that the primitive state is likely to be widespread within a group and to be
associated with other primitive states. Posterior replacement is
found in the families Hynobiidae, Dicamptodontidae and Ambystomatidae; the Hynobiidae also posses the greatest number of primitive
states for other characters.

    5. High haploid chromosome number (20-31), with several
microchromosomes (5 HM); haploid number remains high, but microchromosomes lost (5H); reduction to haploid number of 19 (5(19));
reduction to haploid number of 14 or 13 (5(14)); reduction to
haploid number of 12 or 11 (5(12)). 5HM → 5(19) → 5(14) → 5(12).

    6. All post-atlantal spinal nerves intervertebral (type I
nerves) (6 I): posterior caudal spinal nerves pass through foramina
in vertebrae (type III nerves) (6 III); all post-sacral spinal
nerves type III (6 III'); all but first three spinal nerves type
III (6 III''); all but first two spinal nerves type III (6 III''').
6 I → 6 III → 6 III' → 6 III'' → 6 III'''.

    7. Four gill slits in larva (7(4)); three gill slits in
larva (7(3)); two gill slits in larva (7(2)); one gill slit in
larva (7(1)). 7(4) → 7(3) → 7(2) → 7(1).

    8. Second epibranchial present in adult (8+); second epi-
branchial absent in adult (8-). 8+ → 8-.

    9. Columella free, operculum present (9a); columella free,
operculum absent (9b); columella fused to epiotic, operculum free
(9c); columella reduced to stilus only, fused to operculum (9d);
columella absent, operculum free (9e).

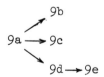

    This character was also coded incorrectly in Edwards
(1976). Two bones, the columella and operculum, are typically
found in the ear region of amphibians. In salamanders, the columella forms early in ontogeny, while the operculum usually forms

only at metamorphosis.  In the primitive condition (Monath, 1965) both columella and operculum are present as free elements in the ear (9a).  This condition is found in some Hynobiidae, Dicamptodon (family Dicamptodontidae), and some Ambystomatidae.  If develop- ment is terminated before metamorphosis occurs, then the operculum does not form, and only the free columella is present (9b); this condition is found in the perennibranchiate families Sirenidae (Larsen, 1963), Proteidae and Necturidae, as well as the partially metamorphosing Cryptobranchidae and Amphiumidae.  The operculum is also absent in some hynobiids and in Rhyacotriton (family Dicamptodontidae).  In the Salamandridae and some of the Amby- stomatidae the columella secondarily fuses with the ear capsule, leaving only a free operculum (9c).  Finally, in plethodontids only one element is found in the ear, but this is usually con- sidered a compound structure formed by the stilus of the columella fusing to the operculum (9d).  In many neotropical plethodontids of the tribe Bolitoglossini the stilus is also lost, leaving only the operculum (9e).  Larsen (1963) questioned the identify of the plate-like portion of the plethodontid ear bone.  This structure is usually considered to be the operculum because it bears the insertion of the "opercularis" muscle, which always attaches to the operculum in other salamander families.  However, it has long been known (Dunn, 1941; Monath, 1965) that the so-called "oper- cularis" muscle in plethodontids is not homologous with the muscle of the same name in other families, the non-plethodontid "oper- cularis" being the levator scapulae while the plethodontid muscle is a portion of the cucullaris.  Larsen also noted that the plethodontid "operculum" actually forms in a manner and at a time more equivalent to the foot plate of the columella.  This subject certainly needs more study, and until such time, we will continue to follow the conventional naming of the plethodontid ear bone with reservations.

10.  Lacrimal present (10+) or absent (10-). 10+ → 10-.

11.  Ypsiloid cartilage present (11+) or absent (11-). 11+ → 11-.

12.  Maxilla present (12+) or absent (12-). 12+ → 12-.

13.  Nasolabial groove absent (13-) or present (13+). 13- → 13+.

14.  Breeding individuals lack gills (14-); breeding indivi- duals perennibranchiate (14+). 14- → 14+.

15.  Opisthotic fused into epiotic (15-); opisthotic separate (15+). 15- → 15+.

16.   Fourth branchial arch present (16+) or absent (16-).
16+ → 16-.

17.   No columellar process on squamosal bone (17-); columellar
process on squamosal bone (17+). 17- $\overset{?}{\rightarrow}$ 17+.

18.   Jugularis branch of hyomandibular nerve passes below
squamosal-columellar ligament (18-); jugularis branch of hyomandibular nerve passes above squamosal-columellar ligament (18+).
18- $\overset{?}{\rightarrow}$ 18+.

In Figure 4 all 18 characters are placed on the phylogeny of
Edwards (1976) without applying any weighting method.  This scheme
requires 57 parallelisms among these 18 characters (see Table II),
and some parallelisms occurred as many as six times.  However,
most of the parallelisms involve loss character states, and as we
argue above should therefore have little weight.

The phylogeny in Figure 5 is more parsimonious, requiring
only 45 parallelisms (Table III), and thus by the rule of maximum
parsimony this tree should be preferred over that of Figure 4.
Figure 5 was constructed by linking groups which showed a high
number of parallelisms in Figure 4, and thus relies heavily on
grouping according to loss character states.  Except for the
placement of the Cryptobranchidae, the phylogeny is very similar
to that of Herre (Fig. 2a).

Edwards (1976) found that the tree in Figure 4 was the most
parsimonious, given the 13 characters he was analyzing.  Figure 5
shows that, when accepting phylogenies according to the rule of
maximum parsimony, the addition of a few characters may materially
change the outcome.  However, as we shall argue below, the additional characters (numbers 14-18) used to construct Figure 5 are
all suspect in that they are due to loss, reduction or neoteny or
have been incompletely analyzed.

When all 18 characters are weighted according to the method
outlined above, only four characters belong to the high weight
groups IV and V (Table IV); these are characters 2 (internal
fertilization), 4 (vomerine tooth replacement - is high weight for
the adult state only), 6 (spinal nerve pattern) and 13 (nasolabial
groove).  Seven characters are due to loss (weighting group I),
three are the result of reduction or fusion (weighting group II),
and three (including the larval vomerine tooth pattern) are due to
neoteny or are clearly larval states (weighting group III).  The
remaining two characters, numbers 17 and 18, need further analysis.
A columellar process on the squamosal bone (character 17) has been
found in adults of _Proteus_ and _Necturus_ and a large specimen of the
neotenic species _Ambystoma mexicanum_ (family Ambystomatidae).  The

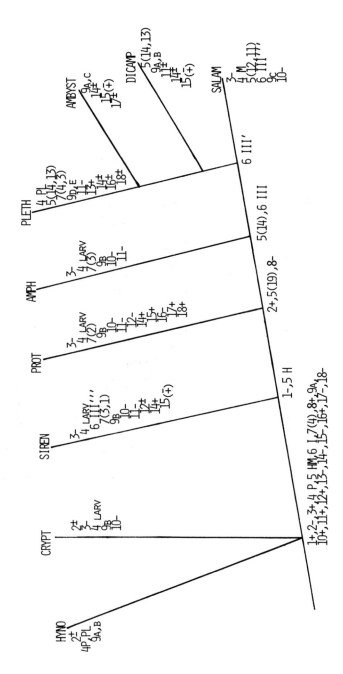

Figure 4. Phylogeny of Edwards (1976) with character states for all 18 characters analyzed in the present study. Parallelisms in this diagram are summarized in Table 2.

TABLE 2

Parallelisms Required by Figure 4

| Character | Number of Times Evolved | Taxa Evolving Character in Parallel |
|---|---|---|
| 2± | 2 | Hyno, Crypt |
| 3- | 5 | Crypt, Siren, Prot, Amph, Salam |
| 4 larv | 4 | Crypt, Siren, Prot, Amph |
| 4 PL | 2 | Hyno, Pleth |
| 5(13) | 2 | Dicamp, Pleth |
| 6 III''' | 2 | Siren, Salam |
| 7(3, 2 or 1) | 4 | Siren, Prot, Amph, Pleth |
| 9b | 6 | Hyno, Crypt, Siren, Prot, Amph, Dicamp |
| 9c | 2 | Ambyst, Salam |
| 10- | 6 | Crypt, Siren, Prot, Amph, (Pleth + Ambyst), Salam |
| 11- | 5 | Siren, Prot, Amph, Pleth, Dicamp |
| 12- | 2 | Siren, Prot |
| 14+ | 5 | Siren, Prot, Ambyst, Pleth, Dicamp |
| 15+ | 4 | Siren, Prot, Dicamp, Ambyst |
| 16- | 2 | Prot, Pleth |
| 17+ | 2 | Prot, Ambyst |
| 18+ | 2 | Prot, Pleth |
| | 57 | |

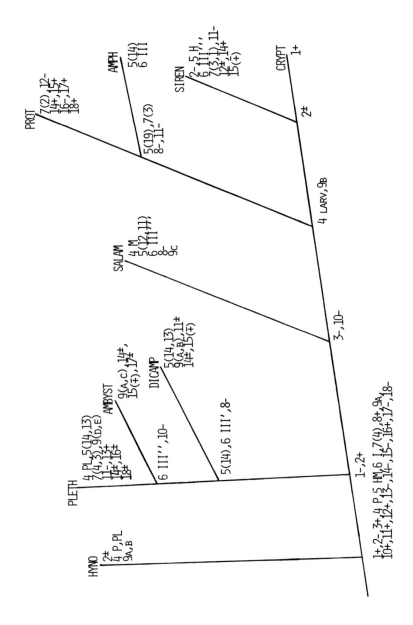

Figure 5. A more parsimonious phylogeny using all 18 characters. This cladogram was constructed by uniting families which showed high degrees of parallelism in figure 4. Parallelisms and reversals are summarized in Table 3.

TABLE 3

Parallelisms or Reversals Required by Figure 5

| Character | Number of Times Evolved | Taxa Evolving Character in Parallel or Undergoing Reversals |
|---|---|---|
| 1+ (reversal) | 1 | Crypt |
| 2± (reversal) | 1 | Crypt |
| 4　PL | 2 | Hyno, Pleth |
| 5　(reductions) | 5 | Dicamp, Pleth, Salam, (Amph + Prot), Siren |
| 6　III''' | 2 | Salam, Siren |
| 7　(3, 2 or 1) | 3 | Pleth, (Amph + Prot), Siren |
| 8- | 3 | (Dicamp + Ambyst + Pleth), Salam, (Amph + Prot) |
| 9b | 3 | Hyno, Dicamp, (Amph + Prot + Siren + Crypt) |
| 9c | 2 | Ambyst, Salam |
| 10- | 2 | (Ambyst + Pleth), (Salam + Amph + Prot + Siren + Crypt) |
| 11- | 4 | Dicamp, Pleth, (Amph + Prot), Siren |
| 12- | 2 | Prot, Siren |
| 14+ | 5 | Dicamp, Ambyst, Pleth, Prot, Siren |
| 15+ | 4 | Dicamp, Ambyst, Prot, Siren |
| 16- | 2 | Pleth, Prot |
| 17+ | 2 | Ambyst, Prot |
| 18+ | 2 | Pleth, Prot |
|  | 45 |  |

process has not been found in other species, but Larsen (1963:86) reports "incipient columellar processes" in <u>Onychodatylus</u> (family Hynobiidae) and <u>Cryptobranchus</u> (family Cryptobranchidae). Similarly, the condition in which the jugularis nerve passes above the columellar ligament (character 18) has been found only in <u>Necturus</u>, <u>Proteus</u> and the troglobitic <u>Typhlomolge</u> (family Plethodontidae), but all salamanders have not been systematically examined for this character state. Therefore, both of the characters need further analysis before they can be weighted correctly.

Hecht and Edwards (1976) have argued that only characters belonging to weighting groups IV or V should be used in determining phylogenies, since the information content of lower groups is so small that true parallelisms cannot be discerned from false ones. Using the four characters from these high weight groups, it is possible to construct two phylogenies (Fig. 6) which each contain two instances of parallelism or reversal. Both schemes require that posterior and lateral vomerine tooth replacement be involved in parallel in the Hynobiidae and Plethodontidae. Figure 6a requires in addition that the two most derived spinal nerve patterns (6 III''') must have evolved in parallel, while figure 6b requires that the spermatophore and all cloacal glands must have been lost in the Sirenidae.

On the other hand, the embryological specialization leading to the derived spinal nerve patterns may be considered simple. Since the spinal nerves form from anterior to posterior, a slight alteration in the spacing between two nerves at one point in development would cause all the more posterior nerves to change their position with respect to the vertebrate. If this alteration occurred late in development, only the more posterior nerves would be type III (exiting through foramina in the posterior halves of the vertebrae), while an alteration earlier in development would result in more type III nerves. This hypothesis suggests that the stimulus leading to incorporation of the nerves occurs very early in development in the families Salamandridae and Sirenidae, so that all but the first two nerves are type III, and that such a stimulus has evolved independently in each family.

The only difference between Figures 6a and 6b is the placement of the family Sirenidae. Figure 6a indicates that sirenids are the earliest originated or most primitive of all salamander families. This is due to their method of fertilization and the total lack of any cloacal glands. By this scheme the sirenids must have evolved the most derived of spinal nerve patterns (6 III''') in parallel with the Salamandridae. In addition, numerous low weight characters (1-, 3-, 4 larv, 5H, 7(3,1), 9b, 10-, 11-, 12±, 14+ 15($\frac{-}{+}$)) must have been evolved in parallel with noncryptobranchoid salamanders; however, all of these, with the

TABLE 4

Weighting of Derived Character States

| Character State | Type of Character | Weighting Group |
|---|---|---|
| 1. Angular | loss | I |
| 2. Internal fertilization | unique complex | IV |
| 3. Septomaxilla | loss | I |
| 4. Vomerine tooth replacement | unique (adult), or neoteny (larval) | V III |
| 5. Chromosome number | reduction | II |
| 6. Spinal nerve pattern | unique | V |
| 7. Number of larval gill slits | reduction | II |
| 8. Second epibranchial | loss | I |
| 9. Columella | reduction, fusion | II |
| 10. Lacrimal | loss | I |
| 11. Ypsiloid apparatus | loss | I |
| 12. Maxilla | loss | I |
| 13. Nasolabial groove | unique | V |
| 14. Perennibranchiate condition | larval | III |
| 15. Separate opisthotic | larval | III |
| 16. Fourth branchial arch | loss | I |
| 17. Columellar process | ? | ? |
| 18. Jugularis nerve above columellar ligament | ? | ? |

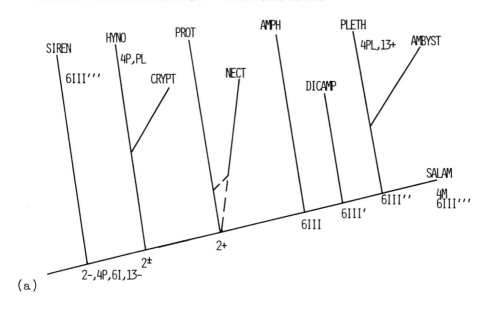

(a)

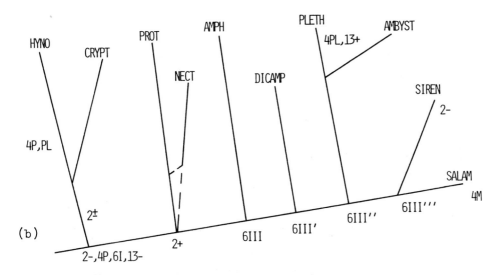

(b)

Figure 6.  The two most parsimonious cladograms using only the
four high weight characters belonging to weighting groups IV and
V.  Each cladogram requires two parallelisms or reversals.  (a)
Parallel evolution of character states 4PL and 6III'''.  (b)  Parallel evolution of character state 4PL and reversal in the Sirenidae to character state 2-.

exception of 1- and 5H, can be ascribed to paedomorphosis (see
below). Loss of the angular (1-) and loss of microchromosomes
with retention of a high haploid number (5H) are both loss characters which belong to the lowest weighting category (group I). The
karyotype of sirenids appears relatively primitive in retaining
high chromosome numbers and in having a mixture of acrocentric,
metacentric and telocentric chromosomes (a condition that
Morescalchi, 1975, calls an asymmetrical karyotype).

Figure 6b suggests that sirenids are among the most derived
of salamanders. This scheme requires that sirenids have lost internal fertilization and all three pairs of cloacal glands. All
of the parallelisms in low weight characters involved in paedogenesis (3-, 4 larv, 7(3,1), 9b, 10-, 11-, 12±, 14+, $15_{(\frac{-}{+})}$) are
also required. In addition, since sirenids have a high chromosome
number under this scheme either all other salamanders have reduced their chromosome numbers in parallel or sirenids have
secondarily increased their chromosome number. Morescalchi and
Olmo (1974) suggest that the sirenids are actually tetraploid, a
hypothesis which agrees well with their placement on Figure 6b.
However, as noted above, the sirenid karyotype is primitive in its
retention of asymmetry. The karyotypes of the amphiumids, dicamp-
todontids, plethodontids, ambystomatids and salamandrids are all
symmetrical (possessing only metacentric chromosomes). Therefore
if the placement of sirenids in Figure 6b is correct, then either
the symmetrical karyotypes have been evolved in parallel four
separate times or the sirenids have re-acquired an asymmetrical
karyotype. In view of these difficulties, it seems to us that
Figure 6a is the more probable phylogeny. However, more characters, especially those relating to fertilization in sirenids, will
have to be examined to test this assertion.

Figure 6a is very similar to the phylogeny of Edwards (1976).
As shown above, this is not the most parsimonious tree, if all 18
characters are used. However, as we have argued repeatedly, it is
not surprising that closely related organisms respond similarly to
similar environments. Sirenids, proteids, necturids and amphiumids
have all evolved many characteristics in parallel (compare Fig. 6
and Table I), including loss of septomaxilla (3-), retention of
larval pattern of vomerine tooth replacement (4 larv), reduction
of number of gill slits (7(3, 2 or 1)), loss of columella (9b),
loss of lacrimal (10-), and loss of ypsiloid apparatus (11-). In
addition, sirenids, necturids, and proteids have lost the maxilla
(12-) and retain external gills throughout life (14+). All of
these character states can be tied to the process of paedomorphosis
in animals which remain aquatic all their lives. It is only by
using character weighting that these characters, which Edwards
(1976) calls the "paedomorphic character suite," can be sorted
from the more important characters.

Character weighting leads to another important conclusion. Necturus and Proteus have usually been considered closely related and have been placed in a single family, the Proteidae, implying a monophyletic origin for the two genera, although Hecht (1957) suggested that they actually evolved in parallel. In a recent study using the weighting method outlined here, Hecht and Edwards (1976) found no high-weight characters linking the two genera, further questioning their monophyletic origin. They concluded that the common ancestor of Proteus and Necturus was probably not a perennibranchiate salamander, and presented two alternate hypotheses: (1) the ancestor was an unknown metamorphosing terrestrial form which would have been in the same monophyletic family; (2) the metamorphosing ancestors belonged to two different families.

Fossil salamanders also present some problems. Paedogenesis has occurred many times in the Amphibia (e.g., Dvinosaurus and the branchiosaurs among the labyrinthodonts), and several elongate, presumably perennibranchiate fossil salamanders are known (Opisthotriton, Hylaeobatrachus, Palaeoproteus, Prodesmodon, Habrosaurus). Since all salamanders go through a similar sequence of development, it is no surprise that such fossils share many characteristics in common with modern elongate perennibranchs such as Proteus. Herre (1935) felt that Palaeoproteus was a fossil proteid, but later work (Estes et al., 1967) suggested that it is really a paedogenetic salamandrid. In a recent study, Estes (1975) analyzed 14 characters in Opisthotriton kayi and found that eight of them could be attributed to paedomorphism; of the remaining six characters, he felt that one of them in particular (columellar process on squamosal bone) indicated a close relationship with the proteids. However, as noted above, Hecht and Edwards (1976) found this character state in Ambystoma mexicanum, and Larsen (1963) reported small columellar processes in Onychodactylus and Cryptobranchus. It is only by identifying characters of high weight (our weighting group IV and V) that true relationships can be inferred. This has been possible with the fossil paedogene Habrosaurus dilatus, which possesses two high weight characters {lower jaw specializations (Estes, 1964) and probably spinal nerve pattern (Edwards, 1976)} which ally it to the modern sirenids. Until such characters can be identified for other fossil paedogenes, their exact phylogenetic placement should remain in doubt.

This exercise shows that the mere multiplication of poorly analyzed morphoclines will not lead to correct phylogenies. Since parallelism is to be expected in closely related organisms and is impossible to detect in low weight characters, which possess little information content, phylogenies should be constructed with well analyzed, high weight characters, in which parallelism should be easier to detect. However, even among such high weight characters, hidden parallelism may occur, as shown by salamander spinal nerve

patterns and adult vomerine tooth replacement patterns.  Thus the
phylogeneticist should constantly re-evaluate organisms in a search
for high weight characters with which to test previously proposed
phylogenies.  All phylogenies constructed using only low weight
characters or one or two high weight characters should be considered suspect.  We must not allow our desire for some answer to
lead us to accept suspect relationships; rather we should admit
that at present some relationships (such as those of Proteus and
Necturus and of many fossil groups) cannot be reliably determined.
Such an admission of uncertainty should lead to renewed efforts to
find high weight characters or to re-examine low weight characters
to increase their information content.

## EVALUATING PHYLOGENIES

In this section we will examine some previous cases of phylogenetic inference and test the validity of the resulting phylogenies.

## CASE I - THE GEKKOTA

The relationships of the lizards of the family Gekkonidae,
a group representing 82 genera and more than 650 species, have
recently been surveyed by three different students (Underwood,
1954, 1955; Kluge, 1967; Moffat, 1973).  Underwood (1957) demonstrated that the closest related group of the Gekkonidae was the
family Pygopodidae and that the two families represent a natural
group, the infraorder Gekkota.  Underwood (1954) in his original
analysis of the Gekkonidae correctly recognized the major groupings
within the family based primarily on characters involving the eye.
In this work, he developed a hypothesis that the most primitive
group within this complex were the eublepharines and that the
concordance between the morphoclines in eye structure and vertebrae
indicated that procoelous vertebrae were primitive and that amphicoelous vertebrae were derived.  This concept was new since
classically, amphicoely had been considered primitive (Camp, 1923).
In 1955 Underwood reversed his 1954 system of relationships because he believed that amphicoely had to be a primitive state in
lizards.  This conclusion was based on the widespread occurrence
of amphicoely in the earliest fossil lizards or lizard ancestors.
This interpretation was supported by Romer (1956), Robinson (1962),
Hoffstetter (1955, 1962, 1964) and Hoffstetter and Gasc (1969).
Kluge (1967) in a review and analysis of the family arrived at the
conclusion that procoely was primitive in the Gekkonidae.  Kluge
was one of the first to use morphoclines and a simplified weighting
system.  Unfortunately, his weighting was limited primarily to only
two value states.  Furthermore, he was influenced in his interpre-

tation of primitive states by extrinsic characters such as temporal
distribution, strict application of the Rule of Parsimony (and
therefore simplified morphoclines) and Darlington's concept of
zoogeography.  Despite these handicaps, he correctly concluded the
primitive nature of the character states generally associated with
the Eublepharinae.  Moffat (1972, 1973) redescribed and reanalyzed
the relationships of the families and subfamilies of the Gekkota
by comparisons within and outside the group.  In making these
comparisons, she examined several morphoclines (Holder, 1960;
Moffat, 1972, 1973) and reexamined the evidence presented by
Underwood (1954, 1955, 1957) and Kluge (1967).  Her major contribution has been the study of the development and phylogeny of
vertebral form and articulation and the reevaluation of the polarity
of this particular morphocline.  As a result of her study, Moffat
(1973) rejected the theory of relationships originally proposed
by Underwood (1954) and Kluge (1967).  She further reexamined and
reanalyzed Kluge's basic contentions of phylogenetic inference.

To illustrate the problems of phylogenetic inference in the
Gekkota, Hecht (1976) had abstracted from Moffat (1973) most of
the pertinent characters used by her and a large portion of those
used by Kluge (1967) (Table V).  The simplification should not
lead to falsification.  It is important to realize that many shared
and derived characters mark the Gekkota and only one has been
listed.  The remaining characters illustrate the problems of trends
such as reduction, fusion, loss, and reorganization.  These characters are only a sample of the characters used by Underwood (1954),
Kluge (1967) and Moffat (1973).

To facilitate the reanalysis and the discussion of the phylogenetic reconstruction of the Gekkota, Hecht (1976) presented six
diagrams (Hecht, 1976, Figs. 1a-1f) which include the conclusions
of Moffat, Kluge and the earlier theory of Underwood.  These diagrams represented a reasonable interpretation of the data available
assuming different axioms.  The distributions of the taxa are provided as subsidiary data to the character analysis.  Since Underwood and Kluge were both influenced by zoogeographic patterns, a
summary of the distributions of the groups is abstracted from
Kluge (1967).  The Pygopodidae has an Australasian distribution.
The Eublepharinae has a disjunct distribution and is known from
southern North America and Central America, West Africa, Ethiopia,
India and Iran, Indonesia and Taiwan.  The Sphaerodactylinae is
known from Antilles, Central and South America.  The Gekkoninae is
circumglobal in both subtropics and tropics.  The Diplodactylinae
is found in Australia, New Zealand, New Caledonia and Loyalty
islands.

There is no question that the Gekkota represent a natural
group of lizards that includes at least two families, Pygopodidae

TABLE 5

Distribution of Character States in the Gekkota Modified from Moffat (1973)

| | P | E | G | D | S |
|---|---|---|---|---|---|
| 1. Eyelids present +; spectacles present − | − | + | − | − | − |
| 2. Four limbs present +; four limbs absent − | − | + | + | + | + |
| 3. Rectus superficialis muscle present +; rectus superficialis muscle absent − | + | − | − | − | − |
| 4. Sacral pleuropophyseal processes present +; sacral pleuropophyseal processes absent − | − | + | + | − | + |
| 5. Postcloacal bones present +; postcloacal bones absent − | + | + | + | + | − |
| 6. Circa 14 scleral ossicles +; reduced number − | + | + | + | + | − |
| 7. Paired premaxillae in adults +; fused premaxillae in adults − | − | + | − | + | − |
| 8. Paired frontals in adults +; fused frontals in adults − | + | − | + | − | − |
| 9. Paired parietals in adults +; fused parietals in adults − | + | − | + | + | + |
| 10. Supratemporal bone present +; supratemporal bone absent − | − | + | − | − | − |
| 11. Splenial present +; splenial absent − | + | + | + | + | − |
| 12. Angular present +; angular absent − | − | + | + | − | − |
| 13. Continuous 2nd visceral arch +; incomplete 2nd visceral arch − | − | + | − | + | − |
| 14. Amphicoelous articulation −; procoelous articulation + | + | + | − | − | −,+ |
| 15. Extreme development of cochlear limbus + | + | + | + | + | + |
| | | | | | |
| Distribution continuous +; distribution disjunct − | + | − | + | + | + |

P − Pygopodidae; E − Eublepharinae; G − Gekkoninae; D − Diplodactylinae; S − Sphaerodactylinae

and Gekkonidae.  Four subfamilies are here included in the family
Gekkonidae:  Eublepharinae, Gekkoninae, Diplodactylinae and
Sphaerodactylinae.  Some of these subfamilies, such as the
Eublepharinae, are recognized by some as families.  No one doubts
that these represent four natural groups, but they are not equi-
distantly related to each other.

Several shared and derived characters mark this group of
lizards but the listing in Table V includes only one such character, number 15.  The alternative to this character state, which is
its absence, is found in all other lizards and lepidosaurians.
The positive statements in characters 1,2,3,6,7,8,9,10,11,12, and
13 are primitive for all lizards and probably lepidosaurians in
general.  The presence of character 5 occurs in one other lizard
family, the Xantusiidae which is regarded by some students as
primitive and also possibly closely related to the Gekkota.  The
presence of character 4 is probably primitive for the Gekkota and
is totally absent in the Pygopodidae because of limb loss.

Character 14 represents the most interesting problem in
morphocline analysis among the entire set.  Underwood (1955) and
Moffat (1973) were heavily influenced by the fossil record
(Robinson, 1962) and the analysis of the Jurassic lizards of the
families Bavariasauridae and Ardeosauridae (Hoffstetter, 1964;
Hoffstetter and Gasc, 1969).  The former was considered a gekkotan
with amphicoelous vertebrae and the latter a close relative with
procoelous vertebrae.  Most of the earlier fossil lizard-like forms
(such as Kuhneosaurus) as well as Sphenodon and other Rhynchocephalia are amphicoelous and Hoffstetter and Gasc (1969) believed
that the condition was primitive for the group.  It should be
pointed out that the gekkota-like fossils are placed within and
near this group on general similarity rather than by any distinctive shared and derived characters.  Moffat (1972, 1973; Holder,
1960) reanalyzed the formation of amphicoely and procoely in
Sphenodon, fossil lizards, and frogs and concluded that the
procoely in the Gekkota was unique and non-synovial, a condition
found in only one species of another relatively primitive family,
the Xantusiidae (which also has cloacal bones and spectacles).
Furthermore, she demonstrated the great similarity in the development of amphicoely in the amphicoelous Gekkonidae and Sphenodon.
In determining the primitive states of the vertebral centrum
morphocline, Moffat (1973) points out that in the Sphaerodactylinae
there are several states which run from amphicoely to partial
procoely.  Her conclusions are that in the Sphaerodactylinae
procoely is derived.  A single gekkonine species of the genus
Pristurus has a similar condition.  Her conclusion simply establishes a polarity which has a primitive state of amphicoely and
develops into two or three parallel, derived states of procoely.
In her two possible phylogenies (Hecht, 1976, fig. 1a and 1b) there

remain three to six other parallelisms among the characters studied. In the analysis of the vertebral centra morphocline, there remains another alternative interpretation, basically accepted by Kluge (1967) and Underwood (1954), which is that procoely is the primitive state (Hecht, 1976, fig. 1c, 1e, 1f). It should be pointed out that Moffat (1973) recognized three but has actually described four character states for vertebral centra: true amphicoely (Gekkoninae, Diplodactylinae), partial amphicoely (Sphaerodactylinae), non-synovial procoely (Eublepharidae and Pygopodidae), and synovial procoely (among all other modern lizards). In this analysis Hecht (1976, fig. 1c, 1e, 1f) merely inverted the polarity and therefore the Gekkota have a primitive form of procoely. If it is accepted that amphicoely is probably primitive for the Lepidosauria and that advanced lizards developed synovial procoely, either polyphyletically or monophyletically, then the condition within the Gekkota is a primitive form of procoely. If this model of polarity is accepted, then there will be greater concordance in the morphoclines of the eyes, eyelids, and vertebral column. This is illustrated by Hecht (1976, fig. 1c, 1e or 1f) with amphicoely as a reversed state. Sphaerodacty-lines represent an intermediate state in the reversal. Further-more, this postulate has a possible functional explanation for the reversed morphocline. All the amphicoelous lizards are the most modified in all other characters and have the most widespread and least relictual distribution. In addition, they are the most derived types within the group in eye structure and with the most modified and varied climbing adaptations. The redevelopment of the amphicoely is, in a biomechanical sense, the redevelopment of a ball-bearing universal type of joint in the vertebral column which may be an adaptation for the secretive climbing habit.

An examination of the six phylogenies indicates that only those schemes which accept the primitive state of the eublepharine eye and eyelids, reduce the number of parallelisms of higher weight characters. In Hecht (1976, fig. 1d), the eublepharid eye and eyelids are considered derived, and are simply a reversal of all polarities in Hecht (1976, fig. 1c), yet it presents a low number of parallelisms. This scheme of reverse polarities defies logic because it requires the redevelopment of the primitive eyelids and the reorganization of the eye. It is highly improbable that such a reversal, although apparently parsimonious, represents reality. Accepting the concordance of the polarity and distribution of the morphoclines of both eye and vertebral column, there can only be one conclusion that the phylogenies based on amphicoely as a primitive state, are not probable. There are three more probable phylogenies as represented by Hecht (1976, fig. 1c, 1e, 1f). Differences between these hypotheses are small and their fundamental axioms are similar.

The above two exercises demonstrate several important points. Firstly, the most important step in phylogenetic inference is the analysis of the morphocline and the recognition of the proper character states. Secondly, parsimony can only be applied after due consideration has been given to a weighting system. Thirdly, taxon by taxon analysis cannot be made until the homologies are clear and the primitive states recognized in the morphoclines. Fourthly, chorological data (distributed in space and time) cannot be used as primary data in determining the primitive or derived taxa in a clade. It is this error which caused Underwood (1955) to reverse his original analysis of evolutionary trends within the Gekkonidae. Amphicoely is still undoubtedly primitive for Lepido-sauria. It is the interpretation of the morphocline which has been slightly erroneous because Moffat (1973) did not recognize that she had described another character state, a primitive form of procoely. Furthermore, Moffat (1972, 1973) demonstrates how close parallel structures can be, since similar morphogenetic patterns will produce similar morphologies.

## HOMINID PHYLOGENY

Eldredge and Tattersall (1975), in an article entitled "Evolu-tionary Models, Phylogenetic Reconstruction, and Another Look at Hominid Phylogeny," have attempted to analyze the taxa generally associated with hominid evolution by using cladistic analysis. They advocated a methodology of phylogenetic reconstruction that is very similar to our own. They recognized both the necessity of comparing forms morphologically without prior judgment as to their biostratigraphy and also the necessity of determining the extent of the morphocline in order to recognize parallelism. The only major differences between their method and our own is that Eldredge and Tattersall do not use character weighting and they insist that all speciation occurs by the method of punctuated equilibria (Eldredge and Gould, 1972), in which evolutionary change occurs only by means of isolated populations during the actual process of speciation. Essentially, Eldredge and Gould (1972) and Eldredge and Tattersall (1975) have been beguiled by one of the models of Mayr (1963), in which the major genetic changes take place only at speciation by the process Mayr calls the "genetic revolution." Recent studies on proteins by such workers as Ayala (1975, 1976) have clearly demonstrated that no particular genetic change is associated with the speciational event. The criticism by Hecht (1974) of punctuated equilibria clearly had this in mind and demonstrated phyletic evolution by the example of morphological geographic lines and their counterparts on the time axis. Behind the interpretation of Hecht (1974) was the observation that direc-tional or orthoselection was one of the primary factors in the changes of gene frequency in response to overriding physical aspects

of the environment.  He furthermore pointed out that phyletic
evolution (in the form of changes within and at the species level)
would be important in the invasion of new adaptive zones such as
the invasion of the aquatic or aerial milieu.  Organisms invading
such a new medium or adaptive zone (such as bipedal terrestrialism)
must respond to the constraints of biomechanics required by this
new environment.  All such organisms may demonstrate parallelisms,
particularly if they are closely related.  Furthermore, phylogenetic
inference using morphoclinal or taxon analysis to determine relationships should not be influenced a priori by the evolutionary
mechanism.

King and Wilson (1975) have demonstrated that the chimpanzee
and human are more than 99 per cent identical in amino acid
sequence within the 44 loci compared.  The differences between the
different forms of Homo and Australopithecus must have been smaller.
The morphological differences between Homo and Pan may be great
but they are mainly the result of allometry, paedogenesis, and
differential morphogenesis.  In these processes, maximal morphological differences can be attained by means of minimum genetic
differences.  Pilbeam and Gould (1975) indicate this utilizing
scaling techniques of comparisons.  Forms with similar genotypes
and similar developmental patterns are most likely to develop
parallelisms especially when their evolution is constrained by
directional selection in the form of the committment to bipedalism
(McHenry, 1975).  At the earliest stages of the hominid evolution
the committment to bipedalism dominated morphological evolution
because of the biomechanical restrictions of possible adaptations.

Secondly, Eldredge and Tattersall (1975) made the assumption
that their analysis was based on evolving units, if not actual
biological species.  (In fact, they treated Homo erectus and Homo
sapiens as sister groups because their prior assumption is based
on a punctuated equilibrium model where there can be no phyletic
evolution.)  However, the level of discrimination in most human
fossils makes it difficult to recognize the morphological limits
of species, let alone the parameters of biological species.  Therefore, in order to organize the data for a cladistic analysis, one
must have a realistic set of taxa, which in this case is the fossil
species.  This is a most optimistic goal.

Furthermore, Eldredge and Tattersall (1975), following the
method of the cladistic systematic approach, searched for shared
and derived characters in a taxon by taxon analysis.  In a proper
analysis, one must first determine the primitive state in a morphocline.  But our present state of knowledge makes this difficult
because of the fragmented materials of the taxa being compared.
If we accept the morphocline as presented, then Eldredge and
Tattersall conclude that Australopithecus africanus is a morphotype

for an hominid ancestor.  Although their methodology denies the existence of <u>actual</u> ancestors in the fossil record, Eldredge and Tattersall also deny that <u>A</u>. <u>africanus</u> is the ancestral group because the chronocline data are in conflict with this interpretation. However, if phyletic evolution takes place, then morphological changes can appear in a lineage without speciation, while in their model a dichotomy must result.

Lastly, the level of resolution allowed by the hominid material as well as the methodology does not allow us to interpret morphoclines and chronoclines with the precision needed to recognize parallelism at the species level.

## CONCLUSIONS

The determination of relationships among organisms is one of the fundamental problems in comparative biology.  Such relationships can be elucidated by phylogenetic reconstruction using the methodology outlined above which presents a set of guidelines.  Our method presents a framework for the whole process of phylogenetic reconstruction from the identification of a character through erection of a cladogram, thereby reducing the ambiguity of the process.  We are in agreement with those who believe that phylogenetic reconstruction should not resemble an art, but instead a scientific procedure.

The basic problem in phylogenetic reconstruction has been the recognition of homology and parallelism and the use of parsimony. Homology can be considered as the monophyly of characters and is the basic concept behind the morphocline and its polarity.  The analysis of several morphoclines allows us to recognize parallelism which is the major cause of error in the determination of the relationships of taxa.  It should be recognized that parallelism is more common than most previous workers have thought.  The use of poorly interpreted or unweighted characters and the universal indisciminant application of parsimony results in the failure to recognize parallelism.  However, by use of our proposed character weighting system, it is possible to determine the level of confidence in a given phylogeny.  The use of high weight characters in phylogeny also results in a more falsifiable scheme.  All phylogenies should be continually tested by the analysis of new data such as newly discovered fossils and living organisms and experimental data such as biochemistry, karyology, and morphogenesis.  In addition, old characters should be reanalyzed to evaluate their information content.

It is obvious that phylogenetic or evolutionary systematics must be based on correct phylogeny.  We believe that the above method will aid in the attainment of this goal.

REFERENCES

Avise, J.C. and Ayala, F.J., 1975, Genetic change and rates of
    cladogenesis.  Genetics 81:757-773.
Ayala, F.J., 1975, Genetic differentiation during the speciation
    process.  Evol. Biol., 8:1-78.
Bannikov, A.G., 1958, Die Biologie des Froschzahnmolches Ranodon
    sibiricus Kessler.  Zool. Jahrb., Abt. f. Syst., 83:245.
Bock, W.J., 1965, The role of adaptive mechanisms in the origin
    of higher levels of organization.  Syst. Zool., 14:272-287.
Bock, W.J., 1973, Philosophical foundations of classical evolu-
    tionary classification.  Syst. Zool., 22:375.
Bonde, N., 1975, Origin of "higher groups": viewpoints of
    phylogenetic systematics, Prob. Actuels Paleo., Evol. Vert.,
    Colloques Inter. C.N.R.S., 218:293.
Camin, J.H. and Sokal, R.R., 1965, A method for deducing branching
    sequences in phylogeny.  Evol., 19:311.
Camp, C.L., 1923, Classification of the lizards.  Bull. Amer. Mus.
    Nat. Hist., 48:289.
Cracraft, J., 1974, Phylogenetic models and classification.  Syst.
    Zool., 23:71.
Dunn, E.R., 1941, The "opercularis" muscle of salamanders.  J.
    Morph., 69:207.
Edwards, J.L., 1976, Spinal nerves and their bearing on salamander
    phylogeny.  J. Morph., 148:305.
Eldredge, N. and Gould, S.J., 1972, Punctuated equilibria:  An
    alternative to phyletic gradualism, in "Models in Paleobiology"
    (T.J.M. Schopf, ed.), pp. 82-115, Freeman, Cooper, San
    Francisco.
Eldredge, N. and Tattersall, I., 1975, Evolutionary models, phylo-
    genetic reconstruction, and another look at hominid phylogeny.
    Contrib. Primat., 5:218-242.
Estes, R., 1964, Fossil vertebrates from the Late Cretaceous Lance
    Formation, eastern Wyoming, Univ. Calif. Publ. Geol. Scis.,
    49:1.
Estes, R., 1975, Lower Vertebrates from the Fort Union Formation,
    Late Paleocene, Big Horn Basin, Wyoming.  Herpetologica 31:
    365-385.
Estes, R., Hecht, M. and Hoffstetter, R., 1967, Paleocene amphibians
    from Cernay, France.  Amer. Mus. Novitates., 2295:1.
Fitch, W.M. and Margoliash, E., 1970, The usefulness of amino acid
    and nucleotide sequences in evolutionary studies.  Evol.
    Biol., 4:67-109.
Gingerich, P.D., 1976, Paleontology and phylogeny: patterns of
    evolution at the species level in Early Tertiary mammals.
    Amer. J. Sci., 276:1.
Goin, C.J. and Goin, O.B., 1962, "Introduction to herpetology,"
    W.H. Freeman and Co., San Francisco.
Harper, C.W., Jr., 1976, Phylogenetic inference in paleontology.
    J. Paleo., 50:180.

Hecht, M.K., 1957, A case of parallel evolution in salamanders. Proc. Zool. Soc., Calcutta, Mookerjee Memor., Vol:283.

Hecht, M.K., 1974, Morphological transformation, the fossil record, and the mechanisms of evolution: A debate, Part I. Evol. Biol., 8:295-308.

Hecht, M.K., 1976, Phylogenetic inference and methodology as applied to the vertebrate record. Evol. Biol., 9:335-363.

Hecht, M.K. and Edwards, J.L., 1976, The determination of parallel or monophyletic relationships: the proteid salamanders - a test case. Amer. Natur., 110:653.

Hennig, W., 1966, "Phylogenetic systematics," University of Illinois Press, Urbana.

Herre, W., 1935, Die Schwanzlurche der mitteleocanen (oberlute-tischen) Braunkohle des Geiseltales und die Phylogenie der Urodelen unter Einschluss der fossilen Formen. Zoologica, 33:1.

Heyer, W.R., 1975, A preliminary analysis of the intergeneric relationships of the frog family Leptodactylidae. Smithsonian Contribution to Zoology, 233:1-29.

Hoffstetter, R., 1955, Squamates de type moderne, "Traite de paleontologie, 5. Amphibiens, reptiles et oiseaux," (J. Piveteau, ed.), pp. 606-662, Masson, Paris.

Hoffstetter, R., 1962, Revue des recentes acquisitions concernant l'histoire et la systematique des squamates. Colloques Inter. C. N. R. S., 104:243.

Hoffstetter, R., 1964, Les Sauria du Jurassique superieur et specialment les Gekkota de Baviere et de Mandchourie, Senskenbergiana Biol., 45:281.

Hoffstetter, R. and Gasc, J.P., 1969, Vertebrae and ribs of modern reptiles, in "Biology of the Reptilia, vol. 1" (C. Gans, A. d'A. Bellairs and T.S. Parsons, eds.), pp. 201-310, Academic Press, London.

Holder, L.A., 1960, The comparative morphology of the axial skeleton in the Australian Gekkonidae. J. Linnean Soc. Zool., 44:300.

Inger, R.F., 1967, The development of frogs, Evol., 21:369.

Jardine, N. and Jardine, C.J., 1967, Numerical Homology. Nature 216(5112):301-302.

King, M.C. and Wilson, A.C., 1975, Evolution at two levels in humans and chimpanzees. Science, 188:107.

Kluge, A.G. and Farris, J.S., 1969, Quantitative phyletics and the evolution of anurans. Syst. Zool., 18:1.

Kurten, B., 1963, Return of a lost structure in the evolution of felid dentition. Soc. Scienti. Fenn. Comment. Biol., 26:4.

Larsen, J., Jr., 1963, The cranial osteology of neotenic and transformed salamanders and its bearing on interfamial relationships, unpublished Ph.D. thesis, Univ. of Washington.

Marx, H. and Rabb, G.B., 1970, Character analysis: an empirical approach applied to advanced snakes. J. Zool., London 161:525.

Marx, H. and Rabb, G.B., 1972, Phyletic analysis of fifty characters of advanced snakes, Fieldiana, Zool., 63:1.

Mayr, E., 1963, Animal species and evolution, The Belknap Press, Harvard University Press, Cambridge, Mass., 797 pp.

Mayr, E., 1965, Numerical phenetics and taxonomic theory. Syst. Zool., 14:73.

McHenry, H.M., 1975, Fossils and the mosaic nature of human evolution. Science 190:425.

McKenna, M.C., 1975, Toward a phylogenetic classification of the Mammalia, in "Phylogeny of the primates" (W.P. Luckett and F.S. Szalay, eds.), pp. 21-46, Plenum, New York.

Michener, C.D. and Sokal, R.R., 1957, A quantitative approach to a problem in classification. Evol., 11:130-162.

Moffat, L.A., 1972, The phylogenetic significance of notochorsal centra in amphibians and reptiles: a comparative study of vertebral morphology and development in Leiopelma (Amphibia: Anura) and the Gekkota (Reptilia: Lacertilia), unpublished Ph.D. thesis, Univ. of Sydney, Australia.

Moffat, L.A., 1973, The concept of primitiveness and its bearing on the phylogenetic classification of the Gekkota. Proc. Linnean Soc. New South Wales 97:275.

Monath, T., 1965, The opercular apparatus of salamanders. J. Morph., 116:149.

Morescalchi, A., 1975, Chromosome evolution in the caudate Amphibia. Evol. Biol., 8:339.

Morescalchi, A. and Olmo, E., 1974, Sirenids: a family of polyploid urodeles? Experientia 30:491.

Nelson, G.J., 1971, Paraphyly and polyphyly: redefinitions. Syst. Zool., 20:471.

Noble, G.K., 1925, An outline of the relation of ontogeny to phylogeny within the Amphibia. II, Amer. Mus. Novitates 166:1.

Noble, G.K., 1931, "The biology of the Amphibia," reprint by Dover Publications, New York, 1954.

Orton, G.L., 1957, The bearing of larval evolution on some problems in frog classification. Syst. Zool., 6:79.

Pilbeam, D. and Gould, S.J., 1974, Size and scaling in human evolution. Science 186:892.

Regal, P.J., 1966, Feeding specializations and the classification of terrestrial salamanders. Evol., 20:392.

Robinson, P., 1962, Gliding lizards from the Upper Keuper of Great Britain, Proc. Geol. Soc., 1601:137.

Rogers, D.J., Fleming, H.S. and Estabrook, G., 1967, Use of computers in studies of taxonomy and evolution. Evol. Biol., 1:169.

Romer, A.S., 1956, "Osteology of the reptiles," Univ. of Chicago Press.

Schaeffer, B. and Hecht, M.K., 1975, Introduction and historical background, symposium on higher levels of organization. Syst. Zool., 14:245.

Schaeffer, B., Hecht, M.K. and Eldredge, N., 1972, Phylogeny and paleontology. Evol. Biol., 6:31.

Simpson, G.G., 1961, "Principles of animal taxonomy," Colombia
     Univ. Press, New York.
Simpson, G.G., 1975, Recent advances in methods of phylogenetic
     inference, in "Phylogeny of the primates" (W.P. Luckett and
     F.S. Szalay, eds.), pp. 3-19, Plenum, New York.
Sneath, P.H.A. and Sokal, R.R., 1973, "Numerical taxonomy,"
     W.H. Freeman and Co., San Francisco.
Sokol, O., 1975, The phylogeny of anuran larvae:  a new look.
     Copeia 1975:1.
Starrett, P.H., 1975, Evolutionary patterns in larval morphology,
     in "Evolutionary biology of the anurans" (J.L. Vial, ed.),
     pp. 251-271, Univ. of Missouri Press, Columbia.
Throckmorton, L.H., 1968, Concordance and discordance of taxonomic
     characters in Drosophila classification.  Syst. Zool., 17:355.
Underwood, G., 1954, On the classification and evolution of
     geckos.  Proc. Zool. Soc. London 1954:46.
Underwood, G., 1955, Classification of geckos.  Nature 175:1089.
Underwood, G., 1957, On lizards of the family Pygopodidae.  A
     contribution to the morphology and phylogeny of the Squamata.
     J. Morph., 100:207.
Wake, D.B., 1966, Comparative osteology and evolution of the
     lungless salamanders, family Plethodontidae.  Mem. South
     Calif. Acad. Sci., 4:1.
Wilson, E.O., 1965, A consistency test for phylogenies based on
     contemporaneous species.  Syst. Zool., 14:214.

# COMMENTS ON PHYLETIC ANALYSIS OF GEKKOTAN LIZARDS

Garth UNDERWOOD

Dept. of Biological Sciences, City of London Polytechnic

London E17NT, England

I arrange Hecht's and Edward's character transformations according to their weighting categories:

1.  Eyelids +, Spectacle -    ... ... ... ... ...    IV

7.  Premaxillae: paired +, fused - ... ... ... ...    III

8.  Frontals: paired +, fused -    ... ... ... ...    III

9.  Parietals: paired +, fused -    ... ... ... ...    III

6.  Scleral ossicles: 14 +, >14 -  ... ... ... ...    II

13.  Visceral arch II: complete +, incomplete - ... ...    II

    2, 3, 4, 5, 10, 11, 12    ... ... ... ... ...    I

I leave the vertebral character (14) in abeyance as that is in question. I add:

16.  Type C double visual cells (Underwood, 1970):
Absent +, present - (P+, EDGS-) ... ... ... ...    V

17.  Egg shell: soft when laid hardens on exposure +,
hard when laid - (P?, ED+, GS-) ... ... ... ...    IV

Type C double visual cells are unique to geckos which are also the only vertebrates in which any second type of double cells are found (Underwood, 1970). Other Squamata (lizards and snakes) lay a soft shelled egg (save _Dibamus_; Smith, 1935).

Character 1

E $\frac{1}{-}$ PGDS $\frac{1}{-}$

Stem    +  /   =

Character 7

E $\frac{1\ 7}{+\ +}$ D $\frac{1\ 7}{-\ +}$ PGS $\frac{1\ 7}{-\ -}$

Stem    + +    = +  /  - =

Character 8

E $\frac{1\ 7\ 8}{+\ +\ =}$ D $\frac{1\ 7\ 8}{-\ +\ =}$ PG $\frac{1\ 7\ 8}{-\ -\ +}$ S $\frac{1\ 7\ 8}{-\ -\ -}$

Stem    + + +    = + +     - = + /  - - =

Character 9

E $\frac{1\ 7\ 8\ 9}{+\ +\ =\ +}$ D $\frac{1\ 7\ 8\ 9}{-\ +\ =\ =}$ P $\frac{1\ 7\ 8\ 9}{-\ -\ +\ +}$ G $\frac{1\ 7\ 8\ 9}{-\ -\ +\ -}$ S $\frac{1\ 7\ 8\ 9}{-\ -\ -\ -}$

Stem    + + + +    = + + +     - = + + /  - - + =    - - - =

Figure 1

Character 16

P $\frac{16}{+}$ EDGS $\frac{16}{-}$

Stem    +   /    =

Character 17

P $\frac{16\ 17}{+\ ?}$ ED $\frac{16\ 17}{=\ +}$ GS $\frac{16\ 17}{-\ -}$

Stem    +  ?       = + /  - =

Character 1

P $\frac{16\ 17\ 1}{+\ ?\ =}$ E $\frac{16\ 17\ 1}{-\ +\ +}$ D $\frac{16\ 17\ 1}{-\ +\ -}$ GS $\frac{16\ 17\ 1}{-\ -\ -}$

Stem    + ? +    = + + /  - + =     - = -

Character 8

P $\frac{16\ 17\ 1\ 8}{+\ ?\ =\ +}$ E $\frac{16\ 17\ 1\ 8}{-\ +\ +\ =}$ D $\frac{16\ 17\ 1\ 8}{-\ +\ -\ =}$ G $\frac{16\ 17\ 1\ 8}{-\ -\ -\ -}$ S $\frac{16\ 17\ 1\ 8}{-\ -\ -\ -}$

Stem  + ? + +    = + + +    - + = +    - = - + /  - - - =

Figure 2

Phyletic analyses of Gekkotan lizards.  A slash (/) indicates a
transformation in the stem stock which separates some of the derived groups.  A double line (=) indicates a derived state, immediately following a character transformation, in the stem stock or
one of the derived groups.     P, Pygopodidae; E, Eublepharinae;
G, Gekkoninae; D, Diplodactylinae; S, Sphaerodactylinae.

Using the first four characters on the weighted list (1, 7, 8 and 9) in that order we successively separate E, D, S, P and G; thus obtaining complete resolution (fig. 1). Four character transformations are required for the resolution but two additional transformations of character 8 are implied and one of character 9 i.e. a total of seven transformations. If procoely be primitive then we have three transitions to amphicoely: in D, G and some S. If amphicoely be primitive we have three transitions to procoely: in E, P and some S. There is no obvious basis on which to prefer one to the other.

If now we take characters 16, 17, 1 and 8 in that order we obtain complete resolution (fig. 2) with, again, a total of seven character transformations. This corresponds with Hecht (1976, fig. 1E) ("Kluge's hypothesis"). Now if procoely be primitive we have two transformations (to amphicoely between E & D, back to procoely within S). Alternatively if amphicoely be primitive we have three transformations to procoely (in P, E and within S). Assuming procoely to be primitive is more parsimonious. It may also be noted that the amphicoelous members of the Sphaerodactylinae (Gonatodes) show a more primitive condition of the digits than do the other members of the group; digits are, however, so varied within groups that they cannot be used at the subfamily level of analysis.

REFERENCES

Hecht, M.K., 1976, Phylogenetic inference and methodology as applied to the vertebrate record. Evol. Biol., 9: 335-363.
Smith, M.A., 1935, Sauria in fauna of British India. Taylor & Francis, London.
Underwood, G.L., 1970, The eye in Biology of the reptilia. Eds. C. Gans & T. Parsons, Academic Press, London.

# ADAPTATION AND THE COMPARATIVE METHOD

Walter J. BOCK

Dept. of Biological Sciences, Columbia University

New York, New York  10027 U.S.A.

## INTRODUCTION

Central of the interests of evolutionary biologists is the concept of adaptation.  Indeed the task that Darwin had before him in writing the "Origin of Species" was to provide a scientific explanation for the origin and modification of adaptations; the concept of biological adaptation was appreciated by humans long before the beginnings of scientific inquiry.  Thus, as evolutionary biologists, we are interested in the adaptive nature of features and in the mechanisms of adaptive evolutionary change. Before it is possible to inquire into the mode of adaptive evolution, or into the possible nonadaptive evolution of certain features, or the adaptive evolutionary history of new groups of organisms, it is necessary to determine the exact adaptive nature or significance of particular features.  Thus we come at once to the questions:  By what methods of analysis can a given feature be judged to be an adaptation?  And do any of these methods possess limitations or even be valid?

Two main methods have been used by biologists to ascertain the adaptiveness of biological features.  The first is by direct analysis in which the properties of form and of function of the feature are studied first followed by observation of its biological roles.  The latter can be done only with the study of the organism living in its natural environment.  Comparison of the biological role with the selection forces arising from the environmental factors acting on the organism provides the basis for ascertaining the adaptive nature of the feature.  This method of direct analysis, together with a means of judging the degree of goodness of the adaptation, has been discussed by Bock and von Wahlert (1965) and will not be considered further in this paper. The second method is by comparison of determination by analogy. The comparative method, which enjoys extremely widespread use in biology and associated sciences, has never been properly scrutinized; it is the subject of this paper. Before the comparative or analogous method of determining adaptations can be analyzed, it is necessary to examine two other topics, namely (a) the comparative method in biology, and (b) the meaning of biological adaptation.

## THE COMPARATIVE METHOD FOR JUDGING ADAPTATIONS

The basic question to be examined in this paper is whether the adaptiveness of a feature in one organism can be ascertained with any degree of certainty by comparison with known adaptations in other organisms. Most biologists would agree, and indeed this approach is a widespread method used in all areas of biology. My thesis is that this method is not permissible, rather that the nature of the adaptiveness of features in every organism must be ascertained independently of known adaptations in other organisms. Any comparisons must be made only _after_ the adaptiveness of features have been determined.

The comparative method for judging adaptations can be stated, as follows, in the form of a general principle. The adaptive nature of a particular feature in one taxon can be determined by comparison of the morphological form of this feature with the morphological form of features of known adaptiveness in other taxa. Agreement in morphological form means that the feature being compared has the same adaptive significance as the feature used as the standard for comparison. Disagreement in morphological form means that the feature being compared does not have the same adaptive significance as the feature being used as the standard of comparison. (For simplicity, I will discuss this principle only in terms of the morphological form of features. It can be put in a general form by considering all properties of form and of function of the features which can be done easily because the properties of function are defined in terms of the properties of form, see below.) The comparative method depends upon the additional principle that given a particular environmental factor, or a selection force (the two are not the same, but both must be mentioned because of vagueness in much of the literature), there is a single optimal adaptation to it. Further, all adaptive evolutionary change under the control of this environmental factor will be toward this single optimal adaptation. In order to ascertain adaptations by comparison, these two basic principles must have a law-like quality, that is, they must

describe a definite cause-effect relationship.  In this case,
the particular environmental factor (or selection force) is the
cause and a definite adaptation is the effect.

If the cause-effect relationship between environmental
factor and adaptation is not law-like, then the comparative
method of ascertaining adaptations cannot be a law-like principle.
The nonlaw-like nature of the principle of a single adaptive
optimum can be demonstrated by showing a number of exceptions to
it.  Moreover, it is necessary to demonstrate two types of exceptions to the principle that the same adaptation always evolves in
response to the same selection force.  The first type of exception is that given a particular selection force, then there are
generally a number of morphologically different adaptations -
multiple pathways of evolution - which could evolve in response
to that selection force (Bock, 1959).  The second type is that
morphological convergence need not be adaptive convergence.  In
other words, convergent features in several taxa could have
evolved in response to different selection forces acting on these
several taxa.  Invalidity of the law-like nature of the principle
of a single optimal adaptation for each selection force will invalidate the law-like nature of the principle by ascertaining
adaptations by analogous comparison.

It must be noted that the demonstration that different adaptive answers may evolve in response to the same selection force
or that convergent features may evolve in response to different
selection forces does not mean that this always occurs.  I do not
claim that these cause-effect relations have a law-like nature.
I claim only that they occur and that they are sufficiently common
to invalidate the principle of a single adaptational optimum for
each selection force.

The principle of the single adaptive optimum has been held
since Darwin and is of extremely widespread use in biology,
serving as the foundation of many important ideas.  The diversity
of studies in which this principle has had a central role can be
shown by a few examples.  These were picked from studies that I
believe are especially important and not to indicate that the
cited workers are less aware of the pitfalls than other biologists.

   (a)  Comparative morphology.  The determination of adapta-
        tions by direct analysis is difficult or impossible in
        many recent organisms because of problems in observing
        them under natural conditions and hence the comparative
        method is used.  In his study of the morphology of the
        passerine family of three-climbing woodhewers
        (Dendrocolaptidae), Feduccia (1973) ascertained climb-
        ing adaptations in the hindlimb and tail by comparisons

with the woodpeckers (Picidae), a group of birds highly
specialized for climbing.  If the morphology of features in the two groups was similar, he concluded that
the structure in the woodhewers was an adaptation for
climbing.  If they differed, then the woodhewer feature was not a climbing adaptation.  Difficulties with
this method are illustrated by examination of the toe
arrangement in the two groups, anisodactyl in the
woodhewers and zygodactyl in the woodpeckers, but both
are clearly adapted for climbing on vertical surfaces
(Bock and de W. Miller, 1959).

(b)  Phylogenetic studies.  Explanation of the evolutionary
history of a taxon should include consideration of the
adaptive origins of the major features of that group.
Cartmill (1974), in discussing the origin of the
primates, argues that the basic features of this order
are not adaptations to the selective demands of arboreal life.  His thesis is based upon a comparison of
the morphology of the arboreal adaptations of primates
with that of the arboreal adaptations of a broad
assemblage of arboreal mammals, including opossums,
tree shrews, palm civets, squirrels and so forth.  Because the morphology of the compared features differs
in these two "groups" of mammals, Cartmill concluded
that those in primates are not adaptations to arboreal
life, but adaptations to other selection forces.  Yet,
it would be as reasonable to have concluded that the
features in the squirrels and other mammals were not
adaptations to arboreal life.  Cartmill did not attempt
to determine the adoptive nature of the primate features by direct analysis, but was content with the results of the comparative method in spite of their being
in conflict with the selective demands of their environment on the primates from the time of their origin.

(c)  Paleontology.  The adaptiveness of a feature cannot be
ascertained by direct methods in fossil organisms and
the paleontologist is forced to use comparative methods.
For example, Ostrum (1974, and elsewhere) suggests that
the morphological adaptations of small ground-dwelling
dinosaur carnivors ancestral to Archaeopteryx could be
determined by comparison with the morphology of recent
cursorial predaceous birds such as the secretary bird
(Sagittarius), or the road runner (Geococcyx). The
problem of determining adaptations in fossils is most
important because fossils offer us the best material
for studying the actual course of adaptive evolutionary
change.  But such studies depend upon prior accurate

ssessment of the adaptations.  Unfortunately the comparative method is used without realization of its
inherent shortcomings.

(d)   Ecological comparisons.  Comparative ecologists and
biogeographers have been interested in the composition
of the faunas and floras in different parts of the
earth and have considered such topics as "ecological
equivalents", "eco-morphological configurations", and
"convergent evolution in species and communities."
Keast (1972) and Karr and James (1975) have, for
example, compared the composition of the avifaunas in
the same general habitat in different parts of the
world, e.g., tropical forests in South America,
Africa and Asia.  The basic assumption used was simply
that ecological equivalents - those organisms occupying
the same part of the habitat in different continents -
will possess the same morphological adaptations.  Fly-
catchers will possess a similar flattened bill, finches
a conical bill, shrikes a stout, hooked bill, nectar-
feeders a long thin bill and so forth.  Hence, examination of the morphology of the organisms is sufficient
to ascertain their ecological position.  These studies
may be valid because the features chosen are ones of
the external morphology that are readily modified
under the action of selection, and that have a narrow
range or possible adaptive answers because of their
form-function possibilities.  And, quite possibly, the
authors actually ascertained the adaptations by direct
analysis, albeit a most superficial one, in spite of
presenting their method as the comparative analysis.
Yet if these discussions of ecological equivalents is
correct, the basic idea should apply equally well to
all morphological features - i.e., aspects of the internal morphology.  In some of these avian groups details of the internal morphology (e.g., aspects of the
bone-muscle system) of the feeding apparatus differ
greatly (pers. obs.).  Hence, either the approach of
designating ecological equivalents by comparative
analysis is invalid or a qualifying assumption must be
made that at least two classes of morphological features
exist - those that permit recognition of eco-morphological equivalents and those that do not.  And methods
must exist by which these two classes of features can
be recognized and separated.  This appears rather unreasonable which suggests that the comparative approach
of recognizing adaptations in these ecological studies
is not valid.

Many other examples may be found in laboratory studies of biology in which parts of a general mechanism, such as the metabolic pathways or the mechanism of muscle contraction, are worked out in different organisms and must be integrated by some comparative method.  This method is essentially the same as that used in the determination of adaptation by analogy as it depends upon the principle that a single adaptive optimum exists for each selection force.  Such a method may be necessary as a practical approach, in the initial stages in the clarification of any biological mechanism, but this does not mean that it is valid for more sophisticated comparative studies once the basic mechanism is understood.

## BIOLOGICAL COMPARISON

Comparison is an integral part of many biological studies, especially in the areas of systematics and evolutionary biology. Indeed it is so ubiquitous in our studies, that many workers speak of "The comparative method" in biology with the implication that all biologists know what it is and that all agree on the general (let alone the precise) methodology of comparisons and on the interpretations that are permissible from a particular comparative study.  Moreover, workers in neighboring fields, such as sociobiology and psychology, speak of the value of applying to their studies the method of comparisons taken from biology.  I simply do not know what "The comparative method" is and know of no discussion in the literature which analyzes in any detail the principle and limitations of comparative methods in biology.  A number of questions come to mind immediately.  Does one method exist?  If so, what are the exact procedures of study?  And what kinds of answers can it provide?  Are there restrictions to extrapolations that may be made from one comparison to another?  Are there different types of comparisons? And so forth.  Consideration of these questions suggests at once that comparisons in biology are not as simple as implied by the allusion to "The comparative method."

Comparisons must be linked with some theory and I will consider them with respect to the theory of organic evolution.  Hopefully comparative studies will permit us to explain more fully the consequences of organic evolution, such as the bases for the similarities and differences between features in diverse organisms which arose during the course of their past evolutionary history. Evolutionary change may be understood as the result of two basic mechanisms - that of phyletic evolution and that of speciation. Phyletic evolution is change with respect to time in a single phyletic lineage resulting from the production of genetically based phenotypic variations and the action of natural selection.

Speciation is the splitting of a single phyletic lineage into two
or more lineages by the origin of intrinsic isolating mechanisms
in the new phyletic lineages.  Phyletic evolution may occur without speciation, but speciation requires phyletic evolution in at
least one of the two separate lineages for the origin of the intrinsic isolating mechanisms.  Phyletic evolution is adaptive in
that it requires the action of natural selection and by definition
those evolutionary changes resulting from the action of selection
are considered to be adaptive.  But this statement is only true
when considering the whole organism within an evolving species.
It is not necessarily true when considering individual features of
an organism separate from other features.  Different features may
be interrelated genetically or developmentally, and one feature
may evolve simply as a consequence of the adaptive evolution of
another feature (= pleiotrophy).

     A species and a phyletic lineage are not the same thing
although they are very frequently interchanged.  A species is
composed of potentially or actually interbreeding populations at
one point in time (Mayr, 1963).  It may be though of those individuals which will share a future descendent.  Mayr is careful to
point out that the species concept is nondimensional as it loses
meaning as one moves away from the species at one locality on
earth both in space and in time.  A phyletic lineage is the consequence of a species reproducing itself generation after generation in time; it is the projection of a species though time.
Thus, a species is the cross-section of a phyletic lineage at any
particular point in time.  Such cross-sections of the same phyletic
lineage at different times do not represent the same species.
This particular dilemna is the consequence of the species concept
being nondimensional.  Different species are cross-sections of
different phyletic lineages, and speciation - the multiplication
of species - is the splitting of a phyletic lineage into two or
more separate lineages.  Phyletic evolution within the same
phyletic lineage never leads to a new species regardless of how
extreme the evolutionary change may be.  Continued phyletic
lineages by speciation and termination of some lineages by extinction will lead to adaptive radiations.  The origin of new major
features and new taxa by adaptive evolution - macroevolution - can
be explained by mechanisms of microevolutionary change if care is
taken to place the events into a proper chronological sequence
(see Bock, 1959, 1963, 1965, 1967, 1969, 1970, 1972a, 1972b;
Bock and de W. Miller, 1959; Bock and von Wahlert, 1963, 1965).
The attributes of features seen in diverse species are the result
of the several evolutionary processes acting during their adaptive
radiation.  Explanation of these attributes - what they are and
their history - can be done only with proper comparative studies,
and these must be based on the theory and mechanisms of evolution.

Several types of comparisons exist in biological studies and the interpretations allowable in each type may differ from the others; interpretations may not be simply extrapolated from one type of comparison to another. Consider a series of phyletic lineages radiating from a single ancestral species by a sequence of speciations so that a number of species exist at the same time period. The comparison between different species - cross-sections of different phyletic lineages - is a horizontal comparison. Such comparisons are between different species regardless of whether the species exist at the same or at different times. The comparison between different members of the same phyletic lineage - cross-sections of the same phyletic lineage at different times - is a vertical comparison. Such comparisons are not between different species but between an ancestral and a descendent phyletic cross-section - ancestral and descendent populations. Thus it is clear that not all comparisons in biology are between species. And it is clear that the differences being examined in these two types of comparisons are not the result of the same evolutionary mechanisms. Many differences seen in horizontal comparisons represent paradaptations (Bock, 1967) - different adaptations to the same selection force. Many of the differences seen in vertical comparisons represent adaptations to the selection force controlling phyletic evolution in a particular lineage. I must note, however, that it is incorrect to assume that all differences observed in horizontal comparisons are automatically paradaptations and those seen in vertical comparisons are adaptations. Moreover, paradaptation and adaptation are not mutually exclusive concepts. However it is clear that the conclusions and interpretations reached in a horizontal comparison cannot be applied automatically to a vertical comparison and vice versa. The conclusion that the differences between features in several species are not adaptive, but rather paradaptive, does not permit the conclusion that these features did not evolve by adaptive evolution.

A second division of comparisons, one that does not correspond simply with horizontal and vertical comparisons is that of non-historical versus historical comparisons. The difference does not depend on whether one includes fossil sequences or possible phylogenies, but on the basic assumptions and approaches used in the comparisons. The fundamental assumption used in nonhistorical comparisons is that no genotypic differences exist for the features being compared in the several species under consideration. This is used by biologists when determining the basic mechanism of some general biological phenomenon, such as contraction of striated muscles, in which the attributes in common are more significant than the differences. This approach is used when comparing the particular characteristics exhibited by different species, as for example the action of the muscle-bone system of the feeding apparatus in a group of birds, against those of a particular mechanical model.  In historical comparisons, it is essential to include the assumption that the genotype of individual species differ and to consider the mechanism by which the genotypes modify.  Thus in a nonhistorical comparison only the mechanism of natural selection is considered while in historical comparisons, both natural selection and the mechanisms of the formation of genetical variation are fundamental to the analysis.  Most comparisons in biology, including evolutionary studies are nonhistorical.  Such comparison is an essential step in the understanding of any biological phenomenon and its importance must not be discounted.  However, it must be remembered that a complete explanation in biology is not possible without undertaking a historical comparison.

In any comparative study, one has available the specimens (representing members of a species) and their attributes.  One does not know whether a vertical of a horizontal comparison can be made aside from the obvious case of those between recent species.  The procedure of study is to make the comparison and by the nature of the possible interpretations that may be reached, decide whether one is engaged in a vertical or a horizontal comparison and hence whether one is comparing members of the same or of different phyletic lineages.

On the basis of these comparisons, one can conclude whether the divergence being examined might represent several different multiple pathways of evolution (Bock, 1959 and see below).  Or one might conclude that the similarity in features is convergent.  By convergent, I follow the general meaning of features that are more similar in two species than in ancestral stages of these species.  That is, during the evolution of two different phyletic lineages, the features have become more similar.

## BIOLOGICAL ADAPTATIONS

The ideas of the form-function complex of features and of adaptation advocated by Bock and von Wahlert (1965) will be accepted here.  Adaptation has always been used for the close correlation between a feature of an organism and its environment. This concept long predates ideas of biological evolution and indeed, the attempt to provide a scientific explanation for adaptation had led to the formulation of the concept of biological evolution by natural selection by Darwin.  Adaptations are parts of the organism; hence it is necessary to characterize first the parts of an organism and their attributes.

For purposes of study, the organism must be divided into separate parts which are features.  Decision on the limits of

features is left to the investigator.  It is important only to
remember that pieces of an animal do not work alone and that the
part chosen by the investigator may not comprise the adaptation
by itself.  Features interact functionally and developmentally;
these interactions are usually very complex, involving many different features.  This somatic interaction is critical to
understanding biological adaptation, but must not be mistaken
for adaptation itself or for "internal selection", a concept
mentioned by a number of workers.

The form of a feature are those properties of material composition and arrangement of this material.  Levels of organization must be included because a living organism is a tiered
organizational system.  The form of a feature is not a single
unchanging property, but generally changes during the life of
the individual.  Sometimes the change is slow, such as wear in
teeth, and sometimes it is rapid, such as change in the shape in
a muscle as it contracts and shortens or in the shape of the
lens of the eye.  The function of a feature are those physical-
chemical properties arising from the form.  If the form of a
feature alters, so does its function.  Thus the force developed
by a muscle decreases as it shortens during contraction.  Or the
bending of light rays will alter as the curvature of the lens
changes.  A feature will generally have a number of functions
because of the diversity of physical and chemical properties
associated with a single form.  Thus a bone will have inertia,
hardness, density, strength against compression, tension and
shear, a mineral reserve, and other functions.  A muscle will
have a maximum tension, distance of shortening and lengthening,
speed of shortening, a particular tension-length curve, rigidity,
heat production and other functions.  The existence of many
separate functions of a form is essential to the thesis that
adaptations cannot be ascertained by comparison.

The combination of the form and a function of the feature -
the form-function complex - is designated as the faculty.  This
concept stresses that adaptation of features is not dependent
only upon the form, but upon the form and the function.  And
further, any statements that either the form alone or the
function alone of a feature is an adaptation or that the form is
adapted to the function are in error.

The form and function of features can, and are usually best,
described in the laboratory.  But they do not indicate the
possible adaptiveness of the feature or even if the feature is
adapted for which studies of the ecological relationships are
needed.  The biological role is how the organism uses a particular
faculty in the course of its life in the normal environment of
the organism.  Determination of biological roles depends upon

observations of the organisms under conditions of captivity and often cannot be ascertained with assurance when the organism is living in a disturbed habitat (Lack, 1965). These studies fall under the heading of ecological or biological morphology. Studies of the biological role which show how an animal uses its locomotory apparatus - whether it is used to escape a predator or to catch prey, whether it serves for long distance migration or for high acceleration and short bursts of speed over small distances. A particular form-function complex may have several biological roles, and generally any feature has a number of biological roles involving different functions. Those functions involved in a biological role are utilized ones, those not associated with a biological role are nonutilized ones. Generally a feature has a large number of nonutilized functions. A feature may have no biological roles, as perhaps a vestigial feature. But a feature cannot be functionless because that indicates that the feature has no form and hence the feature does not exist.

The concept of function and of biological role are generally not distinguished from one another. Regardless of the terms applied to them, these concepts must be kept separate. Otherwise, the concept of adaptation and ideas of adaptive evolutionary change become hopelessly confused. Most biologists, e.g. physiologists, biochemists and other laboratory workers, use function in the sense to which it was restricted by Bock and von Wahlert. Care must be exercised in ascertaining the use of the term "function" by any worker, but this can be done reasonably easily in most cases except when the author is using this word in several senses in the same paper.

A biological role of a faculty interacts with a selection force of an environmental factor (part of the umwelt); this interaction is termed a synerg to indicate that the interaction between biological role and selection involved feedback loops. A particular environmental factor does not automatically place a definite selection force on the organism, but the exact nature of selection depends on how the organism reacts to the environment. This distinction between the nature of the environmental factor and the resulting selection force is not made by most evolutionary biologists, hence the difficulty in formulating the principle of determining adaptations by comparisons in the introductory remarks.

A particular faculty may interact with more than one selection force, and since a feature may have a number of faculties with biological roles, any single feature usually interacts with a number of different selection forces.

An adaptation is a feature of an organism that has at least one biological role interacting with a selection force. The

feature, to be an adaptation, must have properties of form and
function that permit the organism to interact successfully with
the environment - that is, to maintain the synerg between the
biological role and the selection force successfully.  By
successful, I mean that the individual organism survives as an
individual and reproduces to leave progeny in the next generation.
Adaptations must be judged on a probability basis and against
past and present conditions, never against future factors.  Success
is a relative term and some measure of success or the relative
degree of goodness of the adaptation is needed.  Bock and von
Wahlert (1965) suggested the use of the amount of energy required
by the organism to maintain the synerg - a sort of efficiency
judgement - as the means of judging the degree of goodness of a
particular adaptation.  Adaptations must be judged always in
terms of a specific stated environment.  If the environment
changes, then the nature of the adaptation and the basis of
judging its degree of goodness changes.

Adaptation is defined in terms of selection and hence in
terms of environmental factors in the umwelt of the organism.
Thus, all notions of adaptive differences, adaptive evolutionary
changes and so forth must be considered in terms of selection.
It is not sufficient to speak of selection in a general way,
rather one must specify particular selection forces as exactly
as possible.  Selection and environment are always external to
the organism.  Concepts as the internal environment and internal
selection, and features of the organisms which are adaptations to
these internal selection forces of the internal environment,
simply cause unnecessary confusion and misconception in our
understanding of adaptation and of evolutionary change.  These
concepts of internal selection are best understood under the
general principle of functional somatic interactions between
individual features which is a consequence of an organism being
a tiered organizational system of parts.

Not all features or all attributes of features can be
labelled as adaptations, nor can all differences between features
be labelled as adaptive differences, nor can all evolutionary
change of individual features be considered as adaptive evolution.
This stems simply from the relationship between adaptation and
selection and from the concept that phyletic evolution is the
result of a continuous interaction between the production of
genetically based phenotypic variants.  A particular selection
force acting on an organism has no causal influence on the
genetically based phenotypic variants which may appear and be
favored by this selection.  The genetic events, current or past,
that give rise to phenotypic variants acted upon by selection
are strictly chance-based with respect to present and future
selection forces acting on a particular phyletic lineage.  The

attributes of a feature, whether it is an adaptation or not, that
result from the mechanisms leading to the particular genetically
based phenotypes are term paradaptations (Bock, 1967) from the
prefix "para" (= besides) meaning besides or aside from adaptation.
It should be obvious that all features and all variants of
features having arisen with a genetical basis will be a para-
adaptation, and if they have been favored by selection will be
an adaptation.  The concepts of paradaptation and adaptation are
not exclusive as most features or their attributes may be
designated simultaneously as adaptations and paradaptations.  The
concept of paradaptation is closely linked to horizontal comparisons and multiple pathways of adaptation (Bock, 1959, 1967).

## ADAPTATION AND COMPARISON

With the concepts of biological comparison and of biological
adaptation set forth above, it is possible to inquire into the
validity of the principle of determining adaptations by comparisons as a general covering law.  If this principle is to be
law-like - that there is an absolute or close correlation between
cause (selection) and effect (resulting adaptation) - several
associated principles must also have a law-like quality.  These
are:

(a)  That a tight correlation exists between a particular
     environmental factor and the resulting adaptation.
     This means that the environmental factor alone de-
     termines the nature of the selection acting on the
     organism.  The falseness of this idea has been shown
     by Bock and von Wahlert (1965) and von Wahlert (1965)
     who demonstrated that a particular factor of the
     umwelt has a number of possible selection forces and
     that the determination of exactly which one of these
     possible selection forces will actually act as
     selection on an organism depends largely upon the
     organism.  Given an environmental factor acting on
     it, an organism generally has a wide latitude of
     choice of how to react to it and this reaction
     affects the nature of selection.  This interaction
     may be regarded as a feedback between the biological
     role and the potential or actual selection force -
     the synerg (Bock and von Wahlert, 1965).  This
     principle need not be discussed further except to
     point out that adaptations should be discussed with
     respect to selection forces and not the environmental
     factor from which selection arises.

(b) That a single optimum adaptation exists for any given selection force. I will use selection rather than environmental factor because this expression forces consideration of the most demanding comparison. This idea is associated with the concept of multiple pathways of evolution (Bock, 1959) and will be discussed below under that heading. The existence of multiple pathways of evolution represents one class of exceptions that negates the principle of a single adaptive optimum to a selection force and hence the principle of judging adaptations by comparisons.

(c) That convergence is always the result of adaptive evolution to the same selection force. Again I use selection instead of environmental factor to force the most demanding comparison. This idea is associated with the existence of multiple functions and biological roles of a single feature and with the relationship between morphological convergence and adaptational convergence; it will be discussed under the latter heading. The existence of morphological convergences to different selection forces represents the second class of exceptions that negates the principle of a single adaptive optimum to a selection force and hence the principle of judging adaptations by comparison.

It should be noted that these two classes of exceptions comprise the total set of possible exceptions to the principle of a single adaptive optimum to a selection force. The existence of a large number of exceptions in each class provides the test needed to invalidate this principle and with it the principle of determining adaptations by comparison (or by analogy) as a covering law. One can, of course, explain away each exception by invoking ideas of very precise differences in the selection forces acting on the several phyletic lineages, by inserting a concept of functional demand between biological role and function and by claiming that the functional demands are not exactly the same, and so forth. However, this procedure simply adds more and more ad hoc assumptions and secondary hypotheses to the general principle with the consequence being that its law-like nature has been explained away.

A.    Multiple pathways of adaptation: A particular selection acting on a species must favor any phenotypic variants that appear and enable the organism to cope better with the environment (reduces the amount of energy needed to maintain the synerg), but the selection force has no causal influence on the genetical

mechanisms underlying the particular phenotype that appeared.  If
the same selection force acts on several different species, then
one could expect that different historical patterns of genetical
events will (or had) occur in these several lineages, and hence
expect that different phenotypical variants will appear and be
exposed to selection.  The different phenotypes would represent
different paradaptations (Bock, 1967) and, if favored by
selection, would also represent different adaptations to that
selection force.  The result will be different patterns of
evolutionary change - multiple pathways of evolution (Bock, 1959)-
each of which would represent adaptive evolution to the selective
force.  The differences between the different phyletic lineages
would be paradaptive differences with respect to the selection
force controlling the adaptive evolution of each lineage.  Hence
adaptive evolutionary change from one multiple pathway to another
cannot occur under the action of this selection force because
the intermediate morphological stages between the different
adaptive pathways would generally be less adaptive with respect
to that selection force.  Paradaptations representing different
multiple pathways of evolution to the same selection force are
equally adaptive with respect to that selection force.  It is
necessary always to state the selection forces when discussing a
set of multiple pathways and to state the confidence limits
around the "same" selection force and around "equal" adaptations
to this selection force.

   Certainly different paradaptations which are multiple pathways to one selection force have different selective values with
respect to other selection forces.  Adaptive evolutionary change
from one multiple pathway to another can occur under the action
of these other selection forces, but again it must be emphasized
that it is essential to specify the selection forces at all
times.

   Examples of multiple pathways of evolution are extremely
common, but little attention has been given to them.  The arrangement of a right dorsal aorta in birds and of a left dorsal aorta
in mammals in comparison with the double dorsal aorta in their
reptilian ancestors is a good example of multiple pathways.  Evolution from a double to a single dorsal aorta may be considered
as an adaptive evolutionary change (although the details of this
adaptive change are unknown, at least to me), but the loss of the
left aorta in birds and of the right aorta in mammals depended
upon the exact morphology of the heart in the different reptilian
ancestors of birds and of mammals.  And these different heart
morphologies (i.e., how the ventricle was subdivided) depended
upon different historical patterns of genetical events in the
reptilian lineages leading to bird and to mammals.

A second example well known to most vertebrate zoologists
is the structure of the leg in even-toed and odd-toed ungulates.
The difference in the structure of the legs in these large running
mammals depended upon whether the central axis of the leg lay
within the third digit or between third and fourth digits.  The
foot is equally well adapted for "ungulate" locomotion, especially
when considering the broad and widely overlapping types of locomotion of these two groups of mammals.

Within birds, the arrangement of toes in the feet of perching
and in climbing groups represents a pattern of multiple pathways
of evolution (Bock, 1959; Bock and de W. Miller, 1959).  The
ancestral foot is presumed to have a reversed but short hallux
as present in Archaeopteryx (Ostrum, 1976).  Selection for a good
perching foot would favor an arrangement of strong opposing
digits that grasp the perch.  Morphological arrangements of the
toes that would be favored by selection for a good perching foot
would be (1) elongation of the hallux (anisodactyl or syndactyl
foot), (2) reversal of the fourth toe (outer anterior toe =
zygodactyl foot) and (3) reversal of the second toe (inner
anterior toe = heterodactyl foot).  All of these arrangements
have evolved in birds and the first two have evolved several
times (Bock and de W. Miller, 1959).  Evolution from the first
type - the anisodactyl foot - to the second type - the zygodactyl
foot - has occurred in owls (Strigiformes) and in the osprey
(Pandionidae) under the action of selection for a more efficient
prey-catching foot.  The further evolution of climbing feet from
perching feet depends upon the arrangement of toes in the perching
foot.

The avian mandible articulates with the quadrate which itself articulates freely with the brain case.  Anteriad rotation
of the quadrate places a force on the bony palate which in turn
pushes on the ventroposterior corner of the upper jaw; the result of this action is dorsad rotation of the kinetic upper jaw.
Lowering of the upper jaw is achieved by posteriad rotation of
the quadrate and a pull on the base of the upper jaw.  This
mechanism suffers from the defect that it is difficult for the
bird to hold the mandible braced in position and simultaneously
move the upper jaw up and down because the mandible is braced
against the quadrate which interferes with rotation of the
quadrate (see Bock, 1964).  For certain feeding methods, a
mandibular brace other than against the quadrate would be
advantageous.  In many groups of birds, the tip of the internal
process of the mandible abuts against the base of the brain case
to form a brace; in some groups (e.g., Rynchops) a fully developed
diarthrosis is present (Bock, 1960).  In several genera of the
passerine family Meliphagidae, the dorsal edge of the mandibular
ramus abuts against the ectethmoid plate in front of the eye,

forming an ectethmoid articulation and brace (Bock and Morioka, 1971). The two new articulations between the mandible and the brain case - the basitemporal brace and the ectethmoid brace - have evolved at the only two places where a part of the mandible approaches the brain case. They represent two paradaptations to the same selection force associated with feeding, in spite of their strikingly different morphologies.

Many birds extend their tongue far out of their mouth during feeding. This is seen in woodpeckers which capture wood-bording insects with their tongue and in nectar-feeding birds which probe into deep flowers with their tongue to obtain nectar. Both the protractor muscle and the retractor muscle of the tongue must elongate greatly to accomplish the long movement of the tongue out and back into the mouth. The protractor muscle is quite constant in its mode of elongation, but the retractor muscle varies greatly. Considering only the nectar feeding birds to restrict the nature of the selection force to that associated with feeding on nectar in deep flowers, the retractor muscle demonstrates a number of paradaptations (Bock, in preparation). In all cases, the adaptive change is an increase in the length of the muscle fibers. The usual retractor muscle is the M. stylohyoideus originating from the posterior end of the mandibular ramus and inserting onto the basihyale of the tongue. In the hummingbirds (Trochilidae), the sunbirds (Nectariniidae) and the genera Promerops and Toxorhamphus (Meliphagidae), the elongated M. stylohyoideus extends around the back of the cranium to originate on the midline of the dorsal surface of the brain case. In the genus Coereba (Parulidae), the elongated muscle extends along the lateral side of the cranium to originate dorsal to the M. depressor mandibulae. In a number of genera of tanagers (Thraupidae) and warblers (Parulidae) and some species of Myzomela (Meliphagidae), the muscle extends to the base of the occipital plate of the skull - a short but decided elongation of the muscle. In the genus Oedistoma (rudimentary) and in Moho (well developed) and somewhat differently in some species of Myzomela (all Meliphagidae) part or all of the M. stylohyoideus is attached to the insertion of the M. serpihyoideus on a midventral raphe, forming a digastric muscle with the serpihyoideus, and thereby extending its length to the origin of the M. serpihyoideus from the base of the occipital plate. In Melithreptes, some species of Meliphaga and some other genera of the Meliphagidae, the M. stylohyoideus becomes vestigial or disappears entirely. Retraction of the tongue is taken over completely by the elongated M. thryeohyoideus arising from the head of the trachea. (In woodpeckers, the M. thryeohyoideus also takes over retraction of the tongue; the fibers of the elongated paired muscle wrap around the trachea in opposite directions for several turns.) These different configurations of the retractor muscles are all paradaptations as the exact

morphology of the muscle with respect to its origin from the
skull or from the trachea does not affect its adaptive significance
of being able to retract the tongue into the mouth over a long
distance.

Further examples can be added with ease from avian morphology,
and from other groups.  They show that the evolution of different
adaptive answers to the same selection force is commonplace and
hence that the principle that a single adaptive optimum exists
for a given selection force is not valid.

B.    Morphological convergence without adaptive convergence: Evolutionists have assumed almost universally, that morphological
convergence indicates adaptive convergence, namely that the features undergoing convergent evolution were under the control of
the same selection force.  Again this belief is based upon the
notion that one adaptation can evolve in response to one selection force and the idea that features have only a single or
limited number of "functions"; "function" was used to designate
the concept of biological role as defined above as well as the
concept of function.  Realizing that a particular form of a
feature may have a number of functions and biological roles - and
thereby may have a number of possible interactions with different
selection forces - it is reasonable to postulate that different
selection forces could favor the convergent evolution of the
same morphological form of a feature.  The synergs of convergent
features can differ.  It is necessary to specify the selection
forces carefully; it is not sufficient to designate selection as
simply feeding.  Feeding may involve different food times and
different ways of obtaining them, hence to label feeding as the
same selection force is too vague.  Certainly one should attempt
to designate selection forces of approximately the same magnitude
of those used in examples of multiple pathways if the two sets
of exceptions are to have equal relevance.

Examples of morphological, but not adaptive, convergent
features are more difficult to find than those of multiple pathways.  This may result from fewer attempts to identify this
category of convergent features, but more likely it may represent
a true situation.  There are simply fewer examples of convergent
evolution than divergent evolution, and since only a part of the
total number of convergent features are ones of not adaptive
convergent, then the source of examples is reduced.  The examples
cited are all from birds and generally represent simple morphological features.  Most important is the diversity of biological
roles and of the associated selection forces that are postulated
to have controlled the evolution of these convergent features.

The first example is the hooked "shrike-like" bill in the
small passerine predators Lanius and other shrike-like genera
(Laniidae) and the similar shaped bill in the Australian genus
of shrike-tits Falcunculus (Pachycephalinae:Muscicapidae).  The
shrike bill is convergent to a typical hawk or owl bill, being
strong, deep dorsoventrally and somewhat compressed laterally
with a pronounced hook at the top of the upper jaw and usually
a notch of "tooth" behind the hook.  The jaw muscles are strong,
providing the bird with a powerful bite.  Shrikes are predators,
feeding on large insects and small vertebrates; their hooked
bill can be shown to be an adaptation to their predatory feeding
habits.  Falcunculus has a bill morphology and jew musculature
very similar to the shrikes - hence part of its name of shrike-
tit - but it is a member of a different family as shown by a
number of features.  The bills of the shrikes and shrike-tits are
convergent features.  Observation of feeding habits of Falcunculus
shows that they are not passerine predators, rather they creep
around the trunks and large branches of trees in the manner of
tits (Parus) and tear off bark to obtain their food of bark-
dwelling insects.  The noise of this bark-tearing activity is so
loud that one can easily locate shrike-tits in a woodland by
listening for it.  Their strong hooked bill and heavy jaw muscles
are adaptations to the selection forces associated with their
feeding habits of tearing off bark of trees.  Hence this morphological convergence is not an adaptational convergence because
different selection forces have controlled the evolution of the
"shrike-like" bill in Lanius and its relatives, and that in
Falcunculus.

Many birds have a vertical hinge or region of bending in the
mandibular ramus usually located at the corner of the mouth or
approximately midway between the tip of the lower jaw and its
quadrate articulation.  This intramandibular hinge permits the
mandibular rami to bend outwards and hence results in a widening
of the intermandibular space.  The functions associated with the
hinge and the broader intermandibular space are associated with
many different feeding habits and selection forces; hence this
feature, when present as a convergent feature, is often not an
adaptive convergence.  It is present in herons (Ardeidae),
gannets (Sulidae), cormorants (Phalacrocoracidae) and many other
fish-eating birds that have a narrow bill.  The narrow bill is
essential for catching fish, but a broader intermandibular space
is needed to swallow the fish.  A similar hinge is present in the
pelicans (Pelecanidae) and permits the birds to change the shape
of their mandible and attached through pouch to a typical fish-
net used to catch their piscine prey.  Nightjars (Caprimulgidae)
have a similar mechanism (Buhler, 1970) which transforms their
bill into an insect net.  Barn owls (Tyto) possess an intramandibular articulation which permits them to broaden their bill

to swallow their prey and to cough-up pellets of hair and bone. Other owls (Strigidae) have a broader bill and lack the articulation. Lastly, pigeons (Columbidae) possess this articulation which permits the adults to widen their mouth; this allows the young to insert their head into the mouth and esophagus of the adult and feed upon the "crop-milk" produced in the crop.

Extreme protrusion of the tongue is associated with at least two different feeding habits in birds and hence with at least two different selection forces. One is feeding on nectar of flowers with deep corollas as seen in hummingbirds (Trochilidae), sunbirds (Nectariniidae), honeyeaters (Meliphagidae) and several other passerine families. The second is feeding on wood boring insects as seen in woodpeckers (Picidae). In both cases the protractor and the retractor muscles are elongated as an adaptation to the long distances the tongue is moved in association with these different feeding habits. Clearly the functions of the muscle-bone systems are very similar, but the biological roles and the selection forces associated with the two feeding habits are different. Considering only the elongation of the hyoid horns (ceratobranchiale and epibranchiale) and of the protractor muscle (the M. branchiomandibularis) in woodpeckers and in nectar feeding birds, they have elongated similarily and would be morphological convergences. But they are not adaptive convergences because of the different selection forces.

Sickle-shaped bills are seen in many nectar-feeding birds, hummingbirds, sunbirds and honeyeaters in which the shape of the bill is adapted to flowers with curved corollas. A similar bill is found in many birds that creep on the bark of large branches and trunks of trees and probe into crevices for insects, such as in the Certhiidae, the Dendrocolaptidae, the genus Astrapia of the Paradisaeidae and numerous others. This morphological convergence of sickle-shaped bills is clearly not adaptive convergence in these two assemblages of birds. In the Hawaiian honeycreeper genus Hemignathus (Drepanididae), however, the sickle-bill is an adaptation to both nectar-feeding and to probing into deep crevices for insects and would represent an adaptive convergence to both types of sickle-billed passerine birds.

The zygodactyl foot in birds is an adaptation to perching and to catching prey. Clearly in groups such as parrots (Psittacidae), it is an adaptation to perching. In owls (Stigiformes) and in the osprey (Pandionidae), it is an adaptation to catching prey. Hence the convergence in toe arrangement in these groups is not an adaptive convergence because of the different selection forces involved.

Many birds have enlarged mucous-secreting salivary glands for which the mucous is associated with strikingly different biological roles and hence selection forces so that the convergent enlarged mucous glands cannot be adaptive convergences. In woodpeckers (Picidae) the mucous coats the tongue and permits the bird to snare its insect prey. In swifts (Apodidae) the mucous is used to glue twigs and leaves together to build the nest which is glued onto a vertical surface. The extreme specialization is seen in some species of swiftlets (Collocalia) which construct their nests solely of spun fibers of mucous — these are the nexts of the Chinese delicacy of "birds-nest soup." Gray Jays (Perisoreus:Corvidae) have enlarged mucous glands (Bock, 1961) in which the mucous serves for food storage during the winter. Dow (1965) has shown in an elegant study that Gray Jays form food boli by gluing particles of food together and store these boli by gluing them onto branches of trees. The stored food is used during periods of inclement winter weather (e.g., snow storms) whem the birds are unable to find other sources of food. No support could be found for Bock's speculation that the mucous allowed the jays to obtain food with a sticky tongue from crevices, arguing by analogy with woodpeckers. Clearly these uses of mucous involve radically different biological roles and that the convergent enlarged mucous gland in these three groups of birds is not an adaptive convergence. And indeed the original argument that I postulated for the adaptive significance of the enlarged gland in Perisoreus (Bock, 1961) using the comparative method proved to be wrong.

Other avian examples could be given but these are sufficient to show that a number of exceptions exist to the principle that morphological convergence is always adaptive convergence and hence that this principle cannot be used as a covering law. It would be possible to show on the basis of theoretical discussions of functional morphology, how one could look for other exceptions. For example, elongation of muscle fiber permits the muscle to shorten over a greater distance and permits the muscle to shorten with greater speed, assuming in each case a constant load on the muscle. Examples can be shown of muscles elongating as adaptations associated with greater speed of shortening over the same distance. It is possible, at least theoretically, to have convergent elongation of the same muscles for increased speed in one group and for increased distance of shortening in another, and hence the convergence would not be adaptive convergence. Such a discussion would, however, go beyond the scope of this paper which is only to inquire into the comparative method of determining adaptations.

CONCLUSIONS

A.    Expressions such as "The comparative method" in biology are
naive and vague; no single method exists, nor does a definite set
of procedures exist.  The taking over of the idea of "the
comparative method of biology" by other sciences such as sociobiology and psychology, could lead to serious misinterpretations
in these sciences.  Several different types of comparison exist
in biology with different basic assumptions and with different
possible interpretation.  A distinction must be made between
horizontal (between phyletic lineages) and vertical (within a
phyletic lineage) comparison, and between nonhistorical and
historical comparisons.  Most biological comparisons are
horizontal and/or nonhistorical.  Such have been used to ascertain
adaptations by analogy, however it is necessary, at the minimum,
to use a historical comparison to ascertain adaptations by
analogy.  This means the inclusion of the mechanisms by which
genetical variation is generated.  Genetical studies have shown
that these mechanisms are strictly chance-based with respect to
the selection forces acting on species and determining the evolution of adaptations.  Inclusion of a chance-based mechanism has
serious, and usually disastrous, consequences for a covering law
in science.

B.    The concept of biological adaptation is a relationship between a feature of the organism and a factor of the environment;
the feature is the adaptation.  Features have properties of form
and of function, and it is the form-function complex (the faculty)
which is the adaptation via its biological role and the synergeal
interaction between the biological role and a selection force of
the environment.  It was shown that the principle that a single
adaptive optimum exists for a given environmental factor is invalid as a covering law because: (1) the same environmental
factor may exert very different selection forces depending upon
how different species reacted to the demands of the environmental
factor; and (2) the multiple functions and biological roles that
are actually or potentially possessed by a feature with a single
form.  Hence, a single adaptive optimum does not exist for each
selective force.

C.    The concept of multiple pathways of evolution is based upon
the principle that selection has no causal effect upon the
genetical mechanisms that control the production of genotypical
variants underlying the phenotypical variants (paradaptations)
exposed to selection.  And further, different paradaptations may
be favored by selection, leading to the evolution of morphologically
different adaptations to the same selection force in different
phyletic lineages.  These paradaptations corresponding to multiple
pathways of evolution to the same selection force represent the

first class of exceptions to the principle of a single adaptive
optimum to one selection force and hence the principle of determining adaptations by comparison or analogy.

D.    The concept of morphological convergences that are not
adaptive convergences is based upon the demonstration that the
same form of a feature has a number of functions and/or biological
roles.  Thus different selection forces acting on the feature in
different phyletic lineages could result in convergent evolution,
in which the feature in different lineages becomes more similar
morphologically, without the convergent feature in these lineages
having the same adaptive significance.  That is, the principle
that convergence occurs only by the action of the same selection
force acting on different species and resulting in the evolution
of morphologically similar adaptations is not law-like.  These
morphological, but not adaptive, convergences represent the
second class of exceptions to the principle of a single adaptive
optimum to one selection force and hence the principle of determining adaptations by comparisons or analogy.

E.    The principle of determining adaptations by comparison or
analogy is based upon the agreement of morphological form of the
feature in different taxa possessing the same presumed adaptation.
If the morphological form is the same then the features represent
the same adaptation, if the morphological form differs, then the
features are different adaptations.  The law-like quality of this
principle is dependent strictly upon the law-like quality of the
principle that a single adaptive optimum exists for each selection
force and that adaptive evolutionary change under the control of
this selection force is always toward this adaptive optimum.  The
existence of the two classes of exceptions discussed under the
heading of multiple pathways of adaptation and that of morphological, but not adaptive, convergences invalidate the law-like
quality of the principle of a single optimal adaptation to a
selection force.  And hence they invalidate the law-like quality
of the principle of ascertaining adaptations by analogy or
comparison.  Comparison may be used only as a rough working
approximation to narrow down the scope of the analysis needed to
determine the adaptive significance of a feature.  Its use involved the potential difficulty that the analysis may be mis-
directed as mentioned above in the original ideas on the adaptive
significance of the enlarged salivary glands in the Gray Jays
(Perisoreus).

F.    The only valid method by which the adaptive significance of
a feature can be determined is by direct analysis which includes
observing the animal in its natural environment and direct determination of the biological roles and selection forces.

G.    Comparative studies should be undertaken after direct determination of adaptations to ascertain the possible correlations
between particular types of adaptations and their properties of
form and function.   Quite possibly, a close correlation exists
between the form-function complex and the adaptation for some
types of features and/or adaptations.   Other types of adaptations
may not possess any correlation between the adaptation and its
morphology.   Determination of these correlations is necessary if
there is to be any hope of ascertaining adaptations in fossil
organisms and in those Recent organisms for which direct study of
the biological roles, etc. of the organisms in their normal
environment is not possible.   The importance of determining these
correlations between directly analyzed adaptations and form-
function properties of features lies in the extreme value of
observing the historical sequences of changing adaptations in the
fossil record.   Lastly, such correlations are needed to reexamine
some of the comparative studies mentioned at the onset of this
analysis.   It is quite likely that the ideas developed in some of
the comparative ecological studies, for example, are sound even
though they were based upon an invalid method of determining
adaptations.

## ACKNOWLEDGEMENTS

The ideas expressed in this paper were developed over many
years during which time numerous friends provided valuable imput
by their criticisms and suggestions.   I would like to express my
appreciation to all, and to thank especially Professors Ernest
Nagel and F.E. Warburton, and Mr. Arthur Caplan, all of
Columbia University, for their help in many ways from clarifying
concepts in the philosophy of science to pointing out minor
technical errors.   The morphological work upon which my theoretical
ideas were developed has been done with the assistance of financial
support from the National Science Foundation, Washington; this
study was done with support of a research grant BMS-73-06818 from
the N.S.F.

## REFERENCES

Bock, W.J., 1959, Preadaptation and multiple and evolutionary
       pathways.  Evolution, 13:194-211.
Bock, W.J., 1961, Salivary gland in the Gray Jays (Perisoreus).
       Auk, 78:355-365.
Bock, W.J., 1963, Evolution and phylogeny in morphologically
       uniform groups.  Amer. Nat., 97:265-285.
Bock, W.J., 1964, Kinetics of the avian skull.  J. Morph.
       114:1-42.

Bock, W.J., 1965. The role of adaptive mechanisms in the evolution of higher levels of organization. Syst. Zool., 14:272-287.

Bock, W.J., 1967, The use of adaptive characters in avian classification. Proc. XIV Internat. Ornith. Cong., pp. 61-74.

Bock, W.J., 1969, Comparative morphology in systematics, in "Systematic Biology," Nat. Acad. Sciences, pp. 411-448.

Bock, W.J., 1970, Microevolutionary sequences as a fundamental concept in macroevolutionary models. Evolution, 24:704-722.

Bock, W.J., 1972a, Species interactions and macroevolution, in "Evolutionary Biology," (Dobzhansky, Hecht and Steer, eds.), vol. 5, pp. 1-24, Appleton-Century-Crofts, New York.

Bock, W.J., 1972b, Examination of macroevolutionary events by sequential species analysis. Preprints, 17th Internat. Zool. Cong., vol. 2, 17pp.

Bock, W.J. and Miller, de W., 1959, The scansorial foot of the woodpeckers, with comments on the evolution of perching and climbing feet in birds. Amer. Mus. Novitates, #1931, 45 pp.

Bock, W.J. and Morioka, H., 1971, Morphology and evolution of the ectethmoid-mandibular articulation in the Meliphagidae (Aves). J. Morph., 135:13-50.

Bock, W.J., and Wahlert, von G., 1963, Two evolutionary theories - a discussion. British Journ. for the Phil. of Science, 14:140-146.

Bock, W.J., and Wahlert, von G., 1965, Adaptation and the form-function complex. Evolution, 19:269-299.

Buhler, P., 1970, Schadelmorphologie und Kiefermechanik der Caprimulgidae (Aves). Z. Morphol. Tiere, 66:337-399.

Cartmill, M., 1974, Rethinking primate origins. Science, 184:436-443.

Dow, D.D., 1965, The role of saliva in food storage by the Gray Jay. Auk, 82:134-154.

Feduccio, A., 1973, Evolutionary trends in the neotropical oven-birds and woodhewers. Ornith. Monographs, No. 13, pp. iii + 69.

Karr, J.R. and James, F.C., 1975, Ecomorphologic configurations and convergent evolution, in "Ecology and Evolution of Communities," (Cody and Diamond, eds.), Harvard Univ. Press, Cambridge, Mass., pp. 258-291.

Keast, A., 1972, Ecological opportunities and dominant families, as illustrated by the Neotropical Tyrannidae (Aves)., in "Evolutionary Biology," (Dobzhansky, Hecht and Steere, eds.), vol. 5, pp. 229-277, Appleton-Century-Crofts.

Lack, D., 1965, Evolutionary ecology. Ecology, 53:237-245.

Mayr, E., 1963, "Animal species and evolution." Harvard Univ. Press, Cambridge, Mass., XIV + 797.

Ostrum, J., 1974, Archaeopteryx and the origin of flight. Q. Rev. Biol., 49:27-47.

Ostrum, J., 1976, Archaeopteryx and the origin of birds.  Biol.
    J. Linn. Soc., 8:91-182.
von Wahlert, G., 1965, The role of ecological factors in the
    origin of higher levels of organization.  Syst. Zool.,
    14:288-300.

# FUNCTIONAL MORPHOLOGY AND EVOLUTION

P. DULLEMEIJER and C.D.N. BAREL

Zoologisch Laboratorium, University of Leiden

The Netherlands

## INTRODUCTION

The fact that the historical transformation of plants and animals concerned living, active organisms, is for many students sufficient reason to apply functional morphology to evolutionary studies (Dullemeijer, 1974; Gans, 1960; Gutmann, 1966; Gutmann and Peters, 1973; Maglio, 1972). However, many functional morphologists feel obliged to ask how and when historical and dynamic aspects are to be included in their methods and ideas (Crompton, 1963; Dullemeijer, 1970; Gans, 1960, 1969; Liem, 1970, 1973; Liem and Stewart, 1976). Since the two positions have completely different objectives, their approaches to the relation between functional morphology and evolution will be essentially different. In the first approach functional morphology is used as a method to explain evolution. The reverse holds for the second approach: functional morphology includes evolutionary aspects in its explanation hence evolutionary study becomes a method. This means that the same phenomenon (e.g. evolution) is to be explained in one approach and serves as an explanation in the other. It is difficult to synthesize both approaches, because a synthesis would mean replacing the explanation with the explaining theory. In other words, a synthesis between the two approaches would be tautological.

From the foregoing it is clear that the relation between functional morphology and evolution may be analyzed in two ways:

1) The impact of evolutionary studies on functional morphology.

2) The impact of functional morphology on evolutionary studies.

83

We shall approach these problems by presenting a short analysis of the methodology of functional morphology and of evolutionary biology, followed by an analysis of the relation between the two, illustrated with examples from our morphological research on cichlid fishes and snakes.

## METHODOLOGY

### Introduction

The aim of natural science is the explanation of natural phenomena. As it is important to understand what is meant by "explaining," a short definition of explanation is given in this section. Much confusion and unnecessary debate is based on comparing explanations obtained by different methods. It is essential to understand that different methods will lead to different explanations of the same phenomena. To compare these explanations without regard to the methods by which they were obtained is a logical fallacy.

The central concepts of functional morphology - e.g. form, function and relation - have many meanings. To avoid ambiguity an introduction to our approach to functional morphology must include definitions of these concepts. This is preceded by what we see as contained in the current methodologies of evolutionary studies.

### Explanation

Explaining is generally defined as the connection of the phenomenon to be explained, the explanandum, to a known fact or phenomenon by means of a set of initial conditions called the explanans (Hempel and Oppenheim, 1953). Instead of the usual formalistic way in which this deductive explaining is presented, we shall write:

$$F = f(S) \tag{1}$$

in which S is the phenomenon to be explained, f the theory or the set of initial conditions and F the known phenomenon. In the case of functional morphology we can read for S form, for F biological function.

## Disciplines in Evolutionary Studies

Our title requires definitions of disciplines both in functional morphology and in evolutionary studies. Formula (1) can also be used to describe a discipline. Generally, investigators call themselves after the phenomenon which has to be explained; e.g. a morphologist is someone who attempts to explain "form," an evolutionary biologist is someone who tries to explain "evolution." A further specification is obtained when the known phenomenon is added in as a descriptive adjective. Thus a functional morphologist is someone who explains "form" by relating it to "function." Sometimes the discipline is named after the relating factor or theory; e.g. when laws taken from physics are applied the discipline will be called biophysics.

Evolutionary biologists are divisible into: (a) those who try to explain evolution by applying different explananses and known phenomena; (b) those who use evolution as the known phenomenon and use various other explananses; (c) those who explain other phenomena in which the evolutionary theory serves as the explanans with a variety of known phenomena. It is essential to understand that in each category evolution or evolutionary theory plays logically a completely different role. Evolution is the phenomenon to be explained in category (a); in (b) it is the known phenomenon; whereas evolution (theory) serves as explanans in category (c).

Among category (a) are those who try to reduce evolution to genetics, biochemistry, biophysics, etc. They apply various theories among which are parts of functional morphology, as we shall demonstrate in the present paper. Often the investigation is restricted to the construction of phylogenies in the sense of family trees.

Among category (b) we mainly find taxonomists, morphologists and also molecular biologists. They try to connect their aspects of interest to evolution by applying general rules from various fields, such as mechanics and biology, which govern the change and the origin of the aspects they are studying (Mayr, 1963). Among the theories that of natural selection stands out but it is not and need not be the only one. It must be stressed that these adherents have the historical change as the known phenomenon in contrast to those of category (c). This latter group (c) uses another known phenomenon and components of what is called evolutionary theory as the explanatory theory. The known phenomenon is generally, the phylogenetic tree, particularly in systematics. If other aspects are studies we sometimes find geographical distribution, genetics, thermodynamics, etc. An example is Szalay's paper in the present volume. He used the tarsal skeleton as the phenomenon to be explained; taxonomy and phylogenetic tree as the known phenomena; functional morphology and comparative anatomy form the theory. The result relates to the historical succession in a descriptive sense. When a causal factor is looked for, other theories are applied such as natural selection.

We find frequently that authors do not make a sharp distinction between category (b) and (c), particularly when they equate the phylogenetic tree with evolution. This can lead to considerable confusion and renders difficult the derivation of a consistent methodology from the work of evolutionary biologists. Let us suppose that all agree that evolutionary biology should be described as the study in which evolutionary theory is the set of initial conditions (category {c}). The problem then becomes what is the content of these initial conditions and how can they be obtained. Several authors (Mayr, 1963; Simpson, 1961) sum up what evolutionary theory should contain - "In essence it is a two-factor theory, considering the diversity and harmonious adaptation of the organic world as the result of a steady production of variation and of the selective effects of the environment." (Mayr, 1963:1). How we obtain these factors Mayr does not make clear. The evolutionary theory is actually a large hypothesis. As such an hypothesis is obtained by inductive methods, effects other than natural selection and variation cannot be excluded from playing a role. In fact functional morphology makes it highly probable that directed changes, canalization, and non-random variation may be more important than natural selection.

Moreover, the theory is clearly multifactorial, which is supported by Mayr's listing of factors: ectogenetic random response, endogenetic mutational limitations, epigenetic limitations, random events, spontaneous mutations, natural selection (Mayr, 1963:2). The multiplicity of factors involved and their unclear logical position makes it difficult to apply the theory deductively and thus to falsify any model.

Yet we can imagine that it can be done. Take for example the classic case of natural selection on industrial melanism in moths. We can imagine that from the ecological circumstances and the evolutionary theory, taken as the initial conditions, a model for the ratio of melanic versus "normal" moths could be developed. The actual distribution could be used to prove the theory.

The most extensive deductive method is probably shown in Gutmann's hydrocoel theory (see Gutmann, 1972 and in this volume). If this theory is correct then it can be said that most of the factors which play a role come from disciplines other than evolutionary biology.

## Functional Morphology

Functional morphology explains the relation between form and function.  Essential to holistic functional morphology is the concept that the relation between all forms and functions of the animal during its whole life-cycle will result in an explanation of form at any point of its life-cycle.  In practice the functional morphologist is forced to study a limited part.  Often only adult stages of the life-cycle are included and these are studied for a restricted length of time.

Form.  A form is defined as any material area which can spatially be delimited.  Each form in an organism possesses at least five morphological features:  presence, position, shape, size and structure (Dullemeijer, 1974).  The shape is the outline of a form; the structure is the internal arrangement.  The choice of the form to be considered depends upon the functions which the morphologist wants to study; e.g. when the main functional interest concerns mechanical activities, the choice may be the muscle-skeletal-ligament system.  The subdivision of this system into elements (muscular, skeletal and ligamentous) is again based on the morphologist's choice of functions, the elements are distinguished on the basis of their mechanically different properties (Anker, 1974).  This choice is potentially hazardous for the further course of the investigation.  A deficient or poor choice may lead to defective or incorrect conclusions.  Other parts of the organism may be more important in understanding the form-function relationship, or other aspects of the chosen forms may be relevant for the investigation.  Unfortunately the investigator has no means of telling beforehand how the choice has to be made, it is only when discordancies are encountered in his conclusions that he can try other choices.

Function.  The concept of function is complexly related to the fact that each organism has to fulfill its own set of major functional demands in order to survive; e.g. feeding, breathing, locomotion.  These major demands are so widely defined that they hardly suggest any form-substrate.  Any subdivision of a major functional demand is called a function when it covers a number of morphological features.  The concept of function comprises three subconcepts:  activity, demand and biological role (see for definitions Dullemeijer, 1974; other discussions of these concepts are found in Bock and Von Wahlert, 1965; Beckner, 1968; Jeuken, 1958; Driesch, 1909; and Bock in this volume).

Functional component and apparatus. The complex of a function and its accompanying form or its morphological features is called a functional component (Van der Klaauw, 1948-1952; Dullemeijer, 1974).  A functional component should be distinguished from an

apparatus.  An <u>apparatus</u> is a collection of elements of which
certain morphological features form a functional component (cf.
Barel et al., 1976).  The explanatory formula (1) can now be
expanded:

$$F = f\,(a_1,\ a_2,\ b_3,\ c_1,\ c_4,\ ....) \tag{2}$$

a, b, c, etc., stand for elements and the numerical indices for the
various morphological features (e.g. $a_1$ = presence of element a).
This formula is the mathematical representation of the functional
component.  The elements a, b and c form the apparatus which per-
forms function F.  The following example may illustrate the differences between functional component and apparatus.  The gill-
cover is an element in the mandibular depression apparatus of
fishes (see later).  Its shape-features consist, among others, of
two ridges, an articulation-socket and an extensive lamellar part
(see Barel et al., 1976: figs. 18, 19).  One ridge serves for insertion of the m. levator operculi, the other ridge for connection
between socket and interoperculum.  Whatever functional demand
depends upon the necessity to depress the lower jaw, it is improbable that the lamellar part as a whole will be a shape feature
in the concomitant functional component.  Although the gill-cover
is part of the mandibular depression apparatus, only a part of its
form-feature figures in the functional component of the depression
function.

Formula (2), describing the relation between form and function in a functional component, leads to the following observations:

1)  The historical process which produced a form-function
complex does not enter into the formula.  The method may be illustrated by analogy with an engine.  The functional relation between
activity and mechanical parts is only based on the shape, position,
mechanical qualities, etc., of these parts and on generally valid
mechanical laws.  Neither the manufacture of the engine nor its
previous models are relevant in this context.  Every process
leading to the same engine results in the same relation between
activity (function) and constituent parts (forms).  The rules
governing the relation between form and function are generally
valid and do not change in the course of evolution.

2)  It should be noted that function is not the same as
biological role.  An activity may serve in two different major
functional demands.  In this case the function has two different
biological roles; e.g. the force developed by a closing mouth
(function) may be used in fighting and in crushing prey (biological

roles). In the course of history the biological role of a function may change (e.g. with changing environmental factors). It is legitimate to ask whether the differentiation between function and biological role can be maintained on closer examination. The differentiation is probably a matter of abstraction. When biological roles are precisely described it is doubtful whether different roles could be performed by the same functional component.

Constructional morphology. Functional morphology usually only designates the relation between form and function in a functional component. As an organism is a coherent whole of numerous functional components, the interrelation between these components should be studied as well. This interrelation constitutes the object of constructional morphology which forms an essential part of functional morphology in a wider, holistic sense.

THE RELATION BETWEEN FUNCTIONAL
MORPHOLOGY* AND EVOLUTIONARY STUDIES

The Impact of Evolutionary Studies on Functional Morphology

Functional morphology utilizes two principal methods to determine the relation between form and function: viz. induction (= comparison) and deduction (Dullemeijer, 1974). Both methods and the presumed impact of evolutionary studies on them need to be discussed.

Inductive method. Before expounding the formal presentation and before drawing any conclusions we shall first present an example of functional morphological analysis.

Cichlid fishes of the genus Haplochromis from the East African Lakes George and Victoria provide an example of a group of closely related species (Greenwood, 1973, 1974). In each lake this group demonstrates a wide spectrum of feeding types. Comparison shows that the overall arrangement of the parts in the heads, described as presence and positional features, are the same for all the investigated cichlids. Besides this they fulfill certain conditions which facilitate comparison:

a) functional components involving homologous apparatuses demonstrate a wide variety of functions;

*It should be clearly understood that the concept functional morphology in this section is limited to the relation between form and function within a functional component.

b)   lacustrine <u>Haplochromis</u> species are derived from a group of riverine <u>Haplochromis</u>.  The extant insectivorous riverine <u>Haplochromis</u> are thought to be similar to the ancestral group.

Under these conditions one may put forward the hypothesis that:  morphological differences between functional components of riverine and lacustrine species are a necessary accompaniment to the shift in function of these components (Fig. 1).

The procedure followed is that of comparison, the inductive method, based on a classification of forms and functions.  It will be clear from the foregoing that such a classification and subsequently the derived relation between form and function is theoretically independent of historical aspects.  In practice, however, evolution (or better phylogeny) does enter the picture because the relation between form and function is more easily gleaned from the comparison of closely related recent organisms (Fig. 1).  The procedure is very sensitive to differences in judgement of similarity.

One can of course ask questions about development and the historical process, but then the type of explanation is different from the functional morphological one.  When one asks such questions one has another objective which means that one acts in the realm of another discipline.

The inductive method has a disadvantage.  It is theoretically impossible to falsify a supposed relation, because every observation which does not fit, does at best turn the relation into a statistical or stochastic rule.  To reach a conclusion by means of the inductive method, one hundred per cent agreement is not compulsory.  Furthermore any other relationship cannot be excluded; it may hold as well.  The inductive method hides a number of serious pitfalls:

a)   Not all observed differences in morphological features are parameters in the functional component.  In a compact coherent whole such as a fish head, the change of shape, size and position of an element will often affect the morphological features of surrounding elements.  Although the latter changes are a necessary accompaniment to the functional change, they are not pertinent to the relation between form and function under investigation.  For instance, comparing piscivorous and pharyngeal mollusc-crushing <u>Haplochromis</u>, it was demonstrated that the increase in size of the external and posterior levator muscles of the branchial apparatus in mollusc-crushers constrict the space available for the adductor muscle of the gill-cover.  The adductor muscle does not serve in the pharyngeal jaw apparatus (Witte and Barel, 1976).

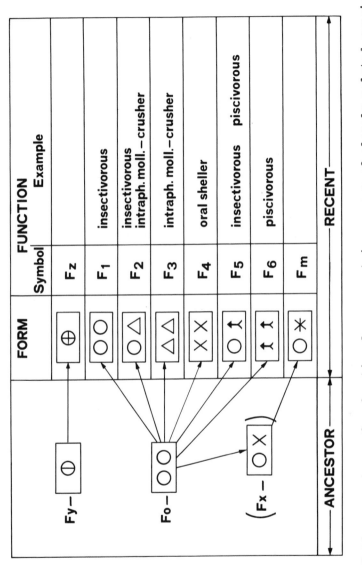

Fig. 1. Evolutionary scheme of a functional component in a group of closely related species such as the Haplochromis species of Lake Victoria. Accepting the component $F_1$ as the ancestral type, the morphological differences between the components $F_1$ and $F_{2-6}$ may be looked upon as morphological transformations necessarily accompanying the functional shifts from $F_1$ to $F_{2-6}$. The inclusion of the components $F_z$ and $F_m$ in the comparison illustrates two possibilities leading to a wrong interpretation. Intraph.moll. = intrapharyngeal mollusc. See text for further explanation.

b) Features that did not change may still be necessary to perform the new function. A comparative method based on the classification of differences does not necessarily hint at all the features to be included in a functional component:

$$F_I = f \, (a_1, \, a_2, \, a_3, \, c_1, \, c_4 \, \dots) \qquad\qquad (3)$$

$$F_{II} = f \, (a_1, \, a_2^*, \, a_3^*, \, c_1^*, \, c_4 \, \dots) \qquad\qquad (4)$$

* indicate altered features of (3)

Especially when considering new adaptations, students are often inclined to concentrate on the altered features and call these restricted parts the adaptation. In fact the whole set of features is the new morphological adaptation to the new function. Leaving out any one of the form-features would make the formula invalid. It cannot be argued which of the features is the more important one.

c) In comparing an ancestor-type (O) with a derived type (M), the functional morphologist attributes all form-differences to the shift in function (Fo) of the ancestor to the new function (Fm) in the derived species (Fig. 1). But if an unknown type (x) with its own function (Fx) occurred between ancestor and descendant, the differences between O and M are no longer necessarily attributable to the shift Fo → Fm only!

Apparently, the usefulness of the inductive method for studying form-function relations is limited. Its greatest power lies in the demonstration of the overall effect of a functional change and, practically, in screening which parts of the organism are to be studied in relation to certain functions. Comparative investigations often mark the initial phase of a functional morphological project.

Deductive method. The inductive method leads to a suggestion as to what kind of relation exists. After such a suggestion the complementary method of deduction can be applied (Dullemeijer, 1974).

In the deductive method the form is derived from functional demands in the first place. These functional demands are formulated as specifications for a model or paradigm. The model is compared to the actual form for which criteria of agreement have to be introduced. These criteria are a free choice of the investigator, nevertheless, in practice various degrees of comparisons are

made (Dullemeijer, 1974). However, the number of form-models obtained in this way is generally so large that further restrictions
must be introduced; e.g. minimum principle (= optimal design),
holistic concept, influence of other functional components, influence of other functions, etc. In the following section the
various steps in the deductive process will be analyzed especially
with regard to (presumed) historical elements in the method.

Deriving form from functional demands - Leaving aside the
problem of determining functional demands, the immediate question
is: What are we allowed to model? The functional morphologist
cannot select construction-materials; they are given data. In
practice he is often even obliged to start with a large part of
the "bauplan" of a group and he only explains "smaller" variations
of the actual form. A number of methodological problems are
created in using a bauplan as a starting point. The bauplan is
obtained by comparison. It is a derived typological concept of
unification and it is highly questionable whether this bauplan
should in turn be used to explain those details of the morphology
which were removed in the preceding methodological step, in order
to obtain a generally valid construction-scheme (viz. the bauplan)
for the group studied. The application of the "bauplan" concept
in functional morphology approaches circular reasoning. Further,
it is doubtful whether the bauplan approaches the ancestral form.
It was the ancestor that was transformed. In closely related
groups with extant ancestral-like types, this type should be used
as a starting point. This has the advantage that only the transformation is to be explained deductively. As has been argued
before, this approach implies the introduction of a phylogenetic
method. One should, however, realize that the ancestral type is
itself a living organism which can be analyzed in terms of deductive functional morphology and that it is a practical limitation
to accept it, as it appears.

The principle of optimal design - Many models can be constructed if the only requirement is that certain materials should perform a function. A further limitation is obtained by introducing
the principle of optimal design or minimum principle. The application of this principle is independent of evolutionary factors. The
deductive steps (viz. model-formulation and optimal design principle) may be illustrated by the studies on the opercular mandibular
depression apparatus in fishes (Fig. 2) by Anker (1974), Barel et
al. (1975) and Barel et al. (in press). Depression of the lower
jaw may be caused by rotation of the gill-cover. Through the
interoperculum and a ligament, gill-cover rotation is transmitted
to the lower jaw. This opercular mandibular depression apparatus
consists of 5 elements, the fifth being the suspensorium to which
lower jaw and operculum are movably articulated (Fig. 2). For a
kinetic analysis of the form of the apparatus is reduced to a

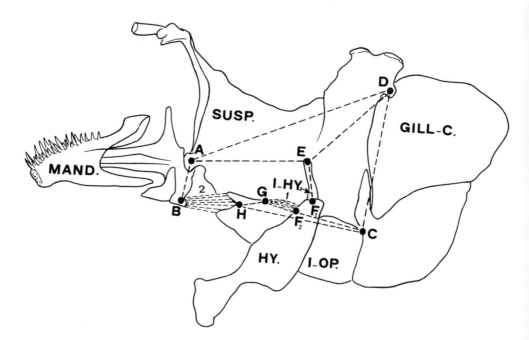

Fig. 2.   The mandibular depression apparatuses of <u>Haplochromis</u>
          <u>elegans</u> in medial view.  A,B,H,C, and D mark the rotation
          points of the bar model of the opercular mandibular de-
          pression apparatus; A,B,H,G,$F_2$,$F_1$, and E are the rotation
          points of the bar model of the hyoid mandibular depression
          apparatus.  Abbreviations:  Mand.=mandible; Susp.=suspen-
          sorium; I-op.=interoperculum; I-hy.=interhyal; Hy=hyoid;
          Gill-c.=gill-cover; l=ligament connecting the inter-
          operculum with the hyoid; 2=ligament connecting the inter-
          operculum with the mandible. (From: Barel, Van der Meulen
          and Berkhoudt, in press.)

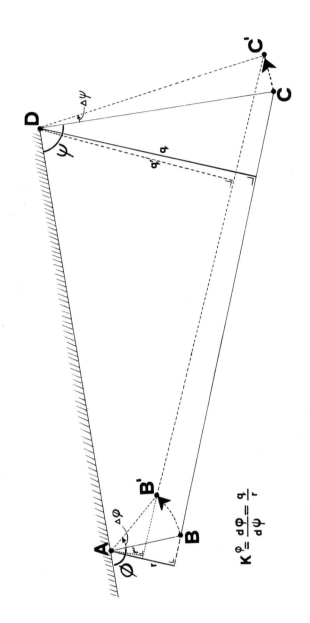

$$K^\phi = \frac{d\Phi}{d\psi} = \frac{q}{r}$$

Fig. 3.  The opercular mandibular depression apparatus reduced to a four bar model.  Capital
letters are the same as in Fig. 2.  The figure illustrates the relation between input-
rotation (CC' and Δψ of the gill-cover) and output-rotation (BB' and Δφ of the mandible).
Kφ is the kinematic transmission coefficient which is a measure of the ratio of output/
input rotation at a position ψ of the bar-system.  The formula is mathematically proved
in Anker (1974).  (Adapted from Barel, Van der Meulen and Berkhoudt, in press.)

four-bar system (Figs. 2, 3).  Of the original form-features of
all elements only a size-feature (lengths AB, BC, CD, DA in Fig.
3) of the elements and the position of one element (the angle ψ
between the gill-cover bar {CD} and the suspensorial bar {AD} is
retained (Barel et al., in press).  The function is mandibular
depression and is measured as the angle (φ) between the suspensorial bar (DA) and the lower jaw bar (AB).  φ can be expressed as
a mathematical function of the (morphological) parameters AB, BC,
CD, DA and ψ (see Hartenberg and Denavit, 1964) of which a short
representation is sufficient here:

$$\phi = f \ (AB,BC,CD,DA,\psi) \tag{5}$$

In the next step an optimalization principle is introduced.  The
m. levator operculi operates the opercular depression apparatus.
It is argued that the best position of the system should be the
one with the highest instantaneous ratio between the torques of
the depressing lower jaw and the rotating gill-cover.  Anker (1974)
demonstrated that the torque-transmission is mathematically the
reverse of movement-transmission, formulated as the kinematic
transmission-coefficient defined by Barel et al. (in press): $\frac{d\phi}{d\psi}$.

The new, mathematical relation between form and function reads:

$$\frac{d\phi}{d\psi} = \frac{df(AB,BC,CD,DA,\psi)}{d\psi} \tag{6}$$

If the principle of optimal design holds, the physiological range
of the depression apparatus should be around a certain φ for which
$\frac{d\phi}{d\psi}$ is minimal.  At this minimum the torque-transmission is maximal
(Anker, 1974).  For Haplochromis burtoni this could indeed be
demonstrated (Barel et al., in press).  In this limited deductive
approach all form-parameters, except ψ are taken for granted.  The
physiological range of rotation was deductively proven.

An example is provided by the shape of the ectopterygoid in
viperid snakes (Fig. 4).  From observations of the movement and
the activity of the muscles a relative force diagram can be constructed.  From the force diagram a model for the shape of the
ectopterygoid can be designed on the basis of mechanical rules
and, it is assumed, mechanically homogenous bony material.  This
model is then compared to the shape of the real ectopterygoid presented in cross-sections.  It is also possible to compare physical
properties derived from the model and from the real element.

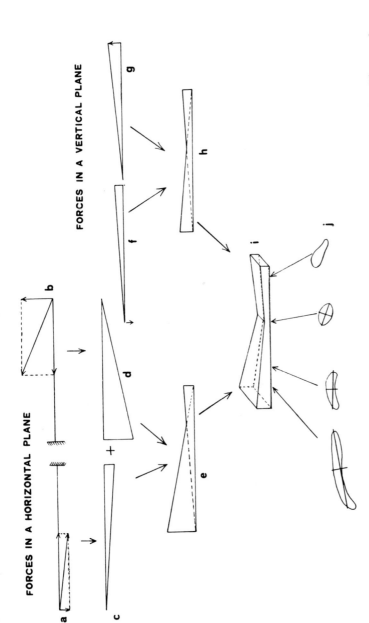

Fig. 4.  Mechanical model for the explanation of the shape of the ectopterygoid in Vipera berus. The construction of the model starts from a series of relative force diagrams. The activities of the muscles in the horizontal plane are shown in retraction (a) and pro-traction (b). In (c) the necessary shape for resisting the bending and compression due to the retractor (m. adductor pterygoideus), in (d) that for the protractors (m. pro-tractor et levator). As both systems work alternately, in the addition both figures can overlap (e). In (f), (g) and (h) the same in the vertical plane. Combining (e) and (h) results in the three-dimensional model (i), which is compared to the actual shape represented by cross-sections (j). After Dullemeijer (1956).

It must be remarked here that in both the bar model and the ectopterygoid model a rigorous selection and reduction of the form-features and the function-features was made initially.  Further investigations are needed to explain the form in more detail. Discordance between model and real shape could have meant that one or more of the steps in the procedure were incorrect or invalid.  This is particularly important for the minimum principle. If the minimum principle was the only factor which caused the disagreement then we would have had an interesting case of sub-optimal adaptation.

The problem of compromises - The fact that many elements contribute to several functional components is a third factor to be taken into account when constructing form-models.  The problem becomes complicated when functions impose contrasting demands on the form of an element.  In such cases the relative importance of the functions must be determined but the process is, in principle, the same as if one were dealing with only one function and does not involve historical factors.

Interaction of spatially separated functional components - (See section on constructional morphology.)

Features without a function - The question remains whether features exist which are only explainable on the basis of their evolutionary history.  Many topographical features seem to belong to this class; e.g. functionally it is difficult to explain why the central nervous system is ventral in vertebrates, although Bonik et al. (to be published) present deductive arguments from constructional requirements which may explain the topography.  But evolution as a theory does not answer this problem either and probably awaits an explanation from functional morphology.  The observation that something is a historical fact is still not an evolutionary explanation but merely a restatement of the problem. Many so-called "functionless" features can be shown to have a function when more factors are considered.  It seems nowadays that no functionless forms are left, at best one of the form features can be termed functionally irrelevant or disadvantageous (Dullemeijer, 1974).

The Impact of Functional Morphology on Evolutionary Studies

Functional morphology is a discipline whose objective is the explanation of the relation between form and function.  This relation is non-causal and is determined with the method described above.

Evolution, however, is not a discipline but a more or less

presupposed process defined as the transformation of organisms
through the interaction between accidental alterations and natural
selection.  These alterations are expressions of individual variability in a population through genetic differences.  The concept
of evolution includes many components.  Logically it is not
obvious whether absence of one of the components results in a
deficient definition.  From the definitions given by various
authors one is inclined to the belief that this is not the case.
It follows therefore that we should have to consider the impact of
functional morphology on each component separately.

However, the illogical connection of a discipline and a
phenomenon forces us to look for other interrelations.  Logically
we can relate a phenomenon only to another phenomenon or an
explanation to another explanation.

The conclusions arising from functional morphology are
phrased in rules and insights which determine the relation between
form and function.  These conclusions can be treated as a new
field for questions and, aiming at a historical explanation, can
be related to evolution.  The new problem may be formulated as
follows:  What significance do the data in these conclusions contain for the study of evolution and how does an explanation of
functional morphology influence an explanation of evolution?  At
least at the level of macroscopial elements, there is some reason
to conclude that no major contribution to evolutionary studies is
to be expected from functional morphology (taken as the form-
function relation in a functional component) because:

a)  Evolution is the historical change in form and function.
During the change the relation between form and function does not
alter, neither does it determine the change in any way.  Factors
outside the form-function relation govern the transformation; the
rules on which the relation is based are independent of evolution.

b)  Until now the observations, viz. the description of
phenomena and conclusions from functional morphology and evolution,
have been associated fairly loosely.  As yet, this inductive
classification has not yielded a theory relating the two.

c)  The deductive method cannot be applied at all to arrive
at an explanation of the relation between functional morphology
and evolution.  The logical basis for this impossibility is the
fact that transformation and form-function do not represent independent aspects.  Although form and function can logically be
separated (albeit related to the same substrate), transformation
always involves form, function, or both.  To establish a relation,
similar, logically independent positions are demanded.  It is,
however, possible to describe the transformation of the functional

components and to set rules for it.

However, functional morphology may possibly contribute to the
explanation of the evolutionary process when "adaptedness and
adaptability" (Dobzhansky, 1968) are considered. The investigations on the optimum principle may be looked upon as first steps
in this direction. If the principle could be successfully applied
in many cases, functional morphology would indeed contribute an
important key to the explanation of evolution. As yet the principle of optimal design (minimum principle) has mostly been used as
a method. The principle is introduced in the construction of
paradigms or models. It is said that the principle holds if
sufficient agreement between model and real form is observed.
Many cases are discussed in Dullemeijer (1974). If the principle
is generally valid, this would imply that evolution should go only
in the direction of optimalization (Bonik et al., to be published).
However, the principle is restricted because: 1) it can only be
applied to functional components, 2) on the molecular level
examples of redundant structures seem to exist, 3) up till now the
principle can only be applied in a relative sense; e.g. the form-
feature "proportion" could be calculated but the absolute magnitude
of the form-feature could not, 4) in compromise structures it is
difficult to apply an optimal principle without knowing what
relative bearing each of the functions has.


THE RELATION BETWEEN CONSTRUCTIONAL MORPHOLOGY AND EVOLUTION

Much more significance to evolutionary biology is gained when
the explanations of functional morphology are extended to the
interaction of functional components (constructional morphology).
Philosophically, the situation can be formulated as follows: the
phenomenon to be explained is evolution; the explaining theory is
functional morphology taken as a set of rules (the initial conditions). The functional components and their interrelations fulfill
the role of the known phenomenon. For an analysis some examples
will be presented which show the methodology and the conclusions
in this kind of morphology. From these examples we shall derive
the principles which can be of use to evolutionary studies.

The 'bauplan' of a rattlesnake may be derived from the
'bauplan' of a common viper by adding a small sense organ (the
pit-organ) to the latter. The consequences of adding this organ
are far-reaching for nearly all parts of the head (Dullemeijer,
1959), and many details are still to be studied. Important for
our elaboration is that the explanation was obtained through
model construction in which the presuppositions also generate
further scientific questions (Dullemeijer, 1974). The construction was obtained by calculating the influence of one functional

component on another, very much in the same way as the function-
form relationship was established. The effect of the influential
component was taken as the demand on the other functional component. This model was compared to the actual situation.

In this way it was possible to 'predict' the head of a
rattlesnake by adding the pit-organ to the head of a viper (Fig. 5).

The mutual influence of the various functional components can
now be depicted symbolically. Such a scheme is called a pattern.
In such a pattern necessary interdependent relations and mutually
exclusive possibilities can be distinguished (Dullemeijer, 1974).
These relations in a pattern can theoretically be used to support
explanations of evolutionary phenomena. In this field many
examples are known; e.g. as the relation between respiratory and
circulatory systems. Cuvier (1805) already stressed the importance
of these relations and pointed out the inseparable connection between the functional components.

Our insight into the rules governing the relations between
functional components is still very patchy. Many current investigations concentrate on this question and the ideas resulting from
these investigations will provide functional morphology with details, probably also data and conditions pertinent to the explanation of evolution.

For a better use of constructional morphology in evolutionary
studies the following set of questions should be asked: (1) Which
functional components are the more changeable and which are the
more stable? What conditions do functional components and their
relations impose on changes? What are the causes of change?

In answering the first question especially, the concepts of
plasticity and flexibility will turn up. The second question
deals with problems such as the threshold value of changing form-
features in a certain pattern, and fitting changed form-features
into a pattern. For the third question concepts such as mutation,
transformation, internal and external selection may be considered.

Concerning functional morphology, as yet only plasticity has
been investigated to a certain extent and therefore deserves
further elaboration.


Plasticity

In cichlids, the mm. levator externus and internus of the
branchial apparatus originate on a concave process (hyomandibulad
shell, Barel et al., 1976; Fig. 6). The shell is similarly shaped

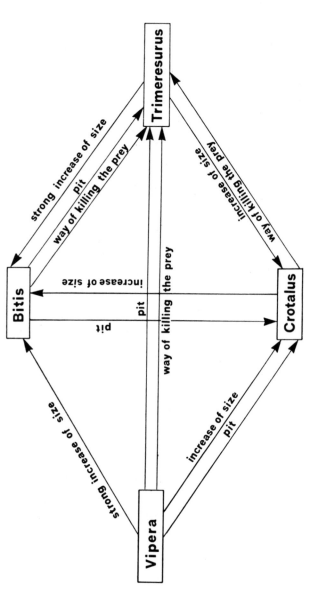

Fig. 5.   Diagram of an idealistic transformation in viperid snake heads.  The snake genera are named in the rectangles.  The arrows run in the direction of the transformation if a particular property is changed or added.  The relationship allows for transformations in any direction.  Other, preferably direct fossil, evidence is necessary to read the transformation in one way (after Dullemeijer, 1974).

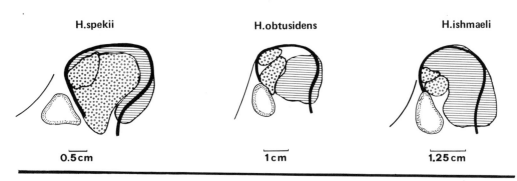

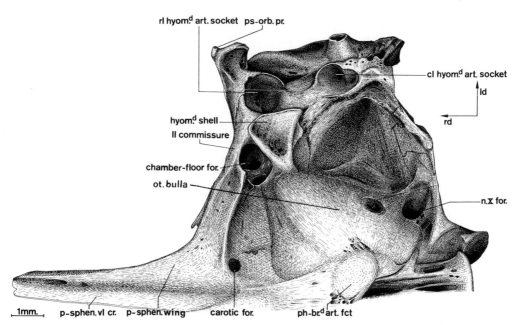

Fig. 6. Attachment areas of the external levator muscle (stripes)
and the internal levator muscle (circles) on the hyomandibular
shell of the neurocranium. The heavy line represents the outline
of the hyomandibular shell; the marginally dotted figure is the
chamber-floor foramen. The topology of the shell and the foramen
can be gleaned from the lower figure (Witte and Van Oijen, 1976:
Fig. 10) which represents a ventral view of half of the otic region
of the neurocranium in Haplochromis elegans. Haplochromis spekii
is piscivorous; H. obtusidens feeds mainly on insects and bivalves;
H. ishmaeli feeds almost exclusively on molluscs (Greenwood, 1960).
Abbreviations: carotic for. = carotic foramen; chamber-floor for. =
chamber-floor foramen; cl hyom.$^d$ art. socket = caudal hyomandibulad
articulatory socket; hyom.$^d$ shell = hyomandibulad shell; ll awning =
lateral awning; ll commissure = lateral commissure; n.X for. =
vagus nerve foramen; ph-br.$^d$ art.fct. = pharyngobranchiad articulation facet; p-shen, vl crest = parasphenoid ventral crest.

in different species, but the relative size of the attachment areas of both muscles differs (Witte and Barel, 1976; Fig. 6). In piscivores the internus muscle occupies the larger part of the shell area; in intrapharyngeal mollusc-crushers the external muscle dominates. A large nerve foramen and an articulation socket for the suspensorium border the hyomandibulad shell. It is tentatively suggested that these, and probably other external factors, determine the shape of the shell so that the area available for muscle attachment is limited. Within this limitation a plasticity in the relative size of the attachment areas of both muscles is observed.

In the lateral aspect of the head of cichlid fishes an area can be distinguished by three elements: the eye and two muscle complexes (adductor muscles and levator arcus palatini + dilatator operculi muscles; Fig. 7). The form-features of these elements are restricted in their size; which is the (relative) measurement of the area which they occupy in lateral view. By comparing 37 Haplochromis species from Lake George and Lake Victoria it was demonstrated that the ratio of largest to smallest absolute size of the occupation area is 12 for the eye, 68 for the adductor complex and 35 for the levator dilatator complex (Barel and Voogt-Kokx, in preparation). These ratios may be used as "an expression of plasticity." A feature with little plasticity is called a dominant feature and an accompanying feature with high plasticity is called subordinate (Dullemeijer, 1974). In our examples the pit-organ in snakes and the size of the eye in fishes are dominant features. In the fishes the adductor complex may be looked upon as subordinate. The dominance of the eye is also demonstrated in fishes with relatively large eyes. In these cases the structure of the adductor muscle "allows" them to be folded around the eye (Dullemeijer, 1974).

The occupation pattern in fishes is correlated with the trophic function. In this example a relatively small area available for the adductor muscle is found in insectivores, feeders on benthic crustaceans and feeders on plankton. A relatively large adductor muscle is correlated with a piscivorous habit. In between the algal grazers, paedophages and oral shellers are located (Barel and Voogt-Kokx, in preparation; Fig. 8, Table I). Also in the snakes the patterns combine with different feeding habits (Dullemeijer, 1959).

For phylogenetic considerations a second important hypothesis arises from the analysis of the occupation pattern (Fig. 5, Table I) - a change in the dominant feature will be "followed" by the subordinate feature. If fishes could do with smaller eyes, more space should become available for the adductor muscle and, at a size, trophic functions other than the existing ones could probably be realized.

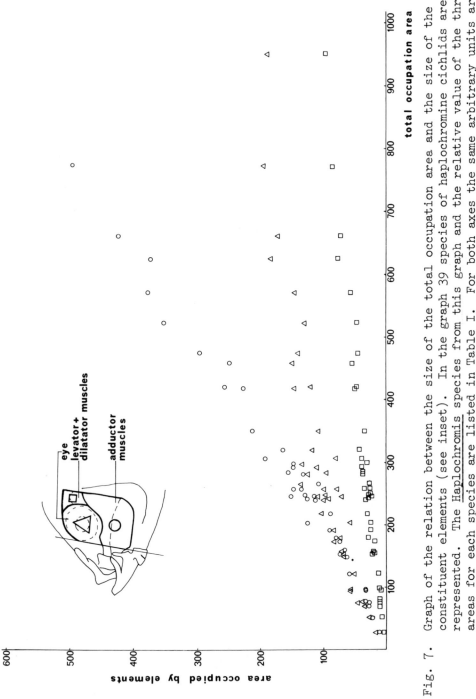

Fig. 7. Graph of the relation between the size of the total occupation area and the size of the constituent elements (see inset). In the graph 39 species of haplochromine cichlids are represented. The Haplochromis species from this graph and the relative value of the three areas for each species are listed in Table I. For both axes the same arbitrary units are used. (From Voogt-Kokx, unpublished.)

TABLE I.  Various _Haplochromis_ species of Lakes Victoria and George, arranged according to increasing relative size of the occupation area of the adductor complex. The sum of the relative sizes of the two muscle-areas and the eye does not equal one hundred per cent because of a narrow zone between the eye and the muscles. This zone is partly filled with fat-tissue and partly caused by shrinkage during preservation. For conclusions see text.

| Species | Lake | Food; Feeding Habit | % of Total Area | | |
| --- | --- | --- | --- | --- | --- |
| | | | Adductor complex | Eye | Lev. + dil. complex |
| H. eduardianus | LG | ? | 32,4 | 53 | 13 |
| H. oregosoma | LG | phytoplankton | 34 | 56 | 8 |
| H. macropsoides | LG | insects | 34 | 47 | 9 |
| H. pappenheimi | LG | zooplankton | 35 | 48 | 10 |
| H. elegans | LG | insects | 38 | 47 | 13 |
| H. "Speke crusher" | LV | phar. moll.-cr. | 40 | 47 | 10 |
| H. nigripinnis | LG | phytoplankton | 40 | 40 | 11 |
| H. martini | LV | small fishes | 40 | 39 | 12 |
| H. barbarae | LV | recently fertilized eggs | 42 | 38 | 15 |
| H. cryptogramma | LV | benthic cruts.+ insects | 42 | 45 | 10 |
| H. pharyngomylus | LV | phar. moll.-cr. | 43 | 37 | 16 |
| H. mylodon | LG | phar. moll.-cr. | 44 | 41 | 13 |
| H. maxillaris | LV | paedophage | 45 | 40 | 13 |
| H. pharyngomylus | LV | phar. moll.-cr. | 45 | 39 | 12 |
| H. "Speke crusher" | LV | phar. moll.-cr. | 45 | 41 | 10 |
| H. microdon | LV | paedophage | 46 | 39 | 13 |
| H. taurinus | LG | paedophage | 46 | 40 | 10 |
| H. limax | LG | "aufwuchs" | 48 | 36 | 10 |
| H. guiarti | LV | fish > insects | 50 | 26 | 13 |
| H. parvidens | LV | paedophage | 51 | 32 | 13 |

| | | | | | |
|---|---|---|---|---|---|
| H. prodromus | LV | oral sheller | 51 | 32 | 11 |
| H. granti | LV | oral sheller | 51 | 35 | 11 |
| H. cronus | LV | paedophage | 52 | 32 | 10 |
| H. squamulatus | LV | fish; paedophage? | 54 | 28 | 12 |
| H. apogonoides | LV | ? | 54 | 32 | 12 |
| H. obesus | LV | paedophage | 54 | 29 | 13 |
| H. prognathus | LV | fish > insects | 56 | 26 | 13 |
| H. xenostoma | LV | fish | 59 | 26 | 12 |
| H. squamipinnis | LG | fish + insects | 60 | 29 | 9 |
| H. plagiostoma | LV | fish | 61 | 26 | 11 |
| H. arcanus | LV | ? | 61 | 23 | 9 |
| H. dentex | LV | fish | 61 | 25 | 12 |
| H. macrognathus | LV | fish | 61 | 20 | 11 |
| H. longirostris | LV | fish + insects | 62 | 18 | 13 |
| H. serranus | LV | fish | 63 | 16 | 11 |
| H. decticostoma | LV | fish | 64 | 16 | 11 |
| H. bartoni | LV | fish | 65 | 20 | 10 |
| H. gowersi | LV | fish | 66 | 20 | 9 |
| H. spekii | LV | fish | 71 | 19 | 10 |

Abbreviations: LV = Lake Victoria; LG = Lake George; phar. moll.-cr. = intrapharyngeal mollusc-crusher.

Information on the food and feeding habits of the formal named species is from Greenwood's papers on the cichlid fishes of Lakes George and Victoria, summarized by Greenwood (1974, 1975). H. "Speke crusher" is a cheironym for a new mollusc-crushing Haplochromis species from the Speke Gulf of Lake Victoria (Greenwood and Barel, to be published). (From Barel and Voogt-Kokx, in preparation.)

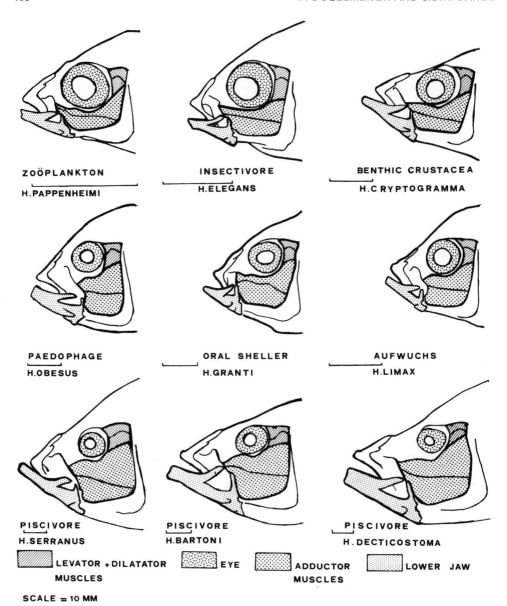

Fig. 8. Occupation pattern of the eye, the adductor complex and
the levator-dilatator complex. For all fishes the total area
available for the elements has been drawn to the same size. The
figure illustrates a reverse relation between the relative size of
the eye and the adductor complex. For the correlation between
the relative size of the adductor complex and the feeding habit
(or food), reference must be made to table 1. H. pappenheimi =
Haplochromis pappenheimi, etc. (From Voogt-Kokx, unpublished).

The preceding analysis of the occupation pattern is a strictly morphological method.  Only morphological phenomena are involved and the conclusions (dominance, subordinate) are, as yet, only based on these phenomena (see, for example, Zangerl {1948} who advocated a methodologically clear separation between methods and inferences, especially where the relations between morphology and phylogeny is concerned).

In the examples given here the constituent elements vary within an available area of (approximately) the same form (Figs. 6 and 8).  For evolutionary considerations the examples bear a number of important implications.  In constructing phylogenies it is often admitted that characters belonging to the same functional component (mostly called functional unit or character complex in phylogenetic studies) have low weight (e.g. Mayr, 1969).  From the foregoing it may be clear that characters belonging to completely different functional components (viz. the eye and the mandibular levation apparatus) demonstrate very strict morphological relations. For instance, one cannot argue that a well developed jaw apparatus and a small eye compared to a less well developed jaw apparatus and a large eye involve two independent character complexes.  In constructing phylogenies not only functional components, but the total pattern should be taken into account (Dullemeijer, 1970).

In the examples of patterns, the plasticity of form is closely related to changes in function.  From an evolutionary standpoint it would be interesting to know whether form could be changed without change in a specific function.  The four-bar model described earlier provides an example of morphological plasticity without significant change in this particular function.  It can be mathematically demonstrated that, within certain limits, the elongation of the horizontal bars CD and DA does not significantly influence the transmission-coefficient.  This means that, in evolution, fishes could elongate their heads without affecting the efficiency of the opercular mandibular depression apparatus.  In extant species, such an elongation is realized in the series Gasterosteus-Spinachia (Fig. 9).

For the foregoing, the evolution of the functional components, expressed in formula (6), may be symbolized as follows:

$$F_I = f\ (AB,BC,CD,DA,\psi) \rightarrow F_I = f\ (AB,BC^*,CD,DA^*,\psi) \tag{7}$$

in which * indicates the elongated bars.

**Gasterosteus aculeatus**

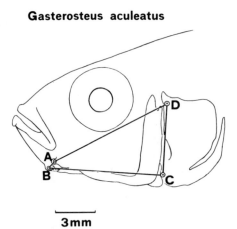

3mm

**Spinachia vulgaris**

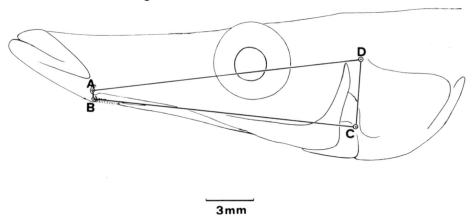

3mm

Fig. 9. The opercular mandibular depression apparatus in two
species of sticklebacks demonstrating the elongation of
the apparatus in <u>Spinachia</u>. The symbols are the same as
in Figures 2 and 3. Transformations such as this hardly
affect the kinetic function of the system.

The authors consider the example to be rather exceptional.
Generally the modification of a morphological feature x of an
element is indicative of modification in all other features of the
functional component(s) to which x belongs.  The following example
from the investigations of Van Oijen (in preparation) on paedophagous Haplochromis species may illustrate this statement.  It was
observed that the reinforced zone of the interoperculum (see
Barel, et al., 1976; Fig. 20) was relatively well developed in
paedophages as compared to this zone in the ancestral type
Haplochromis elegans. The interopercular reinforced zone is a
structural feature in both the hyoid and the opercular mandibular
depression apparatuses.  The elements of the hyoid depression
apparatus involve the chain lower jaw-ligament-interoperculum-
ligament-(hyoid + interhyal)-ligament-sternohyoid muscle-ventral
body musculature (Figs. 2, 10).  For all elements involved, the
features belonging to a depression apparatus could be shown to be
relatively more developed in a paedophage than in H. elegans
(Fig. 10).

In our Haplochromis research, the mutual relations between
elements in a functional component was further demonstrated by
analysis of the pharyngeal jaw apparatus of intrapharyngeal
mollusc-crushers, piscivores and insectivores (Witte and Barel,
1976; Barel et al., 1976).  Comparison of the shape of the upper
pharyngeal jaw (pharyngobranchial 3,4) resulted in many small but
consistent differences between the trophic types.  After including
other skeletal elements and the muscles, the observed differences
could be related to differences in all other elements of the
apparatus and - what is more important - could be explained in
terms of different functional demands imposed on the pharyngeal
jaw apparatus of piscivores and mollusc-crushers (Witte and Barel,
1976).  Apparently different functions do affect all elements involved.  This rigid relation between form and function can be
further demonstrated by comparing similar trophic types, which
evolved independently in different lakes.  Pharyngobranchial 3,4
of mollusc-crushing Haplochromis species from the Zambesi-system,
Lake George and Lake Victoria all exhibit the same organization of
shape-features related to the crushing function (Barel et al.,
1976; Witte and Barel, 1976; Witte, pers. comm. concerning the
Zambesi river).

All comparisons made above (rattlesnake-viper; occupation
patterns in the cheek of cichlids; Gasterosteus-Spinachia;
paedophages-H.elegans; mollusc-crushers-piscivores) can be described as series of idealistic transformations which may be treated
as models for evolutionary processes.  They demonstrate that:
(1) Changes are often reflected in many components.  They include
even the nervous system, as was demonstrated by Molenaar (1974),
who describes a tract and a nucleus in the trigeminal system of

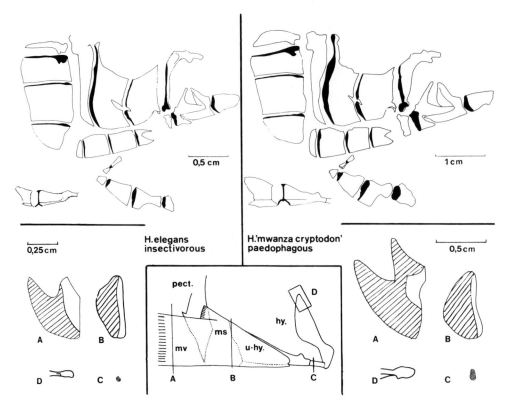

Fig. 10.   The opercular and hyoid mandibular depression apparatus
is of an insectivorous and a paedophagous Haplochromis species
compared.   The figure demonstrates the overall reinforcement of the
elements in the paedophagous hyoid depression apparatus.   The upper
figures represent 'exploded' medial views of the sectioned, skeletal
elements of the apparatuses.   The perspective suggestion is illusory.
The sections are made at right angles to the sagittal plane of the
elements and are figured in the plane of the drawing.   A, B and C
represent cross-sections, and the levels at which they were taken,
through the ventral body musculature, the sternohyoid muscle and the
urohyal-hyoid ligament respectively.   D:  the dorsal view of the
caudal part of the hyoid (marked by the square in the schematic
central figure) where the ligament between hyoid and interoperculum
inserts.   The magnification of the elements is such that the areas
occupied by the suspensoria are approximately equal.   H. "mwanza
cryptodon" is a cheironym of a presumed new, paedophagous Haplochromis species from the Mwanza Gulf of Lake Victoria, closely resembling H. cryptodon.   Abbr.:  hy. = hyoid; ms = sternohyoid muscle;
mv = ventral body musculature; pect. = pectoral element; u-hy. =
urohyal.   (From Van Oijen, unpublished).

certain snakes with infrared-receptors. This part of the trigeminal system is absent in <u>Vipera</u> <u>ammodytes</u> which does not possess such a receptor. Dominant and subordinate elements should be distinguished in such a coherent set of changes. (2) The variability and the freedom of elements is limited. The narrow interrelation largely reduces the variability of the elements and suggests canalization during ontogeny in the sense of Waddington (1966). A comparable canalization can be imagined in phylogeny. This narrow phylogenetic variability implies a minor role for natural selection. That natural selection mostly cannot affect single features but should affect a whole functional component, was demonstrated above and is variously argued by many evolutionary biologists (e.g. Hecht, 1965; Schaeffer et al., 1972). But the concept of dominant and subordinate functional components further limits the possibilities of selection. Subordinate components have to follow the dominant components, when this does not happen, the result is an abnormal or lethal ontogeny. It follows that natural selection should affect the variability of dominant components. That organisms with these variabilities did really exist must be proved.


CONCLUSIONS

The foregoing paper has considered phenomena, factors and conclusions both of evolutionary theory and of functional morphology. Logically evolutionary theory would not be expected to play an essential role in the explanation of functional morphology. When the explanation of functional components is based on evolutionary theory no reference point is available, because explanation takes place independently of evolution. Moreover no known variable can be obtained from the theory of evolution as is demanded by a deductive explanation.

Further it can be seriously doubted whether "evolutionary theory" can be called a theory at all. There is a presumed phenomenon - evolution - which is to be explained using various disciplines. In this explanation natural selection and mutation as essential concepts do play an important part, but they are by no means the only concepts. What is generally called "evolutionary theory" is a mixture of several components, involving phenomena, presuppositions, hypotheses and some explanatory principles of transformation. Only the last ones can be components of a theory. To refer to the mixture of all these components as a theory, causes confusion. When the phenomenon to be explained - the change of species - is indicated with the same word as the explaining principle(s) and when many such principles are concerned, a tautology is hard to exclude. To redirect an unexplained phenomenon from functional morphology to evolution becomes in this way inadmissible.

It should be stressed once more that explaining principles from evolutionary studies plays a role in questions concerning history and that many disciplines including functional morphology (Vogel, 1975; Reif, 1975) can profit from such principles, but they are not essential for the discipline. Conversely, evolutionary biology applies nearly all biological disciplines to the explanation of evolution as well as in the formulation of an evolutionary theory.

The contribution of functional morphology to evolutionary studies stems mainly from its ideas about the transformation of patterns. A pattern is a coherent system of various functional components. In comparing patterns in groups with related organisms, it could be demonstrated that the constituent functional components of a pattern are not equally plastic. Arranging these patterns in a morphological series, relatively unchanging (= dominant) functional components can be distinguished beside more plastic (= subordinate) ones. A change in these dominant components is followed by the plastic ones and therefore determines the transformation (evolution) of the pattern. For phylogenetic analysis the identification of the dominant components in a pattern is of crucial importance. These components should be treated as characters of high weight.


## ACKNOWLEDGEMENTS

The authors are indebted to the following people for their critical readings of the manuscript: Mrs. Thera Voogt-Kokx, Dr. Gerrit Ch. Anker, Dr. Herman Berkhoudt, Dr. Maarten Jan van Hasselt, Martien J.P. van Oijen and Dr. Peter Goody. A special word of thanks goes to Dr. P.H. Greenwood who spent much time in reading and discussing the various draughts of this paper.

The Lake Victoria Haplochromis studied in this paper were obtained from the British Museum (Natural History) and from material collected personally by Anker and Barel from the Lake. The authors are grateful to Dr. P.H. Greenwood for making available his collections and to the Netherlands Foundation for the Advancement of Tropical Research ("WOTRO") which financially supported the collection-trip of Anker and Barel.


## REFERENCES

Anker, G.Ch., 1974, Morphology and kinetics of the head of the stickleback, Gasterosteus aculeatus. Trans. zool. Soc. Lond., 32:311-416.

Barel, C.D.N., Berkhoudt, H., and Anker, G.Ch., 1975, Functional
    aspects of four bar systems as models for mouth-opening
    mechanism in teleost fishes.  Acta Morph. Neerl.-Scand.,
    1975:228-229.
Barel, C.D.N., Witte, F., and Van Oijen, M.J.P., 1976, The shape
    of the skeletal elements in the head of a generalized
    Haplochromis species: H. elegans. Trewavas 1933. Neth. J.
    Zool., 26:163-265.
Barel, C.D.N., Van der Meulen, J.W., and Berkhoudt, H., in press.
    Kinematische Transmissions-Koeffizient und Vierstangensystem
    als Funktion-Parameter und Formmodell fur mandibulare
    Depressions-Apparate bei Teleostiern. Ergeb. exp. Medizin.
Beckner, N., 1968, The biological way of thought.  Cal. Univ.
    Press. Berkeley, Los Angeles, 200 p.
Bock, W.J. and Von Wahlert, G., 1965, Adaptation and the form-
    function complex.  Evolution, 19:269-299.
Bonik, K., Gutmann, W.G., and Peters, D.S. (to be published),
    Optimierung und Okonomisierung im Kontext der Evolutions-
    theorie und phylogenetischer Rekonstruktionen.  Acta
    Biotheoretica.
Crompton, A.W., 1963, On the lower jaw of Diarthrognathus and the
    origin of the mammalian jaw.  Proc. zool. Soc. Lond., 140:
    697-753.
Cuvier, G., 1805, Leçons d'anatomie comparée, Baudoin, Paris.
Dobzhansky, T., 1968, On some fundamental concepts of darwinian
    biology.  Evolution Biol., 2:1-34.
Driesch, H., 1909, Philosophie des Organischen.  Leipzig,
    Engelmann.
Dullemeijer, P., 1956, The functional morphology of the head of
    the common viper, Vipera berus (L.).  Arch. neerl. Zool.,
    11:386-497.
Dullemeijer, P., 1958, The mutual structural influence of the
    elements in a pattern.  Arch. neerl. Zool., 13:suppl. 174-188.
Dullemeijer, P., 1959, A comparative functional-anatomical study
    of the heads of some viperidae.  Morph. Jahrb. 99:881-985.
Dullemeijer, P., 1970, Evolution of patterns and patterns of
    evolution.  Forma et Functio, 3:223-232.
Dullemeijer, P., 1974, Concepts and approaches in animal morphology.
    van Gorcum, Assen.
Gans, C., 1960, A taxonomic revision of the Trogopophina and a
    functional interpretation of the amphisbaenial pattern.
    Bull. Am. Mus. Nat. Hist., 119:129-204.
Gans, C., 1969, Functional components versus mechanical units in
    descriptive morphology.  J. Morph., 128:365-368.
Greenwood, P.H., 1960, A revision of the Lake Victoria Haplochromis
    species (Pisces, Cichlidae).  Part IV.  Bull. Br. Mus. Nat.
    Hist. (Zool.), 6:227-281.
Greenwood, P.H., 1967, A revision of the Lake Victoria Haplochromis
    species (Pisces, Cichlidae).  Part VI.  Bull. Br. Mus. Nat.
    Hist. (Zool.), 15:29-119.

Greenwood, P.H., 1973, A revision of the Haplochromis and related
    species (Pisces, Cichlidae) from Lake George, Uganda. Bull.
    Br. Mus. Nat. Hist., 25:141-242.
Greenwood, P.H., 1974, The cichlid fishes of Lake Victoria, East
    Africa: The biology and evolution of a species flock. Bull.
    Br. Mus. Nat. Hist. suppl. 6.
Gutmann, W.F., 1966, Funktionsmorphologische Beitrage zur
    Gastraeacoelom-theorie. Senck. Biol., 47:225-250.
Gutmann, W.F., 1972, Die Hydroskelett-Theorie. Aufsatze und
    Reden der Senckenbergischen Naturforschenden Gesellschaft,
    No. 21:1-41.
Gutmann, W.F., and Peters, D.S., 1973, Das Grundprinzip des
    wissenschaftlichen Procedere und die Widerlegung der phylo-
    genetisch verbramten Morphologie, in "Phylogenetische
    Rekonstruktionen-Theorie und Praxis," Aufsatze und Reden der
    Senckenbergischen Naturforschenden Gesellschaft, No. 24:
    7-25.
Hartenberg, R.S., and Denavit, J., 1964, Kinematic synthesis of
    linkages. McGraw-Hill.
Hecht, M., 1965, The role of natural selection and evolutionary
    rates in the origin of higher levels of organization. Syst.
    Zool., 14:301-317.
Hempel, C.G., and Oppenheim, P., 1953, The logic of explanation,
    in "Readings in the philosophy of science," (H. Feigl and
    M. Brodbeck, eds.), Appleton Century-Crafts, New York,
    319-352.
Jeuken, M., 1958, Function in Biology. Acta Biotheoretica,
    13:29-46.
Klaauw, C.J. van der, 1948-1952, Size and position of the
    functional components of the skull. A contribution to the
    knowledge of the architecture of the skull, based on data
    in the literature. Arch. Neerl. Zool., 9:1-558.
Liem, K.F., 1970, Comparative functional anatomy of the Nandidae
    (Pisces: Teleostei). Fieldiana: Zoology Memoirs, 56:1-166.
Liem, K.F., 1973, Evolutionary strategies and morphological
    innovations: cichlid pharyngeal jaws. Syst. Zool., 22:425-
    441.
Liem, K.F., and Stewart, D.J., 1976, Evolution of the scale-eating
    cichlid fishes of Lake Tanganyka. A generic revision with a
    description of a new species. Bull. Mus. Comp. Zool.,
    147:319-350.
Maglio, V.J., 1972, Evolution of Mastication in the Elephantidae.
    Evolution, 26:638-658.
Mayr, E., 1963, Animal species and evolution. Harvard University
    Press, Cambridge, Mass.
Mayr, E., 1969, Principles of systematic zoology. McGraw-Hill.
Molenaar, G.J., 1974, An additional trigeminal system in certain
    snakes possessing infrared receptors. Brain Research, 78:
    340-344.

Reif, W.E., 1975, Lenkende und limitierende Faktoren in der
    Evolution.  Acta Biotheoretica, 24:136-162.
Schaeffer, B., Hecht, M., and Eldredge, N., 1972, Phylogeny and
    paleontology.  Evolutionary Biology, 6:31-46.
Simpson, G.G., 1961, The major features of evolution.  Columbia
    University Press. New York. 434 p.
Vogel, K., 1975, Funktionsmorphologie als Hilfsmittel
    palaontologischer Evolutionsforschung.  Franz Steiner Verlag.
    Germany. 19 p.
Waddington, C.H., 1966, Principles of Development and Differentia-
    tion.  Mac Millan, New York, X + 115 p.
Witte, F. and Barel, C.D.N., 1976, The comparative functional
    morphology of the pharyngeal jaw apparatus of piscivorous
    and intrapharyngeal mollusc-crushing Haplochromis species.
    Proc. Second European Ichthyological Congress (to be published).
Zangerl, R., 1948, The methods of comparative anatomy and its
    contribution to the study of evolution.  Evolution, 2:351-374.

# EMBRYOGENESIS, MORPHOGENESIS, GENETICS AND EVOLUTION

Michel DELSOL

Ecole Pratique des Hautes Etudes (Sorbonne-Paris)
Faculte Catholique des Sciences de Lyon
25 rue du Plat, 69288 Lyon, France

## INTRODUCTION

For a long time biologists have known that comparisons can be made between the succession of species in a lineage (phylogeny) and the succession of embryonic stages of a species situated at the end of this line (ontogeny). But it was Haeckel (1866, 1875a, 1875b) who first codified and explained these comparisons, allowing their integration within the setting of the mechanisms and modalities of evolution.

Haeckel divided the embryonic structures of living beings into two categories: palingenetic and caenogenetic. He said that in most cases ontogeny is only a recapitulation of phylogeny. The stages of embryonic development are a condensed and abbreviated repetition of the history of the species; e.g. the branchial openings which appear in the mammalian embryo reproduce the gill pouches of the ancestral fish. These embryonic organs which recall the ancestral state are called palingenetic. From this point of view, the succession of palingenetic structures can be schematized in the following way: the embryogenesis of a group B, supposed to be derived from group A, reproduces, partly at least, the embryonic form of ancestral group A before reaching its own structure.

The characteristics or organs called caenogenetic are present to ensure the survival of the embryo and so they disappear with embryonic life. The mammalian placenta linked with the embryo by the umbilical cord is a typical caenogenetic organ. These characteristics are generally recently acquired (caenogenesis - of recent

119

origin) in the history of the line.

It was soon recognized that Haeckel's theses required modification and several authors studied these problems. Some were very critical of Haeckel's work especially de Beer in his work "Embryology and Evolution" (1929) and in the successive editions of "Embryos and Ancestors" (1940, 1951, 1958). In the french edition of "Embryology and Evolution" (translated by J. Rostand in 1932), he wrote that his own aim was: "d'edifier sur les ruines de la theorie recapitulationniste une synthese plus adequate de notre savoir, touchant les rapports entre l'embryologie et l'evolution."

It is obvious that Haeckel's theories corresponded only to a first approximation, but the judgement put forward by de Beer is, in our opinion, too severe. Moreover de Beer's study was not sufficiently analytical. He merely offered comprehensive definitions of caenogenesis, deviation, neoteny, reduction, adult variation, hypermorphosis, acceleration, paedomorphosis and repetition, one after the other. In this paper we are going to try and analyze these phenomena more fully. This study represents an extension of work begun and partly published in collaboration with H. Tintant (Delsol and Tintant, 1971).

ONTOGENY AND PHYLOGENY:  ATTEMPT AT A GENERAL ANALYSIS

The distinction between embryo and adult is artificial. A being, during its existence, passes through various morphological stages but its life cannot be divided into two parts:  a phase of transformation - embryogenesis, and a phase when the being is finished and does not change any more - adult life. Indeed, even in old organisms, morphological transformations may occur. Nevertheless, bearing in mind the artificial nature of it, the succession of two phases in the existence of most living beings can be admitted: (1) a phase of active morphogenesis called embryonic development; (2) a phase of relative stability, called the adult phase, which can be characterized by the completion or destruction of some structures, and occasionally by the acquisition of a few new structures. The phase of relative stability is also generally characterized by the living being's capacity for reproduction. This division into two stages is mainly to be found in the more highly evolved forms and especially in vertebrates.

In order to define the relations between phylogeny and ontogeny logically and completely, a systematic analysis of the different possibilities could be attempted in two ways: (1) the different ways in which the new hereditary features become visible in ontogeny should be studied and the relations between these phenomena and phylogeny established; (2) the way in which the new hereditary features become visible in phylogeny should be analyzed and the relations between these phenomena and ontogeny established. However, this subdivision proved inconvenient so that it was thought better to divide the hereditary modifications which affect the individuals of a line, and which therefore can modify the direction of phylogeny, into two categories, whatever their age: (1) new features which affect embryonic development or adult life, i.e. the whole of final morphogenesis - readily recognizable in a phylogenetic series; (2) new features which only affect embryonic development and which cannot be seen in adults - difficult to identify in a phylogenetic series. In addition to these two categories of hereditary modifications we will also devote a paragraph to the effects attributable to allometries.

An important point must be made clear before beginning this study. The fact must be stressed that since each feature can generally evolve independently, it will be studied separately here. Sometimes in batrachians, neoteny is observed in only one organ. Nevertheless the fact must be kept in mind that nearly always the evolution of a given feature is more or less linked with the evolution of other features. So in batrachians, groups of features become neotenic because they are linked together by physiological relations which make this common evolution inevitable; e.g. in the axolotl, gills are neotenic but their vascularization has disappeared.

In order to elucidate the facts, we will follow the scheme set out in Figure 1. This general table is divided in two parts, each of which corresponds to the two types of hereditary modifications that we have defined above. In the figure, we start from an early original species of which the development is simplified. In this species we shall distinguish the following: an embryonic stage E characterized by four embryonic features, E1, E2, E3, E4; an adult stage A consisting of juvenile or pre-adult stage Pa and a true adult stage A, characterized by genital glands G.

### New Features which Permanently Affect Embryo
### Or Adult and Their Connection with Phylogeny

A feature can become permanently fixed in a new subject in five ways: four of these ways make the ontogenetical cycle longer or more complex, even if this longer cycle corresponds to the disappearance of a feature; the fifth actually shortens the cycle, development stopping at a stage before that which characterized the ancestor.

Features Appearing at the End of Embryonic
Development: Palingenesis or Recapitulation of Embryonic Stages

When the appearance or addition of new features takes place
at the end of embryonic development, the new organism B or C
develops by adding features B or C to the features A or B of the
form that preceded it.  It is therefore natural for organism B or
C before completion, to go through the stage of the organism A
that immediately preceded it, since B corresponds to A + B, or
C to A + B + C.  In the case of organism C, we omit E2 to illustrate
the frequent condensation of the embryonic stages in palingenesis.
Ancestral adult features then become embryonic features of the new
adult, such embryonic features correspond to Haeckel's palingenetic
features.  Palingenesis disturbs both ontogeny and phylogeny, the
relation between the two is expressed as recapitulation.  There
are three types of recapitulation: (1) recapitulation due to
embryological necessity - true palingenesis; (2) recapitulation
through retrieval for a new use; (3) recapitulation through
heritage.

Recapitulation due to embryological necessity.  The embryonic
development of an adult includes its ancestor's stages either
because the structures of the present adult depend on the embryonic
inductions produced by the ancestor's organs or it uses organs
which existed in the ancestor as basic material.  Reference to
renal and genital structures illustrate these phenomena.  The
mesonephros can appear only after the pronephros because it is induced by the primary ureter which is the duct of the pronephros.
The pronephros - mesonephros succession appears as typically re-
capitulatory because of the process of embryonic induction.  The
Mullerian duct can develop only after the formation of the primary
ureter because, in some species at least, it originates from the
degenerating pronephric duct.  In both of these cases, recapitulation is a necessity.

Recapitulation through retrieval for a new use.  In some cases,
the recapitulated organ acquires a new function and is therefore
retained.  In Nectophrynoides, a respiratory role has been attributed to the tail of the embryos on account of its abundant peripheral vascularization  (Lamotte and Xavier, 1972).  In some
snakes, reduced limbs offer secondary sexual characters.

Recapitulation through heritage.  In cases where recapitulation corresponds merely to the presence of an organ which seems
useless, this organ is said to take shape through heritage.  Many
examples are known in vertebrates but in all these cases the
"uselessness" of the organ will be open to debate.  It is generally
accepted by comparative anatomists that six aortic arches were
present in primitive fish, but in teleosteans, batrachians and

dipnoans, only four aortic arches are known (numbers III, IV, V and VI). In very young <u>Xenopus</u> embryos, aortic arch I and rudiments of aortic arch II form (Millard, 1941, 1945), but degenerate very quickly. However, Terhal (1941) pointed out that he could find no vessels recalling the first two aortic arches in the Urodela he had studied, especially <u>Salamandra maculosa</u>. In anuran tadpoles, the dorsal aortae send out two vessels which seem to correspond to the efferent vessels (mandibular and efferent hyoidean) of the first two arches (Marshall, 1951). They do not develop because of the absence of gills at these stages and there are no afferent vessels.

The degeneration of the eyes has been the object of many studies in fishes and subterranean Urodela. In <u>Proteus anguinus</u>, the optic vesicle evaginates normally from the prosencephalon; the optic cup takes shape while the crystalline lens and cornea differentiate. After hatching, the eye retains its embryonic nature but it only grows slowly, so that the number of retinal cells increase from 2,000 to only 8,000 in ten years. Moreover, the cornea becomes opaque and re-forms into epidermis from which it was derived originally, while degenerative processes appear in the crystalline lens (Durand, 1976).

The regression of limbs in tetrapods has been studied by many authors (see Gans, 1975 for a recent analysis). In a comparison of the development of reptilian limbs, Raynaud (1972) was able to observe several types of rudimentary development. Reptiles with forelimbs which are well developed in the adult have eight somites involved in the constitution of the limb. It is probably the apical crest which induces the growth of the underlaying mesoblast, because in the chicken the ablation of this crest stops differentiation of the early form of the limb (Raynaud, 1972). In <u>Anguis fragilis</u>, the blind-worm, both fore and hindlimbs take shape, remain rudimentary for some time and then regress. Moreover in this species, there are only four somites in the constitution of the early form. In the reticulated python, only the hindlimb takes shape, while in the embryonic grass-snake, <u>Tropidonotus</u>, only the sides of the posterior part of the trunk swell out. According to Raynaud's observations on the blind-worm, the spontaneous degeneration of the epiblastic apical crest early on in limb formation could explain why limb development stops.

It is obvious that in the facts we have just described, the development of the limb, the eye or the anterior aortic arches, we have no reason to think that such early developments have any embryological part to play. We are therefore led to the conclusion that, at genome level, genes are present which induce the formation of useless organs. A series of normal embryological stages seem to develop but apparently get "jammed". In this case,

the impression is that selection has acted only on the disappearance of the adult stage and "has found it unnecessary" to cause
its early form to disappear, because the latter has become a
neutral organ.  It is also likely that selection may not have had
time to act on the early form.  As we do not know under what influences the genes come into action sequentially during embryonic
development, we have no way of knowing why the process stops but
we will insist again on the fact that in all cases described here,
one could question the real usefulness of the organ (Gasc, 1970).

The phylogenetic interest of recapitulation is obvious which-
ever aspect recapitulation takes, "true embryological necessity"
or "through habit".  The study of batrachian larvae and of their
metamorphosis offers several examples.  In the tadpole, as ammonia
excretion precedes that of urea, the constitution of the aortic
system, the branchial system, the kidneys and the main axis of the
tail obviously recall the structure in the ancestral fish.  In
some cases embryological comparisons offer real sequences of development which are so similar that it is very much easier to
recognize homologies than in comparative anatomy.  A good example
is supplied by the transformations of the skin during metamorphosis.
In batrachian tadpoles, the skin is composed of four layers: (1) a
layer of epidermal cells, two or three cells deep; (2) a thin
basic sub-epidermal membrane hardly visible through the light
microscope, the "basement membrane" or "adepidermal membrane"
(Salpeter and Singer, 1960); (3) a layer mainly composed of
collagen fibers, 3 to $6\mu$ thick, the "basement lamella" or "basic
lamella"; (4) a single-celled layer of mesenchymatous dermal cells.
During metamorphosis the skin of the body is seen to undergo
changes.  Initially the basement lamella is invaded by mesenchymatous dermal cells of the fibroblastic type.  Subsequently,
the basement lamella detaches itself from the epidermis and dis-
sociates, the fibroblasts and melanophores multiply in the sub-
epidermal space to form the stratum spongiosum. Then the basement
lamella is reformed beneath this stratum spongiosum and will become the deepest layer of the skin, or stratum compactum (Kemp,
1961).  The phenomena are brought about in exactly the same way
in Anura and Urodela and we can infer that the various skin layers
are certainly homologous in these two groups of batrachians.


Features Appearing in the Adult Stage:  Adult Variation

When the appearance or the addition of a new feature (V) is
visible only at an adult stage, the new organism undergoes embryological development like the one that preceded it.  Several
examples of this process can be seen particularly in features that
correspond to the appearance of new behavior patterns (modifications in nuptial parade).  Allometries which in adults results in

hypertely can also be mentioned.  They form the "bizarre" structures that have been studied for a long time and recently so by
Gould (1974) (Megalocerus giganteus); we shall come back to this
point later.

In such cases it can easily be understood that, in theory,
adult variation modifies the line and therefore disturbs phylogeny
but will not be visible in the embryo.  Nevertheless it should be
kept in mind that the genotype of individuals affected by adult
variation is necessarily modified.  This genotypic variation can
result in clandestine cytological modifications during embryonic
development.

## Features Appearing During Embryonic
## Development and Persisting in the Adult:  Divergence

In this category are included new permanent features which
appear in an embryo which has not completed its development.  In
Figure 1, stage E 4 will become E 4 + D and the adult features
which follow will be disturbed by the new features which appeared
while the embryo was developing.  The subject to be born will be
a new type.  This phenomenon therefore first disturbs ontogeny
and then phylogeny.  Phylogenetic modifications result from disturbances which take place in the course of embryonic development.

## Proterogenesis

Some paleontologists, especially Schindewolf (1936, 1950),
have described a phenomenon which is connected with divergence
and is worth studying separately.  Like divergence it is quite
different from palingenesis, in fact almost the reverse.  In some
fossil series, new features can be observed that first appear in
the embryo but develop no further in the adult (Pr in Figure 1).
Yet if we look over the descendants of these lines, the feature
which appeared progressively in the embryo remains in the course
of growth and is finally found in the adults of the most evolved
forms.

Ammonites provide a good example.  Their shell is composed of
a conical tube coiled up in one plane, the animal only occupying
the end of this tube.  As it grows the animal gradually moves
away from its initial housing situated at the center of the coil
as it partitions off empty spaces, behind it, which serve as
floats.  Thus an ammonite shell that is preserved in good condition allows the characteristics of the successive stages of the
animal to be observed simultaneously from youth to adult age.  For
instance, modifications of the ornamentation can be followed in

the course of its ontogenic development.  In some lines new features appear such as reinforced ribs or tubercles, not in the
adult but in the young animal in the internal windings of the coil.
The descendants of this species will show these ribs or tubercles
on the following windings and little by little, after several
generations, this ornamental feature will be found at the outer-
most coil of the adult shell on the home-place.  With time, it has
become an adult feature.

We are therefore inclined to think that in the early stages
of the lineage the genes responsible for the ornamentation function in early ontogeny but are then subsequently inhibited.  In
the descendants the action of these genes takes place at later and
later stages.  This phenomenon suggests both divergence and
caenogenesis:  it differs from divergence because the feature remains throughout life; it differs from caenogenesis because it
tends to remain in the adult.  This process, called proterogenesis
by Schindewolf, is radically opposed to recapitulation, or palingenesis, the only process considered by Haeckel.

Proterogenesis seems not to have been appreciated by most
embryologists.  This may be explained by the uneven distribution
of this phenomenon in the different zoological groups.  Although
very frequent in invertebrates, it seems to be absent in vertebrates, or at least we know of no convincing examples.  Perhaps
the increasing complexity of organisms has, in the course of their
evolution, caused some details, frequent in simpler forms of life,
to be eliminated.


Features not Appearing at the End of Embryonic Development
and Development Stopping Before the Final Stage: Neoteny

Neoteny is principally characterized by the following facts:
"persistence of organs or dispositions special to larvae, non-
appearance of organs or dispositions special to adults" (Vachon,
1944).  In the new organism, features which could have been
embryonic are seen to appear in the adult.  For some of them the
embryo does not develop beyond a definite embryonic stage and yet
acquires genital organs which is the essential feature characterizing the adult stage.  This therefore represents real shortening
of the cycle.  From the genetic point of view, this phenomenon is
interesting because it is an example of species, at least the
facultative neotenic ones, where there are genes which would
normally have acted at the end of embryonic development and which,
in fact, have remained inactive.

### Features Appearing in the Embryo but not Apparent in the Adult:  Temporary Embryonic Features

Caenogenesis.  Caenogenesis is when a new feature (C in Figure 1) appears in an embryo and then disappears without leaving any traces at the end of embryonic life.  Dollander and Fenart (1970) suggest differentiating in the embryo between those features which correspond to the organization of the body, ontosomatic features, and those features responsible for the life and protection of the embryo, embryotrophic features.  On principle, an embryotrophic character brings no evolutionary novelty to the adult animal.  Caenogenetic organs differ from the other protective structures of the egg or the embryo because they come from the fertilized egg itself.  They are parts of the egg which, instead of producing adult organs, only protect and maintain the life of the embryo.  The appearance of a caenogenetic feature demonstrates that the pool of genes in individuals in the species studied has been changed.

This term "caenogenetic" should be used in a very broad sense. Many caenogenetic organs or organelles can be recognized in vertebrates: e.g. the egg yolk and the organs ensuring larval life (the cornified denticles, the suckers or the tentacles, the elongated structure of the intestine, the branchial filters, the spiracular system, etc.) in batrachians.  All such structures which are often called adaptive structures in larval forms and which exist only in larvae must be considered as caenogenetic. The adaptive structures described by Wassersug (1975) in batrachian tadpoles belong in the same category.

At the end of embryonic development, caenogenetic structures can simply disappear or be replaced.  In batrachians, for example, the cornified denticles and the internal and external gills disappear, the intestine is transformed, etc.  The case of the intestine in batrachians is very interesting.  In the anuran larva, this system is composed of a very simple single-celled epithelium with very small cells recognizable only from their position in the angles between the epithelial cells.  During metamorphosis, these tiny cells multiply actively and form a second epithelial ring which replaces the first, degenerating, one.  In Urodela, the embryonic cells form buds, or minute accumulations of three or four cells.  During metamorphosis, they simply multiply more actively because there is no real metamorphosis in this group but merely a quicker replacement of the intestinal epithelium.  These cells do not disappear completely but form accumulations which ensure the regular regeneration of the functional intestinal cells.  They differ from epithelial cells because they have a large nucleus and very little cytoplasm.

In invertebrates there are morphogenetic sequences characterized by the presence of successive forms. In some cases, a temporary form develops which will completely disappear in the course
of metamorphosis to be replaced by a collection of cells which
have remained embryonic; here we really have very important
caenogenesis.

In protochordates, there are types of budding which can be
connected with caenogenetic phenomena. If, therefore, all larval
structures that last only for a time and then disappear after ensuring the transmission of cells which will produce the adult are
being considered, it must be recognized that caenogenesis is much
less developed in vertebrates than in some other branches of the
animal world. That may be partly due to the fact that, in this
group, latent embryonic cells cannot take shape any longer; the
only known exception is the intestine of batrachian tadpoles.
Thus in vertebrates, cellular types would differentiate earlier
and in a more obligatory way than in invertebrates.

Embryonic dischrony. Dischrony (de Beer's heterochrony) is
when an embryonic feature that used to evolve at a particular
stage in a given subject appears at a different stage in the
development of its descendants. Here is an example borrowed from
de Beer (1932); the frog heart develops only when the embryo has
reached its final form whereas the chicken heart appears considerably earlier. Its appearance compared with that of the rest of
the organism has been accelerated. These phenomena do not necessarily have any effects in the adult. However, it is easy to see
that in the case of dischrony, as for caenogenesis, these modifications can reach the adult indirectly since they are a sign or
changes in gene relations. It should also be noted that if embryonic dyschronies are prolonged into the adult stage, they become divergences.

The phenomenon of differentiation dyschrony should also be
mentioned. A scale of differentiation could be established for
many organs or organ systems which would theoretically reach
their maxima in the adult and would be at a minimum when the
organs appeared. This would, for instance, be easy for the
batrachian's skin which, from the moment when the embryonic epidermis is in place until the adult stage, undergoes a series of
slow transformations, the rhythm of which is, however, quicker
during metamorphosis. If we could give each organ of the body a
differentiation scale, it would probably be noticed that dyschronies are much more frequent than is thought. If such scales
were followed, dyschrony could become neoteny in extreme cases.
Indeed, the slowing down of morphogenesis would be such that the
organ would reach adult age without completing its differentiation;
it would then be neotenic.

### The Particular Case of Allometries

The problem of allometries has already been referred to.  It is important to come back to it since, for some years, various authors have been insisting on the important part played by these phenomena in specific transformations and hence in evolution.  It should be noted in particular that allometries can in some cases end in really new structures (d'Arcy Thompson, 1942; Stahl, 1962; Szarski, 1964; Gould, 1966a, 1966b, 1971, 1974, 1975).  The appearance of a variation in the speed of growth of an organ can occur in nearly all the kinds of relations between embryonic development and evolution described above.  In the most typical and simplest cases, allometry results in hypertely which will be detectable only in adults.  These "bizarre" structures are adult variations.  If allometry appears directly embryonic development is at an end, it corresponds to the standard phenomenon of palingenesis.  If it appears in the early stage of embryonic development and persists into the adult, it is a case of divergence.  It is a well known fact that the proportions of the human fetus and adult show wide differences due to variations in relative growth rates.  Increasing or decreasing allometric growths which appear in embryonic development are prolonged into the adult stage:  they are typical divergences.

Allometries may also appear as proterogeneses.  We know of no examples of them, but the case is theoretically possible.  In other cases, allometry can lead to neoteny.  Artificially accelerating genital development in anurans results in partial neoteny, corresponding to creating increasing allometry of the genital glands.  Some allometries may also result in caenogenesis.  When the intestine of batrachian tadpoles grows inordinately long at the beginning of embryonic life, it obviously shows marked positive allometry which results in a caenogenetic organ.  Lastly, embryonic dyschronies are often only allometric phenomena.

## EPIGENESIS, EMBRYOGENESIS AND EVOLUTION

For some years now, the study of the part played by the action of external factors on embryonic development has been resumed.  These factors have been referred to under the old term of epigenesis, the meaning of which has, however, been altered.  The origin of the word epigenesis is well known.  Some naturalists claimed that the embryonic development of a living being was nothing but the progressive growth of a complex germ which, from the beginning, is constituted like the future adult.  This theory of preformation was generally accepted until the 19th century.  Other naturalists believed that the complex structure of the final living being developed progressively from a very simple form, the

epigenetic theory.  In 1759, Wolff demonstrated in chickens that
the organs developed progressively from simple forms constituted
by "lames et replis," that is to say through epigenesis.  The old
controversy was at an end.  Nevertheless it might be acceptable to
say that there were some correct ideas in the notions upheld by
the partisans of preformation since the embryo, though not "pre-
formed," is genetically "predetermined," which is an idea close
to preformation.  Vodop'janova and Kremjanskij (1954) have even
maintained that the notions of preformation and epigenesis correspond to a dialectical unity.

The term epigenesis has reappeared in the course of recent
decades with an apparently different meaning which, in fact, is
derived from its primitive sense.  For Waddington (especially
1969), an epigenetic factor corresponds to an environmental factor
which disturbs the normal function of a gene inducing it to build
up an unusual structure.  Waddington stressed the fact that the
milieu could modify the genes only within certain limits:  the
internal limits to this action of the gene he called "epigenetic
space."  It is only within this space that the gene can pursue its
normal development, following the ways this author calls "chreodes."
It is obvious that, from this stand-point, epigenetic factors can
only modify the sequence of phenomena set into action by the gene
in small proportions.  When an experimental embryologist drastically interferes with embryonic development causing the production of
abnormal beings, he is producing subjects whose development takes
place away from the epigenetic space.  These individuals cannot
therefore result in normal beings; they are lethal.  It is important to note that these views are in complete agreement with those
proposed by authors who support the synthetic theory of evolution.
Indeed, for Waddington the following points are obvious:  the
cytoplasmic substratum does not alter the structure of the gene;
the substratum modifies the action of the gene only and can influence that action for the present generation only; the characters of the substratum itself were formed during oogenesis under
the influence of the nucleus.

Thus, Waddington's views lead us basically to insist on the
special part played by the environment in the mode of action of
the genes.  Further, there is an important point to be noted here.
When we emphasize the role of the cytoplasmic environment, we must
not forget that the cytoplasm (just like the DNA itself) may be
disturbed by external factors: e.g. temperature, light or hormones.
In this respect, all disturbances that the external environment
may impose upon a living being without at same time altering the
genome in a permanent way must be looked upon as epigenetic.  To
illustrate these facts one may mention all the disturbances that
the students of causal embryology have provoked in eggs, embryos
and larvae in the course of the last century.

From a few examples, we shall be able to understand better those epigenetic disturbances.  An <u>Ambystoma</u> can reproduce before or after metamorphosis according to the environment in which it develops.  In this case, the epigenetic space being very wide, a failure to metamorphose does not mean an end to life or reproduction.  I demonstrated more than twenty years ago (Delsol, 1952, 1953) that in anurans the development of genital glands was independent of that of the rest of the organism and even could, in some cases, be quicker if the tadpole did not undergo metamorphosis.  Such experimental neoteny does not reach its final development for the ducts evacuating sexual products can build up only after metamorphosis.  This is a typical case of a transformation of epigenetic origin that has no result since the transformed individual is doomed to die without descendants.  The experiment caused normal embryonic development to leave the allowed epigenetic space (Delsol, 1953).  Another kind of experiment also allows these phenomena to be defined.  It has long been known from the work of various authors (Helff and Clausen, 1929; Lindeman, 1929, Clausen, 1930) that the tail skin of the batrachian tadpoles grafted on to the body of the same individual (self-grafting) degenerates at metamorphosis in the same way as if it had remained in its proper place.  We showed (Delsol and Flatin, 1969) that this phenomenon did not take place in <u>Alytes obstetricans</u> tadpoles if the tail skin is grafted under the skin of the back.  In this case the grafted tail skin, instead of degenerating at metamorphosis, is completely transformed and even shows the characteristic glands of adult skin.  Zouzouko (1970) confirmed these experiments in <u>Discoglossus</u>.  Here is therefore a typical case of the part played by the environment in embryonic development.  Under normal conditions, the tail skin degenerates at metamorphosis; grafted in the place of normal skin on the surface, it degenerates; grafted under the body skin, it is transformed.

All this shows that Waddington's chreodes can sometimes have a very wide epigenetic space at their disposal.  The action of external features on development therefore reveals possibilities in the modification of ontogeny that are much more extensive that could be believed.

Recently two further authors have considered epigenesis. Brien (1962, 1969, 1974a, 1974b) has widened the meaning of the word recalling that the genetic activities governing ontogeny are under the control of external factors which can be intracellular or intercellular.  Research in causal embryology at the tissue level has shown fully the part played by the interaction of other tissues in embryonic development.  Brien maintains that an embryonic tissue does not develop according to programmed information imposed in an absolute way by a genetic code.  In order to develop normally the tissue must receive inductions from the neighbouring

developing areas.  He also then calls epigenetic factors such inter-
cellular and tissue interactions which act on genic activities and
can alter their course, a real return to Lamarckism.  Indeed
Brien suggests that it is in the course of embryonic development
and according to an epigenetic process that the environment acts
on the genome and alters it directly.

Lovtrup (1974) is a second author who has recently considered
epigenesis.  His ideas and his definition of epigenesis recall
those of Brien.  Epigenesis is seen as a series of events external to the genome represented by information originally located
in what he calls the embryonic substrate.  This information either
in the form of real but non-genetic information, or in the form of
events corresponding to embryogenesis, may be altered, which could
then modify some embryogenetic processes.  Lovtrup considers the
mechanisms that can alter embryonic development:  mutations on the
one hand, and changes in the cytoplasm or in the cortex of the
fertilized egg on the other.  For Lovtrup: "Variation in the
spatio-temporal distribution of the morphogenetic events, as dictated by the nature of the substrate, seem to be at least of as
great, if not greater, importance for phylogenetic progression
than changes in the predetermined mechanisms."  He considers that
it becomes difficult to distinguish between epigenesis and ontogeny
for, if ontogeny is merely the description of the steps through
which the embryo develops, epigenesis appears as the causal explanation of these developmental steps.  Having analyzed many
epigenetic phenomena, Lovtrup sums up Darwinism and Neo-Darwinism
and suggests a new theory which he calls "the complete theory."
For instance, he thinks that epigenesis can explain the phenomena
of divergence and hypertelies and above all some little known
stages of phylogeny corresponding to the passage from one higher
taxon to another.  For students of the "synthetic theory of
evolution" causal embryology is nothing more than phenogenetics.
Epigenesis then represents only the external factors that can more
or less modify what the genes should have produced under normal
conditions.  Brien's and Lovtrup's theories are unacceptable.


GENETICS, EMBRYOGENESIS AND EVOLUTION

In this paper we have considered characters that are being
studied independently of all explanatory theories of evolution.
In our views, the characters which we analyze originate from
mutations and the new species which emerge according to the
processes that we have described will be subject to selection and
accordingly they will survive or disappear depending on the circumstances.  Investigations concerning the relations between embryology and evolution aim only at understanding the causes of the
observed facts of recapitulation and analyzing the possibilities

of relations between phylogeny and ontogeny.  However from the
genetic standpoint, the facts described can shed new light.

As already stressed, they first of all confirm the opinion of
geneticists that at every stage of life in any living being only
a part of the genome is functional.  If the fact is taken into
account that there are many palingenetic organs or transitory
stages and many caenogenetic organs in embryonic development, one
is indeed led to think that the genes that have ensured the existence of such organs become inactive in the adult, or that, although they are pleiotropic, they act only partially.  Besides,
facts representative of neoteny confirm the established idea that
the egg is not a mechanism set up to give birth to only one type
of adult, but which can produce several types within the same
species.  Phenomena of this kind are still more easily recognizable
when the theory of metamerization is being examined.  It is a
known fact that in primitive forms bodily metamerism often produces segments that resemble one another, while, in their descendants, certain segments specialize.  In the simplest cases, this
specialization is a loss.  In the primitive insect Machilis
(de Beer, 1958), the female's reproductive organs are repeated in
the seven abdominal segments; in the most evolved insects, they
exist in one segment only.  Segmentation is generally easily
recognizable in the embryo of a species whose ancestor was meta-
merically segmented.  Obviously, in such cases, the development
of these segments merely represents a typical phenomenon of recapitulation.  But these facts show better still that at each
level of the organism only some of the genes act.

These facts make research on genetic distance much more complicated.  King and Wilson (1975) have established that the genetic
distance between man and chimpanzee is almost non-existent:  90%
of man's polypeptides are identical to the ape's; electrophoretic
comparisons between protein of 44 loci show great similarities;
etc.  Such evidence led King and Wilson to confirm the ideas of
several authors (Ohno, 1972; Wilson et al., 1974a, 1974b; Prager,
1975) who suggest that many anatomical changes could result from
mutations affecting gene expression only; e.g. changes in their
activation rate or their regulation.  It is logical to think that
when differences between two species seem partly due to suspension
of embryonic development, the differences are only modifications
in the expression of genes between them.  The genetic distance
between two groups of the genus Eurycea (Urodela), one neotenic
and the other artificially made adult by thyroxine, is of course
non-existent.  It is therefore not illogical to think that in
some cases mutations of very little importance can have caused
important transformations in the adult form of a species either
because they had activated normally inactive regions of the
genome or because they had rendered inactive usually active regions.

Thus from the point of view of evolutionary mechanisms, works on allometry such as d'Arcy Thompson (1942) on growth and form may not have been given enough importance. Such works suggest that modifications in growth correlations are sufficient to thoroughly transform an adult form and cause a new species to appear. If this is so, it means that in the genome structure very small modifications are sufficient to cause the appearance of new species. It also means that, as for forms that differ only in neotenies, the genetic distance between groups that seem to differ mainly in allometric relations must be very small. Waddington (1962) has already drawn attention to these problems; Bolk (1926) long ago suggested that man and ape differed mainly in delayed growth which he called foetalization and which actually corresponded to neotenies. De Beer (1932, 1958) has stressed the importance of these phenomena but Dullemeijer (1975) has offered recent criticism of these works. Thanks to more precise research, Delattre and Fenart (1960) have shown that many of these foetalizations are really divergences. Now it just happens that in both cases, foetalization or divergence, there might merely be differences in gene expression. Would this be enough to explain King and Wilson's observations? It would of course be interesting to calculate genetic distances between species that differ only in allometric relations and are seemingly very close to one another. This, to our knowledge, has not yet been done.

Research concerning the relations between embryonic development and evolution has at last presented excellent examples of somatic inductions that copy mutations and which Goldschmidt (1935) called phenocopies. It is commonly admitted that facultative neoteny in Ambystoma is a somatic induction that resembles associations of mutations which, in some batrachian species, result in final and hereditary neoteny. The first phenomenon can therefore be considered as a phenocopy compared with the second one. It will also be noted that a phenocopy here is certainly much more complex than the ones usually described by geneticists.

CONCLUSION

It is interesting to note that when Haeckel recognized caenogenesis and palingenesis he cleverly detected the two most important phenomena in the relations between embryonic development and evolution. Among other phenomena, adult variation and neoteny appear to be of minor importance; proterogenesis is difficult to estimate; while divergence and dyschrony are certainly very frequent but perhaps less so than real palingenesis.

From the genetic standpoint, much work needs to be done in the use of data supplied by relations between embryonic development and evolution in order to obtain a better knowledge of genetic mechanisms. It seems to us that the phenomena we have described could, one day, be very useful to geneticists.

## REFERENCES

Beer de, G., 1930, "Embryology and Evolution." Oxford Univ. Press.

Beer de, G., 1932, "Embryologie et Evolution," A. Legrand, Paris, 147 p.

Beer de, G., 1958, "Embryos and Ancestors," Oxford University Press, Third edition, 197 p. (1st ed., 1940; 2nd ed., 1951)

Bolk, L., 1926, "Das Problem der Menschwerdung," Gustav Fisher Verlag, Jena.

Brien, P., 1962, Etude de la formation de la structure des ecailles des Dipneustes actuels et de leur comparaison avec les autres types d'ecailles des Poissons. Ann. Mus. Roy. Af. Centr. Tervuren, Belgique, ser. 8, Sc. Zool. 108:55-128.

Brien, P., 1969, Polymorphisme intraspecifique et Evolution epigenetique. Bull. Soc. Roy. Sci. Liege, 38:718-734.

Brien, P., 1974a, "Propos d'un zoologiste. Le Vivant. Epigenese. Evolution epigenetique," Editions de l'Universite de Bruxelles, 155 p.

Brien, P., 1947b, La Forme et l'Epigenese," Annee Biol., 13, 1-2, pp.3-8.

Clausen, H.J., 1930, Rate of histolysis of Anuran tail, skin and muscle during metamorphosis. Biol. Bull., 59:199-210.

Delattre A. and Fenart, R., 1960, "L'hominisation du crane," Edit. du Centre National de la Recherche Scientifique, Paris, 418 p.

Delsol, M., 1952, Action du thiouracile sur les larves de Batraciens: Neotenie experimentale. Role de l'hypophyse dans ce phenomene. Archives de Biol., 63:279-392.

Delsol, M., 1953, Action du benzoate de dihydrofolliculine sur les canaux de Muller de quelques Batraciens Anoures et Urodeles a l'etat de tetards: phenomenes de neotenie partielle. C.R. Soc. Biol., 147:1895-1898.

Delsol, M. and Flatin, J., 1969, Metamorphose experimentale de la peau de queue du tetard d'Alytes obstetricans Laur. normalement destinee a degenerer. Experientia, 25:392-393.

Delsol, M. and Tintant, H., 1971, Discussion autour d'un vieux probleme: les relations entre Embryologie et Evolution. Revue des Questions Scientifiques, 142:85-101.

Dollander, A. and Fenart, R., 1970, "Elements d'Embryologie," Ed. medicale Flammarion, Paris, tome 1, 366 p.

Dullemeijer, P., 1975, Bolk's Foetalization Theory. Acta. Morphol. Neerl. Scand., 13:77-86.

Durand, J.P., 1976, Rudimentation des Yeux chez les Poissons et Urodeles souterrains. Bull. Soc. Zool. Fr., suppl. No.1, vol. 101, pp.13-21.

Gans, C., 1975, Tetrapod limblessness evolution and functional corollaries. Amer. Zool., 15:455-467.

Garstang, W., 1922, The Theory of Recapitulation: a critical restatement of the biogenetic law. J. Linn. Soc. London Zool., 35:81-101.

Gasc, J.P., 1970, Reflexions sur le concept de regression des organes. Revue des Questions Scientifiques, 141:175-195.

Goldschmidt, R., 1935, Gene und Ausseneigenschaft. I, Zeits.ind. Abst. Vererb., 69:38-69.

Goldschmidt, R., 1940, "The material basis of evolution," Yale University Press, New Haven.

Gould, S.J., 1966a, Allometry in Pleistocene land snails from Bermuda: the influence of size upon shape. J. Paleontol., 40:1131-1141.

Gould, S.J., 1966b, Allometry and Size in Ontogeny and Phylogeny. Biol.Rev.Cambridge Phil.Soc., 41:587-640.

Gould, S.J., 1971, Geometric similarity in allometric growth: a contribution to the problem of scaling in the evolution of size. Amer. Natur., 105:113-136.

Gould, S.J., 1974, The origin and function of "bizarre" structures: antler size and skull size in the "Irish elk" Megaloceros giganteus. Evolution, 28:191-220.

Gould, S.J., 1975, On the scaling of tooth size in the Mammals. Amer. Zool., 15:351-362.

Haeckel, E., 1866, Generelle Morphologie der Organismen, Berlin.

Haeckel, E., 1875a, Die Gastrula und die Eifurchung der Thiere. Jen.Zeit.fur Natur., 9:402.

Haeckel, E., 1875b, Ziele und Wege der heutigen Entwicklungs-geschichte, Jena.

Helff, O.M. and Clausen, H.J., 1929, Studies on amphibian meta-morphosis - 5 - The atrophy of anuran tail muscle during metamorphosis. Physiol. Zool., 2:575.

Kemp, N.E., 1961, Replacement of the Larval basement lamella by adult-type basement membrane in Anuran skin during meta-morphosis. Develop. Biol., 3:391-410.

King, M.C. and Wilson, A.C., 1975, Evolution and two levels in Humans and Chimpanzees. Science, 188 No.4184, pp. 107-116.

Lamotte, M. and Xavier, F., 1972, Recherches sur le developpement embryonnaire de Nectophrynoides occidentalis, amphibien anoure vivipare. I: Principaux traits morphologiques et biometriques du developpement. Ann.d'Embryol. et Morphol., 5:315-340.

Lindeman, V.F., 1929, Integumentary pigmentation in the Frog, Rana pipiens during metamorphosis, with special reference to tail-skin histolysis. Physiol. Zool., 2, 2, pp.255-268.

Lovtrup, S., 1974, "Epigenetics: a Treatise on Theoretical Biology," John Wiley and Sons, Garden City Press, 547 p.

Marshall, A.M., 1951, "The Frog: an Introduction to Anatomy, Histology and Embryology," Twelfth Ed., Macmillan and Co., London, 182 p.

Millard, N., 1941, The vascular anatomy of Xenopus laevis. Trans. Roy.Soc.S.Afr., 28:387-439.

Millard, N., 1945, The development of the arterial system of Xenopus laevis including experiments on the destruction of the larval aortic arches. Trans. Roy.Soc.S.Afr., 30:217-234.

Ohno, S., 1972, An argument for the genetic simplicity of man and other Mammals. Journ. Human Evol., Vol. 1, pp. 651-662.

Prager, E.M. and Wilson, A.C., 1975, Slow evolutionary loss of the potential for interspecific hybridization in Birds: a manifestation of slow regulatory evolution. Proc. Natl. Acad. Sci. U.S.A., 72, 1, pp. 200-204.

Raynaud, A., 1972, Morphogenese des membres rudimentaires chez les Reptiles: un probleme d'Embryologie et d'Evolution. Bull.Soc.Zool., 97(3):469-485.

Schindewolf, O.H., 1950, "Grundfragen der Paleontologie," E. Schweizerbart'sche Verlagsbuchhandlung, Stuttgart, 506 p.

Stahl, W.R., 1962, Similarity and Dimensional Methods in Biology, Science, 137, N° 3525, pp. 205-212.

Szarski, H., 1964, The structure of respiratory organs in relation to body size in Amphibia. Evolution, 18:118-126.

Terhal, H.J.J., 1941, On the heart and arterial arches of Salamandra maculosa Laur. and Ambystoma mexicanum Shaw during metamorphosis. Zool. laboratory of the Government University, Leyden.

Thompson, d'Arcy W., 1942, "On Growth and Form," 2nd edition, Cambridge University Press, Cambridge.

Vachon, M., 1944, L'appendice arachnidien et son evolution. Bull. Soc.Zool.Fr., 69(4):172-177.

Vodop'janova, N.K. and Kremanskji, V.I., 1974, Le principe de l'unite dialectique de la "preformation" et de l'epigenese dans l'embryologie, Filos.Nauki.URSS, 3:42-50. (We have only read an abstract in the Bulletin Signaletique du Centre National de la Recherche Scientifique.)

Waddington, C.H., 1962, "Genetics and Development," Columbia University Press, p. 271.

Waddington, C.H., 1969, The theory of evolution today, in "The Albach Symposium, Beyond Reductionism, New Perspectives in the Life Sciences, 1968, pp. 357-395, N.Y. Macmillan Co.

Wassersug, R.J., 1975, The adaptive significance of the Tadpole Stage with comments on the maintenance of complex life cycles in Anurans. Amer. Zool., 15:405-417.

Wilson, A.C., Maxson, L.R. and Sarich, V.M., 1974a, Two types of molecular evolution - Evidence from studies of interspecific hybridization. Proc. Natl. Acad. Sci. USA, 71, 7, pp. 2843-2847.

Wilson, A.C., Maxson, L.R. and Sarich, V.M., 1974b, The importance of gene re-arrangement in Evolution: Evidence from studies on rates of chromosomal, protein and anatomical evolution. Proc. Natl. Acad. Sci. USA, 71, 8, pp. 3028-3030.

Zouzouko, R.S., 1970, Evolution des greffes cutanees heterotypiques chez la larve et au cours de la metamorphose de _Discoglossus pictus_ Oth. These de 3eme Cycle, Universite de Paris, Faculte Sci. Orsay, 70 p.

# THE DEVELOPMENT OF THE TETRAPOD LIMB:  EMBRYOLOGICAL

# MECHANISMS AND EVOLUTIONARY POSSIBILITIES

Julian LEWIS and Nigel HOLDER

Dept. of Biology as Applied to Medicine
The Middlesex Hospital Medical School
London W1P 6DB United Kingdom

## INTRODUCTION

Professor Delsol (in this volume) has cogently defended Haeckel's approach to the relationship between embryogenesis and evolution, and has developed from it a new and more systematic analysis.  That analysis nevertheless still bears the stamp of Haeckel's times.  For while it is learned in the descriptive anatomy of embryos, and ingenious in the application of evolutionary principles, it treats the mechanisms of embryonic development as an almost impenetrable mystery.  We would not go so far as Oppenheimer (1955), who expresses her horror at Haeckel's influence thus:  "The seduction of embryology by a fanatic who expressed himself even metaphorically in terms of magic represents a darker chapter in its history than any of its earlier or later retreats to mere metaphysics lacking such taint of the mystic."  But it does seem to us that one can now go beyond Haeckel, and beyond de Beer too, and pursue the analysis in terms of specific developmental mechanisms and their evolution.  We can begin, on the one hand, to see how the pattern of the body arises from the rules of cell behaviour in the embryo; and, on the other hand, to see how those rules of cell behaviour may be defined by the genome that lies within each cell (Wolpert and Lewis, 1975).  The nature of this understanding, and its implications for the evolution of vertebrates, can be illustrated by an account of the development of a tetrapod limb.

LIMB DEVELOPMENT IN THE CHICK EMBRYO


Some General Principles

The limbs of the chick develop from small tongue-shaped buds
which start to grow out from the flank at about 3 days of incubation, when the embryo is about 5 mm long.  Each bud is at first
composed simply of an ectodermal jacket, filled with apparently
homogeneous and undifferentiated mesenchyme.  This mesenchyme must
eventually give rise to the skeleton, musculature and other connective tissues of the limb.  As the bud grows, cells in particular
places must differentiate in particular ways, so as to give the
anatomical pattern that is characteristic of the species.  The
problem is to discover how this spatial patterning is controlled,
and so to see how genetic changes may affect it.  We shall suggest
that there is a certain basic mechanism universal among tetrapod
limbs, and characteristic of them, and that the vast multitude of
different forms, from the bird's wing to the elephant's leg, represent merely variations in the quantitative parameters of a
fixed fundamental scheme (Lewis and Wolpert, 1976).  The suggestion in itself is not novel; but some of the details of the basic
mechanism can now be described, and the description throws light
on the types of variation that may occur among tetrapods, and on
the different and more radical mixture of the evolutionary innovation that led to the tetrapods from the fish.

The leg buds appear in the chick embryo at the same time as
the wing buds, and look almost exactly the same; under the microscope, their cells are indistinguishable from those of the wing
buds.  Yet there is an unseen difference.  This is demonstrated by
a profoundly instructive experiment of Saunders, Cairns and
Gasseling (1957).  They cut a small block of mesenchyme from the
region of the prospective thigh, from the early, undifferentiated
leg bud, and grafted it into the tip of the wing bud.  There it
developed neither as thigh (i.e., proximal leg), nor as finger
(i.e., distal wing), but instead as a toe (i.e., distal leg).  We
can conclude (1) that the grafted cells at the time of grafting
had already an intrinsic character peculiar to leg as opposed to
wing; (2) that this specificity did not fade, but was remembered
by them throughout their subsequent growth and differentiation;
(3) that their choice between a proximal and a distal character
was not yet fixed, and so could still be specified by cues from
the host wing; and (4) that their behaviour was thus governed by
a combination of items of information about their position
(Wolpert, 1971), including the early specification as leg rather
than wing, and the later specification according to relative
position within the limb.

These points deserve some comment. First, the leg cells resemble the wing cells not only in their appearance at the time of grafting, but also in the range of tissue types - muscle, cartilage, etc., - to which they will eventually give rise; the difference between them is detected only through the spatial pattern in which those tissue types are later generated. The experiment shows that the eventual difference in patterning is a consequence of an early difference in the intrinsic character of the cells; and it proves that there may be hidden specificities distinguishing cells of the same histological type: cells of the same histological type may be non-equivalent (Lewis and Wolpert, 1976).

We do not know the chemical basis of this non-equivalence; it might, for example, consist in the synthesis of some specific protein, whose sole function is to control the synthesis of other proteins, by activating or repressing transcription of the appropriate genes (Britten and Davidson, 1969). Pursuing this line of speculation, one can see how the cells might be enabled to remember their state of determination: there would need to be a feedback loop in the control system. For example, a controlling protein characteristic of leg cells might promote its own synthesis, by activating transcription of its own gene. Cells lacking that protein would then fail to synthesize it; but if they were some- how primed with a small quantity of that protein, they would be stimulated to produce it in perpetuo: the autocatalytic flame, so to speak, would be lit. Indeed, we might envisage each gene as an element that could be either "on", i.e., being transcribed and translated, or "off", i.e., quiescent. If the product of one gene can turn other genes on or off, the genome as a whole can be con- ceived of as a sort of switching network (Kauffman, 1971), which serves to compute appropriate responses to the morphogenetic signals which the cell receives from its surroundings (Wolpert and Lewis, 1975).

There is some experimental foundation for these ideas, pro- vided, for example, by studies of the lac operon in E. coli (Beckwith and Zipser, 1970), of the control of puffing in Drosophila giant chromosomes (Ashburner, 1974), and of trans- determination in imaginal discs of Drosophila (Kauffman, 1973). Direct evidence as to the molecular biology of development in vertebrates, however, is still extremely scanty.

Although its chemical basis thus remains in many ways a matter for speculation, the phenomenon of non-equivalence in itself is quite well founded in experiment. Just as leg-bud cells are intrinsically different from wing-bud cells, so, for example, the cells that form the humerus are different from those that form the ulna, and those that form the ulna from those that form the radius.

The early differences of cell state give rise to the later differences of organ size and shape.  The grounds for that contention
can best be made clear through an outline of the arguments that
have led to an account of how the cells within the chick wing bud
get the positional information that makes them non-equivalent to
one another.  For more detailed reviews, discussion and references,
see Zwilling, 1961; Saunders, 1972; Summerbell et al., 1973;
Wolpert et al., 1975; Tickle et al., 1975; Stocum, 1975.

The organization of the limb can best be discussed in terms
of its three axes:  proximo-distal, antero-posterior, and dorso-
ventral.  We shall concentrate on the first two of these; the
mechanism of the dorso-ventral patterning is less well understood
(but see Pautou and Kieny, 1973; and MacCabe, Errick and Saunders,
1974).

## The Proximo-Distal Axis

The parts of the wing begin their differentiation in proximo-
distal sequence - the upper arm first, at about 4 days of incubation, the tips of the digits last, about 2 days later (Saunders,
1948).  Meanwhile the mesenchyme at the tip of the bud, in a
region which we have called the 'progress zone' (Summerbell, Lewis
and Wolpert, 1973) remains apparently undifferentiated, and by its
growth serves as a source of tissue for the successively differentiating rudiments.  If the undifferentiated tip of a bud is cut
off and transplanted onto the stump of another bud, it behaves in
a nearly autonomous fashion:  the tip from a young bud (which is,
so to speak, all tip and no stump) generates an almost complete
sequence of limb parts, while the tip from an old bud generates
only the most distal structures (Fig. 1).  If the host stump is of
a different age from the graft, or is cut at different levels,
and so does not contain the complementary set of rudiments, mon-
strous limbs can thus be formed with deletions or reduplications
(Summerbell and Lewis, 1975).  From experiments such as these, it
appears that (1) the undifferentiated tissue at the tip of the bud
is progressively changing its character as it grows older, or, in
other words, it is taking on a more and more distal 'positional
value'; and (2) this positional value is largely independent of
influences from the stump of the bud, where differentiation is
already beginning.

The mesenchyme at the tip, constituting the progress zone, is
kept labile and undifferentiated by its proximity to the apical
ectodermal ridge - a thickened rim of ectoderm at the distal
margin of the bud.  If the ridge is cut off, the progress zone
loses its special character prematurely, and a truncated limb
develops (Saunders, 1948).  If an extra ridge is grafted onto a

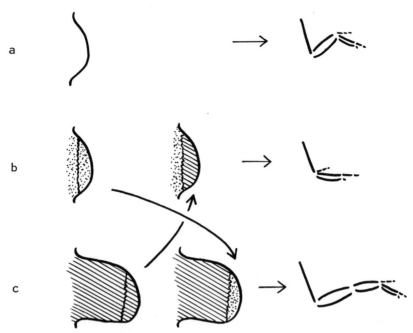

Fig. 1. (a) An undisturbed wing-bud yields a normal wing. (b) The
tip of an old bud (hatched) grafted onto the stump of a young bud
(stippled) gives a limb consisting simply of an upper arm (from
the stump) plus a hand (from the grafted tip). (c) The tip of the
young bud (stippled) grafted onto the stump of the old (hatched)
gives a limb consisting of an upper arm and a forearm (from the
stump) plus a second forearm and a hand (from the grafted tip).

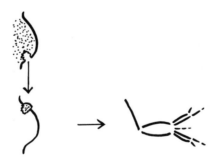

Fig. 2. Polarizing tissue taken from the posterior edge of a do-
nor limb bud and grafted into the anterior edge of a host causes
structures of a posterior character to form from the host tissue
near it, giving a reduplication of the limb pattern.

bud, a supernumerary set of limb parts is generated beneath it
(Saunders, Gasseling and Errick, 1976). The apical ridge thus
acts on limb mesenchyme to make it behave as progress zone tissue.
The ridge does not, however, assign to that tissue its positional
value: the behaviour of the progress zone mesenchyme has been
shown to be independent of the age and origin of the apical ridge
that covers it (Rubin and Saunders, 1972). The cells in the
progress zone evidently change their positional value autonomously
with the passage of time. As growth continues, successive generations of cells are shifted out of range of the apical ridge, and
thus emerge from the progress zone. The change of positional
value thereupon halts, and differentiation of the appropriate limb
parts begins.

There is a strong possibility that the progressive change of
positional value may be coupled to the cell division cycle. From
measurements of cell numbers in the bud, it can be shown that each
segment of the limb is formed by one generation of cells from the
progress zone, i.e., by the cells that emerge from the zone in a
time equal to one cell division cycle (Lewis, 1975). It is note-
worthy also that the segments of the limb when first laid down are
of roughly equal size. The initial rudiment of the wrist, for
example, is as big as that of the forearm (Lewis, 1976). The
eventual differences of size are due mainly to differences in the
subsequent growth rates; and those differences of growth rate, we
would suggest, are due to the differences of positional value, or
in other words, to the non-equivalence of the cells in different
segments.

## The Antero-Posterior Axis

To set up the distinctions between the different digits, or
between the ulna and the radius, the cells of the bud must receive
information about their position along the antero-posterior axis.
The antero-posterior component of positional value, unlike the
proximo-distal, appears to be specified by a signal from a special
region of mesenchyme cells on the posterior margin of the bud -
the polarizing zone. Saunders and Gasseling (1968) found that
when an additional piece of this polarizing tissue was grafted into
the anterior margin, it caused a reduplication of the distal parts
of the limb, with mirror symmetry about the long axis (Fig. 2).

It seems plausible to suggest that the polarizing tissue is
the source of a signal, perhaps a diffusible chemical substance,
which is graded across the bud. The cells may then read off their
positional values from its local intensity or chemical concentration. Tickle et al., (1975) have shown that when additional
polarizing tissue is grafted to various different antero-posterior

levels along the margin of the bud, the resultant pattern of skeletal elements is indeed consistent with such a gradient model.

Summerbell (1974) has demonstrated a connection between the antero-posterior and proximo-distal specification mechanisms, in that only cells in, or close to, the progress zone are responsive to the signal from the polarizing zone. When polarizing tissue is grafted at successively later stages to the anterior edge of the limb, the antero-posterior reduplication is produced from successively more distal levels, corresponding roughly to the levels that were emerging from the progress zone at the time of grafting. It seems that cells in the progress zone are labile with respect to both components of positional value, whereas cells which have left it are fixed with respect to both.

## HOMOLOGY AND THE MECHANICS OF DEVELOPMENT

From this account, it can be seen that the patterning of the limb, or at least of its gross anatomical features, depends on a few simple signals, of which the most notable are those from the apical ridge and from the polarizing zone; but these simple signals govern a rather complicated set of cell responses. We do not know what the signals are in chemical terms; but we have good reason to think that they are the same in different tetrapods. For example, as Tickle et al., (1976) have shown, the leg bud of a mouse embryo contains a zone of polarizing tissue, which can be grafted into a chick wing bud, and can there act upon the chick cells to cause a reduplication in just the same way as does a graft of polarizing tissue taken from another chick.

But if we are to have homology between limbs of different species, it is not enough that the signals should be the same: the cellular responses also must correspond. As we have argued, the different parts of the limb - humerus, ulna, radius, carpals, metacarpals, phalanges, and their attendant soft tissues - are formed from cells in non-equivalent states. Homology of limb pattern among tetrapods must then depend on homology of the set of non-equivalent cell states; and homology of the set of cell states must in all likelihood depend on homology of the genetic control system which defines the cell state in response to the morphogenetic signals. In other words, it seems that in all, or almost all, tetrapods, the limb-forming part of the genetic control system should be formed of corresponding elements, coupled in the same way, and performing essentially the same functions. Quantitatively, of course, the behaviour of the gene control system must vary from species to species; but its qualitative behaviour, as reflected in the topology of the limb, must be the same. Professor Delsol has already mentioned the work of King and Wilson (1975). We would

echo their judgment that the differences between primates lie more
in their control genes than in their structural genes.  But we
would add, that even the differences in the control genes must be
far from radical, and correspond merely to adjustments of parameters - binding constants, rates of synthesis and the like - which
do not disrupt the qualitative performance of the control system.

What then is the explanation of this constancy?  We may
enquire first of all whether mutations can ever occur to upset it.
The answer is that they can and they do.  Numerous mutations have
been observed, in mouse, man, chick and other species (Gruneberg,
1963; Zwilling, 1969; Wolpert, 1976) which cause dramatic dis-
ruptions of the standard tetrapod limb pattern - deletions,
additions, reduplications, etc.  Evidently such changes are not
favoured by natural selection; they are the big leaps which lead
out of the frying-pan into the fire.  The orthodox view seems to
be right:  evolution here proceeds by small changes.  These small
changes cumulatively amount to a quasi-continuous process of
modification, leaving the qualitative behaviour of the gene control
system the same, and the topology of the body invariant.

These continuous modifications are nonetheless a most
marvellous feature of the evolution of tetrapod limbs.  They are
marvellous above all, not for the extremes to which they go, but
for the richness of their variety.  We do not encounter merely
big limbs and small limbs, or broad limbs and narrow limbs; rather,
each element seems capable of modification independently of the
others, as we can see if we contrast, for example, a bat's wing,
a man's arm, and an elephant's foreleg.  This evolutionary freedom
to adapt the parts individually, we suggest, depends on the
individuated characters of the constituent cells; it is made
possible by their non-equivalence.


ACKNOWLEDGMENT

We thank the Medical Research Council for support.


REFERENCES

Ashburner, M., 1974, Sequential gene activation by ecdysone in
    polytene chromosomes of _Drosophila_ melanogaster.  Develop.
    Biol., 39:141.
Beckwith, J.R. and Zisper, D., 1970, "The Lactose Operon," Cold
    Spring Harbor Laboratory, N.Y.
Britten, R.J. and Davidson, E.H., 1969, Gene regulation for higher
    cells; a theory.  Science, 165:349.

Gruneberg, H., 1963, "The Pathology of Development," Blackwell, Oxford.

Kauffman, S., 1971, Gene regulation networks: a theory for their global structures and behaviours. Curr. Top. Develop. Biol., 6:145.

Kauffman, S., 1973, Control circuits for determination and trans-determination. Science, 181:310.

King, M.C. and Wilson, A.C., 1975, Evolution at two levels in Humans and Chimpanzees. Science, 188:107.

Lewis, J.H., 1975, Fate maps and the pattern of cell division: a calculation for the chick wing-bud. J. Embryol. Exp. Morphol., 33:419.

Lewis, J.H., 1976, Growth and determination in the developing limb, in "Vertebrate Limb and Somite Development" (D.A. Ede, M. Balls and J. Hinchliffe, eds.), Cambridge University Press.

Lewis, J.H. and Wolpert, L., 1976, The principle of non-equivalence in development. J. Theor. Biol., in press.

MacCabe, J., Errick, J. and Saunders, J.W., 1974, Ectodermal control of the dorsoventral axis of the leg bud of the chick embryo. Develop. Biol., 39:69.

Oppenheimer, J.M., 1955, Problems, concepts and their history, in "Analysis of Development" (B.H. Willier, P.A. Weiss and V. Hamburger, eds.), Saunders, Philadelphia.

Pautou, M.P. and Kieny, M., 1973, Interaction ectomesodermique dans l'etablissement de la polarite dorsoventrale du pied de l'embryon de poulet. C.R.Acad.Sci.Ser.D., 227:1225.

Rubin, L. and Saunders, J.W., 1972, Ectodermal-mesodermal inter-actions in the growth of limb buds in the chick embryo: Constancy and temporal limits of the ectodermal induction. Develop. Biol., 28:94.

Saunders, J.W., 1948, The proximo-distal sequence of origin of the parts of the chick wing and the role of the ectoderm. J. Exp. Zool., 108:363.

Saunders, J.W., 1972, Developmental control of three-dimensional polarity in the avian limb. Ann. N.Y.Acad.Sci., 193:29.

Saunders, J.W. and Gasseling, M.T., 1968, Ectodermal-mesenchymal interactions in the origin of limb symmetry, in "Epithelial-Mesenchymal Interactions" (R. Fleischmajer and R.E. Billingham, eds.), Williams and Wilkins, Baltimore.

Saunders, J.W., Cairns, J.M. and Gasseling, M.T., 1975, The role of the apical ridge of ectoderm in the differentiation of the morphological structure and inductive specificity of limb parts in the chick. J. Morphol., 101:57.

Saunders, J.W., Gasseling, M.T. and Errick, J.E., 1976, Inductive activity and enduring cellular constitution of a supernumerary apical ectodermal ridge grafted to the limb bud of the chick embryo. Develop. Biol., 50:16.

Stocum, D.L., 1975, Outgrowth and pattern formation during limb ontogeny and regeneration. Differentiation, 3:167.

Summerbell, D., 1974, Interaction between the proximo-distal and antero-posterior coordinates of positional value during the specification of positional information in the early development of the chick limb-bud. J. Embryol. exp. Morphol., 32:227.

Summerbell, D. and Lewis, J.H., 1975, Time, place and positional value in the chick limb-bud. J. Embryol. exp. Morphol., 33:621.

Summerbell, D., Lewis, J.H. and Wolpert, L., 1973, Positional information in chick limb morphogenesis. Nature, 244:492.

Tickle, C., Summerbell, D. and Wolpert, L., 1975, Positional signalling and specification of digits in chick limb morphogenesis. Nature, 254:199.

Tickle, C., Shellswell, G., Crawley, A. and Wolpert, L., 1976, Positional signalling by mouse limb polarizing region in the chick wing bud. Nature, 259:396.

Wolpert, L., 1971, Positional information and pattern formation. Curr.Top.Develop.Biol., 6:183.

Wolpert, L., 1976, Mechanisms of limb development and malformation. Brit.Med.Bull., 32:65.

Wolpert, L. and Lewis, J.H., 1975, Towards a theory of development. Fed. Proc., 34:14.

Wolpert, L., Lewis, J.H. and Summerbell, D., 1975, Morphogenesis of the vertebrate limb, in "Cell Patterning" (Ciba Symposium 29), Associated Scientific Publishers, Amsterdam.

Zwilling, E., 1961, Limb morphogenesis. Adv.Morphogenesis, 1:301.

Zwilling, E., 1969, Abnormal morphogenesis in limb development, in "Limb Development and Deformity: Problems of Evaluation and Rehabilitation" (C.A. Swinyard, ed.), Thomas, Springfield, Illinois.

# PHYLOGENETIC ASPECTS OF KARYOLOGICAL EVIDENCE

Alessandro MORESCALCHI

Institute of Histology and Embryology, Univ. of Naples

Via Mezzocannone 8, 80134 Naples, Italy

## KARYOLOGY AND EVOLUTIONARY CHANGES

Recent progress in the fields of molecular cytology and of genetic differentiation have once more placed the karyotype at the center of evolutionary problems. As has been perceived, especially by White (1973), evolution is essentially a cytogenetic process in which chromosomal rearrangements play an important role. In contrast to changes in other phenotypic characters, they are able to influence directly the properties of the genotype.

### Chromosomal Organization

At the molecular level chromosomes are highly organized structures. The structural (transcriptive) genes represent only a fraction, sometimes a minor fraction, of the chromosomal DNA, generally in the form of long nucleotide sequences, singly or repeated only a few times. In association with this and other DNA fractions one finds the histone and non-histone proteins extremely important in the control of the genome activity and in part specific (Johnson, et al., 1974). The rest of the chromosomal DNA is made up of sequences of different lengths and degrees of repetitiveness; except for some cistrons (t-DNA, r-DNA and histones) these might have regulatory functions on the transcriptional activity (Britten and Davidson, 1969; Georgiev, 1969).

Of particular evolutionary interest might be the non-transcriptional DNA in the form of very short sequences repeated

thousands or millions of times and sometimes separable from the
bulk of genomic DNA as a "satellite" fraction. This fraction is
highly species-specific, which has led to the hypothesis of its
rapid evolutionary turn-over. Moreover, it is usually localized
in the constitutive heterochromatin, perhaps the most dynamic of
the chromosomal structures. Some authors believe that this
highly repetitive DNA may have a structural or functional role,
thus being directly subject to natural selection (Walker, et al.,
1969; Yunis and Yasmineh, 1971; Natarajan and Gropp, 1971).

The different DNA fractions may be distributed along the
chromosomes according to modalities which Davidson et al., (1975)
consider to be of general significance for the metazoans. However, Drosophila differs from this model, and the considerable
differences in the proportions of the various DNA's from one
organism to another (in the amphibians) make it seem likely that
there are differences in the chromosomal organization in
different groups.

The chromosome appears to be, therefore, a heterogeneous
system made up of different molecular structures with different
genetic significance and evolutionary rate (Rice, 1972; Smith,
1974; Yunis, 1974). "Banding" techniques enable one to detect
the different DNA fractions along the chromosomal axis (cf. Schnedl,
1974; Comings, 1975). In species comparisons the "unique"
fractions (appearing as R-bands or interbands) and those of
medium repetitiveness (G- or Q-bands) appear the most conservative
in both size and chromosomal localization, as compared with the
highly repetitive C-bands, which also vary from one population to
another in some mammals. The differences in the G-bands generally
tend to increase in proportion to the phyletic distance (Stock
and Mengden, 1975). Even in the most variable satellite fractions
it appears to be possible to detect, in rodents and in Drosophila,
a certain correlation between their nucleotide compositions and
the phyletic distances between the species under examination
(Flamm, 1972; Blumenfeld, 1974).

By using these techniques therefore it is possible to demonstrate, with a greater degree of resolution than in the past, the
high degree of specificity in karyotype structure and morphology.
Related species with perfectly corresponding karyotypes (homo-
sequential species) are very rare in nature.

## Chromosomal Integration

The differentiation of structural genes, investigated
through amino-acid substitution in proteins, does not seem to
correlate with the rate of evolution of vertebrate organisms as

much as it does with the time of divergence (Wilson, et al., 1974).  The hypothesis of a "molecular clock" is not universally accepted while the above discrepancies could also find some technical explanations (Ayala, 1975; Read, 1975).  However, when allozymic proteins are investigated the degree of genic variability shown by many animals seems to be of adaptive value, as a strategy for various environmental conditions (Levins, 1968; Selander and Kaufman, 1973; Powell, 1975).

Mayr (1975) maintains that a macromolecule is not free to evolve individually but only within a system of continually interacting macromolecules which changes in response to appropriate selective stimuli.  Indeed, the genotype seems to consist of a number of co-adapted gene complexes (supergenes).  The fitness of genes tied up in complexes would be determined far more by the fitness of the complex as a whole than by any functional qualities of individual genes.

At a higher level of complexity supergenes are organized into the chromosomes which provide a mechanism of cohesion between the integrated gene systems and can eventually protect adaptive linkage-groups from free recombination (crossing-over does not equally affect all chromosomal regions and all species - White, 1973; Mayr, 1975 ).  The two opposite evolutionary tendencies - variability and conservativeness of the genetic system - appear to be mediated by the chromosomal phenotype, whose changes can thus severely affect the cohesion and properties of the genotype. Moreover, it is becoming evident that the most important evolutionary changes in higher organisms do not involve the structural genes as much as they do the regulatory (repetitive) fractions of the genotype (Britten and Davidson, 1969; Crick, 1971; Wilson, et al., 1974; see also next section).  In this case also, changes in the chromosomal organization would be able to modify the patterns of gene regulation.

Speciation is normally paralleled by modifications at the chromosomal level.  Following the "stasipatric" model, these changes precede the isolating phase and are one of the more frequent causes of speciation (White, 1973).  In vertebrates, there is good evidence that the rate of structural variation in the chromosomes parallels the rate of anatomical evolution of important groups, such as the amphibians and mammals (Wilson, et al., 1974).  This observation, implying that gene rearrangements may contribute significantly to the adaptive evolution of animals, does not seem in agreement with the views of some authors who maintain that the chromosomal changes in mammals are not especially selective (Hsu and Mead, 1969; Matthey, 1973, 1975; Arrighi, 1974).

Except for some cases of balanced chromosomal polymorphism at intraspecific level, the karyotypic differences seem to reflect others in the genetic constitution, and their history is an integral part of the evolutionary history of the organisms (White, 1973).


## Chromosomes and Evolutionary Rates

It has been known for some time that many organisms demonstrate only a few of the possible types of chromosomal change, with a clear tendency to produce karyotypes which are "symmetrical", that is with chromosomes similar in shape and size (frequent in animals); or "asymmetrical", with chromosomes largely unequal in size and shape (these seem to be achieved in plant evolution: Stebbins, 1971; Sharma, 1974). This phenomenon has been explained by postulating the action of selective factors, of endo- or exocellular origin, which are able to "canalize" the chromosomal differentiation ("karyotypic orthoselection", White, 1973). Following Mayr (1975), these and other orthogenetic trends at the organismal level may be convincingly explained if one considers that a large number of changes are selected against, since they drastically interfere with the cohesion of the genotype.

When analysing the selective factors capable of influencing the chromosome, one must bear in mind that we are dealing with complex and highly organized machinery. As has been said, much, or in fact most, of the nucleoprotein material building the phenotype seems to be engaged in "regulatory" or "structural" functions largely connected with chromosomal metabolism. The heterochromatic areas of the chromosome, rich in repeated sequences and species-specific, seem good candidates to function as "targets" for external selective stimuli. Indeed, the frequency of chromosomal changes seems to depend on the localization and width of these areas (John and Lewis, 1968; Natarajan and Gropp, 1971; Pathak, et al., 1973).

The nature of these selective factors is still conjectural. However, it seems justified to suggest that most of the cytogenetic characters of animals at both meiotic and mitotic levels can be of an adaptive nature. The same chromosomal polymorphism of many eutherian mammals is often related to environmental conditions (White, 1973; Nevo and Shkolnik, 1974). A similar significance has been attributed to the chromosomal variability shown by the higher plants (Stebbins, 1966). The evolutionary changes in the genome size of large groups of plants and animals show a clear correlation with various environmental, ecological and also cytological conditions (Sparrow, et al., 1972; Price, et al., 1973; Morescalchi, 1975; Szarski, 1976).

In conclusion, karyological evolution is largely subject
to natural selection and many aspects of it seem to be adaptive.
In addition, the non-transcriptive areas could be directly
engaged in mediating the action of selective factors of the
chromosomes.  We shall see later that groups of vertebrates
showing different proportions in these areas are also characterized by different patterns of chromosomal evolution, possibly as
adaptive strategies for diverse environmental conditions.

## THE KARYOLOGICAL EVOLUTION OF VERTEBRATES

The most general karyological characters of vertebrates together with the relevant references, are listed in Table I.  It
should be noted that the terrestrial tetrapods have received much
more attention than the aquatic groups.  The next four sub-
sections in the text contain some comments on the data shown in
the Table.

### The Lower Vertebrates

Except for the dipnoans, many teleosts and the polypterids,
the rest of the fish-like vertebrates are usually characterized
by asymmetrical (chromosomes of different shape) and bimodal
(macro- and microchromosomes) karyotypes and a large number of
chromosomes.  Some groups however, have chromosomes which are
hard to compare with those of other fishes (the myxinoids), while
a few species have asymmetrical but unimodal (lacking microchromosomes) karyotypes.  The quantity of nuclear DNA in
agnathans, chondrichthyans and in the lower actinopterygians
varies within relatively small limits (from 2.3 to 7.7 pg/N) with
some exceptions among sharks.  The genome size in teleosts is
slightly lower than the average value in other actinopterygians,
the minimum ones being typical of species which are morphologically specialized.  Whereas other species adapted to stable
(e.g. abyssal) environments tend instead to increase their DNA
(references in Table I).

In chondrosteans, in some teleost families and perhaps also
in lampreys, one can find cases of polyploidy and a few teleosts
seem to have developed heterochromosomes.

The teleosts usually have unimodal or symmetrical karyotypes
and most of the species studied have diploid numbers between 40
and 50.  If one accepts that they are the most recent and
advanced forms among the bony fishes, it is possible to hypothesize that their karyotypes are derived from others more similar
to those of the lower actinopterygians, which are bimodal.  Also,

TABLE I

The Main Karyological Characters of Vertebrates

| GROUPS | 2n(1) | Karyotype(2) | Sex Chromosomes(3) | Polyploidy(4) | 2c(5) | References(6) |
|---|---|---|---|---|---|---|
| **Agnatha** | | | | | | |
| Petromyzonta | 76–168 | Ab | Unknown | Probable | 2.3–4.7 | 1–3 |
| Myxinoidea | 46– 52 | Ab? | Unknown | Unknown | 5.5 | |
| **Chondrichthyes** | | | | | | |
| Holocephali | 58– 86 | Ab | Unknown | Unknown | 3.0 | 3–5 |
| Selachii | 62–100 | Ab | Unknown | Unknown | 5.5–14.7 | |
| Batoidea | 28–104 | Ab | Unknown | Unknown | 5.5 | |
| **Osteichthyes** | | | | | | |
| Actinopterygii | | | | | | |
| Holostei | 46– 88 | Ab | Unknown | Unknown | 2.3–2.8 | |
| Chondrostei | 112–116 | Ab | Unknown | Present | 3.2–5.7 | |
| Teleostei (12 orders) | 18–104 | Au– S | Rare, ♀♂ | Present in some families | 0.8–8.8 | 3,6–14 |
| Polypterini | 36 | S | Absent | Absent | 8.5–13.6 | |
| Choanichthyes | | | | | | |
| Dipnoi | 34– 38 | S | Unknown | Absent | 160–284 | |
| Latimeria | ? | ? | ? | ? | 7.2 | |
| **Amphibia** | | | | | | |
| Apoda | 20– 42 | Ab– S | Unknown | Unknown | 7–50 | |
| Caudata | 22– 70 | Ab– S | Rare, ♀♂ | Rare | 30–190 | 15–16 |
| Salientia | 16– 44 | Ab– S | Very Rare or absent | Many cases | 2–36 | |

| Taxon | 2n | Karyotype morphology | Heterochromosomes | Polyploidy | Genome size (pg/N) | References |
|---|---|---|---|---|---|---|
| **Reptilia** | | | | | | |
| Chelonia | 26–68 | Ab | Rare, ♂ | Absent | 4.6–9.1 | |
| Rhyncocephalia | 36 | Ab | Absent | Absent | 10.6 | |
| Sauria | 20–56 | Ab–Au | Rare, ♀♂ | Rare | 3.9–7.8 | 3,17–22 |
| Ophidia | 24–50 | Ab | Common, ♀ | Absent | 2.8–6.3 | |
| Crocodilia | 30–42 | Au | Absent | Absent | 5.0–5.4 | |
| **Aves** | | | | | | |
| Ratitae | 80–82 | Ab | Rare, ♀ | Absent | 3.09 | 23–29 |
| Carinatae | 52–126 | Ab | Constant, ♀ | Absent | 1.7–3.6 | |
| **Mammalia** | | | | | | |
| Monotremata | 54–64 | Ab | Constant, ♂ | Absent | 5.7–6.0 | |
| Marsupialia | 10–32 | Au–S | Constant, ♂ | Absent | 3.1–9.2 | 30–34 |
| Eutheria | 6–92 | Au–S | Constant, ♂ | Absent | 3.9–11.7 | |

(1) Minimum and maximum diploid chromosome number (2n)

(2) Karyotype morphology: Ab=Asymmetric bimodal; Au=Asymmetric unimodal; S=Symmetric

(3) Heterochromosomes, if present or not. ♂ =male heterogamety; ♀ =female heterogamety; ♀♂ =cases of both male and female heterogamety

(4) Presence or absence of polyploid species or populations

(5) Diploid genome size (2c), in pg per nucleus (pg/N)

(6) More general or recent references: 1. Potter and Robinson, 1973; 2. Robinson, Potter and Atkin, 1975; 3. Ohno, 1970; 4. Nygren and Jahnke, 1972; 5. Donahue, 1974; 6. Fontana and Colombo, 1974; 7. Denton, 1973; 8. Denton and Howell, 1973; 9. Bachmann, Goin and Goin, 1972; 10. Pedersen, 1971; 11. Hinegardner and Rosen, 1972; 12. Ebeling and Chen, 1970; 13. Uyeno and Miller, 1972; 14. Vialli and Casonato, 1972; 15. Morescalchi, 1973; 16. Morescalchi, 1975; 17. Becak and Becak, 1969; 18. Singh, 1972; 19. Gorman, 1973; 20. Bull, Moon and Legler, 1974; 21. Killebrew, 1975; 22. Olmo, 1976; 23. Bachmann, Harrington and Craig, 1972; 24. Ray-Chaudury, 1973; 25. Martin, 1974; 26. Shoffner, 1974; 27. Takagi and Sasaki, 1974; 28. de Boer, 1975; 29. Misra and Srivastava, 1975; 30. Matthey, 1973; 31. White, 1973; 32. Sharma, 1973; 33. Benirschke, 1969; 34. Sparrow, Price and Underbrink, 1972.

among the cartilaginous fishes, the peculiar karyotype of only 28 chromosomes found in <u>Narcine</u> seems isolated within the batoideans, which have more than 80 chromosomes, and could well be of secondary origin.

Both dipnoans and polypterids appear to be differentiated, showing symmetrical karyotypes and a relatively small number of chromosomes (2n= 34-38). On this basis, some investigators consider that the two groups may be related (Denton and Howell, 1973; Capanna and Cataudella, 1973). However, both the chromosome number and the amount of nuclear DNA in polypterids appear to be similar to those of a large number of the teleosts. While dipnoans have much larger chromosomes and much more DNA than polypterids (Table I), it seems likely that the ancestors of the modern lungfishes had less DNA than the living species (Thomson, 1972). The peculiar karyological situation of dipnoans, with a quantitatively hypertrophic genome, finds a parallel in that of the paedogenetic caudates (Morescalchi, 1973, 1975).

Unfortunately, the karyotype of <u>Latimeria</u> has not yet been studied. The quantity of DNA of this "living fossil" is about 7.2 pg/N (Cimino and Bahr, 1974), a value which is not very different from the average value for various primitive bony fishes and various tetrapods.

## Amphibians

In the three orders of living amphibians, various species of the most primitive families have bimodal karyotypes, while most of the species belonging to higher families have symmetrical karyotypes and reduced chromosome numbers (2n= 20-28). The karyological differentiation of these animals seems therefore to have taken the direction of progressively more symmetrical karyotypes with the early "loss" of the microchromosomes.

In anurans, most species studied have DNA values between 4 and 10 pg/N, but some species have definitely higher values (some ranids, <u>Bombina</u>, some tropical leptodactylids) and some have lower values (some deserticolous pelobatids and leptodactylids). As in teleosts, these variations from the mean seem generally correlated with phenomena of adaptive nature (Bachmann, et al., 1972).

In the caudates, the karyotype tends toward a progressively more symmetrical condition, while the genome size varies largely independently of this and its higher levels appear to correlate with metabolic and developmental factors, again clearly demonstrating an adaptive nature (Morescalchi, 1975).

Species of at least 5 families of anurans and a few caudates are polyploid, while <u>Discoglossus</u> (anuran) and some species of salamandrids and plethodontids (caudates) have differentiated sex chromosomes.

## Sauropsids

Most of the families of reptiles show bimodal karyotypes. Among the chelonians only some pelomedusids have symmetrical karyotypes with a reduced number of chromosomes, while the remaining families are more or less conservative in their bimodal formulae.

Most iacertilians and ophidians have bimodal karyotypes of 36 chromosomes although there are differences between the two orders. Matthey (1973) considers that these karyotypes are the result of convergence between the various families of each order, while Gorman (1973) thinks that these represent a primitive condition for lizards and snakes. Among the first, various species groups show karyotypes without microchromosomes.

<u>Sphenodon</u> also has 36 chromosomes, whereas the crocodilians have from 30 to 42 chromosomes, with few or no microchromosomes and a relative constancy in the number of chromosomal arms (=58).

The genome size in reptiles varies from 3 to about 10 pg/N, with relatively small differences between the various orders. Polyploidy seems to be confined to a few parthenogenetic (triploid) species of lizards, while sex chromosomes appear in a single species of turtle (<u>Staurotypus</u>), in species of certain families of lizards (with both male and female heterogamety) and, in a more consistent way, in the higher snakes, all showing female heterogamety though with heterochromosomes of sometimes variable shape (Baker, et al., 1972).

Birds are even more conservative than most reptiles: nearly all species show bimodal karyotypes with a large number of chromosomes; only some <u>Falconiformes</u> seem devoid of microchromosomes. The nuclear DNA amount in birds is on average lower than that of other tetrapods, although similar low levels are sometimes found also in anurans and in snakes. Except for some primitive ratites, the remaining birds show female heterogamety with sex chromosomes showing little variability in different species.

The parallel development of female heterogamety and lower DNA amounts in birds and snakes has often been (and very probably erroneously) interpreted as a proof of relationships between these groups (Ohno, 1967; Ray-Chaudury, 1973; Shoffner, 1974).

## Mammals

Monotremes seem to maintain some reptilian conditions in their bimodal karyotypes, though developing male heterogamety. Marsupials have somewhat lower chromosome numbers, with small departures from two types of karyological formulae (2n=14 or 22) which seem possibly selectively advantageous; it is not known which of these two (if any) is the more primitive.

The sex chromosomes in these mammals, as in the eutherians, are somewhat conservative, being able to vary only in some internal rearrangements, in translocations on the autosomes, or in an increase in the heterochromatic fractions. The X-chromosome of marsupials appears to be structurally and genetically similar to that of eutherians, even though the mechanisms of dosage compensation seem to be different in the two groups (Ohno, 1970, 1972; Sharma, 1973; Pathak and Stock, 1974).

The amounts of nuclear DNA are also similar in the three groups of mammals: this varies from 5 to 9 pg/N, differing from these values only in a few species which are specialized morphologically or karyologically (Green and Bahr, 1975).

The eutherians represent a wide spectrum of interspecific variability in chromosome shape and number. However, more than one-half of the species studied have from 40 to 56 chromosomes (Matthey, 1973). The karyotypes are usually bimodal or symmetrical: in size the Y-chromosome may sometimes appear to be a microchromosome (this is true also for the marsupials, especially Didelphys).

Nevertheless, some families of eutherians are also karyologically conservative (felids, camelids), while others develop only one or a few types of chromosomal change, thus showing rather precise evolutionary trends at the karyological level. A high degree of chromosomal polymorphism is generally found in the groups with restricted vagility (White, 1973). In these karyotypically variable groups the techniques of banding seem to demonstrate that certain linkage groups or entire chromosomes (besides the heterochromosomes) are relatively unchanged at the interspecific level. The greatest variability depends especially on the localization and extension of the C-banded regions. The heterochromatic fractions seem to play important roles in the speciation of eutherians (literature in Natarajan and Gropp, 1971; Fredga and Mandahl, 1973; Ohno, 1973; Pathak, et al., 1973).

CONCLUSIONS

## Macroevolutionary Trends

The presence of bimodal karyotypes in the most ancient groups of aquatic vertebrates has given rise to the hypothesis that these represent the primitive condition. Some more recently developed groups (sauropsids) would preserve this karyotype, others (teleosts, amphibians, most mammals) would tend to acquire unimodal and symmetrical karyotypes. Because the quantity of nuclear DNA varies, except in the living dipnoans and urodeles, between somewhat restricted limits (4-7 pg/N), which can be found in both the older and the Recent groups, it is probable that the phenomena of reduplication of the entire genome (polyploidy) has not significantly contributed to the origin of the major classes as suggested by such qualified investigators as Ohno (1970, 1973).

In fact, only morphologically specialized groups belonging to various classes tend to have DNA values higher or lower than the values cited above. The dipnoans and urodeles have probably massively increased their genomes as a consequence of adaptive processes acting also at cellular level (Morescalchi, 1975; Szarski, 1976).

The minor importance in macroevolution of polyploidy may also be deduced from the irregular distribution of it within the various vertebrate orders (Table I): it is absent in the homeothermic classes; very rare in reptiles; but it is scattered throughout the amphibians (especially in the anurans) and in unrelated groups of teleosts (but its presence here is debatable) and other fishes. A more or less inverse distribution of the heterochromosomes is found: they are established in mammals, carinate birds, the higher snakes, while they appear sporadically in species of some of the other groups, where male and female heterogamety can both be found in the same order.

Evolutionary trends toward the acquisition of symmetrical karyotypes are found especially in the amphibians, in teleosts and more sporadically in other vertebrates. Nevertheless, advanced genera of these groups seem to have evaded this narrow orthoselective passage, showing unexpected "explosions" in chromosomal rearrangements, probably in connection with rapid speciation and radiation. Analogous phenomena are encountered in some groups of reptiles (lizards). However, the karyological conservativeness of most sauropsids might demonstrate the intervention of rigid selective factors acting on new, probably unfit, karyotypes which are displaced from the typical bimodal formulae of these vertebrates.

Among the mammals, many groups of eutherians do not seem to show, on the whole, any kind of canalization in their chromosomal differentiation. However, the most generalized chromosomal changes in this group, apart from the loss or gain in heterochromatic fractions, seem to consist of centromeric fusions/ fissions and of pericentromeric inversions. Neither of these changes appears drastically to alter the composition of the original linkage-groups. Thus, the interspecific karyotype variability of this group seems to be more apparent than real and in fact tends to conceal a relative conservativeness in the main gene clusters (Ohno, 1973).

Wilson, et al. (1975) have suggested that the eutherians show faster rates of karyotypic evolution and speciation than the remaining vertebrates owing to their particular social structure. The high degree of inbreeding typical of these mammals would favor the rapid fixation of new karyotypes and stasipatric speciation. However, other groups with a different social structure are characterized by similar phenomena: for example, the frogs of the genus Eleutherodactylus, actually in a phase of "explosive" radiation in some tropical areas, show modes of karyotypic differentiation which differ according to the population sizes and environmental conditions (De Weese, 1975 and unpublished).

However, contrary to the views of many students of mammals, these modes of karyological evolution are not shared by the majority of vertebrates. The diverse karyological trends encountered in the subphylum can be determined by many factors, the most immediately evident of them being probably of a qualitative (structural) nature.

In amphibians, the rigid evolutionary trends toward symmetrical karyotypes are on a par with considerable interspecific variations in all the genomic fractions of DNA (Table II, essentially from Straus, 1971 and Morescalchi, 1975). The amniotes, on the contrary, show a lesser variability in the total DNA and a relative constancy in the ratios between the various fractions of the genome. The average quantity of highly reiterated DNA seems to be different within the birds (about 15% of the genome) than it is in mammals (sometimes more than 30%: Schultz and Church, 1972; Shields and Straus, 1975). In the eutherians the interspecific differences in the satellite DNA's seem to be highest in the groups showing a large interspecific chromosomal polymorphism, as in the rodents.

Very little is known about reptiles. From unpublished studies by Olmo on the melting temperatures of DNA from some species of chelonians, lizards and snakes, it seems that these groups show a relative homogenity in the proportions of the various genomic fractions, the major variations being, as always, in the more repetitive fractions.

In agreement with these findings, there also seem to exist differences in the extension and localization of the chromosomal C-bands among the various tetrapods, and especially between amphibians and the other classes (Stock, et al., 1973; Hutchison and Pardue, 1975).

Since the chromosomal changes are preferentially localized in the regions rich in highly repetitive DNA, the above-mentioned differences could account, at least in part, for the differing evolutionary rates of tetrapod karyotypes. The genetic signifi-cance of this last phenomenon will probably be more correctly evaluated from a study at a more general level.

TABLE II

Fractions of genomic DNA in some Amphibians
(values in pg/N)

| Anurans | Genome size | Single copy DNA |
|---|---|---|
| Xenopus laevis | 6.3 | 3.8 |
| Scaphiopus couchi | 1.9 | 1.1 |
| Rana clamitans | 11.6 | 2.5 |
| Bufo marinus | 7.0 | 1.8 |

| Caudates | Genome size | DNA with repetition frequency: $10^2$ | $10^6$ |
|---|---|---|---|
| Andrias japonicus | 93 | 10.2 | 26.9 |
| Necturus maculosus | 165 | 16.5 | 46.2 |
| Amphiuma means | 150 | 45 | 25.5 |
| Desmognathus fuscus | 30 | 6.3 | 10.5 |
| Triturus cristatus | 44 | 19.8 | 5.5 |
| Taricha torosa | 56 | 15.1 | 12.3 |
| Taricha rivularis | 60 | 13.2 | 11.4 |
| Ambystoma trigrinum | 55 | 22 | 5.5 |

## Karyology and Adaptive Strategies

During the course of their evolution tetrapods have perfected various systems for the control of internal homeostasis (Bentley, 1971).  The most advanced homeothermic groups are relatively independent of the temporal and spatial patterns of environmental variations.  On the other hand, amphibians, with their complex life cycles, limited mobility and poor development of the systems for osmo- and thermoregulation, appear to be individually very dependent on the degree of environmental "uncertainty" (sensu Levins, 1968).

The higher amniotes and the amphibians therefore tend to "experience" the environment with adaptive strategies which are different.  These differences are also reflected at the genetic level: in fact amphibians show a degree of enzymatic polymorphism which is much greater than that of amniotes (Selander and Kaufman, 1973; Powell, 1975; Nevo, 1976).

The evolution of the tetrapods seems to be characterized also by a progressive "stabilization" of various karyological characters, such as the genome size, the heterochromosomes, the strictly diploid condition, the quantitative relationship between heterochromosomes and autosomes, and the ratios between the different DNA fractions.  These characters appear to be almost constant in homeotherms, while they are very variable within amphibians, sometimes even at subspecific level.

In conclusion it appears that the various modes of karyological differentiation may assume the significance of adaptive strategies towards the different environmental conditions experienced by the main tetrapod groups.  The karyological plasticity of amphibians may constitute, in conjunction with their wide degree of genetic variability, one type of adaptation for environmental situations which are typically "coarse grained" in Levins' terminology.  While the relative stability in many karyological characters shown by the homeothermic groups might constitute the result of a larger degree of independence gained from their genotype against variations in environmental conditions, whose effects tend here to be buffered at the organismal level too.

Like the other morphological, physiological, behavioral and genetic characters of the vertebrates, the karyological characters might then reflect the different adaptive levels reached by the living groups of the subphylum.

REFERENCES

Arrighi, F.E., 1974, Mammalian chromosomes, in "The Cell
    Nucleus" (H. Busch, ed.), Vol.2 pp. 1-32. Academ. Press,
    New York and London.
Ayala, F.J., 1975, Genetic differentiation during the
    speciation process. Evol. Biol. 8:1-78.
Bachmann, K., Goin, O.B., and Goin, C.J., 1972, Nuclear DNA
    amounts in vertebrates. Brookhaven Symp. Biol. 23:419-447.
Bachmann, K., Harrington, B.A. and Craig, J.P., 1972, Genome
    size in birds. Chromosoma (Berl.) 37:405-416.
Baker, R.J., Mengden, G.A. and Bull, J.J., 1972, Karyotypic
    studies on thirty-eight species of North-American
    Snakes. Copeia 1972:257-265.
Becak, W. and Becak, M.L., 1969, Cytotaxonomy and chromosomal
    evolution in Serpentes. Cytogenetics 8:247-262.
Benirschke, K. (ed.), 1969, Comparative Mammalian Cytogenetics.
    Springer-Verlag, Berlin-Heidelberg-New York.
Bentley, P.J., 1971, Endocrines and osmoregulation. Springer-
    Verlag, Berlin-Heidelberg-New York.
Blumenfeld, M., 1974, The evolution of satellite DNA in
    Drosophila virilis. Cold. Spring Harbor Symp. Quant.
    Biol. 38:423-427.
Boer, de, L.E.M., 1975, Karyological heterogeneity in the
    Falconiformes (Aves). Experientia 31:1138-1139.
Britten, R.J. and Davidson, E.H., 1969, Gene regulation for
    higher cells: a theory. Science 165:349-357.
Bull, J.J., Moon, R.G. and Legler, J.M., 1974, Male
    heterogamety in kinosternid turtles (genus Staurotypus).
    Cytogenet. Cell Genet. 13:419-425.
Capanna, E. and Cataudella, S., 1973, The chromosomes of
    Calamoichthys calabaricus (Pisces, Polypteriformes).
    Experientia 29:491-492.
Cimino, M.C. and Bahr, G.F., 1974, The nuclear DNA content and
    chromatin ultrastructure in the coelacanth Latimeria
    chalumnae. Exp. Cell Res. 88:263-272.
Comings, D.E., 1975, Chromosome banding. J. Histochem.
    Cytochem. 23:461-462.
Crick, F., 1971, General model for the chromosomes of higher
    organisms. Nature 234:25-27.
Davidson, E.H., Galau, G.A., Augerer, R.C. and Britten, R.J.,
    1975, Comparative aspects of DNA organization in Metazoa.
    Chromosoma (Berl.) 51:253-259.
Denton, T.E., 1973, Fish chromosome methodology. Charles C.
    Thomas Publ., Springfield, Illinois.
Denton, T.E. and Howell, W.M., 1973, Chromosomes of the
    African Polypterid fishes, Polypterus palmas and
    Calamoichthys calabaricus (Pisces: Brachiopterygii).
    Experientia 29:122-124.

De Weese, J., 1975, Chromosomes in Eleutherodactylus (Anura: Leptodactylidae). Mamm. Chroms. Newsl. 16:121-123.

Donahue, W.H., 1974, A karyotypic study of three species of Rajiformes (Chondrichthyes, Pisces). Can. J. Genet. Cytol. 16:203-211.

Ebeling, A.W. and Chen, T.R., 1970, Heterogamety in teleostean fishes. Trans. Am. Fish. Soc. 99:131-138.

Flamm, W.F., 1972, Highly repetitive sequences of DNA in chromosomes. Int. Rev. Cytol. 32:1-51.

Fontana, F. and Colombo, G., 1974, The chromosomes of Italian sturgeons. Experientia 30:739-742.

Fredga, K. and Mandahl, N., 1973, Autosomal heterochromatin in some Carnivores, in "Chromosome Identification" (T. Caspersson and L. Zech, eds.), pp. 104-127, Nobel Symposium 23, Acad. Press, New York and London.

Georgiev, G.P., 1969, On the structural organization of operon and the regulation of RNA synthesis in animal cells. J. Theor. Biol. 25:473-490.

Gorman, G.G., 1973, The chromosomes of the Reptilia, a cyto-taxonomic interpretation, in "Cytotaxonomy and Vertebrate Evolution" (A.B. Chiarelli and E. Capanna, eds.), pp. 349-424, Acad. Press, New York and London.

Green, R.J. and Bahr, G.F., 1975, Comparison of G-, Q- and EM-banding patterns exhibited by the chromosome complement of the Indian Muntjac, Muntiacus muntjak, with reference to nuclear DNA content and chromatin ultrastructure. Chromosoma (Berl.) 50:69-77.

Hinegardner, R.T. and Rosen, D.E., 1972, Cellular DNA content and the evolution of teleostean fishes. Amer. Natural. 106:621-644.

Hsu, T.C. and Mead, R.A., 1969, Mechanisms of chromosomal changes in mammalian speciation, in "Comparative Mammalian Cytogenetics" (K. Benirschke, ed.), pp. 8-17, Springer-Verlag, Berlin-Heidelberg-New York.

Hutchison, N. and Pardue, M.L., 1975, The mitotic chromosomes of Notophthalmus (=Triturus) viridescens: localization of C-banding regions and DNA sequences complementary to 18 S, 28 S and 5 S ribosomal RNA. Chromosoma (Berl.) 53:51-69.

John, B. and Lewis, K.R., 1968, The chromosome complement. Protoplasmatol. 6A:1-206.

Johnson, J.D., Douvas, A.S., and Bonner, J., 1974, Chromosomal proteins. Int. Rev. Cytol., suppl. 4:273-361.

Killebrew, F.C., 1975, Mitotic chromosomes of Turtles: I. The Pelomedusidae. J. Herpetol. 9:281-285.

Levins, R., 1968, Evolution in changing environments. Princeton Univ. Press, Princeton, New Jersey.

Martin, N.G., 1974, Nuclear DNA content of the Emu. Chromosoma (Berl.) 47:71-74.

Matthey, R., 1973, The chromosome formulae of Eutherian
    Mammals, in "Cytotaxonomy and Vertebrate Evolution"
    (A.B. Chiarelli and E. Capanna, eds.), pp. 530-616,
    Acad. Press, New York and London.
Matthey, R., 1975, Caryotypes de Mammiferes et d"Oiseaux. La
    question des microchromosomes. Quelques reflexions sur
    l'evolution chromosomique. Arch. f. Genetik 48:12-26.
Mayr, E., 1975, The unity of the genotype. Biol. Zbl.
    94:377-388.
Misra, M. and Srivastava, M.D.L., 1975, Chromosomes of two
    species of Coraciiformes. Nucleus 18:89-92.
Morescalchi, A., 1973, Amphibia, in "Cytotaxonomy and
    Vertebrate Evolution" (A.B. Chiarelli and E. Capanna, eds.),
    pp. 233-348, Acad. Press, New York and London.
Morescalchi, A., 1975, Chromosome evolution in the Caudate
    Amphibia. Evol. Biol. 8:339-387.
Natarajan, A.T. and Gropp, A., 1971, The meiotic behavior of
    autosomal heterochromatic segments in hedgehods.
    Chromosoma (Berl.) 35:143-152.
Nevo, E., 1976, Adaptive strategies of genetic systems in
    constant and varying environments, in "Populations
    Genetics and Ecology" (S. Karlin and E. Nevo, eds.),
    pp. 141-158, Acad. Press, New York.
Nevo, E. and Shkolnik, A., 1974, Adaptive metabolic variation
    of chromosome forms in mole rats, Spalax. Experientia
    30:724-726.
Nygren, A. and Jahnke, M., 1972, Microchromosomes in
    primitive Fishes. Swed. J. Agric. Res. 2:229-238.
Ohno, S., 1967, Sex chromosomes and sex-linked genes. Monogr.
    Endocr., Springer-Verlag, Berlin-Heidelberg, New York.
Ohno, S., 1970, Evolution by gene duplication. Springer-Verlag,
    Berlin-Heidelberg-New York.
Ohno, S., 1973, Ancient linkage-groups and frozen accidents.
    Nature 244:259-262.
Olmo, E., 1976, Genome size in some Reptiles. J. Exp. Zool.
    195:305-310.
Pathak, S. and Stock, A.D., 1974, The X chromosomes of Mammals:
    karyological homology as revealed by banding techniques.
    Genetics 78:703-714.
Pathak, S., Hsu, T.C. and Arrighi, F.E., 1973, Chromosomes of
    Peromyscus (Rodentia, Cricetidae). IV. The role of
    heterochromatin in karyotypic evolution. Cytogenet. Cell
    Genet. 12:315-326.
Pedersen, R.A., 1971, DNA content, ribosomal gene multiplicity,
    and cell size in Fish. J. Exp. Zool. 177:65-78.
Potter, I.C. and Robinson, E.S., 1973, The chromosomes of the
    Cyclostomes, in "Cytotaxonomy and Vertebrate Evolution"
    (A.B. Chiarelli and E. Capanna, eds.), pp. 179-203. Acad.
    Press, New York and London.

Powell, J.R., 1975, Protein variation in natural populations of
    animals. Evol. Biol. 8:79-119.
Price, H.J., Sparrow, A.H. and Nauman, A.F., 1973, Evolutionary
    and developmental considerations of the variability of
    nuclear parameters in higher plants. I. Genome volume,
    interphase chromosome volume, and estimated DNA content
    of 236 Gymnosperms.  Brookhaven Symp. Biol. 25:390-421.
Ray-Chaudury, R., 1973, Cytotaxonomy and chromosome evolution
    in Birds, in "Cytotaxonomy and Vertebrate Evolution"
    (A.B. Chiarelli and E. Capanna, eds.), pp. 425-483. Acad.
    Press, New York and London.
Read, D.W., 1975, Primate phylogeny, neutral mutations and
    "molecular clocks". Syst. Zool. 24:209-221.
Rice, N.R., 1972, Change in repeated DNA in evolution.
    Brookhaven Symp. Biol. 23:44-76.
Robinson, E.S., Potter, I.C. and Atkin, N.B., 1975, The nuclear
    DNA content of Lampreys. Experientia 31:912-913.
Schnedl, W., 1974, Banding patterns in chromosomes. Int. Rev.
    Cytol. suppl. 4:237-272.
Schultz, G.A. and Church, R.B., 1972, DNA base sequence
    heterogeneity in the order Galliformes. J. Exp. Zool.
    179:119-128.
Selander, R.K. and Kaugman, D.W., 1973, Genic variability and
    strategies of adaptation in animals. Proc. Nat. Acad. Sci.
    USA 70:1875-1877.
Sharma, G.B., 1973, The chromosomes of non-Eutherian Mammals,
    in "Cytotaxonomy and Vertebrate Evolution" (A.B. Chiarelli
    and E. Capanna, eds.), pp. 485-530, Acad. Press, New York
    and London.
Sharma, A.K., 1974, Plant cytogenetics, in "The Cell Nucleus,
    Vol. 2" (H. Busch, ed.), pp. 263-291, Acad. Press, New York
    and London.
Shields, G.F. and Straus, N.A., 1975, DNA-DNA hybridization
    studies of Birds.  Evolution 29:159-166.
Shoffner, R.N., 1974, Chromosomes of Birds, in "The Cell
    Nucleus, Vol. 2" (H. Busch, Ed.), pp. 223-261, Acad. Press,
    New York and London.
Singh, L., 1972, Evolution of karyotypes in Snakes. Chromosoma
    (Berl.) 38:185-236.
Smith, G.P., 1974, Unequal crossover and the evolution of
    multigene families. Cold Spring Harbor Symp. Quant.
    Biol. 38:507-513.
Sparrow, A.H., Price H.J. and Underbrink, A.G., 1972, A
    survey of DNA content per cell and per chromosome of
    prokaryotic and eukaryotic organisms: some evolutionary
    considerations. Brookhaven Symp. Biol. 23:451-493.
Stebbins, G.L., 1966, Chromosomal variation and evolution.
    Science 152:1463-1469.

Stebbins, G.L., 1971, Chromosomal evolution in higher plants. Edward Arnold (Publs.) Ltd, London.

Stock, A.D. and Mengden, G.A., 1975, Chromosome banding pattern conservatism in Birds and nonhomology of chromosome banding patterns between Birds, Turtles, Snakes and Amphibians. Chromosoma (Berl.) 50:69-77.

Stock, A.D., Arrighi, F.E. and Stefos, K., 1974, Chromosome homology in Birds: banding patterns of the chromosomes of the domestic chicken, ring-necked dove, and domestic pigeon. Cytogenet. Cell Genet. 13:410-418.

Straus, N.A., 1971, Comparative DNA renaturation kinetics in Amphibians. Proc. Nat. Acad. Sci. USA 68:799-802.

Szarski, H., 1976, Cell size and nuclear DNA content in Vertebrates. Int. Rev. Cytol. 44:93-111.

Takagi, N. and Sasaki, M., 1974, A phylogenetic study of bird karyotypes. Chromosoma (Berl.) 46:91-120.

Thomson, K.S., 1972, An attempt to reconstruct evolutionary changes in the cellular DNA content of Lungfish. J. Exp. Zool. 180:363-372.

Uyeno, T. and Miller, R.R., 1972, Second discovery of multiple sex chromosomes among fishes. Experientia 15:223-224.

Vialli, M. and Casonato, P., 1972, Quantita di acido desossiribonucleico e aree nucleari negli eritrociti dei Pesci Brachiopterigi. Rend. Sc. Ist. Lombardo B 106:50-58.

Walker, P.M.B., Flamm, W.G. and McLaren, A., 1969, Highly repetitive DNA in Rodents, in "Handbook of Molecular Cytology" (A. Lima-de-Faria, ed.), pp. 52-66, North Holland publ. Co., Amsterdam-London.

White, M.J.D., 1973, Animal cytology and evolution. 3rd Ed., Cambridge Univ. Press, London.

Wilson, A.C., Maxson, L.R. and Sarich, V.M., 1974, Two types of molecular evolution. Evidence from studies on interspecific hybridization. Proc. Nat. Acad. Sci. USA 71:2843-2847.

Wilson, A.C., Sarich, V.M. and Maxson, L.R., 1974, The importance of gene rearrangements in evolution: evidence from studies on rates of chromosomal, protein and anatomical evolution. Proc. Nat. Acad. Sci. USA 71:3028-3030.

Wilson, A.C., Bush, G.L., Case, S.M. and King, M.C., 1975, Social structuring of mammalian populations and rate of chromosomal evolution. Proc. Nat. Acad. Sci. USA 72:5061-5065.

Yunis, J.J., 1974, Structure and molecular organization of chromosomes, in "Human Chromosome Methodology" (J.J. Yunis, ed.), pp. 1-13, Acad. Press, New York.

Yunis, J.J. and Yasmineh, W.G., 1971, Heterochromatin, satellite DNA, and cell functions. Science 174:1200-1209.

THE PHYLETIC INTERPRETATION OF MACROMOLECULAR

SEQUENCE INFORMATION:  SIMPLE METHODS

Walter M. FITCH

Dept. of Physiological Chemistry, University of
Wisconsin Medical School
Madison, Wisconsin  53706

## INTRODUCTION

The use of specific amino acid and nucleotide sequence information for investigating evolutionary processes is increasing and
will continue to do so as our sequence acquiring and data handling
techniques continue to improve.  This chapter describes simple
methods that enable the beginner to examine sequence information
for himself.

Many interesting genetic and evolutionary events are contained
in macromolecular sequences when viewed in light of the phylogeny*
of the taxa from which they derive.  What are the sources of that
phylogeny?  There are essentially two.  One is to accept current
biological opinion about what are the most probable evolutionary
relationships among the taxa, the other is to utilize the sequence
information itself.  Concerning the reconstruction of ancestral
molecular events (but not the order of branching), the answers
turn out to be very much the same either way (Fitch, 1976).  Because one may in fact wish to discover the phylogeny most consistent
with the data, we shall start by examining methods of reconstructing the phylogenetic relationships after laying some important
groundwork regarding measures of distance.*  This will be followed
by procedures for reconstructing the most probable evolutionary
course of molecular events.  This will permit a second method for
estimating phylogenetic relationships.


* Words or phrases marked with an * are defined in the glossary
  at the end of this chapter.

## MEASURES OF DISTANCE


### Homology Is Not Enough

A distance measure is simply an estimate of the degree of dissimilarity between two taxa as estimated from an examination of one or more phenotypic characters.

There are many measures of distance for biological data both qualitative and quantitative and, for data other than sequences, one should consult other texts such as Sneath and Sokal (1973). For our purposes, the characters* are amino acids or nucleotides in a set of orthologous* sequences from several taxa and the character state* is the particular amino acid or nucleotide in a given location in the sequence of a given taxon.

One must recognize that being homologous* is an insufficient attribute of sequences to permit their use for the purpose of constructing phylogenies of taxa. This arises because gene duplications may give rise, as they have among the hemoglobins, to many genetic loci all homologous to each other. Thus, should we choose to create a phylogeny from the alpha hemoglobins of man and donkey and the beta hemoglobins of chimp and horse we would get a phylogeny based, not on the more recent mammalian divergence of primates and perissodactyls, but rather on the much more ancient gene duplication of the hemoglobin locus. Homologous sequences (and other characters too!) thus come in two varieties. A pair of genes that trace their most recent common ancestry to a gene duplication event are said to be paralogous* because of their common parallel evolution in a single line of descent. The family of hemoglobins (myoglobin, alpha, beta, gamma, delta and epsilon) is typical. A pair of genes that have their most recent common ancestral gene in the most recent common ancestral organism of the taxa in which those genes reside, are said to be orthologous because of the exact correspondence between the branching history of the genes and the branching history of the organisms in which they reside. It is to the orthologous sequences that we must look if we wish to discover a species phylogeny*. It is then that each branch point represents a speciation event (assuming the sequences derive from different species). It is not without interest to examine a set of paralogous sequences where the resulting tree is a gene phylogeny* and each branch point represents a gene duplication event, but from this point on we shall assume that we are using a set of orthologous sequences. The reader should be cautious about literature statements involving homology since they are frequently not true generally but only true of either orthology or, less frequently, paralogy in particular.

## Simple Distance and the Genetic Code

To measure the distance we assume that we can line up the
sequences so that the k-th position is itself strictly homologous
(or conjugately* related, Fitch, 1973).  In practice this is
never a problem for genes that are clearly orthologous since the
need to use gaps* in a sequence to preserve the alignment seems
only rarely to occur, and even when it does, the location of the
gap is seldom in doubt.  The simplest distance between two taxa is
the number of positions that differ between the two orthologous
sequences representing them.  Nor do we need to restrict ourselves
to a single pair of orthologous sequences.  We may continue summing
differences over many pairs of sequences provided that each pair
is orthologous.  The simplest distance is adequate if there are no
positions for which the amino acid is unknown (X) or for which
there is no amino acid at all because a gap has been inserted.
(We shall represent such gap positions by a ', based upon the
orthographical usage of the apostraphe to represent missing
elements and its ease of recognition in a set of aligned sequences.)
This is generally the case but not always.

Before considering the complication posed by gaps and unknown
amino acids, we may note that amino acid sequences may be analyzed
in terms of the nucleotide sequences that, through the genetic
code (Table I), underlie them.  It should be clearly understood
that the number of differences between two sequences is not an
estimate of the number of substitutional* events that has occurred
in their genes since their common ancestor, only a lower bound on
that number. Moreover, as the number of differences increases, the
number of undetected substitutional events also increases, the
latter more rapidly than the former.  It is therefore important
that pertinent information not be wasted.  If the common ancestor
of a position were glycine,  and in one line of descent it remained glycine while in the other it went first to alanine and
very soon thereafter to proline, we might well observe only a
glycine or a proline replacement* in the descendents.  This is
only one amino acid difference.  However, at the level of the
messenger RNA we know from the genetic code where each amino acid
is represented by one or more triplets of nucleotides, that
glycine is encoded by GGX whereas proline is encoded by CCX.  Thus
we can know from a single pair of amino acids that at least two
(and in other, if rare cases, three) substitutions have occurred
since their common ancestor.  The principles that follow are not
dependent upon this transformation from the amino acid to the
nucleotide level.

Assume we have two homologous amino acid sequences, so aligned
(with gaps, if necessary) that the $\underline{k}^{th}$ amino acids of the two

TABLE I.   THE GENETIC CODE FOR THE PARSIMONY PROCEDURE

| Amino Acid | Codon | Amino Acid | Codon |
|------------|-------|------------|-------|
| Ala | GCX | Met | AUG |
| Asx | RAY | Asn | AAY |
| Cys | UGY | Pro | CCX |
| Asp | GAY | Gln | CAR |
| Glu | GAR | Arg | SGX (CGX, SGR) |
| Phe | UUY | Ser | TZX (UCX, AGY) |
| Gly | GGX | Thr | ACX |
| His | CAY | Val | GUX |
| Ile | AUH | Trp | UGG |
| Lys | AAR | Tyr | UAY |
| Lev | YUX (CUX, YUR) | Glx | ZAR |

The unambiguous nucleotides are:  A = Adenine, C = Cytosine,
G = Gaunine, U = Uracil.  The ambiguous nucleotides are: H = A/C/U
(not G), R = A/G (purine), S = A/C, T = A/U, X = A/C/G/U, Y = C/U
(pyrimidine), Z = C/G.  The fully ambiguous codons for arginine,
leucine, and serine contain codons for other amino acids that can
only be eliminated by choosing between two alternatives which are
shown in parentheses.  They are maximally ambiguous subject to the
constraint that no other amino acids be encoded.  Other useful
terms for ambiguous nucleotides that may arise during parsimony
operations are B = C/G/U (not A), D = A/G/U (not C), V = A/C/G
(not U) and W = U/G.

sequences are presumed to share a common ancestor. Let $k_1$ and $k_2$ be the $k^{th}$ amino acids from two sequences. We define their substitution value $SV(k_1, k_2)$, as the minimum number of nucleotides that must be changed to convert the coding from one amino acid to the other. Examples, using the IUB single letter code, are $SV(H,H) = 0$; $SV(H,P) = 1$; $SV(H,T) = 2$; $SV(H,M) = 3$, where H, P, T and M are Histidine, Proline, Threonine and Methionine respectively. A table of these values is given in Table II which also provides the single letter amino acid code.

We define $d_c$ as the (minimum nucleotide substitution) distance over pairs of amino acids and, by inference, therefore excluding pairs where one or both residues is either unknown (X) or a gap ('), i.e., $SV(k,X) = 0$ and $SV(k,') = 0$ for all k. If one does not wish to consider amino acids through theor codons, or if one knows the nucleotide sequence directly, then $SV(k_1, k_2) = 0$ if $k_1 = k_2$ and = 1 otherwise. Then,

$$\text{Eq'n 1} \qquad d_c = \sum_{k=1}^{s} SV(k_1, k_2)$$

where 1 and 2 are the two sequences and s is their sequence length.

We define c as the number of codon (residue) positions for which both amino acids are known and hence $c \leq s$. We define r as the rate of substitution per codon and hence $r = d_c/c$.

We define a common gap* between two sequences as one in which an uninterrupted series of gap residues begin and end at the same residue positions in both sequences. Assuming that the most reasonable explanation of a common gap is an insertion or deletion prior to the common ancestor that diverged to give rise to the two sequences being examined, then that insertion or deletion should not count as an insertion or deletion in determining the distance (amount of change) since their common ancestor. Instead, we treat it as so many residues that, had they been present, would have evolved at the same rate as known amino acid pairs, i.e., they are treated as if they were X's.

We define x as the number of positions where one or both residues of the pair exists but is unknown (X), but neither one is a gap, or where both are members of a common gap. We further define n as the number of residue positions where one or both residues of the pair is a gap not in common. Thus $s = c + x + n$. If there are no gaps, the required distance is $d_u = rs$.

## TABLE II
### MINIMUM NUCLEOTIDE DISTANCE, IN CODONS FOR PAIRS OF AMINO ACIDS

|   | D | C | T | F | E | H | K | A | M | N | Y | P | Q | R | S | W | L | V | I | G | B | Z |   |
|---|---|---|---|---|---|---|---|---|---|---|---|---|---|---|---|---|---|---|---|---|---|---|---|
| D | 0 | 2 | 2 | 2 | 1 | 1 | 2 | 1 | 3 | 1 | 1 | 2 | 2 | 2 | 2 | 3 | 2 | 1 | 2 | 1 | 0 | 1 | ASP |
| C | 2 | 0 | 2 | 1 | 3 | 2 | 3 | 2 | 3 | 2 | 1 | 2 | 3 | 1 | 1 | 1 | 2 | 2 | 2 | 1 | 2 | 3 | CYS |
| T | 2 | 2 | 0 | 2 | 2 | 2 | 1 | 1 | 1 | 1 | 2 | 1 | 2 | 1 | 1 | 2 | 2 | 2 | 1 | 2 | 1 | 2 | THR |
| F | 2 | 1 | 2 | 0 | 3 | 2 | 3 | 2 | 2 | 2 | 1 | 2 | 3 | 2 | 1 | 2 | 1 | 1 | 1 | 2 | 2 | 3 | PHE |
| E | 1 | 3 | 2 | 3 | 0 | 2 | 1 | 1 | 2 | 2 | 2 | 2 | 1 | 2 | 2 | 2 | 2 | 1 | 2 | 1 | 1 | 0 | GLU |
| H | 1 | 2 | 2 | 2 | 2 | 0 | 2 | 2 | 3 | 1 | 1 | 1 | 1 | 1 | 2 | 3 | 1 | 2 | 2 | 2 | 1 | 1 | HIS |
| K | 2 | 3 | 1 | 3 | 1 | 2 | 0 | 2 | 1 | 1 | 2 | 2 | 1 | 1 | 2 | 2 | 2 | 2 | 1 | 2 | 1 | 1 | LYS |
| A | 1 | 2 | 1 | 2 | 1 | 2 | 2 | 0 | 2 | 2 | 2 | 1 | 2 | 2 | 1 | 2 | 2 | 1 | 2 | 1 | 1 | 1 | ALA |
| M | 3 | 3 | 1 | 2 | 2 | 3 | 1 | 2 | 0 | 2 | 3 | 2 | 2 | 1 | 2 | 2 | 1 | 1 | 1 | 2 | 2 | 2 | MET |
| N | 1 | 2 | 1 | 2 | 2 | 1 | 1 | 2 | 2 | 0 | 1 | 2 | 2 | 2 | 1 | 3 | 2 | 2 | 1 | 2 | 0 | 2 | ASN |
| Y | 1 | 1 | 2 | 1 | 2 | 1 | 2 | 2 | 3 | 1 | 0 | 2 | 2 | 2 | 1 | 2 | 2 | 2 | 2 | 2 | 1 | 2 | TYR |
| P | 2 | 2 | 1 | 2 | 2 | 1 | 2 | 1 | 2 | 2 | 2 | 0 | 1 | 1 | 1 | 2 | 1 | 2 | 2 | 2 | 2 | 1 | PRO |
| Q | 2 | 3 | 2 | 3 | 1 | 1 | 1 | 2 | 2 | 2 | 2 | 1 | 0 | 1 | 2 | 2 | 1 | 2 | 2 | 2 | 2 | 0 | GLN |
| R | 2 | 1 | 1 | 2 | 2 | 1 | 1 | 2 | 1 | 2 | 2 | 1 | 1 | 0 | 1 | 1 | 1 | 2 | 1 | 1 | 2 | 1 | ARG |
| S | 2 | 1 | 1 | 1 | 2 | 2 | 2 | 1 | 2 | 1 | 1 | 1 | 2 | 1 | 0 | 1 | 1 | 2 | 1 | 1 | 1 | 2 | SER |
| W | 3 | 1 | 2 | 2 | 2 | 3 | 2 | 2 | 3 | 2 | 2 | 2 | 2 | 1 | 1 | 0 | 1 | 2 | 3 | 1 | 3 | 2 | TRP |
| L | 2 | 2 | 2 | 1 | 2 | 1 | 2 | 2 | 1 | 2 | 2 | 1 | 1 | 1 | 1 | 1 | 0 | 1 | 1 | 2 | 2 | 1 | LEU |
| V | 1 | 2 | 2 | 1 | 1 | 2 | 2 | 1 | 1 | 2 | 2 | 2 | 2 | 2 | 2 | 2 | 1 | 0 | 1 | 1 | 1 | 1 | VAL |
| I | 2 | 2 | 1 | 1 | 2 | 2 | 1 | 2 | 1 | 1 | 2 | 2 | 2 | 1 | 1 | 3 | 1 | 1 | 0 | 2 | 1 | 2 | ILE |
| G | 1 | 1 | 2 | 2 | 1 | 2 | 2 | 1 | 2 | 2 | 2 | 2 | 2 | 1 | 1 | 1 | 2 | 1 | 2 | 0 | 1 | 1 | GLY |
| B | 0 | 2 | 1 | 2 | 1 | 1 | 1 | 1 | 2 | 0 | 1 | 2 | 2 | 2 | 1 | 3 | 2 | 1 | 1 | 1 | 0 | 1 | ASX |
| Z | 1 | 3 | 2 | 3 | 0 | 1 | 1 | 1 | 2 | 2 | 2 | 1 | 0 | 1 | 2 | 2 | 1 | 1 | 2 | 1 | 1 | 0 | GLX |
|   | D | C | T | F | E | H | K | A | M | N | Y | P | Q | R | S | W | L | V | I | G | B | Z |   |

Table is symmetric about the major diagonal.  ASX and GLX imply uncertainty whether amino acid is aspartate (D) or asparagine (N) and glutamate (E) or glutamine (Q) respectively.

## Allowance for Gaps

We shall, following Fitch and Yasunobu (1975), consider three kinds of distances, each dependent upon a different way of treating non-common gaps.  One, $d_u$, treats all residue positions where both amino acids are not known as if they would have evolved at the same rate per codon as those for which both amino acids are known. The second, $d_n$, treats every inserted or deleted codon as equivalent to some number of nucleotide substitutions.  The third, $d_g$, treats every gap, irrespective of its length, as equivalent to some number of nucleotide substitutions.

We define $d_x$ as the minimum distance adjusted for unknown amino acids and assume that if we knew the amino acids present in these positions, they would have the same replacement rate as in those positions where we know both amino acids.  Thus

$$d_x = d_c + rx.$$

If there are no gaps, $d_x$ is the distance being sought and the remainder of this section is unnecessary.

We must now consider non-common gaps and can make three general assumptions regarding the mutation value of gaps.  The first is that the replacement rate should be equivalent to that in the known positions.  This is probably too low a value since insertions and deletions are quite rare compared to nucleotide substitutions and should therefore be weighted more heavily.  Nevertheless, letting $d_u$ be the distance adjusted as if all gap residues were unknown amino acids,

$$d_u = d_x + rn = rs.$$

An alternative assumption is that each residue in a gap should be treated as equivalent to a mutation value of $m_n$ nucleotide substitutions.  One possibility is that $m_n$ be set equal to three since three nucleotides per codon have been inserted or deleted, but any value is in principle acceptable.  We define $d_n$ as the minimum

distance adjusted for the number of residues in non-common gaps.
Then,

$$\frac{d}{n} = \frac{d}{x} + m_n\, n.$$

Still a third alternative assumption stresses the fact that
the insertion or deletion was the genetic event and that we should
evaluate it on the basis of the number of gaps rather than their
total length.  In such a case we may treat each gap, regardless of
length, as equivalent to a mutation value of $m_g$ nucleotide replacements.  If we let $g$ equal the number of non-common gaps and
$d_g$ equal the minimum distance adjusted for the number of non-
common gaps, then

$$\frac{d}{g} = \frac{d}{x} + m_g\, g.$$

If there are no X's, $\frac{d}{x} = \frac{d}{a}$.  If no gaps are longer than 1,
$\frac{d}{n} = \frac{d}{g} +$ a constant.  If there are no non-common gaps, $\frac{d}{u} = \frac{d}{n} =$
$\frac{d}{g} = \frac{d}{x}$.

Although $m_n$ and $m_g$ are in principle arbitrary constants, it
should be clear that in practice there must be some value for these
constants that optimizes the (not necessarily linear) correlation
between $d$ and time since a common ancestor.  The choice between
$d_n$ and $d_g$ as the better estimator is unclear.  The $d_g$ value might
seem better since it is directly proportional to the number of
genetic events.  However, we know that long deletions and insertions are more rare than short ones and should therefore be given
greater weight.  The $d_n$ value can be viewed as a function giving a
weight directly proportional to the length of the gap.  On the
basis of several iron-sulfur proteins (Fitch and Yasunobu, 1975),
a preliminary estimate would suggest $m_n$ and $m_g$ values of approximately 2 and 4 respectively.

The procedures just enumerated provide weighting for gaps,
that is, for characters not present in some taxa.  All other state

changes are unweighted, at least at the level of nucleotide changes. Weighting has been omitted here to avoid problems regarding the objectivity of the weights. It should be noted, however, that procedures for constructing phylogenies do not require unweighted (or, more accurately, equally weighted) characters. This includes parsimony procedures as well as matrix procedures. For those interested in the problem of objective weighting, I would recommend the paper by Farris (1969).

## An Example Computation

In Table III are given three fictitious amino acid sequences that are, for the purpose of computations, to be considered orthologous and conjugately aligned. The sequences were chosen to illustrate the computation. In practice, sequences on which it is reasonable to perform such computations will have relatively fewer X's and gaps, with the result that $d_u$, $d_n$ and $d_g$ will be much more equal.

The details of the computation are shown for the $\alpha,\beta$ pair on the left, the answers for the other two pairs on the right. The table is largely self-explanatory. It should be noted however that $g$ is the number of gaps not in common. Thus, although there are five gaps in $\alpha$ plus $\beta$, the first one of each is a common gap so that $g = 3$. The common gap is treated as an X.

If the distances are expressed as similarities, it may be useful to transform them according to equation 3 of Farris (this volume) prior to trying to build a tree from them.

## Adjustments for Hidden Substitutions

The distances computed here are only differences and there clearly could have been more evolutionary changes to create these differences than simply one for each difference. This is true for every type of distance based upon the simple comparison of two taxa whether the categories of comparison are morphology, ontogeny, physiology, immunology, electrophoresis, DNA hybridization or sequences. Moreover, the expectation is that the total number of evolutionary changes is not a linear function of the number of differences but, at least for the last three categories of comparison, rather a monotonically increasing function of them. This means that most questions that we might wish to ask do not require any correction for the unobserved changes and therefore we shall not include methods for estimating total evolutionary change here. It should be remarked, however, that many questions regarding evolutionary rates do require such an estimation. For those

TABLE III.  EXAMPLE COMPUTATION OF MUTATION DISTANCES

Sequences (s = 12)

| | | | | | | | | | | | | |
|---|---|---|---|---|---|---|---|---|---|---|---|---|
| α | A | ' | B | ' | ' | C | D | X | X | F | X | H |
| β | G | ' | B | E | ' | ' | D | E | X | F | ' | H |
| γ | A | K | B | E | L | C | Q | E | Q | X | E | H |

Let $m_n$ = 2, $m_g$ = 4.  Now, comparing sequences α and β

|  | α,γ | β,γ |
|---|---|---|
| $d_c$ = 1 + 0 + 0 + 0 + 0 = 1 | 2 | 3 |
| $c$ = 5 | 5 | 6 |
| $r$ = $d_c/c$ = 0.2 | 0.4 | 0.5 |
| $g$ = 3 (excludes common gap in position 2) | 2 | 3 |
| $n$ = 4 | 3 | 4 |
| $x$ = 3 (includes common gap in position 2) | 4 | 2 |
| $d_x$ = $d_c$ + rx = 1 + (0.2 x 3) = 1.6 | 3.6 | 4.0 |
| $d_u$ = $d_x$ + rn = 1.6 + (0.2 x 4) = 2.4 | 4.8 | 6.0 |
| $d_n$ = $d_x$ + $m_n n$ = 1.6 + (2 x 4) = 9.6 | 9.6 | 12.0 |
| $d_g$ = $d_x$ + $m_g g$ = 1.6 + (4 x 3) = 13.6 | 11.6 | 16.0 |

wishing to make such estimations for sequence data, they may examine the methods of Dayhoff (1972), Holmquist (1972), Goodman et al. (1974), and Margoliash and Fitch (1968). The assumptions are in each case different and all are oversimplifications. All procedures give improved estimates of total evolutionary change. It is uncertain how good each is or even which is best.

## Discordant Distances

There is another kind of distance that may be calculated that is more closely related to the nature of the most parsimonious* tree and is called the discordant distance*. Consider a taxon and one of its characters where the state of that character is a state possessed by no other taxon. Such a state is said to be singular. Singular character states* cannot give any information about the nature of the most parsimonious tree when every state can give rise to any other state. Thus it is advisable to exclude them from the distance calculation whenever one would hope the tree resulting from operations performed on the matrix of pairwise distances would approach as close as possible the most parsimonious tree. We may therefore revise equation 1 to read:

$$\underline{d}_{\underline{d}} = \sum_{k=1}^{c} SV(\underline{k}_1, \underline{k}_2) \text{ for all } \underline{k} \text{ such that neither } \underline{k}_1 \text{ nor } \underline{k}_2 \text{ are}$$

singular and where $\underline{d}_{\underline{d}}$ denotes the discordant distance. It is particularly important to exclude $\underline{k}$'s that are part of singular gaps. The reasoning behind this useful concept can be found in Fitch (1975, 1977). It is only of value if there are more than three taxa being examined.

## CONSTRUCTION OF A TREE FROM A PAIRWISE DISTANCE MATRIX

### The Construction

The tree is constructed by what Sneath and Sokal (1973) call a hierarchical, agglomerative procedure in which the taxonomic* entities (frequently called OTU's) are originally disjoint sets (that is, they are unconnected). The tree is constructed by joining two of them (taxa) together to create a new set (collectively denoted by their ancestor) that replaces them. The process is repeated until all taxa are joined, that is, until all $\underline{t}$ of them are members of a single set denoted by the ultimate ancestor*. This will require $\underline{t}$-1 joinings, the tree depicting the order of the joinings (clustering). If each joining is between two sets of taxa that are, historically, more closely related to each other than

either is to any of the remaining taxa, the result is the true
phylogeny of the taxa.  In the procedure that follows, it is
assumed that that pair of taxa (or sets of them) are most closely
related historically whose pairwise distance is least.  The method
is that of Fitch and Margoliash (1967).

The matrix of pairwise distances is examined to find the
smallest distance.  That value divides the $t$ taxa, for purposes of
the following computation, into three groups A, B, and X, where A
and B are the pair of taxa for which the distance from A to B,
$\underline{d}(A,B)$, is least and X are all other taxa.  We then compute the
average distance from A to all taxa in X by $\underline{d}(A,X) = \{\Sigma\ \underline{d}(A,X)\}/$
$(\underline{t} - 2)$ for all X $\neq$ A or B, of which there are $\underline{t} - 2$.  Similarly,
we compute $\underline{d}(B,X)$ by substituting B for A in the formula.

In the portion of the tree under construction, an ancestral
node, call it C, becomes the immediate ancestor of A and B and can
be represented as in Fig. 1, left.  Our present task is to assign
lengths to the legs descending from C to A and B.  We let $\underline{d}(A,C) =$
$\underline{a}$, $\underline{d}(B,C) = \underline{b}$ and $\underline{d}(C,X) = \underline{x}$.  Then $\underline{d}(A,B) = \underline{a} + \underline{b}$, $\underline{d}(A,X) = \underline{a} + \underline{x}$
and $\underline{d}(B,X) = \underline{b} + \underline{x}$.  With three equations and three unknowns we
may readily solve for $\underline{a}$ and $\underline{b}$ ($\underline{x}$ is not important except at the
final joining).  This gives $\underline{a} = \{d(A,B) + \underline{d}(A,X) - d(B,X)\}/2$ and
$\underline{b} = \underline{d}(A,B) - \underline{a}$.  It is also useful to define a height for node
$C = \underline{h}(C) = (\underline{a} + \underline{b})/2$.  We now replace A and B by C and the tree
construction process above may be repeated with one less taxon.

The replacement of A and B by C requires a change in the
pairwise distance matrix in which we determine $\underline{d}(C,X) = \{\underline{d}(A,X) +$
$\underline{d}(B,X)\}/2$ for all X $\neq$ A or B.  Following this computation, distances
involving A and B are removed from the matrix and replaced by their
average distances represented by C (the former should not be
destroyed however since examination of alternative trees will require their knowledge).  The matrix has now been fashioned so that
the procedure in the preceding paragraph can be repeated.

In repeating the procedure, the next pair of taxa to be joined
need not involve C.  If not, there is no complication.  If it does
involve joining C, say to D, the only computational problem involves the estimate of the length of the leg joining C to its
common ancestor with D which we will call E.  The situation is
shown in Fig. 1, right.  In analogy to the preceding case, $\underline{c} =$
$\{\underline{d}(C,D) + \underline{d}(C,X) - \underline{d}(D,X)\}/2$.  However, $\underline{c}$ is not equal to $\underline{d}(C,E)$
but equal to the average distance from A and B to E.  The distance
from C to E, $\underline{d}(C,E) = \underline{c}' = \underline{c} - \underline{h}(C)$ since $\underline{h}(C)$ is the average
distance of A and B to C.  One need always remember that the dis-
tances in the matrix are average distances between sets of ultimate
descendents and where the set being joined contains more than one
ultimate descendent, the average height, $\underline{h}$, to their own common

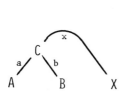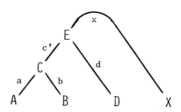

Figure 1. Example Construction of a Tree. This example relates to the data in Table IV. At the first step, taxa A and B are joined (left), all other taxa are assigned to set X, leg lengths a and b are calculated and C becomes the set designation for the set (A,B). At the second step, taxa C and D are joined (right), all other taxa are assigned to set X, leg lengths c' and d are calculated and E becomes the set designation for the set (A,B,D). See text and Table IV for computational details.

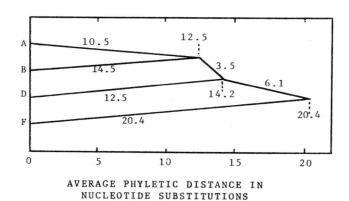

AVERAGE PHYLETIC DISTANCE IN
NUCLEOTIDE SUBSTITUTIONS

Figure 2. Phylogeny of Four Plant Ferredoxins. This is the final result of the tree construction procedure using the data in Table IV. A = Alfalfa, B = Leucena, D = Spinach and F = Scenedesmus. Numbers attached to nodes are peak heights, others are leg lengths in nucleotide substitutions. See text for description of the computation.

ancestor must be subtracted from the computed distance to the new
ancestor formed with the other set to which they are being joined.

## An Example Computation of Tree Construction

The preceding formalisms are more difficult than the actual
process and it seems particularly appropriate to show how easy
the procedure is in practice which will also make the formulae
above more transparent. This is provided by some data from Fitch
and Yasunobu (1975) on plant ferredoxins and shown in the lower
left hand portion of Table IV. The data are $d_u$ values. The
matrix identifies rows and columns not only by the names of the
plants, but by capital letters that will be useful in relating the
analysis to the preceding formalism and Fig. 1. The upper right
hand portion of the matrix should be ignored for the present.

The procedure asks us to find the least value of $d_u$, which is
25, and divide the taxa accordingly. Thus we let A = alfalfa,
B = leucena and X = spinach plus scenedesmus. The problem is
shown graphically in Fig. 1, left. The computation of the distance
from A and B to X is shown in Table IV together with the solution
of the values for $a$ and $b$. One can see the simple algebra on the
right. On the left is shown the more formal terms used in the
previous section.

Following the computation of $h(C)$, the average distance from
A and B to C, the average distances of the set C(A and B) to X are
computed. The distance, $d(C,D)$, proves least and thus C and D
become the sets for the second joining and the remaining set, now
a new and different X, is this time only F (note that X contains
one fewer set after each joining). The procedure is repeated to
get values of $c$ and $d$ (Fig. 1, right) but $c$, being the average
distance from A and B up to E must be reduced by their average
distance to C to obtain $c'$. One then calculates $h(E)$.

Since there remains only one taxon in the set X to joint to
the tree, we need not recompute the distance matrix but need only
estimate how to apportion the leg lengths to the root of the tree
from the two joining sets. The height of the root is obtained
directly from the preceding computation and is simply $(h(E) + x)/2$
where $x = d(C,X) - c$. The lengths of the legs ascending to the
root from the sets E and F are $e' = h(root) - h(E)$ and $f' = h(root)
- h(F)$. Since F is, in this case, an ultimate descendent, $h(F) =
0$ and so $f'$ equals the height of the root. When all these values
are placed upon the completed tree, the result is as shown in
Fig. 2 which gives both leg lengths and peak heights.

TABLE IV. SAMPLE CONSTRUCTION OF A PHYLOGENETIC TREE

|   | Al | Le | Sp | Sc |   |
|---|----|----|----|----|---|
| A | ** | 25.0 | 26.5 | 40.5 | Alfalfa |
| B | 25 | ** | 30.5 | 44.5 | Leucena |
| D | 27 | 30 | ** | 39.0 | Spinach |
| F | 40 | 45 | 39 | ** | Scenedesmus |
|   | A | B | D | F |   |

$$\underline{d}(A,B) \qquad\qquad\qquad = 25.0 = \underline{a} + \underline{b} \quad \text{Since this is smallest,}$$
$$\underline{d}(A,X) = (27 + 40)/2 \qquad = 33.5 = \underline{a} + \underline{x}$$
$$\underline{d}(B,X) = (30 + 45)/2 \qquad = 37.5 = \underline{b} + \underline{x}$$
$$\underline{d}(A,X) - \underline{d}(B,X) \qquad = -4.0 = \underline{a} - \underline{b}$$
$$\underline{d}(A,B) \qquad\qquad\qquad = 25.0 = \underline{a} + \underline{b}$$
$$\underline{d}(A,B) + \underline{d}(A,X) - \underline{d}(B,X) = 21.0 = 2\underline{a}$$
$$\underline{a} = 10.5$$
$$\underline{b} = 14.5$$
$$\underline{h}(C) = (\underline{a} + \underline{b})/2 = 12.5$$
$$\underline{d}(C,D) = (27 + 30)/2 \qquad = 28.5 = \underline{c} + \underline{d}$$
$$\underline{d}(C,X) = (40 + 45)/2 \qquad = 42.5 = \underline{c} + \underline{x}$$
$$\underline{d}(D,X) \qquad\qquad\qquad = 39.0 = \underline{d} + \underline{x}$$
$$3.5 = \underline{c} - \underline{d}$$
$$28.5 = \underline{c} + \underline{d}$$
$$32.0 = 2\underline{c}$$
$$\underline{c} = 16.0$$
$$\underline{d} = 12.5$$
$$\underline{h}(E) = (\underline{c} + \underline{d})/2 = 14.25$$
$$\underline{x} = 26.5$$
$$\underline{c}' = \underline{c} - \overline{\underline{ab}} = 3.5$$

Since only one distance left
$$\underline{h}(\text{root}) = (14.25 + 26.5)/2 = 20.375$$
$$\underline{e}' = 26.5 - 20.375 = 6.125$$

An alternative procedure for constructing a tree from such a
matrix has been given by Farris (1972). It is only slightly more
complicated and possesses the advantage of giving a direct estimate
of the extent of homoplasy* in that tree and of having a closer
relationship to the problem of finding the most parsimonious tree.

### Goodness of Fit

The procedure just given yields a result not unlike other
phenetic procedures in that nothing that has been done guarantees
that the result is the best evolutionary hypothesis for the data
being analyzed. Such a guarantee may not even be possible but
certainly criteria may be set up such that the best evolutionary
hypothesis is the tree that best meets the criteria. I shall
present two such simple criteria now ("%SD" and least homoplasy)
and another later (maximum parsimony).

This procedure (Fitch and Margoliash, 1967) requires that one
first determine the pairwise distance* which is the amount of
change between two species as determined by summing the amounts of
change assigned to the legs connecting them on the phylogenetic
tree constructed. If the values on the legs connecting pairs of
species in Fig. 2 are summed, the results are those shown in the
upper right half of the matrix in Table IV. The best tree may be
considered as the one in which the reconstructed phyletic distances
most closely match the original data. Standard deviation is a
common way of measuring closeness of data pairs.

Let $i$ and $j$ be the column and row indices respectively of a
matrix such as that in Table IV. If $i < j$, then $d_{ij}$ is an original
distance and $d_{ji}$ is a phyletic distance. The test statistic then is

$$"\%SD" = \frac{1}{n} \sum_{i < j} \{100(d_{ij} - d_{ji})/d_{ji}\}^2$$

where $n$ is one less than the number of distance pairs. The factor,
$100/d_{ji}$, is designed to make the differences into per cent differences so that pairs involving long distances are not more important
than those involving short distances. The quotation marks around
%SD designate that the result is not a true standard deviation,
one, because of the percentage feature, and two, because the distances are not independent. As a result, the "%SD" cannot be used
to make a probability statement but it nevertheless remains true
that the smaller the "%SD", the better the tree accounts for the
data.

## An Example Computation of "%SD"

For the matrix in Table IV, there are six pairs of distances, hence $\underline{n}$ = 5. For two of the six pairs, the difference is zero, for the other four it is 0.5 which, times 100, equals 50 and which, when divided by $\underline{d}_{ij}$, gives values of 1.85, 1.67, 1.25 and 1.11. When squared these give values of 3.43, 2.78, 1.56 and 1.23, respectively. The sum of the squares is 9.00 which, divided by 5, gives 1.8. The "%SD" is thus the square root of 1.8 which equals 1.34.

The procedure of Farris (1972) is of such a nature that all pairwise phyletic distances are greater than or equal to the initial pairwise distances. They would all be equal only in the absence of homoplasy and the calculated distances summed over all legs of the tree would be the least possible. The extent to which the sum exceeds this lower bound is, in Farris' procedure, a measure of the extent of homoplasy. Thus, an alternative criterion becomes: That tree is best for which the sum of the leg lengths of the tree is minimal. This is the least homoplasy criterion and, as such, is analogous to the maximum parsimony criterion to be presented further on. The criterion could of course be applied to trees with leg lengths assigned by other methods but the relationship between the criterion and the extent of homoplasy would no longer necessarily obtain.

## Testing Alternative Trees

The procedure for constructing a tree given under "The Construction" section, is really a procedure for fitting the matrix distance data to a tree although it also formulates a rational estimate of what a good phylogeny might look like. In order that we not simply have a phenetic method, it is essential to see whether a better tree than the original one can be found. In this case, better will mean possessing a smaller "%SD". The problem then is to provide an efficient means of finding better trees. I shall mention two.

The first is called branch swapping. Consider any tree that is best so far as is known. Obviously the tree obtained in the section "The Construction" qualifies if no other trees have yet been tested. Consider any 4 nodes, say A, B, C and D that are consecutively adjacent*, that is, A is adjacent to B is adjacent to C is adjacent to D (A-B-C-D). Branch swapping changes the connections such that now D-B-C-A is the adjacency relation. All other relationships in the best tree are preserved and this new tree is tested to see if it is better. If it is, it replaces the

(now second) best tree.  The process is repeated until a best tree
is found for which there is no set of 4 consecutively adjacent
nodes that, if swapped in the approved manner, give a better tree.

The branch swapping procedure has behind it the implicit
assumption that the first tree is sufficiently close to the very
best tree that we shall certainly be able to find it by a series
of small adjustments.  It is clearly not true.  We know that with
real biological data the "surface" of criterion values, whether
%SD, parsimony or whatever, has many local minima and that it is
extraordinarily easy to start close to a local rather than a
global minimum.  There is no way of guaranteeing that the global
minimum will be found short of very thorough (semi-exhaustive)
searches that are impractical for even moderate (say 20) numbers
of taxa.

An alternative procedure for producing alternative trees for
examination is that used by Fitch and Margoliash (1967) but it can
be used with any agglomerative procedure.  At each joining of two
chosen taxa, there may be other pairs of taxa, one of which is a
member of the chosen pair, for which the distance is sufficiently
close as to suggest that the optimal choice is not clear.  For
example, in Table IV we chose to join taxa A and B for which
$\underline{d}(A,B) = 25$ but it is possible that joining A to D since $\underline{d}(A,D) =$
27 would be a better choice in the context of a criterion for the
best complete tree.  One therefore simply examines all alternative
joinings for which the pairwise distance does not exceed the distance for the originally chosen pair by a specified amount, say
five per cent.  The best value of that amount is not determinable
and is usually chosen so that the number of alternative trees
examined is kept to a manageable number, say 100 trees.  It is also
important to keep in mind that five per cent of the smaller chosen
distances may rule out alternatives that should be examined.  If
several taxa differ by only a few nucleotides, one might well wish
to examine all alternative pairs whose distance is within two or
three nucleotide differences of the chosen pair whenever the per
cent limit proves to be less than two or three.

Since one is choosing alternatives on the basis of the degree
of similarity of competing alternatives, this should be a better
search procedure than the branch swapping algorithm.

### Bifurcating Trees Are Maximally Informative

The title of this section is intended to make clear that any
analysis that does not first analyze the strictly bifurcating tree*
cannot be a thorough analysis of the data since trifurcating nodes
(and those of higher degree*) are, mathematically, simply special

cases of adjacent bifurcating nodes for which the internodal distance is equal to zero.  Only when the data do give a distance
equal to zero should the two adjacent nodes be collapsed into a
single node, thereby creating a trifurcating node.

It may of course, occur that a non-zero distance is less than
one feels confident in taking with great seriousness with respect
to indicating the order of speciation.  There are, however, several
means of clearly informing the reader of the analyst's judgement
(such as connecting the adjacent nodes by a dotted rather than a
solid line) that do not thereby hide the analytical result.  To
do otherwise is to assume the reader is not competent to handle
such information.  If, however, a trifurcation is shown because
the distance between two nodes was zero, that fact should be
clearly stated simply so the reader knows you are not among the
arrogant many who won't allow the reader to make his own judge-
ments.

## MAXIMUM PARSIMONY PROCEDURES

### Counting the Number of Required Changes of State

There is a special case in which it is possible to know the
most parsimonious number of changes of state (e.g., nucleotide
substitutions or amino acid replacements) without concern for the
shape of the tree, and that is when there are no parallel or back
substitutions (or replacements), i.e., no homoplasy.  In such a
case, each character will change exactly as many times as there
are character states less one.  Thus the most parsimonious tree
will have as many changes as there are character states, summed
over all characters, minus the number of characters examined.  The
shape of a tree for which the data possess no homoplasy is trivially
easy to find and will not concern us further.  This minimum number
of changes of state can be calculated for any data set, however,
and represents a lower bound to the number of changes in the most
parsimonious tree.  We shall denote this value as $m$ .

There is no simple algorithm that finds the branching tree
structure that is most parsimonious for the data.  There is, however, a procedure that determines the fewest number of changes for
any given tree structure.  This was developed by Fitch (1971) and
proven correct by Hartigan (1973).

This procedure determines the most parsimonious relationship
for a set of sequences given a phylogenetic relationship for
either (1) the taxa from which they derive (if they are orthologous
sequences) or (2) their genes (if they are paralogous sequences).

There must be no doubt from the alignment that a given position in all the taxa have the same common ancestor, i.e., are conjugately aligned.

Step 1. Treat each ancestor in turn, working from present day taxa back to the ultimate ancestor. An ancestor may be treated only if its immediate descendents have been determined. Set a substitution counter to zero. Treatment for a single nucleotide position is as follows: Put into the ancestor all nucleotides common to both immediate descendents. If there are no nucleotides in common, increase the substitution counter by one and put into the ancestor every nucleotide present in either descendent. Mathematically, the ancestral nucleotide set is the intersection of the descendent nucleotide sets unless the intersection is empty, in which case the ancestral nucleotide set is the union of the descendent nucleotide sets. When the nucleotide(s) of the ultimate ancestor has been determined, the number of substitutions in the most parsimonious solution is indicated by the substitution counter and is equal to the number of set unions required. This process is mathematically summarized in Table V.

The above process must be repeated for each position in the sequence. If one wishes only to know the total number of substitutions and the ultimate ancestral sequence, nothing further need be done. If one wishes to determine where in the phylogeny the substitutions occurred and for the precise nature of the nucleotide substitutions, a second and final step must be undertaken.

An example of this procedure is shown in Figure 3. The tree at the top shows the initial condition with a given or proposed tree showing the ancestral relationships among six taxa and one conjugate nucleotide for each taxon. The operation of Step 1 finds the intersection empty at each of the five ancestral nodes except the penultimate node where U is the only nucleotide in common to both its descendents. Thus the union of the descendent sets is performed four times and the most parsimonious accounting of these character states on this tree will require four nucleotide changes.

## Defining the Ancestral Character States

Step 2. In this step one works backward through the nodal sets from the ultimate ancestor in an order precisely opposite to the order in which those nodes were formed. The ultimate ancestor is left unchanged. For each of the ancestral sets, the treatment is as follows: If the nucleotide set of the present node contains all the nucleotides in its immediate ancestor, remove any other

TABLE V.   ANCESTRAL RECONSTRUCTION


Step 1.   Given that $\underline{a}$ is the immediate ancestor formed from descendents $\underline{d}_1$ and $\underline{d}_2$ and N { } represents the set of nucleotides assigned to the designated entities (nodes of the tree), then

$N \{\underline{a}\} = N\{\underline{d}_1\} \cap N\{\underline{d}_2\}$ if and only if the intersection is not
empty, otherwise

$N \{\underline{a}\} = N\{\underline{d}_1\} \cup N\{\underline{d}_2\}$.

The step must be performed for every ancestral node.   Minimum number of substitutions is the number of times a union was performed in Step 1.

Step 2.   Given $\hat{N}$ { } is the corrected nucleotide set assigned to a node and $\underline{e}$ and $\underline{f}$ are the immediate descendents of $\underline{d}$.

a.   $\hat{N}\{d\} = \hat{N}\{a\}$ if and only if $\hat{N}\{d\} \supset \hat{N}\{a\}$; otherwise

b.   $\hat{N}\{d\} = \hat{N}\{d\} \cup N\{a\}$ if $N\{d\}$ was formed by a union; otherwise

c.   $\hat{N}\{d\} = \hat{N}\{d\} \cup (N\{a\} \cap (N\{e\} \cup N\{f\})$.

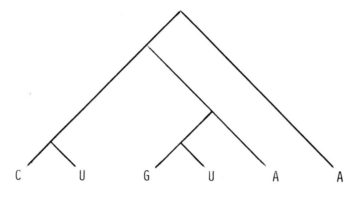

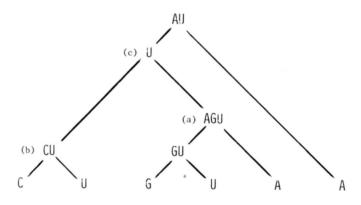

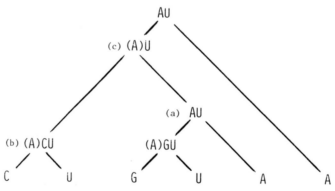

Figure 3.   Reconstruction of Ancestral Nucleotides.   The upper tree shows the initial state with a given tree structure and known character states (amino acids or nucleotides) at the tips.   The middle tree shows the ancestral character states after step 1.   The bottom tree shows the ancestral character state after step 2.   See text for discussion of the steps.

nucleotides not present in its immediate ancestor and proceed to
the next node; otherwise, if the node under consideration was
formed by a union, add to it every nucleotide it does not possess
provided that that nucleotide is also present in its immediate
ancestor and then proceed to the next node; otherwise, add to it
every nucleotide not already present provided that that nucleotide
is also present in both its immediate ancestor and at least one of
its two immediate descendents.  This step is also summarized
mathematically in Table V.  The treatment is repeated until no
further nodes, including those of the starting sequences, remain.
However, the starting sequences may only be reduced in ambiguity,
not expanded.  Any nucleotide added to an ancestral set in the
final step should be specifically noted, parentheses being used
here for that purpose.  The ancestral nodal sets now contain every
nucleotide that could be present in a most parsimonious tree, and
no other nucleotides.

For an example of Step 2 we may examine Figure 3.  The middle
tree shows the tree after Step 1.  The bottom tree shows the tree
after Step 2.  There are three possible cases under Step 2 and the
figure was designed to show an example of each, parenthetically
labelled a - c in accordance to their listing in Table V.  The
penultimate node does not contain all the nucleotides present in
its (the ultimate ancestor nor was it formed by a union so the
ancestral A, which appears in both its ancestor and its right-hand
descendent, must be added (rule 2c).  The penultimate node's
right-hand descendent contains all the nucleotides in its ancestor
so any extra ones (in this case the G) must be removed (rule 2a).
The penultimate node's left-hand descendent does not contain all
the ancestral nucleotides but it was formed by a union.  Hence rule
2 b applies and the ancestral (A) must be added.  The remaining
ancestor is finished under rule 2b also.


Defining the Paths of Most Parsimonious Descent

It is now only necessary to indicate which combinations of
ancestral and descendent nucleotides can be part of the same most
parsimonious tree.  The rule for this is that a nucleotide must
remain unchanged from ancestor to immediate descendent except as
follows:  A nucleotide may change only in two cases, one of which
is that the descendent set does not contain the ancestral nucleotide, the other is that the descendent set contains the ancestral
nucleotide but it is in parentheses.  Where a nucleotide is
permitted to change it may change to any nucleotide in the descendent set that does not possess parentheses around it.  This is
summarized in Table VI.  Figure 4 shows the ancestral character
states of the preceding problem arranged so that all possible
pathways of parsimonious descent are shown by connecting lines

TABLE VI.   COMPUTING FREQUENCIES OF ANCESTRAL NUCLEOTIDES

AND NUCLEOTIDE SUBSTITUTION

1.  All nucleotides in the ultimate ancestor are equally frequent.

2.  Given an ancestral nucleotide of frequence $\underline{n}$ and that it can
    descend over any one of $\underline{p}$ possible paths to the next node, it
    follows each path with frequence $\underline{n}/\underline{p}$.

3.  The sum  of the frequencies on all paths from all nucleotides
    in an ancestral node to all nucleotides in any one descendent
    node must equal one.

4.  The frequency of a nucleotide in a descendent node is the sum
    of all frequencies on all paths that descend to that nucleotide.

5.  The sum of the frequencies of all nucleotides at a given node
    must equal one.

TABLE VII.   NUCLEOTIDE SUBSTITUTIONS BY CHARACTER STATE CHANGE

| | | | | |
|---|---|---|---|---|
| A → C | 0.17 | | G → U | 0.08 |
| A → G | 0.17 | | U → A | 1.25 |
| A → U | 0.58 | | U → C | 0.83 |
| G → U | 0.08 | | U → G | 0.83 |

An "A Solution" by the procedure of Goodman et al., (1974).

| | | | | |
|---|---|---|---|---|
| A → U | 2 | | A → G | 1 |
| A → C | 1 | | | |

PATHWAYS THROUGH ANCESTRAL NUCLEOTIDES

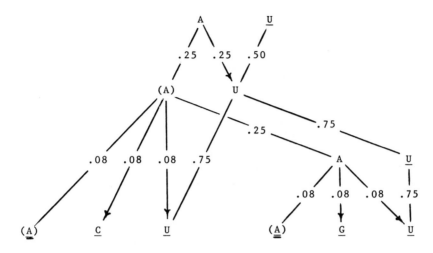

NUCLEOTIDE SUBSTITUTIONS BY INTERVALS

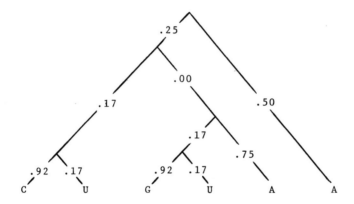

Figure 4.   Paths of Descent Through Ancestral Character States.
The upper figure shows those character states of the five ancestors
present in the lower tree of figure 3.  The states are connected
by lines showing all possible paths of descent, and only those
paths, consistent with the maximum parsimony solution.  Lines
possessing arrows involve a change of character state.  Where an
ancestor is the immediate ancestor of an ultimate descendent, its
character state is underlined if a change of character state will
be required in the descent to that ultimate descendent.  The fraction of the ancestral character states descending along each path
is also shown.  See text for method of determining the possible
paths and calculating their relative usage.  The lower part shows
the total number of changes of state (amino acid replacements or
nucleotide substitutions) in each interval of the complete tree.

which, when there is a nucleotide substitution (mutation fixed),
have an arrow head on them. Nucleotides that are underlined require an (or two) additional nucleotide substitution(s) in reaching
an ultimate descendent, that is, a taxon that is not itself an
ancestor.

Assigning Character State Changes to Intervals on the Tree

The existence of more than one most parsimonious solution
raises the question of how one should enumerate the various possible replacements as to their location in the tree and the nature
of the nucleotide replacement. There are several possibilities,
all somewhat arbitrary but only two are commonly employed, an
average result and an extreme result.

One form of average would take each possible tree and average
over them all. It is easily shown that this leads to extreme
cases. For example, this requires that the ratio of A:U in the
ultimate ancestor of our hypothetical example in Figure 3 to be
10:1. This is also a more difficult average to find.

An easier and more appropriate average is that of Fitch (1971)
and simply assumes that every nucleotide in the ultimate ancestor
is equally frequent. In our example (Fig. 4) A and U would be
0.5 each. We then assume that in the descent from one node to the
next, all possible paths are taken with equal frequency. Thus, in
our example, all 0.5 of the U in the ultimate ancestor must descend
to the U in the penultimate ancestor while the ultimate ancestral A
can descend by way of two paths so that each is traversed with frequency 0.25. Since one of these represents a nucleotide change,
we note that we have 0.25 of a mutational change from A to U.
There is also 0.50 of a mutational change from U to A in the
descent from the ultimate ancestor to the original A in the rightmost taxon of the original set. This is not shown but is understood from the underlining of the U in the ultimate ancestor.

The expected frequencies of an ancestral nucleotide set can
be found by summing the values on the incoming connecting lines
from its ancestor and must, of course, sum to one. These in turn
determine the distribution of frequencies along the outgoing lines
to the descendents of that node. These rules are summarized in
Table VI.

The sums of the frequencies along lines with arrows plus those
of nucleotides that are underlined (times 2 if there are two under-
lines) must equal the total number of substitutions. In our example this number is four. They are divided over the parts of the
tree as shown in the lower portion of Figure 4. They are also
divided among the various possible nucleotide interchanges as
shown in Table VII.

The other procedure is simply to select arbitrarily one of the
many possible trees.  Goodman et al. (1974) have examined two kinds
of extreme cases.  In one case that tree is chosen in which the
nucleotide changes are placed preferentially in legs as close to
the branch tips (i.e., the original sequences or ultimate descendents) as possible.  This is called the A solution.  In our
present hypothetical example, there are two such trees.  One has
only A's, the other only U's at all ancestral nodes.  The choice
between the two trees is arbitrary.  The results for the former
are also shown in Table VII.  These are then the ultimate by this
criterion since no changes occur between ancestral nodes.  Goodman
et al. (1974) have argued that this criterion of preferring the
location of the nucleotide substitutions at the extremities of the
tree causes their preferential distribution among intervals that
span the longer time periods of the tree.  This would be generally
useful because it would tend to compensate for the fact that the
longer an interval is, the smaller the fraction of all substitutions therein that are in fact observed.


Handling the Six-fold Degeneracy of the Genetic Code

The preceding methods are general for any kind of sequences
and, for example, apply just as well to amino acid sequences.  One
can also use nucleotide sequences inferred by using the genetic
code to back-translate amino acid sequences (see Table I).  The
presence of ambiguity as to which of several nucleotide triplets
encode a single amino acid, with one exception, causes no difficulty in the application of this procedure.  The exception is the
treatment of the amino acids serine, leucine and arginine because
of the presence of codons for each of these amino acids that are
different in more than one nucleotide.  Thus AGC and UCC are both
codons for serine and there is no simple way of representing the
choice in the first positions (A or U) independently of the
choice in the second (G or C) without admitting the possibility
that the most parsimonious tree might prefer the choices G and U
in the first and second position.  The difficulty is that GUC
codes for cysteine, not serine.  In other words, nucleotide ambiguity is tolerable only to the extent that the amino acid encoded
remains unambiguous.

There are two approaches to this problem.  One, adopted by
Moore et al. (1973) is not to treat the three codon positions
independently.  This has the disadvantage that much more time is
required whether done by hand or by computer.  The second approach
is that of Fitch and Farris (1974).  Their procedure, while complicated in its details is simple in concept.  For amino acid
positions containing arginine, leucine or serine, the parsimony
procedure just outlined is performed twice rather than once for

the tree nucleotide positions representing the amino acid.

The first time the procedure is performed using complete ambiguity in all positions as if the problem just described did not exist.  After the final step is performed the tree nucleotides representing the arginine, leucine or serine codons are reexamined.  There are three possibilities.  One possibility is that the nucleotides have been reduced in scope to the point that while some uncertainty of the nucleotides may remain, all possible choices now code only for the amino acid intended.  In that case, the remaining nucleotides are the starting point for the second performance of the parsimony procedure.

A second possibility is that the nucleotides have been reduced by the first performance of the parsimony procedure to the point that while some uncertainty of the nucleotides may remain, all possible choices code only for a single amino acid but not the one intended, i.e., not arginine, leucine or serine.  In that case, a codon for the intended amino acid is chosen that is only one nucleotide different from those of the unintended amino acid produced by the first performance of the parsimony procedure.

In both of the preceding cases, the second performance of the parsimony procedure will give the correct most parsimonious result without further concern.  The only problem arises with the third possibility when the result of the first performance of the procedure does not reduce the nucleotides to representatives of a single codon.  In this case one can only try each of two possible codons for the intended amino acid that are given in Table I in parenthesis.  Fortunately such cases are rare and easily resolved by inspection.

It should be clear that keeping track of which nucleotide positions yield what changes permits the knowledge of these changes not only to be located on various intervals of the tree but to various regions of the gene and various nucleotide positions of the codons.


                          Testing Alternative Trees

Since we have a procedure that counts the minimum number of character state (nucleotide or amino acid) changes for any given tree, we may choose from among several potential phylogenies the one requiring the fewest such changes.  This, then, represents the third criterion that we earlier said we would come back to.  It is the most parsimonious tree criterion.  It does not depend upon any assumption that natural organisms evolve parsimoniously.  It does depend upon the assumption that the simplest explanation of the

facts at hand is to be preferred to more complicated ones in precisely the same way that it is simpler to assume that the wings of
bats and birds arose independently rather than assume that a
chiropteran bird independently acquired the entire suite of
mammalian characters.  Parsimony simply says that phylogenetic
explanation is simplest that requires the fewest number of parallel
and back mutations to be fixed.

The number of such parallel and back mutations fixed is easily
determined.  Let $\underline{m}$ be the number of mutational changes in the most
parsimonious tree as found from the section "Counting the Number
of Required Changes of State" (suitably adapted through the
section "Handling the Six-fold Degeneracy of the Genetic Code"
where necessary).  But the section "Counting the Number of Required
Changes of State" also showed us how to determine $\underline{m}$  , the number
of changes that would be necessary in the absence of any homoplasy.
Thus the extent of homoplasy is simply $\underline{h} = \underline{m} - \underline{m}$  .

Clearly phyletic distances can be calculated for parsimonious
trees just like the others.


## SUMMARY

Procedures are detailed that permit one to obtain the following information for sets of orthologous macromolecular (amino acid
or nucleotide) sequences:  1, Measures of genetic distance
(nucleotide differences); 2, Best estimates of the underlying
phylogenies; 3, Determination of the total number of mutational
changes required for the most parsimonious accounting of the data
for any given prospective phylogeny; 4, Determination of which
evolutionary intervals the various changes occurred in; 5, Determination of most parsimonious phyletic distances; 6, Determination
of where in the gene (in which codons) various mutations were
fixed; 7, Determination of where in codons (in which of the
nucleotide positions) those mutations were fixed; 8, Determination
of the frequency with which any character state changed to any
other character state.


## PRACTICE PROBLEMS

1.  What is the value of $\underline{d}_c$ , $\underline{d}_x$ , $\underline{d}_u$ , $\underline{d}_n$ and $\underline{d}_g$ for the following
    pair of sequences:  Assume the values of $\underline{m}_n$ and $\underline{m}_g$ are 2 and
4 respectively.

```
A C E ' F G H X K L M N P
A N D V W A P E R ' ' D T
```

2. Draw the phylogeny for the $\frac{d}{g}$ data calculated in Table III.

3. You are given the phylogeny and the character states for two (2) characters as shown in Figure 5. How many substitutions are required? What are the possible ancestral character states consistent with parsimony? What are the paths of nucleotide descent through the tree?

## GLOSSARY

Adjacent - see tree.

Analogous - similarity arising by virtue of convergent evolution from unrelated ancestors; compare with homologous.

Ancestor, ultimate - the most recent ancestor common to all taxa under consideration; the root of the phylogenetic tree for these taxa.

Branch - see tree.

Character - any attribute of an organism; specifically herein, any position in a macromolecular sequence.

Character state - a particular form that the character may assume; specifically herein, one of twenty amino acids or one of four nucleotides.

Character state, singular - a state in a given character possessed by only one taxon.

Cladogram - an hypothesis about the true historical (evolutionary) relation among taxonomic entities; a putative phylogeny.

Conjugate - said of two codons (or nucleotides) that share a common ancestral codon (or nucleotide). It is for the purpose of aligning two sequences conjugately that gaps are placed in them.

Connected - see tree.

Discordant - the relationship between two unlike states of a character from two taxa when both states are present in other taxa. Thus, given states A,A,C,C for taxa α,β,γ,δ, respectively, the states of β and γ are discordant. The implication of the property is that asserting that β and γ are more closely related to each other than either α is to β or γ to δ requires an extra gene substitution to account for the descent of this character.

Discordant, unavoidably - said of two characters where the avoidance of an extra gene substitution in one character is necessarily at the cost of an extra substitution in the other.

Distance - any measure of dissimilarity, usually used as a basis to infer divergent tree structure but also potentially inferrable from the structure. Similarities have also been used as a basis for tree construction. They are related to dissimilarity distance measures by a constant value for their sum.

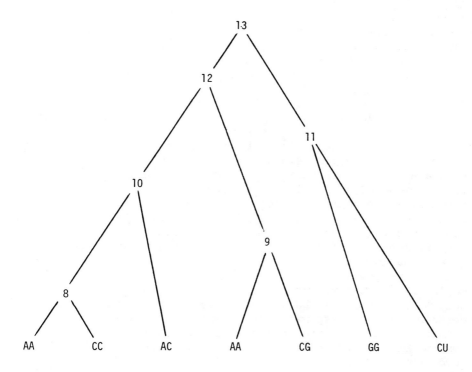

Figure 5. Practice Problem. The tree shows the states for two successive characters in the taxa shown at the branch tips. The taxa are assumed to have the ancestral relationship shown. Data are for problem 3 of the practice problem.

Distance, discordant - a pairwise measure based upon the definition
    of a discordancy.  It is the minimum pairwise distance less all
    differences arising where one or both characters are singular.
Distance, pairwise - any distance measure obtainable by the direct
    comparison of the character states (sequences) of only two taxa;
    where distance is unmodified by some adjective such as discor-
    dant or phyletic, pairwise and the measure will be the number of
    nucleotide (or amino acid) differences between two sequences.
Distance, phyletic - any distance measure between two taxa obtained
    by counting the changes of state along those legs of a tree that
    connect them.
Gap - non-existent positions inserted into homologous sequences
    (and indicated by an apostrophe,') to maintain their characters
    in conjugate alignment.  The implication must be, for two
    sequences one of whose characters is aligned with a gap in the
    other, that since their common ancestor either the character was
    inserted into the one or deleted from the other sequence.
Gap, common - gaps in two or more sequences that begin and end at
    the same positions in both sequences.
Homologous - similarity arising by virtue of (or, better, remaining
    in spite of) divergent evolution from a common ancestor (compare
    with analogous).  There are two distinct forms.  Where gene
    duplication has given rise to separate but homologous loci,
    these loci are called paralogous (para = in parallel) because
    they evolve side by side in a single genome.  Those gene products
    whose most recent common ancestral gene came from the most recent
    ancestral animal of the taxa from which the products were
    obtained are called orthologous (ortho = exact) because of the
    exact correspondence between the history of the gene and its taxa.
Homoplasy - the occurrence of parallel and/or back mutations.
Incompatability - 1, a character change which, if postulated at a
    certain point in a specific phylogeny, would require that change
    to appear again (or its reversal) elsewhere in the phylogeny
    (Camin and Sokal, 1965); 2, more recently, an unavoidable dis-
    cordancy.
Intersection - a mathematical operation in which a new set is
    created composed of only those elements found in both of the two
    sets that are intersected.  The operation is symbolized by a
    Compare with union.
Leg - see tree.
Matrix procedure - any of several procedures for constructing a
    phylogeny on the basis of a matrix of pairwise distances of any
    variety.
Node - see tree.
Orthologous - see homologous.
Paralogous - see homologous.
Parsimony - the explanation of observations with as few assumptions
    as possible; closely related, but not identical, to maximum
    likelihood when applied to sequence data.  They become identical

upon assuming a uniform rate of evolution.  Parsimony does not
assume that events for which the data provide no evidence have
not in fact occurred nor that nature proceeds parsimoniously.

Phenogram - any relation among taxonomic entities, based upon some
collection of their characters (called phenotypes), that can be
expressed diagramatically as a tree.

Phylogeny - the true historical (evolutionary) relation among
taxonomic entities expressed diagramatically as a tree; more
loosely, an estimate thereof, i.e., a cladogram.

Phylogeny, gene - a phylogeny in which the taxonomic entities are
paralogous genes or their products (e.g., myoglobin and the $\alpha$,
$\beta$, $\gamma$, $\delta$, and $\epsilon$ hemoglobins of man) so that the branch points of
the tree denote gene duplications.

Phylogeny, species - a phylogeny in which the characters are all
orthologous; generally assumed when the modifier, species, is
not employed.  Branch points represent speciation.

Replacement - said of amino acids when a descendent does not have
the same amino acid in a given position as its ancestor; compare
with substitution.

Root - see tree.

Similarity - a degree of likeness, generally beyond chance of
intuitive expectations; to be preferred to the terms "homology"
and "analogy" when the origin of the similarity is not known.

Singularity - a singular character state, which see.

Substitution - said of nucleotides when a descendent does not have
the same nucleotide in a given position as its ancestor, compare
with replacement.

Taxonomy - the study of relations among objects or organisms (the
taxonomic entities) for the purposes of classification and/or
understanding their evolutionary relationships.

Tree - a graphical representation of a relation among taxonomic
entities whose terminology is for us slightly confusing because
of the blending of mathematical and biological concepts.  Mathematically, a tree is a graph having points (nodes) and lines
(links, legs, intervals) between pairs of points.  Biological
trees are generally minimally connected, meaning that every point
is connected to every other point by an unspecified number of
successive links (constituting a path) but only one such path
exists between any pair of points.  There are thus no loops
(reticulate evolution and an inbreeding would, of course, require loops).  A subset of a tree that is itself a minimally
connected tree is called a branch.  Two nodes connected by a
path containing only one link are said to be adjacent and have
the relation immediate ancestor and immediate descendent.  The
nodes of a tree are said to have a degree equal to the number of
lines that connect to it.  Our nodes are generally of two types,
those representing the taxonomic entities we observe directly
and those representing their ancestors (there is, of course,
the possibility that a paleontologist might observe two entities

that possess the relation ancestor-descendent).  On our trees,
all observed entities, the <u>ultimate</u> <u>descendents</u>, have degree one.
Tree, strictly bifurcating - the most complete description of a
relationship is a <u>strictly</u> <u>bifurcating</u> <u>tree</u> in which every
ancestral node (except the ultimate ancestor) has degree three,
two lines to its descendents and one line to its immediate
ancestor.  The ultimate ancestor, having no ancestor of its own,
has degree two and is called the <u>root</u>.  Nodes of higher degree
are frequently used when one does not wish to commit oneself to
the order of divergence.
Union - a mathematical operation in which a new set is created
composed of the elements in the two sets that are united.  The
operation is symbolized by a ∪; compare with intersection.

## ANSWERS TO PRACTICE PROBLEMS

1.  $\underline{d}_{\underline{c}}$ = 0 + 2 + 1 + 2 + 1 + 1 + 1 + 1 + 1 = 10; $\underline{c}$ = 9; $\underline{r}$ = 10/9 =

1.11; $\underline{x}$ = 1; $\underline{n}$ = 3; $\underline{g}$ = 2

$\underline{d}_{\underline{x}}$ = $\underline{d}_{\underline{c}}$ + $\underline{rx}$ = 9 + 1.11·1 = 10.11

$\underline{d}_{\underline{u}}$ = $\underline{d}_{\underline{x}}$ + $\underline{rn}$ = 10.11 + 1.11·3 = 13.44

$\underline{d}_{\underline{n}}$ = $\underline{d}_{\underline{x}}$ + $\underline{m}\,\underline{n}_{\underline{n}}$ = 10.11 + 2·3 = 16.11

$\underline{d}_{\underline{g}}$ = $\underline{d}_{\underline{x}}$ + $\underline{m}\,\underline{g}_{\underline{g}}$ = 10.11 + 4·2 + 18.11

2.  The legs to α, β and γ are 4.6, 9.0 and 7.0 respectively and
    look like Figure 1, left with α, β and γ replacing A, B and
    X respectively.

3.  Both positions require four nucleotide replacements.  The
    second position is G in all ancestors except 8 and 9 where it
    is a C.  The first position is more complicated and is shown
    in Figure 6.  There are three possible most parsimonious
    solutions each for A and C in the ultimate ancestor and one
    (1) for G.  Interpretation is like that in Figure 4.

## ACKNOWLEDGEMENTS

    This work was supported by National Science Foundation Grant
BMS 76-20109 and National Institutes for Health Grant AM-15282.

Ancestor

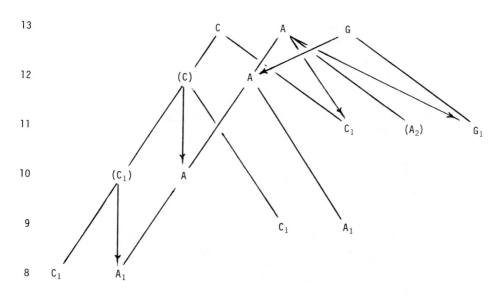

Figure 6.   Answer to Practice Problem 3.   Interpretation as in
figure 4.   Answer is for first nucleotide position.

## REFERENCES

Camin, J.H. and Sokal, R.R., 1965, A Method for Deducing Branching
    Sequences in Phylogeny. Evolution, 19:311.
Dayhoff, M.O., 1972, Atlas of Protein Sequence and Structure.
    Nat. Biomed. Res. Fndn., Washington, D.C.
Farris, J.S., 1969, A Successive Approximations Approach to
    Character Weighting. Syst. Zool., 18:374.
Farris, J.S., 1972, Estimating Phylogenetic Trees from Distance
    Matrices. Amer. Naturalist, 106:645.
Fitch, W.M., 1971, Toward Defining the Course of Evolution:
    Minimum Change for a Specific Tree Topology. Syst. Zool.,
    20:406-416.
Fitch, W.M., 1973, Aspects of Molecular Evolution. Ann. Rev.
    Genetics, 7:343-380.
Fitch, W.M., 1975, Toward Finding the Tree of Maximum Parsimony,
    in "Proc. Eighth Internat. Conf. Numerical Taxonomy" (G.F.
    Estabrook, ed.), W.H. Freeman, San Francisco, 189-230.
Fitch, W.M., 1976, The Molecular Evolution of Cytochrome c in
    Eukaryotes. J. Mol. Evol., 8:13.
Fitch, W.M., 1977, On The Problem of Discovering the Most
    Parsimonious Tree. Amer. Naturalist, in press.
Fitch, W.M. and Farris, J.S., 1974, Evolutionary Trees with
    Minimum Nucleotide Replacements from Amino Acid Sequences.
    J. Mol. Evol., 3:263-278.
Fitch , W.M. and Margoliash, E., 1967, The Construction of
    Phylogenetic Trees - A Generally Applicable Method Utilizing
    Estimates of the Mutation Distance Obtained from Cytochrome
    c Sequences. Science, 155:279-284.
Fitch, W.M. and Yasunobu, K.T., 1975, Phylogenies from Amino Acid
    Sequences Aligned with Gaps. J. Mol. Evol., 5:1-24.
Goodman, M., Moore, G.W., Barnabas, J., and Matsuda, G., 1974,
    The Phylogeny of Humal Globin Genes Investigated by the
    Maximum Parsimony Method. J. Mol. Evol., 3:1.
Hartigan, J.A., 1973, Minimum Mutation Fits to a Given Tree.
    Biometrics, 29:53.
Holmquist, R., 1972, Empirical Support for a Stochastic Model of
    Evolution. J. Mol. Evol., 1:211.
Moore, G.W., Barnabas, J. and Goodman, M., 1973, A Method for
    Constructing Maximum Parsimony Ancestral Amino Acid Sequences
    on a Given Network. J. Theor. Biol., 38:459.
Sneath, P.H.A. and Sokal, R.R., 1973, Numerical Taxonomy, W.H.
    Freeman, San Francisco.

# EVOLUTION OF PANCREATIC RIBONUCLEASES

Jaap BEINTEMA

Biochemisch Lab., der Rijksuniversiteit te Groningen

Zernikelaan, Groningen, The Netherlands

Pancreatic ribonucleases form a group of homologous proteins found in considerable quantities in the pancreas of a number of mammalian taxa and of a few reptiles (Barnard, 1969; Beintema et al., 1973). The ribonuclease activity found in different species varies greatly. Large quantities are found in species which are ruminants or have a ruminant-like digestion, but also in a number of species with coecal digestion. Barnard (1969) proposed that elevated levels of pancreatic ribonuclease are the response to the need to digest the large amounts of ribonucleic acid derived from the microflora of the stomach of ruminants. As rumination and ruminant-like digestion have evolved several times independently during the evolution of mammals (Moir, 1968), pancreatic ribonucleases may well show different solutions to the same evolutionary pressure.

The primary structures of pancreatic-type ribonucleases from 23 species are known. Several species contain identical ribonucleases (cow/bison; sheep/goat), other species show polymorphism (dromedary) or the presence of two structural genes (guinea pig pancreas contains two ribonucleases which differ at 24% of the amino acid positions). 27 different amino acid sequences (including the ribonuclease from bovine seminal plasma, which is homologous to the pancreatic ribonucleases) were used to construct a most parsimonious tree (J.J. Beintema and W.M. Fitch, unpublished) (Fig. 1). The "artiodactyl" part of the tree conforms quite well to the classical biological tree of this mammalian order, but other parts of the tree agree less well with the classical tree, probably as a result of the occurrence of many parallel and back mutations. Bovine seminal ribonuclease was found

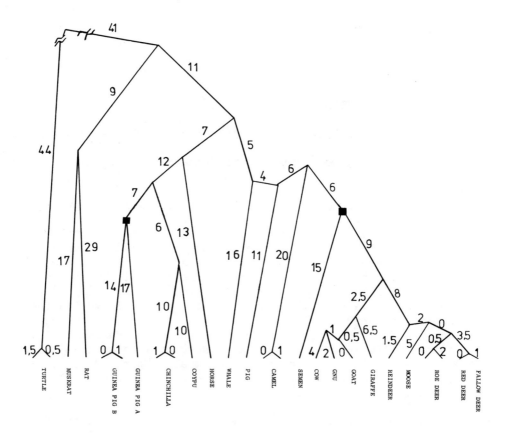

Fig. 1.   Most parsimonious tree of pancreatic-type ribonucleases.
The number of nucleotide replacements in each branch is
indicated.   : gene duplication, resulting in two
paralogous gene products.

to be the result of a gene duplication which occurred before the divergence of the true ruminants, or pecora, but after the divergence of this group from the cameloids.

A number of pancreatic ribonucleases have carbohydrate covalently attached to the surface of the molecule. There is an enzyme system that attaches carbohydrate to the side chain of the amino acid residue asparagine (Asn) if this residue occurs in a sequence -Asn-X-Ser/Thr-, where X stands for any amino acid and Ser and Thr mean the hydroxyamino acids serine and threonine, respectively. The surface positions where carbohydrate attachment occurs belong to the most variable parts of the ribonuclease molecule and so it is not surprising that we find differences in the presence of carbohydrate between closely related species (Fig. 2).

We do not yet know the exact function of carbohydrate on the ribonuclease molecule. However, the presence of carbohydrate is not a neutral characteristic since species with coecal digestion, like pig, horse and guinea pig, produce ribonucleases with large carbohydrate moieties attached to several positions at the surface of the molecule. Another indication of the significance of carbohydrate for natural selection is the strong similarity in the sequences 7 - 40 (about one-third of the whole molecule) of pig and guinea pig ribonuclease B (Van den Berg and Beintema, 1975) as a result not only of conservation of the ancestral structure but of a considerable degree of parallel evolution as well (J.J. Beintema and W.M. Fitch, unpublished). The sequence 7 - 40 forms part of the surface of the molecule and encompasses two carbohydrate attachment sites. We have suggested that the presence of carbohydrate protects ribonuclease from absorption in the gut, enabling it to be transported to the large intestine where it should hydrolyze the ribonucleic acid from the coecal microflora (Beintema et al., 1976). This would be analogous to the function for ribonucleases postulated by Barnard (1969) for ruminants.

We have looked for the presence of Asn-X-Ser/Thr sequences in the ancestral ribonucleases derived from the most parsimonious tree. Position 76, to which carbohydrate is attached in the enzymes from pig and whale, can be taken as example (Fig. 2). Deer have the sequence Asn-Ser-Ala and bovids have Tyr-Ser-Thr at this site and so have no attached carbohydrate.

But the ancestral pecora sequence is Asn-Ser-Thr and may have been glycosidated. In the same way, we looked to the other sites where carbohydrate attachment occurs. Our general impression is that ancestral ribonucleases may have had more carbohydrate attached to them than most present ribonucleases. This is true especially for the pecora and can be correlated with the hypothesis

| Species | position 34 35 36 | Species | position 76 77 78 |
|---|---|---|---|
| cow, sheep, goat, topi, roe deer | CHO* / -Asn-Leu-Thr- | bovids | -Tyr-Ser-Thr- |
| giraffe, okapi, moose | CHO / -Asn-Leu-Thr- | giraffe | -Tyr-Ser-Ala- |
| reindeer | -Asp-Leu-Thr- | deer | -Asn-Ser-Ala- |
| eland | -Asp-Met-Thr- | bovine seminal plasma | -Lys-Ser-Thr- |
| nilgai | -Ser-Met-Thr- | camels | -Ser-Thr-Ser- |
| fallow deer, red deer, whale, bovine seminal plasma | -Lys-Met-Thr- | pig | CHO / -Asn-Ser-Thr- |
| camels, muskrat, guinea pig A | -Glu-Met-Thr- | whale | CHO* / -Asn-Ser-Thr- |
| pig, horse, coypu, chinchilla, guinea pig B | CHO / -Asn-Met-Thr- | horse | -Ser-Ser-Ser- |
| rat | -Gly-Met-Thr- | guinea pig A | -Tyr-Ser-Ser- |
|  |  | guinea pig B | -Tyr-Ser-Arg- |
|  |  | coypu, chinchilla | -Asn-Ser-Asn- |
|  |  | muskrat | -Arg-Ser-Ala- |
|  |  | rat | -Ser-Ser-Thr- |
| **Hypothetical sequences** |  |  |  |
| ancestor pecora | -Asn-Leu-Thr- | ancestor pecora | -Asn-Ser-Thr- |
| ancestor pecora and camels | -Lys-Met-Thr- | ancestor pecora and camels | -Asn-Ser-Thr- |
| ancestor artiodactyls | -Lys-Met-Thr- | ancestor artiodactyls | -Asn-Ser-Thr- |

Fig. 2.    Amino acid sequences of two of the four carbohydrate attachment sites in ribonucleases. Ribonucleases containing carbohydrate and hypothetical ancestor ribonucleases with sequences which obey the specificity requirements for carbohydrate attachment are underlined. CHO = carbohydrate. *, partially glycosidated.

that the ruminants descended from species with coecal digestion (Moir, 1968). This would imply that generally the presence of carbohydrate in ruminant ribonucleases is a relic from the evolutionary history of the enzyme.

REFERENCES

Barnard, E.A., 1969, Biological function of pancreatic ribonuclease. Nature, 221:340-344.

Beintema, J.J., Scheffer, A.J., van Dijk, H., Welling, G.W. and Zwiers, H., 1973, Pancreatic ribonuclease: Distribution and comparisons in mammals. Nat. New Biol., 241:76-78.

Beintema, J.J., Gaastra, W., Scheffer, A.J. and Welling, G.W., 1976, Carbohydrate in pancreatic ribonucleases. Eur. J. Biochem., 63:441-448.

Moir, R.J., 1968, Ruminant digestion and evolution. Handbook of physiology - Section 6: Alimentary canal, Vol. V, (Ed.: Code, C.F.): p. 2673-2694. American Physiological Society, Washington, D.C. (U.S.A.).

Van den Berg, A. and Beintema, J.J., 1975, Non-constant evolution rates in pancreatic ribonucleases from rodents: Amino acid sequences of guinea pig, chinchilla and coypu ribonucleases. Nature, 253:207-210.

THE PHYLETIC INTERPRETATION OF MACROMOLECULAR SEQUENCE

INFORMATION:   SAMPLE CASES

Walter M. FITCH

Dept. of Physiological Chemistry, Univ. of Wisconsin

Madison, Wisconsin   53706

## INTRODUCTION

Methods were provided in the preceding section for the formulation of phylogenies, the reconstruction of ancestral sequences,
and the counting of evolutionary changes in the context of the
phylogeny, all based upon macromolecular sequences such as DNA
and protein.  The results of such an analysis are not facts in the
sense that every detail must be a correct representation of the way
it really happened.  Those results are, however, chosen so that
more of those details are expected to be correct than alternative
formulations.

But while one might be concerned over the correctness of an
individual item, it is clear that one is getting enormously useful information when those details are examined in the aggregate.
For example, we may ask whether mutations are fixed randomly
among the various codons of the gene for a given protein, whether
some nucleotide changes are fixed more often than others, whether
those changes are fixed randomly with respect to the three nucleotide positions of the codons, whether different genes fix mutations at the same rate and whether the same gene fixed mutations
at the same rate in different lines of descent.  Examples of all
these questions will be illustrated in the material that follows.

It must be remembered that these analyses are examining sequences existing in present day populations and that all differences, apart from residual polymorphism, represent mutations that
have been fixed.  This is an extremely small fraction of the total
number of mutations that occurred in the evolving populations.
Those changes that survive the evolutionary process therefore

represent an enormous filtering of the primary mutational events
so that it would be a gross mistake to interpret the results to
be presented as proving anything about the mutation process per se.
As a consequence we shall continually refer to replacements (amino
acids) and substitutions (nucleotides) to avoid mutational implications.

The results also do not depend upon a neutralist or selection-
ist view.  The results recount the changes observed whereas neutral-
ism and selection are explanatory mechanisms about those changes.
There is, however, one exception to the comments of these last
two paragraphs and that involves rates.  Wherever the reader is
willing to assume that the changes observed are predominantly
neutral in character, then estimates of the rates of fixation are
in fact the estimates of the rate of production of neutral mutations.

TREE CONSTRUCTION

The Effect of Gap Weighting

In the section on "Allowance for Gaps" in the preceding
paper, I presented several ways of allowing for the effect of gaps
placed into sequences to optimize the homology among sequences
when computing measures of distance.  It should be clear, if different sequences have different numbers of gaps of differing
length, that the topology of the tree may change as a function of
the weights assigned.  Several excellent examples of this have
been described (Fitch and Yasunobu, 1975) one of which is seen in
figure 1.  Since all gaps are, in this case, of length one, it is
immaterial whether one weights according to the number of gaps or
the number of residue positions in those gaps.  Rather than assigning a specific weight to gaps (since we did not know what
weight was appropriate), we asked what values of the weight would
produce different phylogenies.  We found that we could obtain
five different trees as shown in figure 1.  When the value of the
weight is less than 0.65 nucleotide substitutions per gap, the
leftmost tree is obtained; for values greater than 5.62, the rightmost tree is obtained.  The rightmost tree is the only one that
makes a monophyletic group of the clostridia.  The most important
point, however, is that the effect of the weight given to the gap
value can have an enormous significance upon the nature of the
inferred phylogeny.  Moreover, the problem of the proper weight
cannot be ignored if gaps are present because to ignore the gaps
is equivalent to a conscious decision to give them a weight of
zero or, worse, a weight equal to the average value of non-gapped
residues for each pair and which therefore varies from one pair
to the next.

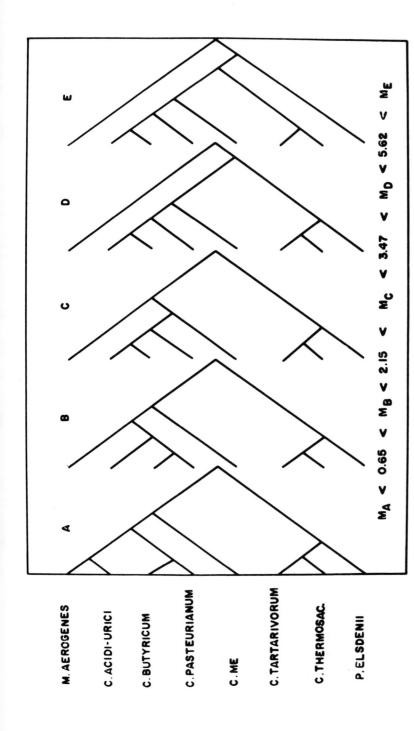

Figure 1. Phylogenies of Bacterial Ferredoxins. Each tree is the best tree for a different matrix of pairwise distances. Those distances vary solely as a consequence of the value, in nucleotide substitutions, accorded to gaps in the homologous alignment of sequences. That value, M, is bounded for each tree (shown by the subscript of M) as indicated in the lower portion of the graph. The branch lengths on the tree have no meaning, only their order of branching. From Fitch and Yasunobu, 1975.

    In the example just given, all gaps were of length one.  In
other cases presented by Fitch and Yasunobu (1975), the gaps were
not all of the same length and, as might be expected in such cases,
the nature of the inferred phylogeny may vary considerably according to whether the weights are on the basis of the number of gaps
or on the number of residues in gaps.

    Also in the example given, there was little doubt as to the
proper location of the gaps.  In the case of the ferredoxins, the
proper alignment between plant and prokaryote ferredoxins is not
unambiguous and again Fitch and Yasunobu showed that the nature of
the inferred phylogeny may vary significantly as a function of how
the gaps are located.

    The problems associated with gaps do not ruin the utility of
molecular sequence data for studying evolution.  They do suggest,
however, an obligation upon authors to advise the reader, when
gaps have been required, about their weights and what changes in
those weights would have changed the phylogeny inferred and/or
other conclusions.  But neither are these problems peculiar to
molecular sequences.  Any set of taxonomic information that is
incomplete, regardless of whether it is because the characters are
absent or just weren't analyzed has the same problems (except that
the length of the gap is not meaningful).  And in a similar manner
it cannot be ignored since to ignore it is, in fact, to give it a
special value.

                         Parsimony Analyses

    Regardless of how trees are constructed and regardless of the
nature of the data, not all possible trees can be investigated,
even for modest numbers of taxa.  Thus every procedure has some
reasonable procedure for finding one or more good trees plus,
usually, some systematic method of altering these good trees to
see if they can be improved upon.  There must, of course, be some
criterion of goodness and, for parsimony procedures, it is the
fewest number of necessary changes of character state in the
evolution of the characters examined or, its equivalent, the tree
of least length.  This is true regardless of the nature of any
weights assigned to the various changes of state, that is, weighting and parsimony are perfectly compatible.  For macromolecular
sequences, the changes of state are amino acid replacements or
nucleotide substitutions.

    Since there are plenty of trees obtained by parsimony pro-
cedure in the literature, I shall present an example that gives
some idea about the distribution of the number of changes of state
required when all possible topologies may be investigated.  For
eight taxa, there are 10,395 possible unrooted trees and I have
a program that will determine the numbers of state changes

required for each of these trees given a set of homologous macromolecular sequences for the eight taxa.

   The data set chosen is that of the prokaryotic cytochromes
c-551 investigated by Sneath et al (1975).  They actually examined
nine taxa but Pseudomonas aeruginosa strain 129B differs by only
one amino acid replacement (and one nucleotide substitution) from
strain P6009 and its presence does not affect the relationships
among the other taxa.  I have therefore removed it from the set to
attain the managable eight taxa.  I examined all 10,395 trees on
the basis of minimizing the number of nucleotide substitutions and
then repeated the analysis on the basis of minimizing the number
of amino acid replacements.

   The most parsimonious tree for these sequences is shown in
figure 2.  The numbers refer to nucleotide substitutions of which
there are a total of 132.  There was one other tree requiring only
132 nucleotide substitutions and that is similar to the one shown
except that fluorescens strains 50 and 181 are interchanged on the
branch tips (and the numbers on the legs also, of course).  A
third tree, with fluorescens strains 18 and 181 interchanged, requires 133 nucleotide substitutions and a fourth tree, with aeruginosa and vinelandii interchanged, requires 134 nucleotide substitutions.  The least parsimonious trees, of which there are 28, require 190 nucleotide substitutions.  The distribution of the number of nucleotide substitutions for the entire collection is shown
in figure 3 by the hatched area.  The median is at 175 nucleotide
substitutions, the mode at 183.  The distribution is clearly non-
Gaussian, possessing a long tail to the left, a desireable property
in a procedure that would like the extreme leftmost value to be as
significantly differentiated from other topological solutions as
possible.  In the absence of any parallel and back substitutions,
the best tree would have required only 102 nucleotide substitutions
so that the analysis reveals that there must have been at least 30
unavoidable discordancies (extra homoplasic substitutions) occurring
in the evolution of these sequences, more if in fact neither of
the two trees requiring 132 substitutions is the true phylogeny.

   When the analysis is performed in terms of amino acid replacements rather than nucleotide substitutions as if we knew nothing
about the genetic code, the number of amino acid replacements for
the two most parsimonious trees is 104 and these two trees are the
same as the two best for the nucleotide substitution analysis.
There are two more trees requiring 105 amino acid replacements
(one of which was third best by the nucleotide analysis) and five
more requiring 106 amino acid replacements (one of which was fourth
best by the nucleotide analysis).  The least parsimonious trees, of
which there are 98, require 144 amino acid replacements.  The distribution of the number of amino acid replacements for the entire
collection is shown in figure 3 by the open areas.  The median is

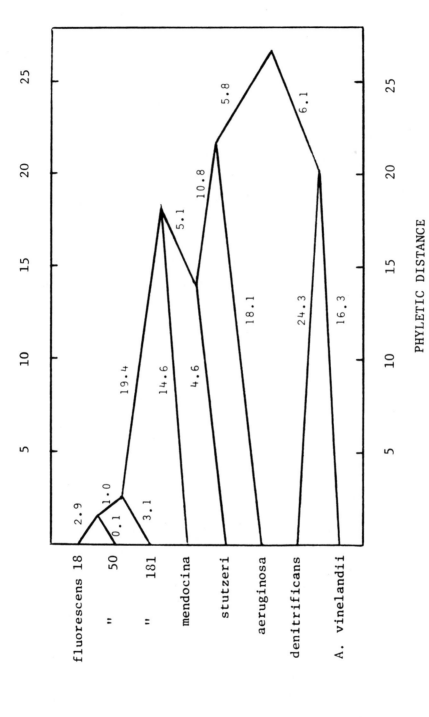

PHYLETIC DISTANCE

Figure 2. Most Parsimonious Phylogeny for Eight Bacterial Cytochromes c-551. Sequence data from Sneath et al., 1975. Numbers on legs are numbers of nucleotide substitutions in the interval, averaged over various equally parsimonious ways of distributing them (see previous chapter). Nodes are placed at a distance equal to the average number of substitutions on the two descendent sides of that node. The total number of nucleotide substitutions in 132.

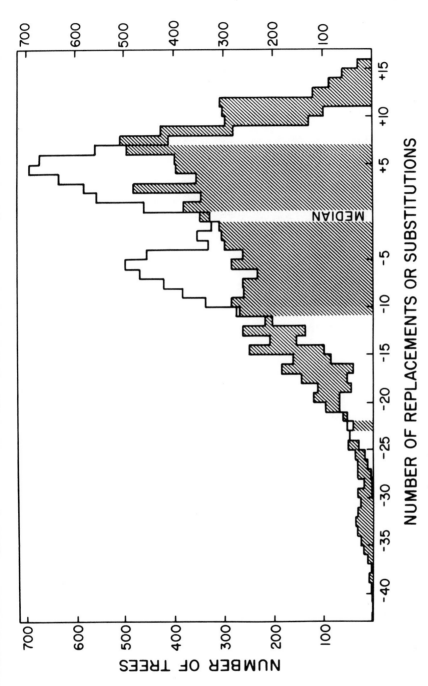

**NUMBER OF REPLACEMENTS OR SUBSTITUTIONS**

Figure 3. Distribution of the 10,395 Possible Unrooted Trees for Cytochrome c-551 According to the Most Parsimonious Number of Changes Required for Each Tree. There are two distributions, one for amino acid replacements (unshaded) and one for nucleotide substitutions (hatched). The median value for the two distributions is 133 and 175 respectively. The two distributions are plotted so as to superimpose their medians. The area immediately below a line has the shading characteristic of the distribution that that line represents except at -24 where the lines for the two distributions are undistinguishable.

at 133 amino acid replacements, the mode at 138.  Again the distribution is non-Gaussian with an extended tail to the left but
the spread is clearly much reduced.  In the absence of any parallel and back replacements, the best tree would have required only
89 amino acid replacements so that the analysis reveals that there
must have been at least 15 unavoidable discordancies (extra homoplasic replacements) occurring in the evolution of these sequences,
more if in fact neither of the two trees requiring 104 replacements is the true phylogeny.

Perhaps of greater interest is the comparison of the two
analyses.  First of all, there is a high correlation between the
results of the nucleotide substitution and amino acid replacement
analysis.  This, of course, ought to be in fact the case since the
latter is an observation dependent upon the former.  Since the
correlation is hardly perfect, the question, then concerns the
relative merits of the two forms of analysis.  First of all there
were at least 27 percent {100 (132 - 104)/104} more genetic events
than the replacement analysis revealed.  It could be higher because the 132 substitutions is also a minimum estimate.  To ignore
so much information is certainly wasteful, probably dangerous, and
possibly foolish.

But ultimately, topologies are differentiated on the basis of
avoiding as much as possible any parallel or backward changes of
state and amino acids readily hide them because there are 20 of
these and only four nucleotides.  For example nine taxa with the
amino acids lys, glu, gln, thr, ala, pro, met, val and leu require
no extra replacements and cannot cause any problems when compared
to other amino acid positions but at the level of the nucleotides
they are, among themselves, variously discordant and either the
first or the second nucleotide position of their codons will be
discordant with respect to every other position in the gene that
has at least two different nucleotides each present in two or more
of the nine taxa.  The result is, in these data, that the replacement analysis can only see 15 unavoidable discordancies, the substitution analysis sees 30.  Thus, in terms of the most crucial
kinds of information which determines the most parsimonious result,
the replacement analysis has missed at least as many crucial events
as it has found, possibly more since 30 is also a minimum estimate.

And finally, any method that truly spreads the distribution
out is, a priori, more likely to permit better discrimination a-
mong alternatives.  The replacement analysis distributes 10,395
trees over a range of only 41 amino acid replacements.  The substitution analysis distributes these over a range of 50 nucleotide substitutions, an increase of 44 percent in the range.  More
importantly, a major part (13 out of 18) of that increase is in
the lower tail of the distribution.  The result is that there are
only 20 trees with not more than five changes of state greater than

the most parsimonious tree on the basis of nucleotide substitutions, but 62 in that same region on the basis of amino acid replacements. The conclusion would seem unavoidable that using a replacement analysis should be avoided.

It might be useful to provide here a simple example of how a most parsimonious tree based on substitutions need not be the same as one based on replacements. Figure 4 shows, in the upper portion, a dipeptide for five taxa and the only unrooted tree that is most parsimonious in terms of amino acid replacements. In the lower portion is shown the nucleotides associated with the codons for the amino acids in the upper figure together with the only unrooted tree that is most parsimonious in terms of nucleotide substitutions. These two trees are as different as two unrooted trees for five taxa can possibly be where differentness is measured in terms of the number of interchanges of neighbor branches required to trans- form one tree into the other.

## ANCESTRAL SEQUENCES

### The Meaning of Ancestral Ambiguity

If two descendants have a different nucleotide (amino acid) in a given position, their ancestor is assigned both nucleotides (amino acids) and we assert that either could be the ancestor and that a mutation in one line or the other led to the difference, the line of descent in which the substitution (replacement) oc- curred being dependent or which is the true ancestor. Thus the presence of ambiguity is generally regarded as a profession of ignorance as to which of two or more alternatives is in fact cor- rect. That, quite possibly, may be correct most of the time, but there is an alternative that should be recognized. The ancestral taxon in which a proposed ancestral sequence is presumably present is in fact a population. Given that a nucleotide substitution actually occurred somewhere in these lineages, there must have been a time when the population was polymorphic. It may well then be that on some occasions the ambiguity in an ancestral form is not a representation of our ignorance but of the polymorphism pre- sent at the time.

There is an interesting example that may represent a case in point. In figure 5 are shown several primates and the amino acid they possess at position 135 of gamma hemoglobin. Ignoring the orang for the moment, it is clear that the hominoids have threo- nine and the monkeys have alanine in this position. The codons for these amino acids are also shown and they necessarily differ only in the first position as A and G respectively. Accepting the tree as the biologists give it to us (and with which the amino acid data agree) we get the parsimony result shown by the solid lines.

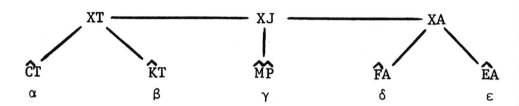

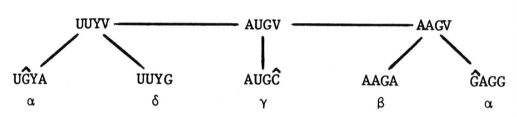

Figure 4. Two Most Parsimonious Trees for Same Data. The upper
tree is the most parsimonious tree if one counts amino acid replacements for the dipeptides shown in the IUB single letter code (see
previous chapter, table II, for their meaning). The lower tree is
the most parsimonious tree if one counts nucleotide substitutions
in the dicodons of the dipeptides shown above. Nucleotides five
and six of the dicodons are omitted because they require no substitutions. Character states appearing only once in a given position
are denoted by the presence of a carat above them. Note that these
two most parsimonious trees are different. The ambiguous amino
acids X = C, E, F, K or M and J = A, P or T. The ambiguous nucleotides are Y = C or U and V = A, C or G. From Fitch, 1977.

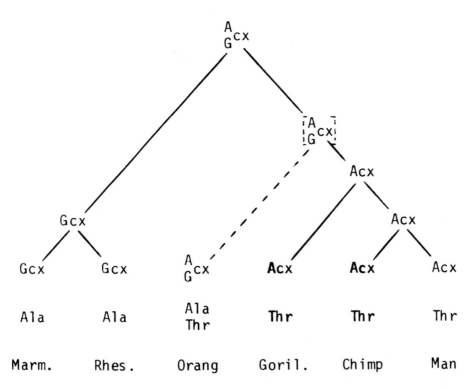

Figure 5. Ambiguity of Reconstruction as an Example of Transition State. The amino acids are those present in position 135 of gamma hemoglobin including the polymorphism in the orang. The codons are also shown with the first nucleotide capitalized as it is the only one of significance to the parsimony procedure. See text for discussion. From Fitch, 1977.

The more recently determined orang data reveals the population to be polymorphic, possessing both alanine and threonine at position 135. If we now add this taxon to the tree as shown by the dashed line, nothing is changed except the insertion of a polymorphic ancestor that <u>must</u> contain the abiguity shown if we are to account for the simultaneous presence of <u>both</u> alanine and threonine in the orang population today. That is to say, the post parsimonious accounting of the data at the tree tips absolutely necessitates the presence of a polymorphic ancestral hominoid population that retained its polymorphism in its descent to the orang but lost the alanine allele in its descent to the others.

## Convergence and Divergence

Consider the problem of whether alpha and beta hemoglobins are as similar as they are because of convergent evolution from ancestrally unrelated genes or because of divergent evolution from a common ancestor. The process of ancestral reconstruction then permits one to examine separately the ancestral sequences of alpha and beta hemoglobin and ask whether they are significantly more alike or less alike than today's alpha and beta sequences. If the ancestral sequences are significantly more alike than today's sequences, the genes have been diverging and we may properly conclude that they are homologous gene products that arose following some event that produced two hemoglobin loci that subsequently diverged to today's forms. If, on the other hand, they are significantly less alike than today's sequences, the genes have been converging and we may properly conclude that they are analogous gene products. The method for establishing significance was given by Fitch (1970). In that paper it was demonstrated that the fungal cytochromes <u>c</u> were homologous to the plant and animal cytochromes <u>c</u>. Since then Langley and Fitch (1973) have shown that alpha and beta hemoglobins are homologous and in 1976b, Fitch and Langley showed that the hemoglobins are homologous to myoglobin. All other statements in the literature about homology are not in fact demonstrated conclusions about ancestry but statements that confusingly use homology as a synonym for similarity. Nevertheless the implicit assumption that significant similarity over the length of a gene is the result of common ancestry may in fact be true because there is as yet no evidence that functional constraints are so severe that if the same function were to arise twice, they would require a very similar sequence. Indeed, if we are careful to distinguish parallelism from convergence, where parallelism means the same set of changes occurring independently in two different lines of descent without concomittant divergent changes, then not only has convergence in protein sequence not been observed, neither has parallelism.

The example that comes closest to illustrating parallelism in

protein sequence evolution is to be found in the pancreatic ribonucleases where in the restricted region of amino acids 9 to 34
there have been 21 nucleotide replacements since the common ancestor of pig and guinea pig B (see figure 6). Seven have occurred
in the pig line since its common ancestor with other artiodactyls
and eight in the guinea pig B line since the gene duplication that
gave rise to the paralogous guinea pig A gene. These latter two
ancestors differ by at least six nucleotides in this region of
their genes. The net result of the 15 ( = 7 + 8) subsequent nucleotide substitutions in these two lines is to leave this region of
the gene still only six nucleotides different. It has been necessary to restrict our view to only 20 percent of the gene to find
such a case but it nevertheless remains very exceptional. It is
also unexplained but may be associated with the independent evolution of carbohydrate binding sites at both positions 21 and 34
in both of these two lines.


## DISTRIBUTION OF EVOLUTIONARY CHANGES

### Are All Nucleotide Interchanges Equally Likely?

It might be assumed, if the frequency of a nucleotide in a
gene is say, 24 percent, that a mutation of that nucleotide would
be fixed 24 percent of the time and that it would be substituted
by each of the other three nucleotides 24/3 = 8 percent of the
time. One can readily test that assumption and Fitch (1976a) has
examined 333 and 234 evolutionary nucleotide substitutions in the
first and second positions respectively of the codons deduced from
the examination of the evolution of 50 eukaryotic cytochromes $c$
from species including protozoans, fungi, plants and animals. The
results are summarized in table I. There are only 4.66 + 5.16 =
9.82 C-U interchanges in the first nucleotide position but there
are 72.93 + 68.74 = 141.67 A-G interchanges. Even making allowance for the fact that there are initially more purines (A's and
G's) than pyrimidines (C's and U's) in the first codon position of
the cytochrome $c$ gene does not alter the greatly significant difference in these two values. Such non-randomness has been long
recognized and thought originally to indicate either a possible
non-randomness in the underlying mutational process or differential
(selective?) recognition of the two complementary DNA strands
(Fitch, 1967). Both ideas seem naive interpretations now although
still espoused by some. Of course both ideas may ultimately prove
to be correct, it is only that such data as these are not capable
of supporting such notions.

One reason is that the data in table I make a simple alternative much more attractive, namely that the gene is in compositional equilibrium which means that our expectations were wrongly computed and that what we wish to know is if the pairs of numbers are,

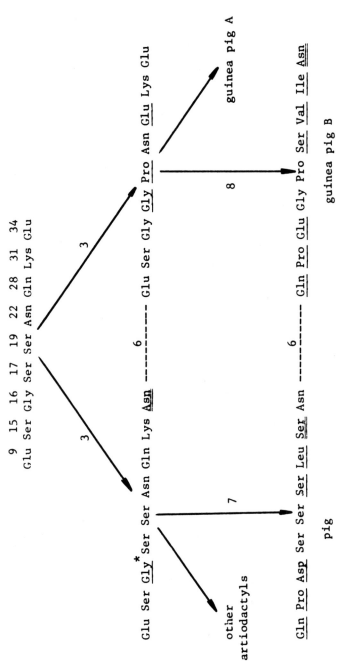

Figure 6. Parallel Evolution in a Part of Pancreatic Ribonuclease. The region shown is from residues 9 to 34 with all positions that had amino acid replacements in this region shown and their position number indicated above the upper ancestral sequence. This region represents a-bout 20 percent of the total. Solid arrows show the phylogenetic relationship. Numbers along the arrows show the number of nucleotide substitutions during the interval, their codon position being indicated by the underlining in the descendent fragment. The dashed line and its number simply indicate the number of nucleotide differences between the codons for this pair of frag-ments resulting from the nucleotide substitutions indicated on the arrows. A silent mutation was located, somewhat arbitrarily, in the descent to the glycine with the asterisk. These data are part of the complete tree for 26 ribonucleases studies by Beintema et al., 1977.

Table 1.  NUCLEOTIDE INTERCHANGE IN EACH DIRECTION IN

THE FIRST TWO CODON POSITIONS OF CYTOCHROME c

|   | U | G | C | A |   |
|---|---|---|---|---|---|
| U |  | 17.29 | 4.66 | 19.62 | F |
|   |  | 26.53 | 5.16 | 15.93 |  |
|   |  |  |  |  | I |
| G | 2.00 |  | 35.28 | 72.93 | R |
|   | 3.00 |  | 20.85 | 68.74 |  |
|   |  |  |  |  | S |
| C | 28.29 | 13.33 |  | 22.28 | T |
|   | 11.04 | 11.77 |  | 23.74 |  |
| A | 22.33 | 23.00 | 53.00 |  |  |
|   | 16.50 | 13.00 | 37.33 |  |  |

S   E   C   O   N   D

    Table shows the frequency of various nucleotide interchanges
in pairs.  For each pair, the upper figure is the number of occurrences wherein the row nucleotide was substituted by the column
nucleotide, the lower figure is the number of occurrences wherein
the column nucleotide was substituted by the row nucleotide.  The
upper right-hand portion of the table contains data for the first
nucleotide position of the codon, the lower left-hand portion for
the second nucleotide position.  From Fitch 1976a.

statistically, equal, whether A substitutes for G as often as G
substitutes for A.  The answer is that they are not statistically
different in the first position and only marginally so in the
second (p = 0.015).  Thus we may conclude that nucleotide substitutions in the gene for cytochrome c behave as if the gene were
close to compositional equilibrium in the first and second nucleotide positions of the codons.  Since that equilibrium is not of
equimolar concentrations of four nucleotides, we may strongly suspect evolutionary constraints.  Indeed, if the amino acid composition of the gene product is constrained, the nucleotide composition
of the gene is also greatly constrained in the first two codon
positions.

     This does not, however, completely explain the non-randomness
with which the various nucleotides are interchanged.  We may note
that that non-randomness was based upon two assumptions, either or
both of which may be false and we may ask about the legitimacy of
each.  Those two assumptions were that each of the four nucleotides
was likely to be substituted in proportion to its frequency in the
gene and that upon substitution each of the other three nucleotides
was equally likely to substitute for it.

     Table II shows the expected frequency with which each nucleo-
tide should have been substituted under the first assumption and
the frequency actually observed.  The three condon positions are
treated independently so that the total expected equals the total
observed in each row.  This eliminates any bias resulting from the
overall substitutional frequencies being unequal among the three
different codon positions.  We may clearly conclude that nucleotides are being substituted in the cytochrome c gene approximately
in proportion to their compositional frequency in that gene.  The
largest chi-square value, 5 for 1 degree of freedom, implies a
p = 0.025 if only one sample were examined, but given that we have
examined 12 (4 nucleotides x 3 positions), there is nothing significant about this extreme.

     Table III shows the frequency that each nucleotide was ex-
pected to be the substituting nucleotide under the second assumption and the frequency actually observed.  Again the three codon
positions are treated independently and for the same reasons.  The
conclusion is even more clear here.  Except for uridine in the
second position, every nucleotide is the substituting nucleotide
with a frequency significantly different from expectation (p <
0.0015) in both of the first two nucleotide positions of the codons.
Thus, in combination with table II, we may conclude that the non-
randomness observed in the substitution of nucleotides is not the
result of a bias as to which kind of nucleotide may be substituted
but rather what kind of nucleotide it may be substituted by.  Therefore it is the second assumption that is false.

Table II.  ARE NUCLEOTIDES SUBSTITUTED IN PROPORTION

TO THEIR COMPOSITION FREQUENCY?

| Nucleotide Position | | A | C | G | U |
|---|---|---|---|---|---|
| 1 | E | 111.00 | 48.80 | 117.61 | 53.59 |
| | | (0.06) | (0.01) | (1.91) | (2.69) |
| | O | 108.40 | 48.28 | 134.74 | 41.57 |
| 2 | E | 90.08 | 63.16 | 40.38 | 40.38 |
| | | (0.76) | (3.95) | (5.00) | (2.40) |
| | O | 98.33 | 78.96 | 26.17 | 30.54 |
| 3 | E | 28.79 | 18.59 | 25.03 | 18.59 |
| | | (1.13) | (1.02) | (0.36) | (1.02) |
| | O | 34.50 | 14.23 | 28.04 | 14.23 |
| All | E | 223.05 | 137.00 | 179.06 | 118.89 |
| | | (1.96) | (0.59) | (0.18) | (8.91) |
| | O | 241.24 | 141.47 | 188.95 | 86.34 |

A total of 658 nucleotide substitutions were observed in the
evolution of 50 cytochromes c with 333, 234 and 91 in the first,
second and third codon positions, respectively.  The expected
substitutions (E) (the upper number of each pair) were determined
from the nucleotide composition of the ancestral cytochrome c gene.
The lower number of each pair is the number of times such a nucleotide was observed (O) to have been substituted.  The numbers in
parentheses are chi-square values for the pair they stand beside.
The third position is given for the sake of completeness only since
the observational data are almost entirely purine-pyrimidine interchanges.  From Fitch, 1976a.

Table III.   WHEN A NUCLEOTIDE IS SUBSTITUTED, IS EACH OF THE
OTHER THREE NUCLEOTIDES EQUALLY LIKELY TO SUBSTITUTE FOR IT?

|     |   | A | C | G | U |
|-----|---|---|---|---|---|
| 1 | E | 74.86 | 94.90 | 66.08 | 97.14 |
|   |   | (21.34) | (10.27) | (25.19) | (25.25) |
|   | O | 114.83 | 63.68 | 106.88 | 47.61 |
| 2 | E | 45.22 | 51.68 | 69.28 | 67.82 |
|   |   | (10.33) | (10.71) | (12.95) | (0.40) |
|   | O | 66.83 | 75.21 | 39.33 | 62.62 |
| 3 | E | 18.83 | 25.59 | 20.99 | 25.59 |
|   |   | (1.61) | (1.19) | (1.46) | (1.19) |
|   | O | 13.33 | 31.11 | 15.46 | 31.10 |
| All | E | 138.92 | 172.18 | 156.35 | 190.55 |
|   |   | (22.64) | (0.29) | (0.18) | (18.40) |
|   | O | 195.00 | 169.99 | 161.66 | 131.34 |

Nucleotide substitutions in the evolution of 50 cytochromes c
numbering 333, 234 and 91 were observed in the first, second and
third coding positions.  The upper number of each pair is the number of times that the designated nucleotide would have been expected
(E) to have substituted for the incumbent nucleotide if all alternatives to the incumbent nucleotide are, on the average, equally
acceptable, i.e. the probability of being the substituting nucleotide equals 1/3.  The number of times each nucleotide was substituted is given in the lower of the pair of numbers in table II,
and one-third of each such number for each nucleotide in a given
codon position was distributed among the other three to get the
expected number of substituting nucleotides.  For example, in cell
(U,1), 97.14 = (108.47 + 48.28 + 134.74)/3.  The observed (O) numbers of substituting nucleotides is the lower number of each pair.
The numbers in parentheses are the chi-squared values for the pairs
they stand beside.  For chi-square values greater than 10.1, the
probability of occurring by chance is < 0.0015.  From Fitch, 1976a.

The preceding shows which assumption is false but does not explain why it is false. When one examines, however, the nature of the amino acid replacements that occur as a result of the most common nucleotide substitutions, one readily notes that the pairs of amino acids so delineated are among those pairs that are the most similar, frequently having the same charge characteristics or differing only by a methylene ($-CH_2-$) group. In other words, the most commonly occurring nucleotide substitutions are those that bring about the smallest changes in the character of the resulting protein.

### Are the Observable Nucleotide Substitutions Equally Likely in the Three Positions of the Codons?

We know that the parsimony procedure for the reconstruction of evolutionary changes will miss some of the changes that actually occurred. For example, if there is an A → C → G set of nucleotide substitutions between two successive nodes on the phylogenetic tree, there are no descendents of the C form to give us any knowledge of that unsuspected intermediate and we infer an A → G substitution. Also, if A → C occurs on each of the two immediate evolutionary intervals of descent <u>from</u> an ancestral node, parsimony will show this as a single A → C event in the descent <u>to</u> that ancestral node. But the "observable" in the title is also intended to remind us, if we are examining amino acid sequences, that we are unlikely to detect nucleotide substitutions that do not alter the amino acid encoded. For example, GGX codes for glycine regardless of the nature of X and hence substitutions of one nucleotide for another in the third position may well be unobservable. As a consequence all computations regarding what we expect to observe must recognize that some changes are silent at the level of amino acid expression and not expected to be observed. (One also calculates expectations under the assumption that mutations to terminating codons are not expected to be observed.) It is particularly important to recognize the meaning of observable in this section where we ask about the randomness of nucleotide substitutions among the three positions of the codons because the problem of third (and occasionally first) position silent changes means that we should expect, a priori, to see fewer substitutions in the third position.

It is instructive to recognize a truism that is pertinent to the case at hand, namely, that significant non-randomness is simply a demonstration of an inadequate (rejectable) model and that at least one assumption in any such model must be false. For the question of this section we must make an assumption about the nature of the "sample space" (i.e. the codons) in which the observable nucleotide substitutions can occur. We may assume: (1), that it is equivalent to the 61 amino acid codons, each in equal

frequency; or (2), that it is equivalent to the codons used in the
(ancestral) gene for cytochrome c and in those frequencies; or
(3), that it is equivalent to the subset of those codons in cytochrome c that are known to have sustained nucleotide substitutions.
One might suggest others.  Suffice it to say that the degree on
non-randomness diminishes greatly as one successively tests models
based on assumptions 1, 2 and 3, showing that the non-randomness,
in the first two cases at least, was largely the result of an improperly described sample space.  Our analyses here will be confined
to assumption 3.  The details of the calculation of the expected
values can be found in Fitch (1973).

When the analysis is applied to the same 658 nucleotide substitutions from the evolution of cytochrome c already examined, we
get the results shown in table IV.  It is expected that one will
observe about 10 percent more substitutions in the second than in
the first nucleotide position of the codon.  In fact there are 42
percent more in the first than in the second nucleotide position.
The differences in both positions are very significant (p < $10^{-3}$)
deviations from expectation.  The number of third position changes
is what is expected.  The same pattern is repeated in the evolution
of alpha and beta hemoglobins but with continually decreasing non-
randomness until, in the case of beta hemoglobin, there is no
significant non-randomness although the general pattern of more
than the expected number of substitutions in the first and fewer
in the second and third positions is again repeated.

As in the previous section, the conclusion of non-randomness
is clear but the explanation is not.  As before, there is the obvious associated fact that mutations in the first position are
somewhat less likely to change drastically the character of the
protein than those in the second position and its implication that
small changes are more likely to be evolutionarily acceptable than
more dramatic changes.

More interesting, however, is the fact that the degree of non-
randomness changes so markedly (three orders of magnitude) as one
progresses from cytochrome c to alpha hemoglobin to beta.  There
are three associations that may be made about this order:  1, cytochrome c evolves more slowly than alpha which in turn evolves more
slowly than beta hemoglobin; 2, cytochrome c (probably) interacts
with more other elements (ligands, proteins, etc.) than alpha hemoglobin which in tern interacts with more other elements than beta;
3, cytochrome c is older than alpha hemoglobin which in turn is
older than beta (assuming that "true" beta of today could only a-
rise after the duplication that enabled the gamma hemoglobin for
fetal function to evolve).  All of this suggests that the opportunities for selective improvement among the three proteins are
increasingly restricted from beta hemoglobin to alpha to cytochrome c and that the more severe the restriction, the more

Table IV.   NUCLEOTIDE SUBSTITUTIONS BY CODON POSITION

|  |  | Codon Position | | | |
|---|---|---|---|---|---|
|  |  | First | Second | Third | Total |
| Cytochrome c | Expected | 268.6 | 296.3 | 93.1 | 658 |
|  | Found | 333.0 | 234.0 | 91.0 | 658 |
|  | $x^2$ | 15.5 | 13.1 | 0.1 | 28.6 |
|  | P | $<10^{-4}$ | $<10^{-3}$ | 0.77 | $<10^{-6}$ |
| Alpha Hemoglobin | Expected | 148.0 | 156.5 | 42.6 | 347 |
|  | Found | 177.0 | 138.0 | 32.0 | 347 |
|  | $x^2$ | 5.7 | 2.2 | 2.6 | 10.5 |
|  | P | 0.02 | 0.14 | 0.11 | $<0.01$ |
| Beta Hemoglobin | Expected | 118.9 | 129.9 | 38.2 | 286 |
|  | Found | 131.0 | 123.0 | 32.0 | 286 |
|  | $x^2$ | 1.2 | 0.3 | 1.0 | 2.5 |
|  | P | 0.28 | 0.60 | 0.32 | 0.28 |
| Overall | Expected | 538.4 | 578.4 | 177.8 | 1291 |
|  | Found | 641.0 | 495.0 | 155.0 | 1291 |
|  | $x^2$ | 21.1 | 12.0 | 2.9 | 36.0 |
|  | P | $10^{-4}$ | $<0.01$ | 0.41 | $<10^{-5}$ |

The degrees of freedom are 1 for any one position and protein, 2 for the total of any one protein, 3 for any one position overall and 6 for the total overall and p if the probability of a chi-square as large as that shown occurring by chance. From Fitch, 1976a (cytochrome c) and 1973 (hemoglobins).

non-random is the nature of the nucleotide substitution relative
to the position in the codon in which they occur.


### Are Nucleotide Interchanges Equally Likely in All Codons?

One expects that some codons will sustain more nucleotide
substitutions than others, the numbers of substitutions per codon
ranging from zero to some upper limit depending upon the total
number of substitutions and the total number of codons among which
the substitutions are distributed. If the substitutions occur
randomly over the protein, we expect them to be Poisson distributed.
A test for such randomness was first given by Fitch and Margoliash
(1967). An example from that early analysis is given in table V.
There it can be seen that the fit is terrible, there being an enormous excess of unsubstituted codons and a few codons that have been
substituted with much greater frequency than expected.

A reasonable biological interpretation of the above result is
that the excess number of unsubstituted codons arises because some
of the encoded amino acids are so important to the structure and
function of cytochrome c that any change in that amino acid would
be vigorously selected against with the result that the zero class
includes two kinds of codons, those that might have been substi-
tuded but, by chance, were not, plus those that could not have
been substituted and survived the selection process. The former
are variable but unvaried, the latter are invariable. I specifically avoid the confusing term "invariant" as a word that, on various occasions has meant both unvaried and invariable.

The significant excess of codons with large numbers of substitutions demonstrates that even among the variable class of codons,
they are not all equally variable. Markowitz (1970) devised a procedure for fitting data such as that in table V to a model that includes a class of codons that is invariable (evolutionarily speaking, not mutationally) plus two other classes of variable codons
that differ only in their probability of sustaining the next substitution. The best fit provides an estimate of the number of
codons in each of the three classes. It appears that a large number of observed substitutions are required before the three-class
model no longer fits well. It does not fit well to the 658 substitutions in the evolution of 50 eukaryotic cytochrome c. The
conclusion, however, should be, not that there are three variable
classes of codons (plus the invariable), but that there are many
and that the analysis is not sufficiently sensitive to detect them
all.

Table V.   POISSON FIT TO NUCLEOTIDE SUBSTITUTIONS PER CODON

IN THE EVOLUTION OF CYTOCHROME c FROM 20 TAXA

| Substitutions/codon | 0 | 1 | 2 | 3 | 4 | 5 | 6 | 7 | 8 | $\geq 9$ |
|---|---|---|---|---|---|---|---|---|---|---|
| Expected | 13.4 | 28.2 | 29.6 | 20.7 | 10.9 | 4.6 | 1.6 | 0.5 | 0.1 | 0.04 |
| Found | 35 | 19 | 24 | 10 | 8 | 3 | 4 | 2 | 0 | 5 |

Each number is the number of codons that would be expected, or was found, to have the number of nucleotide substitutions shown at the top of the column.   There were 231 nucleotide substitutions distributed over 110 codons sustaining n substitutions is $E\{n\} = 110 r^n e^{-r}/n!$  The probability that the above observed distribution would occur by chance is $< 10^{-6}$.   From Fitch and Margoliash, 1967.

Table VI.   RATES OF NUCLEOTIDE SUBSTITUTIONS:   HORSE PIG-COMPARISON

| Protein | Substitutions | $Rate_1$ | Codons | $Rate_2$ | Covarions | $Rate_3$ |
|---|---|---|---|---|---|---|
| Fibrinopeptide A* | 10 | 78 | 19 | 4.11 | 18 | 4.34 |
| Ribonuclease | 41 | 320 | 124 | 2.58 | 34 | 9.42 |
| Insulin C* | 8 | 62 | 31 | 2.02 | 18 | 3.47 |
| Beta Hemoglobin | 30 | 234 | 146 | 1.61 | 39 | 6.01 |
| Alpha Hemoglobin | 22 | 172 | 141 | 1.22 | 50 | 3.44 |
| Cytochrome c | 5 | 39 | 104 | 0.38 | 12 | 3.26 |
| Overall | 116 | 151 | 573 | 1.56 | 181 | 5.00 |

$Rate_1$, substitutions per gene per $10^9$ years; $Rate_2$, substitutions per codon per $10^9$ years; $Rate_3$, substitutions per covarion per $10^9$ years.   It is assumed that the common ancestor of horse and pig occurred $64 \times 10^6$ years ago.   Substitutions are those in both lines of descent since their common ancestor.   From Beintema et al., 1977.

*These are only portions of the complete gene.

# COVARIONS AND RATES OF EVOLUTION

## Concomitantly Variable Codons or Covarions

The Markowitz procedure of the previous section was applied to cytochrome c data that differ from one set to another in the range of eukaryotic taxa whose cytochrome c were examined (Fitch and Markowitz, 1970). At one extreme, the sequences of all known taxa were included, at the other extreme, only those of non-primate mammals. The initially surprising result was that the estimate of the number of invariable codons greatly increased as the range of taxa was narrowed (range was quantified as the average number of nucleotide substitutions in the descent from the most recent common ancestor of all the taxa examined to those descendent taxa). A recent example of this for the 50 cytochromes c is provided in figure 7.

In fact, there is a biologically reasonable explanation for this. There may well be positions in the cytochrome c of plants where the amino acids present are vital for them but which, in the insects, could tolerate change or, alternatively, could not change among the insects but where the cytochrome function was optimized using an amino acid different from the one used in plants. Fungi, protozoans, vertebrates and whatever may all have a number of amino acid positions peculiarly well adapted to their particular group with the result that positions that are invariable in one group are not in another and, as we expand the range of taxa in the sample, the number of amino acid positions simultaneously invariable in all taxa continuously declines, just as the figure shows.

What then is the meaning of the y-intercept in figure 7 where the line is the weighted least squares fit to the data? It must be the fraction of the gene's codons that is invariable in any one taxon (range of species = 0 nucleotide substitutions) and the remainder must be the fraction that is variable; in this case that there are, on average, only 12 variable codons in cytochrome c, in any one taxon at any time. These are the concomitantly variable codons or covarions.

How stable is this estimate of 12 cytochrome c covarions as one changes phyla? That is not exactly clear, but in figure 7 the range narrows toward the mammals for the filled circles and toward the dicotyledonous plants for the open circles. Thus 12 covarions is a reasonable estimate for at least 2 quite different taxa.

## Evolutionary Rate per Covarion

Evolutionary rate is something that can be, and unfortunately

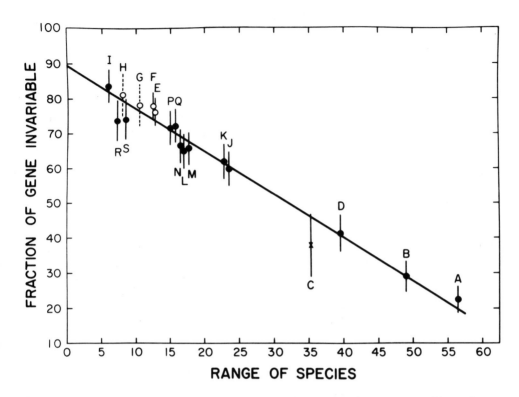

Figure 7.   Fraction of Cytochrome c Gene Invariable as a Function
of the Range of Species.   Each point represents all the taxa descended from a single common ancestor (A represents all 50 taxa,
other representations can be found in Fitch, 1976a).   The point is
plotted along the abscissa according to the average number of nucleotides substituted in the descent from the common ancestor of
the taxa represented by the point.   The point is plotted along the
ordinate according to the fraction of the gene that is computed to
be invariable over the taxa represented by the point.   This fraction is calculated by the method of Markowitz, 1970.   The vertical
bars are the standard deviations for that fraction.   The line is a
weighted least squares fit with the weights inversely proportional
to the variance of the ordinate value.   It has a slope of -0.0123
and an intercept of 89.3 ± 5.9.   Fraction has the units percent of
gene, range has the units nucleotide substitutions.   From Fitch,
1976a.

is, provided in many different units. For molecular data it may
involve either amino acid replacements or nucleotide substitutions;
I shall use the latter. It may involve various time units; I shall
use $10^9$ years. It may involve various DNA size classes including
nucleotide, codon, covarion, gene and genome; I shall use several
but relate them to each other.

If one asks whether proteins evolve at equal rates, the answer
depends upon the units as well as on estimates from paleontological
dating. Errors arising from dating estimates can sometimes be eliminated. In comparing protein evolutionary rates, if one has, as
we do, sequences from both the pig and horse, then the relative
rates must be the same regardless of the date of their common ancestor. Some of these are shown in table VI.

One can see that rates are quite variable. The rate of substitution per gene varies by an order of magnitude and would vary
by at least another order of magnitude if we could include histones
in these data. The rate per codon is as variable as that per gene.
The rate per covarion is the least variable and, until ribonuclease
was examined, the most recently examined of the group, the range of
variation in rate was less than twofold. The lower variability in
rates of substitution per covarion is what one should observe since
codons that cannot change at a given point in time ought not to be
added into the denominator when determining the rate of that change.

### Maximum Likelihood Fit for Evolutionary Rates

In 1968, Kimura proposed his neutral mutation theory in which
he asserted that the vast majority of amino acid replacements that
have occurred in the evolution of proteins were the result of stochastic rather than selective processes. He further demonstrated
that if the proposal were true, it followed that the amino acids
that, selectively, were not significantly more or less beneficial
than the originally encoded amino acid. This special rate, a rate
less than the total mutation rate, is called the neutral mutation
rate. If there are positions where all mutations are neutral,
then, for those positions only, the neutral mutation rate and the
total mutation rate are the same. It has been suggested that mutations in the third codon position that do not change the encoded
amino acid may be neutral.

If the neutral mutation rate is the same from one taxon to
another and from one evolutionary interval to another, and if the
vast majority of amino acid replacements are indeed neutral, then
one can test Kimura's theory as originally proposed by asking
whether the amino acid replacements, or their corresponding nucleotide substitutions, are occurring at a uniform rate over time and
in different lines of descent.

An early attempt to make this test was by the use of the covarion rates shown in table VI. The general equality of these
values was in agreement with the expectations of the neutral theory
but could not prove it. Another more sensitive test of the theory
was developed in Langley and Fitch (1973, 1974) in which they
asked what was the (uniform) rate of nucleotide substitution for
each of several protein genes that would give estimates of the
number of such substitutions in each evolutionary interval that
were in greatest agreement with those actually observed. Said
differently, we wish to find those rates of nucleotide substitution
in each of several protein genes that maximize the likelihood that
any differences between observation and expectation are simply the
result of chance. The result, which requires a computer to obtain,
gives estimates of the rate of nucleotide substitution in each protein, relative times since the common ancestor of any two taxa
whose protein sequences are included in the data set, and estimates
of the goodness of fit to two different kinds of uniformity of
evolutionary rate.

Let us examine the estimated rates first. These shown in
table VII for the seven proteins examined. The format is the same
as for table VI and, for the five proteins common to both tables,
the extent of agreement of the rates is considerable. The average
substitutional rate per codon per $10^9$ years is 1.53 and 1.56 for
tables VI and VII respectively or, averaged, 0.52 observable substitutions per nucleotide per $10^9$ years where, as before, observable implies a change of the amino acid encoded.

Before examining the uniformity of rate test, it is important
to note that there are two kinds of uniformity. In one kind (total rates), the total number of substitutions per unit time remains
constant but there may be great differences in the number of substitutions in any one protein gene during various intervals, i.e.,
the relative rates of substitution in the different protein genes
can vary considerably. In the other kind (relative rates) the
ratio of the number of substitutions in any two genes remains constant but there may be great differences in the total number of
substitutions per unit time, i.e., the rates of substitution in
all genes speeds up or slows down together.

In Kimura's neutral mutation theory, both rates should be
constant and the appropriate tests are shown in table VIII. It
can be seen that both relative and total rates of amino acid
changing nucleotide substitutions are significantly non-uniform.
We may therefore conclude that the nucleotide substitutions do not
occur in conformity to the proposition that the vast majority are
neutral. It is not at all clear, however, how large a fraction
could not be neutral if the observed results are not to deviate
significantly from expectation. Moreover, if time were measured
in generations rather than in years, there would be some improvement

Table VII.  RATES OF NUCLEOTIDE SUBSTITUTION:  MAXIMUM LIKELIHOOD

| Protein | $Rate_1$ | Codons | $Rate_2$ | Covarions | $Rate_3$ |
|---|---|---|---|---|---|
| Fibrinopeptide B | 114 | 21 | 5.40 | | |
| Fibrinopeptide A | 68 | 19 | 3.55 | 18 | 3.76 |
| Insulin C | 93 | 31 | 3.01 | 18 | 5.17 |
| Beta Hemoglobin | 263 | 146 | 1.80 | 39 | 6.73 |
| Alpha Hemoglobin | 224 | 141 | 1.58 | 50 | 4.47 |
| Myoglobin | 140 | 153 | 0.91 | | |
| Cytochrome c | 40 | 104 | 0.38 | 12 | 3.34 |

$Rate_1$, substitutions per gene per $10^9$ years; $Rate_2$, substitutions per codon per $10^9$ years; $Rate_3$, substitutions per covarion per $10^9$ years.  It is assumed that the common ancestor of marsupials and placentals occurred $120 \times 10^6$ years ago.  From Fitch and Langley, 1976a,b.

Table VIII. TESTS OF UNIFORMITY OF RATES HYPOTHESES

| | $\chi^2$ | df | p |
|---|---|---|---|
| Among proteins within legs (relative rates) | 166.3 | 123 | $6 \times 10^{-3}$ |
| Among legs over proteins (total rates) | 82.4 | 31 | $4 \times 10^{-6}$ |
| Total | 248.7 | 154 | $6 \times 10^{-6}$ |

Initial maximum likelihood fit gave the results shown after correction for the non-linearity.  $\chi^2$ is chi-squared, df is degrees of freedom, and p is the probability that results this far removed from expectation would arise by chance if the null hypothesis were true, i.e., if amino acid changing, nucleotide substitutions are accumulating uniformly over time.  From Fitch and Langley, 1976a,b.

in the total rate test. This change could not affect the relative
rate test since one is only comparing genes in the same line of
descent and they necessarily share the same amount of time whatever the units in which it is measured.


## Silent Substitutions

The recent advent of messenger RNA sequencing has permitted
us to estimate the frequency of nucleotide substitutions that do
not change the amino acid encoded. This involves 22 third codon
position silent substitutions out of 57 codons compared and leads,
as shown in table IX to an overall silent substitution rate of 4.03
per third position nucleotide per $10^9$ years, a rate eight times
greater than that for the amino acid changing variety. If these
silent substitutions are indeed neutral, then the mutation rate
must be 5.6 nucleotide changes per nucleotide per $10^9$ years.

It is instructive to compare the two kinds of substitutional
rates in the context of a structural gene since the third position
silent substitutions must be averaged over the other two positions.
This is shown in table X. From this it can be seen that the number of silent nucleotide substitutions is about 2.5 times that of
the amino acid changing variety. Said differently, the total number of nucleotide substitutions, in a gene whose protein product
is evolving at an average rate, is about 3.5 times the number
detected from an examination of the amino acid sequences.

Also shown in table X are two other rates based upon DNA
hybridization techniques. These rates are at the high end of
interpretation since the conversion factor from the melting point
lowering varies among authors from 1.5%/1° to 1%/1.5° with 1%/1°
the most common choice. I used the largest factor here to emphasize that although this makes the substitution rate as large as
possible the larger 1.5 rate is still 16 percent below the substitution rate in genes coding for proteins. Thus, if we conclude
that there is a restriction on the number of ways one can change a
structural gene without decreasing the fitness of the product, we
must also conclude that there is at least as much restriction for
single-copy (non-repetitive) DNA.

There is yet a third conclusion derivable from the data of
tables IX and X. Since amino acid changing nucleotide substitutions proceed at nine percent (= .52 x 100/5.63) of the mutation
rate, we must conclude one of two alternatives: 1, if the silent
substitutions are neutral, then at least nine percent of the non-
silent mutations are not deleterious; or 2, silent substitutions
are not all neutral and the mutation rate, therefore, is greater
than estimated.

Table IX.   RATES OF SILENT NUCLEOTIDE SUBSTITUTION

|   |   | $\alpha$ | $\beta$ | H | Total |
|---|---|---|---|---|---|
| A | Codons (third position) | 11 | 19 | 27 | 57 |
| B | Number of A unchanged | 5 | 12 | 18 | 35 |
| C | Substitutions/site (= -ln (B/A)) | .788 | .460 | .405 | .488 |
| D | Myrs of evolution | 122 | 122 | 120 | 121 |
| E | Rate$_1$ (= C/D x $10^{-3}$) | 6.46 | 3.77 | 3.38 | 4.03 |
| F | Rate$_2$ (= E/0.716) | 9.02 | 5.27 | 4.72 | 5.63 |

The column headings stand for alpha and beta hemoglobins and histone H4.  Messenger RNA's were partially sequenced for alpha and beta hemoglobins from the human (B. Forget et al., 1974, and personal communication), and from the rabbit (W. Salser, 1975, and personal communication) and for histone from two species of sea urchin, Lytechinus pictus and Strongylocentrotus purpuratus (M. Grunstein, 1976).  Row A gives the number of codons whose third position nucleotide is known in both species.  Row B gives the number of third position nucleotides that are identical.  It does not necessarily follow that they are unchanged but the mathematical treatment operates as if they were unchanged.  Row C is the average number of changes per site (B/A) corrected for sites containing multiple changes using the Poisson distribution where $B/A = e^{-r}$ and r is the corrected value wanted.  Row D is twice the number of years (in millions) since the common ancestor of the two species.  Row E is the rate in silent nucleotide substitutions/third position nucleotide/$10^9$ years.  Row F is a lower bound of the mutation rate in mutations/nucleotide/$10^9$ years assuming that all third codon position silent mutations are neutral and 0.716 of all third position mutations are silent.  From Fitch and Langley, 1976a,b.

Table X.   EVOLUTIONARY RATES OF NUCLEOTIDE SUBSTITUTION

| Type | Rate | Remarks and Assumptions |
|---|---|---|
| Amino Acid Changing | | Marsupial-placental split 120 million years ago |
| $\alpha$, $\beta$ hemoglobin | 0.6 | |
| fibrinopeptide B | 1.8 | |
| overall | 0.5 | Including cytochrome <u>c</u>, insulin C-peptide and myoglobin |
| Silent Substitutions | | Corrected for multiple substitutions in any one site |
| $\alpha$, $\beta$ hemoglobin | 1.6 | From human and rabbit messenger RNA; split 61 million years ago |
| overall | 1.3 | Including also two sea urchin histone messenger RNAs; split 60 million years ago |
| Total | | |
| $\alpha$, $\beta$ hemoglobin | 2.2 | |
| overall | 1.8 | |
| Single-Copy DNA Hybridization | 1.5 | 1.5% base change/1°C $\Delta T_m$; Rhesus-man split 34 million years ago |
| 28s Ribosomal DNA Hybridization | 0.8 | 1.5% base change/1°C $\Delta T_m$; Baboon-man split 34 million years ago |

Rates in nucleotide substitutions/nucleotide position/$10^9$ years. The amino acid changing rates are the rates per codon of Table VI divided by 3.  Silent substitution rates are from Table IX and were divided by 3 since they apply to only the third nucleotide position, whereas we wish the rate for all positions on average.  This ignores those silent substitutions that do not occur in the third position. From Fitch and Langley, 1976b.

## On Evolutionary Clock

The seven proteins examined in the previous sections were
not used to determine the phylogeny of the species from which they
derive.  Instead, it was assumed that the proper phylogeny is that
of current biological opinion.  The maximum likelihood solution
gives relative times of divergence and in figure 8 is shown the
phylogeny of the seventeen taxa used in this study with the ancestral nodes located along the abscissa as the number of amino acid
changing, nucleotide substitutions occurring, on average, in all
seven genes.  For comparison one can treat the number of such
nucleotide substitutions as if they were directly proportional to
time.  If we assume that the ancestral mammal at the marsupial-
placental divergence occurred 120 million years ago, then the scale
along the top of the figure is applicable and becomes an estimate
of other dates of divergence.

How good are these estimates?  That cannot be readily answered
because it is not clear how good the paleontological dates for
these divergences are.  Nevertheless a comparison of the two is
provided in figure 9.  The line is arbitrarily drawn between the
origin and point 16, the marsupial-placental divergence, and it is
apparent that, apart from the primates, the points fall reasonably
close to the line.

There are two fundamental kinds of clocks, metronomic and stochastic.  In the former, the intervals between ticks are uniform;
in the latter the intervals vary and are Poisson distributed.
Radioactive decay is an example of the latter.  In the previous
section we noted that there was significant non-randomness in the
rate of nucleotide substitution, which means that nucleotide substitutions are not as well behaved as radioactive decay for use
as ticks to be counted in a molecular clock.

But we can ask how bad is it?  It is clear from figure 9
that if we average over enough substitutions in enough protein
genes and enough taxa, the result is rather good.  In this case
there are a total of 687 nucleotide substitutions in the data, which
represents an enormous sequencing effort by biochemists.

One estimate of how bad the non-randomness is comes from the
observation that the chi-square values in table VIII are roughly
twice the number of degrees of freedom indicating that the variance
in the time interval between nucleotide substitutions is roughly
twice that for a truly stochastic process.  Therefore we must
count twice as many nucleotide substitutions as radioactive dis-
integrations to achieve the same degree of accuracy.  Nevertheless,
it remains true that, given a sufficient sample of nucleotide substitutions, one can get reasonable dates of evolutionary divergence;
that is, there really is a clock, its just unusually erratic.

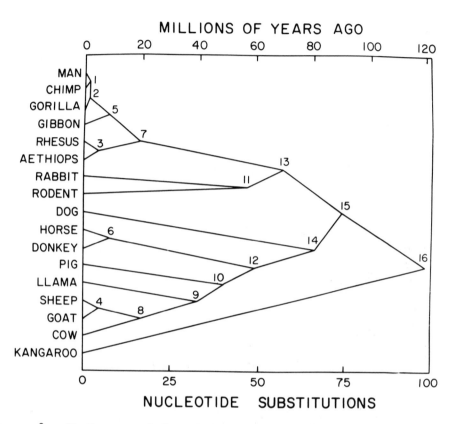

Figure 8.   Phylogeny of Seventeen Mammals.   The branching order is
understood to be current biological opinion.   The nodal "heights"
along the abscissa are according to the maximum likelihood fit to
the nucleotide substitutions observed for the most parsimonious
accounting of the evolution of seven proteins using this tree.   The
names of the proteins examined are given in table VII.   The axis
at the top of the figure is a simple linear interpolation that results by assuming that the marsupial-placental ancestor occurred
120 million years ago and that nucleotide substitutions are directly proportioned to time.   From Fitch and Langley, 1976a,b.

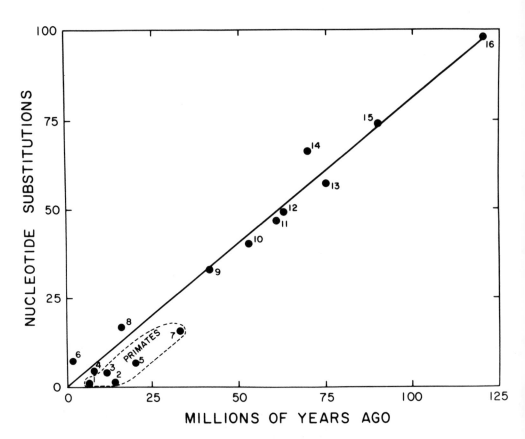

Figure 9.  Relation of Nucleotide Substitutions from the Maximum
Parsimony Fit of Figure 8 and Paleontologically Estimated Times of
Divergence.  Numbers beside the points identify the common ancestor
with the same number in figure 8 and with the abscissa of that
figure the same as the ordinate of this figure.  The abscissa of
this figure are values estimated by Leigh VanValen without knowledge of our results (except points 1, 2, 3, and 5 which were provided by Morris Goodman who attributed them to Malcolm McKenna).
For each point, VanValen presented what he described as "subjective
standard deviations" of his estimate.  For the marsupial-placental
divergence (16) this was ± 15 million years.  All others averaged
close to plus or minus four (largest six) million years with a tendency for the larger deviations to be for the larger estimates of
divergence times.  Line was arbitrarily drawn through the origin
and point 16.  From Fitch and Langley, 1976a,b.

Table XI.  COMPARATIVE DISTRIBUTIONS OF RESIDUES
WITH EVOLUTIONARY REPLACEMENTS OF AMINO ACIDS
AND HUMAN VARIANTS IN $\alpha$ AND $\beta$ HEMOGLOBIN

|            |          | Varied | Unvaried | Total |
|------------|----------|--------|----------|-------|
| Replaced   | Expected | 72.6   | 113.4    | 186   |
|            | Found    | 70     | 116      |       |
| Unreplaced | Expected | 39.4   | 61.6     | 101   |
|            | Found    | 42     | 59       |       |
| Total      |          | 112    | 175      | 287   |

$\chi_1^2 = 0.43$; $p > 0.5$.  Data were for 112 variants.  From Fitch, 1974.

A few caveats should be inserted before rushing to calibrate evolution with our molecular clock.  It is clear that it requires large amounts of sequence information.  It is also clear that special groups (taxa) may pose special problems.  The primates, as a group, all fall below the line.  Is there a special cause operating here, perhaps longer than average generation times?  And if certain groups can cause problems, we must certainly be careful regarding groups not present in these data.  It is possible that proteins not among those examined could behave differently.  One would expect that the presence of paralogous gene products in a collection thought to be orthologous could play havoc in estimates of divergence times.  Agreement here results from interpolation between the marsupial-placental divergence.  Extrapolation beyond the reference divergence time carries additional sources of error.  Finally, as detailed in an evaluation of molecular clocks (Fitch, 1976b), other methods, even though basically molecular, have additional sources of error present and hence are necessarily even less reliable.

## Variants Versus Substitutions

In the case of human hemoglobins, there are more than 230 abnormal alpha and beta chain variants known to exist in the population.  The future will, presumably, see similar data for other proteins.  It is presumably from a pool of such variants that the next evolutionary amino acid replacement will come.  Is there some relationship then between the amino acid positions that are variant in human hemoglobins and those that were replaced during the evolution of hemoglobin?  Table XI shows the distribution of 287 alpha and beta hemoglobin positions according to whether they possess one or more variants and whether they sustained one or more amino acid replacements.  A high degree of relationship between variants and replacements would lead to significantly higher than expected numbers of positions that were both varied and replaced or were both unvaried and unreplaced with correspondingly reduced numbers that were varied but unreplaced or were replaced but unvaried.  The result, however, is that the two distributions behave ($p > 0.5$) as if they were completely independent.

## Acknowledgements

This work was supported by National Science Foundation grant BMS 76-20109 and National Institutes of Health grant AM-15282.

REFERENCES

Beintema, J.J., Gaastra, W., Lenstra, J.A., Welling, G.W. and
    Fitch, W.M., 1977, The Molecular Evolution of Pancreatic
    Ribonuclease, submitted for publication.
Fitch, W.M., 1967, Evidence Suggesting a Non-Random Character to
    Nucleotide Replacements in Naturally Occurring Mutations,
    J. Mol. Biol., 26, 499-507.
Fitch, W.M., 1970, Distinguishing Homologous From Analogous Pro-
    teins, Syst. Zool., 19, 99-113.
Fitch, W.M., 1973, Is the Fixation of Observable Mutations Distri-
    buted Randomly Among the Three Nucleotide Positions of the
    Codon?, J. Mol. Evol., 2, 123-136.
Fitch, W.M., 1974, A Comparison Between Evolutionary Substitutions
    and Variants in Human Hemoglobins, Annals N.Y. Acad. Sci.,
    241, 439-448.
Fitch, W.M., 1976a, The Molecular Evolution of Cytochrome c in
    Eukaryotes, J. Mol. Evol., 8, 13-40.
Fitch, W.M., 1976b, An Evaluation of Molecular Evolutionary Clocks,
    in Molecular Study of Biological Evolution, F.J. Ayala, ed.,
    Sinauer Associates, Sunderland, Mass., 160-178.
Fitch, W.M., 1977 , On the Problem of Discovering the Most Parsi-
    monious Tree, Amer. Natur., in press.
Fitch, W.M. and Langley, C.H., 1976a, Protein Evolution and the
    Molecular Clock, Fed. Proc., 35, 2092-2097.
Fitch, W.M. and Langley, C.H., 1976b, Evolutionary Rates in Pro-
    teins, Neutral Mutations and the Molecular Clock, in Progress
    in Molecular Anthropology, M. Goodman and R.E. Tashian, eds.,
    Plenum Press, N.Y., in press.
Fitch, W.M. and Margoliash, E., 1967, A Method for Estimating the
    Number of Invariant Amino Acid Coding Positions in a Gene
    Using Cytochrome c as a Model Case, Biochem. Gen., 1, 65-71.
Fitch, W.M. and Markowitz, E., 1970, An Improved Method for Deter-
    mining Codon Variability in a Gene and Its Application to
    the Rate of Fixations of Mutations in Evolution, Bioch. Gen.,
    4, 579-593.
Fitch, W.M. and Yasunobu, K.T., 1975, Phylogenies from Amino Acid
    Sequences Aligned with Gaps, J. Mol. Evol., 5, 1-24.
Forget, B.G., Marotta, C.A., Weisman, S.M., Verma, I.M., McCaffrey,
    R.P. and Baltimore, D., 1974, Nucleotide Sequences of Human
    Globin Messenger RNA, Ann. N.Y. Acad. Sci., 241, 290-309.
Grunstein, M., Schedl, P. and Kedes, L., 1976, Sequence Analysis
    and Evolution of Sea Urchin (Lytechinus pictus and Strongylo-
    centrotus purpuratus) Histone H4 Messenger RNAs, J. Mol. Biol.,
    104, 351.
Kimura, M., 1968, Evolutionary Rate at the Molecular Level, Nature,
    217, 624.
Langley, C.H. and Fitch, W.M., 1973, The Constancy of Evolution:  A
    Statistical Analysis of the ∝ and β Hemoglobins, Cytochrome c

and Fibrinopeptide A, in Genetic Structure of Populations, Newton E. Morton, ed., Univ. Press of Hawaii, Honolulu, 246-262.

Langley, C.H. and Fitch, W.M., 1974, An Examination of the Constancy of the Rate of Molecular Evolution, J. Mol. Evol. $\underline{3}$, 161-177.

Markowitz, E., 1970, Estimation and Testing Goodness-of-Fit for Some Models of Codon Fixation Variability, Bioch. Gen., $\underline{4}$, 595-601.

Salser, W., Bowen, S., Browne, D., El Adli, F., Federoff, N., Fry, K., Heindell, H., Paddock, G., Poon, R., Wallace, B. and Witcome, P., 1975, Investigation of the Organization of Mammalian Chromosomes at the DNA Sequence Level, Fed. Proc., $\underline{35}$, 23-35.

Sneath, P.H.A., Sackin, M.J. and Ambler, R.P., 1975, Detecting Evolutionary Incompatibilities from Protein Sequences, Syst. Zool., $\underline{24}$, 311-332.

# ZOOGEOGRAPHY AND PHYLOGENY: THE THEORETICAL BACKGROUND

# AND METHODOLOGY TO THE ANALYSIS OF MAMMAL AND BIRD FAUNA

J. Allen KEAST

Department of Biology, Queen's University

Kingston, Canada K7L 3N6

## INTRODUCTION

Biogeography is one of the fundamental divisions of evolutionary biology. It is the objective here to review its basic premises, concepts, methodology and its major contributions to our understanding of the evolution of faunal structure and its phylogenetic implications. Perspective can best be gained by using the historical approach which will also enable shortcomings to be highlighted. Current approaches to "ecological" biogeography will also be briefly reviewed and the question asked what has its contribution been to phylogenetic understanding.

Zoogeography defined basically as the study of the geographic distributions of animals has gone through a series of phases: (a) descriptive (the documenting of distributions and distributional patterns); (b) historical (interpreting these with the aid of constantly improving fossil record); (c) ecological (incorporating environmental analyses); and (d) predictive. Vuilleumier (1975) provides an alternative grouping into descriptive, analytical, and predictive. Either way it must be stressed that whilst those phases may have appeared sequentially one did not replace the other and all continue to be actively pursued today. (It must be pointed out here that some writers have recently been critical of the lack of a unified methodology; e.g. Vuilleumier {1975}). The descriptive phase still, of necessity, largely dominates in the case of the lesser known animal groups. One school of thought still argues that zoogeography should remain essentially descriptive and limit itself to the analysis of distributional patterns. Recently Briggs (1974) has offered

separate definitions of biogeography and ecology; happily in his
analyses he does not maintain the separation.  The narrow approach,
of course, hardly accords with the spirit of the discipline as
seen by Darwin and Wallace.  A recent development in vertebrate
biogeography is the mathematical analysis of distribution patterns.  At the other end of the 'analytical' scale predictive
zoogeography is still much more of a pipe dream than a reality.
We have not yet even attempted it with the more complex continental faunas.  Thanks, however, to the fascinating work on island
faunas in recent years we now have a tentative predictive base
for understanding how species numbers relate to island area and
distance from the nearest major faunal source.


THE HISTORY OF BIOGEOGRAPHIC THEORY

Although the basic ideas long preceded them, zoogeography as
a discipline stems from the books of Darwin (1859) and Wallace
(1876), two major works which are conceptually mature, even by
today's standards.

Many of our ideas on zoogeography relative to phylogeny stem
directly from Darwin.  The 'dilemma' of the Southern Hemisphere
cold temperate disjuncts was freely debated by Darwin and Hooker
(1860); the former arguing that they were the relics of dispersals from the North and the latter that they indicated a
southern dispersion via Antarctica.  Only in the last few years
has the new sea floor spreading data confirmed a Mesozoic clustering of continents in the far South.  Likewise long distance
dispersal was debated with great clarity.  Significant also was
Darwin's demonstration that the faunas of islands like the Galapagos are derived from those of the nearest continent, that island
faunas represent only a filtered sample of these, and that archi-
pelagos (with differentiated populations on the different islands)
may provide beautiful living demonstrations of evolution in
action.

The introductory chapters of Wallace (1876) contain the basic
tenets of zoogeography:  plants and animals are not uniformly
distribution - they show 'habitat specificity'; distributional
barriers such as mountain ranges, rivers, and arms of the sea,
break up distributions and lead to differentiation; the more
complete the barrier the more different forms on either side of it
are likely to be; there are distinct Arctic, forest, desert biotas;
the various continents contain distinct biotas including equivalent but independently derived forest, desert, etc., assemblages;
dispersal is fundamental to life but species differ greatly in
their dispersal capacities; distributions have changed in geological time as part of the evolutionary process; the fossil

record can tell much about faunal histories as well as about past
climates and distribution of land on the globe; glaciations
greatly affected animal distribution patterns; changes in land
area and distribution must have affected climate; a proper taxonomy is fundamental to zoogeographic analysis; and broad conclusions about the distributions and histories of biotas can only be
drawn from the simultaneous consideration of large numbers of
distribution patterns.


### The Concepts of Zoogeographic Regions and Life Zones

Sclater (1958) in his analysis of bird distribution discovered that the world was broadly divisible into six zoogeographic
regions:  Palearctic, Nearctic (together forming the Holarctic),
Neotropical, Ethiopian, Oriental, and Australian.  The Divisions
were supported by comprehensive documentation.  These zoogeographic regions are separated by distributional barriers and each has
been the center of origin for a series of groups.  Degree of difference (and endemism) has been a function of the completeness
and duration of the isolating barrier in geological time.  Thus,
Australia has been the most isolated and has the most distinctive
fauna.  The Ethiopian and Oriental regions on the one hand, and
Nearctic and Palearctic on the other have been intermittently
connected, as well as being in the same latitudes.  Hence their
faunas bear (and have borne) a considerable measure of similarity
to each other.  Sclater thus initiated the treatment of faunas at
the regional level, an approach that is still very prominent today.

Despite various alternative classifications in the next 30
years, including ones like that of Allen (1878) that created
additional zones to allow for the climatic belts within continents,
Sclater's scheme is the one in general use today (see the review
of Serventy {1960}).

An additional way of considering distributions is, of course,
in terms of life zones and climatic belts; e.g. Merriam (1892)
with respect to North America.  Life zone divisions do not correspond to faunal subregion ones, as Darlington (1957) stresses.
Figure 1 shows the clearcut zonation of the world's vegetation
belts.


### The Area and Climate Concepts of Matthew

Matthew (1915) advanced the view that the Holarctic Region
(which in his definition included Africa as well as Asia, Europe,
and North America), as the largest land-mass had been the center
of origin for most of the major mammalian stocks.  These under the

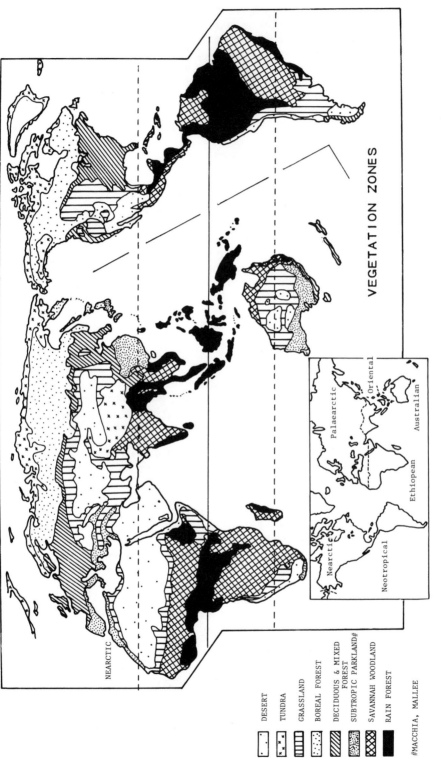

Fig. 1. The world's major vegetation regions reflect the climatic belts and have evolved their own adapted biotas. In the case of the desert regions of Africa and Asia, equivalent vegetation belts are in close proximity and facilitate colonization from outside regional biotas which are usually locally derived. (See Fig. 7). The desert biotas of Australia and the Neotropical region are secondarily derived from the faunal assemblage of that continent. Inset shows the

influence of climatic changes, had expanded outwards (southwards) in successive waves. Figure 2 shows a series of Matthew's range maps which would seem to be persuasive evidence in support of his stand. The concept long commanded considerable respect in zoogeographic circles and in later years has been reassessed by Darlington (1957, 1959).

### Continental Drift and the Influence of Wegener

Biologists had long postulated hypothetical land connections to explain puzzling distributional disjuncts. Hence when Wegener (1915, 1924) advanced his arguments that the world's continents had changed position in geological time it had a profound impact on biogeographic thought (Fig. 3). (Wegener's maps should be compared with more recent maps (Fig. 4) in order to understand the nature of the original hypothesis and the limitations incorporated in it.) It particularly found favour with Southern Hemisphere workers who had long seen the need to postulate a direct far southern interchange for such forms as marsupials, their associated trematode, cestode and mallophagan parasites, parastacid crayfishes, the crayfish ectoparasite Temnocephala, and leptodactylid and hylid frogs (Harrison 1924 is an example of a typical paper of this type).

By the 1930's the biological literature on continental drift numbered some 350 titles (Wittman, 1934). It had also acquired some very powerful opponents who saw no need to invoke continental movements to explain distributional patterns; e.g. the oligochaete specialist Michaelson (1922). Northern Hemisphere geologists generally came to oppose the idea, arguing that there was no known mechanism whereby continents could move. Simpson (1940) criticized the rationalizations of those workers who invoked continental connections without considering what the patterns advanced would require of other groups. He systematically worked through a number of cases that had been advanced as proving southern junctions (marsupials, ratites, meiolanid turtles, galaxiid fishes), and dismissed each on various grounds. For example, the marsupials could have entered Australia from the North; some galaxiids spawn in salt water and hence could have dispersed through the sea; the Australian meiolanid turtles were Pleistocene and the South American ones Eocene. He concluded by suggesting that no known biotic fact demands an Antarctic land-migration route for its explanation. It must be borne in mind that virtually all geological thought in the 1930's - 1950's was solidly against drift so that these interpretations were based on what was then regarded as overwhelming probability. Not only Simpson but virtually all paleontologists of this era shared these views (Simpson, 1943). They do not necessarily indicate a

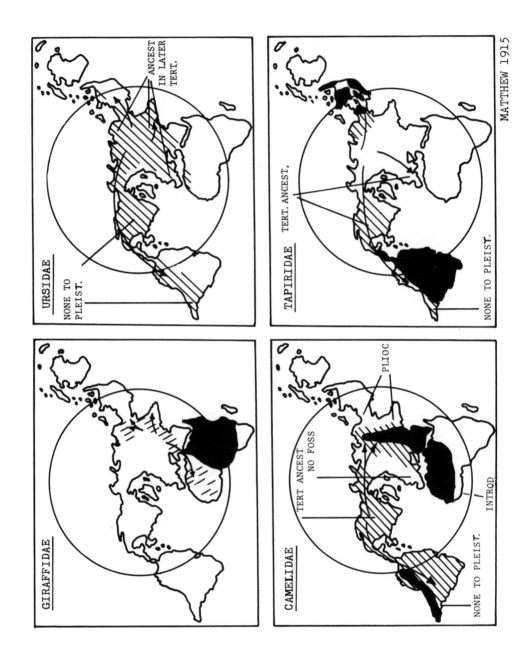

URSIDAE

NONE TO PLEIST.

ANCEST IN LATER TERT.

TAPIRIDAE

TERT. ANCEST.

NONE TO PLEIST.

GIRAFFIDAE

CAMELIDAE

TERT ANCEST NO FOSS

NONE TO PLEIST.

INTROD

PLIOC

MATTHEW 1915

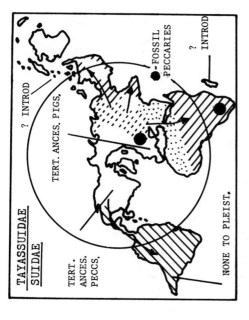

Fig. 2. The contemporary and fossil distributions of five mammalian groups (giraffes, bears, camels, tapirs, and pigs and peccaries) according to Matthew (1915), to show the kinds of distribution patterns that persuaded him that most "dominant" forms originated in the major land-mass of Holarctica (which in his definition included Africa). Most of the patterns have not been significantly changed by later knowledge. Fossil giraffes, however, are now known from throughout Africa. Note that bears only reached South America after the formation of the Panamanian land-bridge at the end of the Pliocene and that this generalized group has failed to penetrate Ethiopian Africa. Camels first appear in the fossil record in North America, suggesting that they had a Nearctic origin. The group only penetrated South America with the formation of the Panamanian land-bridge. Tapirs are limited to South America and Asia today but were formerly widespread in North America (from where they colonized South America during the Pleistocene), and in Asia. His map of peccary distribution is now known to be wring and hence is modified here. The latter were formerly thought to be the American counterpart of the suids but new fossil discoveries now show them formerly to have occurred in Europe and Africa (Hendey, 1976).

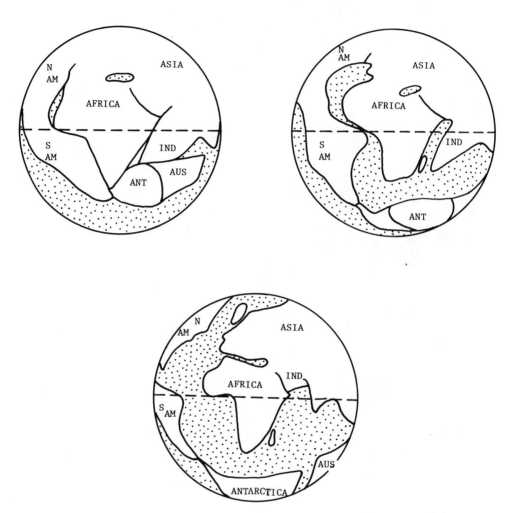

Fig. 3.  Continental drift according to Wegener (1915, 1924), reconstructions based on the matching of shoreline configurations,
corresponding geological strata, etc.  Early biological data allowed for lingering similarities between South America and Australia
via Antarctica:  this has been amply confirmed by latter-day data.
Note also the continuing connection across the North Atlantic now
confirmed by marked similarities in the Eocene fossil mammals of
Europe and North America (McKenna, 1975).

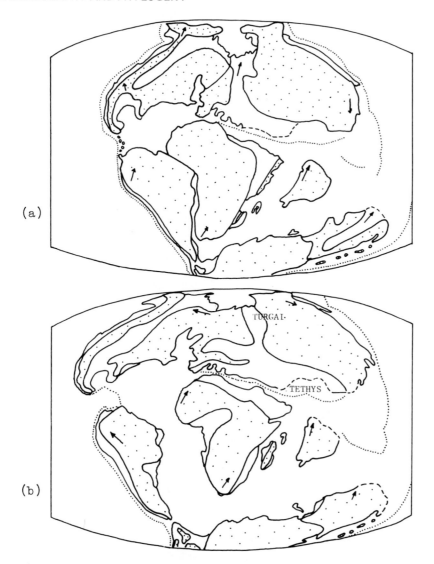

(a)

(b)

TORGAI

TETHYS

Fig. 4.  Distribution of the world's land and epicontinental seas
in (a) the Late Jurassic, about 135 million years before present,
when it was clustered into a northern Laurasia and southern Gondwana; and (b) Late Cretaceous, about 65 m.y.b.p., based on Dietz
and Holden (1970), and other sources.  The lower illustration bears
surprising resemblance to the postulated land distribution of
Wegener, made 50 years before sea floor spreading was discovered.
Note the late connection between Australia and Antarctia, and continuing proximity of South America and New Zealand to Antarctica
which explains similarities in their cold temperate biotas.  The
African temperate elements are rather different.  Redrawn from
Tedford (1974).

fundamental flaw in paleontological methodology as some recent
workers have asserted (e.g. Croizat et al., 1974).  At the same
time one cannot help but be alarmed at the recent rush of some
biologists to embrace drift since its geological proof in the
1960's, initiating massive reinterpretations without any new,
original taxonomic work.


                        The Contribution of Vertebrate
                     Paleontology to Biogeographic Theory

     Fossil histories of evolutionary lines and changing faunal
assemblages in geological time have long been fundamental in
demonstrating migratory pathways and helping disentangle why
contemporary distribution patterns are as they are.  Darwin and
Wallace were already drawing liberally on contemporary paleontological knowledge and the book by Matthew (1915) is one of
several major syntheses of "historical" zoogeography.  Some later
workers have sought to discount the fossil record because of its
incompleteness and because they feel far too much reliance has
been placed on it.  On the other hand its positive achievements
are great:  (1) it has revealed in fairly complete detail the
fossil and distributional histories of various groups, including
the well documented equids and proboscidians; (2) it emphasized
how misleading deductions made purely on contemporary distribution
patterns can be, e.g. lungfishes and osteoglossomorphs now exclusively southern (Fig. 5) were formerly widespread in the
Northern Hemisphere; camels (now confined to Asia and South
America) appear first in the North American fossil record and
peccaries (now almost completely Neotropical) formerly occurred
in Africa and Europe (Hendey, 1976), a fact not known in Matthew's
time (Fig. 2); (3) the discovery of Lystrosaurus and an associated
Triassic fossil fauna in Antarctica (Elliot et al., 1970) has recently helped confirm the clustering of the southern continents
at that time.

     Paleontologists as a group have contributed greatly to bio-
geographic theory, and vertebrate paleontology will remain the
final arbiter of many matters of past history, notwithstanding the
expansion of the new cladistic, and comparable, approaches.  Criticism that vertebrate paleontologists make interpretations far
beyond what their data merits is hardly uniformly true.  Simpson
(1947:628) in his comprehensive paper on Palearctic-nearctic
Tertiary mammal interchanges notes that: "The problem of direction
of migration would seem to be relatively simple.  It might be
supposed that a given group or its ancestors would appear earlier
on one land mass than on another, and that the direction of migration would then be obvious.  As a matter of fact, there are relatively few groups for which the direction of migration can be

given with any strong assurance. This is in most cases one of the
greatest uncertainties regarding the geographical history of
Holarctic mammals."


## Latter Day Work Emphasizing
## The "Classical" Approach to Zoogeography

Classical zoogeography could be said to have reached a peak
with the publication of Hesse et al., (1951); Ekman (1953); and
Darlington (1957). The last-named, which can be read in conjunction with such other papers of the writer as Darlington (1948,
1959) is a catalogue of the distributions of all the families of
land vertebrates, and faunas of the major continents and islands.
Darlington sought to generalize on, and interpret distributional
patterns and the evolution of these patterns, and suggested a set
of "working principles" to help overcome errors of interpretation.
These are generalized and necessarily inexact, but of great historical interest. Their widespread use partly explains the im-
patience of some recent writers to develop a more exacting methodology.

A variety of other prominent works in biogeography include
Croizat (1956, 1964) in which the emphasis is on deducing general
distribution patterns ("multiple tracks") by the analysis of a large
number of individual distribution patterns; Hubbs (1958), a regional work mainly devoted to the zoogeography of western North
America; Udvardy (1959); Banarescu (1970); Hallam (1973); Cox
et al., (1973); Muller (1973); and Briggs (1974). None of these
works really lift us into the analytical era, nor do they satisfy
all the requirements of an advanced text in biogeography. There
are also a range of regional works in biogeography; e.g. the
series on different continents and regions of the world published
by Junk of the Hague; Meggers et al., (1973).

Phytogeographic texts include Cain (1944); Good (1964); and
Carlquist (1974).


## Geographic Speciation

Wide acceptance of speciation as a geographic phenomenon in
sexually reproducing species had its beginnings with Mayr (1942).
Mayr developed his data and quantified approach to the study of
island evolution from extensive studies of the birds of the southwest Pacific region. In the course of these studies raciation,
clines, barriers, hybrid zones, and faunal origins were compre-
hensively examined as well as geographic variation, the polytypic
species concept (of which Mayr has certainly been the major

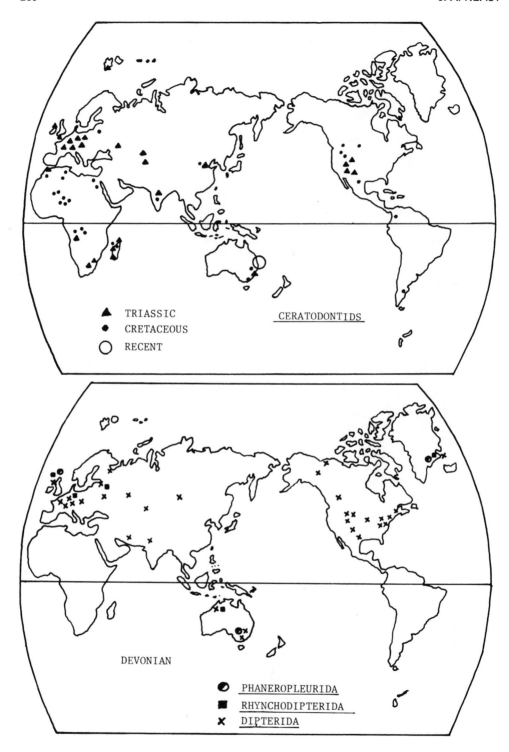

Fig. 5.  The lungfish (Dipnoi), because of their limitation to
Africa, Australia, and South America today, were once held to be
a Gondwana group that originated in the Southern Hemisphere.  The
fossil record, however, showed this view to be false and that the
group had had a world-wide distribution for much of its history,
thus providing another example of how misleading it is to try to
derive distributional histories from living forms alone.  The
top figure shows the distribution of Ceratodontids in the Triassic,
Cretaceous, and Recent.  The group survives today only as the
Australian Neoceratodus, confined to two small Queensland rivers.
The bottom map shows the wide distribution of three groups of
Dipnoi in the Devonian:  the Dipterida (Lower and Upper Devonian,
Phaneropleurida (Upper Devonian), and Rhynchodipteridae (Middle
Devonian.)  Seeming absences from Africa and South America are
presumably due to lack of rocks of the right age.  Even "peculiar"
forms like Griphognathus (Rhynchodipterida) has been found in
localities as distant as Australia, Germany, and Latvia also emphasizing the very different distribution of land in the Devonian;
on the other hand there is no fossil record of Lepidosiren and
Protopterus before the Tertiary and outside of South America and
Africa (H.P. Schultze, personal communication).  The Devonian map
is drawn after compilation of H.P. Schultze, Gottingen, who kindly
made it available.

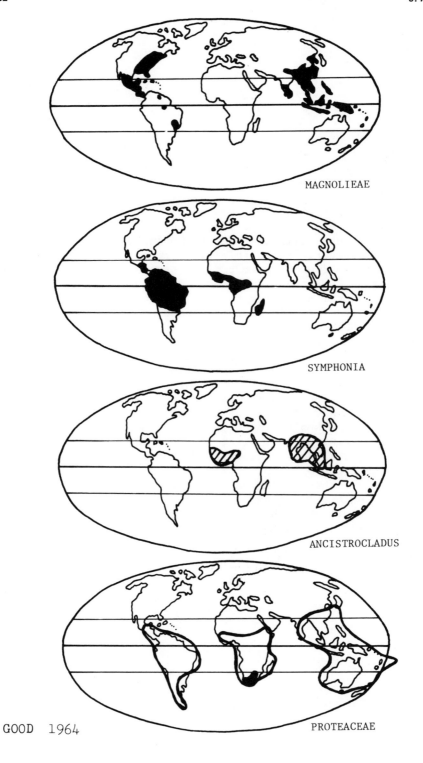

MAGNOLIEAE

SYMPHONIA

ANCISTROCLADUS

GOOD 1964

PROTEACEAE

Fig. 6.  A series of maps showing four different kinds of disjunct
plant distributions and redrawn from Good (1964), chosen because
each (a) can be held to be explained by different primary geographic events and, (b) is equally capable of alternative explanation.
The Magnolieae, an old group, have a strongly fractured distribution in the Americas and east Asia.  The pattern could, theoretically, represent the remains of a formerly continuous pan-tropical
and warm temperate one stemming from as far back as the Eocene.
Alternatively they may never have occupied western Asia and Europe,
former links having been between the Americas and Asia direct.  The
genus Symphonia has a West African-South American distribution, a
pattern that in older groups has been held to reflect the South
Atlantic land link that persisted to the Middle Cretaceous (or
continuing spatial proximity into the early Tertiary (e.g. Raven
and Axelrod, 1974).  The genus Ancistrocladus, with African - Oriental rainforest disjuncts could, in view of the great biotic
similarities of these regions and repeated mammal interchanges
between them (Cooke, 1972) represent a contraction from a later
continuous Pliocene or Pleistocene range.  The Proteaceae are an
old southern temperate group that extends as fossil pollen back
to the Cretaceous:  the tropical extensions are secondary (see full
discussion in Johnson and Briggs, 1963).  Note that in each case
the pairs, or series, of disjuncts occupy the same climatic belts.
Although each might be held as evidence for former direct land
connections in the second and third especially, the second region
could equally have been colonized by long distance dispersal.  Note
that there are counterparts to these distribution types in the
mammals, the first by tapirs, the second by characin fishes, the
third by elephants, rhinoceroses, extinct and living giraffids,
and the bottom by various families of mayflies, Odonata, stoneflies,
etc.  In cases where a range of species have a common distribution
pattern a common distributional history is not necessary.  Some
groups may be relictual there; others recent colonizers of a climatic zone that has the right attributes.  Hence, in advancing a
postulate great care has to be given to the age of the line and
whether or not specific routes were available to it.  If a line is
only known as far back as the Oligocene it is hardly appropriate
to try to explain its distributional history in terms of Cretaceous
land connections.

proponent), and the biology of speciation.  In other works (Mayr,
1940, 1946, 1963; Mayr and Phelps, 1967) significant analyses of
faunal histories were made, using the techniques of centers of
abundance, levels of differentiation, and levels of species diversity, approaches necessary when there is no supporting fossil
history.

Outgrowths of Mayr's ideas on geographic speciation and avian
biogeography are to be found in a series of studies on the avifaunas of the different continents, e.g. Australia (Keast, 1961);
Africa (Moreau, 1966; Hall and Moreau, 1970); South America
(Haffer, 1967, 1969, 1974; Vuilleumier, 1969, 1970, 1972; Short,
1975) and North America (Mayr and Short, 1970).


## The Theoretical Aspects of Land Biogeography

Islands, supporting small, relatively simple faunas, the age
of which can sometimes be dated, have long contributed dispropor-
tionately to evolutionary knowledge and biogeographic theory.  The
importance of the Galapagos to the original formulation of
Darwin's ideas on evolution have already been noted.  Studies on
this fauna have continued with Lack (1947), Bowman (1961), and
others, analyzing the avifauna in ever greater detail.  Hawaii,
with its spectacular examples of adaptive radiations from an
initially small, highly unbalanced series of colonizations is
unique (Zimmerman {1948} on the insects; Amadon {1950} and Berger
{1972} on the birds; Fosberg {1948} and Carlquist {1970} on
plants).  Recent years have seen a series of symposia on island
faunas (e.g. those organized by the Institute of Tropical Biology
in Puerto Rico in 1971; International Congress of Ornithology,
Canberra, 1974):  in these such individual facts of island faunas
ecological 'release' have been extensively expounded.  See also
Grant (1966, 1972); Keast (1968, 1971); Diamond (1970, 1973, 1975),
and Terborgh and Faaborg (1973).

MacArthur and Wilson (1967) for the first time, put into a
theoretical and mathematical framework such facets as the relationship between area and species number, area and diversity, colonization, species turnover rates, and extinction.  The ideas have
since been considerably expanded by Wilson and Simberloff (1969),
Simberloff (1969, 1970), Simberloff and Wilson (1969), Schoener
(1968), and others.


## CONCEPTS AND PROBLEMS OF ZOOGEOGRAPHIC ANALYSIS

Modern biogeography has inherited several concepts and problems of analysis that merit specific comment.

## Current Taxonomic State

That any zoogeographic analysis is dependent on the quality and state of the group systematics being used is obvious.  Inadequate data will bias results and one cannot draw generalities from poorly known groups.  Excessive splitting at the generic level is bad, relationships being thereby lost; excessive lumping hides relationships.  Notwithstanding that many workers believe that an 'absolute' taxonomy is possible there is going to remain a degree of arbitrariness.  Accordingly, the biogeographer has to be very cautious in comparing numbers of families, genera, and species, in different groups, and in making numerical comparisons between different continents. Species are evolving units and hence all levels of evolutionary differentiation can be expected in any fauna.  Another problem is that classificatory levels are commonly related to rapidity of evolution.  Thus the 'species groups' in recent Nothofagus had already differentiated in the Cretaceous when the marsupials (that have now achieved ordinal rank) were only starting.

A test that is commonly used to compare different taxonomies is to calculate ratios of numbers of species to genus, and genera to family.  If these are reasonably similar the taxonomies should be approximately equivalent.  They commonly are.  Thus, in a recent set of comparisons between the mammal faunas of Africa, South America, and Australia, the figures shown below emerged (Keast, 1972a):

|  | Africa | South America | Australia |
|---|---|---|---|
| Numbers of species per genus | 3.24 | 2.91 | 3.11 |
| Numbers of genera per family | 4.46 | 5.56 | 6.50 |

Notwithstanding that it is "comforting" to find similar ratios it must be stressed that there is no reason why these have to be so.  A region could well have a high proportion of distinctive relict genera.

Greatly improved numerical comparisons between continents are possible using the polytypic species concept.  Here allopatric counterparts are counted only once.

Contemporary Distribution
Patterns Relative to Past Histories

In view of (a) the passage of 120 million years since Africa
started to separate from South America, and 43 million since
Australia left Antarctica; (b) the vast climatic fluctuations
since then; and (c) gross changes in competitive interrelationships stemming from the radiation and extinction of a great many
evolutionary lines, it is questionable how much one can expect to
learn of past distribution patterns from contemporary ones.  The
writer's guardedness in this regard conflicts with the basic
thesis of Croizat (1964).

Continental Area and the Origin of Dominant Groups

Darwin (1859), and Matthew (1915, 1928) advanced the idea
that the large northern land-masses (Holarctica) had produced the
dominant groups of animals which spread out from there to occupy
much of the world.  Darwin was particularly impressed with the
seeming superiority of European plants and animals over those of
New Zealand and Australia.  After noting that these "versatile and
aggressive invaders" were likely to exterminate many of the New
Zealand endemics he offered the explanation that groups that
evolve in the large land-masses are at an advantage because they
developed in a more competitive background.  Matthew attributed
the success to their having evolved during the demanding arid and
cool phases, a time of stringency, scarcity of resources, and
enhanced competition.

Darlington (1957) has re-examined the whole question of
dominant groups.  He accepted that the Holarctic had been active
in producing these.  He saw them both as the product of evolution
in a large continental area and climatic moulding, and as being
effective because they were the survivors of continued cycles of
extinction, faunistic replacements, and range spreads.

These discussions seem curiously archaic today.  We no longer
look for such simple all-embracing answers.  The Holarctic, because
of its sheer size would be expected to produce a high proportion
of new lines.  Horton (1974a) makes this point, noting that if two
areas, both with all their niches full, are exchanging species,
then more movement will take place from the area with the larger
number of species, simply because it has the larger number of
potential immigrants.  Horton also suggests that smaller areas are
likely to have unoccupied niches (in contrast to the larger ones),
making it easier for them to be colonized.  I suspect, however, in
the New Zealand and Australian cases noted by Darwin it was not a
case of unfilled niches but recently changed ones (as a result of

human activity) that helped give the European open country and
farmland invaders a marked advantage.

The Holarctic "superiority" concept may thus be challenged on
several grounds, as it has long been by the Gondwanalanders like
Harrison (1924). It can now be appreciated that a large land mass,
per se, will only give rise to new forms if it also has isolating
barriers permitting initial differentiations and sufficient consistency of conditions to enable the new forms to consolidate.
Thus some of the Holarctic's "dominant" forms apparently arose in
Africa (proboscideans, probably man) and North American (camels).
These are regional parts of Holarctica which have been intermittently semi-isolated or isolated.

Well documented cases of "world-fauna" forms being more
successful colonizers of island continents than the reverse include (a) the disproportionate colonization of South America by
North American mammals following completion of the Panamanian
landbridge and, (b) the one way interchange between Asia and
Australia. At the more restricted level Australia has contributed
more birds to New Zealand by over-water colonization than the reverse, just as have the forests of southeastern Australia to the
isolated ones of the southwest corner of that continent. Horton
(1974b) suggests that the American example can be explained by
differing climatic and habitat adaptations of the two sets of
forms. The question that can be asked, however, if it really was
harder for the rain forest - tropical savannah adapted Neotropical
forms to penetrate and establish in North America than for the
temperate North American ones to colonize the tropics or extend
through the tropical highlands to temperate South American beyond
if potential competitors were absent. Since "pioneer" species of
porcupines, tanagers, hummingbirds, and wrens have done so the
answer is possibly no.

The question of dominant forms cannot be said to be resolved.
It is difficult to completely discard the proposition that forms
that evolve in a highly competitive, faunistically saturated environment do not have some competitive advantage over those that
evolved in a smaller, peripheral area.


The Centers of Origin Concept

The centers of origin concept is basic to contemporary biogeography. It is based on the following:

(1) Many families and sub-familial groups are limited to
single parts of the World and there is overwhelming
evidence that they have been limited to that area

throughout their history; e.g. the various families of
Australian marsupials. There is no fossil record from
anywhere else, and the continent was isolated from the
Eocene onwards.

(2)   Patterns of radiation within such groups are spelled out,
      sometimes in excellent detail, in the fossil record and
      they lie wholly within the confines of that continent;
      e.g. the history of South American mammals (Patterson
      and Pascual, 1972).

(3)   There are many examples of such groups having later in
      their history, and as derived forms, secondarily colon-
      ized other continents; e.g. North America by ground
      sloths, armadillos, and porcupines from South America.

(4)   The further stage in the process, where the colonizing
      group has undergone radiation in their new home is
      amply illustrated by rodents in Australia, meliphagids
      in New Zealand, drepanidids in Hawaii, and so on.

(5)   Faunas are characteristically built up by waves of
      colonization from an outside source superimposed on
      original stocks, e.g. the contemporary Nearctic mammal
      and bird faunas, the South American mammal faunas, and
      the contemporary Australian bird and mammal faunas.
      Allowance should be made here, of course, for the fact
      that outward colonization will take place equally from
      areas of divergence, or survival, in addition to origin.

(6)   Allopatric replacement of faunal assemblages, genera,
      species, and races, is one of the fundamental facts of
      animal geography. Geographic speciation, now univer-
      sally accepted, demands differentiation in isolation,
      and secondary range spread from there. Monophyly re-
      quires that each form arise somewhere, all cannot arise
      everywhere.

(7)   The isolation of faunas by continental movements, clim-
      atic expansions and contractions, and physical barriers
      provides ample opportunity for the origin of new forms.
      Different histories and degrees of isolation explain
      how the various continents and regions developed strong-
      ly endemic biotas at various times in their histories.

Recently Croizat et al., (1974:276) have challenged the
centers of origin concept: "The majority of naturalists today
accept concepts such as center of origin as foolproof fundamentals
of biogeography without having much understanding of their history

and real meaning, and without any awareness of the conflict in
Darwin's own views (vicariance versus center of origin and dispersal).  Having failed to dissect these concepts (center of
origin, vicariance) to their core, contemporary zoogeographers
founder in a self-created morass of chance hops; great capacities
for, or mysterious means of, dispersal; rare accidents of over-
sea transportation; small probabilities that with time become
certainties; and other pseudo-explanations."  These criticisms are
based on the observation that biogeographers are frequently unable to find centers of origin for groups and sometimes spend far
too much time worrying about origins of groups.  These are perfectly valid criticisms.  There is also ample evidence that faunas may
become fragmented and many evolutionists express doubt about their
materal; e.g. Olson (1971:758) stated that "The concepts of
centers of origin and dispersal are deeply ingrained in biogeographic thought and supported by so much evidence in the best
known cases that other concepts have received little attention.
Yet in many specific cases the nagging questions of what the
center was or whether there was a center do arise.  What is commonly seen in the fossil record seems to suggest that evolution
always occurred somewhere else."  Examples introduced by Croizat
et al. (1974) to illustrate their points include the inability of
Mayr (1946) to suggest centers of origin for 30% of the North
American bird families in his analysis of that fauna.  It would
seem to this writer, by contrast, remarkable that he was able to
do so for 70% of them.

     The work of Croizat (1958, 1964) is based on the concept that
contemporary (and fossil) distribution patterns result largely
from the geological fragmentation of ranges, supplemented by some
dispersal.  As will be shown later such a concept has applicability
in some cases only.

     Earlier biogeographers debated such subjects as to whether
"successful" lines and new morphological types were more likely to
arise in the center, or periphery, of a group's range.  Since
isolation, rather than just selective pressures alone, is the
important factor initiating differentiation, the arguments have
little relevance today.  The limited gene pools of founder populations may lead to the rapid evolution of bizarre types:  such
"founder populations," however, have little future if subsequently
exposed to mainland stocks.


                   'Numbers Clues" in Biogeography

     Darlington (1957:31) notes that a commonly used pattern in
biogeographic analysis is to "find the place where the largest
numbers of genera or species of a given group of animals now occur

and take that as the place of origin of the group.  In doing this
they tacitly assume that genera and species usually increase in
numbers at about the same rate everywhere."  The concept must be
used in conjunction with degree of differentiation achieved in an
area.  If a solitary but well differentiated genus occurs well
away from the rest of the group it suggests an origin, or early
colonization there.  This could reverse the 'numbers clues'.
Additional clues as to area of origin are the extent of area
occupied by the group today, it being inferred that the area
occupied by a group increases proportionately with its age, continuity of area occupied, and the distribution of related, or
competing, families.  The rationale behind all this stems from
the innumerable cases of families being known from one continental
area and having radiated extensively there (e.g. Antilocapridae in
North America, Timaliinae in southeast Asia), and other cases
where groups have, obviously, only in later times colonized
another land-mass and set up a bridgehead there; e.g. motacillids,
alaudids, and sylviid warblers, in Australia.  By contrast, the
less the disparsity in numbers of species between two continents
and the older a group, geologically speaking, the less reliable
deductions will be.

    Where a group is equally developed on two, or more, land
areas the worker is not justified in trying to apply this technique.  The fossil record, as noted, shows just how misleading
contemporary distribution patterns can be in determining origins.

    The 'numbers clues technique' for determining areas of origin
obviously has its limitations.  On the other hand, used in modera-
tion, and caution it can be and has been successful (Mayr {1964}
on the origins of the Nearactic avifauna) especially where there
is such accessory data as habitat specializations.  In groups
lacking any sort of fossil record it may be the only approach
possible.

ORIGINS OF CONTINENTAL BIOTAS

Disjunct Biotas

    These have been a basis for discussion and speculation since
the beginnings of biogeography.  Were they to be explained by
former land connections, long distance dispersal, or the shifts
of climatic belts?  The acceptance of continental drift has led
to a reconsideration of pantropical disjuncts (e.g. Axelrod,
(1970); Raven and Axelrod, (1974); distributional links between
Europe and North America in the Eocene (e.g. McKenna, 1975),
southern disjuncts (Keast, 1973); and the relationships of disjunct families and genera of vertebrates (Cracraft, 1974b, and

elsewhere).  At the same time many disjuncts are not attributable
to land fragmentation alone and in large measure must be explained
by climatic change, range expansion and range shrinkage.  See
many maps of plant disjuncts in Good (1964), Thorne (1963, 1972)
and Fig. 6.  The same obviously underlies the disjunct distribution patterns of certain beetle genera and their food plants on
either side of the North Pacific, or between southeast Asia and
the southeastern United States (maps in Linsley, 1963, and Fig.
7).  Note also the many cases of isolated Andean and Rocky Mountain distributions described in Raven (1972), and elsewhere, and
isolation on African mountain tops (Hedburg, 1971).  Most of these
cases can be attributed to long distance dispersal (Fig. 7).


Continental Drift and Biogeography

There has been a tremendous literature on this in recent
years (for reviews see Keast, 1972c, 1973; Cracraft, 1973, 1974b;
Jardine and McKenzie, 1972; Raven and Axelrod, 1972, 1974).  These
have been devoted to reinterpreting Mesozoic origins and relationships, and thence Tertiary and contemporary distributions.  Many
conclusions necessarily remain tentatively at speculative and there
is a great need for new phyletic studies.

Families that originated after the Eocene have not been influenced by continental movements, except incidentally.  Early
mammal evolution took place pre-drift.  Apart from the trans-North
Atlantic dispersals described by McKenna (1975) for the Eocene,
and the presumed trans-Antarctic distributions of marsupials prior
to the breakaway of Australia about 47 million years ago (McGowran,
1973; other references in Keast, 1976) continental movements have
not been important to later mammalian distributions and development.  Note, however, the arguments that ancestral caviomorph
rodents and primates may actually have reached South America
across a steadily widening South Atlantic in the late Eocene
rather than from North America (Lavocat, 1969; Hoffstetter, 1972;
Hershkovitz, 1972).

Drift may, after all, help explain the close relationships
of the world's rain forest floras.  Raven and Axelrod (1974) make
the interesting point that although Africa and South America
separated in the middle Cretaceous by the latest Cretaceous the
steadily widening gap would only have been about 800 kms, equal to
that separating Jamaica from North America, an island that has
built up a comprehensive flora by trans-oceanic dispersal.  However, the lower sea levels of the Pleistocene would have reduced
the isolation of Jamaica to 300-400 kms; this was not true of
Africa relative to South America.

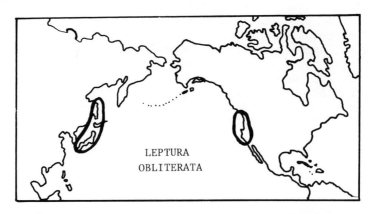

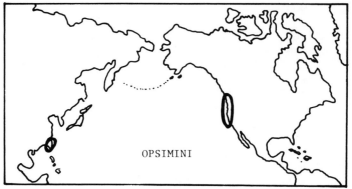

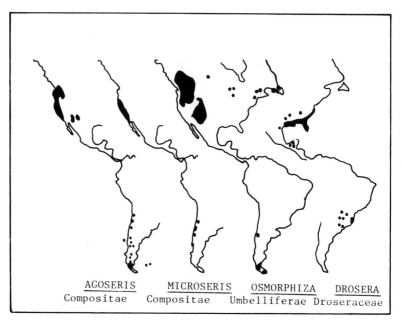

Fig. 7. The top two figures provide examples of trans-Pacific disjuncts taken from the paper by Linsley (1963) on the cerambicid beetles and their food plants. These, many of which involve south-eastern, rather than western North American, forms are examples of trans-Bering colonizations and then southwards shrinkage with the appropriate climatic belts. (Note the evidence of Solbrig, 1972, that some such relicts formerly had a continuous distribution through Asia and that a trans-Bering hypothesis does not apply to all such cases.)

The Pacific was formerly a much larger ocean and progressively shrank through the latter half of the Mesozoic as a result of the westward drift of the Americas and eastward drift of northern Asia (Larson and Chase, 1972). Alaska and Siberia achieved their present proximity about 63 million years ago (Pitman and Talwani, 1972).

The bottom diagram shows four pairs of amphitropical disjuncts that have differentiated at the species level. Raven (1963), in a compilation, found that there were 30 species of bipolar or high-latitude plant disjuncts and about 130 temperate, plus a substantial number of desert, ones. Since the floras of these two regions must have been distinct since the Middle Cretaceous and are very different at present it was concluded that the great majority of these plants must have reached their disjunct areas by long-distance dispersal relatively recently, and that the Pleistocene was the most likely time. Long distance dispersal is more characteristic of plants than animals, since they can disperse by a single propagule. Redrawn from Wood (1972).

## Corridors, Filters, and Sweepstakes Routes

In his discussions of paths of faunal interchange Simpson
(1940, 1965) noted that the probability of spread of a group of
animals from one region to another may have any level from nearly
impossible to nearly certain, depending on the geographic and
other environmental conditions between the two regions.  Stressing
that his scheme did not involve sharp distinctions he allowed
three pathways and levels of probability:  corridors, filters,
and sweepstakes routes.  A corridor is a route along which the
spread of many or most of the animals of one region to another is
probable (e.g. the Tertiary 'corridor' that permitted a common
mammal fauna from Europe to China); a filter is a route which
allows the passage of certain adaptive types (e.g. the Bering and
Panamanian landbridges); a sweepstakes route is one across which
spread is highly improbable for most animals (e.g. the trans-
oceanic colonization of Hawaii and Madagascar).  The last is a
formidable barrier that is only sometimes crossed.

The division remains an important one in contemporary biogeography.  It has recently been subjected to a modern analysis
by McKenna (1973) who has examined it in the new plate tectonics-
continental drift framework.  He provides a list of corridors,
filters, and sweepstakes routes, as they apparently existed from
the late Mesozoic onwards.

## Long Distance Dispersal and Chance

All living forms have a dispersal phase in their life history.  Some species (especially flightless ones, rain forest
species, etc.) are extremely local but one of the basic adaptations of island forms is an exceptional adaptation for dispersal
(at least initially) - see Carlquist (1974) and Diamond (1975).
Along with the ability to cover long distances equally important
is establishment potential, whether or not the recipient area is
already faunistically saturated, and so on (Fig. 8).

Botanists have on the whole been more concerned with long
distance dispersal than zoologists (Cain, 1944; Good, 1964;
Carlquist, 1974; and the various papers in Gressitt, 1963).  Much
of the disagreement, particularly amongst zoologists, surrounds
just what is, and what is not, possible.  Simpson (1952) suggests
that given enough time almost anything is possible while Axelrod
(1952) noted that changing climates would vary probabilities.  A
recent worker who feels as does Simpson is Carlquist (1974:12)
who has written "Among organisms for which long-distance dispersal is possible, eventual introduction to an island is more
probable than nonintroduction."  Commonly invoked to help explain

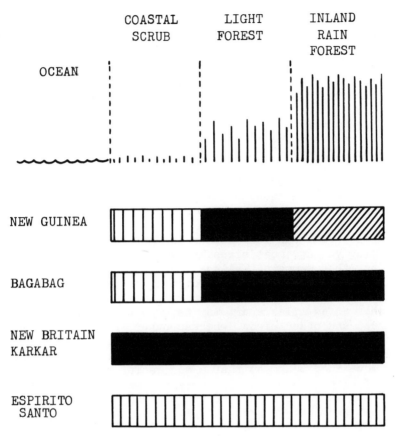

Fig. 8. Dispersal and island colonization, indirect evidence of the effect of the presence/absence of a competitor on habitat oc- cupation. Three species of ground doves occur in different habi- tats in New Guinea as follows: Chalcophaps indica (vertical bar) in coastal scrub; C. stephani (solid shading) in light forest, and Gallicolumba rufigula (diagonal bars) in inland rain forest. The last of these is absent from the islands. On two of the groups single species occupy the full habitat spectrum but on Bagabag Island two of them divide it. Since from their distributions the two Chalcophaps are obviously equally good colonizers these patterns are evidence that the presence of an efficient competitor can slow and presumably prevent the successful establishment of an immigrant on an island. Being able to establish is a basic component of dis- persal. Redrawn from Diamond (1975).

specific problems are rare hurricanes and equivalent phenomena;
e.g. Hedburg (1971) considering the colonization of isolated
African mountain tops.  Transport of seeds on the feet of birds
and passage on the surface of the sea have also been documented.

Careful analysis of the faunas of such islands as Hawaii and
the sub-Antarctic islands in recent years has given us a fairly
good idea of which insect groups can, and which cannot establish
themselves across a thousand miles of ocean (Zimmerman, 1948;
Gressitt, 1963).  Of particular importance here are studies
whereby insects were netted from the jetstreams and by ship and
air far out over the Pacific (Gressitt and Yoshimoto, 1963)
although, unfortunately, the netting procedure used did not discriminate between living and dead insects (Gressitt, personal
communication).  The fauna of the sub-Antarctic islands south of
New Zealand must have been built up by long-distance dispersal
since these were glaciated (Fleming, 1962; 1975).

Bird dispersal has long been a subject for discussion.
Despite being volant, most birds are as subject to distributional
barriers as other mammals (Wallace, 1876).  Nevertheless, North
American migratory birds are periodically swept across the
Atlantic (the reverse is rare).  Mayr (1942) notes that Pacific
Islands only a few miles apart may have quite different species
of Zosterops and other birds, whereas other species have very
wide ranges through group after group.  He noted three common
characteristics (adaptations) of these successful colonizers:
(1) they move in flocks (so that a successful propagule does not
find itself without a mate); (2) they are generalists so far as
food and habitat is concerned; (3) they tend to move into the
air (so that they become prone to being caught up in hurricanes,
etc.).  Diamond (1975) in the course of distributional studies on
the islands off New Guinea and the Solomons has documented large
numbers of cases where ecologically equivalent pairs of species
replace each other on adjacent islands, thus emphasizing how a
competitor may prevent successful colonization (Fig. 8).  Ricklefs
and Cox (1972) illustrate how island colonizing birds may go
through a series of stages whereby, following an initial one when
they have high colonizing ability and are ecologically versatile,
they may develop the characteristics of extreme sedentarinesses,
restrict themselves to the central forests, and develop into
endemic species.

These discussions of the potential efficiency of plant, insect, and bird dispersal does not, of course, get away from the
fact that whole groups of animals have only the minimal capacity
for overseas dispersal.  Salamanders and "primary division"
freshwater fishes, and to a lesser degree snakes are examples
amongst the vertebrates, as their absence from islands shows.

Thus it can be stated that the primitive frog Leiopelma and the rhynchocephalian Sphenodon must have reached New Zealand at a time when it was joined to Antarctica. (The same is presumably true of Nothofagus, whose seeds cannot survive brief immersion in salt water - Preest, 1963) (see Fig. 9).

Obviously, notwithstanding continuing argument long distance dispersal accounts for many contemporary distributions. It is one of the major ways in which faunas are built up. The New Zealand biotas, for example, is a combination of old, original, archaic elements (Fig. 9), and forms that have successively arrived by colonization from elsewhere (Fleming, 1975).

### ATTEMPTS AT A BETTER QUANTIFICATION OF HISTORICAL BIOGEOGRAPHICAL KNOWLEDGE

#### The Measurement of Faunal Resemblances

Measurement of faunal resemblances was introduced by Simpson (1960, 1965) as a means of calculating the degree of relationship between different faunas. A widely used, very simple, formula (Simpson, 1960) for comparing two regional biotas is:

$100 \frac{C}{N}$ which equals the % of members of the smaller fauna also present in the larger fauna.

Extensions of the approach involve the construction of similarity matrices (Cheetham and Hazel, 1969); these authors also comment upon the various coefficients given by Simpson. Important in the introduction to similarity coefficients in the determination of natural areas is the paper of Webb (1950). Peters (1971) reviews this and other literature and recommends an improved method that involves ranking similarity coefficients and adding additional information to provide data on the relationships of each unit with all other units involved in the analysis.

#### Numerical Analysis of Regional Faunas

In recent years several authors have introduced numerical analysis to distributional studies. Hagmeier (1966) and Hagmeier and Stults (1964) have developed printouts of the distribution patterns of North American mammals based on data in Hall and Kelson (1959). They divided the continent into 2,490 fifty mile square blocks, developed indices of faunal change between these, and then attempted to discover the degree of relationship that

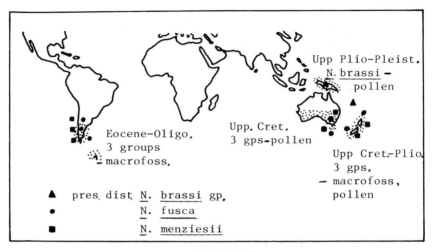

NOTHOFAGUS

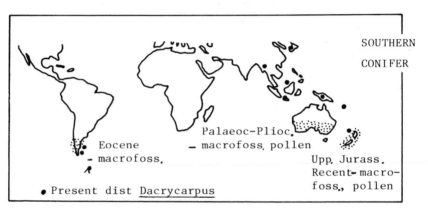

DACRYCARPUS

Fig. 9.  Contemporary, and fossil distributions of two groups of
southern cold temperate disjunct groups, Nothofagus (above), and
the southern podocarps (Dacrycarpus section).  The evidence indicates that these groups have been southern throughout their history and records of fossil Nothofagus pollen in Russia are generally discounted.  They are hence taken as examples of the various
groups that had a continuous southern range prior to drift.  Redrawn from Keast (1975) and taken from the various sources listed
therein.  For data on the Galaxiidae see McDowall and Witaker
(1975).

existed between the resultant series of mammal provinces (Fig. 10).
Interesting distributional patterns emerged from this work.
Peters (1971) has criticized this approach on the grounds that the
provinces so derived were used for subsequent work regardless of
their "validity as biogeographic entities."

Kikkawa (1968), and Kikkawa and Pearse (1969) investigated
the ecological associations of bird species and habitats in eastern Australia by means of similarity analysis.  Subsequently
avian distributional patterns throughout the continent were
studied using divisive informational analysis.  They selected
their 121 study sites so as to include all niches available in
each area.  Their technique permitted the combination of sites
into groups to calculate the heterogeneity of each.  The work
helped confirm distributional limits of arid and forest-adapted
avifaunas and the location of such distributional barriers as the
Hammersley-Kimberley arid gap indicated by distributional and
speciation data (Serventy and Whittell, 1962; Keast, 1961).

Holloway and Jardine (1968) used cluster analysis and non-
metric multidimensional scaling to study the butterflies, birds,
and bats of the Indo-Australian area.  Faunal dissimilarity coefficients were developed for the various intermediate islands
and primary areas were defined (Fig. 11, 12).  Peters (1971) and
Vuilleumier (1975) comment on these interesting attempts to re-
fine and quantify descriptive biogeography.

Deductions from Replicated<br>Distribution Patterns - Croizat's Approach

Croizat (1958, 1964) has been read fairly widely by botanists
but largely ignored by such leading zoogeographers as Simpson,
Darlington, and Mayr.  In recent years Nelson (1975, 1974a) and
Rosen (1974a), both ichthyologists, have advocated Croizat's views
(see also Croizat et al., 1974).  Croizat basically analyzes
large numbers of individual distribution patterns ("tracks"),
which, he suggests, fall into a series of basic types ("multiple
tracks") revealing common ancestral distribution patterns that
antedate contemporary distributional barriers.  The concept is
thus partly a "static" one and he sees major groups as having had
wide ranges to continental fragmentation ("vicariism").  Croizat
insists that by patient and impartial mapping and superimposition
of distribution patterns the origins of groups and faunas reveal
themselves.  The concept of originally widespread faunas that
have become fragmented contrasts strikingly with the Darwinian
and post-Darwinian ideas of groups arising at point sources (i.e.
on individual continents or areas thereof), and from there colonizing outwards.

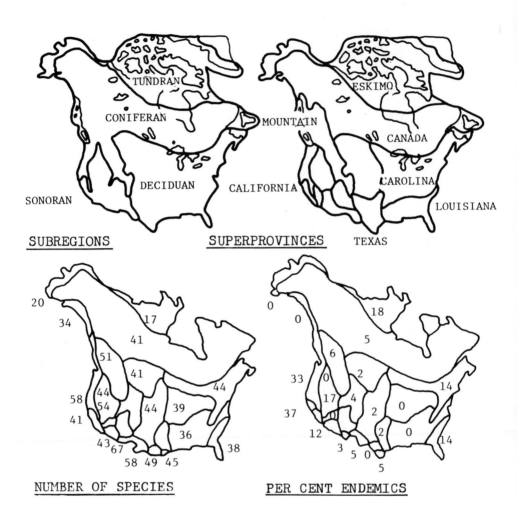

SUBREGIONS          SUPERPROVINCES

NUMBER OF SPECIES          PER CENT ENDEMICS

Fig. 10. Distributional patterns of North American mammals redrawn from Hagmeir and Stults (1964), and Hagmeir (1966) based on distributional data in Hall and Kelson (1959). The printouts are based on indices of faunal change between 2,490 fifty mile square blocks. The maps, which are selected from a series show:  A, subregions; B, superprovinces; C, numbers of species and; D, percent of fauna endemic to the provinces. "Subregions" and "superprovinces" represent decreasing levels of difference, and are defined mathematically. The work was one of the first attempts to develop a numerical analysis of regional faunas.

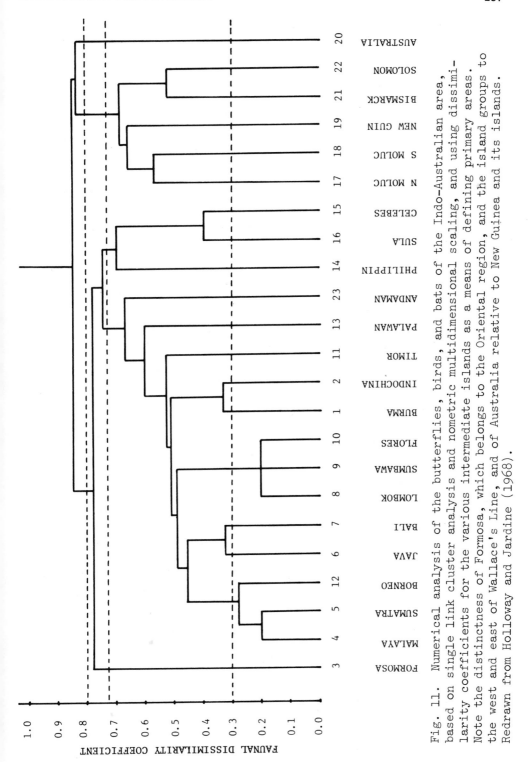

Fig. 11. Numerical analysis of the butterflies, birds, and bats of the Indo-Australian area, based on single link cluster analysis and nometric multidimensional scaling, and using dissimilarity coefficients for the various intermediate islands as a means of defining primary areas. Note the distinctness of Formosa, which belongs to the Oriental region, and the island groups to the west and east of Wallace's Line, and of Australia relative to New Guinea and its islands. Redrawn from Holloway and Jardine (1968).

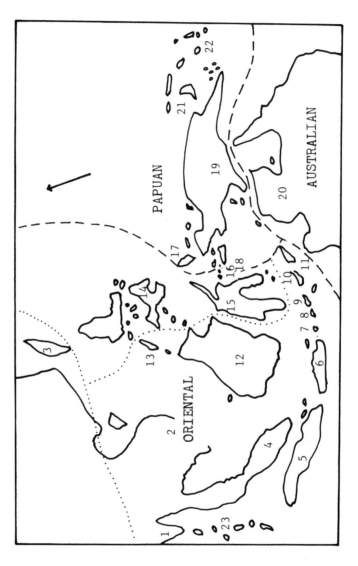

Fig. 12. The main biogeographic divisions and distribution of the island groups relative to this, as per the data in the previous figure.

   The strength of the Croizat method, it is suggested, lies in
its impartial avoidance of assumptions, and the fact that conclusions are based only on "averages".  Rosen (1974a) states that
the results are testable and Nelson (1973) that they are potentially amenable to mathematical procedures.  This, it is stressed,
contrasts with the lack of exact methodology in the case of
"dynamic biogeography" and the inability of its proponents to deduce generalizations.

   I agree with Cracraft (1975) that the Croizat approach does
have the capacity for improving historical biogeographic analysis,
and that it has, and will, force biogeographers to think in terms
of biotas rather than individual species, as well as original
histories.  There are, however, both errors in Croizat's analysis
of individual cases and basic flaws in his method.

   Figure 13 provides an example of Croizat's approach.  It involves the origin and history of two groups of finches, the
Fringillidae and Estrildidae and is taken from Croizat (1970).
The Fringillidae today occur in the Americas, Eurasia, and Africa,
the Estrildidae in Africa and Australia (twin radiations) and have
a range of species in the intermediate Southeast Asia area.
Croizat (1970:284) suggests that Africa is the "Keystone of the
distribution" of these groups where a "common arch-ancestor
occurred" not later than about mid-Cretaceous" when Africa was
part "of the as yet unsundered Gondwanic landmass."  Subsequently
the ancestral stocks of the Fringillidae "streamed northwards
veering east to eventually reach the New World, avoiding warmer
Asia and Malaysia to Australia."  He suggests that the Carduelinae,
79% of whose species are today North American" 'departed from the
Cape' in the Eocene.  The Estrildinidae, by contrast early in
their history moved eastwards.  Madagascar was not colonized because it had already become isolated in the mid-Jurassic.  In each
case descendent forms evolved and radiated in situ, vide in the
fringillids there are three "centers" represented by <u>Serinus</u> in
Africa, <u>Carpodacus</u> in southern Asia, and <u>Carduelis</u> in the
Americas.  These basic types extend back to the Eocene by which
time they were already deployed along an arch "which would take in
Capetown, Irkutsk, and Mexico City" today.

   Several weaknesses in this analysis are obvious.  The fringillids could equally have evolved in Asia or North America: in
fact some workers suggest a New World ancestor and believe that
the African members represent a later arrival and radiation.  The
Estrildines could equally have originated in Australia, early or
later.  The fact that most avian colonization has been <u>into</u>
Australia would argue against this.  However, it is a reasonable
supposition that the two families arose on different continents.
If one accepts the newer drift data that Australia only approached

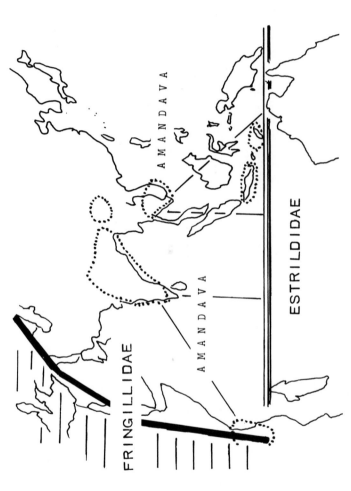

Fig. 13. The main trends of dispersal in the fringillid and estrildid finches derived from "multiple track" analysis (redrawn from Croizat, 1970). The former show one kind of "track", the other a different one. The distribution of _Amandava_ (Estrildidae) is shown as it is the lone genus to be "dispatched" beyond Africa. Since Croizat rigidly excludes "chance dispersal" and "storm driven immigrants" as a dispersive mechanism it is not clear how _Amandava_ reached its present range. He suggests alternatively that "...Pemba, Zanzibar and Mafia 'open the way' to a link between East Africa and India" (p. 288), or that "an alternative bond may be considered with a link directly across the Indian Ocean to a point between Java and Timor" (p. 289). Whether or not this is seen as a former land connection is not stated but presumably it is. However, no such existed in later geological times.

Asia in the Miocene the Australian occurrence and radiations are
of Pliocene age, or younger.  The marked similarity of the
Australian and African estrildids make it impossible that the two
could independently be derived from a proto-ancestor that occurred
prior to the mid-Cretaceous isolation of Africa from Gondwanaland.
The second major problem is the time-scale.  By over-emphasizing
the "disjunction" aspect Croizat is forced to postulate an ancestor far back in the Middle Cretaceous.  In reality, an overview of
the avian fossil record suggests that these small finches go back
no further than the Middle Tertiary, certainly much of the radiation is very late.

    One must be fair to Croizat and stress that much of the
interpretation is in the form of a series of suggestions.  There
is also the possibility of distortion where, as here, a reviewer
elaborates on a single example.  (Various other examples in this
1970 paper do actually have equal shortcomings).  The real point
is that this particular example of the use of the method (a) involves conclusions that are arbitrary, without substance, and
almost certainly wrong, (b) lacks quantification, (c) is based on
a time scale that is wrong, (d) fails to consider and assess valid
alternative hypotheses.

    Weakness at the individual case level need not, of course,
weaken the method but here a number of general criticisms must be
levelled:

    (1)  Similar distribution patterns may result from quite
    different sets of factors; e.g., the distribution of a
    group could equally indicate habitat and climatic zone pre-
    ferences as early history.

    (2)  Except possibly in the case of specialized peripheral
    relicts like the "southern cold temperate disjuncts" (e.g.,
    Nothofagus, Fig. 9) it is doubtful if contemporary patterns
    can be expected to reflect late Mesozoic historic patterns in
    more than a few cases.  In 60 million years of evolution
    since the late Cretaceous, for example, ranges must have been
    repeatedly disrupted by climatic shifts and the origin, radia-
    tion, and extinction, of new competitors.

    (3)  The whole question of dispersal is inadequately resolved.
    Much of the data throughout is presented strictly from the
    viewpoint of vicariances and whilst the term dispersal is
    freely used no real effort is made to balance one against the
    other.

    (4)  There are problems in the interpretations of the "gener-
    alized tracks" as Croizat himself acknowledges.

The major point of disagreement with Croizat must continue to be, of course, the basic concept that dispersal is of lesser importance than disjunction.

If the Croizat approach, then, does not supply the ultimate method for analyzing distributional patterns is it still possible to develop such? Any new approach must obviously fit dispersal into a proper perspective. Yet this is extremely difficult to do since not only do different groups vary greatly in their dispersal characteristics but ease of dispersal between different areas obviously changes in geological time. This writer argues hence, that it is not justifiable to try to come up with all-embracing generalizations. Rather analysis must proceed at the familial and generic level giving equal weight to original historic patterns, dispersal, distributional changes associated with climatic change, evolutionary changes in situ, and interrelationships with the associated biota. Note that even animals like oligochaetes can disperse far more effectively than their basic anatomy and physiology might suggest (Michaelson, 1922; Jamieson, 1973).

## The Cladistic Approach to Biogeography

Stemming from the initial concepts of Henning (1965, 1966), with elaboration by Brundin (1966) the cladistic ("phylogenetic") approach to classification has been extended to biogeographic analysis of the past histories of continents. The technique, that entails deriving classifications and histories by means of successively derived "suites" of characters, has, if widely applicable, a natural extension into biogeography. This will, of course, be particularly true if it offers the ultimate classification, as has been argued. The method has now been the subject of a considerable literature with successive authors discussing its philosophical basis, advantages, and limitations. Later in this volume Bonde will deal with the cladistic approach to vertebrate classification, but see, in addition, Mayr (1965, 1974), Winterbottom (1971), Darlington (1970, 1971), Nelson (1972, 1974b), Rosen (1974b, 1974c; and Cracraft (1974c, 1975).

Brunden (1966), Cracraft (1972), and others, suggest that the cladistic approach is superior to the classic evolutionary one in biogeography, it being based not on relative and difficult to define levels of similarity and difference but on the basis of precise descendent branch points. There remain, however, the problem of setting rank levels. Henning (1966) has introduced the convention of basing rank on relative age within the system. Fossils, as noted, may provide an absolute minimum age.

Several authors have now used the phylogenetic approach in biogeographic analyses of southern "cold temperate" groups (Brundin, 1966, with chironomids: Illies, 1966, stoneflies; Edmonds, 1972, 1975, Ephemeroptera; and Cracraft, 1974a, 1974c, with ratite birds).  In many of these cases it has been possible to correlate specific branch points and "character suites" with land separation events.  When, however, it came to analyzing evolution of groups within continents (e.g., Edmonds, 1975, with the Neotropical Ephemeroptera) the results were less tidy.

Cracraft's analysis of the phylogeny of the ratites (1974c) may be reviewed here as an example of the application of the cladistic method to historical ornithogeography.  It is a bold effort to develop an orderly history for this interesting group, but at the same time highlights, in almost "classic" form, the difficulties of being limited to incomplete fossil material and of trying to fit limited geological and other facts into a framework.

The contemporary ratites fall into 6 major groups (families sensu Cracraft): "primitive" Tinamidae (that alone retain the powers of flight), South America; Apterygidae (kiwis), New Zealand; Dromicidae (emus), Australia; Casuaridae (cassowaries), Australia; Struthionidae (ostriches), Africa, fossil in Asia; and Rheidae (rheas), South America.  In addition there are two ex- tinct groups: Dinornithidae (moas), New Zealand; and Aepyorni- thidae (elephant-birds), Madagascar; the latter having a much wider earlier range.  The ratites have long been held to be a case of southern dispersal.  The newer continental drift data makes this hypothesis both plausible and likely.

Figure 14 shows Cracraft's reconstructions of the phylogeny of the ratites in terms of descendent series of "character suites". A simple series of cladistic branchings is indicated.  The group is seen as having a monophyletic origin (see also Bock, 1963) with the Tinamidae being the most primitive members, followed by the New Zealand Apterygidae and Dinornithidae, than the Aepyornithidae, Dromicidae and Casuariidae, whilst the Rheidae and Struthionidae are the most advanced members.  Beyond the Tinamidae flight is lost and the wings are markedly but variously reduced with some secondary enlargement in the rhea and ostrich where they are used in display.

Cracraft's scheme is logical but nevertheless, has its pro- blems:

(1)  It is limited to the postcranial skeleton and poten- tially useful skull data is not introduced.

(2)  Some of the "character suites" separating the groups

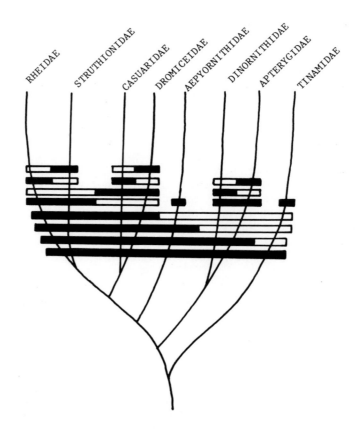

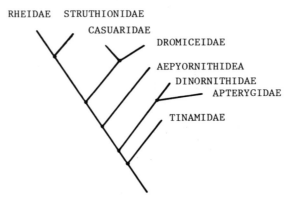

Fig. 14.   Theory of the relationships of the ratites, redrawn from
Cracraft (1974c).  Each monophyletic lineage is defined by a suite
of shared derived character-states whereas the coordinate sister-
group is characterized by primitive character-states (Cracraft).
The former is shown in black, the latter in white.  The bottom
figure shows a phylogeny derived from the above.

appear weak and are certainly difficult to evaluate.  Thus
the Aepyornithidae are characterized by being morphologically
very primitive and retaining a number of generalized features, e.g., the broad pelvis, and tibiotarus with the base
of the enemial crests only moderately constructed, the crests
not projecting greatly, and the posterior side of the internal condyle not strongly excavated to form a groove. This
is not a true diagnosis.

(3)  Only 7 of the 15 "character suites" separating the
various groups are described; here the reader has to look
through series of illustrations of segments of tibiae,
tarsometatarsals, etc., to confirm the author's deductions
from the positions and shapes of condyles, crests, ridges,
etc., a difficult task.  Just what weight should be given to
these characters, moreover, is uncertain in the absence of
knowledge as to their possible functional significance.

(4)  There are "philosophical" difficulties, in the absence
of any standard of reference, such as an outgroup comparison
(Hecht and Edwards, in this volume), in deciding which characters
are primitive.

(5)  Hennig (1966) stresses the great difficulty of trying
to use the cladistic approach with fossils from which the critical
parts are commonly absent.  This handicap comes through in the
present study.

(6)  Whilst the proposed order of differentiation of the different lineages partly accords with acknowledged separational
events (e.g., earlier separation of New Zealand than Australia) it
does not in other cases.  How, for example, did two such distinct
groups as the kiwis and moas arise on the small islands of New
Zealand, or did differentiation occur elsewhere, for example on a
larger Antarctic land-mass, prior to the separation of these
islands?  If the "initial step" in ratite evolution was flight-
lessness how could they have got from South America to Antarctica
in the first place?  Elliot (1972) first suggested that junction
may have persisted through to the Late Cretaceous or Early
Tertiary; subsequently, however, Dalziel et al. (1973) have suggested that the initial break may have occurred during the Late
Jurassic.  This latter date would have been well ahead of ratite
differentiation.  Where did the ostriches and elephant birds
arise?  How did the elephant-isolated in the mid-Jurassic to mid-
Cretaceous?  Presumably dispersal of the New Zealand, Australian
and Madagascar ratites must all have involved crossings under
ocean gap or, at least, archipelago conditions.  Again, if the
very close relationship of the ostrich and rhea is true (postcranial, and other morphology, as noted indicates that these two

were the last to separate), how does it come about that they
occur on continents that separated in the Middle Cretaceaous?

Workers such as Brundin, Illies, and Edmonds, using extant
material with its mass of available characters are at a distinct
advantage over Cracraft.  Likewise the capacity of these insects
to fly across water-gaps simplifies the dispersive picture.  However, various recent workers using the cladistic approach have
confined their analyses to single organ systems or areas of the
body.  Ferris et al. (1976), for example, working with nematodes
attempt to construct a phylogenetic history, involving an early
division of ancestral Gondwanan and Laurasian types on the basis
of the distribution of contemporary forms with constricted verses
unconstricted oesophagi.  Admittedly nematodes have only so many
good characters but it would still be comforting to have the
argument supported by more than one character, even if it is a
very fundamental one.  It should be noted that Henning (1965) has
stressed that the greater the range of characters used the less
the chances of error.

Finally, at the present stage most phylogenetic writings are
either preoccupied with theory (but not the methodology of
character analyses), or are directed only at the highest, and
therefore most distinct, taxonomic categories.  It remains to be
seen how effective the approach will be in resolving problems at
the family - genus level and some of the more interesting biogeographic ones such as the relationships and history of the Neotropical and Australian marsupials or the history of the lemurs.

## ECOLOGICAL BIOGEOGRAPHY, FAUNAL
## COMPOSITION AND ADAPTIVE ZONES

Faunal origins are only one facet of biogeography.  Several
others are of as great, if not greater, interest.  For instance
the analysis of mechanisms whereby species, groups, and faunas
maintain themselves, what combinations of morphological and
ecological types are present and why, and how faunas evolve as
interrelated assemblages of forms.  This, the ecological dimension
to biogeography has, as yet, only attracted a fraction of the
attention of the historical aspects.

## THE SCOPE OF THE PROBLEM, AND CURRENT APPROACHES

Many of the facets of ecological zoogeography have already
been noted.  There has been much work on local assemblages with
workers variously emphasizing the trophic, habitat, or social
aspects.  Studies of this type have been made on African herbivores,

bats, rodents, grassland birds, and freshwater and tropical
reef fish.  Considerable attention has been given as to how related species divide up living space (vertical feeding zonation,
habitat specialization), separate out in terms of foods eaten,
and invoke other strategies for minimizing interspecific
"competition."  There have been relatively few attempts at
looking at entire faunas and determining how different continents
compare in terms of species numbers, percentages of species
occupying the major adaptive zones, and the influence of vegetation types and history on faunal composition.  (See books by
MacArthur {1972} and especially that of Cody and Diamond {1975}
and papers of Cody {1968} and Diamond {1970, 1975} for comparative
data at the community level).  Keast, Erk, and Glass (1972) have
made comparative analyses of the mammal faunas of the three
southern continents, attempting to answer such questions as
whether the numbers of families, genera, and species are a function of area, and whether continental faunas are in a state of
'balance'.


                    The Simple Island Fauna

     That the associated fauna, along with the physical environ-
ment, has a major role in moulding evolutionary lines and the
ecological channelling of species is obvious although surprisingly
little attention has been given to it (Bock, 1970, 1972; Keast,
1972).  Clear demonstrations of the production of morphological
types to fill the roles of absent parrots, honeyeaters, creepers,
etc., are to be found in the Hawaiin avifauna.  However here, as
well as in the Galapagos and elsewhere, this radiation has not
impinged upon the adaptive zones of the few other groups that
have made the water crossing.  Island archipelagoe show that,
given enough time forms will evolve to fill vacant niches.  An
example is the development of a long-billed "nectarinid-like"
Zosterops, and a thrush-like bulbul in Mauritius to fill the
niche of absent sunbirds and thrushes.  Moreover, isolated islands
are obviously characterized by an equivalent range of basic
morphological types:  one or two parrots, a couple of pigeons, a
swift, a flycatcher or two, a couple of warblers, a thrush, a
long-billed nectar-feeder and so on.  The pattern has not yet
been subject to quantitative analysis but it is true for islands
as far apart as the Mascarenes, Samoa and Tonga, and St. Lucia
and the other Less Antilles.  For a better background to this kind
of problem see the discussion of "assembly rules" and "incidence
functions" as determinants of species occurrences on islands near
New Guinea (Diamond, 1975).

     Later in this volume Sondaar will consider insular evolution
in large mammals which is very different to that found in birds.

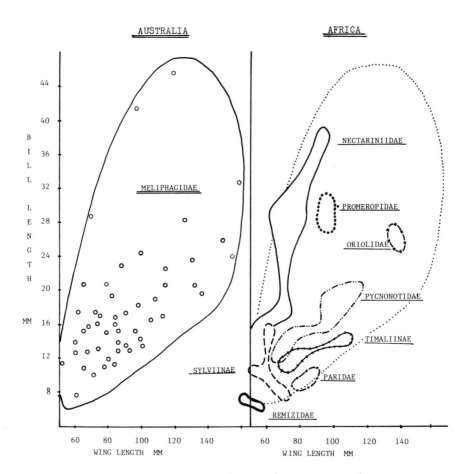

Fig. 15. The Australian Melophagidae (honeyeaters), a group that,
to judge from the unusually wide range of adaptive zones and niches
occupied, was an early inhabitant of this insular continent and was
able to radiate the absence of closely-related competitor groups.
Comparative ecological and behavioral observations show that it
occupies, or partly occupies, the adaptive zones of several African
families.  The inference is that there has been a continuing interchange between Africa and the North with newer specialized families
arriving and leading to continued resorting and finer division of
opportunities at the familial level.  This figure "contours" the
Meliphagidae and a series of African counterparts and part counterparts in terms of two adaptive characters, wing length (a general
indicator of size), and bill length which linked to kind and sizes
of food taken.

### The Eco-Morphological Approach To Faunal Analysis

The more complicated continental faunas present a wide range
of analytical problems.  One that merits fuller study is just how
morphologically different species have to be for coexistence to be
possible.  This is based on the appreciation that there is likely
to be close correlation between morphological structures and way
of life and that this extends not only to the major guilds (e.g.
in the case of birds, glycatchers, as against warblers or finches),
but down to the species level.  Species analyses gained impetus
from the observation of Schoener (1965) that sympatric species of
birds differed in bill size by a minimum of about 10% and Bock
(1966) demonstrated the functional importance of bill shape.  It
should also be noted that island populations of species very
commonly show increased bill and tarsus length.  The reason for
this has defied explanation.  Sometimes, however, it is linked to
ecological shifts (Keast, 1971).  Recent years have been attempts
not only to identify morphology with adaptive zone and niche but
to compare faunas in terms of numbers of species with particular
kinds of adaptive morphological attributes (e.g. Keast, 1972
(Fig. 15); Lein, 1972; Wilson, 1973; Karr and James, 1975; see
Fig. 16).

"Morphological distance" is of considerable interest as a
possible indicator of how close the morphology of sympatric species
can be.  The decision has to be made at the beginning of the
analysis as to whether the emphasis is to be on the multivariate
approach and a series of characters in combination or whether the
investigator wishes to emphasize single, highly adaptive characters such as the bill or dentition.

### Habitat Utilization and Separation in Continental Faunas

The individual group or species will interact with the other
members of a continental fauna only to the extent that it is in
physical contact with these.  In Africa in particular, however,
not only are the various mammal species linked with specific
habitat types (i.e. there are desert, savannah, and rain forest
assemblages) but whole faunas may be partly duplicated to the
North and South of the continent.  Examples are the southern
African grassland-desert assemblage (characterized by Antidorcas
marsupialis, Damaliscus pygargus, Pelea capreolus; the east
African assemblage (characterized by Gazella granti and G. thomp-
soni, and the Somali arid assemblage (characterized by Ammodorcas
clarkei and Gazella pelzelnii. Habitat specializations may be
looked upon by the ecologist as a 'device' permitting a greater
number of species to occur.  A faunal evolutionist will note that
many species are now confined to a restricted area of specialized

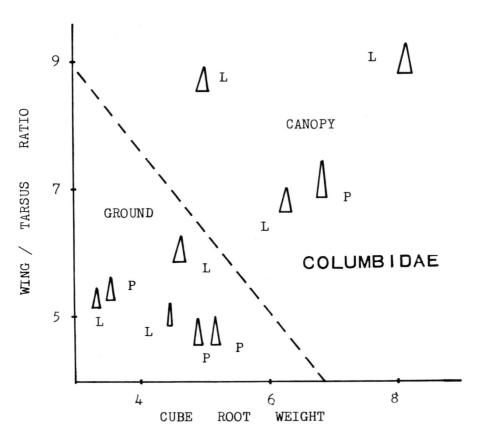

Fig. 16. Ground and canopy-feeding pigeons of Liberia (indicated
by "L") and Panama ("P"), compared. Each triangle refers to one
bird species. The position of the triangle gives the wing/tarsus
ratio and the cube root of the weight; the base and height of the
triangle are proportional to bill width and bill length, respectively. Note the occurrence of two pairs of morphological counterparts
in the ground - and one in the canopy-feeding assemblages, but with
some differences in bill shape. Redrawn from Karr and James (1975).

(and commonly shrinking) habitat, and that habitat and regional
faunas may mean partial duplications of whole assemblages.  Except
in the case of generalists habitat separation it may be stressed,
means that a species or group at any time is under the "moulding"
influence only of the associated biota of that habitat type.

Whilst there are cases of whole "evolutionary lines" being
restricted to a limited region or habitat (e.g. rain forest) this
is rarely true of the more successful lines.  Here genera and
species are evolving simultaneously in the different habitats.
Accordingly, here interactions are with the whole spectrum of
faunal types in the continental assemblage.  In the longer-term
evolutionary (geological) context habitat specializations can be
looked upon as a temporary expedient for a more efficient use of
ecological opportunities.  Species interactions in habitat assemblages can be relegated to a temporary feature of longer term
adaptive interactions.

Species Separation, Feeding and Different Feeding Strategies

The existence of a series of basic ecological opportunities
(carnivore, large herbivore, anteater, etc.) is reflected in different feeding 'guilds' within biotas.  In the longer term lineages
are groups that can be defined in terms of the adaptive zones they
occupy.

Diet separations and "devices" serving to minimize interspecific food competition within communities and faunas have been
treated by many authors.  The hundred-odd species of large herbivores in Africa, for example, "separate out" as follows:  browsing
at different heights; being predominantly grazers against browsers;
undertaking different degrees and kinds of seasonal feeding movements; and in terms of habitat.  Stomach and fecal analyses, plus
the visual identification of plants being eaten show, however, only
partial separation in terms of the species of food plants eaten.
Fish studies in small Canadian lakes, by contrast, show a high
degree of dietary separation, moderate degree of habitat separation, and differences in diurnal feeding periodicity.  Major
structural differences, moreover, channel the species towards different foods and ways of life (Keast, 1970).  The Great Lakes fish
fauna has a dual origin, post-glacially being a mixture of cold-
adapted forms from the north and glacial front lakes and warm
water species from the Mississippi-Missouri system to the south.
These different adaptations permit a degree of "temporal" separation, especially in terms of optimal temperatures for feeding and
growth spawning (Keast, 1970).

Detailed feeding studies show, however, the feeding interactions of cohabiting fish species to be, in reality, enormously

complex with a whole series of different strategies being used
simultaneously.  Since they are cold-blooded animals, limited to
3-4 months of growth a year in Ontario, the longer lived species
have several size cohorts simultaneously present in the system.
Intraspecific competition between these is a potential threat to
the species, hence "adaptational effort" must be directed to
minimize this.  When the feeding of size (age) groups is added to
the picture it is found that, in addition to the basic separations
in food type referred to above, the following "ecological categories" of fish occur:  (1) generalists, species that at any time
take a wide range of foods (Perca flavescens);  (2) specialists
that are limited to 3-4 major items (Notropis heterodon);
(3) species that change their diet strikingly as they grow, e.g.
Ambloplites rupestris undertakes three shifts, each of which is as
distinct from the other as are those separating species;
(4) species that begin life as generalists and end up as specialists, (Lepomis gibbosus; Micropterus salmoides); and (5) opportun-
ists that go from one "superabundant" resource to another, a
successional pattern of feeding that greatly minimizes the chances
of both inter- and intra-specific competition.

    Thus feeding specializations and separations within biotas
and communities can be both highly refined and complex.  Strategies are closely linked to the opportunities of the particular
seasonal and biotic environment as well as the "threat" of potential competitors.  In the longer term evolutionary sense there is
far more to group and species interactions, diversifications and
extinctions, than a crude appraisal of the geological record and
successional occurrence of different sets of morphological types
might suggest.


        Continental Faunas and Different Levels of Maturity

    Consideration of the mammal faunas of different continents
shows that complexities are not limited to subdivisions into
faunal assemblages and the interactions between species within
these, but to the entire faunas.  Faunas may, for example, be
simple or complex and be at different stages of evolution.  This,
as will be seen, has wide implications both in terms of moulding
the phylogenics of contemporary lines and in highlighting the kind
of 'ecological bottleneck' that evolutionary lines must have passed
through in their history.  Accordingly the subject of faunal matur-
ity or "sophistication" merits treatment in some detail.  It could
be illustrated in various ways but an effective approach might be
to concentrate on the large carnivore guild since this represents
the apex of the ecological pyramid.

    Figure 17 shows the types of large mammalian carnivores and

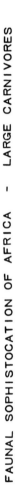

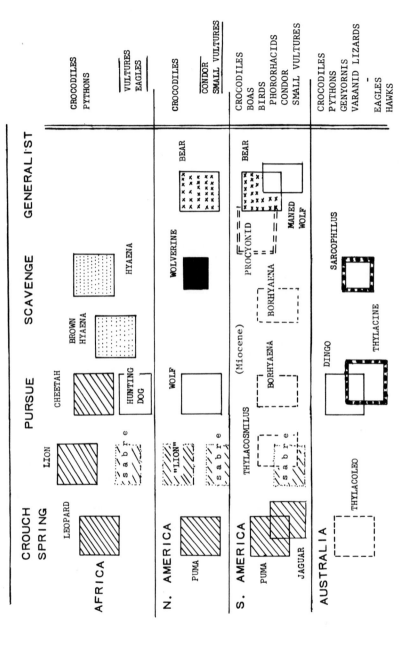

Fig. 17. The suite of large mammalian carnivores, and their role division in Africa, North America, South America, and Australia. Extinct forms are shown in broken outline. The different mammalian families are indicated as follows: felids (cross-hatching), canids (white squares), hyaenids (spotting), mustelids (black), ursids (crosses), procyonids (double-line border), and marsupials (checker border). See text.

their role division, contemporaneously in four continents, Africa, North America, South America and Australia. Some extinct forms are also shown to illustrate that continents now impoverished formerly had comprehensive and diversified faunas. Where the large carnivore guild is or was partly occupied by non-mammals this is also shown.

Several relevant facts are indicated in the figure:

(1)  Large carnivores are divisible into four categories: (a) crouch and spring hunters; (b) pursuit (run down) hunters; (c) scavengers (that usually do a little hunting); (d) generalists. This division, incidentally, also applies in birds and fish so would seem to be fundamental.

(2)  All four continents have evolved the first three types, which have clear-cut morphologies. These morphologies will not be reviewed here except to state that, for example, the optimum characters of a scavenger represent adaptations towards a massive head, enlarged premolars for bone-crushing, a locking device at the canine-incisor junction for grabbing flesh; powerful neck and massive foreparts associated with pulling, etc. Coyotes and jackals are excluded as large scavengers both because of their small size and because they lack these adaptations.

(3)  If a continent has nothing else it tends to have a crouch and spring carnivore (a felid), a pursuit one (a canid, except in the 'marsupial continents'), and a scavenger (a hyaenid in Africa, a mustelid (Gulo) in North America, and a marsupial (Sarcophilus) in Australia {Keast, 1972b}). Extinction of major types has, however, been prominent in the "marsupial continents" Australia and South America that remained isolated throughout the Tertiary. The latter lost the scavenging and (?) hunting borhyaenids in the Miocene; it then evolved bear-sized procyonids and subsequently lost these (Marshall in this volume). In Australia Thylacoleo, that became extinct in the Pleistocene could have represented the crouch-and-spring guild in this continent.

(4)  Sabre tooths evolved on all continents except Australia; but all are now extinct. Interestingly South America evolved a Mid-tertiary sabre-tooth marsupial, Thylacosmilus (Marshall in this volume) that antedated (and became extinct before the arrival of)the machaerodonts from North America at the end of the Pliocene.

(5)  All continents have supplementary non-mammalian carnivores in specialized habitats with rather different strategies

in the form of crocodiles (tropical) and (except for North
America) pythons or boas.  South America which apparently did
not evolve a large scavenging mammal (and which has lost its
"running" borhyaenids) had, in the Miocene and Pliocene
heavy-billed scavenging birds, the phoronhacids.  Australia,
which apparently also had some "difficulty" evolving large
mammalian predators had a Pleistocene varanid lizard,
Megalania prisca (Hecht, 1975), up to 7 metres in length,
which must surely have been a top level carnivore - scavenger.
(Note here the pig and deer eating abilities of Varanus
komodoensis.)

(6)  The scavenger guild gets some help from vultures, hawks,
etc., which have the advantage in that they can cover great
distances in search of carcasses.

(7)  When a continent cannot or can no longer support specialized types it may, so to speak compromise by emphasizing
generalized types (e.g. bears amongst the large carnivores)
or leaving an adaptive zone unfilled.  A nice feature about
large carnivores, of course, is that they are generally
continent-wide (or nearly so) in their ranges and are not
narrowly limited by habitat.  Certainly this is true of
Africa, largely true of North America and apparently formerly
also applied in Australia.  In the case of secondarily impoverished South America, however, the bear (Speoarcthos) and
the maned wolf (Chrysocyon) are narrowly habitat-specific.
Since both are descended from late Pliocene-early Pleistocene
invaders it could be that these two lines have always been
morphologically specialized or have not yet had time to
adapt more widely ecologically.

The great diversity and "sophistication" of the African large
mammal carnivore fauna relative to that of the other continents is
apparent in the following aspects:  (a) It has the most massive
forms (except for the jaguar in South America).  (b)  A cat
(Acinonyx) has evolved into the pursuit carnivore role.  (The
cheetah formerly also widespread in Eurasia has the long legs but
not the elongated muzzle of the pursuit canids.)  (c)  The scavengers (Hyaenidae) are not only larger but show more advanced adaptations for this role than those of the other continents.  (d)  Intermediate forms occur between the three categories, such as lion
and brown hyaena.  (e)  The complex of areal scavengers (vulture
fauna) is also much more elaborate in Africa than in the other
continents.  (f)  Africa lacks "generalist" large carnivores.  It
would appear that this type is characteristic of the 'biologically
deficient' continents, that is those that cannot, or can no longer,
support a good fauna of specialized types.

The conclusion that results from the above is that these
four continents are at very different levels of 'faunal evolution'.
The African one is at a peak of sophistication and complexity;
those of North America and South America are secondarily degenerate; that of Australia is primitive and never got very far in
evolving a large mammalian carnivore fauna.  It is a mixture of
primitive and structurally unspecialized types (the thylacine and
borhyaenids are both structurally inferior to the wolf - Keast,
unpublished), and mammals never completely excluded reptiles and
birds from the carnivore-scavenger guilds.  It seems likely,
moreover, that the thylacine and Sarcophilus never inhabited the
very exposed desert habitats.  Notwithstanding this the scavenging
eagle and hawk avifauna of Australia had not produced very large-
bodied forms:  it is a modest counterpart of this avifauna elsewhere.

## PHYLOGENY AND BIOGEOGRAPHY, AN INTEGRATION

If phylogeny is the evolutionary history of a lineage and
biogeography is the geography of distribution a close interrelationship of the two is obvious.  Geographic (spatial and
climatic) events shape faunas and species from the time of their
origin.  The origin of species (by geographic speciation) involves the spatial isolation of stocks and their subsequent re-
uniting (necessary to prove that the speciation process has been
completed).  Likewise the spatial dimension dominates faunas.  The
early histories of many of today's evolutionary lines were shaped
by continental separations and junctions in the Mesozoic and Early
Tertiary.  Subsequently the lines have been moulded by physio-
graphic events such as mountain building, which have drastically
altered climates, climatic oscillations, and the distributions of
habitats (vegetation formations).  On a wider global scale continental movements have changed climatic patterns and oceanic
currents.  These have, of course, been superimposed on long-term
climatic trends such as the Eocene warming and cooling thereafter,
climaxing in the glacial periods.

The changing geography of the World, and the associated
climatic changes and oscillations are one form of "pressure" on
evolutionary lines.  The biotic environment is another.  Here a
fluctuating assemblage of plants, prey, and parasites impinges on
the fauna.  Competitor, or potentially competitor species also
exercise a "pressure".  The evolving and radiating animal groups
must respond, in turn, to these.  Immigrant forms appear and
groups become extinct, all as part of the normal and long-term
evolutionary process.

These interactions between an evolving line and the geographic and biological environment are shown, diagrammatically in Figure 18.

## SUMMARY

Phylogeny is the evolutionary history of a lineage and biogeography is the geography of distribution - a close interrelationship between the two is obvious. The history of biogeographic studies is broadly divisible into 4 successive phases: descriptive, historical, ecological, and predictive. In the first two phases the origin of continental biotas was a dominant theme. The centers of origin concept is still basic to contemporary biogeography - groups arise in particular areas and spread outwards. Dispersal is considered in the light of continental drift, land bridges, long distance dispersal, disjunct distributions, etc. The conflicting view, based on the concept that contemporary distribution patterns result largely from vicariance, fragmentation of a former continuous range, supplemented by some dispersal, was presented. Many problems remain in the methodology of biogeographical analysis, especially the fact that any results are dependent on the quality of the taxonomy being used. Caution must be used in comparing numbers of families, genera and species in different groups and in making numerical comparisons between different continents. It is also questionable how much one can expect to learn of past distribution patterns from contemporary ones only. Paleontological data is of great value in elucidating such matters, as it is in suggesting areas of origin of major groups. Attempts have been made at better quantification of biogeographic knowledge, initially simply by measurements of faunal resemblances. In recent years numerical and cluster analyses have been introduced with qualified success.

The ecological dimension to biogeography has, as yet, only attracted a fraction of the attention of the historical aspects. The associated fauna, along with the physical environment, has a major role in moulding evolutionary lines and the ecological channeling of species. There is a correlation between morphological structures and way of life so that it is possible to compare faunas in terms of numbers of species with particular kinds of adaptive morphological attributes. Faunas may themselves be simple or complex and may be at different stages of evolution. The changing geography of the World and the associated climatic changes and oscillations are one form of 'pressure' on evolutionary lines. The biotic environment is another, a fluctuating assemblages of plants, prey, parasites, competitors, etc., all impinge upon the fauna. Immigrant forms appear and groups become extinct, all as part of the normal and long term evolutionary process.

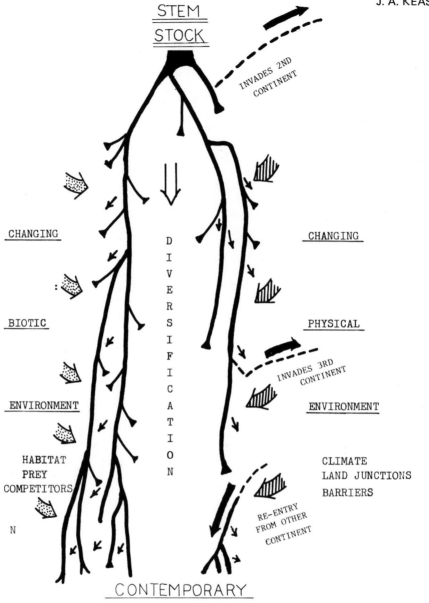

Fig. 18.  The interrelationships of phylogeny and biogeography in
moulding an evolutionary line.  The basic gene pool, that is constantly changing as radiation progresses and new lines are produced,
is subject to constant and changing "pressure" from the physical
environment (climate especially), the biotic environment (habitat,
prey organisms), and competitor lines.  Geographic changes divide
and reunite stocks and facilitate (or prohibit) the arrival of new
potential competitors, etc.  The scheme is hypothetical but is
characteristic of that facing continental faunas.

## ACKNOWLEDGEMENTS

The preparation of the present manuscript was carried out whilst the writer was on a research grant from the National Research Council of Canada.  I should like to thank this body for their support.

## REFERENCES

Allen, J.A., 1878, The geographical distribution of the Mammalia. Bull. U.S. Geol. and Geogr. Survey of the Territories, 4: 313-377.

Amadon, D., 1850, The Hawaiian honeycreepers (Aves, Drepaniidae), Bull. Amer. Mus. Natur. Hist., 95(4):153-262.

Axelrod, D.I., 1952, Variables affecting the probabilities of dispersal in geologic time, in "The Problem of Land Connections across the South Atlantic, with Special Reference to the Mesozoic," (E. Mayr, ed.), Bull. Amer. Mus. Nat. Hist., 99:177-188.

Axelrod, D.I., 1970, Mesozoic paleogeography and early angiosperm history. Bot. Review, 36:277-319.

Banarescu, P., 1970, "Principii si Probleme de Zoogeografie," Acad. Repub. Soc., Romania, Bucharest.

Berger, A.J., 1972, "Hawaiin Bird-Life," University of Hawaii Press, Honolulu.

Bock, W.J., 1963, The cranial evidence for ratite affinities. Proc. 13 Internal Ornith. Cong., Vol. 1:39-54.

Bock, W.J., 1966, An approach to the functional analysis of bill shape, Auk 83:10-51.

Bock, W.J., 1970, Microevolutionary sequences as a fundamental concept in macroevolutionary models. Evolution, 24:704-722.

Bock, W.J., 1972, Species interaction and macroevolution, in "Evolutionary Biology," (T. Dobzhansky, M. Hecht, and R. Steere, eds.), 1-24.

Bowman, R.I., 1961, Morphological differentiation and adaptation in the Galapagos finches, Univ. Calif. Publ. Zool., 58:1-302.

Briggs, J.C., 1974, "Marine Zoogeography," McGraw-Hill, New York.

Brundin, L., 1966, Transantarctic relationships and their significance as evidenced by chironomid midges. Kungl. Svenska Ventenskapsakademiens Handlingar, 11:1-472.

Cain, S.A., 1944, "Foundations of Plant Geography," Harper Bros., New York.

Carlquist, S., 1970, "Hawaii, a Natural History," Natural History Press, New York.

Carlquist, S., 1974, "Island Biology," Columbia University Press, New York.

Cheetham, A.H. and Hazel, J.E., 1969, Binary (presence-absence) similarity coefficients.  J. Paleontology, 43:1130-1136.

Cody, M.L., 1968, On the methods of resource division in grass-
    land bird communities. Amer. Natur., 102:107-147.
Cody, M.L. and Diamond, J.R., 1975, "Ecology and Evolution of
    Communities," Harvard Univ. Press.
Cooke, H.B.S., 1972, The Fossil Mammal Fauna of Africa, in
    "Evolution, Mammals, and Southern Continents," State Univ.
    New York Press, Albany:89-140.
Cox, C.B., Healey, I.M., and Moore, P.H., 1972, "Biogeography an
    ecological and evolutionary approach," John Wiley, New York.
Cracraft, J., 1972, The relationships of the higher taxa of
    birds: problems in phylogenetic reasoning. Condor, 74:379-392.
Cracraft, J., 1973, Continental drift, paleoclimatology, and the
    evolution and biogeography of birds.  J. Zool., 169:455-545.
Cracraft, J., 1974a, Phylogenetic models and classification.
    Syst. Zool., 23:7190.
Cracraft, J., 1974b, Continental drift and vertebrate distribu-
    tion.  Annual Rev. Ecol. Syst., 5:215-261.
Cracraft, J., 1974c, Phylogeny and evolution of the ratite birds,
    Ibis, 116:494-521.
Cracraft, J., 1975, Historical biogeography and earth history:
    perspectives for a future synthesis.  Ann. Missouri Botan.
    Gard., 62:227-250.
Croizat, L., 1958, "Panbiogeography," published by the author,
    Caracas.
Croizat, L., 1964, "Space, time, form: the biological synthesis,"
    published by the author, Caracas.
Croizat, L., 1970, A selection of notes on the broad trends of
    dispersal mostly of Old World Avifauna.  Natural History
    Bull. Siam Society, 23:257-324.
Croizat, L., Nelson, G., and Rosen, D.E., 1974, Centers of origin
    and related concepts.  Systematic Zool., 23:265-287.
Dalziel, I.W.D., Lowrie, W., Kingfield, R., and Opdyke, N.C.,
    1973, Paleomagnetic data from the southermost Andes and
    Arctandes, in "Implications of Continental Drift to the
    Earth Sciences," (Tarling and Runcorn, eds.), Vol. 1,
    Academic Press, London: pp. 87-101.
Darlington, P.J., 1948, The geographical distribution of cold-
    blooded vertebrates.  Quart. Rev. Biol., 23:1-26, 105-123.
Darlington, P.J., Jr., 1957, "Zoogeography," John Wiley, New York.
Darlington, P.J., Jr., 1959, "Area, climate, and evolution,"
    Evolution, 13:488-510.
Darlington, P.J., Jr., 1970, A practical criticism of Hennig-
    Brundin "Phylogenetic Systematics" and Antarctic biogeography.
    System. Zool., 19:1-18.
Darlington, P.J., Jr., 1971, Interconnected patterns of bio-
    geography and evolution.  Proc. Nat. Acad. Sci. U.S., 68:
    1254-1258.
Darwin, C., 1859, 1950, "On the Origin of Species," London,
    Murray reprinted 1950, Watts & Co., London.

Diamond, J.M., 1970, Ecological consequences of island coloniza-
    tion by south-west Pacific birds, I.  Types of niche shifts.
    Proc. Nat. Acad. Sci. U.S., 67:529-536.
Diamond, J.M., 1973, Distributional ecology of New Guinea birds.
    Science, 179:759-769.
Diamond, J.M., 1975, Assembly of species communities, in "Ecology
    and Evolution of Communities," (M.L. Cody and J.L. Diamond,
    eds.), pp. 342-444, Harvard Univ. Press.
Dietz, R.S. and Holden, J.C., 1970, Reconstruction of Pangaea:
    breakup and dispersion of continents, Permian to present.
    J. Geophys. Res., 75:4939-4956.
Edmonds, G.F., Jr., 1972, Biogeography and evolution of
    Ephemeroptera. Ann. Rev. Entomol., 17:21-42.
Edmonds, G.F., Jr., 1975, Phylogenetic biogeography of mayflies.
    Ann. Missouri Botan. Gard., 62:251-263.
Ekman, S., 1953, "Zoogeography of the Sea," Sidgwick and Jackson,
    London.
Elliot, D.H., Colbert, E.H., Breed, W.J., Jensen, J.A., and
    Powell, J.S., 1970, Triassic tetrapods from Antarctica;
    evidence for continental drift.  Science 169:1197-1201.
Elliot, D.H., 1972, Aspects of Antarctic geology and drift
    reconstructions, in "Antarctic Geology and Geophysics,"
    (R.J. Adie, ed.), Universitetsforlaget, Oslo: pp. 849-858.
Ferris, V.R., Goseco, C.G., and Ferris, J.M., 1976, Biogeography
    of free-living soil nematodes from the perspective of
    plate tectonics.  Science 193:508-509.
Fleming, C.A., 1962, New Zealand Biogeography, a paleontologists
    approach.  Tuatara 10(2):53-108.
Fleming, C.A., 1975, The geological history of New Zealand and
    its biota, in "Biogeography and Ecology in New Zealand,"
    1-86, W. Junk, The Hague.
Fosberg, F.R., 1948, Derivation of the flora of the Hawaiian
    islands, in "Insect of Hawaii, vol. 1: Introduction,"
    (E.C. Simmerman, ed.), pp. 107-19, University of Hawaii
    Press, Honolulu.
Good, R., 1964, "The Geography of the Flowering Plants,"
    Longmans, London.
Grant, P.R., 1966, Ecological comparibility of bird species on
    islands. Amer. Natur., 100:451-462.
Grant, P.R., 1972, Convergent and divergent character displace-
    ment.  Biol. J. Linn. Soc., 4:39-68.
Gressitt, J.L., 1963, Insects of Antarctica and the sub-Antarctic
    islands, Chapter in Pacific Island Biogeography (J.L.
    Gressitt, ed.), 435-442.
Gressitt, J.L. and Yoshimoto, C.M., 1963, Dispersal of animals
    in the Pacific, "Pacific Island Biogeography," (J.L.
    Gressitt, ed.), pp. 283-292, Bishop Museum Press, Honolulu.
Haffer, J., 1967, Speciation in Colombian forest birds west of
    the Andes.  Amer. Mus. Nov., 2294:1-57.

Haffer, J., 1969, Speciation in Amazonian forest birds.  Science, 165:131-137.

Haffer, J., 1974, Avian speciation in tropical South America.  Publ. Nuttall Ornithol. Club, No. 14, pp. 1-390.

Hagmeir, E.M., 1966, A numerical analysis of the distributional patterns of North American mammals, II.  Re-evaluation of the provinces.  System. Zool., 15:279-299.

Hagmeir, E.M. and Stults, C.D., 1964, A numerical analysis of the distributional patterns of North American mammals.  System. Zool., 13:125-155.

Hall, B.P. and Moreau, R.E., 1970, "Atlas of Speciation of African Birds," British Museum (Natur. Hist.), London.

Hall, E.R. and Kelson, K.R., 1959, "The Mammals of North America," Ronald Press, New York.

Hallam, A. ed., 1973, "Atlas of Palaeobiogeography," Elsevier, Amsterdam.

Harrison, L., 1924, The migration route of the Australian marsupial fauna.  Austral. Zool., 3:247-263.

Hecht, M.K., 1975, The morphology and relationship of the larger known terrestrial lizard, Megalania prisca avian from the Pleistocene of Australia.  Pro. Rov. Soc. Victoria, 87:239-350.

Hedberg, O., 1971, Evolution of the Afroapline flora, in "Adaptive Aspects of Insular Evolution," (W.L. Stern, ed.), pp. 16-23, Washington State University Press, Pullman, Wash.

Hendey, Q.B., 1976, Fossil peccary from the Pliocene of South Africa.  Science, 192:787-789.

Hennig, W., 1965, Phylogenetic systematics.  Ann. Rev. Entomol., 10:97-116.

Hennig, W., 1966, "Phylogenetic Systematics," University of Illinois Press, Urbana, Illinois.

Hershkovitz, P., 1972, The recent mammals of the Neotropical region; a zoogeographic and ecological review, in "Evolution, Mammals, and Southern Continents," (A. Keast, F.C. Erk, and B. Glass, eds.), State University of New York Press, Albany, pp. 311-431.

Hess, R., Allee, W.C., and Schmidt, K.P., 1951, "Ecological Animal Geography," 2nd ed., John Wiley, New York.

Hoffstetter, R., 1972, Relationships, origins, and history of the ceboid monkeys and caviomorph rodents:  a modern re-interpretation.  Evol. Biol., 6:323-347.

Holloway, J.D. and Jardine, N., 1968, Two approaches to zoogeography: a study based on the distributions of butterflies, birds and bats in the Indo Australian area.  Proc. Linn. Soc. London, 179:153-158.

Hooker, J.D., 1860, On the origin and distribution of species: Introductory essay to the flora of Tasmania.  Amer. J. Sci. Arts, ser. 2, 29:1-25; 305-326.

Horton, D.R., 1974a, Species movement in zoogeography. J. Biogeography, 1:155-158.

Horton, D.R., 1974b, Dominance and zoogeography of the southern continents. Systematic Zoology, 23:440-445.

Hubbs, C.L., 1958, "Zoogeography," Publication No. 51, Amer. Assoc. Adv. Science, Washington.

Illies, J., 1966, Phylogeny and zoogeography of the Plectopera. Ann. Rev. Entomol., 10:117-140.

Jamieson, B.N., 1973, The Oligochaeta of Tasmania, in "Biogeography and Ecology in Tasmania," (W. Williams, ed.), Dr. W. Junk, The Hague.

Jardine, N. and McKenzie, O., 1972, Continental drift and the dispersal and evolution of organisms. Nature (London), 235:20-24.

Johnson, L.A.S. and Briggs, B.G., 1963, Evolution in the Proteaceae. Aust. J. Bot., 11:21-61.

Karr, J.R. and James, F.C., 1975, Eco-morphological configurations and convergent evolution in species and communities, in "Ecology and Evolution of Communities," (M.L. Cody and J.M. Diamond, eds.), Harvard Univ. Press.

Keast, A., 1961, Bird speciation on the Australian continent. Bull. Mus. Comp. Zool., Harvard 123:305-495.

Keast, A., 1968, Competitive interaction and the volution of ecological niches as illustrated by the Australian honeyeater genus Melithreptus (Meliphagidae). Evolution, 22:762-784.

Keast, A., 1970, Trophic interrelationships of fish faunas in some small Ontario waterways, in "Marine Food Chains," (J.H. Steele, ed.), Oliver and Boyd, Edinburgh, pp. 61-96.

Keast, A., 1971, Adaptive evolution and shifts in niche occupation in island birds, in "Adaptive Aspects of Insular Evolution," (W.L. Stern, ed.), pp. 39-53. Washington State University Press, Pullman, Wash.

Keast, A., 1972a, Comparisons of contemporary mammal faunas of southern continents, in "Evolution, Mammals and Southern Continents," (A. Keast, F.C. Erk; and B. Glass, eds.), pp. 433-502.

Keast, A., 1972b, Australian mammals, zoogeography and evolution, in "Evolution, Mammals and Southern Continents," (A. Keast, F.C. Erk and B. Glass, eds.), pp. 195-246.

Keast, A., 1972c, Continental drift and the evolution of the biota of southern continents, in "Evolution, Mammals, and Southern Continents," (A. Keast, F.C. Erk and B. Glass, eds.), pp. 23-87.

Keast, A., 1972d, Ecological opportunities and dominant families, as illustrated by the Neotropical Tyrannidae (Aves). Evolutionary Biology, 5:229-277.

Keast, A., 1973, Contemporary biotas and the separation sequences of the southern continents, in "Implications of Continental

Drift to the Earth Sciences, Vol. 1," (D.H. Tarling and
S.K. Runcorn, eds.), Academic Press, London, pp. 309-343.

Keast, A., 1976, Historical biogeography, antarctic dispersal,
and Eocene climates, in "The Biology of Marsupials,"
(B. Stonehouse and D. Gilmore, eds.), pp. 69-95.

Keast, A., Erk, F.C., and Glass, B.J., 1972, "Mammals, Evolution,
and Southern Continents," State University of New York
Press, Albany.

Kikkawa, J., 1968, Ecological association of bird species and
habitats in eastern Australia; a similarity analysis.
J. Animal Ecol., 37:143-165.

Kikkawa, J. and Pearse, K., 1969, Geographical distribution of
land birds in Australia - a numerical analysis.  Aust. J.
Zool., 17:821-840.

Lack, D., 1947, "Darwin's Finches," Cambridge University Press,
Cambridge.

Larson, R.L. and Chase, C.D., 1972, Late Mesozoic evolution of
the western Pacific Ocean.  Bull. Geol. Soc. Am., 83:3627-
3644.

Lavocat, R., 1969, La systematique des rongeurs hystricomorphs
et la derive des continents. Comptes Rend. Acad. Sci.
(Paris), 269:1496-1497.

Lein, M.R., 1972, A tropic comparison of avifaunas.  Syst. Zool.,
21:135-150.

Linsley, E.G., 1963, Bering Arc relationships of Cerambycidae and
their host plants, in "Pacific Basin Biogeography," (J.L.
Gressitt, ed.), Bishop Museum Press, Honolulu.

MacArthur, R.H., 1972, "Geographical Ecology," Harper and Row,
New York.

MacArthur, R.H. and Wilson, E.O., 1967, "The Theory of Island
Biogeography," Princeton University Press, Princeton, N.J.

McDowall, R.M., 1971, Generalized tracks and dispersal in
biogeography more comments on Leon Croizat's biogeography,
in press.

McDowall, R.M. and Whitaker, A.M., 1975, The freshwater fishes,
in "Biogeography and Ecology in New Zealand," (G. Kuschel,
ed.), pp. 277-299, W. Junk, The Hague.

McGowran, B., 1973, Rifting and drift of Australia and the migration of mammals.  Science, 180:759-761.

McKenna, M.C., 1973, Sweepstakes, filters, corridors, Noah's arks,
and beached viking funeral ships in palaeogeography, in
"Implications of Continental Drift to the Earth Sciences,
Vol. 1," (D.H. Tarling and S.K. Runcorn, eds.), pp. 295-308,
Academic Press, London.

McKenna, M.C., 1975, Fossil mammals and early Eocene North
Atlantic land continuity.  Ann. Missouri Bot. Gard., 62:335-
353.

Marshall, L.G., 1977, Evolution of the Borhyaenidae, extinct
South American predaceous marsupials. Univ. Calif. Publ.

Geol. Sci., in press.

Matthew, W.D., 1915, 1939, "Climate and Evolution," Special
    Publ. New York Acad. Sci. I:1-223, Republished 1939.

Mayr, E., 1940, The origin and history of the bird fauna of
    Polynesia. Proc. Sixth Pacific Sci. Congre., 4:197-216.

Mayr, E., 1942, "Systematic and the Origin of Species,"
    Columbia University Press, New York.

Mayr, E., 1946, History of the North American bird fauna.
    Wilson Bull., 58:3-41.

Mayr, E., 1963, "Animal Species and Evolution," Harvard
    University Press, Cambridge, Mass.

Mayr, E., 1965, Classification and phylogeny. Am. Zoologist,
    5:165-174.

Mayr, E., 1974, Cladistic analysis or cladistic classification.
    Z. Zool. Syst. Evol. forsch., 12:94-128.

Mayr, E. and Phelps, W.H., 1967, The origin of the bird fauna of
    the South Venezuelan highlands. Bull. Amer. Mus. Natur.
    Hist., 136:269-335.

Mayr, E. and Short, L.L., 1970, Species taxa of North American
    birds: a contribution to comparative systematics. Publ.
    Nuttall Ornithol. Club, Cambridge, Mass. No. 9:1-127.

Meggers, B.J., Ayensu, E.S., and Duckworth, W.D., 1973, "Tropical
    forest ecosystems in African and South America: a Comparative
    Review," Smithsonian Institution, Washington.

Merriam, C.H., 1892, The geographic distribution of life in
    North America. Proc. Biol. Soc. Washington, 7:1-64.

Michaelson, W., 1922, Die verbreitung der Oligochaten im lichte
    er Wegener's chen Theorie der kontinentverschiebung.
    Verh. Ver. Naturw. Unterh. Hamburg, (3) 29:325-344.

Moreau, R.E., 1966, "The Bird Faunas of Africa and its Islands,"
    Academic Press, New York.

Muller, P., 1973, "The Dispersal Centres of Terrestrial Verte-
    brates in the Neotropical Realm," Dr. W. Junk, The Hague.

Nelson, G.J., 1972, Comments on Hennig's "Phylogenetic Systematics"
    and its influence on ichthyology. Systematic Zool., 21:
    364-374.

Nelson, G., 1973, Comments on Leon Croizat's biogeography.
    Systematic Zoology, 22:312-320.

Nelson, G., 1974a, Historical biogeography: an alternative
    formalization. System. Zoology, 23:555-558.

Nelson, G., 1974b, Darwin-Hennig classification: a reply to
    Ernst Mayr. System. Zool., 23:452-458.

Olson, F.C., 1971, "Vertebrate Paleozoology," Wiley-Interscience,
    New York.

Peters, J.A., 1971, A new approach to the analysis of Bio-
    geographic data. Smithsonian Contributions to Zoology,
    107:1-28.

Patterson, B. and Pascual, R., 1972, The fossil mammal fauna
    of South America, in "Evolution, Mammals and Southern

Continents," (A. Keast, F.C. Evk, and B. Glass, eds.),
State University of New York Press, Albany, pp. 247-310.

Pitman, W.C., III and Talwani, M., 1972, Sea-floor spreading
in the North Atlantic. Bull. Geol. Soc. Am., 83:619-646.

Preese, D.S., 1963, A note on the dispersal characteristics of
the seed of the New Zealand podocarps and beeches and
their biogeographical significance, in "Pacific Basin
Biogeography," (J.L. Gressitt, ed.), pp. 415-424, Bishop
Museum, Press, Honolulu.

Raven, P.H., 1963, Amphitropical relationships in the floras
of North and South America. Quart. Rev. Biol., 38:151-177.

Raven, P., 1972, Plant species disjunctions: a summary. Ann.
Missouri Bot. Gard., 59:234-246.

Raven, P. and Axelrod, D.I., 1972, Plate tectonics and
Australasian paleobiogeography. Science, 176:1379-1388.

Raven, P. and Axelrod, D.I., 1974, Angiosperm biogeography and
past continental movements. Ann. Missouri Bot. Gard., 61:
539-673.

Ricklefs, R.E. and Cox, G.W., 1972, Taxoncycles in the West
Indian avifauna. Amer. Natural, 106:195-219.

Rosen, D.E., 1974a, Review "Space, Time, Form; the Biological
Synthesis: by Leon Croizat, System. Zoology, 23:288-290.

Rosen, D.E., 1974b, Phylogeny and zoogeography of salmoniform
fishes and relationships of Lepidogalaxias salamandroides.
Bull. Amer. Mus. Nat. History, 153:269-325.

Rosen, D.E., 1974c, Cladism or gradism: a reply to Ernst Mayr.
System. Zool., 23:446-451.

Schoener, T.W., 1965, The evolution of bill size differences
among sympatric congeneric species of birds. Evolution,
19:189-213.

Schoener, T.W., 1968, The Anolis lizards of Bimini: resource
partitioning in a complex fauna. Ecology, 49:704-726.

Sclater, P.L., 1858, On the general geographical distribution
of the members of the Class Aves. J. Proc. Linnean Soc.
(London), Zool., 2:130-145.

Serventy, D.L., 1953, Some speciation problems in Australian
birds. Emu, 53:131-145.

Serventy, D.L., 1960, Geographical distribution of living birds,
in "Biology and Comparative Physiology of Birds," (A.J.
Marshall, ed.), vol. 1, pp. 95-126, Academic Press, N.Y.

Serventy, D.L. and Whittell, H.M., 1962, "Birds of Western
Australia," Paterson Brokensha, Perth.

Short, L.L., 1975, A zoogeographical analysis of the South
American chaco avifauna. Bull. Amer. Mus. Nat. Hist.,
154:163-352.

Simberloff, D.S., 1969, Experimental zoogeography of islands,
A model for insular colonization. Ecology, 50:296-314.

Simberloff, D.S., 1970, Taxonomic diversity of island biotas.
Evolution, 24:23-47.

Simberloff, D.S., and Wilson, E.O., 1969, Experimental
    Zoogeography of islands. The colonization of empty
    islands. Ecology, 50:278-296.
Simpson, G.G., 1940, Mammals and land bridges. J. Washington
    Acad. Sci., 30:137-163.
Simpson, G.G., 1943, Mammals and the nature of continents,
    Amer. J. Sci., 241:1-31.
Simpson, G.G., 1947, Holarctic mammalian faunas and continental
    relationships during the Mesozoic. Bull. Geol. Soc. Amer.,
    68:613-387.
Simpson, G.G., 1952, Probabilities of dispersal in geologic
    time. Bull. Amer. Mus. Natur. Hist., 99:163-76.
Simpson, G.G., 1960, Notes on the measurement of faunal
    resemblance. Amer. J. Science, 258:-A:300-311.
Simpson, G.G., 1965, "The Geography of Evolution," Capricorn
    Books, New York.
Solbrig, O.T., 1972, Disjunction in plants: a symposium. Ann.
    Missouri Botanical Garden, 59:105-106.
Tedford, R.H., 1974, Marsupials and the new paleogeography, in
    "Paleogeographic Provinces and Provinciality," (C.A. Ross,
    ed.), Soc. Econ. Paleontol. Mineral. Spec. Pub. no., 21:109-
    126.
Terborgh, J. and Faaborg, J., 1973, Turnover and ecological re-
    lease in the avifauna of Mona Island, Puerto Rico. Auk.,
    90:759-779.
Thorne, R.F., 1963, Biotic distribution patterns in the
    tropical Pacific, in "Pacific Basin Biogeograph," (J.L.
    Gressitt, ed.), pp. 311-354, Bishop Museum Press, Honolulu.
Thorne, R.F., 1972, Floristic relationships between Tropical
    Africa and Tropical America, in "Tropical Forest Ecosystems
    in Africa and South America: a Comparative Review," (B.J.
    Meggars Ayensu, E.S., and Duckworth, W.D., eds.), pp. 27-47.
    Smithsonian Institution, Washington, D.C.
Udvardy, M.D.F., 1969, "Dynamic Zoogeography with Special
    Reference to Land Animals," Van Nostrand-Reinhold,
    Princeton.
Vuilleumier, F., 1969, Pleistocene speciation in birds living
    in the high Andes. Nature (London), 223:1179-1180.
Vuilleumier, F., 1970, Insular biogeography in continental
    regions, I. The northern Andes of South America. Amer.
    Natur., 104:373-388.
Vuilleumier, F., 1972, Speciation in South American birds: a
    progress report. Acta. Congr. Latinoamer. Zool., 1968,
    Vol. 1., pp. 239-255.
Vuilleumier, F., 1975, Zoogeography, in "Avian Biology, Vol. 5,"
    (D. Farner and J.R. King, eds.), Academic Press, pp. 421-496.
Wallace, A.R., 1876, "The Geographical Distribution of Animals,"
    London, Macmillan, 2 vols.
Webb, L., 1950, Biogeographic regions of Texas and Oklahoma.
    Ecology, 31:426-433.

Wegener, A., 1915, Die Enstehung der Kontinente and Ozeane, Brunswick. Friedr. Wieweg und Sohn.

Wegener, A., 1924, "The Origin of Continents and Oceans," (translated from 3rd German ed.), London, Methuen.

Wilson, D.E., 1973, Bat faunas: a trophic comparison. Syst. Zool., 22:14-29.

Wilson, E.O. and Simberloff, D.S., 1969, Experimental Zoogeography of islands, Defaunation and monitoring techniques. Ecology, 50:267-278.

Winterbottom, R., 1971, The familial phylogeny of the Tetradon-tiformes (Acanthopterygii, Pisces) as evidences by their comparative mycology. Ph.D. thesis, Queen's University, Kingston, Ont., 445 pp.

Wittmann, O., 1934, Die biogeographischen beziehungen der Sudkontinente. Zoogeographica, 11:246-304.

Wood, C.E., Jr., 1972, Morphology and phytogeography; the classical approach to the study of disjunctions. Ann. Missouri Botanical Garden, 59:105-106.

Zimmerman, E.C., 1948, "Insects of Hawaii, vol. 1: Introduction," University of Hawaii Press, Honolulu.

# PART 2

# Macroevolutionary Trends among Vertebrate Taxa

# PHYLOGENETIC RELATIONSHIPS AND A

# CLASSIFICATION OF THE EUTHERIAN MAMMALIA

Frederick S. SZALAY

Hunter College, City University of New York

695 Park Avenue, New York, N.Y.   10021

## INTRODUCTION

It is remarkable that the enormous increase in the fossil evidence, new facts from the study of living mammals, and the promising rise of molecular studies have in fact had less than the usually declared effect in modifying a number of basic ideas on methods of phylogenetic inference which have been in practice since the beginning of this century.  Similarly, one finds that many of the early (late 19th and early 20th century) hypotheses of relationships are being "rediscovered" or have remained un-falsified when evidence already known to these workers is re-discovered.  Growth in the data base and the increasing number of students addressing themselves to phylogenetic problems neverthe-less have resulted in an enormous increase in the literature and many very sound phylogenetic and morphological studies.

In this review of early eutherian relationships I will attempt to corroborate a number of previously proposed hypotheses of re-lationships, falsify some others by presenting new data and inter-pretations, and present some new speculative hypotheses (Fig. 1). I will also present a classification of eutherian higher taxa and some reasons for or against specific classificatory practices.

## THE NATURE OF THE EVIDENCE

A large number of studies conducted in the past decade on fossil mammals have attempted to show that many of the early hypotheses of eutherian phylogenetic relationships were false

315

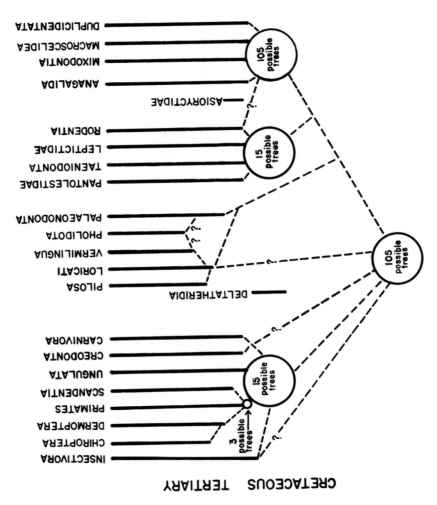

Fig. 1. An hypothesis of phylogenetic relationships of various monophyletic groups within the Eutheria. Circles represent presently unresolved polychotomies. Numbers in the circles depict the possible number of phylogenetic trees given the number of the taxa sharing a common ancestor within that circle. Question marks indicate doubtful but possible derivations.

and offered new solutions.  Unfortunately, the majority of these
studies have concentrated on dental morphology and have simply
failed to seriously consider other already known and new available fossil evidence, thus inadvertently divorcing taxon phylogeny
from character genealogy.  This practice is prevalent in the study
of fossil mammals and, although unavoidable in the study of many
groups with no adequate neontological or fossil postcranial evidence, it is clearly not so in many other taxa.

Given that most mammal species can be recognized by cheek
teeth alone, it follows that crown morphology is recognizably
unique (autapomorphous) if not in all then in most species or at
least from genus to genus.  As molar crowns are often "finger-
prints" of mammal species, the elements by themselves often have
no more "pure" phyletic meaning than information of mechanical
significance.  Study of crown morphology, the basis of much
assessment of mammal relationships, is therefore, by necessity
often based on patristic evidence rather than on clearly identified
derived characters.  In spite of genuine efforts to surmount this
problem, when only cheek teeth are available often only a grade of
similarity rather than a specific synapomorphy can be ascertained.
Some paleomammalogists would dispute this conclusion, but nevertheless, a molar tooth is more often than not a heritage dictated
and mechanically-functionally transformed single character complex.
Teeth, only if most carefully scrutinized for heritage and habitus
aspect of their characters, permit justification of polarity hypotheses.  Teeth, by and large have not been so studied and therefore often studies on molars alone, called cladistic in principle,
are operationally strictly patristic inasmuch as they measure
total similarity or difference rather than a hierarchy of recency
of specific dental characteristics.  It should be said however,
that these patristic similarities are often so uncannily many-
faceted that extreme subtleties are possible in determining which
two taxa are closer although not necessarily most recently related
to each other.  For reasons that cannot be adequately treated here
teeth of closely related lineages are notoriously prone to parallelism and convergence.

The degree to which existing postcranial evidence has been
ignored in paleomammalogical hypotheses is remarkable.  The only
way one may explain this curious phenomenon is that some assumptions about bone morphology must be widespread.

The bias and the misunderstanding of bone evolution of most
paleomammalogists is reflected in Gingerich's (1976) comment on
tooth as opposed to (implied) bone heritability.  According to
this author, whose views are quite representative of the widespread assumptions in the literature (Gingerich, 1976, p. 21),
"Most other skeletal characters in vertebrate and invertebrate

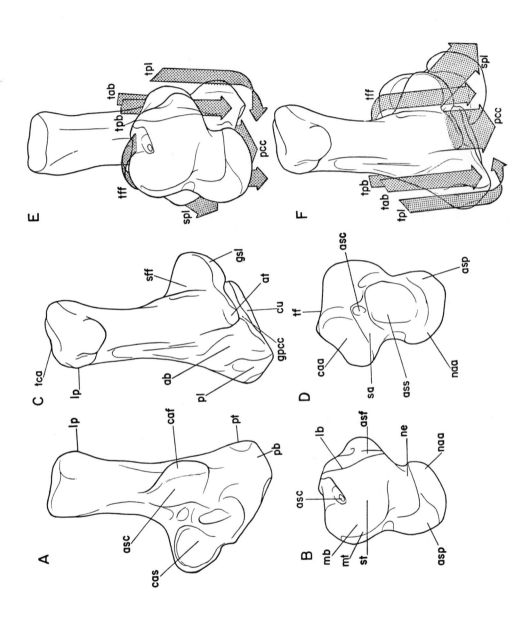

Fig. 2. Morphology and some muscle-tendon relationships of the primitive eutherian astragalocalcaneal complex, exemplified by the late Cretaceous Protungulatum. Dorsal (A, B, E) and ventral (C, D, F) views of the calcaneum (A, C), astragalus (B, D), and astragalocalcaneal complex (E, F).

Abbreviations on the calcaneum: ab, groove for m. abductor digiti quinti?; asc, astragalocalcaneal facet; at, anterior plantar tubercle; caf, calcaneal fibular facet; cas, calcaneal sustentacular facet; cu, cuboid facet; gpcc, groove for plantar calcaneocuboid ligament; gsl, sustentacular groove for attachment of "spring" ligament; lp, lateral process of tuber calcanei; pb, groove for tendon of m. peroneus brevis; pl, groove for tendon of m. peroneus longus; pt, peroneal tubercle; sff, sustentacular groove for tendon of m. flexor digitorum fibularis; tca, tuber of the calcaneum.

Abbreviations on the astragalus: asc, astragalar canal; asf, astragalar fibular facet; asp, astragalar "spring" ligament facet; ass, astragalar sustentacular facet; caa, calcaneoastragalar facet; lb, lateral border (crest) of trochlea; mb, medial border (crest) of trochlea; mt, medial tibial facet of trochlea; naa, naviculoastragalar facet; ne, neck of astragalus; sa, sulcus astragali; st, superior tibial facet of trochlea; tf, trochlear groove for tendon of m. flexor digitorum fibularis.

Abbreviations on the astragalocalcaneal complex: pcc, (long) plantar calcaneocuboid ligament; spl, "spring" ligament (calcaneonavicular ligament); tab, tendon or fleshy fibers of abductor digiti quinti or other muscles; tff, tendon m. flexor (digitorum) fibularis; tpb, tendon m. peroneus brevis; tpl, tendon m. peroneus longus.

animals undergo significant changes in ontogeny, <u>making many of
these characters unsuitable for detailed analysis in the fossil
record</u>...." (emphasis mine).  The implied lack of heritability of
bone as compared with teeth is of course an arbitrary statement,
and ontogenetic change in bony features (grossly overestimated
for mammals, and usually based on quite irrelevant experimental
manipulations or pathological readjustments in rare specimens)
does not in any way negate or contradict heritability as high as
for tooth morphology, kidney histology, or specificity for inner-
vation of muscular tissue, to cite a few of the possibly thousands
of examples.[1]

It appears to me that studies of fossil mammals often ignore
postcranial evidence on the basis of the notion that because bone
is much more "plastic functionally" (i.e. ontogenetically) than
teeth, this fact must be true phylogenetically, so that anything
may be expected to be due to "convergent functions."  The faulty
syllogism therefore assumes that phylogenetic intertia is more
characteristic of teeth than of other components of the skeleton.
This scheme (also erroneously assuming that teeth are phylogenetically less responsive to functional demands than bone) over-
looks the fact that evolutionary change (in bones) from species
to species is not the result of ontogenetic change but the result
of changes in some regulator gene frequencies.  It further fails
to realize that, were bone changes as ontogenetically plastic as
the assumption purports them to be, then intraspecific variation
should be much greater in postcranials than in dental parameters.
This, of course, is simply not true for any sample of postcranials
when compared to the dental variance of the same animals.  Convergence is not any more or less difficult to detect in bone and
joint morphology than in tooth crown patterns (see especially
Fig. 3 and 4).  Historically, one line of evidence has been
favored.  The dubious necessity to continue this methodological
trend, solely because of the greater preponderance of teeth, was
perhaps justified in the past but it is no longer tenable at
present.

## PRECISION OF TAXONOMIC DIAGNOSES AND THEIR
## INFLUENCE ON THE CONSTRUCTION OF PHYLOGENETIC HYPOTHESES

It is indisputable that whenever large numbers of taxa represented by good dental records show gradations of morphology, i.e.
morphoclines with easily recognizable polarities, and corresponding
chronoclines also exist, then dental characters, by and large,
faithfully mirror a phylogeny.  Skulls and associated postcranial
remains, usually discovered more rarely, tend to confirm such
phylogenetic hypotheses initially based on teeth.  Some groups of
rodents, perissodactyls, elephants, and a large number of other

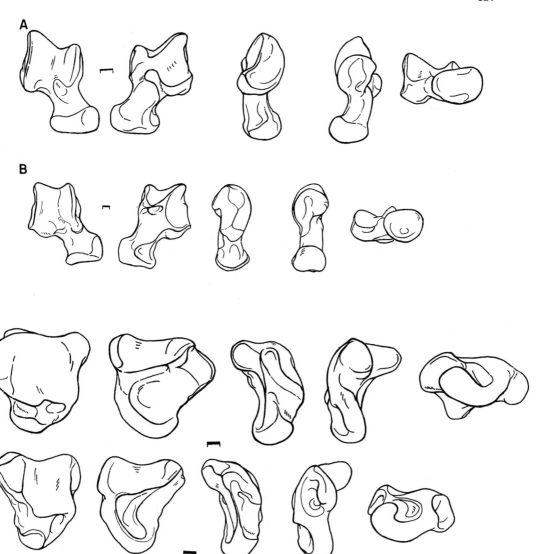

Fig. 3. Comparison of the right astragali of: A. the claw climbing sciurid rodent, Sundasciurus sp. (AMNH 2140); B. the gliding
sciurid, Petaurista sp. (AMNH 112975); C. the claw climbing phalangerid, Pseudocheirus archeri (AMNH uncatalogued specimen); D.
the gliding phalangerid, Schoinobates volans (AMNH 65364). From
left to right: dorsal, ventral, lateral, medial, and distal views,
respectively. In spite of similar locomotor habits of A and C
(climbing only) and B and D (gliding and climbing) the morphological simularities are compelling along the lines of heritage rather
than habitus. Scales represent one mm.

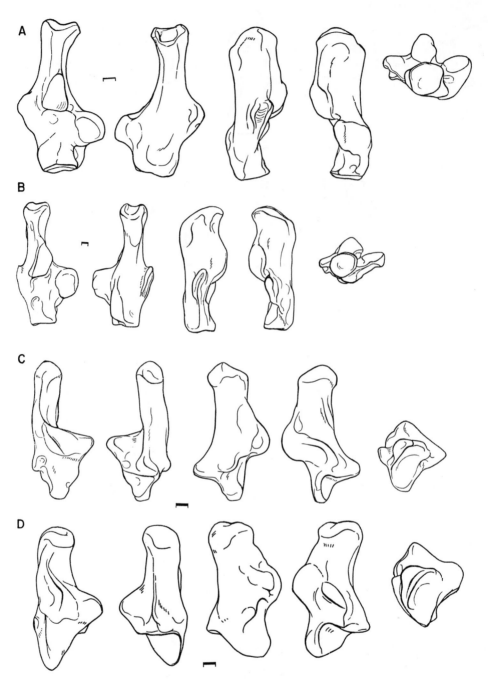

Fig. 4. Comparison of the right calcanea of: A. the claw climbing sciurid rodent, Sundasciurus sp. (AMNH 2140); B. the gliding
sciurid, Petaurista sp. (AMNH 112975); C. the claw climbing phalangerid, Pseudocheirus archeri (AMNH uncatalogued specimen); D.
the gliding phalangerid, Schoinobates volans (AMNH 65364). From
left to right: dorsal, ventral, lateral, medial, and distal views,
respectively.

well differentiated groups show the above view to be often correct.

Perhaps subsequent to these generalities acknowledged by students of fossil mammals, it is a common practice in paleomammalogy that taxa are exclusively defined on dental criteria. Even when attempts are made to delineate derived aspects of the dentition, known postcranials are rarely considered in the diagnosis. Somehow, the postcranial evidence is often not assessed to be sufficiently notable or unique and therefore diagnostic, or the usefulness of combining more than one character complex of a fossil taxon in a diagnosis is not considered important. I believe that this is the direct result of a traditional view which regards teeth as panacea and a subsequent slighting of other evidence.

It may seem unnecessary, but it should be pointed out that the sharing of any distinctive, uniquely derived postcranial adaptation will have much greater significance in assessing relationships among taxa than the sharing of a particular common, but nevertheless primitive (ancient) molar pattern present in the common ancestry of all of the taxa under investigation. In order to arrive at highly specific, precisely stated, and therefore (because of their vulnerability) most useful hypotheses, we should define a taxon on all of its known derived conditions (autapomorphies). For example, leptictids or didelphodontine eutherians are not only a set of dental conditions (largely primitive), but also specific recognizable tarsal characters. Similarly, we may not call taxa anagalidan, primate or rodent if they do not share diagnostic features on a level restricted to their respective groups. By the same token, no matter how similar in ancestral molar features various eutherian dental taxa may be to leptictids, for example (known by an adequate postcranial record), we should not refer to these as "leptictids." If we do, we simply dilute a specific taxonomic concept where the use of "primitive eutherian" or a more restrictive other concept would have been more precise and instructive.

Our phylogenetic hypotheses will become ambiguous, not precise, and therefore untestable if we diagnose selected anatomical regions of phena rather than taxa. Using all homologies which display a number of character states is the only hope for untangling phylogeny befuddled by mosaic evolution. Dental or other features of compared taxa may reflect a range of relative ages of similarities, another set of homologies may show that, within the dentally defined group, several other subgroups exist, exhibiting their most recent similarities in cranial or postcranial features. It is, therefore, a prerequisite to sound phylogenetic analysis that fossil groups be delineated in all known parts of their anatomy by the combination of all their advanced features. Without this procedure we cannot diagnose, in a strict sense, or construct useful morphotypes, and therefore cannot construct precise phylogenetic hypotheses.

## SYSTEMATIC PROCEDURES FOR THE
## ESTABLISHMENT OF HISTORICAL RELATIONSHIPS OF TAXA

It is important to have a clearly defined understanding where decisions about geneology must be made, and, subsequently, areas where disagreements should be resolved.  In conducting character analyses, the bases of phylogenetic hypotheses, the following levels of observation and decision, and identifiable areas of disagreement, exist:  1) problems and potential disagreements on the existence of a homology; 2) problems and potential disagreements on the polarity of a number of homologous states; 3) problems and potential disagreements in weighting similarities which suggest contrasting hypotheses of relationships (i.e. sorting out relative recency of shared and derived characters).

With these areas of disagreement in mind, the following is an attempted summary of the operations involved in character analysis and in choosing one of several competing phylogenetic schemes of taxa:

1.  Observations are made (as an indirect result of a host of unexpressed assumptions and hypotheses) and a particular set of circumstances is stated to be present in two or more taxa.  In other words, characters are recognized and delineated.  We can refer to this as data gathering.

2.  If these similarities, as originally recognized, can also be recognized by others (i.e. if they are repeatable), then it may be said that we have an empirical data base.

3.  The hypothesis may now be advanced that the similarity is either the result of homology and more specifically, that it is the sharing of an ancient (primitive, ancestral) or less ancient (advanced, derived) character, or, if not an homology, that it is convergence.  This hypothesis is arrived at when alternative character states are compared by an examination of both ontogenetic and adult states, as well as by mechanical analysis of the character.  This pivotal phase of analysis requires the use of the biologically most sophisticated methods, techniques, and interpretive schemes.  Decisions on this level profoundly affect what is commonly called "testing" of phylogenetic hypotheses.

One should have sound biological reasons to hypothesize a character state to be primitive when it appears late in the fossil

record, and taxa with what is identified subsequently as advanced
states abound in earlier strata.  For obvious reasons (see
Simpson, 1975) character states of a given feature are less variable at the onset of evolution.  It is therefore expected that the
primitive, rather than advanced, character states should occur
with greater frequency at earlier dates.  Character states become
progressively more variable and diversified as the record gets
younger and the opposite is more than likely when the taxa we
study are older.  Working hypotheses (to be rigorously screened
by morphological criteria), which are based on biostratigraphic
evidence, therefore are important starting points for the establishment of morphocline polarities.

4.  "Testing" of polarities should proceed beyond character
analysis when possible, by comparing the hypothesized polarities
of character clines to one another, a method referred to by
Hennig as "reciprocal illumination."

5.  With the known or suspected polarities of as many character clines as possible, using shared and derived characters, and
by weighting the phylogenetic valency of biologically different
kinds of shared and derived characters (see especially Hecht and
Edwards, 1976), a phylogenetic hypothesis is constructed (using
both "sister group" and "ancestor-descendant" concepts, depending
on the temporal nature of the evidence) about taxa into which one
should attempt to place the investigated homologies in a relative
time framework.  When possible attempts should be made to arrange
a phylogenetic hypothesis in an absolute time framework, using all
the available rockstratigraphic and biostratigraphic evidence.  It
is desirable that this phylogenetic hypothesis should postulate
the least number of possible derivations for unique and functionally highly integrated features.  That theory of relationships of
taxa which accounts most parsimoniously for all the postulated
polarities of the known and weighted characters (Hecht and
Edwards, 1976) is to be preferred.  Should this call for parsimony
be not heeded then nothing prevents one from postulating any
phylogenetic hypothesis, because without this methodological limitation independent evolution of characters may always be postulated.

When one considers all these factors in hypothesis formulation then one conforms most closely to the notion of striving for
low probability, high information content, and easy falsifiability.
Contrary to some statements on the alleged untestability of
ancestor-descendant hypotheses (Cracraft, 1974) consideration of
the fossil record and its temporal data assures an even greater
information content and opportunity for contradiction when the
fossil record and the biology of characters are known.  This results in a more desirable hypothesis.  Students who choose to
ignore and attempt to be independent from the fossil record by

offering hypotheses which are formulated in such broad and ambiguous ways that new discoveries do not affect them are of questionable value for the study of evolutionary history) both its geneological and transformational aspects).

## SOME PREVIOUSLY SUGGESTED HYPOTHESES
## OF EUTHERIAN HIGHER TAXON RELATIONSHIPS

Were we to choose a dozen or so synthetic works on mammalian relationships published in this century and at the same time rich with information on fossils as well as living taxa then perhaps the works of Winge (1941, but written between 1887 and 1918), Schlosser (1887-1890), Weber (1904), Bensley (1903), Gregory (1910, 1951), Stehlin (1912-1916), Matthew (1909, 1937), and Simpson (1937, 1945) would be among these. This is not an attempt to single out the "best" but merely to point to a guide to a vast subject matter which requires extensive acquaintance with morphological evidence of both fossils and living species. Few students in this century have had the combination of biological savoir faire, dedication, opportunity, and ability to absorb and synthesize mammalian evolutionary history as did W.K. Gregory, W.D. Matthew, and G.G. Simpson.

The following review of various hypotheses of some ordinal and supraordinal relationships of eutherian mammals is far from complete. It will serve, however, to bring into focus major outlines of thought and the nature of the evidence used by various students of placental relationships at least during this century, with an emphasis on the views of the past two decades.

Winge's studies, although published in English in 1941, were written between 1887 and 1918. His conclusions on early eutherian relationships (written in 1917) were, as far as I can ascertain, based entirely on cranial morphology. All eutherian orders were derived from the "Insectivora." The relationships of the group are shown in a phylogenetic tree (p. 145) in which he had the galeopithecids give rise to the Leptictidae on the one hand and the Tupaiidae on the other. From the leptictids he derived the tillodonts, perypitichids, and taeniodonts, and from the tupaiids he derived the macroscelidids and talpids; the latter was shown as giving rise to the centetids and the soricid-erinaceid group. Winge's hypothesis of early eutherian relationships, his way of diagnosing groups, and his practices for determining relationships are esoteric and, with all possible objectivity, appear to have been thoroughly falsified.

Max Weber's (1904) synthesis of relationships of the Insectivora, following the works of Dobson, Edwards, Haeckel, Leche,

Gill, Mivart, Peters, and A. Milne-Edwards, divided the group into
the Menotyphla (tupaiids and macroscelidids) and Lipotyphla.  The
latter group was split into one comprising of talpids, soricids,
and erinaceids, and another which included potamogalids, centetids,
and chrysochlorids.

Gregory's (1910) monumental synthesis of the mammals resulted
in a classification specifying a number of superordinal categories
which reflected his idea of early eutherian relationships at that
time.  The Therictoidea included the Insectivora (essentially the
Lipotyphla) and the Ferae Linnaeus (1758), the Archonta included
the Menotyphla (Tupaiidae and Macroscelididae), Dermoptera,
Chiroptera, and Primates.  He considered the Rodentia (both the
rodents, present sense, and lagomorphs) and the Edentata (with
a questioned inclusion of taeniodonts, tubulidentates, and pholidotans) to have been independently derived from an undiscovered
ancestral stock.  It appears that Gregory considered primarily
cranial and dental evidence.  Most importantly, the "Insectivora"
of past grouping was placed into two supraordinal categories,
underscoring Gregory's attempts to separate genealogy from level
of organization whenever the evidence clearly permitted it.

Matthew (1909, 1937) extensively utilized the morphology of
the teeth in his assessment of evolutionary diversity of mammal
faunas at a given time (e.g. early and middle Paleocene mammals),
but he also attempted to diagnose constellations of tarsal characters.  His efforts to group taxa and relate various types of feet
to one another unfortunately failed, largely because he neglected
to consider various similarities in terms of relative recency of
these features.  His assessment of similarities was unquestionably
patristic - considering total shared and inherited morphological
resemblance.  He nevertheless recognized general similarities
(which are now recognized as advanced) shared between taxa.  For
example, Matthew (1937, p. 317) showed that the foot of Onycho-
dectes, an early taeniodont, is essentially like that of leptictids and rodents!  He did not, however, employ this information
in hiw own review of the phylogenetic ties of Paleocene-Eocene
eutherians (Matthew, 1937, fig. 83).

Simpson's (1945) synthesis is well known, so I will only note
that within his cohort Unguiculata he included the Insectivora as
a waste-basket order, the Dermoptera, Chiroptera, Primates, Tillo-
dontia, Taeniodonta, Edentata and Pholidota.  He grouped the remaining eutherians into the cohorts Glires, Mutica (whales), and
Ferungulata.

McDowell's (1958) study of insectivoran cranial anatomy led
to several hypotheses of relationships.  He firmly united members
of Haeckel's 1866 Lipotyphla.  In addition, he believed that the

leptictids should be closely associated with the tupaiids, and
placed into either the Primates or Menotyphla. The Dermoptera was
derived from stem menotyphlans. McDowell noted that the zalamdalestids are closely related to the leptictids, and that the
macroscelidids, as held before, are close relatives of tupaiids
and leptictids. He set aside the Palaeoryctidae, based on basicranial features of Palaeoryctes, from the Lipotyphla, and included
in that group the Deltatheridiidae.

Van Valen's 1966 view of eutherian relationships was based in
its entirety on dental characters alone. The subsequent synthesis
was stimulating and in many instances frankly speculative. Lipotyphla was again reaffirmed to be monophyletic, with some possibility
that zalambdodont lipotyphlans may have been palaeoryctid derived,
and the Palaeoryctidae (including in this the Deltatheridiidae)
and their alleged descendants, nominally the creodonts, were given
ordinal recognition as the Deltatheridia. Again in 1967, Van
Valen commented on a whole host of eutherian groups, but even those
taxa extensively known by postcranial morphology were noted by the
relevance of their molar characters alone. In essence, Van Valen,
presumably on the basis of a figure of relationships lacking a
time axis (but not a cladogram or a phylogenetic tree), included
a host of groups in the Insectivora. However, he included the
tenrecoids and chrysochlorids within the Deltatheridia and, as in
1966, considered the palaeoryctoids to be hyaenodontan deltatheri-
dians.

In 1968 Szalay presented a speculative phylogeny of eutherians.
That schema was unfortunately also entirely based on dental charac-
ters. I had at that time grouped the Lagomorpha, Anagalidae, and
Didymoconidae as a monophyletic taxon.

McKenna's 1969 review of mammal differentiation maintained
that a leptictid-insectivoran-primate group was separate from a
palaeoryctoid-ungulate-carnivoran-creodont group by the late
Cretaceous. Lillegraven (1969), like McKenna, has palaeoryctids
(most specifically Cimolestes spp.) give rise to the creodont,
carnivoran, and ungulate radiations. Lillegraven postulated
origins of late Cretaceous leptictids (e.g. Gypsonictops) from
the Asiatic Kennalestes (i.e. quite independent from North American
palaeoryctids such as Cimolestes and Procerberus) and the derivation of other Insectivora and the Primates from late Cretaceous
representatives of the Leptictidae. The Rodentia was derived
from the Primates.

Van Valen (1971) recently published a phylogeny of the therian
mammals in which he grouped the Carnivora, Hyaenodonta, Taeniodonta, Lagomorpha, Edentata, Chiroptera, and Primates (and the
alleged derivatives of the Primates, the Rodentia) along with

ancestral Insectivora, as the Unguiculata. All other orders were
derived from the Condylarthra, and in turn all these were grouped
within the Ungulata. All the unguiculate orders, except the
Rodentia, were shown to be independently evolved from Insectivora.

Following the hypotheses of Broom (1926, 1927, 1936)
Hoffstetter (1972, 1973) has suggested summarily that eutherians
early divided into two groups, the Henotherida and Neotherida.
The division is based on the alleged number of bony elements in
basicranial axes as reported by Broom.

Szalay and Decker (1974) presented a brief resume of what
they assessed to be eutherian interrelationships. In addition to
dental evidence, the postcranial anatomy, particularly the tarsus
of several fossil groups, was taken into consideration. The
characters judged to be most recently shared between groups,
irrespective of whether cranial or postcranial, were employed for
this analysis. In their schematic phylogeny, palaeoryctoids (including leptictids and derivatives), edentates, rodents, and taeniodonts were shown to be monophyletic, whereas the condylarths and
derivatives, lipotyphlans, primates, dermopterans, and chiropterans
were shown to be another natural group. The Lagomorpha-Anagalida
(shown as a monophyletic group), Hyaenodonta (=Creodonta), and
Carnivora were shown to have been derived prior to the last common
ancestry of the palaeoryctoid and condylarth-lipotyphlan groups.

A recent synthesis, that of McKenna (1975), presented a detailed discussion of mammal relationships and attempted a largely
dichotomous classification of the Mammalia based on what he considered to be the most plausible hypotheses of relationships.
Although a large number of soft anatomical and postcranial features were employed in describing some taxa, I judge that the
majority of dichotomous splittings postulated by McKenna were
based on dental aspects of the fossil record and living species.

The phylogenetic hypotheses McKenna held were expressed with
a precision previously unattempted in mammal classifications.
McKenna's (1975, pp. 40-42) detailed holophyletic classification
of the Mammalia precisely reflects his hypotheses of genealogical
interrelationships. His schema postulated that the first split
of the eutherians involved the Edentata and the ancestor of all
others, the latter group being referred to as Epitheria. The
latter further split into the Ernotheria, which is characterized
by the loss of $P^3_3$, and the Preptotheria, which in turn is diagnosed by the retention of $dP^5_5$. The Ernotheria was said to have
divided into a number of genera grouped together (Kennalestes,
Asioryctes, palaeoryctines) and the Leptictida, which is characterized by the inferred loss of $dP^3_3$ and astragali and calcanea
which are lagomorph-like. The Leptictida was postulated to have

split into the Anagalida and the Tertiary leptictids, which have
an inflated sternum, multi-cuspate incisors, and a fused tibia-
fibula.  The Anagalida, characterized by somewhat prismatic cheek
teeth, divided into the anagalid-macroscelidid group and into the
Lagomorpha, diagnosed by a deepened jaw.  The furcation of the
Lagomorpha into the Pseudictopidae and the zalambdalestid-lagomorph
(sensu stricto) group is based on the shared enlarged $I_1$ and $I^2$ of
the latter group in contrast to the multicuspate incisors of the
former.

The Preptotheria were said to have split into the Deltatheridia (Deltatheroides-Deltatheridium group), diagnosed by the inferred loss of $dP_2^2$ and $P_2^2$ or $dP_3^3$ and $P_3^3$ and a shortened snout, and
the Tokotheria (Ferae, Insectivora, Archonta, Ungulata) which he
postulated to share in a derived manner the loss of $M_3^3$ and the reduction of the incisor formula to $I_3^3$ .

## SUGGESTED MONOPHYLETIC GROUPS WITHIN THE EUTHERIA

### Critique of Some Previously Suggested Hypotheses

Before I express my current views on eutherian interrelationships, corroborate several previously proposed hypotheses, and
propose some new ones, I will first present some arguments which
I believe falsify a number of hypotheses proposed in the past.

In order to explore eutherian relationships, partly based on
dental criteria, it is important that some aspects of therian and
metatherian dental attributes be discussed.

The dental formula of the therian morphotype recently suggested by McKenna (1975) is endorsed here.  The hypothesized dental formula for this ancestor is as follows:

$$I \; \frac{1,2,3 \quad d1,2,3}{d1,2,3 \quad 1,2,3} ; \quad C \; \frac{1 \quad d1}{d1 \quad 1} ; \quad P \; \frac{1,2,3,4,5 \quad d1,2,3,4,5}{d1,2,3,4,5 \quad 1,2,3,4,5} ; \quad M \; \frac{d1,2,3}{d1,2,3} .$$

McKenna's (1975) arguments that "$P^4$" and "$P^3$" of the Asiatic
Cretaceous eutherians Zalamdalestes and Kennalestes are really
$P^5$ and $P^4$ respectively, are not improbable.  This view is based on
morphology and taxa representing stages of an evolutionary sequence,
rather than on mixed criteria which would also employ the sequence
of eruption as a clue to serial homology.  Along the same line of
reasoning, the hypothesis that the marsupial $M_1^1$ is really $dP_5^5$ of

the morphotype therian dentition (Archer, 1974) is also plausible.

McKenna's schema of mammal relationships is based on his interpretation of cheek tooth homologies. Unfortunately, the pivotal arguments for such a schema, namely the reasons for specific tooth homologies, are skimmed over. The recent literature dealing with the relevant ontogenetic evidence reflects a complete lack of consensus about "homologies" of therian cheek teeth or a common method whereby these serially repetitive structures may be treated. McKenna did not mention Ziegler's (1971) study in which that author concluded that premolar loss in mammals proceeds from front to back and after the replacement tooth is lost, the deciduous homologue in the same tooth family is the next tooth to be lost. Archer (1974), however, in his detailed account of cheek tooth development in Antechinus flavipes, noted that he could find neither paleontological nor ontogenetic evidence for this rule in metatherians.

Archer's (1974) study of cheek tooth development in Antechinus flavipes clearly shows that there are eight independent tooth families. In applying the terminology of Thomas (1887) to these (P1, P3, P4, dP4, M1, M2, M3, M4), he notes that (p. 56): "Application of these does not mean that I imply any successional relationships between any of the teeth or believe a P2 family is lost in the dentition of Antechinus flavipes. Thomas' (1887) nomenclature is used simply because it is familiar and widely accepted in connection with Australian metatherians." He furthermore states that (p. 59): "It is clear that in Antechinus flavipes teeth established posterior to the C position are separate tooth families and each has only one generation. There are, therefore, no true successional postcanine teeth in the sense of milk and permanent teeth of succeeding generations such as are believed to occur in most eutherians."

Different students continue to use different criteria in establishing either serial homology or whether the tooth in question is a replacement or a deciduous one. The determination of the homology of successive teeth in different species by Schwartz (1975) for example, is sometimes based on mixed and therefore unacceptable criteria. Recognition of a specific tooth in a fossil jaw as being small, i.e. "reduced" (a topographical-homology criterion) is not comparable to the criterion whereby the homologies are determined by the sequence of eruption in developing jaws. In the first case the assumption is that successive teeth represent homologous loci whereas in the second case it is assumed that an inferred eruption sequence is fixed with no differences between species, which in turn would justify its use for a phyletically relevant "marker" role.

The replacement of teeth is the only developmentally asses-
sable criterion of homology in fossil and most living forms and it
is well known that the same number of deciduous teeth, having the
same number of replacement teeth, can have widely different eruption sequences among mammals.  If this is so then either the
eruption sequence or the sequence of teeth in a jaw is not a
proper criterion to identify tooth homologies.  Because the exact
number of deciduous teeth in the usually small samples of the
fossil record cannot be determined, the "size sequence" of teeth
in fossil jaws has often come to replace actual knowledge of replacement as the criterion for homology.  This operational criterion correlates closely with homologies determined by replacement
criteria in living mammals, but not with criteria based on eruption sequences.

One of the most useful operational methods of determining
the serial homology of mammal teeth is occlusal relationships of
tooth pairs.  Before this is rejected, evidence need be shown that
eruption sequence is a "more permanent" feature of the organism
(i.e. genotype is less apt to allow change in the eruption sequence than in the occlusal relationships of upper and lower teeth).
Undoubtedly, this field of inquiry requires extremely cautious,
non-circular exploration before it is used as an explanatory
hypothesis of the facts of the number and kinds of teeth encountered.

It is clear that methods of determining dental homologies
which can incorporate fossil evidence have not been carefully delineated.  So perhaps new dental phylogenies and subsequent cladistic classifications based on dental criteria alone should be
preceded by biological inquiries into dental morphogenesis.

McKenna (1975, pp. 30-31), apparently satisfied with the reliability of his criteria of dental homologies, postulated that
"...the number of teeth at premolar loci later having been reduced to 4 in several ways in adult eutherians but to 2 nonreplaced
or nonreplacing premolar loci, a replacing premolar $(P^4_4)$, and a
retained last milk premolar in primitive marsupials."  Several
points must be clarified.  A number of living eutherians replace
four teeth.  Let us, for the moment, assume (although this is
open to debate) that the deciduous premolars of, e.g., carnivorans
or insectivorans are serial homologs of the teeth replacing them
(for example, as suggested by McKenna, those of $P^{1,2,4,5}_{1,2,4,5}$ for

leptictids or anagalidans).  The loss of a lower and upper tooth
from the therian morphotype zahnreihe of five replacing premolars
by the vast majority of eutherian ordinal morphotypes may be interpreted as a synapomorphy, occurring only once, or, as McKenna
suggests, convergently, several times.  What is crucial is

knowledge of whether these teeth were deciduous, or replacement
or both.  These facts are lacking as yet.  The evidence published
by Clemens (1973) shows Gypsonictops to have five, as opposed to
four premolars, and I assume, as McKenna (1975) did, that this is
concrete evidence for the existence of five as opposed to four
ancestral premolar loci.  From this point on, however, I disagree
with McKenna's hypotheses of convergent reduction of premolar
numbers (given a total of four lost teeth!) in eutherians he calls
non-ernotheres.  Rather than considering the independent reduction
and loss of one tooth (deciduous and permanent) in a smattering of
taxa at the same locus, is it not more parsimonious to regard this
reduction and loss as representative of a common ancestral stage
leading to the ancestral condition of all other eutherian formulae?
The tooth at the P3 locus of Gypsonictops may represent either a
reduced $P\frac{3}{3}$ with a deciduous precursor or a $dP^3_3$.  If the latter is
proved to be correct (the replacement tooth having been suppressed
in a common eutherian ancestor) then four premolared and three
molared eutherians, regardless of how we want to express the loss
of the P3 locus in our formulae, would have serially homologous
teeth.  Work by Kindahl (1957, 1958, 1967) has shown that reduction of loss of a locus often manifests itself on the deciduous
tooth before affecting the replacement.  This would mean that the
holophyly of the Ernotheria, if a natural group, would have to
rest on characters other than premolar homologies.

Even though, as McKenna (1975) boldly and, I believe, justi-
fiably suggests, most early therians had five premolar loci, the
presence of probably even more premolars in Kuehneotherium and
ancestry (McKenna, 1975) and therefore in an all inclusive concept of the Theria should caution us about a reordering of relationships based on newly interpreted formulae, particularly where
corroborative evidence seems to be missing.[2]

Sharing of possibly primitive, rather than any demonstrably
derived, characters between the morphotype of Tertiary leptictids
and the Anagalida sensu McKenna, 1975 weakens or obviates the
hypothesis on which the superorder Leptictida is based.  If the
loss of $P^3_3$ will prove to be the universal therian or eutherian
method of premolar reduction, then this primitive feature, again,
shared between the superorder Kennalestida McKenna, 1975 and superorder Leptictida McKenna, 1975 is not, as noted above, a good
character for the erection of the Ernotheria McKenna, 1975.  Furthermore, McKenna's suggestion that ernotheres are united by,
among other things, lagomorph-like foot structure is sharply contradicted by the well known astragalocalcaneal complex of leptictimorphs (comprised of Leptictidae, Pantolestidae, and Taeniodonta,
see below).

The Preptotheria are suggested to have been derived from an

ancestor with a dental formula of 5/4 incisor count; $C^{1}_{1}$; $dP^{1}_{1}$;

$P^{2,3,4,5}_{2,3,4,5}$. $M^{1,2,3}_{1,2,3}$ a pattern suggested by McKenna (1975) to have been
inherited from a <u>Peramus</u>-like form in the Jurassic. As noted
above, however, no convincing evidence suggests, whether premolar
reduction from five to four in number was accomplished (a) more
than once; (b) several times, the reduction being from the same
locus independently; or (c) several times, premolar loss occurring
at different loci. It is therefore difficult to understand why
McKenna (1975) considers all non-ernotheres to have departed independently from a five premolar locus ancestry when corroborating
evidence, i.e. other shared and derived features, are lacking.
Characters which unite the Lagomorpha, Anagalida, and Macroscelididae are valid, but do not include, as a uniquely derived feature, the loss of teeth at the $P^{3}_{3}$ loci. At present, <u>Deltatheroides</u>
and <u>Deltatheridium</u> are probably not relevant to the loss of $P^{3}_{3}$.
As McKenna (1975) states, no one has observed actual premolar replacement in deltatheriids. The reduction of the last molar
further complicates the significance of deltatheriids to the problem of tooth elimination at the $P^{3}_{3}$ locus (and therefore the enigma
of non-ernothere relationships) beyond any reasonable future
solution.

The Tokotheria, suggested by McKenna (1975) to be the sister
group of the amended concept of the Deltatheridia, is stated by
him to be holophyletic, although, as he admits, the internal relationships are ill understood. In my view, the Tokotheria, in
which McKenna ties together the Ferae, Insectivora, Archonta, and
Ungulata through four respective taxa suggested to be ancestral,
the genera <u>Cimolestes</u>, <u>Batodon</u>, <u>Purgatorius</u>, and <u>Protungulatum</u>,
is not based on any unambiguous similarities which may be inter-
preted as shared derived. I believe that his hypothesis is entirely based on the recognition of "grade" similarity of primitive
eutherian tribosphenic molar morphology, its interpretation as a
monophyletically attained characteristic of the tokothere clade,
superficial convergence, and in a number of assumptions about premolar adaptations (see below). It is unfortunate that no attempt
was made to consider other than dental hard anatomy, particularly
in the affinities of <u>Cimolestes</u>, and subsequently in the erection
of the new order Cimolesta. Similarly, the inclusion of the
Pantodonta in the Cimolesta and within the Tokotheria is explic-
able only by the dismissal of postcranial evidence by McKenna and
the use of grade similarities of molars as synapomorphies shared
with <u>Cimolestes</u>.

McKenna states that the "...tokotheres share or further
modify a postcanine dental formula consisting of $dP^{1}_{1}P^{2}_{2}P^{3}_{3}P^{4}_{4}dP^{5}_{5}M^{1}_{1}M^{2}_{2}$.

$dP\frac{5}{5}$ is the tooth usually called $M\frac{1}{1}$; the last premolar is primi-

tively a nonmolariform $P\frac{4}{4}$, the same tooth as the penultimate pre-

molar in ernotheres. These homologies are fundamental" McKenna
presents his hypothesis clearly, and the logic in the application
of his assumptions to his interpretation cannot be faulted. However, it may be worth restating why McKenna considers the tokothere
morphotype formula to be as given by him. He believes that the
last premolar (in conventional terms) of the tokothere group is
(a) nonmolariform and (b) therefore homologous with the ernothere
$P\frac{4}{4}$. I believe that these interpretations are the bases of his
hypothesis (or conclusion) for the tokothere morphotype dental
formula, and furthermore that they rest on some fundamental assumptions about teeth. There are two reasons why I disagree with
McKenna's assumptions and, therefore, interpretations of tokothere
monophyly.

First, I believe that the overwhelming evidence from many
groups of mammals (fossil as well as living) is that the last premolar (conventional concept) will often vary from non-molariform
to molariform, all within one clearly recognized and tightly knit
(monophyletic) group. I also believe that too much emphasis is
placed on the "type" of last premolar that "groups" have, or, to
put it another way, often "groups" are created because the con-
formation of the past premolar supposedly displays "adaptation-
free" qualities. Another facet of the "last premolar stasis"
assumption is that the dynamic role of mechanical solutions to
constantly shifting diets, resulting in shifting the areas of
activities where various roles are performed, is considered to be
somehow biologically subordinated to the phyletically determined
morphology of the premolars. In my experience this is not a very
reliable rule.

Second, I simply disagree with McKenna's assessment of what
constitutes morphotype conditions of the last premolar in the
Ferae, Insectivora, Archonta, and Ungulata.


Phylogeny of the Major Stocks of Eutheria

Figure 1 summarizes my present views of eutherian interrelationships, based almost exclusively on interpretations of
osteological evidence. Given what I believe to be unresolved
problems of origin or branching (in this figure represented by
circles and broken lines) I see no justification for the coining
of new names to express several of the proposed speculative
hypotheses. Perhaps a three-partite division of the Eutheria, as
shown, is warranted, but new names for these divisions would

TABLE 1

| Number of taxa | Rooted trees |
|---|---|
| 2 | 1 |
| 3 | 3 |
| 4 | 15 |
| 5 | 105 |
| 6 | 945 |
| 7 | 10,395 |
| 8 | 135,135 |
| 9 | 2,027,025 |
| 10 | 34,459,425 |
| 11 | 654,729,075 |
| etc. | etc. |

The number of possible phylogenetic trees given the number of taxa left.

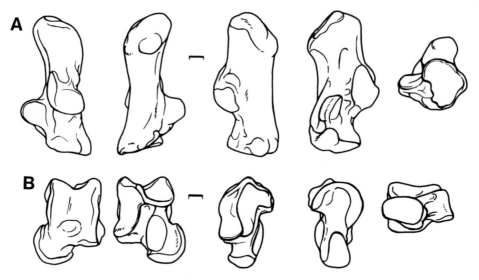

Fig. 5.  Left calcaneum (above) and left astragalus (below) of
Erinaceus europaeus.  From left to right:  dorsal, ventral, lateral,
medial, and distal views, respectively.  The presence of a large
fibular facet, a clearly discernable cavitation, judged to be
homologous with the dorsal astragalar foramen, renders the astragalocalcaneal complex of this hedgehog one of the most primitive
among living eutherians.

clearly clutter the literature without adding science to it, as the number of equally possible trees, as I see it, is still 15-10395.

Group 1.
  Cohort Archonta Gregory, 1910.

Gregory's (1910) hypothesis of tupaiid-primate-dermopteran and chiropteran relationships was so genuinely based on phylogenetic reasoning that later students, using a more patristic approach to phyletics, often found it impossible to accept (see for example the comment by Simpson, 1945, p. 173). In 1974 I published a brief note on the special similarity of the tarsus in Paleocene primates and colugos, and recently noted pedal similarities (see below) which I consider almost certainly derived features shared by Ptilocercus, Paleocene primates, and dermopterans. The Archonta of Gregory (sans the Macroscelidae, see McKenna, 1975), including the Scandentia (Tupaiiformes), Primates, Dermoptera, and Chiroptera, appears to be a sound monophyletic, and probably also holophyletic category.

Although the Archonta is judged to be monophyletic, branching relationships between the Scandentia[3] (tupaiids and mixodectids), primates, and colugos is not clearly understood (see Fig. 1). As noted by Simpson (1937), Van Valen (1966), and Rose (1975), the mixodectids and the earliest known dermopterans, the Plagiomenidae, may be closely related and I suspect that the ties of tupaiids and paromomyiform primates to these families was also close but genealogically still unclear.

I find that the single, most clearly discernable special similarity of the tarsus of Ptilocercus, outside the Tupaiidae, is with known paromomyiforms and dermopterans (Figs. 6, 7). In spite of the numerous statements in the literature that tupaiids are largely primitive eutherians postcranially (Jenkins, 1974), they are not any more so than are the murid rodents, for example. The postcranial morphology of a laboratory rat is in fact probably more similar character for character to a eutherian morphotype than are the features of tupaiids. The similarities of the Ptilocercus astragalocalcaneal complex to that of Plesiadapis gidleyi, for example, are unique derived similarities, present only in colugos and primitive adapids. The following combination of characters are archontan features in the known tarsals. The tibial trochlea is proximo-distally elongated while traces of the superior astragalar foramen are still retained; the body of the astragalus is laterally high and sharply crested, whereas its medial border is rounded; the astragalar sustentacular facet tends to be confluent with the naviculoastragalar facet medially; on the calcaneum the astragalocalcaneal facet, aligned proximodistally,

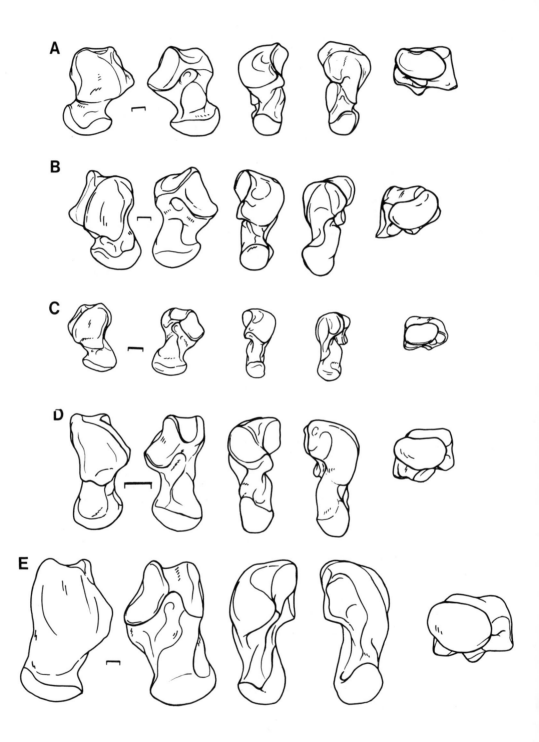

Fig. 6. Comparison of astragali of some representative archontans.
A. Plesiadapid primate from the Torrejonian Paleocene of Bison
Basin (AMNH 89533). B. The primate Plesiadapis gidleyi from the
Torrejonian Paleocene of Gidley Quarry (AMNH 17379). C. The paromomyid primate Phenacolemur sp. from the early Eocene East Alheit
Quarry (AMNH 88812 and 29131). D. The recent tupaiid Ptilocercus
lowii (CNHM 76855). E. The recent dermopteran Galeopithecus volans (UMZ Cambridge, England E4023I). From left to right: dorsal,
ventral, lateral, medial, and distal views, respectively. Scales
represent one mm.

A combination of anteroposteriorly elongated tibial trochlear
facet, rounded medial crest and sharp lateral crest of the tibial
trochlea, medial articular surface bypassing the dorsal astragalar
foramen, ventral astragalar foramen, sustentacular facet nearly
confluent medially with the navicular facet, and medially relatively deep navicular facet are interpreted to be a shared and derived
character constellation of archontan common ancestor. This primitive archontan character complex is most closely approached among
known Archonta by Plesiadapis, Phenacolemur, and Ptilocercus.

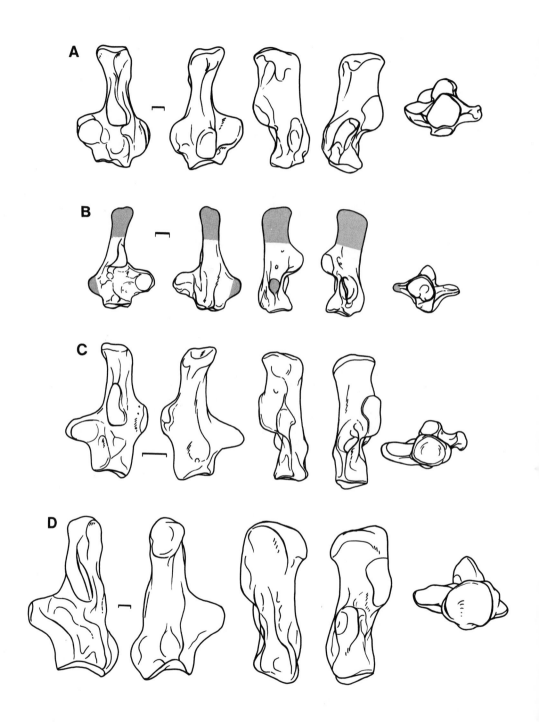

Fig. 7. Comparison of calcanea of some representative archontans. A. The primate Plesiadapis gidleyi from the Torrejonian Paleocene of Gidley Quarry (AMNH 17379). B. The paromomyid primate Phenacolemur sp. from the early Eocene East Alheit Quarry (AMNH 88815). C. The Recent tupaiid Ptilocercus lowii (CNHM 76855). D. The Recent dermopteran Galeopithecus volans (UMZ Cambridge, England E4023I). From left to right: dorsal, ventral, lateral, medial, and distal views, respectively. Uniform stippling indicates reconstruction. Scales represent one mm.

The compination of a screw-like astragalocalcaneal facet (Permitting rotation as well as translation), a well defined sustentacular groove for the tendon of m. flexor digitorum fibularis, and a nearly circular, slightly concave cuboid facet are interpreted to be a shared and derived character constellation of the archontan common ancestor. This primitive archontan character complex is most closely approached among known Archonta by Plesiadapis, Phenacolemur, and Ptilocercus.

allows both rotation and at least a slight amount of translation;
the fibular facet is almost completely eliminated; the cuboidocal-
caneal facet is founded with a slight depression and is aligned
transversely to the long axis of the calcaneum.

This constellation of special similarities represents the
base for tupaiid-primate-dermopteran ties, although Gregory did
not consider the pedal evidence, some of which (from the Paleocene)
was unavailable to him.  Nevertheless, the significance of tupaiid-
primate similarities, as special similarities, was strongly
impressed on Gregory (1910, p. 274): "Thus in the constitution and
arrangement of the auditory region, the Tupaiidae, as observed by
all authorities, differ radically from the other Insectivores and
approach the non-Malagasy Lemurs (Weber, 1904, pp. 366, 745).  If
as Van Kampen suggests (1905, p. 452) the entotympanic represents
the expanded and secondarily independent tympanic wing of the
petrosal, which is seen in many Marsupials and Insectivores, then
the difference in the bulla...becomes a difference in degree rather
than of kind."  The path of descent of bulla morphology nevertheless remains obscure when all the groups in the Archonta are considered.  In spite of the unity of the primates in possessing a
uniquely derived condition, the petrosal bulla (Szalay, 1975b), it
is conceivable, as Gregory suggests, that (a) the tupaiid entotympanic is a secondary feature from an archontan ancestry with a
petrosal bulla, or (b) that the entotympanic, ectotympanic, and
petrosal solutions to cover the bulla ventrally were evolved independently from an ancestral archontan with either a cartilaginous
or composite bulla.  In addition, there stands McDowell's (1958)
suggestion that the petrosal bulla is the result of progressively
earlier ontogenetic ossification of the entotympanic and its
fusion to the petrosal center of ossification.

The hypothesis of colugo relationships is derived from a very
old idea inasmuch as in 1758 Linnaeus referred to one of the dermopteran species as Lemur volans. In spite of early suspicions,
views fluctuated widely, as reviewed by Gregory (1910, pp. 316-317),
and some discussions by Simpson (1937, pp. 127-131), Van Valen
(1966), and Rose (1975) added new ideas, interpretations and,
admittedly, some confusion to the understanding of the relationships of the strange colugos.  I recently published (Szalay, 1975a)
an abstract indicating my interpretation of some postcranial characters of some specific primates and dermopterans as shared derived
characters.

Fossil postcranials will undoubtedly be of crucial importance
for further testing the hypothesis of bat origins.  As Gregory
(1910, p. 317) noted, Leche discovered that the patagium of
dermopterans and fruit bats are strikingly similar in pattern and
therefore most likely homologous.  This character alone is of

greater significance for relating the Chiroptera to the Dermoptera than any other feature which may be cited for the "insectivore" origin of bats, based on vague and convergent grade similarities of molars. Unlike those of the gliders of the Sciuridae, Anomal-uridae, and Phalangeridae, the colugo patagium surrounds the body, is attached to the tail, extends between the long fingers, and is supplied by the same muscles and nerves as those of bats.

If the archontans are holophyletic, then which other known groups are their most recent relatives? Much of the literature pertaining to this question has focused on the origin of the Primates. Most recently, papers by Szalay (1975b), Luckett (1975) and Goodman (1975) have probed this question but adequate answers have yet to be found. As in 1975, for the Primates specifically, I would venture to hypothesize the origin of the Archonta, an arboreal to volant group, from some unknown stocks of terrestrial adapisoricids. In addition to this speculative view of archontan descent, I would like to note that the earliest archontans, forms probably not unlike Ptilocercus, were arboreal and that this mode of life, contra Jenkins' (1974) views on the origins of primate arboreality and Kay and Cartmill's (1974) notions of primate origins, was a derived condition not only for the earliest primates but also for the Scandentia. The postcranial morphology of the combined primate and tupaiid (i.e. Ptilocercus) evidence points to an ancestry which was arboreal in contrast to its ancestral stock. For some detailed arguments on joint mechanics which pertain to this problem see Szalay and Decker (1974) and Szalay, Tattersall, and Decker (1975).

Cohort Ungulata Linnaeus, 1766.

There is no clear cut consensus on the exact genealogical ties of the early representatives of this cohort, although most students of Cretaceous and early Tertiary ungulates appear to agree on the monophyly of what has been previously grouped as the Condylarthra and derivatives (see especially Van Valen, 1971, and McKenna, 1975). It should be added here that, as far as I know, no substantive evidence exists which may be interpreted to contradict the concept Ferungulata formally expressed by Simpson (1945), although no shared and undisputably derived characters are known to me which would strongly warrant that hypothesis.

McKenna (1975) recently pointed out that the concept Condy-larthra originally did not include the Arctocyonidae, now generally recognized as the stem group of the Ungulata. Similar problems surround the concept Creodonta.

As the rules governing higher category names are neither clear nor rigid, and as many of these names have not been clearly defined

when erected, one may attempt to use the consensus of past
decades to tighten up diagnoses or amend original names (see, for
example, McKenna, 1975). There are several compelling arguments
why the name Condylarthra should, even though in a modified way,
be retained in the biological, and not nomenclatorial, sense used
by Marsh, Ameghino, Weber, and Simpson, and more recently by Van
Valen (1966) and many others (see especially Simpson, 1945, pp.
233-238). Yet removal of the hyopsodontids and periptychids to
the Artiodactyla and the phenacodontids to the Perissodactyla, as
done by McKenna, would leave no remaining taxon for Cope's 1881
concept of the Condylarthra. The association of the Arctocyonidae
with the base of the great ungulate radiations has been one of the
substantial advances of paleomammalogy. The arctocyonids have
become widely accepted as the most primitive of the known Ungulata.
It is for this reason, stimulated by McKenna's (1975) arguments,
that Van Valen's usage of Arctocyonia is employed here to include
the Arctocyonidae and Mesonychidae.[4]

Specific derived features which unite the arctocyonids with
their many descendants are numerous and detailed on skulls, teeth,
and postcrania. All of these would be very difficult to delineate,
so this task is well beyond the ambitions of this paper. The
overall patristic dental as well as tarsal similarity of the
primitive hyopsodontids, periptychids, phenacodontids, and mesonychids to such genera as Protungulatum (Figs. 8, 9) and other
arctocyonids makes this hypothesis of relationships a strong one,
whatever the settlement on nomenclatorial conventions.

The character which still validates the concept Ungulata is
the presence, except when lost, of hooves. If it can be shown
that arctocyonids and the ungulates evolved hooves convergently
then perhaps the concept is not a viable one. Until then, however, the Ungulata appear to share a derived character state in
contrast to the antecedent condition.

Cohort Ferae Linnaeus, 1758.

The most convincing special similarity shared between primitive Carnivora and primitive Creodonta lies in the construction of
the tarsus (Figs. 10, 11). The following combination of features
are shared between my concept of the carnivoran and creodont
morphotypes, and as far as I know this combination is unique to
the ancestry of these hypothetical taxa. On the calcaneum, the
primitively well developed fibular facet is present but it is
modified into two facets which meet along a lateral crest. The
anterior plantar tubercle appears relatively far back, giving the
lateral and medial profiles of the calcanea their characteristic
appearance. The peroneal process is reduced. Although the astragalar foramen is retained, the tibial trochlea is elongated with

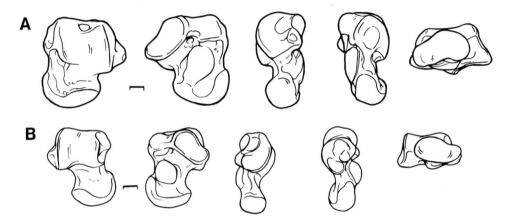

Fig. 8. Comparison of astragali of the Cretaceous ungulate <u>Pro-
tungulatum</u> <u>donnae</u> (UMVP 1914) and the Cretaceous lepitictimorph
<u>Procerberus</u> <u>formicarum</u> (UMVP 1821). From left to right: dorsal,
ventral, lateral, medial, and distal views, respectively. Scales
represent one mm.

The astragalus of <u>Protungulatum</u> is distinctly more primitive
than that of <u>Procerberus</u> in lacking sharp trochlear crests, in
having a large astragalar canal and a broad, short neck and head.
Both are primitive in having a broad, shallow and short trochlea
limited to the body. The most notable specialization of the primitive leptictimorph is the posteriorly extensive tibial trochlea
and the slight relative enlargement of the lateral half of this
articulation.

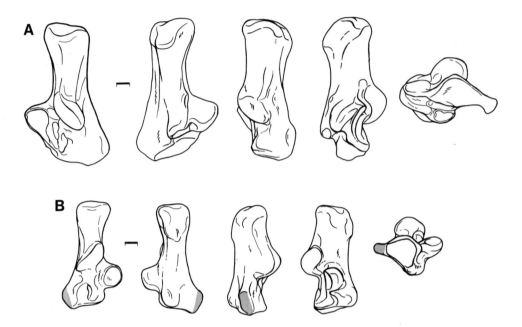

Fig. 9. Comparison of calcanea of the Cretaceous ungulate Protungulatum donnae (UMVP 1823) and the Cretaceous leptictimorph
Procerberus formicarum (UMVP 1883). From left to right: dorsal,
ventral, lateral, medial, and distal views, respectively. Uniform stippling indicates reconstruction. Scales represent one mm.

The calcaneum of Protungulatum is distinctly more primitive
than that of Procerberus in having a fully functional fibular facet
and a groove between the anterior plantar tubercle and the cuboid
facet.

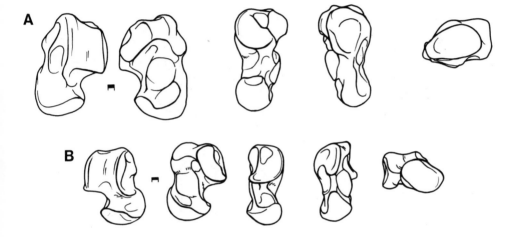

Fig. 10. Comparison of astragali of representative Ferae. A. The late Eocene miacoid carnivoran Didymictis sp. (AMNH 2855). B. The Bridgerian Eocene Sinopa grangeri (AMNH 13142). From left to right: dorsal, ventral, lateral, medial, and distal views, respectively. Scales represent one mm.

The astragalus of primitive Ferae may be characterized by the combination of a medially longer tibial trochlear facet and a tendency of the sustentacular facet to be laterally confluent with the navicular facet.

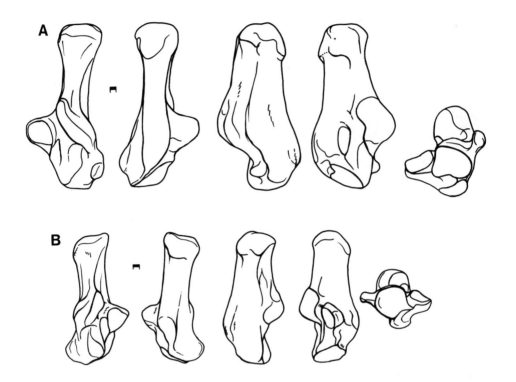

Fig. 11.  Comparison of calcanea of representative Ferae.  A. The late Eocene miacoid carnivoran _Didymictis_ _sp_. (AMNH 2855).  B. The Bridgerian Eocene _Sinopa_ _grangeri_ (AMNH 13142).  From left to right: dorsal, ventral, lateral, medial, and distal views, respectively. Scales represent one mm.

   The calcaneum of primitive Ferae may be characterized by a fibular facet modified into two which meet along a lateral crest, and an anterior plantar tubercle which is positioned more posteriorly than in other primitive eutherians.

a characteristically more extensive mediodorsal than laterodorsal surface. This difference in proportion of the astragalar body contrasts the primitive feraeans with leptictids and rodents, for example.

As is the case in numerous other instances in the study of mammalian relationships, disassociation of the Carnivora and Creodonta by Van Valen (1966) and the ensuing chaos in classification were the result of the exclusive use of dental features.

Group 1, incertae sedis.
    Order Insectivora Illiger, 1811.

The concept Insectivora, as a monophyletic unit, is restricted to Haeckel's Lipotyphla of 1866, and I follow McKenna's (1975) arrangement of the included families in the Erinaceomorpha and Soricomorpha. Erinaceomorphs would be comprised of the erinaceids and the nebulously defined early Tertiary Adapisoricidae, whereas the soricomorphs would include all the families listed by McKenna (1975, p. 38). It should be emphasized that at present it is quite impossible to ascertain whether or not the Insectivora is a paraphyletic or holophyletic group. It is likely to be paraphyletic, as the archontans were probably derived from some unknown adapisoricid. The known fossil record simply does not permit a holophyletic grouping of the known Insectivora (sensu stricto).

Morphotype conditions of the dentition and skull have been discussed in a large numbers of papers dealing with the Insectivora, sensu lato (see in particular Novacek, in press). With the exception of Matthew's (1937) comments, analytical discussions of the tarsus, rich in character states, are non-existent.

Group 2
    Cohort? Edentata Cuvier, 1798.

1. Palaeanodont-pholidotan ties. In view of the extremely primitive characteristics of the astragalus referred to Teutomanis by Helbing (1938), serious doubts exist about the close ties between pholidoyans and palaeanodonts advocated by Emry (1970). The astragalus referred to Teutomanis has an anterioposteriorly extremely short trochlear facet and a large, unreduced astragalar foramen. Combined with the primitive features, however, there are indications of the presence of a flat to concave navicular facet on the head of the astragalus, a derived similarity or convergence shared with extant manids. This type of astragalus sharply contrasts with that of the earliest well known palaeanodont, Palaeanodon (see Figs. 12, 13), from the early Eocene. The latter has the advanced trochlear morphology shown in leptictids and a primitive astragalar head morphology. A careful assessment of foot

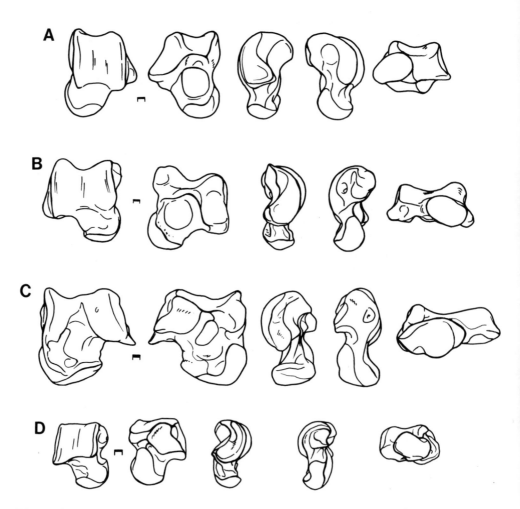

Fig. 12. Comparison of astragali of representative palaeanodonts
(A, B), Loricati (C) and leptictimorph pantolestids (D). A. The
Wasatchian early Eocene Palaeanodon ignavus (AMNH 15137). B. The
medial Wasachian palaeanodont Metacheiromys sp. (AMNH 18666). C.
The Recent armadillo Dasypus novemcinctus (private collection).
D. The Bridgerian Eocene Pantolestes nanus (AMNH 12152). From left
to right: dorsal, ventral, lateral, medial, and distal views,
respectively. Scales represent one mm.

    The most striking special similarity is apparent between astra-
gali of Metacheiromys and armadillos. Pantolestid astragali resemble those of primitive palaeanodonts perhaps only in a primitive or
parallel manner, yet the possibility of special relationships cannot be ruled out.

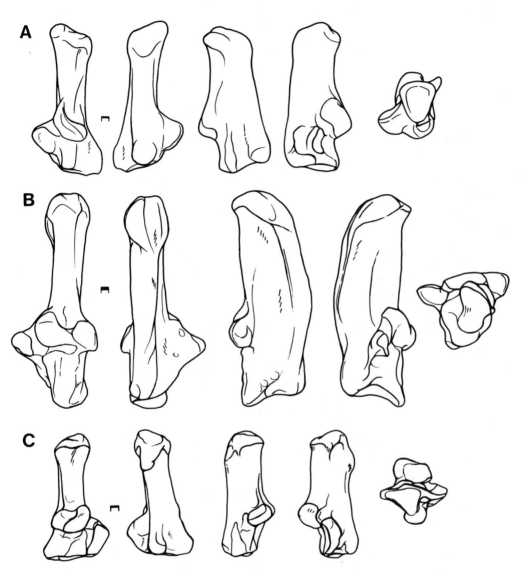

Fig. 13.   Comparison of calcanea of representative palaeanodonts
(a), Loricati (B), and leptictimorph pantolestids (C).   A. The
Wasatchian early Eocene Palaeanodon ignavus (AMNH 15137).   B. The
Recent armadillo Dasypus novemcinctus (personal collection).   C.
The Bridgerian Eocene Pantolestes nanus (AMNH 12152).   From left
to right:  dorsal, ventral, lateral, medial, and distal views,
respectively.   Scales represent one mm.

bones is clearly in order.  Helbing's referral of the astragalus
to Teutomanis  may well be incorrect; if so, the objection to a
palaeanodont-pholidotan tie could be removed.  Nevertheless,
Emry's (1970, p. 500) statement that, "...the metacheiromyids
lacked the specializations seen in the feet of Manis, but so do
the primitive manids, which are more like the metacheiromyids in
this respect" is incorrect.  To the contrary, primitive manids,
i.e. Teutomanis (if they are manids) are not more like metacheiromyids than extant ones.  The pedal evidence is stressed here
because Simpson's (1931) and Emry's (1970) careful reviews of
palaeanodont characters indicate (at least to me) that all other
lines of evidence are unequivocally ambiguous as to the alleged
shared special non-convergent similarities of palaeanodonts and
pholidotans.  If the tarsals allocated to Teutomanis  are indeed
pholidotan, then the tarsal evidence is distinctly contrary to
palaeanodont (as we know them from Palaenodon and Metacheiromys)
and pholidotan monophyly.

     2.  Palaeanodont-edentate ties.  In contrast to what I per-
ceive to be a lack of special similarity in the astragalocalcaneal
complex of the Palaeanodonta and Pholidota, the palaeanodont and
dasypodid cingulatan derived similarities of these two bones are
extremely suggestive of a common ancestral pattern.  As I judge
the dasypodid pes to be the most primitive one among known edentates (yet characteristically different from "primitive" eutherians), and as I hypothesize the palaeanodont and edentate morphotype similarities of the astragalocalcaneal complex to be shared
derived, I subsequently hypothesize sister group or perhaps
ancestor-descendant relationships between the Palaeanodonta and
Edentata.  Unclear notions of what may be the characteristic features of the morphotype of this group because of the numerous
autapomorphies of both groups, make the procedure of reciprocal
illuminations by other characters extremely difficult.

     3.  Palaeanodont origins.  I may venture a hypothesis about
the origin of the palaeanodonts.  The astragalocalcaneal complex
of paleanodonts and pantolestids are astonishingly similar in
some ways.  The differences are restricted to the greater inter-
locking of the astragalus by the crus in the former, but at the
same time allowing the original great fore and aft mobility characteristically present in leptictimorphs.  Whether the similarities
are ancestral or parallel, from a leptictid ancestral condition,
or shared and derived cannot be answered at present, but in either
case their phylogenetic significance is tantalizing.

     4.  Pholidotan-edentate ties.  If the astragalocalcaneal
complex allocated by Helbing (1938) to Teutomanis  is manid pholidotan, then the synapomorphies hypothesized below are probably
convergent.  Nevertheless, it is worthwhile to note that relatively

primitive myrmecophagids such as <u>Tamandua</u> share a similar condition
of astragalar head-navicular articulation (not a true pivot, but
rather a ball and socket articulation) as well as an extremely
similar lower ankle joint.  The astragalonavicular ball and socket  joint unites the Myrmecophagidae, Megalonychidae, Mylodontidae,
and Bradypodidae in the suborder Pilosa of Flower, 1883.  Whether
or not the Pholidota may be derived from a myrmecophagid-like
species is suggestive but still largely uncorroborated and certainly does not qualify for a new taxonomic concept uniting the
myrmecophagids with pholidotans.

   Cohort Glires Linnaeus, 1758.
     Order Leptictimorpha, new.

   Within this new order I unite the leptictids with groups I
consider to be derivatives of them.  Gregory wrote of the leptictids in 1910 (p. 260) that "The most primitive representatives of
this group (i.e. the Erinaceoidea of his sense) are the Eocene and
Oligocene Leptictidae."  This notion, based on the morphology of
molar teeth, continued throughout the ensuing 60 years and Cretaceous mammals with a tribosphenic (i.e. primitive) molar morphology continue to be characterized (necessarily by the patristic
features of molar crowns) as leptictids or "close" to them.  Even
if the postulated five premolars of ancestral therians (Metatheria
and Eutheria) are preserved in <u>Gypsonictops</u> and <u>Kennalestes</u>, the
almost certainly derived astragalocalcaneal complex of the Cretaceous Leptictidae (<u>Procerberus</u>, <u>Gypsonictops</u>, and <u>Cimolestes</u>)
bars this group from being ancestral to erinaceotans, archontans,
ungulatans, or feraeans. If <u>Asioryctes</u> and <u>Kennalestes</u> had the
apomorph tarsus of the Leptictidae then they were leptictid; if
they did not, then the Leptictidae, defined here on the basis of
<u>possession of synapomorphies</u>, should not include <u>Asioryctes</u> or
<u>Kennalestes</u> even if one of the latter was the ancestor of leptic-
timorphs.

   The ordinal concept Leptictimorpha is necessary to unite the
Leptictidae, Pantolestidae, and Taeniodonta.  There is no  other
concept which includes the leptictids and their undoubted closest
relatives with shared and derived similarities, in particular
<u>Cimolestes</u> and derivatives, the pantolestids and the taeniodonts.
The superorder Leptictida McKenna, 1975, included the grandorder
Ictopsia McKenna, 1975, as well as the orders Macroscelidea and
Lagomorpha.  The order Cimolesta McKenna, 1975, was thought to
have been allied with the Creodonta and Carnivora rather than the
leptictids.  The evidence presented here leaves little doubt that
bona fide leptictids such as <u>Leptictis</u>, <u>Prodiacodon</u>, and <u>Myrmeco-</u>
<u>boides</u> share the derived astragalocalcaneal morphology of the late
Cretaceous Bug Creek eutherians other than the arctocyonid <u>Protun-</u>
<u>gulatum</u> of that deposit.

The assessment of the leptictimorph morphotype is based on astragali and calcanea of <u>Cimolestes</u> <u>incisus</u>, <u>C</u>. <u>magnus</u>, <u>Procerberus</u> <u>formicarum</u>, and <u>Gypsonictops</u> <u>hypoconus</u> from the Bug Creek Anthills local fauna from the Hell Creek Formation (Sloan and Van Valen, 1965). These two bones, with the exception of the astragali of the arctocyonid <u>Protungulatum</u> <u>donnae</u>, which are easily identifiable by comparison to other associated arctocyonid tarsals, are the only eutherian tarsals in that fauna and occur in great abundance, covering the size range of the non-arctocyonid species listed above. These astragali and calcanea (Figs. 8, 9) although differing in size, are essentially identical, and they also closely resemble their homologues in <u>Prodiacodon</u> and <u>Leptictis</u>. It is beyond reasonable doubt that in spite of the identifiable generic differences of the dentition of such taxa as <u>Gypsonictops</u>, <u>Leptictis</u>, <u>Prodiacodon</u>, <u>Procerberus</u>, and <u>Cimolestes</u>, these genera share a more recent similarity in their astragalocalcaneal complex than any other similarity which they may share with other groups in their molar morphology. Thus, irrespective of greater molar similarities of either of the North American leptictimorphs ("palaeoryctoids") to <u>Kennalestes</u>, <u>Asioryctes</u>, or <u>Zalambdalestes</u>, these probably only reflect a more ancient similarity. The overwhelming aspect of the molar similarity of these taxa, as well as that of creodonts, carnivorans, and erinaceomorphs are probably ancestral similarities as well as some vague parallelisms.

The leptictimorph morphotype may be characterized by an upper ankle joint modified for extremes of plantar-flexion, judged by the great increase of the trochlear arc of the astragalus and the complete obliteration of the astragalar foramen, as well as by increased lateral stability, suggested by increased sharpness of the lateral border of the tibial trochlea and the reduction or loss of the fibular facet of the calcaneum. The primitively extensive groove for the plantar calcaneocuboid ligament is essentially obliterated. At the same time the very extensive navicular facet of the astragalus (a derived eutherian condition) hints at an unusually great range of mobility of the fore pes. It is not unlikely that, in a squirrel-like fashion, the ancestral leptictimorph was a capable scansorial animal. Perhaps it was even adapted for the characteristic hyperinversion seen in squirrels, a function probably also present in the earliest rodent.

I should add that I find McKenna's (1975) derivation of the Pantodonta from <u>Cimolestes</u> unacceptable. The foot structure of pantodonts (see especially <u>Pantolambda</u>) is clearly similar to those of arctocyonid "condylarths" and is unlike that of <u>Cimolestes</u> or <u>Procerberus</u>, for example. Admittedly this similarity is presently recognized as only patristic. Nevertheless the unmistakable diagnostic specializations of the <u>Cimolestes</u> foot are not present in early pantodont tarsi. Inclusion of the Pantodonta within the

recently erected order Cimolesta McKenna, 1975 appears to be
reasonably predicated on derivation of primitive pantodont dental
morphology from a Cimolestes-like ancestry.  The same reasoning
was apparently applied in the allocation of the recently described
Hypsilolambda,[5] suggested to be a pantodont (Chow, 1975).  The
wide molar stylar shelves of this genus have been assessed to resemble the sweeping ectolophs of pantodonts; the relatively large
size of this animal has influenced this conclusion.  The molar
ectoloph of pantodonts is a derived feature, and whenever this
character evolved as a probable response to some form of browsing
in later ungulates, the stylar shelf is invariably wider as a
mechanical necessity for this arrangement.  The foot of Pantolambda,
an early pantodont, was exceptionally well documented by Matthew
(1937).  The calcaneum does not have some of the specific characters, such as a tripartite superior facet of creodonts and carnivorans.  In all details, the pantodont foot is arctocyonid-like.
It is important to stress that the taeaniodonts, a Paleocene-Eocene
group of leptictimorph eutherians which also attained large size,
clearly exhibit shared derived tarsal similarities when compared
to the leptictimorph morphotype.

    Lack of any unambiguously shared derived characters between
leptictimorphs and the Ferae prohibits hypotheses which would
claim their close relationships.  The advanced foot morphology of
the Leptictimorpha (including Cimolesta McKenna, 1975) shares no
agreed upon derived similarities with the Ferae and the dental
similarities may well be nothing more than the perniciously per-
vasive primitive tribosphenic characteristics found throughout
Cretaceous Mammalia.  Leptictimorphs must be diagnosed on the
basis of their total known characters.  The rather general primitive similarity of the teeth and postcranial autapomorphies of
this group make it doubtful that they are either a sister group or
ancestral taxon of the Ferae.

    Order Rodentia Bowdich, 1821.

    As McKenna (1975) noted* in his classifications, the rodents
are as yet very much an incertae sedis group when compared to
other Mammalia.  As stated in 1974, I still view the basicranial
as well as skeletal evidence as possibly indicating leptictimorph
relationships.

*He has mistakenly quoted Szalay and Decker, stating that these
authors implied the origin of the Rodentia from a miscellaneous
ancestral stock of eutherians including as diverse groups as the
Pantodonta, Cimolestes, and Carnivora.  Szalay and Decker (1974,
Fig. 2) showed the Rodentia originating from "Palaeoryctoid insectivorans" (here referred to as leptictimorphs) along with taeniodonts (Van Valen, 1966) and the edentates.

I find no supporting evidence for the derivation of the rodents from primates as suggested by McKenna (1961) and Wood (1962), and embraced by Van Valen (1966, 1971) and Szalay (1968).

The pedal morphology of early (as well as many Recent) rodents (Figs. 14, 15) is, in spite of the clearly derived condition of the "split navicular" (probably a medial sesamoid), very similar to the characteristic, not at all primitive eutherian foot bones (particularly astragali and calcanea) of leptictids. The possibility that this striking similarity between the rodent and leptictimorph morphotype astragalus and calcaneum is due to convergence cannot be ruled out. Dental evidence is of no avail, however, because leptictimorphs are relatively primitive whereas the earliest rodents are clearly autapomorphous as far as is known. As noted under the Lagomorpha, the special similarities in the morphology of fetal membrane and placenta ontogeny (Mossman, 1937, W.P. Luckett, pers. comm., and this volume) support the concept of Glires.

Judged from pedal evidence, the morphotype rodent was a scansorial, perhaps somewhat squirrel-like, animal.

Order Lagomorpha Brandt, 1855 (emended).

As early studies by Evans (1942) hinted, but particularly as McKenna (1975) suggested, the macroscelidids are close relatives of the duplicidentate lagomorphs. In addition, the Zalamdalestidae, Eurymylidae, Anagalidae, and possibly the Didymoconidae are members of an essentially closely knit group of mammals. Although the exact genealogy of the various undisputed monophyletic groups included within the concept Lagomorpha is not understood, the order itself probably evolved from a species whose specializations can be assessed from known taxa. The earliest recognizable specialization was probably a characteristic restriction of upper and lower ankle joint movements (see Figs. 16, 17) as well as cheek tooth hypsodonty for some form of terrestrial herbivory. I do not believe that incisor enlargement in the eurymylids, zalamdalestids, and duplicidentates is necessarily synapomorphous and I am skeptical about the use of tooth enlargement alone (see Hecht and Edwards, 1976) as a significant derived character to determine relationships within the Lagomorpha.

All known representatives of this group are small, were likely to have been herbivorous, terrestrial, and specially adapted for a cursorial-saltatorial locomotor behavior. I believe that their monophyly (pending outcome of rodent relationships), as well as the range of their feeding and locomotor adaptations, warrant their inclusion in one order, the Lagomorpha. The previous solution of having three orders for the known species of this group

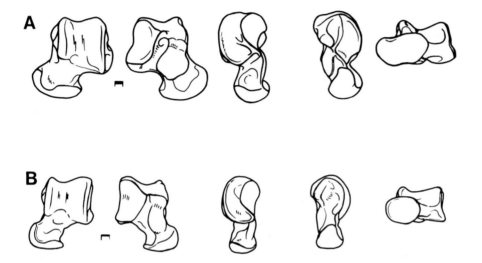

Fig. 14.  Astragali of rodents.  A. The late Wasatchian Eocene
Paramys copei (AMNH 2682).  B. The Recent Rattus norvegicus (personal collection).  From left to right:  dorsal, ventral, lateral,
medial, and distal views, respectively.  Scales represent one mm.
    Comparison of these, particularly Paramys, with Procerberus on
Fig. 8 reveals close special similarity.  Also compare closely
with Fig. 3A and B to appreciate the constancy of the rodent character complex.

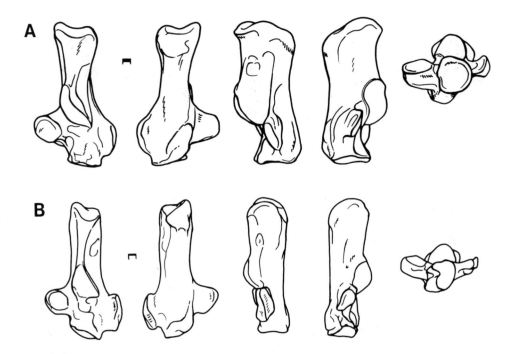

Fig. 15. Calcanea of rodents. A. The late Wasatchian Eocene *Paramys copei* (AMNH 2682). B. The Recent *Rattus norvegicus* (personal collection). From left to right: dorsal, ventral, lateral, medial, and distal views, respectively. Scales represent one mm.

Comparison of these particularly *Paramys*, with *Procerberus* on Fig. 8 reveals close special similarity.

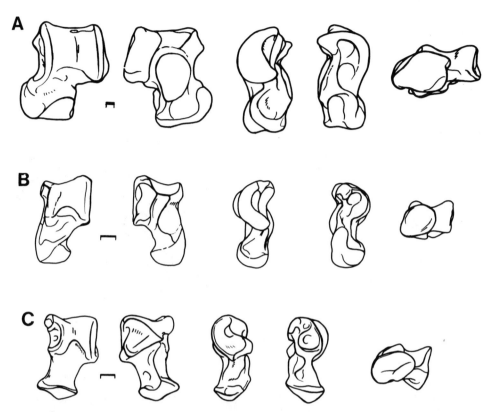

Fig. 16.  Comparison of astragali of Lagomorpha.  A. The Asiatic late Paleocene lagomorph Pseudictops lophiodon (AMNH 21755).  B. The Chadronian Oligocene Paleolagus haydeni (AMNH 6275).  C.  The Recent macroscelidean Petrodromus tetradactylus (AMNH 203336). From left to right:  dorsal, ventral, lateral, medial, and distal views, respectively.  Scales represent one mm.

The most primitive published lagomorph astragalus of Pseudictops is unlike that of any known leptictimorph inasmuch as the dorsal astragalar foramen is retained, the trochlear facet is relatively short, yet the navicular facet is transformed into a saddle-like joint indicative of cursorial adaptations.  The macroscelidean Petrodromus is more advanced that the duplicidentate lagomorph Paleolagus in having developed a peculiar navicular pivot or projection.

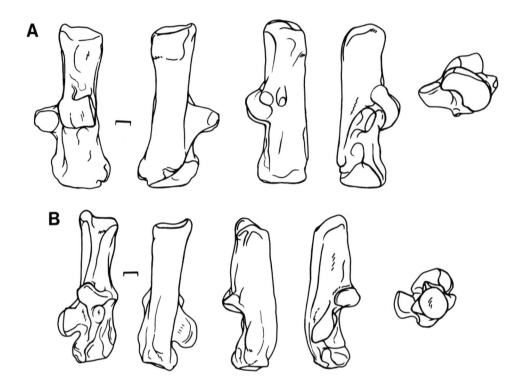

Fig. 17.  Comparison of calcanea of some Lagomorpha.  A. The Chadronian Oligocene Paleolagus haydeni (AMNH 6275).  B. The Recent
macroscelidean Petrodromus tetradactylus (AMNH 203336).

adds nothing to their understanding, and inflates the image of eutherian diversity for non-specialist students. This extended concept of the Lagomorpha is not any broader or more heterogeneous than that of the Carnivora or the Primates.

Duplicidentate, macroscelidean (see McKenna, 1975), anagalid, pseudictopid, and eurymylid (Szalay and McKenna, 1971; Szalay and Decker, 1974) lagomorphs (Kielan-Jaworowska, 1975) share unequivocal special similarities which unite the Order Lagomorpha. Leptictids, however, as argued above, are not considered to share any of the advanced "ernothere" features and therefore are excluded from this group. Within the Lagomorpha the macroscelideans may be specially related to duplicidentates, in spite of the presence of enlarged incisors in zalamdalestids. Unique cranial similarities, such as the crestlike sculpturing of the orbital edges, for example, are shared specifically between duplicidentates and macroscelidiids. As noted under Rodentia, the concept Glires Linnaeus 1758, employed by Simpson (1945) to unite the Rodentia and Lagomorpha has not been falsified. Development of fetal membranes and placental morphology in fact appear to support such a grouping (P. Luckett, this volume).

The morphotype tarsus of the Lagomorpha was somewhat specialized for restrictive, fore and aft mechanics in the upper ankle, lower ankle, and mid tarsal joints, unlike known leptictid feet. The astragalar foramen, unlike in the leptictids, persists even in clearly cursorial primitive anagalidans like pseudictopids. In addition to pedal specializations the ancestry of this group appears to have had molariform posterior premolars, and relatively hypsodont cheek teeth probably used for grazing. In summary, the lagomorph morphotype was probably a terrestrial, cursorial, herbivorous small mammal.

Subclass Theria, _incertae_ _sedis_.
Order Deltatheridia Van Valen, 1966 (sensu Kielan-Jaworowska, 1975).

Kielan-Jaworowska's (1975) recent account of this Asiatic Cretaceous group (restricted to Deltatheridium and Deltatheroides) is the most complete, being based on excellent new specimens. As Kielan-Jaworowska's and McKenna's (1975) discussions reflect, there are two equally plausible interpretations for the homology of the cheek teeth replacement. Without evidence of tooth replacement the dental formula of deltatheriids cannot be categorically stated to include three premolars (Kielan-Jaworowska) or four premolars, the latter given in the formula:

$dP\frac{1}{1}$, $P\frac{2}{2}$ or $\frac{3}{3}$, $P\frac{4}{4}$, and $dP\frac{5}{5}$, as well as $M\frac{1}{1}$, $M\frac{2}{2}$, and $M\frac{3}{3}$ (McKenna).

The presence of four premolars (by the replacement criterion) and
three (and two) molars cannot be ruled out, and the only undisputed
fact is the presence of seven postcanine cheek teeth in deltatheriids.  At present I cannot recognize any similarities between
deltatheriids and any known therians which can be shown with any
degree of reliance to be shared and derived.


CLASSIFICATION AND THE CLASSIFICATION OF THE EUTHERIA

A highly constructive furor has been raging for the past two
decades or more, most particularly within the pages of the journal
of Systematic Zoology, over how to classify.  Before I express my
specific preferences for a classification of the Eutheria, given
my present stand on the phylogeny of this taxon, I will briefly
attempt to give a few reasons why cladistic, Hennigean classification is not, on operational as well as theoretical (biological)
grounds, acceptable to me.  I will not even attempt to strive for
completeness on this topic as such papers as Mayr (1974), Sokal
(1975), and Simpson (1975) have covered most of the pertinent
points exhaustively and astutely.

A.    Cladistic classification as advocated by Hennig (1950, 1965,
1966) and his followers is the outgrowth of the now vigorously
pursued practice of dichotomous cladistic analysis.  The efforts
to resolve phylogeny into cladistic dichotomies resulted in such
absurd notions as the virtual denial of descent by ignoring the
concept of ancestors, phyletic evolution and evolutionary stasis
of features, lineages, as well as the difference between the
species concept of a given instant and lineage.  All phena simply
are not the result of dichotomous speciation.  From this semantic,
"logical" stance of Hennigean cladistics emanate the views that
ancestors cannot be found or that they are not worthwhile recognizing, and that all phylogenies can be theoretically resolved
into dichotomies.  Regardless of the operational problems in resolving phena, species, or other categories into dichotomies, the
theoretical simplification of ancestor-descendants (see Eldredge
and Tattersall, 1975, for example) into dichotomous splittings
and the failure to recognize lineages in instances when the fossil
record is excellent and (therefore the dichotomous divergence
model lowers the information content of possible hypotheses) are
the major theoretical objections to a rigidly "cladistic" method
of analysis.

Although I am, as others, firmly convinced that classification should and will become increasingly more phylogenetic (but
not cladistic) as we continue to gain a better understanding of
phylogeny, the manner of constructing evolutionary classifications
by using paraphyletic groups is not any less arbitrary than an

avowedly Hennigean one which enforces holophyly, sistergroup relationships but not ancestor-descendant ties and the subsequent
procedure of subordination.

The threat of a practice of holophyletic, dichotomously
nested formal classifications, irregardless of the uneven strength
of the research results on which they may be based threatens the
usefulness of classifications as summaries of evolutionary histories.

In biostratigraphic studies formal naming of faunal divisions
or biostratigraphic zones when evidence is less than satisfactory
is a broadly unapproved practice.  It is often pointed out that
many localities, detailed measured sections, and a variety of taxa
are prerequisites for a truly useful biostratigraphic zonation
(McKenna, 1976).  Perhaps similarly the usefulness of formal
classifications should be judged by comparable criteria.  In phylogenetic studies these criteria may be well researched and several
character complexes of groups, tested morphocline polarities, and
the inclusion of more than single genera in higher categories.

If we cannot unequivocally reject alternate phylogenetic
hypotheses but still favor one of the several options, then what
justification is there for classifying purely dichotomously and
cladistically?  In failing to falsify alternate hypotheses with
unequivocally delineated and interpreted evidence, one cannot
sufficiently corroborate a choice of phylogenies or erect a classification which expresses dichotomous relationships.  Consequently,
such a classification (see that of McKenna, 1975, for example) will
be short-lived and will fail to fulfill the need for significant
summaries of reliable, or well tested, hypotheses of relationships.
Polychotomies, when based on several well understood characters
which are unquestionably shared and derived, non-convergent similarities, are likely to be lasting and instructive groupings inasmuch as major evolutionary splittings are involved.  Further recognition of special relationships within these groups can then proceed without frequent complete rearrangements of higher category
classifications.

The number of taxonomic categories (and therefore supraspecific names) necessary to express the cladistic relationships
of all existing species in a pure hennigean system is n-1, n representing the number of species that ever existed.  Hopefully every-
one will agree that a practical use of the available categories is
absurd.  But how will a consensus or agreement result?  "Art" and
"caprice," despite the claims of hennigean classificators, are
potentially rampant in that system.  Just recently, for example,
McKenna's (1975) scheme of mammalian classification (although
some of his phylogenetic conclusions are corroborated in this

paper) offered a rather large number of new categories in addition
to new taxonomic names.  If we put these together with those traditionally used in taxonomy the list appears uncomfortably longer
and complicated; and although such a list may seem long now, it is
only the beginning of a potentially unmanageable and frightening
avalanche of new supraspecific categories, each of which is a new
opportunity for an enormous number of new names.  It is my belief
that the message of such classifications gets lost in the medium.
Pictorial phylogenetic hypotheses are far superior to <u>complete</u>
classifications (and completeness will be striven for by full
time taxonomists whose primary interest is classification and not
evolutionary relationships of taxa as genealogies of supraspecific
taxa will be worked out in greater and greater detail).

The danger of creating new higher categories and new taxa on
these levels becomes real after we consider the growth of possible
new taxa with the use of each new category.  The appetite of the
monster of a dichotomous (Hennigean) cladistic system of classification has the potential to devour classification itself (a reduc-
tionist activity which is mainly useful to summarize organic diversity).  Let us suppose that we have decided on the holophyly of
ten taxa.  The rooted trees possible to express the different
cladistic relationships and, therefore, classification of the ten
taxa is 34,459,425 (see Table I).  Given the fact that these ten
taxa are of high categories (orders, superorders, or cohorts), an
increasingly greater ambiguity exists in the assessment of morphocline polarities on which their hypothesized phylogenetic ties are
based and subsequently a great potential exists for legitimate,
conflicting research results as to phylogenetic ties.  Given mosaic
evolution, as research on a greater and greater number of features
will increase so will the diversity of views.  Is it wise and helpful to create new categories and new taxa, forcing all subsequent
students to continue obfuscating the taxonomic literature with
taxonomic concepts that may be extremely short lived? Does it add
to the biological (phylogenetic included) knowledge of mammals,
for example, to know, by a formal name, what some highly hypothetical higher category may be?  It does, perhaps, if some unequivocal evidence was marshalled to back the holophyly of these
taxa.  But even if we accept the use of dichotomous expression of
holophyletic groups as a desirable procedure in classification,
there will be some who will still remain arbitrary in their choice
of the number of categories they wish to employ.  Those, however,
who reject the latter arbitrariness of a Hennigean system or have
their own ideas on where to formalize alleged dichotomous branchings with names, have, of course, at their disposal n-1 categorical levels to choose from.

Mathematicians are not compelled to communicate in verbal
language and there is no reason why systematists should want to

formally express each of their genealogical hypotheses, regardless
of how tested they may be, in new names.  Phylogenetic trees are
precise and they are just as publishable as formal classifications.
They create none of the ghost towns lined with the tombstones of
names commemmorating falsified phylogenetic hypotheses.  Phylogenetic trees of considerable complexity can be rapidly grasped
by the mind, but the same cannot be said for most taxonomic concepts quagmired in the history of systematic research.  I believe
that the complexity of biological history of organisms is sufficiently awe inspiring without additional man-made complexities;
so should we not, for this reason, avoid adding to a history of
nomina?  Rather than stimulating new research, this new trend of
creating new formal categories of classification that are likely
to give rise to an enormous number of new taxa, may confuse,
obfuscate, and continue to add "authority" rather than format for
flexibility and testing in systematics.  Finally, given that a
Hennigean system of classification is not any less arbitrary than
any other, what will be the result of the shifting sands of higher
category concepts of major groups on the students, teachers, and
non-specialists?  If we choose to care at all.

This should not be taken as a defeatist commentary, but it
appears that phylogeny of all species can never be completely
assessed largely because much of the information necessary to a
purely phylogenetic analysis is just not available.  The latitude
between genotype and phenotype is great, parallelisms are often
undetectable, and analysis of several supraspecific taxa does not
result in the same phylogeny as would be gleaned from a species
level phylogenetic analysis.  We cannot hope to reflect phylogeny
with any accuracy closely approaching the actual history as our
theoretically impeccable intentions must be compromised to become
operational procedures based on characters that will remain
patristic despite our valiant efforts to separate the plesiomorph
from the apomorph.

Judicious use of phenetics and evaluation of adaptations offer
hope for a new evolutionary classification which will make heuristic use of paraphyletic taxa.  The purpose of a classification
is to describe a finite organic world, a finite history with a
beginning and an end today, rather than to be a logical system.  I
fail to see logic in biological history, or in the attempts to
reflect organic diversity in cladistically nested dichotomies.
Even if we were theoretically satisfied with dichotomies and had
operational procedures to decipher all dichotomous branchings, the
"logic" of subordinating all species, and subsequent taxa, according to the rules of cladistic classifications would result in an
edifice that is best appreciated by one's imagination.  Clearly,
classification will continue to be more and more phylogenetic.  But
is it really necessary that each hypothesized (from suspected to

well corroborated) sistergroup relationship should be given
dichotomous taxonomic (formal) recognition?

In addition to views against a pro forma, nominally constructed Hennigean classification which attempts to express everything
in a dichotomous format, I would like to note that this classification can show greater precision than evidence warrants. A
quantity (in this case a phylogenetic hypothesis) may be expressed
to a greater precision than is justified by the accuracy of the
information (morphoclines with ascertained polarities) from which
that quantity was derived. Just as excessive decimals in figures
are meaningless and possibly misleading as to level of accuracy,
so a classification, particularly a dichotomously conceived one,
may be also unjustified.

Because we cannot, I believe, attain a nearly complete knowledge of one aspect of organic evolution, namely the history of
splittings, derivations, and duration of lineages, then perhaps we
should not sacrifice the use of one of our great heuristic devices
in biology, i.e. classification, for the exclusive use of mirroring
genealogical hypotheses. We may, however, continue to teach with
the use of evolutionary classifications, with due regard to monophyly, total biological history, combining genealogy with adaptive
peaks and adaptive valleys, a modestly synthetic reflection of the
evolutionary events. The details of genealogy and adaptational
history lost in such efforts are favorably balanced by the synthetic picture gained. If we make classification an expression of
genealogical hypotheses only then will we have closed a window
through which our students may have begun to view evolutionary
history, an essential part of which is grasped in the transformation of one adaptation into another.

B.   The actual sequence of splitting among early eutherians is
still poorly understood. Most views, including those presented in
this paper, are working hypotheses which need to be scrutinized
from the vantage point of new features.

If the edentate groups prove to be monophyletic, and if this
cohort is more recently related to the Leptictimorpha, then perhaps a formal bipartite division of the Eutheria may become possible. Meanwhile, however, division into Group I and II should
suffice. I believe that only a formalization of false precision
of our as yet inaccurate grasp of higher taxon phylogeny would
result if we were to further subordinate the orders and Cohorts in
the Eutheria. For obvious reasons therefore, and because I do not
believe that a formal classification is the proper red cape of
challenge for the improvement of phylogenetic knowledge, the
classification proposed (Table II) is a conservative one. To
paraphrase G.G. Simpson, it is, I believe, consistent with the
phylogenetic hypotheses (and doubts) presented in Figure 1.

TABLE II

A HIGHER CATEGORY CLASSIFICATION OF THE EUTHERIA

Infraclass Eutheria Gill, 1872
    Group I
        Cohort Archonta Gregory, 1910
            Order Scandentia Wagner, 1855 (incl. Tupaiidae and
                Mixodectidae)
            Order Primates Linnaeus, 1758 (suborders not listed
                here)
            Order Dermoptera Illiger, 1811
            Order Chiroptera Blumenbach, 1779
        Cohort Ungulata Linnaeus, 1766
            Superorder Protungulata Weber, 1904
            Order Arctocyonia Van Valen, 1969 (including the
                Arctocyonidae, Mesonychidae, and Tillodontidae)
          Suborder Procreodi Matthew, 1915
          Suborder Acreodi Matthew, 1909
          Suborder Tillodontia Marsh, 1885
          Order Pantodonta Cope, 1873
            Superorder Cete Linnaeus, 1758
          Order Cetacea Brisson, 1762 (suborders not listed
             here)
            Superorder Paraxonia Marsh, 1884
          Order Artiodactyla Owen, 1848 (suborders not listed
             here)
            Superorder Mesaxonia Marsh, 1884
          Order Perissodactyla Owen, 1848 (suborders not listed
             here)
          Order Dinocerata Marsh, 1873 (suborders not listed
             here)
          Order Embrithopoda Andrews, 1906
          Order Hyracoidae Huxley, 1869
          Order Proboscidea Illiger, 1811 (suborders not
             listed here)
          Order Sirenia Illiger, 1811
          Order Desmostylia Reinhart, 1953
          Order Meridiungulata McKenna, 1975
            Suborder Litopterna Ameghino, 1889
            Suborder Notoungulata Roth, 1903
              Infraorder Notioprogonia Simpson, 1934
              Infraorder Toxodontia Owen, 1858
              Infraorder Typotheria Zittel, 1893
              Infraorder Hegetotheria Simpson, 1945
            Suborder Astrapotheria Lydekker, 1894
            Suborder Trigonostylopoidea Simpson, 1967
            Suborder Xenungulata Paula Couto, 1952
            Suborder Pyrotheria Ameghino, 1895

Cohort Ungulata, _incertae_ _sedis_
      Order Tubulidentata Huxley, 1872
Cohort Ferae Linnaeus, 1758
      Order Creodonta Cope, 1875
      Order Carnivora Bowdich, 1821 (suborders not listed
            here)
Group I, _incertae_ _sedis_
      Order Insectivora Illiger, 1811
        Suborder Erinaceomorpha Gregory, 1910
        Suborder Soricomorpha Gregory, 1910
        Suborder Zalambdodonta Gill, 1885 (incl. Tenrecidae
            and Chrysochloridae)

Group II
      Cohort? Edentata Cuvier, 1798
            Order Palaeanodonta Matthew, 1918
            Order Xenarthra Cope, 1889
              Suborder Pilosa Flower, 1883
              Suborder Loricati Vicq d'Azyr, 1792
              Suborder Vermilingua Illiger, 1811
            Order Pholidota Weber, 1904
      Cohort Glires Linnaeus, 1758
            Order Leptictimorpha, new (incl. Leptictinae and
                  Palaeoryctinae as Leptictidae, Pantolestidae,
                  Taeniodontidae and possibly the Microsyopidae;
                  _Didelphodus_ may belong to one of several
                  other orders)
            Order Rodentia Bowdich, 1821 (suborders not listed
                  here)
            Order Lagomorpha Brandt, 1855
              Suborder Anagalida Szalay and McKenna, 1971
              Suborder Mixodontia Sych, 1971
              Suborder Macroscelidae Butler, 1956
              Suborder Duplicidentata Illiger, 1811
            Order Lagomorpha, _incertae_ _sedis_
                        Family Zalambdalestidae Gregory and Simpson,
                              1926
                        Family Didymoconidae Kretzoi, 1943
Infraclass Eutheria, _incertae_ _sedis_
                        Family Asioryctidae
Subclass Theria Parker and Haswell, 1897, _incertae_ _sedis_
            Order Deltatheridia Van Valen, 1966 (_sensu_ Kielan-
                  Jaworowska, 1975)

SUMMARY

1) Most past studies of eutherian phylogenetic relationships which considered both fossil and extant taxa heavily favored cranial and dental evidence. Besides the greater abundance of fossil dentitions the major reason for overlooking bone morphology in favor of teeth is believed to be based on a mistaken view of ontogeny and phylogeny. It is widely, and erroneously, held by paleomammalogists that because bone is sometimes ontogenetically more plastic than teeth, it is phylogenetically more prone to change and therefore convergence. This view ignores the fact that evolutionary modification is the result of change in gene frequencies and not of ontogenetic transformation of occasional individuals.

2) Paleomammalogists commonly define taxa on dental criteria alone. Taxa, however, should be diagnosed by all their known derived character states. Dental, cranial, postcranial, and other anatomical character states of taxa compared may reflect a range of relative ages in their similarities. It is a prerequisite for setting up phylogenetic hypotheses of taxa that following extensive biological investigations of available characters we employ the most recently acquired character states shared in order to diagnose a particular taxon. Without attempting such a procedure when the available data permits we cannot hope to construct phylogenetic hypotheses which stand up to the scrutiny of further character analyses.

3) Some previously proposed hypotheses of eutherian phylogeny are evaluated.

4) The hypothesis presented on Fig. 1 is discussed primarily from the perspective of osteological evidence of fossil and extant forms.

5) An evolutionary classification is presented employing paraphyletic taxa using a minimum number of orders and supraordinal categories. The major reason for this practice is the tenuous nature of the relationships between most ordinal taxa in spite of the fact that large groups (the orders) themselves can be most often delineated.

6) Strictly cladistic classifications (i.e., those based on cladograms, expressing phylogenetic hypotheses only in terms of sister group relationships) are based on incomplete evolutionary theory and subsequently peculiar operational criteria. Whereas evolutionary classifications will employ paraphyletic groupings as dictated by data and weighting of evolutionary changes, the "logical" practice of subordinating taxa according to dichotomously

resolved sister group relationships (i.e., not phylogenies) requires the use of n-1 taxonomic categories.  Because the use of an
inordinately large number of categories renders such a logical
system useless, the necessary choices to limit the number of
categories (or usually, to increase them) introduces "art" and
"caprice."  The use of authority, therefore, is potentially more
rampant in a hennigean classification than in an evolutionary one.
Contrary to an evolutionary classification in which biological
information may be gleaned from the arrangement of groups, a
strictly cladistic classification offers no information beyond
hypotheses of furcations.  As hypotheses of minor branching relationships are falsified and new ones are posed drastic alterations become necessary in a dichotomous system of classification.

## ACKNOWLEDGEMENTS

I am indebted to Drs. Sidney Anderson, Malcolm C. McKenna,
Robert E. Sloan, Donald E. Savage and Richard H. Tedford for making
specimens in their care available for study.  Dr. Niels Bonde provided useful advice on translating phylogenetic trees into classifications.  The illustrations were skillfully drawn by Ms. Anita
J. Cleary.

## APPENDIX

[1] One need not go far to find evidence for genetic influences on
bone.  Not only the shape, but relative size and structure of bones
(not considering drastic surgical experiments where every tissue can
be subject to the mechanical influences of interference) are strongly affected by heredity.  It is well known, for example, that bones
of monozygotic twins are nearly identical and that anomalies such
as bipartite scaphoid are genetically determined (inasmuch as it is
also present in the other twin).  In spite of the obfuscating theoretical influence of genetic or experimental effects upon other
organs such as the thyroid, for example, as these affect skeleton
secondarily, bony morphology is not any "less" genetically determined than is the dentition.

[2] It may be unwise to rewrite the dental formulae of most mammals
based on only assumptions that some of the premolars may or may not
have been replaced.  In the vast majority of cases of fossil mammals
evidence for replacement is lacking.  Furthermore, if we accept the
proposition that the therian ancestor had five premolars (i.e. five
sets of deciduous-replacement pairs) then I find the current system
(i.e. the format of expression) illogical for maintaining premolar
vs. molar designations when the only phyletically and developmentally
significant criteria between these categories is replacement vs. lack
of it, yet insisting that the first molar of therians is the fifth

deciduous premolar.  The logical solution is to determine the total number of teeth in a morphotype zahnreihe and simply number them from front to back.  Each replacement wave should receive its special designation, of course,  The following format is proposed to designate serial homologies, giving the hypothesized dental formula for the common ancestor of all therians (I stands for incisors, C for canines and PC for postcanines):

$$I\frac{\genfrac{}{}{0pt}{}{1,2,2,?,?}{dl,2,3,?,?}}{\genfrac{}{}{0pt}{}{dl,2,3,?,?}{1,2,3,?,?}}; \quad C\frac{\genfrac{}{}{0pt}{}{1}{dl}}{\genfrac{}{}{0pt}{}{dl}{1}}; \quad PC\frac{\genfrac{}{}{0pt}{}{1,2,3,4,5}{dl,2,3,4,5,6,7,?}}{\genfrac{}{}{0pt}{}{dl,2,3,4,5,6,7,?}{1,2,3,4,5}}$$

[3] A number of Paleogene genera (e.g. Tupaiodon, Messelina, Litolestes, etc.) usually referred to as "Insectivora" may be early Tertiary tupaiids or other archontans.

[4] McKenna's (1975) usage of the Arctocyonia is restricted to include arctocyonids only, while he employed ordinal ranking for the Acreodi (mesonychids) to show their suspected close ties to whales.  I believe that the same information distribution is accomplished either by saying that cetaceans maybe derived from the cursorially specialized mesonychids or by the use of a phylogenetic tree.

[5] Hypsilolambda has a primitive molar dentition, which does not appear pantodontan to me in features which I consider diagnostic of that group.

## REFERENCES

Archer, M., 1974, The development of the cheek-teeth in Antechinus flavipes (Marsupialia, Dasyuridae).  J. Royal Soc. of Western Australia, 57:54-63.

Bensley, B.A., 1903, A theory of the origin and evolution of the Australian Marsupialia.  Amer. Naturalist, 25:245-269.

Broom, R., 1926, On the mammalian presphenoide and mesethmoid bones.  Proc. Zool. Soc. London, pp. 257-264.

Broom, R., 1927, Some further points on the structure of the mammalian basicranial axis.  Prox. Zool. Soc. London, pp. 233-244.

Broom, R., 1935, A further contribution to our knowledge on the structure of the mammalian basicranial axis.  Ann. Transvall Mus., 18:33-36.

Chow, M., 1975 (Artilce without English title).  Wert. Palasiatica, 13:154-162.

Clemens, W.A., 1973, Fossil mammals of the type Lance Formation, Wyoming.  Part III.  Eutheria and summary.  Univ. Calif. Publ. Geol. Sci., 94:1-102.

Eldredge, N. and Tattersall, I., 1975, Evolutionary models, phylogenetic reconstruction, and another look at hominid phylogeny, in "Approaches to primate paleobiology," (F.S. Szalay, ed.), Contributions to Primatology, 5:218-242.  S. Karger, Basel.

Emry, R.J., 1970, A North American Oligocene pangolin and other additions to the Pholidota. Bull.Amer.Mus.Nat.Hist., 142:459-510.

Evans, R.S., 1942, The osteology and relationships of the elephant shrews (Macroscelididae. Bull.Amer.Mus.Nat.Hist., 80:85-125.

Gingerich, P.D., 1976, Paleontology and phylogeny: patterns of evolution at the species level in early Tertiary mammals. Amer. J. Sci., 276:1-28.

Good, M., 1975, Protein sequence and immunological specificity: their role in phylogenetic studies of primates, in "Phylogeny of the primates: a multidisciplinary approach," (W.P. Luckett and F.S. Szalay, eds.), Plenum, New York and London, pp. 219-248.

Gregory, W.K., 1910, The orders of mammals. Bull. Amer. Mus. Nat. Hist., Vol. 27.

Gregory, W.K., 1951, Evolution emerging, a survey of changing patterns from primeval life to amn. Macmillan Co., N.Y., Vol. 1.

Hecht, M.K. and Edwards, J., 1976, The determination of parallel or monophyletic relationships: The proteid salamanders--A test case. Amer. Natur., 110:653-677.

Helbing, H., 1938, Nachweis manisartiger Saugetiere im Stratifizierten europaischen Oligocaen. Eclog. Geol. Helvetiae, 31:296-303.

Hennig, W., 1950, Grundzuge einer Theorie de phylogenetischen Systematik. Deutscher Zentralverlag, Berlin.

Hennig, W., 1965, Phylogenetic systematics. Ann. Rev. Entomol., 10:97-115.

Hennig, W., 1966, Phylogenetic systematics. Univ. Ill. Press, Chicago

Hoffstetter, R., 1972, Relationships, origins, and history of the ceboid monkeys and Caviomorph rodents: a modern reinterpretation. Evol. Biol., 6:323-347.

Jenkins, F.A., Jr., 1974, Tree shrew locomotion and the origins of primate arborealism, in "Primate locomotion," (F.A. Jenkins, Jr., ed.), Academic Press.

Kay, R.F. and Cartmill, M., 1974, Skull of Palaechthon nacimienti. Nature, 252:37-38.

Kielan-Jaworowska, Z., 1975, Results of the Polish-Mongolian palaeontological expeditions. Part VI. Preliminary description of two new eutherian genera from the late Cretaceous of Mongolia. Palaeont. Polonica, 33:5-16.

Kindahl, M.E., 1957, On the development of the tooth in Tupaia javanica. Ark. Zool., 10:463-479.

Kindahl, M.E., 1958a, Notes on the tooth development in Talpa europaea. Ark. Zool., 11:187-191.

Kindahl, M.E., 1958b, On the tooth development in Soricidee. Acta Odont. Scand., 17:203-237.

Kindahl, M.E., 1967, Some comparative aspects of the reduction of premolars in Insectivora. J. Dent. Res., 46:805-808.

Lillegraven, J.A., 1969, Latest Cretaceous mammals of upper part of Edmonton Formation of Alberta, Canada, and review of marsupial-placental dichotomy in mammalina evolution. Univ. Kansas Paleont. Contrib., article 50 (vertebrata 12).

Luckett, W.P., 1975, Ontogeny of the fetal membranes and placenta: their bearing on primate phylogeny, in "Phylogeny of the primates: a multidisciplinary approach," (W.P. Luckett and F.S. Szalay, eds.), Plenum, N.Y. and London, pp. 157-182.

Matthew, W.D., 1909, The Carnivora and Insectivora of the Bridger Basin, middle Eocene. Mem. Amer. Mus. Nat. Hist., 9:291-567.

Matthew, W.D., 1937, Paleocene faunas of the San Juan Basin, New Mexico. Trans. Amer. Phil. Soc., 30:1-510.

Mayr, E., 1974, Cladistic analysis or cladistic classification. Z. Zool. Syst. Evol.-forsch., 12:95-128.

McDowell, S.B., 1958, The Greater Antillean insectivores. Bull. Amer. Mus. Nat. Hist., 115:115-214.

McKenna, M.C., 1961, A note on the origin of rodents. Amer. Mus. Nov., no. 2037:1-5.

McKenna, M.C., 1969, The origin and early differentiation of therian mammals. Ann. N.Y. Acad. Sci., 167:217-240.

McKenna, M.C., 1975, Toward a phylogenetic classification of the Mammalia, in "Phylogeny of the primates: a multidisciplinary approach," (W.P. Luckett and F.S. Szalay, eds.), Plenum, N.Y. and London, pp. 21-46.

McKenna, M.C., 1976, Esthonyx in the upper faunal assemblage, Huerfano Formation, Eocene of Colorado. Jour. Paleont. 50(2):354-355.

Mossman, H.W., 1937, Comparative morphogenesis of the fetal membranes and accessory uterine structures. Contr. Embryol., 26:129-246.

Nelson, G., 1973, Classification as an expression of phylogenetic relationships. Syst. Zool., 22:344-359.

Popper, K.R., 1959, The logic of scientific discovery. Basic Books, N.Y.

Rose, K.R., 1975, Elpidophorus, the earliest dermopteran (Dermoptera, Plagiomenidae). J. Mamm., 56:676-679.

Schlosser, M., 1887-1890, Die Affen, Lemuren, Chiropteren, Marsupialier, Creodonten und Carnivoren des europaischen Tertiars. Beitr. zur Palaont. Oestereich-Ungarns, pt. 1, 6:1-224; pt. 2, 7:1-162, pt. 3, 8:1-106.

Schwartz, J.H., 1975, Re-evaluation of the morphocline of molar appearance in the primates. Folia primat., 23:290-307.

Simpson, G.G., 1931, Metacheiromys and the Edentata. Bull. Amer. Mus. Nat. Hist., 59:295-381.

Simpson, G.G., 1937, The Fort Union of the Crazy Mountain Field, Montana, and its mammalian faunas. U.S. Nat. Mus. Bull., 169:1-287.

Simpson, G.G., 1945, The principles of classification and a classification of mammals. Bull. Amer. Must. Nat. Hist., 85:1-350.

Simpson, G.G., 1975, Recent advances in methods of phylogenetic inference, in "Phylogeny of the primates: a multidisciplinary

approach," (W.P. Luckett and F.S. Szalay, eds.), Plenum,
    N.Y. and London, pp. 3-19.
Sloan, R.E. and Van Valen, L., 1965, Cretaceous mammals from
    Montana.  Science, 148:220-227.
Sokal, R.R., 1975, Mayr on Cladism - and his critics.  Syst.
    Zool., 24:257-262.
Stehlin, H.G., 1912-1916, Die Saugetiere des schweizerischen
    Eocaens.  Critischer Catalog der Materialen.  Abh. schweiz.
    Pal. Ges., 38 and 41:1165-1152.
Szalay, F.S., 1968, The beginnings of primates.  Evolution,
    22:19-36.
Szalay, F.S., 1975a, Early primates as a source for the taxon
    Dermoptera (Abs.).  Amer. J. Phys. Anthro., 42(2):332-333.
Szalay, F.S., 1975b, Phylogeny of primate higher taxa:  the
    basicranial evidence, in "Phylogeny of the primates:  a
    multidisciplinary approach," (W.P. Luckett and F.S. Szalay,
    eds.), Plenum, N.Y. and London, pp. 91-125.
Szalay, F.S. and Decker, R.L., 1974, Origins, evolution and
    function of the pes in the Eocene Adapidae (Lemuriformes,
    Primates), in "Primate locomotion," (F.A. Jenkins, Jr., ed.),
    Academic Press, N.Y., pp. 239-259.
Szalay, F.S. and McKenna, M.C., 1971, Beginnings of the age of
    mammals in Asia.  Bull. Amer. Mus. Nat. Hist., 144:269-318.
Szalay, F.S., Tattersall, I., and Decker, R.L., 1975, Phylo-
    genetic relationships of Plesiadapis - postcranial evidence,
    in "Approaches to Primate Paleobiology, Contributions to
    Primatology," (F.S. Szalay, ed.), Karger, Basel, Vol. 5,
    pp. 136-166.
van Kampen, P.N., 1905, Die Tympanalgegend des Saugetierschadels.
    Morphol. Jahrbuch, 34:321-722.
Van Valen, L., 1966, Deltatheridia:  a new order of mammals.
    Bull. Amer. Mus. Nat. Hist., 132:1-126.
Van Valen, L., 1967, New Paleocene insectivores and insectivore
    classification.  Bull. Amer. Mus. Nat. Hist., 135:217-284.
Van Valen, L., 1971, Adaptive zones and the orders of mammals.
    Evolution, 25:420-428.
Weber, M., 1904, Die Saugetiere. Einfuhrung in die Anatomie und
    Systematik der recenten und fossilen Mammalia.  Gustav
    Fischer, Jena.
Wiley, E.O., 1975, Karl R. Popper, systematics, and classifica-
    tion:  a reply to Walter Bock and other evolutionary taxono-
    mists.  Syst. Zool., 24:233-243.
Winge, H., 1941, The interrelationships of the mammalian genera
    translated from Danish by E. Deichman and G.M. Allen.
    Kobenhavn C.A. Reitzels forlag., pp. 1-412.
Wood, A.E., 1962, The early Tertiary rodents of the family
    Paramyidae.  Trans. Amer. Philos. Soc., n.s., 52:1-261.
Ziegler, A.C., 1971, A theory of the evolution of therian dental
    formulas and replacement patterns. Quart. Rev. Bio., 46:226-249.

# WING DESIGN AND THE ORIGIN OF BATS*

P. PIRLOT

Departement des Sciences Biologiques, Univ. de Montreal

Montreal 101, Canada

with an appendix on

## SIZE LIMITS IN FLYING ANIMALS

by

R.J. TEMPLIN

## INTRODUCTION

Anyone watching bats on the wing, at least in tropical countries and especially in South America, soon becomes aware that not all of them fly in the same manner. Some are faster, some slower, some make extremely swift turns or loops after braking with an astonishing accuracy one foot away from a mistnet, others cruise along relatively straight routes and, although never bumping into anything solid even in a jungle, blindly fall into a black thin mesh stretched across a trail. In general, large bats are heavy flyers while many small species appear to be expert acrobats in their nimble movements.

One would expect that such differences in flight pattern could be partly accounted for by differences in wing structure, in particular, relative width and length. Also the proportions of each section of the wing (arm, forearm, metacarpals, and phalanges) with respect to total wing area may vary between species. In

*With the financial assistance of the National Research Council of Canada (grant A 0778 to the first author).

fact, it is to be expected that wing build is related in some way
to the type of movements a bat has to perform in order to find
its food. These are particular to each dietary group in the
order Chiroptera. It is obviously a widely different practical
problem if a bat has to catch at full speed a rapidly moving insect, or approach a flower to lap nectar from inside its corolla
in a state of stationary flight, or delicately land near or on a
sleeping mammal to feed upon its blood, or scan the water surface
in search of small fishes swimming one inch below it, or fly
down and up again under the overhanging giant leaf of a banana
tree to take a bite at a fruit underneath. Whatever the help of
sonar control and the accuracy of fast reflexes through the central nervous system, the flying structures must be adjusted to a
number of situations. Does each show its particular adaptation
in its overall shape? Can one describe such adjustments with the
aid of a few biometrical data? Such data should pertain: a) to
body weight; b) wing-span, wing-breadth, wing-outline and total
lifting area; c) to muscular power (this being in turn correlated with metabolic level). Finally, questions related to flying
ability in animals may up to a point be compared with the technical problems encountered in aeroplane design.

Several other papers have dealt with the structural characteristics of bat wings. However, to my knowledge none has considered wings from the three specific viewpoints taken here. Indeed, the purpose of the present work is: (1) to investigate the
relation between wing geometry and type of flight with special
reference to the search for food; (2) to deal with the related
problem of size limits in flying animals so that the particular
position of bats among these animals is clearly described; (3) to
see if the grouping of bat families according to wing structure in
this paper corresponds to the groupings proposed on the basis of
brain organization in an earlier paper, and also with reference to
feeding behaviour, and to interpret such correspondence, if any,
from a phylogenetic viewpoint. Point (2) requires competence in
aeronautical engineering and is treated by Templin in the appendix. For a study of the structure of bat wings using other
approaches, the reader is referred to such papers as Lawlor (1973),
Norberg (1970, 1972), Davis (1969), Vaughan (1966, 1970), Bader
and Hall (1960), Findley et al. (1972). A number of other papers
deal with the flying ability of particular bat species, but since
their viewpoint is quite different from mine, I have not made
reference to each of them. I have also not used all of the
measurements published by other authors since my own data was
sufficient. I was also not sure that these other data were collected in exactly the same way as mine.

## THE WORLD OF BATS

I will briefly remind the reader of the subdivision of Chiroptera into seven groups according to diet. The first and very large group comprises all the insectivorous species. These are present almost everywhere in the world and with the help of sonar they hunt for insects catching and swallowing them on the wing. The second group is much more restricted in number consisting of a few tropical nectarivorous species. These feed inside flowers, drinking the nectar and also apparently absorbing pollen and insects found therein. The third group, exceedingly diversified but restricted in geographical distribution, consists of all the New World frugivorous bats, tropical species feeding on various kinds of fruit. The fourth group is also frugivorous but is distributed solely in the Old World. It includes some very large species (the so-called flying foxes) for which the name Megachiroptera is applied. However, the group also contains small species but all megachiropterans lack sonar. The fifth group is very small, consisting of a few carnivorous species; some semi-carnivorous bats may be included here but their status is not clear. The sixth group comprises the rare piscivorous bats. The seventh group is that of the vampires, three South American genera only. In this classification, we therefore have six distinct diets but one of these, the frugivorous, is represented by two widely divergent groups. One of these, the Megachiroptera, separated from the rest very early in the history of the order; the second one, the South American phyllostomatids, being presumably younger.

The diversity of modes of life among the order Chiroptera is striking. Indeed, no other order of placental mammals shows such a variety of diets and accordingly none is as likely to possess so many morphological, physiological and behavioral adaptations. Nevertheless, Chiroptera are at the same time a homogeneous group from the viewpoint of locomotion, being the only truly flying mammals, possessing real wings not just parachute membranes after the fashion of the cobegos and a few squirrels.

One should refrain from establishing tempting parallels between bats and birds. The occupancy of the aerial niche is about the only feature that bats and birds share in common. Apart from that bats have nothing to do with birds: they are true mammals, their wings are membranes, not feathers, their metabolism is different and so is their nervous system. Each group has taken up aerial locomotion in its own way although they may have solved in a comparable way certain problems related to the mechanics of flight. We shall confine ourselves to the study of wing proportions from that viewpoint.

PHYLOGENETIC ASPECTS OF THE PROBLEM

As suggested above, one may expect wing structure in bats to be related to the particular behaviour displayed by each species or group of species in their search for food. An approach to structural adaptations of bats based on a dietary classification is not new. It has been used for studies of relative brain size (Pirlot, 1969, 1970; Pirlot and Stephan, 1970), for quantitative analysis of brain composition (Stephan and Pirlot, 1970; Pirlot and Schneider, 1974; Pirlot and Pottier, in press), and for comparative investigations of cochlear, vestibular and optic centers (Baron, 1973). It was found that particular features of the central nervous system are associated with feeding habits. The functional significance of morphological organization at the neuro-sensory level was thus demonstrated.

Furthermore, on the same functional morphological basis, phylogenetic relations could be tentatively proposed. A reference type for brain structure was selected from among the least advanced, presumably archetypal, living mammals, i.e. from the order Insectivora, and the chiropteran dietary groups were scaled with respect to it. It is important to note that such a dietary group may include more than one bat family. The adaptive significance of the structures under consideration was brought to light above the specific or even the generic levels only. The general results of that approach can be summarized as follows: a) The insect-catchers appear to be the most primitive group among bats with minimum encephalization and neocorticalization, maximum specialization for acoustic tracking of prey on the wing and a relatively great development of the cerebellum and some sub-cortical vestibular centers. b) There is clear indication that the next group up the evolutionary scale is that of the nectar-lappers; these also possess their own feeding and flying specializations. c) The widest variations are observed among the frugivorous bats. The neotropical types, apparently a relatively recent offshoot within the order, are superior to the insect- and nectar-feeders as far as encephalization and neocorticalization are concerned and they seem to have various levels of efficiency with respect to echolocation. The paleotropical fruit-eaters, certainly a very old branching line from early Chiroptera, have very high encephalization and neocorticalization with a well-developed visual system but no provision for sonar navigation. d) Two semi-carnivorous and one truly carnivorous genera were studied. In this group too the trend is toward very progressive higher brain components (neocortex, cerebellum). e) Finally, two other highly evolved types with certain behavioural specializations are the fishing <u>Noctilio</u> and the blood-sucking <u>Desmodus</u> and <u>Diaemus</u>. These bats display high encephalization and neocorticalization; the first one is equipped with powerful sonar

which aids in spotting fish in the water; the last two are endowed
with an exceptionally developed neocortex and cerebellum corresponding to the elaborate strategy and movement control required
for their approach to relatively enormous targets.

Thus, on the basis of brain structure, we can see that the
phylogeny of bats entails:  a) an early divergence between Megachiroptera and Microchiroptera with a strong development of the
higher brain centers and vision in the former; b) a sequence from
insectivorous to nectarivorous to frugivorous types, showing progression in the higher brain centers and predominance of audition
with very specialized offshoots in the carnivorous, piscivorous
and blood-sucking directions.  The exact points of origin of these
side-lines are unknown.

Given the above results obtained from neuro-anatomical
studies, it is interesting to ask whether some close relationship
could also be found between morphological features of the wings
and locomotory requirements involved in the search for each kind
of food.  In other words, could the classification of bats into
5 or 6 dietary types be matched with aeronautical characteristics
of the flying equipment?  At the same time, one can possibly check
if the phylogenetic implications reported for brain morphology
still hold true or if new ones must be substituted or added.

One would expect bat wings to show some basic features related to the original necessity for catching insects, under the
assumption that the first bats were insect-hunters, as is clearly
suggested by brain organization.  One may imagine that the first
bats were in fact flying insectivores, distinguished from the
strictly terrestrial species by the possession of sustaining
membranes along the body and the fore-limbs.  We know that flying
equipment emerged several times in vertebrate history through
mutation and selection.  It is likely that distinct aeronautical
characteristics appeared in the flying reptiles, in birds and in
bats, according to the special needs of each group.  In all cases,
however, the equipment must have been ecologically and behaviourally
adapted to a way of life not yet taken up by previously existing
types.  It is usually assumed that birds started to occupy the
aerial niche by gliding from tree to tree.  Whether they were, as
descendants of the diapsids, insectivorous, or frugivorous, may
be debatable.  Among mammals, the limited aerial locomotion of
Dermoptera and some squirrels seems to be related to searching for
plant foods.  It was one niche as yet unoccupied by other mammals
and it probably reflected nothing more than a safe way of moving
from tree to tree without running the risk of locomotion on the
ground.  But the diet of an animal that flies or glides from
branch to branch is not necessarily vegetarian.  Such an animal
may feed on insects found on bark, leaves, holes in the wood, etc.

or even on flying insects.  As an example from the Primates, I
can mention the case of the bush-baby Galago demidovi that I observed at night in a well-lit cage in an African forest.  These
nimble creatures could spot large insects entering the cage
through the mest and, jumping from one wall to the opposite one,
were able to swallow them during their jump.  It is hard to say
whether they could do that in a dark jungle.  Obviously, these
Primates relied on their eye-sight for that hunting technique.
Other animals such as insectivores may have to rely upon acoustic
tracking instead.  This may very well have been the case in the
immediate insectivore ancestors of bats, animals with moderate or
poor visual ability.  The first bats presumably inherited from
their forerunners an acoustic equipment, the sonar, and were thus
able to hunt in the dark.  It is in this context that we shall
examine the design of the wings in bats.

## WING-SHAPE IN BATS

As an approach, it is instructive to have a look at the outlines of the wings in a few bats (Fig. 1).  A first glance reveals
some differences but none of them is very striking or in obvious
relation to any particular type of flight.  The proportion of
width over length seems to be the most likely source of variation
but it is not easy to interpret it functionally (Table I).  Shape
either more rounded or more angular also supplies some information
but its exact significance is not too clear either.  The wing of
Eumops (Fig. 1) is probably the most distinctive of the ten shown
here.  Figure 2 illustrates, by way of geometrical simplification,
the difference between the two extreme values of Table I (Sphaero-
nycteris and Taphozous).

A better way of looking at shape is to borrow from the aeronautical engineer the concept of "aspect ratio".  By definition,
this is the ratio of the wing-span to the mean wing-chord.  The
chord is the distance between leading edge and trailing edge of
the wing.  Due to the festooned outline of the wing, it is somewhat difficult to estimate the aspect ratio in bats.

We can, as a first step, content ourselves with two linear
measurements that are very easily made:  one is the wing-span
(E) and the other is the length of the fifth digit (L) which is
taken here as an approximate substitute for the width of the wing
or chord.  These two data are given in Table I for 50 species of
bats, mostly tropical.  They suffice for a simple comparison and
provide graph Figure 3 in which the E is plotted against L.  The
mean E and L values for each species were used.  It can be seen
that almost all of the points are grouped along a straight line.
The four values for the molossids, a family with longer and

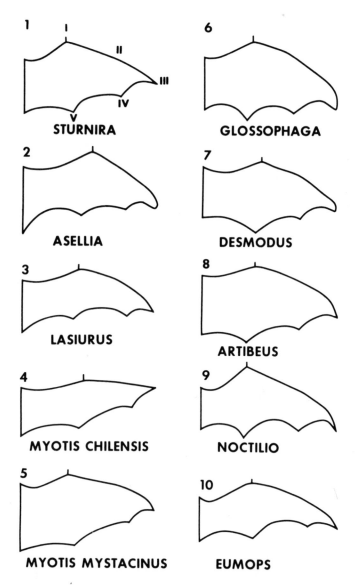

Fig. 1. Outlines of wings in nine genera (ten species) of bats. List of species in Table I. The roman figures I to V on wing 1 indicate the tips of the digits.

TABLE I

WING MEASUREMENTS AND PROPORTIONS

| SPECIES | N | E (mm) | L (mm) | L/E (%) | E/L |
|---|---|---|---|---|---|
| Phyllostomatidae | | | | | |
| Artibeus jamaicensis | 139 | 447 | 85 | 19.0 | 5.26 |
| Artibeus lituratus | 30 | 518 | 99 | 19.2 | 5.23 |
| Glossophaga soricina | 34 | 281 | 53 | 18.8 | 5.30 |
| Carollia perspicillata | 175 | 326 | 63 | 19.4 | 5.17 |
| Uroderma bilobatum | 44 | 324 | 62 | 19.2 | 5.23 |
| Sturnira lilium | 48 | 317 | 60 | 19.0 | 5.28 |
| Mimon crenulatum | 4 | 386 | 70 | 18.1 | 5.51 |
| Micronycteris megalotis | 4 | 275 | 53 | 19.3 | 5.19 |
| Vampyrops helleri | 4 | 311 | 56 | 18.1 | 5.55 |
| Phyllostomus discolor | 3 | 438 | 78 | 17.8 | 5.61 |
| Pteronotus davyi | 1 | 334 | 61 | 18.2 | 5.47 |
| Artibeus nanus | 2 | 272 | 53 | 19.6 | 5.13 |
| Choeronycteris mexicana | 2 | 328 | 59 | 18.1 | 5.56 |
| Sphaeronycteris toxophyllum | 1 | 330 | 67 | 20.3 | 4.92 |
| Lonchophylla robusta | 1 | 261 | 47 | 18.0 | 5.55 |
| Lonchorhina aurita | 1 | 368 | 73 | 19.8 | 5.04 |
| Phyllostomus hastatus | 2 | 629 | 107 | 17.0 | 5.88 |
| | | | | Aver. 18.76 | 5.34 |
| Pteropodidae | | | | | |
| Cynopterus brachyotis | 18 | 390 | 75 | 19.2 | 5.20 |
| Cynopterus horsfieldii | 33 | 489 | 93 | 19.0 | 5.26 |
| Cynopterus sphinx | 14 | 457 | 87 | 19.1 | 5.25 |
| Eonycteris spelaea | 27 | 434 | 74 | 17.1 | 5.86 |
| Balionycteris maculata | 3 | 311 | 61 | 19.5 | 5.10 |
| Macroglossus lagochilus | 2 | 330 | 66 | 19.9 | 5.00 |
| Rousettus aegyptiacus | 2 | 498 | 98 | 19.6 | 5.08 |
| Pteropus intermedius | 3 | 940 | 172 | 18.2 | 5.46 |
| Pteropus giganteus | 4 | 1085 | 209 | 19.2 | 5.19 |
| | | | | Aver. 19.98 | 5.26 |

| | N | E | L | E/L |
|---|---|---|---|---|
| Desmodontidae | | | | |
| Desmodus rotundus | 17 | 423 | 18.2 | 5.49 |
| Diaemus youngi | 8 | 419 | 17.7 | 5.66 |
| Noctilionidae | | | Aver. 17.95 | 5.57 |
| Noctilio leporinus | 8 | 624 | 17.7 | 5.62 |
| Noctilio labialis | 6 | 443 | 16.5 | 6.07 |
| Vespertilionidae | | | Aver. 17.10 | 5.84 |
| Myotis nigricans | 7 | 221 | 18.9 | 5.27 |
| Tylonycteris robustula | 6 | 221 | 17.0 | 5.81 |
| Myotis mystacinus | 2 | 265 | 19.0 | 5.30 |
| Myotis macrodactylus | 1 | 249 | 19.2 | 5.19 |
| Kerivoula sp. | 1 | 330 | 19.6 | 5.08 |
| Scotophilus temminckii | 15 | 331 | 17.5 | 5.71 |
| Nyctalus noctula | 25 | 351 | 15.9 | 6.27 |
| Rhinolophidae | | | Aver. 18.16 | 5.51 |
| Rhinolophus sp. | 1 | 356 | 19.6 | 5.09 |
| Hipposideridae | | | | |
| Hipposideros bicolor | 30 | 292 | 19.5 | 5.12 |
| Hipposideros cinereus | 1 | 245 | 19.1 | 5.21 |
| Hipposideros sp. | 5 | 326 | 18.6 | 5.34 |
| Emballonuridae | | | Aver. 19.06 | 5.22 |
| Rhynchiscus naso | 4 | 257 | 17.2 | 5.84 |
| Peropteryx macrotis | 1 | 318 | 16.6 | 6.00 |
| Saccopteryx bilineata | 1 | 340 | 18.5 | 5.40 |
| Taphazous melanopogon | 4 | 438 | 13.5 | 7.42 |
| Emballonura monticola | 3 | 264 | 16.2 | 6.14 |
| Molossidae | | | Aver. 16.40 | 6.16 |
| Molossus major | 81 | 296 | 14.0 | 7.21 |
| Molossus ater | 1 | 365 | 13.8 | 7.30 |
| Eumops perotis | 7 | 513 | 14.1 | 7.12 |
| Tadarida plicatus | 12 | 343 | 13.8 | 7.30 |
| Total... | 848 | | Aver. 13.87 | 7.23 |

N: number of specimens; E: wing-span; L: length of fifth digit; E/L: aspect ratio.

narrower wings, are located separately above the main group also
on a straight line.  The five values for the emballonurids are in
a way in an intermediate position.  They seem to split into two
directions, three species following the main group and the other
two staying closer to the molossids.  The sample is too small to
permit much comment upon this but one may note that Novick (1958)
considered the emballonurids as an old family, possibly ancestral
to many others.

The linear regression equation for the main group (7 families,
41 species) is $E = 4.77 + 5.32 L$, and for the molossids (4 species)
$E = 3.78 + 6.73 L$.  It is not worth considering the regression for
the emballonurids since they are at the same time so few and so
obviously split into two divergent directions.  For both the main
group and the molossids, the correlation coefficient is very high.

The mean ratio E/L is, in increasing order: 5.09 for the
rhinolophids, 5.22 for the hipposiderids, 5.26 for the pteropids,
5.34 for the phyllostomatids, 5.51 for the vespertilionids, 5.57
for the desmodontids, 5.84 for the noctilionids, then rises to
6.16 for the emballonurids and 7.23 for the molossids.  The
gradient is fairly continuous although the emballonurids are a
mixed group with high (7.42) and low (5.40) intragroup values
while all molossids show high ratios.  In fact, taking the two
extreme values for all bats 7.42 (Taphozous) and 4.92 (Sphaero-
nycteris), it can be seen that the former represents an increase
of about 50% over the latter, which is rather impressive from the
viewpoint of structure and function.  Comparing the two extreme
family averages, one gets 7.23 (molossids, 4 species) and 5.09
(rhinolophids, one species), that is a 41% difference.  Even if
one considers only the well represented family with the lowest
figure, that is the pteropodids with their 9 species, in comparison
with the molossids which show the highest figure, the difference
between the aspect ratios (5.26 to 7.23) is still about 37%.

How and when these structural differences occurred in the
evolution of the distinct line remains to be clarified.  Likewise,
the question of what a relatively narrow wing means exactly, from
the point of view of adaptation and behaviour, cannot be answered
at this stage either.  It is worth noting that the rhinolophids
and the hipposiderids are extremely good insect-catchers in spite
of their having relatively short wings.  Those two families
possess ratios comparable to that of the pteropodids, the latter
being rather clumsy on the wing.

It has been reported that the aspect ratio "is commonly about
7 in birds, bats and insects, though there is a good deal of
variation" (Alexander, 1971).  In our consideration the ratio
usually falls between 5 and 6.  The highest values are found in

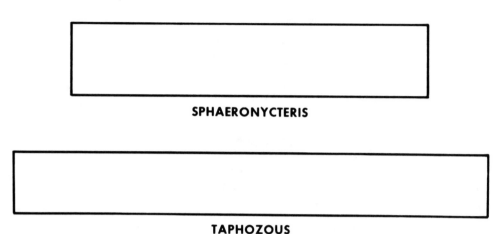

Fig. 2.   Schematized lifting areas, reduced to rectangles, in two extreme types of bats.

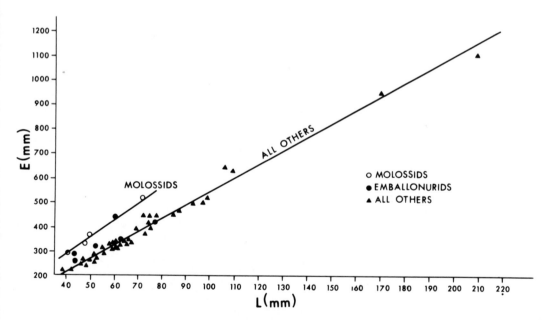

Fig. 3.   Relation between E and L corresponding to the aspect ratio.

molossids (all above 7) then among the emballonurids.  One
vespertilionid and <u>Noctilio labialis</u> also have relatively high
aspect ratios.  One may note that Farney and Fleharty (1969)
reported for North American bats ratios varying from 5.85 (<u>Myotis</u>,
a vespertilionid) to 8.57 (<u>Tadarida</u>, a molossid).  Long narrow
wings result in a high and favourable ratio of lift to drag.  It
would seem that most bats are inferior to the majority of birds
in that respect.


## VARIATION OF WING-SPAN AND WING-AREA IN RELATION TO BODY SIZE

A considerable amount of data has been published on the wing-
span, wing-area and body-weight of insects, birds and bats; e.g.
Greenewalt (1962).  However, since the methods for measuring wing-
span and wing-area are not uniform amongst authors, I suspect that
inaccuracies may affect certain data transcribed by Greenewalt.
This might be the case, for example, when one wing only is measured and the figure multiplied by 2, or when the body and uropatagium
surfaces are completely overlooked in the estimation of the total
sustaining area.  Nevertheless, use can be made of many numerical
data reported by Greenewalt, especially of the abundant measurements for birds (over 200 species) and also of the few figures for
bats drawn from various sources.

The question of how body size should be expressed often
raises difficulties too because linear measurements are not easy
to obtain due to the irregular shapes of animals bodies, and
weights are highly variable individually.  This drawback which I
have already encountered elsewhere (Pirlot, 1969, 1970) does not
preclude describing the general trend of the relation between
weight and another measurement.  We are interested here in <u>large</u>
differences between families and even more between groups with
very distinct modes of life.  Comparable studies have been published by Poole (1936) and Struhsaker (1961) but much remains to
be done before we really understand the modes of flight in bats
in relation to their wing structure.  It should be noted that I
have taken the definition of aspect ratio as provided by an
aeronautical engineer, i.e. wing-span over mean chord, and not
the "ratio of wing-span squared to the surface area of the wings"
used by Struhsaker (1961:153).


### Wing-Span Versus Body-Weight

For 36 species of my Microchiroptera, the relation can be
expressed by $E = 253.85 \ W^{0.299}$ and for 13 species reported by
Greenewalt (1962) by $E = 12.03 \ W^{0.315}$, W being the body-weight.
Only six species on Greenewalt's list also appear on mine.  If

one analyzes the same relation in each of the main groups among my
bats, one obtains the results in Table II.

The figures for bats can be compared with those obtained from
the bird data transcribed by Greenewalt from Magnan. The results
for birds are shown in Table III. The values of b are probably
the most interesting ones. Alexander (1971), in a case of isometry for birds of different sizes, suggested that wing-span
should be proportional to (body-weight)$^{0.33}$. As will be shown
below, my own estimates for birds range from 0.258 to 0.465 with
a median around 0.36. I fail to see clear relationships between
b values and the types of flight or the systematic position of
the birds. In bats, on the other hand, the b values for my
Microchiroptera taken together or for the five groups in Table II
all fall within the same range with almost the same extremes
(0.271 and 0.422) and the same median as in birds (0.35). In this
respect, bats and birds in general do not seem to differ strikingly as far as this short analysis can show.


## Wing-Area Versus Body-Weight

It seems, a priori, that one of the factors involved in
flight pattern must be the total surface upon which the flying
animal is actually borne by air. This will determine in some way
the ease with which the animal flies; the frequency of beats, the
friction and other specific characteristics are bound to be related to it.

There is here a small problem in definition and terminology.
Aeronautical engineers are familiar with the concept of "wing-area".
To the surface designated by this expression can be added in bats,
as functionally sustaining or lifting, the lower (or belly) surface of the body and also the between-leg membrane called the
uropatagium: i.e. all surfaces that make the 'total lifting area.'
Certainly, in many bats, the wing-area represents practically all
of the lifting area. However, the distinction must be kept in
mind because in a number of cases the other components of the
total lifting area are not negligible although perhaps somewhat
difficult to estimate.

Direct measurements are not easy. Stretching and pinning
down the wing of a freshly killed or anaesthesized bat on a board
is not so satisfactory a procedure as one might think, owing to
the very great elasticity of the membrane. Nevertheless, if a
contour-line is drawn and the surface measured, for example with
a planimeter or by counting $cm^2$ or $mm^2$ on ruled paper, it is possible to obtain a reasonably good estimate of the total lifting
area. It is also tempting to simplify the procedure by using some

TABLE II

MORPHOMETRIC RELATIONS IN BATS

| Groups | : N : | Relation E = $_a$W$^b$ | | | | Relation A = $_a$W$^b$ | | |
|---|---|---|---|---|---|---|---|---|
| | | a | b | r | | a | b | r |
| Phyllostomatidae | 17 | 136.08 | 0.330 | 0.93 | | 24.81 | 0.636 | 0.94 |
| Pteropodisdae | 9 | 116.94 | 0.352 | 0.97 | | 18.34 | 0.691 | 0.97 |
| Vespertilionidae | 4 | 148.09 | 0.271 | 0.82 | | 38.79 | 0.408 | 0.70 |
| Emballonuridae | 5 | 170.12 | 0.301 | 0.96 | | 50.42 | 0.424 | 0.87 |
| Molossidae | 4 | 95.10 | 0.422 | 0.95 | | 8.12 | 0.851 | 0.95 |

E: wing-span; W: body-weight; A: wing area

linear measurements easy to take in the field.  Total wing-span is
relatively easy to measure although some standard (i.e. standardly
personal) habit must be acquired in order to stretch out the membranes enough but not too much.  One may then take the fifth digit
again as representing the mean width of the wings.  The product of
these two figures, that is E for wing-span and L for fifth digit
or width gives the surface of a rectangle which one could use if
it were roughly equal to and therefore as good as the actually
desired area.  A look at Figure 1 shows that this is not the case.
Quite evidently, the product of EL is larger than the sum of both
membranes and the body in between.  In reality, L is too great a
quantity for the purpose.  The question is then:  Is that excess
relatively constant or, better, is the true area a fixed proportion of the calculated rectangle?  The answer can only be experimental; it requires tedious work.  I have done it on nine of the
specimens used for Figure 1.  After carefully measuring lengths
and surfaces twice, using once a polar planimeter and once ruled
paper and obtaining results practically identical with both
methods, I came to the conclusion that the true area is about two-
thirds of the rectangle.  Therefore, 0.66 EL would provide a
roughly correct estimate.  In the case of vespertilionids with a
large uropatagium, however, the correction factor should be somewhat higher, probably between 0.72 and 0.75.  The "errors" made
with the corrected estimate-products in comparison with results
obtained by planimetry (assuming the latter to be exact) range
from 0.25% to 5.4%.  On the whole, the approximation is not bad.
It will be noted that in his calculations (see Appendix) Templin
used the formula for an ellipse in order to estimate the wing area.
This amounts to applying the correction 0.7854 ($\pi/4$) to the product
EL, this being taken as the product of the two axes.  With that
method, the A/W estimates would be slightly higher than the ones
I have used but the difference is immaterial.

All other things being equal, one would expect the total
lifting area to be positively correlated with the weight.  Naturally, since weight increases as the third power and surface as the
second power, there will be some relative decrease of the latter.
In Table IV, together with the mean values for surface and body-
weight, the ratio of both is given, i.e. the horizontal area of
body per gram.  For computation, the uncorrected measurements of
E and L were first used (4th column), then the correction was
applied.  The correction factor was 0.66 except in the few cases
indicated in the table.  The A/W ratio is the reciprocal of the
so-called loading coefficient, i.e. $gr/cm^2$ used by other students
(Farney and Fleharty, 1969).  Transforming the figures given by
these authors, I find that their estimates for North American bats
would range from 8 to 17 for the A/W ratio (that is from 0.126 to
0.059 for their loading index).  Our figures agree fairly well
considering that my group of tropical bats is more varied.  I have

used the ratio A/W instead of the loading ratio because I am interested primarily in describing the wing and it is easier to
visualize a concrete surface per gram than an abstract ratio
written with three decimals, although both expressions are of
course correct.

The conclusion to be drawn from Table IV is that the ratio of
the total lifing area to the weight of the airborne animal varies
considerably from 4.2 to 21.8, that is five-fold. As could be
expected, it tends to decrease when absolute body-weight increases,
which may partly explain the heavier and clumsier flight of large
species. The correlation, however, is not quite regular (see
Micronycteris of the phyllostomatids and Tylonycteris of the
vespertilionids) as can be seen from Figure 4 where the corrected
A/W ratios are plotted against body-weights. This provides only
a visual estimate of the relation. It will be noted that the
correction factor for the hipposiderids, the emballonurids and
the noctilionids was taken as 0.72 and for the vespertilionids as
0.75. The choice is partly arbitrary and rests upon experience
with those four families and knowledge of the importance of the
uropatagium in each of them.

The fact that, with closely similar A/W ratios, some molossids
fly so much better than certain pteropodids or large phyllostomatids suggests that the differences between their respective skill
cannot be simply accounted for by the biometric structure, i.e. by
the static morphology of the wings, but that some dynamic factor
comes into play. This could be the efficiency of the muscular
apparatus.

As noted by Alexander (1971:5), if birds were isometric, they
would show wing-areas proportional to (body-weight)$^{0.67}$; but he
notes that these areas actually appear proportional to (body-
weight)$^{0.76}$. For Magnan's birds (in Greenewalt, 1962) I found for
the exponent values ranging from 0.327 to 0.857 with a median at
about 0.59 (Table III). For the five groups of bats in Table II,
I obtained the extremes 0.408 and 0.851 with a median value about
0.63. Again birds and bats in general seem to be comparably
organized with respect to the relation between wing-area and body-
weight although the difference between 0.59 and 0.63 for the exponent is not negligible.

### Significance of the Biometrical Relations
### Between Wing-Span, Wing-Area and Body-Weight

It would be desirable to categorize birds and bats according
to flying patterns and Magnan attempted to do it for birds (see
Table III). It would be particularly helpful if one could establish a similar set of classes for both groups so as to make a

TABLE III

MORPHOMETRIC RELATIONS IN BIRDS (after Magnan's data)

| Groups (see below) | N | Relation $E = \frac{W^b}{a}$ | | | | Relation $A = \frac{W^b}{a}$ | | |
|---|---|---|---|---|---|---|---|---|
| | | a | b | r | | a | b | r |
| I. Long range gliding day predators | 17 | 240.14 | 0.258 | 0.93 | | 50.64 | 0.575 | 0.97 |
| II. Long range gliding webfooted birds | 6 | 116.49 | 0.367 | 0.99 | | 16.26 | 0.670 | 0.99 |
| III. Flapping-gliding waders | 9 | 109.70 | 0.360 | 0.98 | | 20.89 | 0.677 | 0.98 |
| IV. Flapping-gliding night predators | 7 | 131.12 | 0.340 | 0.95 | | 24.79 | 0.665 | 0.96 |
| V. Flapping-gliding day-predators | 9 | 143.52 | 0.306 | 0.96 | | 25.37 | 0.613 | 0.96 |
| VI. Flapping-gliding Corvidae | 12 | 78.76 | 0.393 | 0.93 | | 17.46 | 0.681 | 0.96 |
| VII. Flapping-gliding passerines | 6 | 121.87 | 0.340 | 0.99 | | 15.59 | 0.705 | 0.98 |
| VIII. Flapping-gliding webfooted birds | 7 | 213.30 | 0.265 | 0.93 | | 46.41 | 0.494 | 0.89 |
| IX. Flapping passerines with steady flight | 50 | 118.63 | 0.265 | 0.88 | | 29.91 | 0.425 | 0.66 |
| X. & XI. Flapping passerines with little steady flight plus one hummingbird | 23 | 93.18 | 0.327 | 0.91 | | 10.76 | 0.750 | 0.92 |
| XII. & XIII. Flapping land and shore waders | 26 | 105.49 | 0.334 | 0.97 | | 15.75 | 0.639 | 0.84 |
| XIV. & XV. Flapping pigeons and fowl | 18 | 81.84 | 0.330 | 0.91 | | 14.41 | 0.586 | 0.95 |
| XVI. Flapping and swimming webfooted birds | 19 | 38.34 | 0.465 | 0.96 | | 2.35 | 0.857 | 0.96 |
| XVII to XIX. Flapping and diving webfooted birds, waders and one passerine | 17 | 99.22 | 0.312 | 0.91 | | 68.98 | 0.327 | 0.50 |

Symbols as in Table II. I have translated the French condensed titles given my Magnan as accurately as I could.

TABLE IV

LIFTING AREA AND BODY WEIGHT

| Species | N | A (cm$^2$) | W (gr) | A/W uncorrected | A/W corrected |
|---|---|---|---|---|---|
| Phyllostomatidae | | | | | |
| Artibeus jamaicensis | 94 | 379.5 | 39.1 | 9.7 | 6.5 |
| Artibeus lituratus | 26 | 514.7 | 58.8 | 8.7 | 5.7 |
| Glossophaga soricina | 24 | 155.8 | 10.9 | 14.3 | 9.4 |
| Carollia perspicillata | 50 | 213.5 | 15.6 | 13.6 | 9.0 |
| Uroderma bilobatum | 20 | 206.5 | 16.7 | 12.4 | 8.2 |
| Sturnira lilium | 24 | 192.4 | 17.4 | 11.0 | 7.3 |
| Mimon crenulatum | 4 | 270.4 | 14.8 | 18.3 | 12.1 |
| Micronycteris megalotis | 2 | 155.3 | 8.0 | 19.4 | 12.8 |
| Vampyrops helleri | 1 | 171.6 | 12.0 | 14.3 | 9.4 |
| Pteronotus davyi | 1 | 203.7 | 10.9 | 18.7 | 12.3 |
| Artibeus nanus | 1 | 151.2 | 11.0 | 13.7 | 9.0 |
| Pteropodidae | | | | | |
| Cynopterus brachyotis | 12 | 299.8 | 25.8 | 11.6 | 7.6 |
| Cynopterus horsfieldii | 27 | 463.3 | 52.9 | 8.7 | 5.7 |
| Cynopterus sphinx | 8 | 405.2 | 41.1 | 9.8 | 6.5 |
| Eonycteris spelaea | 27 | 329.6 | 42.8 | 7.7 | 5.1 |
| Balionycteris maculata | 2 | 182.7 | 13.9 | 13.1 | 8.6 |
| Macroglossus minimus | 1 | 236.0 | 23.6 | 9.9 | 6.5 |
| Desmodontidae | | | | | |
| Desmodus rotundus | 10 | 312.1 | 31.2 | 10.0 | 6.6 |
| Diaemus youngi | 5 | 314.8 | 36.1 | 8.7 | 5.7 |
| Noctilionidae | | | | | |
| Noctilio leporinus | 8 | 693.7 | 60.3 | 11.5 | 7.6   8.3 (a) |
| Noctilio labialis | 8 | 324.4 | 39.0 | 8.3 | 5.5   5.9 (a) |

| | N | A | W | | | |
|---|---|---|---|---|---|---|
| Vespertilionidae | | | | | | |
| Myotis nigricans | 7 | 95.1 | 4.6 | 20.6 | 13.6 | 15.4 (b) |
| Myotis mystacinus | 2 | 133.9 | 5.3 | 25.2 | 16.6 | 18.9 (b) |
| Tylonycteris robustula | 6 | 83.4 | 8.0 | 10.4 | 6.9 | 7.8 (b) |
| Hipposideridae | | | | | | |
| Hipposideros bicolor | 21 | 167.2 | 8.3 | 20.1 | 13.3 | 14.5 (a) |
| Hipposideros cinereus | 1 | 115.1 | 3.8 | 30.3 | 20.0 | 21.8 (a) |
| Emballonuridae | | | | | | |
| Rhynchiscus naso | 4 | 113.9 | 4.4 | 25.9 | 17.1 | 18.6 (a) |
| Taphozous melanopogon | 3 | 255.0 | 25.6 | 9.9 | 6.5 | 7.1 (a) |
| Emballonura monticola | 1 | 130.2 | 5.3 | 24.5 | 16.2 | 17.6 (a) |
| Molossidae | | | | | | |
| Molossus major | 80 | 122.4 | 13.0 | 9.4 | 6.2 | |
| Molossus ater | 1 | 182.5 | 28.6 | 6.4 | 4.2 | |
| Eumops perotis | 7 | 373.2 | 46.5 | 8.0 | 5.3 | |
| Tadarida plicatus | 12 | 163.4 | 23.6 | 6.9 | 4.5 | |

N: number of specimens; A: total lifting area; W: body weight - (a) with correction factor 0.72 - (b) with correction factor 0.75. This table does not repeat exactly Table I in its composition because body-weights and/or wing measurements were not available for a part of my collection. For corrections on A/W see text. The mean body-weight of Tadarida was kinly supplied by Dr. A.J. Beck, University of California, and eas estimated by him from 69 specimens.

comparison between the two more meaningful. I cannot do it at the
present time, not even by combining my own field observations with
those reported by other students. Very different flight patterns
are easy to identify but sorting out all the bat species according
to as many distinct types of flight as Magnan thinks he can recognize in birds is much more difficult. However, assuming that the
classifications proposed by Magnan for birds (Table III) and the
family grouping of bats (Table II) are by and large natural and
to that extent legitimate, one may look at the two parameters r
and b involved in the relations here discussed and see if they
show some degree of significance.

As far as the correlation coefficient (r) is concerned, the
values in Table II are statistically significant on the .01 level
for the phyllostomatids and the pteropodids; they are not significant even on the .10 level for the vespertilionids; the value
0.96 is significant at .01 but 0.87 is so at .10 only for the
emballonurids; both values (0.95) for the molossids are significant on the .05 level. In Table III, all r values are significant
at .01 except the last one to the right (0.50) which is so on the
.05 level only. It thus seems that the correlation is generally
quite strong between wingsize and body-weight, especially in birds
and in each flight class. The few lower figures are probably due
to sampling heterogeneity.

Regarding the regression coefficient (b), it is interesting
to ask whether, for all groups, the estimated b values are significantly different from the theoretical values expected in case of
isometry. The data at hand are not abundant for bats and they are
second-hand for birds, therefore, it is not worth carrying out an
elaborate test. Nevertheless a simple consideration of the standard error of b, i.e. of the statistic $s_b$, can be useful. I have
estimated $s_b$ for each of the following groups: (1) bats from
Greenewalt (13 data); (2) phyllostomatids as in Table II (17 data);
(3) pooled Microchiroptera for which satisfactory data were available (36 data); (4) all birds from Magnan grouped as in Table III.
I have computed the value of twice the standard error $s_b$ and
checked if the theoretical (isometry) figure would fall into the
interval between plus and minus that quantity around the observed
b value. As this would cover (theoretically) 95% of a normal
distribution around b, only the values falling outside that range
could be said to differ significantly from b. This may not be
the most sophisticated test but it is a simple method by which at
least important trends or departures from expectations can be
brought to light.

In the case of wing-span, the expected or isometry value for
b is 0.33. I found that this value is within the 2 $s_b$ range for
the bats of group (1) above, for those of group (2), for those of

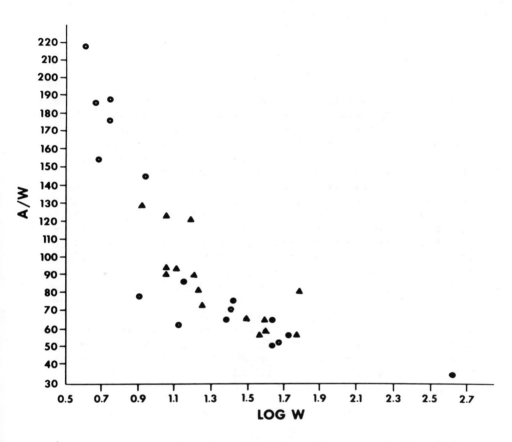

Fig. 4. Graphic illustration of the negative correlation between lifting area (A) and body-weight (W). Circles: various insectivorous bats. Triangles: phyllostomatids. Solid dots: pteropodids.

group (3) and for all birds under (4) except classes I, IX and
XVI from Magnan (although in two of the last three classes the
expected value would be included in a 3 $s_b$ or 99% interval).

In the case of wing-area, the expected or isometry value for
b is 0.67. I found that this value falls within the 2 $s_b$ range
for the bats of group (1) above, for those of group (2) and for
all birds under (4) except classes I, IX, XVI and XVII-XVIII
(although again a 3 $s_b$ interval would include it for these
classes); also, in the case of pooled Microchiroptera (group 3 in
bats), the expected value is a little over 2 $s_b$ but again within
3 $s_b$.

Although this treatment of the data is very crude, it clearly
indicates that the departure from isometry is not great within the
group of animals observed. One may doubt if it is significant at
all from a biological viewpoint. This holds good for bats and
birds alike. What such a general similarity in construction
really means is perhaps not easy to show in detail but it does
point to some general conditions to be fulfilled if the animal is
to fly. Returning to the statements by Alexander (1971) that b
tends toward 0.39 for wing-span and toward 0.76 for wing-area in
birds, I must stress the variations within the b-values. The mean
of all b's for wing-area is about 0.62 for birds and about 0.60
for bats whereas for wing-span it is about 0.33 for birds and bats
alike. In other words, my observations tend to yield figures that
lie rather <u>below</u> the isometry level or on that level at the most,
while Alexander suggests a tendency for the wing to show a development of span and area <u>above</u> that level. Considering that most
flying animals of recent times are small and that even the largest
ones are not very large, relatively speaking, it seems to me that
my figures agree better with the general size limitation and the
"status", so to speak, of those animals.


FLYING ABILITY IN BATS

From the above tables and comments, we may now proceed to
consider the question of how well equipped for flight Chiroptera
are, from the viewpoint of wing morphology. Reference will be
made to some principles applied in aircraft design and construction. Interesting comments on this topic have been proposed by
Templin (1970) who, however, restricted himself to considering the
case of unpowered landing in flying animals (bats not included)
and estimated the maximum permissible weight or size as a function
of aspect ratio.

When the case of powered flight in bats was submitted to him,
Templin reworked the whole problem in the manner shown in Fig. 5

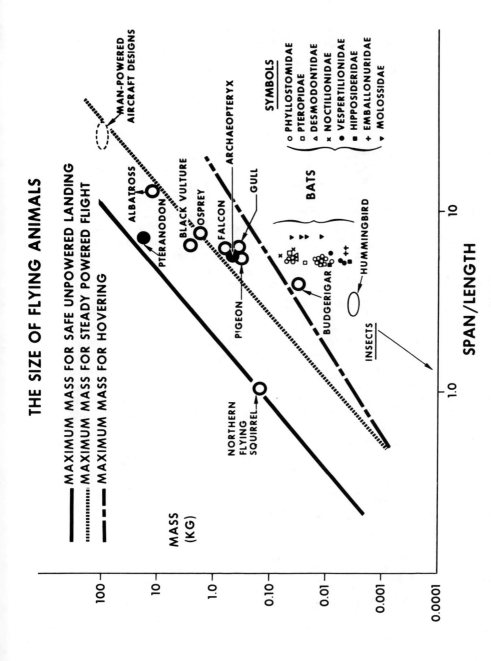

Fig. 5. Relation between maximum mass and aspect ratio in birds and bats, for three modes of flight, as established by Templin.

and explained by him in the Appendix.  He used E/L that is wing-
span over length of 5th digit, the latter being taken as approximately equal to body-length.  This assumption is justified.  Indeed, while in Brazil some time ago, I checked on the magnitude of
the difference between 5th digit and body-length in 108 specimens
from 14 Amazonian species and I found that those two dimensions
were practically equal: the 5th digit is likely to be between
1.6% and 3.2% shorter or longer than the body in fresh animals.
Considering the difficulty of obtaining accurate measurements on
a soft and more or less plastic body, this variation certainly is
quite acceptable.

As can be seen from Fig. 5, the bat data, when plotted on the
graph of maximum permissible weight versus E/L suggest that
Chiroptera are far below the danger level with respect to their
permissible weight as a function of wing design.  The mean specific weights are located very favourably, well under the critical
limit, even for hovering (stationary) flight which is the most
difficult type of flying considered here.  All bats so far studied
are capable of sustaining flight at zero speed and, as Templin
pointed out to me, it is not surprising that the smallest of them
are capable of rapid manoeuvres.  Those species are hipposiderids,
emballonurids, vespertilionids and molossids.

On the graph (Fig. 5), the abscissa is a log-scale and has
the effect of concentrating the E/L ratios for most of the bats
on a narrow vertical column.  Interesting facts can be observed
here.  On the one hand, some phyllostomatids are fairly close to
the mass limit for hovering but the small ones among them are
located far below.  The latter group includes Glossophaga, a
truly hovering bat.  On the other hand, the closest to the upper
permissible limit is Noctilio leporinus, the fishing bat; the
farthest are the swift and nimble hipposiderids.  In fact, few of
these bats may be assumed to hover for a long time.  But one may
wonder if a number of them do not approach a hovering state at
times and for a very brief duration.  For example, although it is
certain from observations both in nature and in captivity that
frugivorous bats do a great deal of climbing after they have lan-
ded on a tree, one may imagine that a brief stop at the moment of
landing would help them choose the best branch and fruit among the
several probably within reach or sight at that instant.  Even the
pteropodids appear on the graph to be constructed within the range
of safety for slow if not zero speed flight and two of them, Eonyc-
teris and Macroglossus probably do a great deal of hovering (for
nectar-feeding in all likelihood).  Also, although we know very
little about the strategy by which a vampire (Desmodus or Diaemus)
approaches its prey, it is reasonable to imagine that there is a
short critical moment during which the predator selects a spot near
or on its victim, where it can land and find suitable vascularization.  It cannot afford to search anywhere and randomly through its

fur or upon its skin.  That strategy probably results in a brief
hovering phase.  As far as the true insect-eaters are concerned,
it would seem that they do not have to hover significantly and, as
already said, their light weight is in favour of skilled manoeuvering.  However, it is now known that they frequently do hover.
Hovering has been reported for a vespertilionid and a rhinolophid
(Allen, 1939).  I can recall here observations that I made on two
occasions in Malaya with the genus _Hipposideros_ and on one occas-
ion in the Sahara with the genus _Asellia_ (also a hipposiderid).  I
then saw clearly that these animals are able to brake with an
extraordinary efficiency less than one foot away from a mistnet,
flap their wings violently during perhaps two or three seconds of
quasi-stationary flight without falling down or touching the net,
then fly away.  Also, one must refer to Norberg's work on _Plecotus_
_auritus_. That author writes: "The long-eared bat, _Plecotus auritus_,
often flies about in the crowns of trees, hunting insects among
the branches. It is very clever in manoeuvering in narrow spaces
and hovers easily.  It can lift off vertically from the ground,
and can move vertically up and down" (Norberg, 1970:62).  Finally,
one should remember that bats do not land the way birds do.  The
latter maintain a normal, i.e. head up position when they land on
a branch or on the edge of a nest.  Bats, on the contrary, have to
turn themselves over, or at least begin to do so, in order to hang
from their claws, i.e. feet up and head down.  I have observed
that manoeuvre many times on the wall of a cave entrance and it
was clear to me that it involved a short moment of stationary
flight (with _Brachyphylla_, for example). This may be general in
bats.

   In all cases, even if some bats seem to be capable of performances for which they have little need, all of them appear to possess a reserve of "aerial freedom" that fits very well with their
general life-habits.  The best constructed of all in this respect
may be the molossids with their high aspect ratio.  In Fig. 5,
they are located far to the right, away from the limiting line.

   One may wonder why bats are not considered as good flying
organisms as birds are.  They give the impression of making pain-
staking efforts when they fly, especially if one compares them
with swiftlets or other good insect-catchers among birds.  Neither
do they offer such spectacular and long-range migratory phenomena
as birds do.  The findings of Templin regarding the position of
the albatros (and also of the vulture and the osprey - see Fig. 5)
in this respect suggest that this weakness may be due to the fact
that Chiroptera, apart from a few exceptions, do not glide over
great distances and therefore get tired easily whereas the above-
mentioned birds have recourse to gliding as an energy-saving type
of locomotion.  This statement is not in contradiction with the
commonly known fact that limited gliding is observed, at times, in

some pteropodids (<u>Pteropus</u>, <u>Rousettus</u>).  Large species of that
family may have recourse to gliding and use thermal currents: this
strategy would compensate for the fact that their weight is critical or subcritical (see also the remark about the possibility of
<u>large</u> bats being located on the graph above the hovering line, at
the end of paragraph iii below).

There would seem to be a limitation of the bat's flying power
quite independent of wing architecture.  Although more observations should be presented in order to substantiate this hypothesis
the following tentative remarks can be made.

i)  Peculiarities of bat metabolism are likely to play a
certain part here.  Chiroptera have a relatively large area for
heat loss.  Indeed, their wings are densely vascularized, very
thin and not thermally insulated with abundant hair that would
act like bird feathers.  One may expect great energy wastage
through loss of calories.

ii)  The A/W ratio has been shown to be much higher in small
than in large species (see Fig. 4).  The energy loss is therefore
relatively more important in the former.  Big bats loose relatively less and produce relatively more calories with their greater
mass, this being also true for other mammals but vastly more so
for bats.  That may partly or perhaps entirely explain why long-
distance flights are difficult for the smallest species:  they
would have to stop too often for eating.  Humming-birds are said
to stop every two hours or so for feeding during their migratory
movements, paying a penalty for being so small, so that they
approach the lowest permissible size for homeotherms with an un-
favourable surface-to-volume ratio.  It is true that certain small
bats do travel some distances in the Fall and in the Spring, for
example in Western Europe and from Eastern Canada to the United
States.  But one may suspect that, in Canada in particular, most
of them do not go very far South (Massachusetts, for example).
Their displacements are not very extensive in Europe either, from
Northern Germany to Belgium or even from the North of France to
the Pyrenees the trips are not very long.  Bats are not likely to
cross vast bodies of water either.  They did reach remote islands,
Australia in particular.  But so did other mammals that are unable to fly, and sweepstakes transportation may have been in some
regions such as the Pacific the commonest means of dispersal for
small bats as it was for rodents.  On the contrary, large pteropodids (up to five feet in span) are known to travel daily and
seasonally very great distances in South-east Asia, in the Indonesian archipelago especially, according to oral reports which I
obtained in that area.

iii) Another feature of the life-habits of large frugivorous

bats is probably related to their stronger although clumsier
flight-pattern. These feed mainly on carbohydrates (bananas,
sugary fruit-pulps) which must give them more metabolic fuel than
the insectivorous diet does to small bats. In fact, one cannot
imagine a large bat living exclusively on mosquitoes or flies:
this occupation would require too much time for a large size
animal unless it is specifically equipped for it as ant-eaters
are. It is striking that the one very large species which is not
frugivorous has turned truly carnivorous (Vampyrum spectrum of
South America, up to four feet in wing-span) and eats the flesh
of other vertebrates, not just tiny insects. In fact, as a result of more elaborate and further investigations, the graph in
Fig. 5 may have to be slightly modified. Indeed, the calculation
of the limiting curve depends on the assumption that one is dealing
with "normal" flying mammals, animals endowed with an energy
efficiency of about 20%. Now, if this figure is valid for gliding
squirrels and also for birds with their good thermal insulation,
it may have to be adjusted for bats if (and we have just seen
that this is likely) the latter are thermo-dynamically at a disadvantage in comparison to birds. Since the ordinate on the graph
shows the mass-equivalent of the energy available (and required),
it may be that the plots for the bats have to be shifted a little
upward. Some of them, especially large size species, may then be
pushed above the hovering line. However, there are enough biologically valid reasons for believing that such a change would not
alter the general conclusion that bats are well designed not only
for powered steady flight but also for certain hovering or close-
to-hovering performances.


## OTHER CHARACTERISTICS OF WING-DESIGN IN BATS

1) One of the fundamental characteristics of wing design is
the disc area ideally covered by both wings ($S_d$) i.e. the circle
the diameter of which is wing-span. That area therefore is: $S_d =
\pi(E/2)^2 = 0.7854 \ E^2$. From this, one can estimate the wing disc
loading, that is the ratio body-weight/wing disc area. For humming
birds, it has been predicted by Hainsworth and Wolf (1975:230) that
"wing disc loading should decrease with increasing body-size."
These authors report finding a significant negative correlation
r = -0.38 for 40 species and they also state: "This suggests that
humming birds in general have compensated somewhat for increased
costs of hovering associated with size by making adjustments in
wing length" (l.c.). In my bats, wing disc loading varies from
1.57 to 2.84 and it increases with body size. Its relation to
body size can be expressed by the equation log y = 0.1267 + 0.0029
log x (y being the wing disc loading and x the body-weight) with a
significant positive r = +0.78 (P < 0.01). Note that one would
expect the loading to increase with size since body-weight increases

by the third power whereas wing surface increases by the second
power only. Therefore there is no compensating increase of wing
length in the larger bats, many of them being nevertheless capable
of hovering.

2) The lift-drag ratio is another interesting characteristic
of wing geometry. Its expression is basically L/D but the "effective ratio" is usually smaller, depending on the ratio $P_o/P_{am}$
between profile power and absolute minimum power (the profile
power being needed because the wings have to be flapped and this
creates some drag). Following Pennycuick's (1972) suggestion
based on studies of pigeon, one can adopt the value 2 for the
ratio of the two powers; then, effective lift-drag can be estimated
from $(L/D)' = 0.4 \, (S_d/P_a)$ where $P_a$ is the equivalent flat plate
area (greatest frontal area of the body). In·my bats, the ratio
varies from 7.640 to 10.576 with a mean at 8.931. Most of the
values fall between 8 and 9. From engineering principles, I
understand that man-made aircraft may have a much more favourable
ratio, well over 10 and up to 30, for example. For my bats, if we
assumed $P_o/P_{am}$ to be 1 instead of 2, we would obtain for the
ratio the extreme values 10.620 to 14.700; if, on the contrary, we
took $P_o/P_{am} = 3$, then the ratio would range from 6.170 to 8.540
only.

3) A third characteristic is the Reynolds number, a dimension-
less number that depends on the length of the body and the speed
of the flying animal. Its expression is $7 \times 10^4$ (length in m) x
(velocity in m/sec) as given·by Alexander (1971). A high Reynolds
number is associated with a lesser effect of drag or a better
lift. I was unable to find detailed data on the speed of all bats
considered here. However, assuming that 10 m/sec is a fair average
for many species, one obtains Reynolds numbers ranging from $2.66 \times 10^4$ to $7.77 \times 10^4$. The lowest values would then be those for insectivorous species (vespertilionids, hipposiderids, emballonurids
and molossids). This is but a very crude representation of
reality because speed may vary from 5 to 20 m/sec in bats. A
velocity of 10 m/sec is 36 km/hr and one may wonder if most bats
could maintain that speed for a long time. Nevertheless, it seems
that some molossids and vespertilionids far exceed that speed,
possibly approaching 100 km/hr (close to 28 m/sec) during short
periods.

Discussion and Conclusions

Looking at my material arranged according to families, one
gets the impression that bats are rather uniformly built as regards the geometry of their flying apparatus. However, the outline of both wings taken together, if reduced to a simple E x L

rectangle, shows a certain amount of variation.  On the other
hand, the relation between total lifting area and body-weight
shows an allometric decrease of the relative magnitude of the area.
This agrees well with the theoretical expectation and with the
observed differences in flight patterns between large and small
bats, the former being much clumsier than the latter.

The ratio E/L (Fig. 3) may be taken as a measure of how slender wings are.  But its mean family-values, when arranged in decreasing order, do not represent an array from good flyers to
poor flyers.  The situation in that respect is more complex.  The
lowest values belong to a few nimble flyers (especially hipposiderids) but also to average ones (phyllostomatids) and to some
bats that are considered the clumsiest but are nevertheless long-
range travellers (pteropodids).  The highest values belong to
excellent flyers (the molossids and emballonurids) which certainly
is not surprising.  This suggests that a relatively broad wing
can be adapted to two different purposes: skilled manoeuvres or
sustained flight.  If a bat has a high E/L ratio and is at the
same time a good insect-hunter, it probably possesses either
great muscular power or outstanding mechanical efficiency of the
transmission gear (muscles and tendons in the forearm).  This
would be the case in the molossids that seem to be the "fast
aircraft" among Chiroptera.  They certainly are in a favourable
position in this respect, as confirmed by the mass-limit graph
(Fig. 5).  They remind one of modern jets as opposed to older
propeller planes.  The emballonurids may be looked upon as intermediate although well specialized for flight.  Novick (1958),
studying their sonar, once expressed the opinion that they are
possibly ancestral forms.  In a recent study, Pirlot and Pottier
(in press) find that emballonurids are characterized by a very
high ratio of mesocortex (or schizocortex) to neocortex, which
can be taken as an indication of primitiveness.  This does not
mean that modern emballonurids are the exact image of the ancestor
to all bats.  It does suggest, however, that they are the closest
to the origin of the order, whatever specialized features they
may have acquired since the early Cenozoic (Stephan and Pirlot,
1970, have shown that specialization in bats should be dissociated
from evolution and primitiveness).

That idea finds support here.  From a first model of a flying
insectivore (i.e. the first bat), it may be that evolutionary
change occurred in two directions:  one produced high-powered and
swift animals that are, in this sense, the most advanced flyers
(molossids); the other one diversified into a variety of adaptations from clumsier with more power (pteropodids although these
may originate from a distinct ancestor) to swifter and more skill-
ful at manoeuvring but with less endurance (many small insect-
eaters).  One could therefore regard the molossids as a relatively

old and very early specialized family but the hipposiderids, for
example, might also be a fairly old type.  The youngest of all
certainly are the phyllostomatids that appear to be the most
difficult to characterize and the most heterogeneous of all.  They
seem to be, still today, in full evolutionary momentum along
several distinct lines of specialization and it would be impossible
to predict how far they will go in those various directions.  This
is of course a speculative suggestion that calls for further confirmation but it is already fairly well supported.  It is independent of earlier conclusions reached through quantitative analysis
of the brain of bats (Pirlot and Stephan, 1970; Stephan and
Pirlot, 1970; Pirlot and Pottier, in press) but it agrees with
them.  In particular, the question of the development of the
cerebellum with respect to skill in aerial manoeuvres is probably
quite distinct from the problem of wing design.  Accurate control
of flight can be achieved through the balance centers for various
types of wings and various types of usage.  It is interesting to
note that, in a study of the vestibular nuclei, Baron (in press)
finds distinctive features that agree fairly well with my hypothesis, especially with respect to the specific control of body
balance which must be achieved by molossids, vespertilionids and
pteropodids respectively.  Taking the vestibular complex as a
whole, Baron reports that _Myobis_ (a vespertilionid) is the most
advanced, followed very closely by the pteropodids, then by
_Desmodus_, _Rhinolophus_, _Noctilio_ and various phyllostomatids, all
about at the same level.  Below the bats just named come the
hipposiderids and one molossid.  These results suggest that
accurate vestibular control is needed not only for acrobatic aerial
manoeuvres but also for delicate climbing in trees and on the body
of large prey.

If I may be permitted to speculate a little further, this study
also leads me to suggest that the history of bats as flying organisms may have been somewhat different from that often pictured for
birds.  It has been said repeatedly that reptiles started gliding
from tree to tree and, one day, were able to fly as birds.  This
is also perhaps the way people look at flying squirrels: a glide
may be a possible intermediary stage toward real flight.  In the
case of bats, I would propose that the most significant selection
has been in favour of a kind of hovering, not gliding, although
gliding may have been important at early stages too.  I am not,
by any means, suggesting that hovering was the only type of movement at those early stages.  We can assume that the first Chiroptera
acquired one or several of the abilities to run, jump, glide,
flap-fly, hover, etc.  But given the fact that a general potentiality to hover is observed in the wing design of modern bats,
and the fact that a number of these bats do actually hover (the
most difficult type of flight, it must be remembered), it is
reasonable to assume that selection acted very early in favour of

hovering.  As a comparison, I would say that when we observe the
general presence of enamel ridges in ungulate cheek-teeth, we conclude that selection favoured such ridges from the early history
of that order.  The above assumption does not preclude that other
types of locomotion (gliding in particular) may have been in use
among bats at the same time as or even before hovering.  In fact,
considering that hovering requires a particular strengthening of
bones and muscles, as was pointed out to me by Clark (see comments
following this paper), I admit that it has probably taken more
time to establish itself than gliding and simple steady flight.
But I also maintain that mutation and selection in favour of
hovering most probably came into play soon after the age of insectivores.  After all, we have no reason for thinking that mutations towards hovering were less likely to occur than mutations
for simple steady flight.  As a matter of fact, hovering flight
is but one type of flapping flight just like ordinary steady
flight.  It is quite possible that both abilities (for ordinary
and for hovering flight) emerged practically together as the result of one sequence of mutations.  I further believe that we can
exclude the probability of an accidental convergence producing
hovering in the various lineages of bats since we are here dealing
with one single order and it is more than likely that bats derive
from one common group of insectivores (at the most one would
accept the possibility of a convergence between Mega- and Microchiroptera under a diphyletic hypothesis).

To summarize, I will compare the probable sequence of events
in birds and in bats.  In birds, running, jumping and gliding as
the first flight led to flap-flying in the form of steady flight.
Gliding is still widely retained in birds whereas stationary
flight (hovering) is an exception.  In bats, even if running,
jumping and gliding were also present at the beginning and also
led to flap-flying, only a few pteropodids seem to have retained
gliding.  The majority of bats took up a flapping flight that
fulfills the conditions for hovering as well as for steady locomotion. The basis of my argument is the generality of gliding and
the rarity of hovering in birds, on the one hand, and the generality of hovering and the rarity of gliding among bats, on the other,
together with the fact that the least evolved species of Chiroptera
(insectivorous and nectarivorous) do exhibit hovering flight.  Indeed, it was shown by Pirlot and Stephan (1970) and by Stephan and
Pirlot (1970), on the basis of brain structure; (1) that insect-
eating bats are the most primitive ones contrary to what has often
been assumed a priori, (2) that they are closest to some terrestrial, little advanced and unspecialized insectivores, and (3) that
the most skilled hoverers of present time, the nectar-feeders, are
the group closest to the insect-eaters.

What was the adaptive value of hovering during the early
evolution of the Chiroptera?  It was probably useful for catching
insects in the easiest position, i.e. resting on plants.  Jumping
at them from the ground or from a branch was possibly the first
technique utilized, after which hovering for two or three seconds
would have been a great improvement and one likely to be "encouraged" by natural selection.  The particular significance of nectar-
feeding must be remembered here.  Nectar-feeders are considered
highly specialized animals, but in which sense is that correct?
In my opinion, only in the sense that they have retained one
special behavioral feature that was probably much more common in
the early history of Chiroptera; that is one special version of
the general insectivorous habit.  Indeed, whereas all bats were,
from their origin or soon thereafter, capable of stationary flight,
some of them evaded strong competition by exploiting that ability
to a maximum and hovering very near the flowers in order to catch
insects inside their corollas, thereby becoming nectar-feeders.  It
is not surprising that some nectar-feeders can be maintained in
captivity on a diet consisting of insects only.  It is not surprising either that they are found in an intermediate position between insect-catchers and frugivorous types as regards their brain
organization (Stephan and Pirlot, 1970; Pirlot and Pottier, in
press).  It is not irrelevant to mention also that members of the
presumably very old Emballonuridae, e.g. Saccopteryx, have the
habit of roosting on the bark of large trees, a few feet above the
ground:  this I observed several times in the Amazonian forest and
on Margarita island; it fits with my hypothesis.

At this stage the question "Which species among the bats
studied is the most advanced with respect to wing structure?" must
be asked.  This is difficult to answer, but one would think that
the molossids might be.  For a given mass, they fall far to the
right on the graph (Fig. 5).  They probably offer among Chiroptera
one of the best or perhaps the best combination of great muscular
power and favourable aspect ratio permitting both reasonably high
wing loading (roughly the reciprocal of A/W in Table IV) and high
speed flight.  But this is just one side of reality, the one that
corresponds best with our customary way of judging the overall performance of an aircraft.  In other respects (e.g. long range
traveling, skilled avoidance of obstacles, accurate landing), other
bats may be more advanced.

One must now return to the question that was asked in the
introduction to this paper: Is it possible to classify bats according to wing design alone into groups that would correspond to the
dietary subdivision?  The answer must be only in part.  The only
clear subdivision on the basis of wing design is into three groups:
(1) molossids; (2) emballonurids; and (3) all the other families.
This means that a relatively great degree of uniformity is observable in the geometry of bat wings.  In other words, the differences

in flight patterns among Chiroptera cannot be accounted for in
terms of geometric differences only. A bat's wing is, from the
point of view of its proportions, adaptable to various types of
dietary habits.  That does not preclude, however, the possibility
that molossids' wings are built for higher speed and emballonurids'
wings are intermediate in design.  Differences in wing design did
appear in the course of evolution even within the group including
all bats other than molossids and emballonurids.  But they were
limited to some geometrical characteristics making for a swift or
a slow flight, for ease or clumsiness in manoeuvring.  The most
interesting feature seen in bats, in that respect, is their
theoretical ability to hover, actually exploited by a few species
but virtually present in all of them, even in those that make
little or no use of it. Of course, one may expect degrees to exist
in this ability but its generality within the order seems to
justify the hypothesis put forward in this paper, an early extensive (although not exclusive) use of stationary flight in the
search for food.

From the above discussion we must admit that the exact origin
of bats remains, by and large, unknown, but we seem to find partial
clues in features shown by their brains as well as by their wing
design.  Whereas wing structure is rather uniform among bats,
brain organization, perhaps being more closely related to the
various life-habits, is more diversified within the order.  For a
proper understanding of both, one must refer to the particular
ecological circumstances that were prevailing at the dawn of the
mammalian era.  At that time there were still in existence a number of dangerous reptiles although the very large types had probably all or almost all disappeared.  Birds had filled most of the
aerial niches.  But both reptiles and birds were mainly diurnal
animals (in spite of the fact that some dinosaurs may have evolved
towards a nocturnal way of life and that some birds such as the
guacharos have been suspected of using sonar for night-flight).
Their most important sense was vision.  It is not surprising that
bats had but little choice: from the moment mutations gave them
sustaining membranes and they were able to flap these, they were
practically restricted to nocturnal aerial life for which the most
useful sensory system was audition.  One must remember that in
fact all primitive and rather small mammals had to be active by
night so as to avoid reptilian enemies.  Audition and olfaction
were their main assets.  In the case of bats, olfaction again was
not very useful, as expected in flying animals, but bats developed
their acoustic power to a degree of specialization and sophistication attained by no other land mammals. Accordingly, while birds
had acquired sizeable optic lobes, bats became endowed with voluminous colliculi posteriores and other well developed auditory relays.  One may thus suggest that the first bats were mammalian insectivores that started hunting acoustically by night in the spaces
between large plants while both the branches and the ground around

were predominantly occupied by shrews, birds, probably some Carnivores, and a number of reptiles. Under this assumption, direct
competition must have been greatly reduced for bats. This suggestion applies to the vast majority of Chiroptera. However, one
group among them evolved differently. Good vision was preserved
in one very ancient group, the pteropodids, which proves that
another orientation of the evolutionary trend, i.e. towards optic
navigation and twilight activity was still possible in the early
evolution of the order. (This suggestion does raise the question,
not to be debated here, of whether modern bats are mono- or
diphyletic.) For pteropodids, a corollary of their good vision
and feeding habits was that sonar equipment was not necessary and
indeed did not develop (see Baron, 1974, for a quantitative analysis
demonstrating that pteropodids have relatively underdeveloped
auditory relays in comparison with other vegetarian and with insectivorous species). Under such circumstances, it is conceivable
that hovering ability carried a selective advantage for all bats,
whether insectivorous or vegetarian. On the hypothesis of such a
general ability for hovering flight, the origin of bats is best
understood in dense forests, probably of tropical type, i.e. in
whatever type of high vegetation existed at that time.

Finally, I must insist that, while comparisons between birds
and bats come readily to mind, they are a hazardous enterprise
because the significance of both similarities and differences is
not easy to evaluate. Welty (1962) remarks that the largest bird
(ostrich) is only 64,000 times heavier than the lightest (a humming
bird) whereas the biggest mammal (a whale) is 59,000,000 times
heavier than the smallest (a shrew). A comparison among bats seems
more meaningful. Among Chiroptera, the extremes are about 1200 g
and 3.5 g so that the heaviest would be about 340 times the lightest one. In other words, even large bats are severely restricted
as far as weight is concerned. Metabolic factors may explain that
restriction; as already said, bat wings are less protected by far
than bird wings against heat loss. The sparse hair covering their
richly vascularized patagium does not seem to be a very efficient
protection against caloric dissipation, especially during nocturnal
flight. So, gigantic bats with proportionate wings were not likely
to emerge in the course of evolution. It is not surprising that
the mass limit of Chiroptera is found within (or near) the range
of values required for hovering flight. This is also in agreement
with the fact that bats, contrary to birds, possess neither hollow
bones nor conspicuous adaptations of the respiratory system (air
sacs). They have a mammalian skeleton, i.e. a solid bony framework which is definitely less favourable for flight than the
pneumatic bird bones. Their skull and jaws are relatively heavy
whereas the avian skull and bill are exceedingly light. All in
all, bats seem to be in a critical or subcritical position as
flying animals, while being at the same time capable of hovering

performances. One feature, however, appears to be a special asset
for bats and that is their having <u>red</u> breast muscles. This feature
gives them more power during flight than is the case with many
birds that possess <u>white</u> breast muscles only. In spite of this,
bats have in general much less flight endurance than birds.


## ACKNOWLEDGEMENTS

My thanks are due to Mr. R.J. Templin who wrote the appendix;
to my colleague in Montreal Professor G. Baron who provided useful
data on the stato-acoustic system of bats; to Professor Max K.
Hecht, Dr. J.D. Smith and Mr. B.D. Clark who offered helpful crici-
cisms during the NATO Advanced Study Institute; and to Mrs. Pottier
who helped me with the final writing and to Mrs. Farid who drew
the figures.


## BIBLIOGRAPHY

Alexander, R.McN., 1971, Size and shape. London, E. Arnold, 59 pp.
Allen, G.M., 1939, Bats. New York, 368 pp.
Bader, R.S. and Hall, J.S., 1960, Osteometric variation and func-
    tion in bats. Evolution, 14:8-17.
Baron, G., 1973, Volumetrischer Vergleich sensorischer Hirnstruk-
    turen bei Fledermausen. Period. Biol., 75:47-53.
Baron, G., in press, The vestibular complex in relation to flight
    behavior among Chiroptera: a volumetric analysis.
Davis, R., 1969, Wing loading in pallid bats. J. Mamm., 50:140-144.
Farney, J. and Fleharty, E., 1969, Aspect ratio, loading, wing
    span, and membrane area of bats. J. Mamm., 50:362-367.
Findley, J.S., Studier, E.H. and Wilson, D.E., 1972, Morphologic
    properties of bat wings. J. Mamm., 53:429-444.
Greenewalt, C.H., 1962, Dimensional relationships for flying
    animals. Smiths. Misc. Collect., 144(2):1-46.
Hainsworth, F.R. and Wolf, L.L., 1975, Wing disc loading: implica-
    tions and importance for hummingbird energetics. Am. Nat.,
    109(966):229-33.
Lawlor, T.E., 1973, Aerodynamic characteristics of some neotropical
    bats. J. Mamm.,
Norberg, U.M., 1970, Hovering flight of <u>Plecotus auritus</u> Linnaeus.
    Bijd. Dierk., 40:62-66.
Norberg, U.M., 1972, Bat wing structures important for aerodynamics
    and rigidity (Mammalia: Chiroptera). Z. Morph., 73:45-72.
Novick, A., 1958, Orientation in paleotropical bats. I. Micro-
    chiroptera. J. Exp. Zool., 138(1):81-154.
Pirlot, P., 1969, Relations ponderales entre l'encephale et le
    corps chez les Chiropteres. I. Especes neotropicales. Rev.
    Can. Biol., 28(2):127-136.

Pirlot, P., 1970, Id. II. Especes paleotropicales et palearctiques.
    Discussion et conclusions.  Bijd. Dierk., 40(2):103-115.
Pirlot, P. and Stephan, H., 1970, Encephalization in Chiroptera.
    Can. J. Zool., 48:433-444.
Pirlot, P. and Pottier, J., in press, Quantitative study of the
    brain in bats.
Poole, E., 1963, Relative wing ratios of bats and birds.  J. Mamm.,
    17:412-413.
Stephan, H. and Pirlot, P., 1970, Volumetric comparisons of brain
    structures in bats.  Z.f.zool.Syst.u.Evolutionsforschung,
    8(3):200-236.
Struhsaker, T.T., 1961, Morphological factors regulating flight in
    bats.  J. Mamm., 42:152-159.
Templin, R.J., 1970, Aerodynamics low and slow.  Can. Aeron. and
    Space J., 16(8):318-328.
Vaughan, T.A., 1966, Morphology and flight characters of Molossid
    bats.  J. Mamm., 47:249-260.
Vaughan, T.A., 1970, Flight patterns and aerodynamics, in "Biology
    of Bats," (W.A. Wimsatt, ed.), New York Academic Press,
    vol. I:195-216.
Welty, J.C., 1962, The life of birds.  Philadelphia and London,
    Saunders, 546 pp.

APPENDIX

SIZE LIMITS IN FLYING ANIMALS

R.J. TEMPLIN

Low Speed Aerodynamics Section, National Research

Council of Canada

## INTRODUCTION

It has been known for a long time that the flying ability of winged animals is related to their size, and that there are size limits beyond which various kinds of flight seem to be impossible. A recent discussion of this problem is given, for example by Alexander (1971), but no quantitative estimates of absolute size limits have been found. Figure 5 presents the results of the present attempt, which must be considered to be tentative because of a number of rather sweeping assumptions and approximations that have been made. It is the purpose of this Appendix to explain, at least in outline, the process by which the three limiting curves in Figure 5 have been calculated.

To begin with, the common assumption has been made that animal mass M is proportional to the cube of some representative linear dimension, and this has been taken to be the body length L, so that

$$\text{mass } M = K_M L^3 \tag{1}$$

While it might be expected that this equation would be correct only for geometrically similar animals, and therefore that variations in the ratio of wing span to body length, E/L, in winged animals would have some effect on the parameter $K_M$, we have found no consistent variation. This may be due to the fact that, even in long-span birds, the wing weight seems to be less than about 20% of total weight (Greenewalt, 1970), and there is in any case a

411

considerable scatter in weight data.  From an analysis of a large number of weight and body length values for flying and non-flying animals, we have taken a mean value of $K_M$ to be 35 kg/m$^3$.

However, if the ratio of span to length is of minor importance in determining animal mass, this is definitely not true when considering aerodynamic and flight performance characteristics.  For example, we have assumed that the maximum available wing area is given by a wing of span E, and of elliptical plan form with its chord length equal to the body length at the body centre line. Hence

$$\text{wing surface area } S = \frac{\pi}{4} EL = \frac{\pi}{4} \frac{E}{L} \times L^2 \qquad (2)$$

In this equation, the common assumption that the wing surface area varies directly with the square of the linear dimension L is modified by the wing "aspect ratio" E/L.  (In aeronautical parlance, the term aspect ratio is reserved for the ratio of wing span to mean wing chord, but the difference is of no account for our present purposes.)

Similarly, we will need to consider another lifting area A, which defines the maximum flow stream tube area affected by the wing system.  It is essential to the calculation of induced drag (drag due to lift) and since it corresponds to helicopter rotor disc area, it also determines the power required to hover.  Its definition is:

$$\text{wing "disc" area } A = \frac{\pi}{4} E^2 = \frac{\pi}{4} \left(\frac{E}{L}\right)^2 L^2 \qquad (3)$$

Again, the aspect ratio E/L is important in the scaling rules, and so it is not surprising that it should appear as a primary parameter in our graph.

In what follows, we distinguish between three main modes of flight:  gliding, continuous hovering flight and continuous level flight.  By continuous flight we mean flight of indefinite duration at a power level such that no oxygen deficit is incurred.

MAXIMUM MASS FOR SAFE UNPOWERED LANDING

The upper curve of Figure 5 was first presented (Templin, 1970) as part of a discussion of the possible evolutionary development of wings.  For animals whose only mode of flight is gliding, or that have insufficient power available for continuous flapping flight, landings must still be possible at flying speeds low enough to prevent injury. Rough considerations indicate that this speed should be not more than about 6 m/sec., which is the velocity after a free fall, with negligible air resistance, from a height of roughly 2 meters. Many large animals can jump to about this height, and even insects are known to be damaged or crushed at impact speeds not much greater than 6 m/sec.

There is a temptation, by analogy with aircraft performance calculations, to assume that this safe landing speed must be equal to or greater than the level flight stalling speed $V_S$. The stalling speed is the minimum speed at which the wing system is capable of generating aerodynamic lift equal to the flying weight, and is defined by the equation

$$\text{Lift} = Mg = \frac{1}{2}\rho V_S^2 \; SC_{Lmax} \tag{4}$$

Where M = animal mass
g = acceleration of gravity
S = wing area as defined by (2)
p = air density
$C_{Lmax}$ = maximum wing lift coefficient

We have not assumed the minimum flying speed $V_m$ to be equal to $V_S$ but have taken

$$V_S = BV_m \tag{5}$$

where B is a factor to be determined by analysis. Many of the birds, and other flying animals such as the flying squirrel, appear to have developed a type of flight trajectory on approach to a landing, which achieves a minimum speed <u>well below</u> stalling speed, so that the factor of B in the above equation may be considerably greater than unity. In order to investigate this possibility further, a computer study was carried out for a set of idealized unsteady flight paths. It was assumed that the flying animal approaches landing in its flattest possible glide (the flatness depending mainly on the wing aspect ratio), at a speed which is well above stalling speed. At some instant prior to landing, the wing incidence was assumed to be increased to just below the stalling angle, and held there for the remainder of the flight. At this instant the lift is increased well above the animal weight, so that the flight path begins to curve upward. Drag also increases greatly and speed drops, due both to the drag increase and eventually to the upward-sloping flight path. When the speed has dropped to the level flight stalling speed, it continues to decrease and the flight path develops downward curvature. Aerodynamic stall does not occur however, because of the fixed wing angle of attack. As speed decreases further, lift

falls and the flight path begins to slope downward.  The drag remains high and speed continues to drop until it passes through a
minimum and then increases with further downward slope of the
path.  It was assumed that this entire manoeuvre has been so well
"judged" that the landing point occurs at the point where the
speed is a minimum.  It is interesting that, because of the S-
shaped curvature of the path, the landing point (minimum speed
point) occurs at a height that is nearly the same as the height
at which the steady glide ended, and the braking manoeuvre began.
The analysis showed that the ratio of the stalling speed to the
minimum speed (the factor B) is independent on aspect ratio over
a range from about 1 to 8.  The value of B was about 1.5 at the
higher aspect ratios, and increased slightly with decreasing $\frac{E}{L}$.

We now have all the information required for the completion
of the "safe landing" mass limit.

Substituting (5) and (2) into (4), and rearranging slightly,

$$\frac{M}{L^2} = \frac{\pi}{8} \frac{\rho}{g} B^2 V_m^2 C_{Lmax} \left(\frac{E}{L}\right)$$

But, from (1),

$$L^2 = \frac{M^{2/3}}{K_M^{2/3}} \quad \text{hence,}$$

$$M^{1/3} = \frac{\pi}{8} \frac{\rho}{g} \frac{B^2 V_m^2}{K_m^{2/3}} \frac{E}{L} C_{Lmax}$$

and, finally,

$$M = \left(\frac{\pi}{8} \frac{\rho}{g}\right)^3 \frac{B^6 V_m^6}{K_m^2} \left(\frac{E}{L}\right)^3 C_{Lmax}^3 \tag{6}$$

It will be noted that the mass given by (6) is proportional
to the sixth power of B and of the assumed minimum speed $V_m$.  This
sensitivity is unfortunate, but the situation is retrieved in part
by the compressed logarithmic mass scale of Figure 5.  It is also
interesting to note, however, that the braking manoeuvre "trick",

which yields a value of roughly 1.5 for the factor B, accounts for an additional order of magnitude in permissible safe mass (because $1.5^6 = 11.4$). Whether or not our idealized description of this manoeuvre is accurate, there can be little doubt that its development by the largest gliding animals has been a factor in their growth. It is a trick not considered to be legal in aeronautical circles.

We now substitute the following numerical values for the constants in the equation:

$\rho$ = 1.23 kg/m$^3$ (standard sea level density)

g = 9.807 m/sec.$^2$

B = 1.5

$V_m$ = 6 m/sec.

$K_M$ = 35 kg/m$^3$

$C_{Lmax}$ = 1.2 (a guess, for fairly highly cambered wings at low Reynolds no.)

The result is the required equation for the safe mass limit:

$$M = 0.09 \left( \frac{E}{L} \right)^3 \qquad (7)$$

In actually plotting the upper curve in Figure 5, allowance was made for the slight increase in B found in the computer analysis as E/L decreases, but apart from this adjustment, equation (7) shows that the safe mass limit for unpowered landings varies approximately as the cube of the span-to-length ratio.

## MAXIMUM MASS FOR HOVERING

A calculation of the mass limits for powered flight requires a knowledge of the continuous power available from the flight muscle system. Many measurements have been made of muscle weight in flying animals, but since we are interested in the mass limits for steady flight (during which no oxygen deficit need be incurred), it is probably more accurate to base estimates of available power on measurements of maximum metabolic rates. Tucker (1969b) has collected and presented data for insects, birds and for a number of non-flying animals, including man. Over a very wide range of weight, from insects weighing about 0.03 gm., to the horse at 700 kg., the maximum metabolic rate data fit closely a power law of the form

$$R = K_R M^{2/3} \tag{8}$$

where R = metabolic rate

$K_R$ = a constant

M = animal mass

When R is expressed in watts (Newton-Meters/sec) and M is in kilograms, we find the numerical value of $K_R$ to be 25 from Tucker's data.

Since only a fraction (about 20 per cent) of the metabolic energy rate is converted to mechanical power, the remainder presumably must be dissipated as heat through the animal surface. For geometrically similar bodies of equal density, the surface area is proportional to $L^2$ and thus $M^{2/3}$, and it may be that this provides an explanation of the form of the empirical equation (8). It should be noted, however, that Alexander (1971) proposes a slightly different power law with an exponent of 3/4 rather than 2/3.

It can be shown that the aerodynamic power required for level flight (either hovering flight, or level flight at minimum power) is proportional, in geometrically similar families, to the ratio

$$\frac{M^{2/3}}{L}, \text{ and hence also to } M^{7/6}$$

Since we are assuming that the power available is proportional to $M^{2/3}$ we obtain the ratio:

$$\frac{\text{power required for steady flight}}{\text{power available}} = \frac{M^{7/6}}{M^{2/3}} = M^{1/2}$$

The ratio increases with increasing mass, and thus, in any geometrically similar family, there will be a mass above which it exceeds unity and steady sustained flight becomes impossible. Since the minimum power required for steady level flight is generally less than that required to hover, the hovering mass limit is the lowest curve on our graph.

The minimum power required to hover at weight W by means of a lifting system with a "disc" area A can be shown from simple momentum and energy considerations to be

$$P_r = \frac{W^{3/2}}{\sqrt{2\rho eA}} \qquad (9)$$

where e is an empirical factor which is unity for an ideal, perfectly efficient lifting system. Its actual value is always less, and the product eA can be considered as an "effective" swept disc area. Some guesswork is involved in assigning a reasonable value to e, and we have chosen a tentative value of 0.5. For helicopters it is considerably closer to unity, but the vibrating or oscillating wings of hovering insects, birds and bats probably sweep out a stream tube area considerably smaller than the area of a circular disc with diameter equal to wing span. The same factor e appears in the equation for induced power (power due to lift) in forward flight as well as in hovering, and in fixed-wing aircraft or gliders its value may be as high as 0.9. However, a previous analysis (Templin, 1970) of wind tunnel measurements made by Tucker (1969a) had indicated that, in level cruising flight, the value of e for birds may be as low as 0.25 in some cases. Since these same birds probably are as efficient as aircraft when in gliding (non-flapping) flight, the average value of e during intermittent flapping-gliding flight may be expected to lie between these two extremes. We have thus taken e = 0.5 as a reasonable guess throughout the powered flight regime.

If we substitute from (2) and (1) into (9) we obtain

$$P_r = \frac{W^{3/2}}{\sqrt{\frac{\pi}{2}\rho e \; \frac{E}{L} \; L}}$$

$$= \frac{g^{3/2} M^{7/6} K_M^{1/3}}{\sqrt{\frac{\pi}{2}\rho e \; \frac{E}{L}}} \qquad (10)$$

We now assume that the maximum metabolic rate determines the continuous power available and that the thermodynamic conversion efficiency is η.  Thus the power available is, from equation (8):

$$P_a = \eta K_R M^{2/3} \tag{11}$$

At the limiting mass, power available and power required are equal.  We therefore equate the right-hand sides of equations (10) and (11) to obtain a new equation.  When it is rearranged so that the mass terms are collected to the left-hand side, we obtain:

$$M^{\frac{1}{2}} = \frac{\left(\frac{\pi}{2}\rho e\right)^{\frac{1}{2}} K_R \left(\frac{E}{L}\right)}{g^{3/2} K_M^{1/3}}$$

and hence, by squaring both sides,

$$M = \frac{\pi}{2} \frac{\rho e \eta^2 K_R^2}{g^3 K_M^{2/3}} \left(\frac{E}{L}\right)^2 \tag{12}$$

This is the equation for the hovering mass limit.  It predicts that the limiting mass is proportional to the square of the span-length ratio E/L.

The thermodynamic efficiency η has been taken to be 0.2 (20 per cent).  This value is based in part on an analysis (Templin, 1970) of Tucker's wind tunnel measurements of the energy consumption of a budgerigar in steady flapping flight, for which he used a tilting wind tunnel to simulate descending, level, or climbing flight.  Tucker measured metabolic rate during flight, and since the differences in power required with change in flight path angle, at any given speed are easily calculated, the conversion efficiency could be deduced.  Dawson and Sills (1966) also give a value of 0.2 for the thermodynamic efficiency of a man when running or cycling.

We now substitute the following numerical values into equation (12):

$$\rho = \text{air density} = 1.23 \text{ kg/m}^3$$

$$e = 0.5$$

$$\eta = 0.2$$

$$K_R = 25 \text{ watts/kg}^{2/3}$$

$$g = 9.807 \text{ m/sec}^2$$

$$K_M = 35 \text{ kg/m}^3$$

We obtain

$$M(\text{kg}) = 0.0024 \left(\frac{E}{L}\right)^2 \qquad\qquad (13)$$

This is the equation for the lower curve plotted in Figure 5.

## MASS LIMIT FOR STEADY POWERED FLIGHT

The calculation of the steady powered flight mass limit curve is more complex than for the other two, mainly because of the necessity to take into account the parasite drag coefficient, which varies with aspect ratio and with the Reynolds number (a function of size, speed, and kinematic viscosity of air). The complete development of the method will not therefore be described here. It can be described in outline, however, by considering a modification of the hovering case.

As previously mentioned, the power required for level flight is generally less than the power required to hover. If the ratio of cruising to hovering power is C, say, then we could rewrite the power equation (10) to include the factor C. Carrying through the remaining algebra of the last section, we find that the mass limit equation (12) remains the same except for a new factor $1/C^2$. In other words, the limiting mass is inversely proportional to the square of the power ratio C. For example, if C = 1/3, that is if the power required for level flight were one-third of the hovering power, the mass limit for steady level flight would be 9 times the hovering limit.

Upon what does the power ratio C depend? It can be shown to be given by the following equation:

$$C = \frac{2^{3/2}}{3^{3/4}} \left( \frac{f}{eA} \right)^{\frac{1}{4}} = 1.24 \left( \frac{f}{eA} \right)^{\frac{1}{4}} \tag{14}$$

The quantity eA has been defined already as the cross-section area of the effective momentum stream tube of air that is captured by the lifting wing system. The greater this area the lower will be the induced drag for a given aerodynamic lift. The quantity f is an area also, and is called the "parasite area" in aeronautical parlance. It is approximately the area of a plate, normal to the flight direction, that would have the same drag as the profile drag of the animal body and wings, and its value is roughly of the same order of magnitude as the body frontal area. Since our limiting mass for level flight is $1/C^2$ times the hovering flight limit, the ratio of the two limiting masses is proportional to $\frac{A}{f}$, and thus varies somewhat like the ratio of wing span to body diameter. This explains why the two mass limit curves come close together at small span-length ratios. The wing span of many insects is rather small in relation to body size, and as a result it is not much more difficult for them to hover than to fly in level flight. At the other end of the span-length scale, animals capable of continuous cruising flight may be much larger than geometrically similar hovering animals.

The estimation of the parasite area f was the complicating factor in the calculation of the steady powered flight curve of Figure 5, and will not be described in detail. The method used, however, included the profile drag of wings and body, and took into account at least approximately the variation of profile drag coefficients with Reynolds number.

## DISCUSSION

In addition to the data for bats, the graph in Figure 5 contains a few points for other flying animals to better display the total range found in nature. At least in a rough way the known flying characteristics of these animals are appropriate to the regions they occupy in the graph. The largest flying animal known in vertebrate history was the Pteranodon, with a wing span of about 7.5 meters. Bramwell and Whitfield (1970) estimate its weight to have been about 40 lb. (18 kg. mass), and if this estimate is anywhere near correct, the theory that it was primarily a glider is confirmed in Figure 5. In some parts of the world enthusiastic groups are now attempting to put man in the air under his own power. The graph shows why the proposed man-powered aircraft designs have large wing spans.

The smallest bats, the hummingbird, and insects lie well below the calculated mass limit for continuous hovering. In this region of the graph there is sufficient excess power available to permit rapid manoeuvres without loss of flight speed. Sharp turns or pull-ups, are coupled with high accelerations normal to the flight path and require the development of aerodynamic lift much in excess of flying weight. Increased power is required to balance the resulting increase of drag due to lift. In aeronautical terminology the normal acceleration (normal to the flight path) is often measured in "g" units, that is, in multiples of the accelera-tion of gravity. Steady level flight is 1-g flight, and thus our middle curve showing the calculated mass limit for sustained level flight can be thought of as a mass limit for 1-g flight with no oxygen deficit. It would be possible to calculate a set of nearly parallel curves on the logarithmic scales of the graph, lying below the steady level flight curve, each one applicable to a mean manoeuvring g-level. One such curve would pass through the region of the graph occupied by the smaller bats. The same curve would also traverse the region occupied by many of the insects. Thus it would appear that the graph confirms the obvious: insect-eating bats, in spite of their size, can out-manoeuvre some insects.

## REFERENCES

Alexander, R.M., 1971, Size and Shape. Studies in Biology No. 29. Edward Arnold Ltd., London.

Bramwell, C.D. and Whitfield, G.R., 1970, Flying speed of the Largest Aerial Vertebrate. Nature, Vol. 225, p. 660.

Dawson, L.G. and Sills, T.D., 1966, Speed is Cheap. Paper delivered to Royal Aeronautical Soc. Glasgow Branch Centenary Celebrations, 11 Oct. 1966.

Greenewalt, C.H., 1962, Dimensional Relationships for Flying Animals. Smithson. Misc. Collns., Vol. 144, No. 2.

Templin, R.J., 1970, Aerodynamics Low and Slow. Journ. Can. Aeronaut. and Space., Vol. 16, No. 8, Oct. 1970, p. 318-328.

Tuckers, V.A., 1969a, The Energetics of Bird Flight. Scientific American, Vol. 220, p. 7-78.

Tuckers, V.A., 1969b, Energetic Cost of Locomotion in Animals. Comp. Biochem. Physiol., Vol. 34, pp 841-846.

ENERGETICS OF HOVERING FLIGHT AND THE ORIGIN OF BATS

Brian D. CLARK

Division of Biological Sciences, The University of
Michigan
Ann Arbor, Michigan  48109 U.S.A.

It has been suggested that the evolution of the Chiroptera involved early selection for brief periods of hovering flight (Pirlot, this volume).  Proto-bats are represented as being insectivorous, first leaping at their prey from the ground or from plants, and later hovering for periods of a few seconds by vigorously flapping their forelimbs.  After entering the aerial niche in this manner, bats are supposed to have undergone an adaptive radiation into the variety of flight habits used by their extant representatives.  A similar scheme has been advanced by Jepsen (1970) who proposes that the mammal that first took to hovering already had webbed hands and "little skin flaps extending from arms to sides of the body" which it used in capturing insects.

This model for the evolution of the Chiroptera is untenable as it requires the capacity for hovering flight in an intermediate form with patagia adapted not as airfoils but as insect nets.  It is unlikely that those anatomical and physiological parameters of the forelimb muscles which were established when these muscles served only for quadrapedal locomotion would allow the power output required in hovering.

Let us consider the hypothetical animal which first leaped into the air and hovered.  The insect-catching structures on the forelimbs (e.g. enlarged, webbed "hands") must have been small enough to avoid serious impairment of quadrapedal locomotion, yet large enough to provide sufficient aerodynamic forces when flapped. The muscles which had been moving the pectoral appendage in a roughly parasaggital plane during quadrapedal locomotion would suddenly be required to yield large power outputs in the novel movements involved in hovering.

Hill (1950) has shown that striated muscles develop maximum power output when they contract against loads around 0.3 times their maximum isometric tension. Alexander (1973) has demonstrated that the muscles involved in movement are set to contract against such loads in many locomotor patterns. This implies that in the course of evolution muscle mass, fiber length, internal architecture, moment, myosin speed and other parameters are matched to body mass, moment of inertia of the limbs, degree of limb excursion and velocity of locomotion. Selection optimizes muscle power output in specific locomotory behaviors.

Consequently the muscles moving the pectoral limb of the proto-bat were probably adjusted to produce a high power output during quadrapedal locomotion. If the animal shifted suddenly to flight, these muscles would be required to contract over an altered range of lengths and at altered velocities, and most importantly, at altered loads. As a result, power output of the forelimb musculature would have been lower in forelimb flapping than in quadrupedal running.

On the other hand, we do know something of the energetics of powered flight in birds and bats (Pennycuick, 1968, 1975; Thomas, 1975; Tucker, 1968, 1973). The curve for power requirement versus forward speed for flying animals is U-shaped. They need to expend more power at zero speed (hovering) than they do at intermediate speeds. At still higher speeds the power requirement again increases.

A power curve of this general shape may also be applied to the flight of the proto-bat. It seems unlikely that this animal, with patagia imperfectly adapted as airfoils and flight muscles adjusted primarily for quadrapedal running, would have invaded the aerial adaptive zone with a flight behavior as strenuous as hovering. It is more reasonable to suggest that bat ancestors were gliders which gradually evolved the capacity for sustained (and controlled) flight at speeds where power requirements were minimal.

These arguments may also have some bearing on the origin of flight in birds. Ostrom (1974) proposes that Archaeopteryx, like the proto-bat of Jepsen and of Pirlot, was a cursorial predator the flight of which consisted of "flapping leaps to catch insect prey." If birds are derived from bipedal archosaurs, and the forelimbs developed strong powers of adduction in their capacity as insect nets, then it is conceivable that the first fliers may have had muscles capable of the power outputs required for hovering or low

speed flight.  However if <u>Archaeopteryx</u> was capable of brief
periods of hovering or of ascending flight (for which power requirements are also quite high) then it should immediately have
been capable of sustained level flight at moderate speeds, as the
power requirements for this are minimal.

Models suggesting that a group of flying vertebrates originated by a sudden transition from a cursorial form to one which
hovered or flew at low speeds should consequently be reconsidered
in the light of what we now know about flight energetics and the
physiology of locomotion.  It is easier for an animal to suspend
itself at moderate speeds than at very low (or very high) speeds,
and any proposed intermediate which could hover undoubtedly could
also fly forward.  It is unlikely that bats (and perhaps birds as
well; assuming a bipedal ancestry) first evolved the capacity for
sustained flight at any speeds other than the least strenuous ones.

## REFERENCES

Alexander, R. McN., 1973, Muscle performance in locomotion, <u>in</u>
    "Comparative Physiology" (L. Bolis, K. Schmidt-Nielsen and
    S.H.P. Maddrell, eds.), pp. 1-21, North-Holland, Amsterdam.
Hill, A.V., 1950, The dimensions of animals and their muscular
    dynamics.  Science Progr., 38:209-230.
Jepsen, G.L., 1970, Bat origins and evolution, <u>in</u> "Biology of
    Bats" (W.A. Wimsatt, ed.), Vol. 1, pp. 1-64, Academic Press,
    New York.
Ostrom, J.H., 1974, <u>Archaeopteryx</u> and the origin of flight.  Quart.
    Rev. Biol., 49:27-47.
Pennycuick, C.J., 1968, Power requirements for horizontal flight
    in the pigeon <u>Columba livia</u>. J. Exp. Biol., 49:527-555.
Pennycuick, C.J., 1975, Mechanics of flight, <u>in</u> "Avian Biology"
    (D.S. Farner and J.R. King, eds.), Vol. 5, pp. 1-75,
    Academic Press, New York.
Pirlot, P., this volume, Wing structure and the origin of bats.
Thomas, S.P., 1975, Metabolism during flight in two species of
    bats, <u>Phyllostomus</u> hastatus and <u>Pteropus</u> gouldii. J. Exp.
    Biol., 63:273-293.
Tucker, V.A., 1968, Respiratory exchange and evaporative water
    loss in the flying budgerigar.  J. Exp. Biol., 48:67-87.
Tucker, V.A., 1973, Bird metabolism during flight:  evaluation of
    a theory.  J. Exp. Biol., 58:689-709.

# COMMENTS ON FLIGHT AND THE EVOLUTION OF BATS

James Dale SMITH

Department of Biology, California State University

Fullerton, California  92634

Inasmuch as bats possess wings, are capable of sustained flight, and have become totally committed anatomically to this mode of existence, an interpretation of chiropteran evolution must first deal with the evolution of chiropteran flight.  Unlike birds which can simply fold their wings, when not in use (and walk about on relatively unspecialized pelvic appendages) bats are essentially incapable of alternative forms of locomotion when not in flight. Of course, all bats can scurry about to some extent and vampire bats (Desmodontinae) are quite agile at walking on their elbows and wrists.  However, by virtue of their anatomical adaptations for flight, bats have largely abandoned their terrestrial (i.e., quadrupedal) locomotory abilities.  No other volant animal has made such a complete and drastic modification in its locomotory style; the only possible exceptions are the reptilian pterosaurs and flightless birds.

It is generally accepted that bats evolved from small, arboreal insectivores that may have possessed gliding membranes (Fig. 1).  The argument for an arboreal ancestor as opposed to a strictly terrestrial ancestor seems obvious in light of the fact that all volant mammals normally launch from trees or heights above the ground.  As shown in Figure 2 (A-C), the patagium, medial to the wrist, could have developed in a fashion similar to most mammalian gliders.  At the same time, that portion of the "wing" lateral to the wrist may have expanded, as a result of digital elongation and the concurrent enlargement of the interdigital membrane panels. Initially, this may have provided an increase in the size of the gliding surfaces and perhaps some degree of maneuverability by abduction/adduction of the manus.  Certainly, such early stages

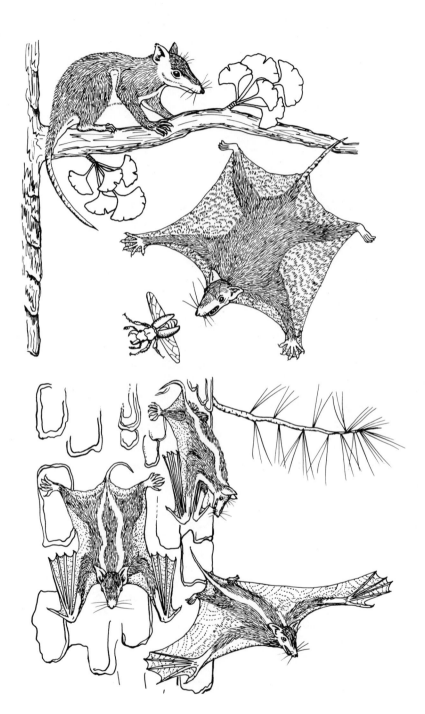

Fig. 1.  Top. - Hypothetical arboreal, insectivorous ancestor of bats showing patagial development (see Fig. 2 A-B).
Bottom. - Hypothetical intermediate ancestor of bats showing further development of the
wing (see Fig. 2 C-D).

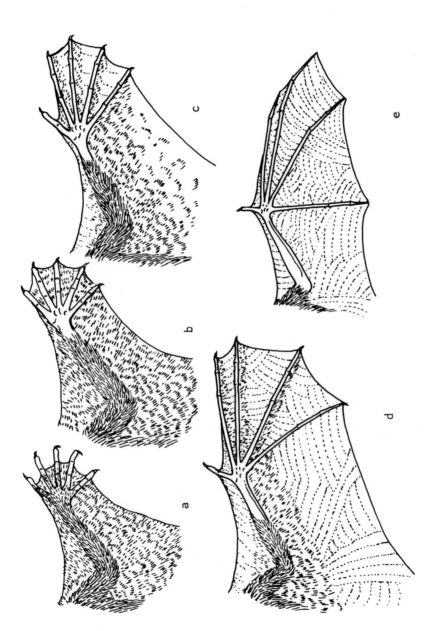

Fig. 2. Hypothetical, progressive stages in the development of the wing of bats. A-C, early stages in which emphasis was on gliding. D, an intermediate stage in wing development. E, a fully developed bat wing.

in the development of the wing would have allowed both volant and relatively normal quadrupedal locomotion. However, the continued development of the wing, in this manner, eventually would have produced an ungainly and clumsy structure (Fig. 2 C-E) that necessitated movement as a wing rather than a fixed gliding device. I suspect that this transition, once initiated, was rather rapid and perhaps accounts for the apparent paucity of "transition forms" in the fossil record.

Dr. Pirlot (this volume) has suggested that "the first selection has been for some degree of hovering rather than for gliding in early Chiroptera....Jumping at insects from the ground was possibly the first hunting technique of those animals and, then, hovering during one or two seconds would have been a great improvement in efficiency, and one likely to be encouraged by natural selection." My investigations with chiropteran wings and bat flight and those of others (Norberg, 1970; Vaughan, 1970; Findley et al., 1972) indicates that hovering requires a rather sophisticated wing (long tip portion and low aspect ratio or E/L ratio of Pirlot) as well as a rapid and precise wing beat cycle. In view of the apparent complexity of hovering flight and as a consequence of how the chiropteran wing may have developed, I doubt that the first bats were capable of hovering flight. On the contrary, I suspect that the first bats possessed short-tipped wings, of rather low aspect, and that the ancestors of these bats progressed through a gliding stage (Figs. 1 and 2).

Having successfully traversed the critical point in wing development, bats were well on their way to occupying an aerial insectivorous niche. Further refinements of the wing probably were related to increased maneuverability and/or speed. Certainly, hovering was developed by some bats such as the nectarivorous glossophagines and by foliage gleaning, insectivorous bats such as some vespertilionids, emballonurids, and rhinolophids.

With regards to the ancestry of bats, Dr. Pirlot has implicated the Emballonuridae as "likely to be the most primitive and therefore the closest to the first bats." A similar ancestry also was implied by Miller (1907) and later followed by Simpson (1945). Actually, Miller (1907:81) regarded the family Rhinopomatidae as "the lowest of the suborder (Microchiroptera)" based largely on the presence of two phalanges in the second digit, free premaxilla, and "strongly" primitive shoulder joint. Closest to the rhinopomatids, Miller (1907:84) aligned the emballonurids, which he noted to have the greatest number of primitive characters, with the least degree of specialization. My investigations (1972, 1976) and those by Hill (1974) of the features discussed by Miller, suggest that he confused "primitive" with "unspecialized". Furthermore, preliminary results of my studies on the structure of bat wings indicate that emballonurids possess many rather specialized features.

Pirlot has suggested two features which have influenced his conclusions concerning the ancestral nature of emballonurids. First, he referred to Novick and Griffin as stating that "on the basis of the characteristics of sonar the emballonurids will be among the very first if not ancestral to the others." I have examined the works of these two early investigators of chiropteran acoustics and find no such reference. Griffin (1958) does refer to the primitive nature of the shoulder of emballonurids (_fide_ Miller, 1907). J.A. Simmons (personal communication) commented that the ultrasonics of emballonurids are not primitive by any measure, but, on the contrary, are rather sophisticated and highly adaptive.

In addition, Pirlot reported a high ratio of Neocortex/Mesocortex; approximately 10 per cent in emballonurids as compared to 4-5 per cent in other mammals. This high ratio, he suggests, is a further indication of the primitive quality of emballonurids. However, Henson (1970:135) noted a large expanse of Schizocortex (= Mesocortex) in _Eptesicus_ (Vespertilionidae) and other insectivorous Microchiroptera which resembles the condition in other lower forms. This suggests to me that proportionately large, mesocortical areas are not restricted to the emballonurids and therefore cannot be used in this argument of ancestry.

In view of the foregoing discussion, what can be said of chiropteran origin and evolution? Bat biologists have only begun to seriously consider this question in the past five to ten years and, although the situation is not as incomplete as before, we still are a long way from constructing hard and fast phylogenies. Among the major problems have been a paucity of fossils (not to be utilized as ancestors, but more as time and zoogeographic reference points in the evolution of the order), and an apparent disregard or lack of understanding of the mosaic nature of chiropteran adaptation. In the last few years, more fossil material has been discovered; particularly in the Old World from Eocene-Miocene faunas (Westphal, 1959; Lavocat, 1961; Butler and Hopwood, 1957; Butler and Greenwood, 1965; Butler, 1969; Bachmayer and Wilson, 1970; Engesser, 1972; see Smith, 1976 for additional references). Comparable material from the New World still remains scarce. Regarding the second problem, I feel it is important to reiterate Hill's (1974) warning against inferring phylogeny from a variety of living forms each of which exhibits a variety of different specializations and modifications in different degrees. Furthermore, chiropteran evolution must be viewed against the background of evolving World-ecosystems and faunal complexity.

I have previously presented (1976) a more detailed assessment
of chiropteran evolution than is possible here.  To summarize that
material, the suborder Microchiroptera may be considered to include
five monophyletic superfamilies (inclusive families in parentheses)
as follows:  Paleochiropterygoidea (Palaeochiropterygidae,
Icaronycteridae {part}; Russell and Sige, 1970); Phyllostomatoidea
(Phyllostomatidae, Noctilionidae, Mormoopidae; Smith, 1972);
Emballonuroidea (Emballonuridae, Rhinopomatidae, Crasseonyteridae;
Hill, 1974); Rhinolophoidea (Rhinolophidae, Nycteridae, Megadermatidae; Winge, 1923); Vespertilionoidea (Vespertilionidae,
Myzapodidae, Natalidae, Thyropteridae, Furipteridae, Molossidae,
Mystacinidae; Winge, 1923).  The microchiropterans appear to be a
monophyletic group (Fig. 3) with a common ancestor in the Paleocene
or perhaps the late Cretaceous.

The Paleochiropterygoidea (all extinct) are represented in
the early to middle Eocene and Oligocene of Europe and North
America.  Five living microchiropteran families (Emballonuridae,
Rhinolophidae, Megadermatidae, Vespertilionidae, and Molossidae)
have been reported from the Eocene-Miocene of the Old World, and
two (Phyllostomatidae and Vespertilionidae) have been reported
from the Miocene of the New World (Smith, 1976).  I have suggested
(1972, 1976) that the Phyllostomatoidea is autochthonous to the
New World with the remaining, living superfamilies originating in
the Old World.  Of these the emballonuroids and rhinolophoids seem
to share a closer common ancestor than either does with the
vespertilionoids (Fig. 3).

The Old World chiropteran faunas apparently have been derived
through the adaptive radiation of the emballonuroid, rhinolophoid,
and vespertilionoid lineages.  The New World tropics may have received an early vespertilionoid stock which differentiated there
into the Natalidae, Thyropteridae, and Furipteridae.  Also,
judging from the degree of morphological specialization, the
Emballonuridae appears to have arrived rather early in the New
World tropics.  Finally, later vespertilionid and molossid stocks
dispersed to the New World, perhaps in the late Miocene, and
speciated there.  The general composition of World chiropteran
faunas appears to have been established by the late Miocene to
early Pliocene.

To this point, I have not discussed the fruit bats of the
family Pteropodidae {Megachiroptera}.  Members of this family,
which show a number of unique features, are restricted to the Old
World tropics of Asia and Africa.  They are generally large bats
with wing spans up to over one meter and weights of a kilogram or
more.  Their eyes are large and their heads and faces resemble
canids; hence the name flying foxes.  The typical microchiropteran
ability to acoustically orient (echolocate) is absent; although

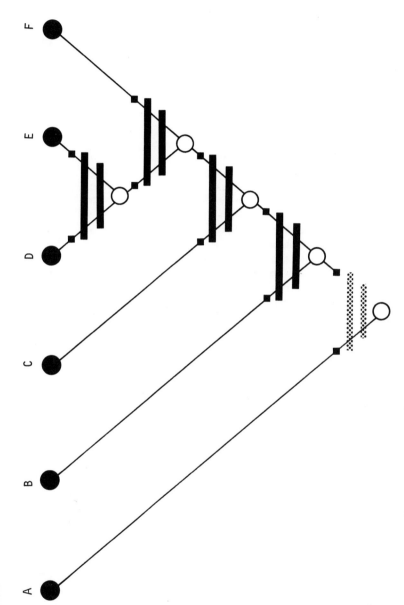

Fig. 3.  Cladogram with suggested phyletic relationships of chiropteran superfamilies.  A, Pteropodoidea; B, Paleochiropterygoidea; C, Phyllostomatoidea; D, Emballonuroidea; E, Rhinolophoidea; F, Vespertilionoidea.  Black bars indicate synapomorphous characters which establish monophyly of groups.  Black squares indicate apomorphous characters of lineages.  Dotted bars indicate synapomorphous features to be established in order to ascertain the phylogenetic relationships of the Megachiroptera (A).

one genus, <u>Rousettus</u>, has apparently developed an effective, but anatomically different, means of producing ultrasonics (Novick, 1958). The wings of megachiropterans have several unique features including: long and highly flexible first digit; second digit with two phalanges and a terminal claw (a similar feature also present in the Paleochiropterygoidea); third digit with a markedly elongated second phalange. Dr. Pirlot (this volume) and others (see Henson, 1970) have commented on the pronounced enlargement of the Neocortex of the brain as well as other unique neural features of the fruit bats. Megachiropterans are present in the Oligocene bat fauna of Northern Italy (<u>Archaeopteropus</u>, Meschinelli, 1903; Dal Piaz, 1937) and Miocene of Kenya (<u>Propotto</u>, Walker, 1969). Russell and Sige (1970) tentatively assigned <u>Archaeopteropus</u> to their Icaronycteridae (Microchiroptera); this arrangement was apparently based on the interpretation of Dal Piaz (1937). I have examined this specimen and find no justification for their arrangement.

A consideration of the phylogenetic relationships of the Megachiroptera poses some interesting problems. Most chiroptologists have recognized the distinctness of the pteropodids and most post-Linnean classifications reflect this with an assignment to some higher taxonomic category; currently the suborder Megachiroptera. Although fruit bats are usually placed within the context of the order Chiroptera, most authors neglect to elucidate or otherwise justify such an arrangement. If any mention of relationship is made, it usually is along the lines of the primitive and/or highly derived nature of the family or the phylogenetics are simply left as a dangling inference. In my consideration of bat evolution (1976), I suggested that the order Chiroptera might represent a polyphyletic group (i.e., not a monophyletic taxon).

Hennig (1966:90) states that monophyly can be established only by means of synapomorphous characters. As I have noted above, there appear to be synapomorphous features (autapomorphies, collectively) which imply monophyly of the Microchiroptera, on one hand, and the Megachiroptera, on the other. The resolution of the problem concerning the relationships of these two monophyletic groups, lies in establishing synapomorphous characters (Fig. 3) which attest to a greater monophyletic group of which these two are a part.

I suspect the possession of wings has been used in this context. It should be noted, however, that wings have evolved independently (convergence) in reptiles, birds, and mammals. In the class Aves, possession of wings (although varied in size and shape) has not been utilized to characterize any particular taxonomic ranking; quite the converse seems to be true. Within the class Mammalia, and further attesting to the convergent nature of volant

adaptations, there is a variety of gliding devices, of similar
structure, which have apparently developed independently in unrelated groups (i.e., gliding marsupials, dermopterans, and
several unrelated species of rodents). Viewed in this light, the
wings of bats may not constitute synapomorphies, but perhaps
represent homoplastic convergences.

In brief, the problem can be simply summarized as follows:
1) if the order Chiroptera is a monophyletic group, the Megachiroptera and Microchiroptera, as sister groups (Fig. 3), should
be found to share a number of synapomorphous characters; 2) if on
the other hand, the order is not monophyletic synapomorphous
characters should be found with another sister group. It is the
search for and critical evaluation of such features which I
suggest is yet to be accomplished.

In closing, I would like to comment on Dr. Pirlot's reference
to neural anatomy. While it is not the main topic of discussion,
he seems to imply a greater evolutionary significance to changes
in neural morphology; particularly, in this case, the relative
size of the Neocortex. It has been shown that the Neocortex is
relatively large in pteropodids, frugivorous phyllostomatids and
sangivorous desmodontines, and the piscivorous Noctilio leporinus.
On the other hand, insectivorous bats tend to have a smaller
expanse of Neocortex, but relatively large hind brains. I suggest
that neural anatomy is functionally plastic and that, at this
time, there is little evidence to justify giving high weight to
such neural features. The enlargement of the Neocortex seems to
be functionally associated with orientation by olfaction and/or
vision and low-speed, highly maneuverable flight, whereas, enlargement of the hind brain seems to involve acoustic orientation
and generally high-speed flight. Mann (1963) and Henson (1970)
have suggested similar interpretations. In conclusion, chiropteran
evolution has progressed as a mosaic of adaptations and no one
particular feature or set of features can be taken out of the
adaptive context and treated as more important than any other.

## REFERENCES

Bachmayer, F., and Wilson, R.W., 1970, Die Fauna der altpliozanen
     Hohlenund Spaltenfullungen bei Kohfidisch, Burgenland
     (Osterreich). Ann. Naturhistor. Mus. Wien., 73:533-587.
Butler, P.M., 1969, Insectivores and bats from the Miocene of
     East Africa. New material, Vol. 1:1-36, Academic Press,
     New York.
Butler, P.M., and Greenwood, M., 1965, Olduvai Gorge 1951-1961,
     Vol. 1:1-15, Cambridge Univ. Press, Cambridge.

Butler, P.M., and Hopwood, A.T., 1957, Insectivora and chiroptera
    from the Miocene rock of Kenya colony. Brit. Mus. (Nat.
    Hist.), Fossil mammals of Africa, 13:1-35.
Dal Piaz, G., 1937, I. Mammiferi dell'Oligocene veneto.
    Archaeopteropus transiens, Mem. Instit. geol. R. Univ.
    Padova, 11:1-8.
Engesser, B., 1972, Die obermiozane Saugetierfauna von Anwil
    (Baselland). pp. 363, Ludin AG, Liestal.
Findley, J.S., Studier, E.H., and Wilson, D.E., 1972, Morphologic
    properties of bat wings. Jour. Mamm., 53:429-444.
Griffin, D.R., 1958, Listening in the dark. pp. xviii+ 413,
    Yale Univ. Press, New Haven.
Hennig, W., 1966, Phylogenetic systematics (Transl. Davis, D.D.
    and Zangerl, R.) pp. vi + 263, Univ. Illinois Press, Urbana.
Henson, O.W., 1970, The central nervous system, in "Biology of
    Bats" (W.A. Wimsatt, ed.) Vol. 2, pp. 57-152, Academic Press,
    New York.
Hill, J.E., 1974, A new family, genus and species of bat (Mammalia:
    Chiroptera) from Thailand. Bull. Brit. Mus. (Nat. Hist.),
    27:303-336.
Lavocat, R., 1961, Le gisement de Vertebres Miocenes de Beni
    Mellal (Maroc). Etude systematique de la Faune des
    Mammiferes et conclusions generales. Notes et Mem. Serv.
    geol. Maroc., 15:121.
Mann, G., 1963, Phylogeny and cortical evolution in chiroptera.
    Evolution, 17:589-591.
Meshinelli, L., 1903, Un nuovo chiroptero fossile (Archaeopteropus
    transiens Mesch.) delle liquiti di Monteviale. Atti. reale
    Instit. veneto Sci. Lett. Arti., 62(2):1329-1344.
Miller, G.S., Jr., 1907, The families and genera of bats. Bull.
    U. S. Nat. Mus., 57: pp. xvii + 282.
Norberg, U.M., 1970, Hovering flight of Plecotus auritus Linnaeus.
    Ark. Zool., 22:483-543.
Novick, A., 1958, Orientation in paleotropical bats. II.
    Megachiroptera. Jour. Exp. Zool., 137:443-462.
Russell, D.E., and Sige, B., 1970, Revision des chiroptereas
    lutetien de Messel (Hesse, Allemagne). Palaeovertebrata,
    Montpellier, 3:83-182.
Simpson, G.G., 1945, The principles of classification and a
    classification of mammals. Bull. Amer. Mus. Nat. Hist.,
    85: pp xvi + 350.
Smith, J.D., 1972, Systematics of the chiropteran family
    Mormoopidae. Univ. Kansas Mus. Nat. Hist., Misc. Publ.,
    56:1-132.
Smith, J.D., 1976, Chiropteran evolution, in "Biology of bats of
    the New World family Phyllostomatidae. Part I" (R.J. Baker,
    J.K. Jones, Jr., and D.C. Carter, eds.). Spec. Publ. Mus.,
    Texas Tech Univ., 10:49-69.

Vaughan, T.A., 1970, Flight patterns and aerodynamics, *in* "Biology of Bats" (W.A. Wimsatt, ed.) Vol. 1, pp. 195-216, Academic Press, New York.

Walker, A., 1969, True affinities of *Propotto leakeyi*. Simpson 1967, Nature, 223:647-648.

Westphal, F., 1959, Neue Wirbeltierreste (Fledermase, Frosche, Reptilien) aus dem obermiozanen Travertin von Bottigen (Schwabische Alb). Neues Jb. Geol. u. Palaont., Abh., Stuttgart, 107:341-366.

Winge, A.H., 1923, Pattedyr-Slaegter. 1. Monotremata, Marsupialia, Insectivora, Chiroptera, Edentata, pp. 360, Copenhagen. (Transl. Deichmann, E., and Allen, G.M., 1941, The inter-relationships of the mammalian genera, Vol. 1. Reitzels ed., pp. 418, Copenhagen).

# ONTOGENY OF AMNIOTE FETAL MEMBRANES

# AND THEIR APPLICATION TO PHYLOGENY

W. Patrick LUCKETT

Dept. of Anatomy, Creighton Univ. School of Medicine

Omaha, Nebraska  68178 U.S.A.

## INTRODUCTION

The extraembryonic or fetal membranes of vertebrates play an important functional role in the nutrition, respiration, excretion, and protection of the embryo and fetus during prenatal life. They are auxiliary structures which develop in continuity with the tissues of the embryo proper, and both embryo and fetal membranes are derived from the same three basic germ layers (ectoderm, mesoderm, and endoderm).  The fetal membranes are transitory structures which persist for only a relatively brief period during the ontogeny of the individual; nevertheless, their functional differentiation is essential for the normal development of the embryo during prenatal life.  The functional life of the fetal membranes is terminated at the time of birth or hatching; they may become partially resorbed into the body of the newborn, or, more commonly, they become disrupted and degenerate.

All vertebrate eggs develop in an aqueous environment, regardless of whether they are laid in water, deposited on land, or retained within the mother's body.  The eggs of all oviparous agnathans, chondrichthyes, osteichthyes, and most amphibians are deposited in water, either before or after fertilization.  A few oviparous amphibians lay their eggs on land, usually in damp places, and the thick jelly envelopes which surround the fertilized eggs apparently provide enough water for the developing embryo, although some water may be absorbed into the jelly envelopes from the surrounding humid environment.

In contrast, the shelled eggs of all oviparous reptiles,

birds, and prototherian mammals are deposited on land, following internal fertilization and a brief but variable period of intrauterine development. A prerequisite for this mode of development was the evolution of an egg which contained an adequate reservoir of nutrients, inorganic ions, and water, so that the newly-hatched young could assume an independent, terrestrial existence, without undergoing an aquatic phase of larval development and metamorphosis. Such a self-contained egg, closed off from the surrounding environment by a rigid shell that functions primarily to retard the loss of water, has been designated as a cleidoic egg (Needham, 1942). Romer (1967) considered the origin of the cleidoic egg in ancestral reptiles to be "the most marvelous 'invention' in vertebrate history."

The cleidoic egg of all extant reptiles, birds, and prototherian mammals are characterized by the development of four fetal membranes: (1) amnion, (2) chorion, (3) allantois, and (4) yolk sac. These same fetal membranes have been retained, although sometimes modified, in viviparous reptiles and mammals. The first three listed fetal membranes are unique, shared, derived characters of reptiles, birds, and mammals, and a super-class Amniota has been erected for these three vertebrate classes, based on their common possession of an amnion (as well as chorion and allantois). In contrast, a yolk sac develops in all vertebrates which possess yolk-rich or megalecithal eggs, including hagfish, elasmobranchs, some teleosts, most apodans, and some urodeles. The developmental pattern and function of the fetal membranes have remained relatively constant within reptiles and their descendants, and it is reasonable to assume that the amniote fetal membranes originated in the last common ancestor of extant reptiles, birds, and mammals. The origin and evolution of the amniote fetal membranes were concomitant with the development of a yolk-laden cleidoic egg in the basal reptilian stock; the oldest known fossil reptile egg dates from the Lower Permian of Texas (Romer, 1957).

The yolk sac is the most primitive of the vertebrate membranes and also the most widespread; it occurs in all amniotes and in those anamniotes which possess a yolk-rich or megalecithal egg. In its simplest form, as in elasmobranchs and teleosts, all three germ layers (ectoderm, mesoderm, and endoderm) spread over the yolk mass and envelop it to form a trilaminar yolk sac (Fig. 1). The yolk sac mesoderm (or its equivalent) is the initial site of hematopoiesis in all vertebrates, and the vascularized yolk sac serves as a fetal nutritive organ in all vertebrates, with megalecithal eggs. In amniotes, the trilaminar yolk sac is only a transitory condition during prenatal life. An exocoelom develops in continuity with the embryonic coelom, and its peripheral expansion separates the trilaminar yolk sac into

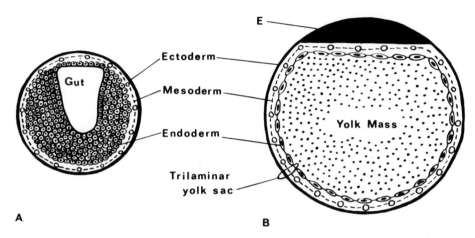

Fig. 1.    Comparison of (A) miolecithal amphibian embryo without
           definitive extraembryonic tissues and (B) megalecithal
           teleost or elasmobranch embryo with extraembryonic tri-
           laminar yolk sac.

an inner splanchnopleuric layer (endoderm plus vascular splanchnic
mesoderm) and an outer somatopleuric layer (ectoderm plus avascular somatic mesoderm).  The yolk sac splanchnopleure becomes
the definitive yolk sac of amniotes, whereas the outer somatopleuric layer becomes part of the chorion (Figs. 2 and 3).

     The amnion provides a fluid-filled environment in which the
embryo can develop without adhesion to the overlying shell membranes and shell, and in which the delicate embryonic tissues are
protected from desiccation and distortion.  The amnion develops
by a fundamentally similar process of upfolding by the extraembryonic somatopleure in reptiles, birds, monotremes, marsupials,
and many groups of eutherian mammals (Figs. 2 and 3).  However,
amniogenesis occurs by a process of cavitation (Fig. 22) within
the preprimitive streak epiblast in some eutherian mammals,
apparently correlated with the mechanism of blastocyst implantation (for further discussion, see Luckett, 1975).

     The somatopleuric chorion develops in continuity with the
amnion as a common chorioamniotic fold when amniogenesis occurs
by folding.  The chorion becomes separated from the amnion secondarily by closure and fusion of the chorioamniotic folds, and by
the expansion of the exocoelom between the two membranes (Figs. 2
and 3).  In this way the entire embryo and the other fetal membranes - amnion, yolk sac, and allantois - come to be completely
surrounded by the chorion.  This developmental relationship establishes a pathway for physiological exchange between the embryo and

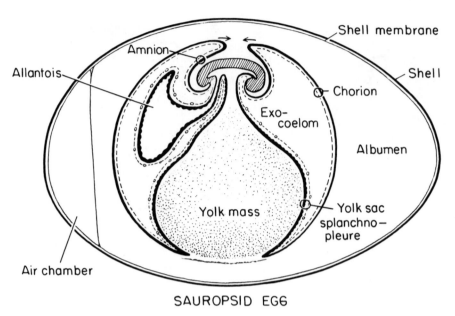

SAUROPSID EGG

Fig. 2.   Generalized scheme of early fetal membrane relation-
          ships in cleidoic egg of reptiles and birds (From
          Luckett, 1975).

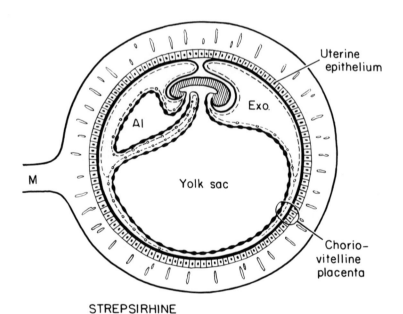

STREPSIRHINE

Fig. 3.   Scheme of early fetal membrane relationships in strep-
          sirhine Primates.   Al=allantois; M-uterine mesometrium
          (From Luckett, 1975).

its surrounding environment. In the cleidoic egg the chorion contacts the albumen, shell membrane, and shell, whereas in viviparous reptiles, metatherians, and eutherians the chorion is directly apposed to the uterine endometrium (compare Figs. 2 and 3).

The splanchnopleuric _allantois_ is essentially an outgrowth of the embryonic hindgut, and subsequently it becomes continuous with the urogenital sinus and urinary bladder. Initially, it may serve as a receptacle for nitrogenous waste products (urea and uric acid) derived from the embryonic mesonephric kidney. During later somite stages, the vascularized allantois expands into the exocoelom (Figs. 2 and 3) and gradually comes to line much of the inner surface of the chorion, fusing with it to form a vascularized chorioallantoic membrane. This membrane plays an important role in the exchange of respiratory gases between the fetus and the external environment, across the shell of the cleidoic egg. Moreover, it is the chorioallantoic membrane that becomes modified to form the fetal component of the definitive or chorioallantoic placenta characteristic of viviparous reptiles and eutherian mammals (Fig. 4).

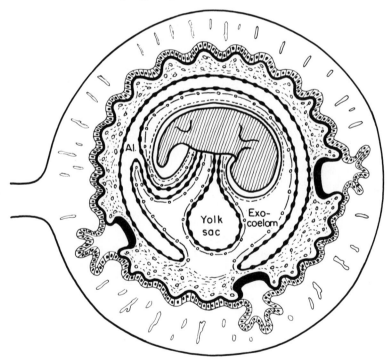

Fig. 4.  Scheme of definitive fetal membrane relationships in strepsirhine Primates.  Al=allantois.  (From Luckett, 1975)

Following a description of the basic patterns of fetal membrane development in oviparous reptiles, birds, and prototherian mammals, I wish to examine the possible effects of the assumption of viviparity on the development and modification of the fetal membranes in reptiles and therian mammals. Such a consideration of the comparative morphogenesis of fetal membranes facilitates a reconstruction of the morphotype of the fetal membranes in the ancestral reptilian stock, utilizing the principles of cladistic analysis elucidated by Hennig (1966) for distinguishing between primitive (ancestral) and derived (specialized) character states. Finally, the fetal membrane evidence will be used to assess the cladistic relationships among the higher taxonomic categories of reptiles and mammals.

The rationale and procedures for utilizing fetal membrane development in studies of mammalian phylogeny have been presented in detail elsewhere (Luckett, 1975, 1976) and will only be summarized here. The fetal membranes of amniotes develop from all three germ layers and comprise an interrelated complex of genetic information; this minimizes the possibility of convergent evolution in their entire ontogenetic pattern. The developmental relationships of the fetal membranes have remained quite constant during the course of amniote evolution (even more so in reptiles and birds than in mammals). Thus, broad comparative studies of the complete ontogeny of a complex organ system such as the fetal membranes, coupled with their conservative nature, serve to offset the absence of these characters in the fossil record and to increase their usefulness in assessing phylogenetic relationships among higher taxonomic categories.

Broad comparative neontological studies can provide evidence of cladistic but not ancestor-descendant relationships. Cladistic relationships are defined in terms of relative recency of common ancestry and are based on the possession of shared derived homologous characters in related "sister groups." Cladistic analysis involves the identification of all alternative states of homologous characters, and the subsequent arrangement of these character states in a sequence or morphocline from most primitive to most derived (Hennig, 1966; Schaeffer et al., 1972). The relative primitiveness or derivedness of character states are determined in part by the distribution of character states in higher categories and by detailed ontogenetic studies when possible. Character states which are widespread in higher taxa (and in more distantly related taxa) are considered to be primitive retentions of the ancestral condition in that category. Conversely, relatively rare and uniquely acquired character states are generally considered to be derived, particularly when their presence in sister groups can be shown to be the result of common ontogenetic pathways. As an example, the development of a cleidoic egg (associated with an amnion, chorion, and allantois) in all oviparous reptiles, birds, and prototherians, coupled with the homologous ontogenetic pathway of the fetal membranes in all amniotes, strongly suggest that the occurrence of this character complex represents the primitive (plesiomorphous) condition in the ancestral reptilian stock. However, within the phylum Chordata the occurrence of a cleidoic egg and its associated fetal membranes in the superclass Amniota is a derived (apomorphous) condition. This emphasizes that character states are relatively primitive or relatively derived depending on the taxonomic categories being compared.

### FETAL MEMBRANES OF REPTILIA

All extant members of the orders Chelonia, Crocodilia, and Rhynchocephalia are oviparous, whereas many families of Squamata contain both oviparous and viviparous species (Bauchot, 1965; Bellairs, 1971). The possible effects of viviparity on fetal membrane morphogenesis in reptiles will be discussed later. The megalecithal eggs of chelonians, crocodilians, and Sphenodon are surrounded by an albumen layer that is secreted by the oviduct, but there is little, if any, albumen about the eggs of Squamata. A thin shell membrane and shell envelop the albumen layer; the shell is calcareous in chelonians and crocodilians, but it is more leathery in oviparous lizards and snakes. A shell has not been detected in viviparous reptiles, although a shell membrane is always present, at least in early stages (Weekes, 1935).

The formation of the three embryonic germ layers during gastrulation in several chelonians and lacertilians has been summarized by Kerr (1919), Pasteels (1957a), and Bellairs (1971) and will not be considered here.

### Order Chelonia

The most complete description of fetal membrane development in chelonians remains that of Mitsukuri (1891), and Fisk and Tribe (1949) have provided additional data on amniogenesis. Newly laid eggs of the snapping turtle Chelydra serpentina (Yntema, 1968) and painted turtle Chrysemys picta (Mahmoud et al., 1973) measure about 32 x 21 mm, and the fertilized egg has already developed to the gastrulation stage with an open blastopore canal in all chelonians examined. The earliest indication of amnion formation is the elevation of a thickened, horseshoe-shaped ridge of extraembryonic ectoderm in front of the anterior margin of the embryonic shield in neural plate-early neural groove stages (about three days incubation in Chelydra and Chrysemys).As emphasized by Fisk and Tribe (1949), this amnionic (or more precisely,

amniochorionic) primordium is a solid ectodermal ridge (ectamnion)
initially, rather than a doubled layered fold, as is frequently
assumed (Fig. 5A). The formation of somatopleuric chorioamniotic
folds is a secondary condition resulting from the expansion of
the exocoelom into the lateral margins of the ectamnion (Fig. 5B).
The ectamniotic ridge continues to grow caudally so that the
entire dorsal surface of the embryo becomes covered by the 7-10
somite stage. Ectamniotic growth continues caudally beyond the
level of the blastopore canal, then narrows considerably to form
an elongated posterior amniotic tube (Fig. 6) which remains open
to the surface of the extraembryonic blastoderm by an oval or
horseshoe-shaped aperature, the amniotic navel. The amniotic
navel gradually becomes closed sometime between the 12-20 somite
stage.

During the initial stages of amniogenesis, the embryonic
head is differentiating as a thickened neural plate and neural
groove, and the anterior end of the neural plate begins to sink
into the underlying yolk (Mitsukuri, 1891). This results in the
formation of a horseshoe-shaped groove between the neural plate
and the thickened ectamniotic primordium, with the anterior wall
of the groove being formed by a mesoderm-free region of extraembryonic ectoderm and endoderm. Further growth and flexure of
the head as the neural folds begin to fuse result in the continued
sinking of the head into the underlying yolk, and extraembryonic
ectoderm and endoderm gradually cover much of the head as a
proamnion (Fig. 5A). As a result, the developing amnion consists
of two distinct but contiguous regions in 2-3 somite embryos of
Clemmys, Chelydra and Chrysemys: (1) a mesoderm-free proamnion
covering the ventrally-flexed head, and (2) a thickened ectamniotic ridge which continues caudally over the dorsal surface of the
embryo to the level of the first somite (Mitsukuri, 1891; Fisk
and Tribe, 1949; personal observations).

At this stage of amniogenesis there are no somatopleuric
amniotic folds; this condition is not attained until there is
expansion of the exocoelom into the anterolateral margins of the
ectamniotic headfold in 4-7 somite embryos (Fig. 5). As emphasized by Mitsukuri (1891), the expanding exocoelom appears to
play an active role in splitting the solid ectamniotic ridge into
somatopleuric amnionic and chorionic layers in the trunk region
of the embryo (Fig. 5B), although a midline remnant of the
original solid ectamnion persists for much or all of gestation as
the chorioamniotic connection. All three regions of the amnion
(ectamnion, proamnion, and somatopleuric amnion) are still evident in 40 somite embryos of the terrapin Malaclemys (personal
observations), and an understanding of the developmental interrelationships between these three phases of amniogenesis is
essential for reconstruction of the possible phyletic origin of
the amnion in the ancestral reptilian stock.

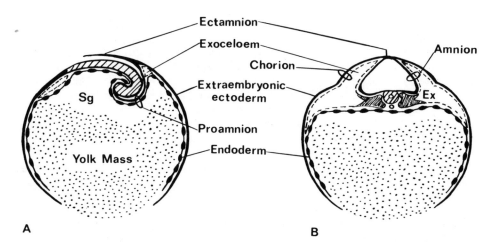

Fig. 5.   Diagram of 7-somite chelonian embryo in sagittal (A)
and transverse (B) sections.  The exocoelom (Ex) expands
dorsomedially into the originally solid ectamnion, resulting in the formation of a somatopleuric amnion (A)
and chorion (C) (Adapted from Mitsukuri, 1891).

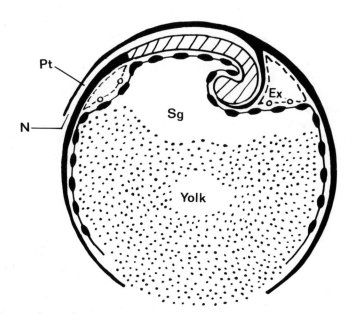

Fig. 6.   Midsagittal diagram of chelonian embryo with 16 somites,
illustrating the extent of the posterior amniotic tube
(Pt) and the patent amniotic navel (N).

Extraembryonic ectoderm and endoderm spread slowly over the surface of the yolk during early somite stages, followed by peripheral spread of extraembryonic medoderm to form the wall of the trilaminar yolk sac.  Peripheral expansion of the exocoelom during early somite stages results in the separation of the tri-laminar yolk sac into an inner splanchnopleuric or definitive yolk sac and an outer somatopleuric chorion (Fig. 5).  The occurrence of blood islands and vessels (Fig. 6) in the proximal wall of the yolk sac mesoderm in 14 somite stages of Chrysemys and Malaclemys suggests the possibility of transport of digested yolk from yolk sac vessels to the developing embryo at this stage. As noted above, dorsal expansion of the exocoelom also separates the ectamnion into inner amnionic and outer chorionic layers. Peripheral spread of the trilaminar yolk sac wall to completely envelop the yolk mass apparently proceeds rather slowly in later stages of development, although this has not been adequately described.  Figures published by Agassiz (1857) suggest that complete envelopment of the yolk mass occurs in late stages of incubation (as it does in birds).

The allantois probably originates from the hindgut in 30-35 somite stages.  A broad flattened allantoic diverticulum extends ventrally from the floor of the hindgut in 39-40 somite embryos (7-8 mm) of Malaclemys, whereas there is no suggestion of either hindgut or allantois in 18-26 somite stages.  The allantois ex-pands into the exocoelom and makes contact with the chorion at sometime after the 40 somite stage in a 6 mm curved embryo.  With continued expansion the allantois comes to line and vascularize much of the inner surface of the chorion to form a fused chorioallantoic membrane.  Mitsukuri (1891) observed that albumen persists until late stages of incubation, although it becomes continually re-duced. The chorioallantoic membrane becomes modified at the abembryonic pole of the egg, adjacent to the persisting albumen. Chorionic epithelial cells in this region are columnar and vacuo-lated and apparently function to absorb albumen; this modification is homologous with the albumen sac of birds.

Order Crocodilia

The eggs of crocodilians, like those of chelonians, are surrounded by an abundant albumen layer and thick, calcareous shell.  The limited available descriptions of the fetal membranes in Crocodylus (Voeltzkow, 1902) and Alligator (Reese, 1908, 1915) are restricted primarily to surface views of the blastoderm and embryo.  In both genera the primordium of the amniotic headfold is first evident in eggs with early neural groove embryos.  In an 8 somite alligator embryo, amnion formation is limited to a pro-amniotic covering of the head (Reese, 1908).  In contrast, the

dorsal surface of the embryo (excluding the proamnion-covered
head) is completely covered by the ectamniotic ridge in 7-8 somite
chelonian embryos; this difference appears to be due to the
absence of an extensive ectamnion in early crocodilian embryos.
In 15-17 somite alligator embryos the chorioamniotic headfold
covers about 2/3 of the surface of the embryo, and the exocoelom
is associated with at least the lateral portion of the headfold.
Much of the trunk region of the amnion is somatopleuric at this
stage and is separated from the chorion by an expanded exocoelom.
An older crocodile embryo (probably 28-30 somites) examined by
Fisk and Tribe (1949) exhibited an extensive exocoelom separating
chorion and amnion over much of the trunk region, whereas a short
ectodermal chorioamniotic connection was limited to the caudal
trunk region in front of the small patent amniotic navel. Although
based on incomplete observations, the available evidence suggests
that amniogenesis in crocodilians is accomplished primarily by
formation of somatopleuric folds associated with exocoelomic expansion, with only limited formation of a solid ectamniotic ridge.

The allantois projects freely into the exocoelom in ± 30
somite alligator embryos (Reese, 1908, 1915), and the amniotic
navel is apparently closed at about this time.  Further expansion
of the allantois and its fusion with the chorion have not been
described, nor is there any mention of the possible occurrence of
an absorptive albumen sac or the stage of complete envelopment of
the yolk mass.

## Order Rhynchocephalia

The order Rhynchocephalia is represented by the single extant
genus Sphenodon which inhabits a few islands off the New Zealand
coast.  Our knowledge of fetal membrane morphogenesis in Sphenodon
is surprisingly extensive, thanks to the studies of Dendy (1899),
Schauinsland (1899, 1903), and Fisk and Tribe (1949).  A horseshoe-
shaped proamniotic headfold is first evident in early neural
plate stages, and by the time neural folds are developed an extensive ectamnion (Fig. 7) extends caudally from the proamnion to
envelop the dorsal surface of the embryo (Fisk and Tribe, 1949).
The ectamnion continues to grow caudally, so that in later neural
fold embryos it extends beyond the blastopore as a narrow elongate
posterior amniotic tube (Fig. 8).  In 12-14 somite stages the
exocoelom has invaded the lateral margins of the solid ectamnion
in the trunk region, forming an extensive region of somatopleuric
amnion and reducing the ectamnion to a thin medial strip of persisting chorioamniotic connection.  As in chelonians, the amnion
consists of three distinct regions at this stage:  proamnion,
ectamnion, and somatopleuric amnion.  In addition, a small somatopleuric tailfold of the amnion develops over the caudal end of the

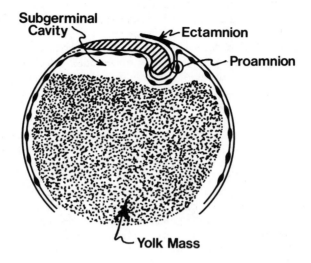

Fig. 7.   Presomite embryo of _Sphenodon_ illustrating proamnion and
          ectamniotic ridge.

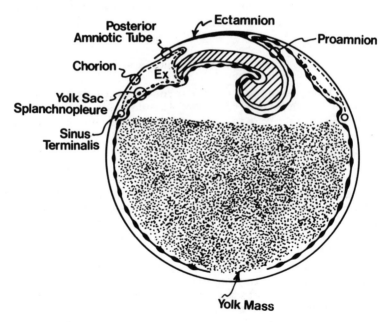

Fig. 8.   _Sphenodon_ embryo with 14 somites, exhibiting posterior
          amniotic tube, proamnion, and ectamnion.

embryo, concomitant with caudal expansion of the exocoelom (Fig. 8).

Between the 14-20 somite stages there is continued growth of the caudal portion of the embryo, and the posterior emniotic tube has disappeared in 20 somite embryos. Further wxpansion of the exocoelom results in the complete obliteration of the ectodermal chorioamniotic connection by the 41 somite stage. The alantois is evident as a short, finger-like ventral diverticulum of the hindgut in an embryo with about 30 somites, and it has assumed the form of a small vesicle projecting into the exocoelom in 41 somite embryos (Fisk and Tribe, 1949). The alantois expands and fuses with the chorion in later embryos to form an extensive chorioallantois as in other reptiles.

Order Squamata

Although squamatans are the most numerous and widespread of extant reptiles, details of their fetal membrane morphogenesis are known for only a few species (see Table I). Incomplete data from other species suggest, however, that a similar developmental pattern occurs in other members of the order. A horseshoe-shaped proamniotic headfold appears initially in neural fold, presomite stages of the lacertids Lacerta agilis and L. vivipara (Peter, 1904; Dufaure and Hubert, 1961) and the iguanids Liolaemus tenius and L. gravenhorti (Lemus and Duvauchelle, 1966; Lemus, 1967). A small somatopleuric amniotic tailfold develops in 8-9 somite Lacerta agilis embryos, and the amniotic navel is closed in 12-17 somite embryos of Lacerta and Liolaemus. A proamniotic headfold develops at the same stage in the gekkonid Platydactylus (Will, 1893), the scincid Mabuyu (Pasteels, 1970), and the colubrids Natrix and Thamnophis (Krull, 1906; Zehr, 1962). There is no report of a posterior amniotic tube in any squamatans.

A different mechanism of amniogenesis has been recorded in the chamaeleonid Chamaeleo chamaeleon by Schauinsland (1903, 1906) and Peter (1934, 1935). The ectamniotic ridge appears precociously in early stages with a flattened bilaminar embryonid disc. In contrast to its initial horseshoe-shaped appearance in other reptiles, the ectamnion develops as a circular ridge which envelops the entire embryonid disc. Centripetal movement of the circular ectamniotic ridge leads to rapid closure of the amniotic naval before the differentiation of the neural groove. According to Peter (1934), the amniotic navel is closed and extraembryonic ectoderm completely envelops the yolk at about the time that the egg is laid. Unfortunately, later developmental stages of the amnion and other fetal membranes are unknown in this species. This precocious differentiation of the amnion has been interpreted as the primitive reptilian condition (Fisk and Tribe, 1949); however, Pasteels (1957b) reported that other chamaeleon species (C. bitaeniatus,

TABLE I

Selected developmental stages of the fetal membranes in reptiles and the chick
(see text for references)

| | Developmental stages at oviposition | Amniotic headfold initiated | Amniotic tailfold initiated | Closure of amniotic navel | Allantoic diverticulum initiated | Fusion of chorio-allantois |
|---|---|---|---|---|---|---|
| CHELONIA Chrysemys, Clemmys, Chelydra, Trionyx, Malaclemys | Neural plate | Presomite | None | 14-19 somites | 30-35 somites | After 40 somites |
| CROCODILIA Alligator Crocodylus | Somite stage | Presomite | None | $\pm$ 30 somites | $\pm$ 30 somites | After 40 somites |
| RHYNCHOCEPHALIA Sphenodon | Pregastrula | Presomite | Small; 14 somites | 14-20 somites | $\pm$ 30 somites | After 41 somites |
| LACERTILIA Lacerta agilis | 50 somites | Presomite | Small; 9 somites | 14-17 somites | 8-14 somites | 38 somites |

LACERTILIA (Con't)

| | | | | | | |
|---|---|---|---|---|---|---|
| L. vivipara | Ovoviviparous | Presomite | Small | 15-18 somites | 15 somites | After 35 somites |
| Liolaemus tenius 58 somites | | Presomite | Small, if present | 12 somites | 9-12 somites | 39 somites |
| L. gravenhorti | Viviparous | Presomite | Small, if present | 16 somites | 8-10 somites | 38 somites |

AVES

| | | | | | | |
|---|---|---|---|---|---|---|
| Gallus gallus | Preprimitive streak | 12 somites | 27 somites | 38 somites | 28 somites | 50 somites |

C. dilepis) differ in their mode of amniogenesis from C. chamaeleon
and instead resemble other lacertilians.

The peripheral spread of extraembryonic ectoderm to completely
envelop the yolk mass is completed at about the 24 somite stage in
Lacerta agilis (Peter, 1934), and the formation of the infolded
yolk navel at the site of final closure was described by Hrabowski
(1926). The relatively small (±3.0mm) egg of the viviparous
scincid Chalcides (=Seps) chalcides is enveloped more precociously;
enclosure is already completed in late presomite-early somite
embryos (personal observations). Viviparity, however, does not
necessarily lead to rapid envelopment of the yolk mass. A small
open yolk navel is still evident abembryonically in a 50 somite
embryo of the viviparous gekkonid Hoplodactylus maculatus (Boyd,
1942), and in comparable stages of the viviparous colubrid snakes
Enhydris dussumieri and Thamnophis sirtalis (Parameswaran, 1963;
Hoffman, 1970).

An endodermal allantoic diverticulum is first evident in
Lacerta agilis embryos with 11-14 somites (Peter, 1904), although
a solid mesodermal thickening associated with the caudal region
of the blastopore has been interpreted as the initial sign of
allantoic differentiation. A small allantoic diverticulum is
also detected at about the same time in Liolaemus (Lemus and
Duvauchelle, 1966; Lemus, 1967). Expansion of the allantoic
vesicle into the exocoelom leads to its initial fusion with the
chorion at about the 38 somite stage in Lacerta agilis (Peter,
1904) and in 38-44 somite embryos of Liolaemus. The further expansion of the allantois within the exocoelom to envelop the
embryo and yolk sac and to form an extensive chorioallantoic
membrane has been described and illustrated in Lacerta agilis,
L. vivipara, and Anguis fragilis by Greil (1914) and Hrabowski
(1926).

Modifications of reptilian fetal membranes associated with
viviparity. The term viviparity has been utilized in different
contexts by various authors to categorize the method of embryonic
development in certain reptiles. This is exemplified by the
varying descriptions of Lacerta vivipara as viviparous or ovoviviparous (Panigel, 1951, 1956), despite the fact that a thinly-
shelled egg is normally laid. Strictly defined, this mode of
development could be considered as oviparous, even though hatching occurs about 24 hours after oviposition, and prenatal development takes place almost completely within the confines of the
mother's reproductive tract. There appears to be little difference in the morphogenesis of the fetal membranes in L. vivipara
and the more clearly oviparous species L. agilis (Peter, 1904;
Hrabowski, 1926). In both species the extensive chorioallantoic
membrane of later stages is thinned and highly vascular and

functions primarily for gas exchange with the surrounding environment, whether this be shell membrane, shell, and external atmosphere, or a thinned shell membrane and highly vascular uterine endometrium.

Some lacertilians and ophidians exhibit a more distinct pattern of viviparity, in that the shell membrane is disrupted during intrauterine life, and a placental relationship is established by the intimate apposition of fetal and maternal tissues. The belief that viviparity has evolved independently within different families of the order Squamata is supported by the fact that every family and most genera which contain viviparous species also contain oviparous forms (Weekes, 1935; Bauchot, 1965). There are two principal regions of potential feto-maternal exchange in viviparous species: (1) the abembryonic pole of the yolk sac, in particular, its avascular bilaminar portion, and (2) the chorioallantoic membrane.

The abembryonic, bilaminar yolk sac exhibits similar specializations for absorption in Lacerta agilis, L. viviparous, and in the more distinctly viviparous species of the families Scincidae, Gekkonidae, Anguidae, Xantusiidae, Colubridae, Elapidae, and Hydrophiidae (see Weekes, 1935; Bauchot, 1965; and Hoffman, 1970 for a summary of the literature). In all species examined, there is a hypertrophy of the closely apposed cells of the uterine epithelium and ectoderm of the bilaminar yolk sac. Peripheral spread of extraembryonic mesoderm proceeds rather slowly, so that the abembryonic pole of the egg remains bilaminar (ectoderm and endoderm) for a considerable portion of gestation. Unfortunately, the terminology used to describe this region of the bilaminar yolk sac placenta has varied considerably among different authors; this is related in part to disagreement concerning the origin and relationships of tissues involved in the formation of a specialized region of the yolk sac - the yolk cleft.

The yolk cleft develops within the yolk mass beyond the margins of the sinus terminalis, during the period of peripheral expansion of the allantoic vesicle and exocoelom. Essentially three different origins of the yolk cleft have been proposed. (1) The cleft is lined by endoderm and may be continuous with the subgerminal cavity of the yolk (Virchow, 1892; Hrabowski, 1926). (2) The cleft develops as a result of splitting of an originally solid strand of "intravitelline mesoderm" which invaginates the yolk from the margins of the sinus terminalis (Weekes, 1927, 1929; Hoffman, 1970); this implies a relationship with the exocoelom. (3) The yolk cleft develops between the solid invasive strand of intravitelline mesoderm and the endoderm associated with the outer wall of the yolk sac (Boyd, 1942); in this case, the cleft would not be related to either the sub-germinal cavity or the exocoelom.

These proposed relationships are incompatible with each other, and it is unlikely that the origin of the yolk cleft varies within the Squamata. Preliminary observations on the develop-mental relationships of the yolk cleft in the viviparous scincid <u>Chalcides chalcides</u> support the hypothesis that the cleft arises by cavitation within a solid strand of yolk sac mesoderm (Luckett, unpublished). However, the yolk cleft never becomes continuous with the exocoelom, although both arise as spaces within the extraembryonic mesoderm (Fig. 9). Although little attention has been given to the possible functional significance of the yolk cleft, it is likely that it plays a role in the absorption of uterine secretions during the prolonged period of intrauterine life in both oviparous and viviparous Squamata.

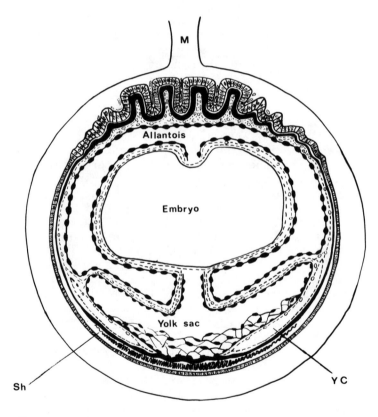

Fig. 9.  Pregnant uterus of the viviparous lizard <u>Chalcides</u>
         <u>chalcides</u>, illustrating the villous epitheliochorial
         placenta at the mesometrial pole.  The expanded allantois
         has fused to the parietal surface of the yolk sac.

Hoffman's (1970) ultrastructural study of the apposed
uterine epithelium and ectoderm of the bilaminar yold sac placenta
in the garter nake, Thamnophis sirtalis, supports the concept
(Bauchot, 1965) that this is a major site of absorptive activity
in most viviparous reptiles, and his autoradiographic evidence
suggests that amino acids or protein may be supplied to the
embryo via this route.

In the majority of viviparous squamatans examined to date,
the chorioallantoic membrane exhibits an attenuated and highly
vascular appearance, similar to that of oviparous reptiles.  The
adjacent uterine epithelium is also thinned and well vascularized,
and this relationship suggests that the chorioallantoic placenta
functions primarily for respiratory gas exchange between mother
and fetus (Weekes, 1935; Bauchot, 1965; Hoffman, 1970).  A number
of authors have reported that the attenuaged chorionic ectoderm
and adjacent uterine epithelium may partially degenerate and
disappear in the region of chorioallantoic placentation.  However,
as emphasized by Hoffman (1970), these observations are questionable and require validation by electron microscopic investigation.
His studies on Thamnophis sirtalis, utilizing superior fixation,
demonstrated the integrity of all layers of the thinned chorioallantoic membrane and uterine epithelium, as well as the persistence of a thin shell membrane between fetal and maternal
tissues throughout pregnancy (similar to the condition in Lacerta
vivipara).

In a few species of scincids (Chalcids chalcides, Lygosoma
entrecasteauxi), the yolk content of the egg is considerably reduced in relation to comparable-sized reptiles, and the chorioallantoic placenta exhibits a specialized elliptical zone in
addition to the more typical and attenuated paraplacental zone
(Giacomini, 1891; Harrison and Weekes, 1925; Cage-Hoedemaker,
1933, Weekes, 1935).  The chorioallantoic placenta of Chalcides
chalcides is the most specialized of any known reptile, and its
histological appearance in late stages has been described by
Cate-Hoedemaker (1933).  The newly ovulated egg of C. chalcides
is one of the smallest (± 3.0 mm) recorded for reptiles, but in
contrast to most viviparous reptiles, the "egg" undergoes considerable increase in diameter during untrauterine gestation,
reaching a maximum size of about 22 mm.  The elliptical portion
(= "placentome") of the chorioallantoic placenta is located
mesometrially in the uterus and consists of interdigitating
villous folds of the chorioallantois and uterine mucosa (Fig. 9).
The intimately apposed cells of the chorionic ectoderm and uterine
epithelium are columnar and bear a well developed brush border on
their apposed apical surfaces.  The chorionic ectodermal cells are
highly vacuolated, and it is likely that uterine secretions are
absorbed by this epithelium.  A shell membrane is absent between
the fetal and maternal tissues of the placentome, and the intimate

apposition of maternal uterine epithelium and fetal chorion is
strikingly similar to the histological relationships of the
epitheliochorial placenta in eutherian mammals such as the pig
or galago.


## FETAL MEMBRANES OF AVES

Following ovulation and fertilization, the chicken zygote
undergoes cleavage and early endoderm (hypoblast) formation during
its passage through the oviduct.  The albumen, shell membranes,
and shell are secreted by the oviduct, and the shelled egg is
laid at about 22-25 hours after ovulation.  Newly laid eggs of
the emu, chicken, turkey, duck, pigeon, and sparrow exhibit comparable stages of preprimitive streak development (Haswell, 1887;
Witschi, 1956; Romanoff, 1960), and it is assumed that this is
true for other birds.  The morphogenetic pattern of fetal membrane
development appears to be virtually identical in all avian orders
which have been studied.  Information is available for some
species of the orders Casuariformes (emu), Procellariformes
(albatross and shearwater), Sphenisciformes (emperor penguin),
Pelecaniformes (frigate bird, booby, tropic bird), Anseriformes
(duck), Galliformes (chicken, turkey), Ralliformes (coot),
Charadriiformes (tern, godwit), Columbiformes (pigeon), and
Passeriformes (sparrow, lapwing) (see Romanoff, 1960 for references).  The following summary is based primarily on development
of the chick (for further details, see Hamilton, 1952; Witschi,
1956; Romanoff, 1960).

The earliest indication of amnion formation is evident in
8-9 somite chick embryos as a horseshoe-shaped ectamniotic thickening near the anterior margin of the mesoderm-free proamnion
during the second day of incubation (about 33 hours).  During the
12-13 somite stage, there is a great expansion of the exocoelom
lateral to the head, concomitant with cephalic flexure and invagination of the head into the underlying yolk.  This results
in the elevation of a proamniotic headfold over the forebrain
region in the chick, lapwing, penguin, and coot.  However, the
major portion of the amniotic headfold is formed by dorsal expansion of the exocoelom; this elevates somatopleuric chorioamniotic folds over the remainder of the head and trunk.  The
junction between future amnionic and chorionic ectoderm in the
folds is marked by the thickened ectamnion, and ectamnion persists
at the site of midline fusion of the lateral wings of the headfold to form a transitory chorioamniotic connection.  This developmental interrelationship between proamnion, ectamnion, and
somatopleuric headfolds differs considerably from the condition in
chelonians and Sphenodon, but it is quite similar to the known
pattern in crocodilians.

A somatopleuric amniotic tailfold is first evident in 27-30
somite chick embryos (50-55 hours incubation).  Chorioamniotic
head- and tailfolds converge and begin to fuse initially at their
lateral extents.  At the 34 somite stage, an oval amniotic navel
is still patent in chick embryos and extends over the level of
the twenty-second to thirty-third somites (Witschi, 1956).  Continued midline convergence results in the complete closure of the
amniotic navel in 38-40 somite embryos of the chick, sparrow,
lapwing, and penguin (Witschi, 1956; Grosser and Tandler, 1909;
Glenister, 1954).

In some aquatic birds, including the tern, godwit, albatross,
shearwater, frigate bird, booby, and tropic bird, a transitory
posterior amniotic tube extends beyond the caudal end of the
embryo proper, similar to the condition in chelonians and Sphenodon
(Schauinsland, 1903, 1906).  This feature is associated with precocious development of the amniotic headfold and with the absence
of a tailfold.  The ectamnion appears to be extensively developed
(as in chelonians and Sphenodon) and plays a major role in the
formation of the posterior amniotic tube.

The allantois appears as a broad ventral diverticulum from
the floor of the hindgut in 28-30 somite chick embryos (52-55
hours incubation), and it expands slowly and protrudes into the
exocoelom during the third day.  Continued expansion of the allantois results in its initial fusion with the chorion in 50-52
somite embryos (96-100 hours incubation) of the chick and sparrow
(Hamilton, 1952; Witschi, 1956).  The expanding allantois continues to spread within the exocoelom and greatly increases the
extent of its fusion with the chorion.  The entire dorsal surface
of the chick embryo is covered by the vesicular allantois by the
end of the sixth day of incubation, and by the end of the ninth
day the inner surface of the chorion is completely lined by the
allantois (Hamilton, 1952).  The distal ends of the fused chorioallantois fold about the residual albumen during the 9th-12th
days of incubation and envelop it to form an extensive albumen
sac.  This specialized region for the absorption of residual albumen occurs in all birds which have been examined and is homologous with the albumen sac of chelonians.

The expansion of the allantois and its subsequent fusion with
the chorion appears to depend on the functional differentiation of
the mesonephric kidney during the fourth day of incubation.  The
caudally growing mesonephric duct unites with the cloaca in 40-41
somite chick embryos at 72-78 hours incubation (Hamilton, 1952),
and both morphological and experimental evidence suggests that
the mesonephros begins to excrete urine at about this time.  If
the mesonephric ducts are obstructed from uniting with the cloaca
during the 29-35 somite stage, embryos examined 1-5 days later

exhibit a marked accumulation of fluid in the proximal remnant of
the mesonephric duct and mesonephros (Boyden, 1924).  Initial
differentiation of the allantois proceeded normally, with formation of a small vesicle which projected freely into the exocoelom.
However, further expansion of the allantois was arrested or
greatly retarded, and the small allantoic vesicle failed to fuse
with the chorion.  Boyden (1924) concluded that normal expansion
of the allantoic vesicle depends on its mechanical distension by
mesonephric excretions.

By the end of the second day of incubation, extraembryonic
ectoderm has spread peripherally to cover about half the surface
of the yolk mass in the chick egg.  In contrast, peripheral
spread of extraembryonic mesoderm lags considerably, due to its
later origin from the primitive streak.  Peripheral expansion of
the exocoelom during early somite stages begins to separate the
mesodermal layer of the trilaminar yolk sac into an outer somatic
layer which becomes part of the chorion, and an inner splanchnic
layer which forms the outer wall of the definitive yolk sac
(Fig. 2).  About three-fourths of the surface of the yolk is
covered by extraembryonic ectoderm at the end of the fifth day,
but further enclosure of the yolk mass proceeds relatively slowly.
As a result, an abembryonic region of yolk remains exposed until
relatively late stages of incubation (16-19 days) as the yolk
navel or yolk sac umbilicus (Fig. 2).  The yolk sac is withdrawn
into the peritoneal cavity shortly before the time of hatching
in the chick and sparrow.  Residual yolk may be an important nutrient source for about one week after hatching in the chick, and
the regressing yolk sac may persist for 2-4 weeks (Romanoff, 1960).
In passeriform birds with small eggs and altricial young, such as
the sparrow, there is less reserve yolk at the time of hatching,
and the young are more dependent on immediate parental feeding
(Witschi, 1956).


                        FETAL MEMBRANES OF MAMMALIA

In contrast to other amniotes, the class Mammalia contains
both oviparous forms with megalecithal eggs and meroblastic
cleavage (Monotremata), and viviparous groups with microlecithal
eggs and holoblastic cleavage (Eutheria).  In some ways the
Metatheria exhibit intermediate conditions between these two extremes and provide valuable evidence for evaluating the morphocline polarity of developmental and fetal membrane characters
within the Mammalia.  The fetal membranes of monotremes and
marsupials exhibit a general pattern of evolutionary conservatism,
whereas those of eutherians present considerable variability in
their morphogenesis.  This appears to be related to the reduction
of intraovular yolk and the acquisition of true viviparity in all
eutherians.

## Prototheria

Unconfirmed reports existed during much of the 19th century concerning the possible oviparous or ovoviviparous manner of reproduction in the echidna (<u>Tachyglossus</u>) and platypus (<u>Ornithorhynchus</u>). It was not until 1884, however, that Caldwell obtained shelled eggs containing embryos of both genera, and he reported his findings to the British Association for the Advancement of Science in a succinct and now famous telegram: "Monotremes oviparous, ovum meroblastic." A historical account of the search for monotreme eggs was subsequently published by Caldwell (1887).

The early stages of fertilization, cleavage, endoderm formation, and primitive streak development in monotremes were described in great detail by J.P. Hill and his colleagues (Wilson and Hill, 1907, 1915; Flynn and Hill, 1939, 1947). In contrast, later stages of intrauterine development are poorly known; the only adequate descriptions available are of 11-18 somite platypus embryos (Hill and Martin, 1894; Wilson and Hill, 1907). The fetal membranes of Monotremata are known solely from the account of their definitive nature in <u>Tachyglossus</u> (Semon, 1894b). It is evident from unpublished notes that the late Professor J.P. Hill had planned to publish the results of his observations on the further embryonic development and fetal membrane morphogenesis in monotremes. Fortunately, Hill's extensive collection of intrauterine and incubated monotreme eggs is now permanently maintained at the Hubrecht Laboratory in the Netherlands, and my observations on the early development of monotreme fetal membranes are based on a study of this material through the courtesy of the Hubrecht Laboratory.

Ovarian eggs of <u>Tachyglossus</u> and <u>Ornithorhynchus</u> at the time of ovulation are megalecithal and have a maximum diameter of about 4.4 mm. This is considerably smaller than in comparable sized reptiles and birds (see Table II), and the ovulated ovum does not contain enough stored nutrients to supply the embryo during the intrauterine and incubation phases of development (Hill, 1910; Flynn and Hill, 1947). Associated with the reduction in stored yolk is the relatively precocious growth and differentiation of the extraembryonic ectoderm. Following meroblastic cleavage, cells at the margins of the blastodisc begin to spread peripherally over the surface of the yolk mass. Peripheral spreading is associated with a thinning of the blastodisc centrally, so that the blastodisc, originally 6-7 cells thick, assumes a unilaminar condition (Fig. 10). Subsequent segregation and internal migration of presumptive endodermal cells leads to the formation of a bilaminar blastoderm. Both layers of the blastoderm continue to spread peripherally and completely enclose the yolk mass in late preprimitive streak stages of the echidna and

TABLE II

Sizes of ovarian and shelled eggs of selected amniotes

| Species | Diameter of ovarian egg | Diameter of shelled egg | Authority |
|---|---|---|---|
| REPTILIA | | | |
| Alligator mississippiensis | 25-30mm | 74 x 43mm | Reese, 1908 |
| Chelydra serpentina | | 32 x 21mm | Yntema, 1968 |
| Sphenodon punctatus, | | 30 x 23mm | Dendy, 1899 |
| Chamaeleo chamaeleon | | 18 x 11mm | Peter, 1934 |
| Lacerta agilis | 7mm | 14 x 9mm | Peter, 1904 |
| Enhydris dussumieri | 20mm | viviparous | Parameswaran, 1963 |
| Hoplodactylus maculatus | 9mm | viviparous | Boyd, 1942 |
| Chalcides chalcides | 3mm | viviparous | Cate Hoedemaker, 1933 |
| AVES | | | |
| Struthio camelus | 80mm | 170 x 135mm | Romanoff and Romanoff, 1949 |
| Gallus gallus | 32mm | 57 x 42mm | Romanoff and Romanoff, 1949 |
| Trochilus colubris | 6mm | 13 x 8mm | Romanoff and Romanoff, 1949 |
| MAMMALIA | | | |
| Tachyglossus aculeatus | 4.3mm | 17 x 15mm | Flynn and Hill, 1939 |
| Ornithorhynchus anatinus | 4.4mm | 17 x 15mm | Flynn and Hill, 1939 |
| Dasyurus quoll | 0.24mm | viviparous | Hill, 1910 |
| Didelphis marsupialis | 0.14mm | viviparous | Hartman, 1916 |

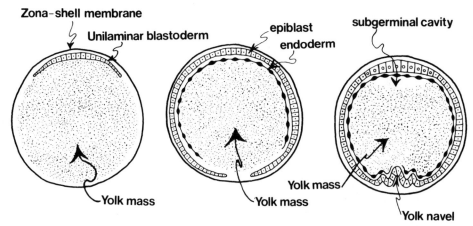

Fig. 10.  Stages in the formation of the bilaminar "blastocyst"
in monotremes, illustrating the peripheral spread of
extraembryonic ectoderm and endoderm to completely
envelop the yolk mass.


platypus (Flynn and Hill, 1947).  The site of final closure of
the yolk is marked temporarily by a thickened, infolded yolk
navel (Fig. 10).

     Following enclosure of the yolk there is rapid distension of
the bilaminar "blastocyst," due to the absorption of uterine
gland secretions.  At the time of enclosure, the bilaminar blastocyst of Tachyglossus measures about 5.4 mm in diameter, whereas
the newly laid egg has a maximum diameter of about 17 mm; similar
intrauterine growth occurs in Ornithorhynchus. Such intrauterine
growth of the egg is unknown in oviparous reptiles or birds.  The
precocious differentiation of extraembryonic ectoderm as a nutritive organ in monotremes foreshadows its increased functional
importance as the trophoblast of therian mammals (Hill, 1910;
Flynn and Hill, 1947).

     Extraembryonic mesoderm spreads peripherally during somite
stages of development, and in 11-18 somite platypus embryos
blood islands and vessels are well developed in the proximal
trilaminar portion of the yolk sac (Hill and Martin, 1894;
Wilson and Hill, 1907).  In the latest stages of intrauterine
development examined (18 somites-platypus, 19 somites-echidna),
a slight crescentic thickening of extraembryonic ectoderm anterior
to the flattened neural plate marks the site of subsequent development of the proamniotic headfold (Hill and Martin, 1894;
personal observations).  The 19 somite stage marks the end of the
intrauterine developmental period in Tachyglossus; such embryos
are found in both intrauterine and recently laid pouch eggs of
the Hill Collection.

In a recently laid pouch egg of echidna with 19 somites the head has begun to invaginate into the underlying yolk sac, and there is a slight indication of a proamniotic headfold with a thickened ectamniotic ridge at its apex (Fig. 11). In 27-30 somite embryos the head and heart region are invaginated into an extensive proamnion, and a somatopleuric tailfold is elevated over the caudal end of the embryo (Fig. 12). An amniotic navel is still open over the trunk region between the level of the 9th-14th somites. The exocoelom extends into the lateral margins of the ectamnion, resulting in the separation of somatopleuric amnion and chorion, but a thickened plate of ectamnion still forms a prominent chorioamniotic connection in the midlongitudinal plane (Fig. 12), comparable to the condition in reptiles. In 35-36 somite platypus embryos the amniotic navel is completely closed, and a prominent chorioamniotic connection persists over much of the middorsal extent of the embryo (Fig. 13). An extensive pro-amnion persists anteriorly, whereas the amnion and chorion are separated completely over the caudal end of the embryo by the ex-panded exocoelom. A portion of the chorioamniotic connection persists throughout the incubation period in both Ornithorhynchus and Tachyglossus (Fig. 14), as reported originally by Semon (1894b).

The bilaminar wall of the yolk sac (bilaminar omphalopleure) is transformed into a trilaminar yolk sac (choriovitelline membrane) by the continued peripheral spread of extraembryonic mesoderm. In an incubated platypus egg with a 35-36 somite embryo, more than half the surface of the yolk sac is vascularized, and the bilaminar omphalopleure is reduced to a small abembryonic region (Fig. 13). The junction between trilaminar and bilaminar omphalopleures is marked by a prominent vascular channel, the sinus terminalis. Cytological differences in the components of the two different regions of the yolk sac reflect differing func-tional specializations. The ectoderm of the bilaminar omphalo-pleure is low columnar and vacuolated, whereas that of the tri-laminar yolk sac is squamous where it overlies the enlarged vitelline simusoids. The relationships of the endodermal lining of the yolk sac exhibit the reverse appearance. The endoderm is squamous in the bilaminar omphalopleure, but columnar and dis-tended with phagocytosed yolk droplets in the trilaminar region. The appearance of the bilaminar omphalopleure is reflective of its function during the intrauterine period for the abosrption of uterine gland secretions.

The functional activity of the choriovitelline membrane is more complex at this stage of development, since it appears to possess both nutritive and respiratory functions. The partially liquefied contents of the yolk sac are phagocytosed and digested by the columnar endodermal cells, and the products of digestion are transferred to the adjacent vitelline vessels for transport

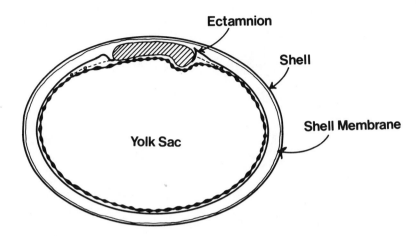

Fig. 11.   Echidna embryo with 19 somites from recently laid
           pouch egg.  Note the early differentiation of pro-
           amniotic and ectamniotic headfolds.

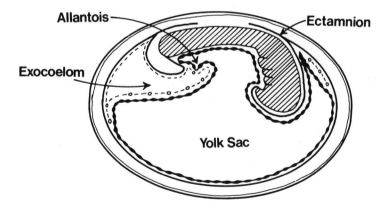

Fig. 12.   Echidna embryo with 27-30 somites from pouch egg.
           Note the extent of development of the amniotic head-
           and tailfolds, and the early differentiation of the
           allantois.

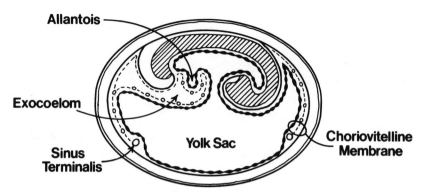

Fig. 13.   Platypus embryo with 35 somites from incubated egg.
           The amniotic navel is closed by an extensive chorio-
           amniotic connection.  A sinus terminalis marks the
           junction between the choriovitelline membrane (=tri-
           laminar omphalopleure) and the persisting bilaminar
           omphalopleure.

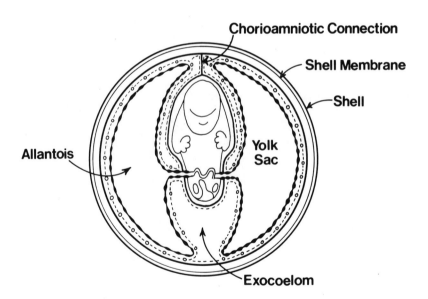

Fig. 14.   Definitive arrangement of the fetal membranes in a
           late incubation stage of the platypus.  Note the ex-
           tensive area of both chorioallantoic and choriovitelline
           membranes, and the persisting chorioamniotic connection.

to the developing embryo.  The intracellular processes which
occur are presumably similar to those described for the chick
yolk sac (Lambson, 1970).  Thinning of the chorionic ectoderm
which overlies the dilated vitelline sinusoids is suggestive of
a role in respiratory gas exchange across the shell membrane and
shell.  The yolk sac provides the only vascularization of the
chorion in 19-40 somite stages, since the allantois does not yet
contact the chorion.

A distinct allantois is first evident in 27-30 somite
echidna embryos as a transversely widened diverticulum projecting
into the exocoelom (Fig. 12), and it is only slightly larger in
a 35-36 somite platypus embryo (Fig. 13).  The latter embryo has
a general length of about 6.5 mm, and the allantois measures
1.4 x 0.8 mm.  In an older 8.5 mm embryo with more than 40 somites,
the allantois has enlarged considerably (9.5 x 8.0 mm) and fused
with the chorion over the right side of the embryo to form an extensive, vascular chorioallantoic membrane.  The definitive nature
of the allantois and yolk sac have been described by Semon (1894b).
Both allantois and yolk sac remain large in late stages of incubation, and each occupies about half the inner surface of the
chorion (Fig. 14).  The persisting chorioamniotic connection
separates yolk sac and allantois in the middorsal plane.  The
yolk sac is completely vascularized in late stages, and both
choriovitelline and chorioallantoic membranes appear to function
in respiratory gas exchange.


## Metatheria

Mature ovarian ova of marsupials are considerably smaller in
diameter (0.15-0.24 mm) than those of monotremes and viviparous
reptiles (Hill, 1910); this reduction in stored yolk necessitates
an extra-ovarian source of nutrition for the developing embryo
and fetus.  This source consists of the uterine endometrium during
the abbreviated period of intrauterine development, and the mammary glands during the "fetal" growth period which occurs in the
pouch.  During its passage through the oviduct the unsegmented
ovum becomes surrounded by an albumen layer and shell membrane
which are homologous with those of monotremes, reptiles, and
birds (Caldwell, 1887; Hill, 1910; Hartman, 1916; Sharman, 1961).

The yolk-poor or microlecithal ovum undergoes holoblastic
cleavage following fertilization, in contrast to the primitive
pattern of meroblastic cleavage in monotremes, reptiles, and
birds.  An event of considerable phylogenetic significance occurs
during the early cleavage stages of the Australian dasyurid,
Dasyurus quoll (= D. viverrinus), and the American didelphid,
Didelphis marsupialis (and presumably in other marsupials).

Cytoplasmic yolk spheres are extruded from the daughter cells
during the first two or three cleavages (Hill, 1910; Hartman,
1916), resulting in a reduction in volume of the daughter cells.
Yolk elimination doubtlessly facilitates the completion of holoblastic cleavage; on the other hand, however, it increases
greatly the dependency of the early marsupial embryo on an extra-
ovarian source of nutrition.

The blastomeres assume a peripheral distribution adjacent to
the zona pellucida by the 16 cell stage in both Dasyurus and
Didelphis, and the eliminated yolk occupies a central position in
the presumptive blastocyst cavity. Further cell divisions lead
to the formation of a unilaminar blastocyst in Dasyurus at the
108-130 cell stage (0.4 mm) (Hill, 1910), whereas blastocyst
formation in Didelphis is completed earlier, in 32-50 cell stages
measuring 0.12-0.15 mm (Hartman, 1916, 1928). Hill (1910) believed that presumptive embryonic and extraembryonic regions are
discernible in Dasyurus at the 16 cell stage; however, Hartman
(1916) could detect no apparent differences in the blastomeres of
Didelphis until the formation of the unilaminar blastocyst. In
the absence of experimental studies which follow the developmental fate of isolated early blastomeres, comparable to those
undertaken by Tarkowski and Wroblewska (1967) on the mouse, Hill's
hypothesis must remain questionable.

Endodermal cells differentiate from the proliferating and
thickened mass of cells at the embryonic pole of the unilaminar
blastocyst, and they spread peripherally to completely line the
yolk sac cavity of the bilaminar blastocyst (Fig. 15). The endodermal lining is completed in Didelphis by the end of the 6th day
of gestation in 0.75-1.0 mm blastocysts (Hartman, 1919; McCrady,
1938), and in a comparable stage in the wallaby, Macropus
rufogriseus (Hill, 1910). At the time of early endodermal differentiation, the unilaminar blastocyst of Dasyurus is considerably larger (± 5.0 mm) than that of other marsupials studied
(Hill, 1910), and the endodermal lining of the yolk sac is not
completed until primitive streak stages (8.5 mm blastocyst).

The outer wall of the marsupial bilaminar blastocyst is
clearly differentiated into two tissue types with different developmental fates: (1) an embryonic pole of columnar epithelial
cells, at this stage comprising the embryonic epiblast, which is
destined to give rise to all the tissues of the embryo proper,
and (2) a more extensive region of extraembryonic ectoderm
(Fig. 15). As emphasized by Hill (1910), the precociously differentiated extraembryonic ectoderm of the marsupial blastocyst
is homologous developmentally with the same layer in prototherians,
reptiles, and birds, and with the trophoblast of eutherian mammals. The demonstrated developmental homology and nutritive

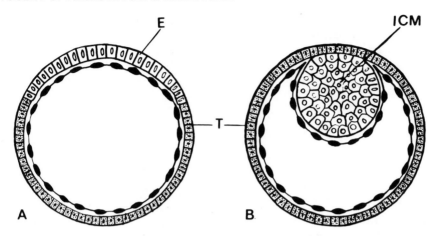

Fig. 15.   Bilaminar blastocysts of (A) metatherian and (B)
           eutherian mammals.  Note the occurrence of the embryonic
           epiblast at the surface of the metatherian blastocyst,
           in comparison to the eutherian blastocyst in which the
           inner cell mass (IC) is overlain initially by a layer
           of polar trophoblast (P).


function of the extraembryonic ectoderm in marsupials and eutherians warrant the use of the term trophoblast for this layer in
both mammalian taxa, as suggested originally by Hill (1910).

     Despite the increased study of marsupial reproductive biology
and embryology during the past 15 years (see Tyndale-Biscoe, 1973,
and Tyndale-Biscoe et al., 1974, for a summary of the literature),
the early stages of fetal membrane development remain inadequately
described for both American and Australian species.  Relatively
complete data on fetal membrane morphogenesis are available only
for Didelphis marsupialis (Selenka, 1887; McCrady, 1938), and
even in this species only a few stages have been illustrated histologically.

     Amniogenesis is initiated relatively late in marsupials (as
it is in monotremes).  Didelphis embryos with 12-16 somites
(McCrady's Stage 25) lack any distinct evidence of amniogenesis,
and the same is true of a 15 somite embryo of the Australian
peramelid, Perameles nasuta (personal observations).  Cephalic
flexure in late 9 day embryos of Didelphis (Stage 26) results in
the formation of a proamniotic headfold (Selenka, 1887; McCrady,
1938) (Fig. 16).  The embryo at this stage measures 6 mm in
length and apparently possesses ± 17-19 somites (counted on Fig.
28 of McCrady, 1938).  A somatopleuric amniotic tailfold becomes
evident in Didelphis during Stage 27 (estimated 20-25 somites),
associated with the caudodorsal expansion of the exocoelom

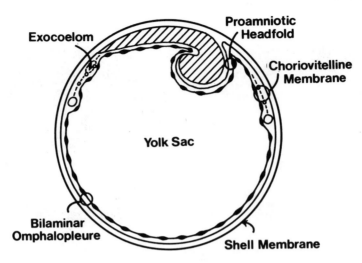

Fig. 16.   Embryo of the opossum Didelphis with 17-19 somites,
           illustrating the onset of amniotic formation.  Note
           the persistence of the shell membrane and the extent
           of the choriovitelline membrane and bilaminar omphalo-
           pleure.

(Selenka, 1887; McCrady, 1938).  The amniotic tailfold has a
similar appearance in a 23 somite (4.7 mm) embryo of Dasyurus
quoll (personal observations).  The development of cervical
flexure of the embryo during this stage in Didelphis results in
the further invagination of the head into the proamnion, and the
proamniotic headfold now covers the entire head and extends to
the level of the first somites (McCrady, 1938).  A comparable
stage in the development of amniotic head- and tailfolds is evident in a 4.8 mm (19 days) embryo of the macropodid Setonix
brachyurus and a 5.4 mm embryo of the phalangeroid Pseudocheirus
peregrinus, described by Sharman (1961).

     During Stage 28 (26-30 somites), the amniotic folds of
Didelphis approach each other, and the exocoelom invades the
caudolateral margins of the proamniotic fold, leading to the
formation of a separate amnion and chorion in the portion anterior
to the open amniotic navel.  Thus, at the time of closure of the
amniotic navel during Stage 29 (late 10th day, ± 30 somites), the
amnion and chorion are completely separated by the exocoelom over
the caudal 2/3 of the embryo, while the proamnion persists over
the cranial 1/3 (Fig. 17).  The chorionic vesicle measures about
16 mm at this stage and is still surrounded by the shell membrane;
the embryo is characterized by club-shaped forelimb buds (McCrady,
1938).  Selenka (1887) and McCrady (1938) reported that the
proamnion of Didelphis persists and increases in extent during
the remaining brief period of intrauterine life (gestation period

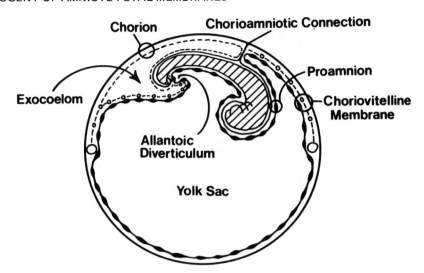

Fig. 17.   30 somite embryo of Didelphis exhibiting early stage of
allantoic differentiation.

= 12.5 days).   In the Australian genera examined by Sharman (1961)
and Renfree (1973), however, the proamnion is only a transitory
structure which is supplanted in late stages by somatopleuric
amnion, associated with the continued expansion of the exocoelom
(Fig. 18).

The yolk sac of all Australian and American marsupials
examined is characterized by the early differentiation and persistence of three distinct regions of differing functional significance.   During late presomite stages, extraembryonic mesoderm
spreads peripherally beyond the margins of the embryo proper.   By
the 12-16 somite stages in Didelphis and Perameles, mesoderm has
covered the proximal third of the yolk sac endoderm.   Development
of the exocoelom leads to the separation of a vascular yolk sac
splanchnopleure in the region of the embryo, whereas a region of
unsplit mesoderm persists peripherally and comprises part of the
wall of the trilaminar omphalopleure or choriovitelline membrane
(Fig. 16).   The sinus terminalis marks the extent of peripheral
spread of mesoderm over the yolk sac, and it is already well developed in 12-somite embryos of Didelphis (McCrady, 1938).   The
abembryonic 2/3 of the yolk sac, beyond the margins of the sinus
terminalis, remains mesoderm-free as the bilaminar omphalopleure.
Whereas an extensive bilaminar omphalopleure is only a transitory
condition during the ontogeny of the yolk sac in monotremes and
reptiles, it persists over the abembryonic half of the yolk sac
(Figs. 16-19) until the end of pregnancy in all marsupials
(Selenka, 1887; Semon, 1894a; Hill, 1900; Pearson, 1949; Sharman,
1961).

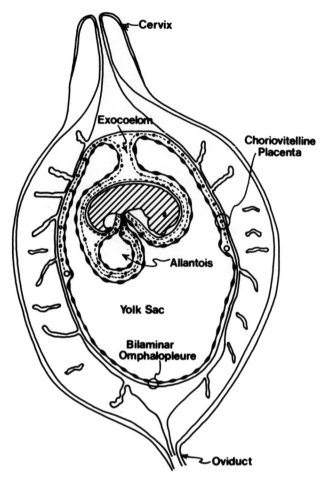

Fig. 18.   Uterus and definitive fetal membranes of the wallaby
           Macropus eugenii. The allantoic vesicle is separated
           from the chorion by folds of the yolk sac (Modified
           from Renfree, 1973).

The histological appearance of the marsupial bilaminar and
trilaminar omphalopleures is similar to that of monotremes during
late intrauterine-early incubation stages.  The trophoblast of
the bilaminar omphalopleure is cuboidal or low columnar, and
ultrastructural features such as the presence of a well developed
microvillous border, associated with pinocytotic vesicles and
larger vacuoles in the apical cytoplasm (Enders and Enders, 1969;
Tyndale-Biscoe, 1973), support the concept that the bilaminar
omphalopleure plays a primary role in fetal nutrition, both before
and after the rupture of the shell membrane.  Renfree's (1973) extensive studies on the composition of uterine and yolk sac fluids

in the tammar wallaby, Macropus eugenii, demonstrated that the
concentration of amino acids in yolk sac fluid is ten times
greater than in maternal serum, and three times greater than in
endometrial exudates.  These data indicate that an active transport mechanism is involved in the uptake of amino acids by the
yolk sac.  Renfree's investigations on the uptake of amino acids,
glucose, and proteins support the concept, originally based solely
on morphological evidence, that the bilaminar omphalopleure is
the major pathway for maternal-embryonic nutrient transfer during
intrauterine life in most marsupials.

A broad, circumferential zone of trilaminar omphalopleure
becomes closely apposed to the endometrium following the rupture
of the shell membrane during late stages of gestation (Fig. 18).
The attenuated layers of the trilaminar omphalopleure and adjacent
endometrium form a choriovitelline placenta between the embryonic
and maternal vascular systems, and this is the principal site of
respiratory gas exchange for the embryo in most marsupial families.
Thus, the morphological differentiation of the marsupial yolk sac
into bilaminar and trilaminar omphalopleures is correlated with
their functional specializations as nutritive and respiratory
organs, respectively.

The allantoic diverticulum is first evident in early limb bud
Didelphis embryos with 26-30 somites (Fig. 17) during the 10th
day of gestation (McCrady, 1938), and in comparable developmental
stages of Dasyurus (personal observations), the phalangerid
Trichosurus vulpecula, and the macropodids Setonix brachyurus,
Macropus rufogriseus, and Potorous tridactylus (Sharman, 1961).
Further expansion of the allantoic vesicle into the exocoelom
during the 11th day in Didelphis is concomitant with the functional differentiation of mesonephric glomeruli and the union of the
mesonephric duct with the cloaca (McCrady, 1938).  The allantois
continues to expand, and by the 12th day it attains a maximum
diameter equal to the length of the embryo.  The allantois functions primarily as a urinary reservoir in Didelphis and most
other marsupials; the concentration of urea in the allantois of
Macropus eugenii is 3-5 times greater than that of the yolk sac
(Renfree, 1973).  In the 12-24 hour period prior to parturition
in Didelphis (12.5 days gestation length), the cloacal membrane
ruptures, releasing urine into the amniotic cavity.  As a result,
there is a partial collapse of the allantois at this time and a
concomitant increase in volume of the amniotic fluid (McCrady,
1938).  The allantois becomes partially enveloped by folds of the
yolk sac in Didelphis and Macropus; this prevents it from contacting and fusing with the overlying chorion (Fig. 18).  Failure of
chorioallantoic fusion suggests that the allantois plays little,
if any, role as a respiratory organ in these genera, and a similar
relationship characterizes all other marsupial families which have

been examined, with the exception of the Phascolarctidae and
Peramelidae, (Semon, 1894a; Hill, 1895; Sharman, 1961; Tyndale-
Biscoe, 1973).

   Implantation and Placentation in Marsupialia. Following the
rupture of the shell membrane during the last third of pregnancy,
the trophoblast enters into an intimate apposition with the
maternal endometrium and establishes a placental relationship in
all marsupials.  Regions of both the bilaminar and trilaminar
omphalopleures become closely apposed to the endometrium to initiate the development of a bilaminar yolk sac placenta and trilaminar
or choriovitelline placenta, respectively.  Hughes (1974) has
emphasized that the rupture of the shell membrane and the subsequent attachment of trophoblast to endometrium do not occur
until late embryonic-early fetal stages of development in marsupials.  However, in discussing the variation in degree of invasive activity of the trophoblast at the time of implantation or
attachment, Hughes concluded that "...marsupial species without
an invasive trophoblast do not exhibit implantation at any stage
of pregnancy."  This statement is misleading in that it implies
that invasive activity is necessary for implantation to occur,
and it is unfortunate that Hughes did not distinguish simply between invasive and non-invasive implantation.  Ultrastructural
analyses of the mechanism of early implantation in a variety of
eutherian mammals (Schlafke and Enders, 1975) have revealed a
sequence of processes involving apposition, adhesion, and subsequent invasion by the trophoblast of some species, and it is
evident that only the first of these processes is involved in the
noninvasive implantation of many marsupials (and eutherians).
Enders and Enders (1969) demonstrated that a well-developed microvillous border occurs on the apposed apical surfaces of the
trophoblast and uterine epithelium in both the bilaminar and trilaminar placentas of the didelphid Philander opossum. The apposed
microvillous borders do not interdigitate, although Tyndale-Biscoe
(1973) has reported such an interdigitation in Macropus eugenii.

   Following Hill's (1900) description of an annular zone of invasive trophoblast in the attached bilaminar omphalopleure of
Dasyurus, limited invasive activity by the trophoblast of either
the bilaminar or trilaminar omphalopleure has been observed in
some species of six families of marsupials:  Dasyuridae, Phascolarctidae, Vombatidae, Phalangeridae, Macropodidae, and Didelphidae (Hughes, 1974; Enders and Enders, 1969).  It is probable that
limited invasive activity has evolved independently in most of
these families; the majority of phalangerids, macropodids, and
didelphids examined to date exhibit no evidence of invasive
trophoblast (Hughes, 1974).

   As noted previously, the moderate-sized allantoic vesicle is prevented from contacting the chorion in most marsupials, as a result, there is no development of a chorioallantoic placenta. However, three different families (Phascolarctidae, Peramelidae, and Dasyuridae) exhibit varying degrees of chorioallantoic fusion (Fig. 19). Semon (1849a) described and illustrated fusion of the allantois and chorion in the koala, Phascolarctos cinereus, and he suggested that its chorioallantois functions principally as a respiratory organ, since it is only loosely attached to the underlying endometrium. The occurrence of a vascular chorioallantois in Phascolarctos was confirmed by Hughes (1974) in a 17 mm full term embryo, and early stages in the growth and fusion of the chorioallantois have been observed in a series of 4-17 mm embryos (Luckett, unpublished). A similar chorioallantoic relationship has been reported (but not described) for the wombat, Vombatus (Pearson, 1949; Hill, 1949).

Hill (1900) reported that the elongated allantoic diverticulum of Dasyurus quoll briefly contacts the chorion; however, the allantoic vasculature was poorly developed, and in later stages the allantois lies free in the exocoelom as a degenerated and avascular structure. In a 3.5 mm limb bud embryo of Dasyurus examined at the Hubrecht Laboratory, the bulbous allantois projects freely into the exocoelom and its distal portion is poorly vascularized. In contrast to other marsupials at comparable developmental stages, the moderately developed mesonephric kidney of this embryo lacks glomeruli, and the same is true for older 4.2-4.7 mm embryos, in which the allantois is even more degenerate and avascular (personal observations). Physical distension of the allantoic vesicle by urine may play an important role in its early differentiation and expansion, and the atypical appearance of the mesonephros may be correlated with abnormal development of the allantois in Dasyurus. It is unlikely that the allantois of Dasyurus plays any role in either embryonic respiration or nutrition, since it is already degenerate and avascular when it temporarily contacts the chorion in a 5.8 mm embryo.

J.P. Hill's (1895, 1897) description of an invasive, discoidal chorioallantoic placenta in the peramelids Perameles obesula (= Isoodon obesulus) and P. nasuta was followed by great controversy as to whether this represented a primitive metatherian (and therian) condition, or whether chorioallantoic placentation has evolved convergently in peramelids and eutherians (see discussion below). Further studies (Flynn, 1923; Hill, 1949) suggested that there is a fusion of chorioallantoic trophoblast and uterine epithelium to form a "conjoint syncytium." This conjoint layer subsequently exhibits some degenerative changes and becomes thinned in later stages, so that maternal and fetal capillaries become more intimately apposed. Amoroso (1952) has interpreted this as an endothelio-endothelial relationship. However, recent

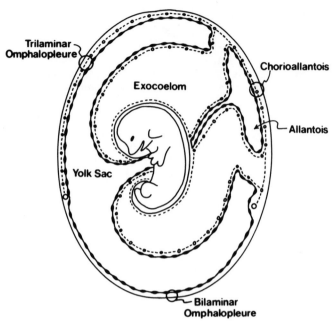

Fig. 19.   Definitive nature of the fetal membranes in the bandi-
           coot <u>Perameles</u>, illustrating the chorioallantoic pla-
           centa and the persisting choriovitelline placenta.
           (Modified from Hill, 1895)

ultrastructural studies of placentation in peramelids reveal that
a tissue layer (maternal epithelial symplasma?) persists between
fetal and maternal capillaries in late stages (Padykula and
Taylor, 1974, in press).  Padykula and Taylor believe that the
trophoblast disappears between fetal and maternal vessels in near
term placentas, a condition unknown in the chorioallantoic placenta
of eutherian mammals.  The possible relationship of trophoblast
loss to the evolution of prolonged intrauterine development in
eutherians will be discussed below.

        The viewpoint that chorioallantoic placentation in peramelids
is developmentally and histologically similar to the condition in
certain eutherians (Flynn, 1922, 1923) is not supported by the
present study.  The lack of formation of vascular "mesodermal
villi" and the apparent degeneration of trophoblast are striking
features of peramelid placentation which differ from the relationships in eutherian chorioallantoic placentation.  Furthermore, the chorioallantoic placenta does not temporarily or functionally <u>replace</u> the choriovitelline placenta in peramelids (as
it does in eutherians).  As noted by Hill (1949), embryonic nutritional and respiratory requirements continue to be met in
great measure by the bilaminar and trilaminar omphalopleures of

peramelids after the establishment of the chorioallantoic placenta, and the latter serves to supplement, rather than replace, the functional activities of the yolk sac placentas.


## Eutheria

All eutherian mammals are characterized by the absence of a shell membrane and by the development of a definitive chorio-allantoic placenta. Furthermore, the intrauterine gestation period is prolonged, in comparison with marsupials, so that the neonate exhibits an advanced state of development. Concomitant with the absence of a shell membrane, the eutherian blastocyst attaches to the endometrium at an earlier developmental stage (presomite embryo) than it does in marsupials. Differences between eutherian and marsupial embryos become evident during early cleavage stages. A solid cluster of blastomeres, the morula, characterizes the 8-30 cell stages of eutherians, whereas the blastomeres are distributed peripherally to form a unilaminar vesicle in comparable stages of marsupials. Fluid-filled inter-cellular spaces develop within the morula, and the enlargement and subsequent coalescence of these spaces lead to the formation of a hollow, spherical blastocyst.

The early eutherian blastocyst differs from that of marsupials in that two distinct cell types are evident at the time of its initial differentiation: (1) a peripheral layer of epithelial cells, the trophoblast, forms the complete outer wall of the blastocyst, and (2) a small cluster of cells, the inner cell mass, is attached to the inner surface of one pole of the blastocyst (Fig. 15). The inner cell mass or embryoblast will give rise to all the tissues of the embryo proper, and it is homologous with the embryonic pole of the unilaminar marsupial blastocyst. In contrast to the marsupial blastocyst, the presumptive embryonic mass of the eutherian blastocyst is completely segregated from the intrauterine environment by a covering layer of polar tropho-blast (Fig. 15). The trophoblast is the first tissue to dif-ferentiate in the eutherian blastocyst, and this reflects its essential role as a selectively-permeable layer for physiological exchange between the embryo and mother.

Following the differentiation of endoderm from the embryonic mass and its peripheral spread to form a bilaminar blastocyst, there is considerable variation in the further development of the fetal membranes and placenta among the various orders and sub-orders of eutherians. Variation occurs most frequently in the (1) fate of the polar trophoblast, (2) pole of the trophoblast involved in the initial attachment to the endometrium, (3) orien-tation of the embryonic mass at the time of implantation, (4) re-gion of the endometrium on which the blastocyst initially attaches,

(5) depth of blastocyst implantation, (6) mechanism of amnion formation, (7) fate of the bilaminar omphalopleure portion of the yolk sac, (8) participation of the yolk sac splanchnopleure in the formation of a temporary choriovitelline placenta, (9) nature of the allantoic vesicle and/or body stalk, and (10) location, shape, and finer morphology of the chorioallantoic placenta.

A detailed survey of the morphogenesis of the fetal membranes and placenta in the different orders of eutherian mammals is beyond the scope of the present study. Such information is available for all orders and suborders, most superfamilies, and many families, and much of this data has been summarized previously (Mossman, 1937; Amoroso, 1952; Starck, 1959). Mossman's (1937) comprehensive survey provided a basis for evaluating the relative evolutionary conservatism of each fetal membrane character, as judged by its developmental constancy within higher taxonomic categories (orders, suborders, and superfamilies). The most conservative characters are: (1) orientation of the embryonic mass to the endometrium at the time of implantation, (2) the nature of the yolk sac and its vascular splanchnopleure, (3) the nature of the allantoic vesicle, and (4) the detailed structure of the definitive chorioallantoic placenta. Mossman (1937, 1953) suggested that the relatively conservative nature of the mammalian fetal membranes (in comparison with cranioskeletal characters) may be the result of their relative isolation from the selective effects of the external environment. There is no evidence that external selective forces, such as diet, habitat preference, or climate, have had any apparent selective effect on the evolutionary development of the eutherian fetal membranes. Nevertheless, as suggested elsewhere (Luckett, 1974, 1975), it seems likely that there have been intense centripetal selective forces acting upon certain fetal membrane characters, resulting in the maintenance of a relatively constant developmental pattern. Selection is considered to be centripetal when there is a tendency to retain a relatively static optimal condition of a character and to eliminate variant forms, and such conservative characters are doubtlessly of great adaptive importance to the organism, perhaps even essential to its survival (Farris, 1966).

In lieu of presenting a detailed survey of the varying patterns of fetal membrane morphogenesis in eutherians, a reconstruction of the morphotype of the fetal membrane pattern in the ancestral eutherian stock(s) is attempted, utilizing the principles of cladistic analysis summarized above. This reconstruction encompassed an evalution of fetal membrane development in all the orders and suborders of eutherians, as well as a consideration of the more uniform patterns found in marsupials, monotremes, and reptiles. The conservative nature of the fetal membranes permits generalizations about their relationships within families and

superfamilies to be based on developmental data from relatively
few species, since significant differences are rarely found within
families (Mossman, 1937). A reconstruction of the morphotype of
primate fetal membranes has been presented elsewhere (Luckett,
1974, 1975), and the primitive and derived character states of
rodent fetal membranes have been discussed by Mossman (1937) and
Luckett (1971). The data utilized in the morphotype reconstruction can be summarized within four general categories: (1) implantation and amniogenesis, (2) yolk sac and choriovitelline
placentation, (3) allantoic vesicle, and (4) chorioallantoic
placenta.

   *Implantation and amniogenesis.* Many of the subsequent developmental differences in eutherian placantation are related to
the mechanism of implantation. Variations may occur in the pole
of the blastocyst involved in the initial attachment, the degree
of invasive activity of the attached trophoblast, and the region
of the uterine endometrium to which the blastocyst attaches. The
orientation of the embryonic disc to the endometrium at the time
of implantation is the most conservative feature of eutherian
placentation. The pattern is constant within most orders (Fig.
20), regardless of intraordinal differences in the degree of invasive activity (Fig. 21). The significance of this constant
pattern of orientation is unclear, although Mossman (1971) has
emphasized that the orientation of the embryonic disc is correlated with the subsequent location and further differentiation
of the yolk sac, amnion, and allantois, and with the site of
development of the definitive chorioallantoic placenta. Although
an antimesometrial orientation of the embryonic disc is the most
common pattern in eutherians (Carnivora, Artiodactyla, Perissodactyla, Pholidota, Edentata, Dermoptera, most Insectivora), it
is unclear whether this should be considered the primitive
eutherian condition. A notable exception to the constancy of
embryonic disc orientation is found within the Insectivora, where
differences in implantation (and other fetal membrane characters)
are common between superfamilies. The evidence of the fetal
membranes strongly supports the concept that the order Insectivora
is a paraphyletic or polyphyletic assemblage based primarily on
the common retention of many primitive eutherian features.

   Based upon ontogenetic and comparative data, as well as a
consideration of the further developmental relationships of the
fetal membranes, the primitive eutherian pattern of blastocyst
implantation is considered to involve superficial, noninvasive,
and central or somewhat eccentric attachment of a relatively expanded bilaminar blastocyst by a broad circumferential or equatorial zone of trophoblast (Fig. 21). This pattern is found in
Artiodactyla, Perissodactyla, Pholidota, strepsirhine Primates,
and presumably Cetacea, and in all these taxa it is associated

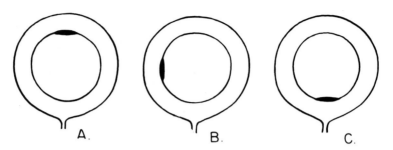

Fig. 20.   Patterns of orientation of embryonic disc to endometrium
           during implantation in eutherians.  (A) Antimesometrial
           orientation, characteristic of most orders, including
           Carnivora, Dermoptera, and most Insectivora; (B) Ortho-
           mesometrial orientation, found in Primates and most
           Chiroptera; (C) Mesometrial orientation, characteristic
           of Lagomorpha and Rodentia. (From Luckett, 1975)

with the subsequent development of an epitheliochorial placenta.
The simplest modification of this pattern is seen in carnivorans,
soricids, and tupaiids, in which the equatorial (paraembryonic)
trophoblast exhibits limited invasive activity.  The most derived
pattern of implantation is found in hystricognath rodents and
hominoid primates.  Implantation occurs precociously at the unilaminar blastocyst stage, and the blastocyst rapidly invades the
endometrium and becomes completely encapsulated within it, a
condition termed interstitial implantation (Fig. 21).  Intermediate conditions between superficial and interstitial implantation are found in geomyoid and muroid rodents (Mossman, 1937).

     Concomitant with the mechanism of implantation is the fate
of the polar trophoblast of the blastocyst, and this in turn influences the process of amnion formation.  Whenever blastocyst
implantation is effected solely by the paraembryonic trophoblast,
the polar trophoblast which overlies the embryonic mass is invariably disrupted and lost during the expansion of the bilaminar
blastocyst (Fig. 21).  As a result, the embryonic mass is secondarily exposed at the surface of the blastocyst, resulting in a
relationship that is identical to the early bilaminar blastocyst
of marsupials.  Amniogenesis subsequently occurs by somatopleuric
folding in all eutherians in which the polar trophoblast is lost
(Figs. 3, 22).  Amniotic folds in eutherians are invariably
associated with the expansion of the exocoelom during early somite
stages; as a result, there is little evidence of an ectamnion.
Amniogenesis occurs by folding in all eutherians with epitheliochorial placentation, and in most taxa with endotheliochorial
placentation (Table III).  This primitive feature is shared with
all marsupials and monotremes (as well as reptiles and birds).

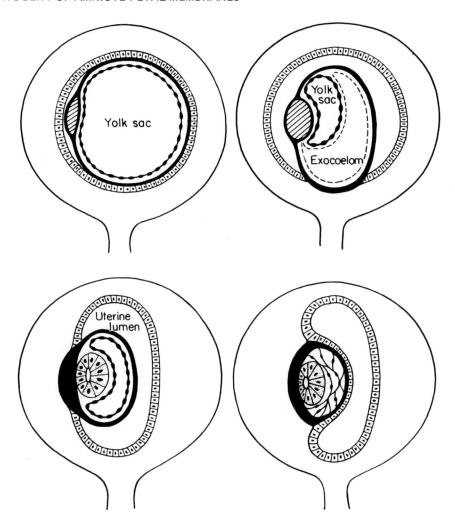

Fig. 21. Differing patterns of blastocyst implantation within the
order Primates. (A) Noninvasive, central implantation
of strepsirhines. (B) Invasive, eccentric implantation
in Tarsius. (C) Invasive, eccentric implantation of
ceboids and cercopithecids. (D) Interstitial implantation of hominoids. (From Luckett, 1975).

Amniogenesis occurs by a process of cavitation within the
embryonic mass in all eutherians in which the polar trophoblast
persists, usually as a result of implantation occurring at the
embryonic pole of the blastocyst (Fig. 21). Macroscelididae,
tenrecid and erinaceid Insectivora, Dermoptera, Hyracoidea, some
Chiroptera, anthropoid Primates, and hystricognathous Rodentia

are characterized by this derived condition.  An intermediate
type of amniogenesis occurs in several taxa, including Artiodactyla, Tupaiidae, and Tarsiidae.  In these, a primordial amniotic cavity develops by cavitation, but the roof of the
primordial cavity ruptures at about the time of implantation.  As
a result, the embryonic mass is secondarily exposed at the surface
of the blastocyst, and definitive amniogenesis occurs subsequently
by folding.  The key factor in this process of amniogenesis
appears to be the involvement of the paraembryonic trophoblast in
the implantation phase, and the failure of the polar trophoblast
to fuse with the overlying endometrium.  As a rule, definitive
amniogenesis occurs by folding in all eutherians in which the
polar trophoblast is lost, regardless of whether a primordial
amniotic cavity develops.  Conversely, amniogenesis occurs by
cavitation when the polar trophoblast persists.

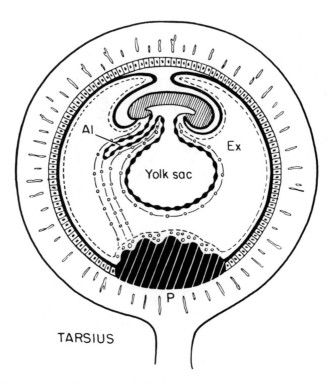

Fig. 22.  Fetal membranes and embryo of the haplorhine primate
          Tarsius. Amniogenesis occurs by folding, and the re-
          duced allantoic diverticulum (Al) projects into a
          prominent mesodermal body stalk.  P=hemochorial pla-
          cental disc; Ex=exocoelom. (From Luckett, 1975)

Yolk sac and choriovitelline placentation. In all eutherians, extraembryonic mesoderm spreads peripherally to convert at least the embryonic third of the bilaminar omphalopleure into a trilaminar yolk sac (Mossman, 1937). Further differentiation of the yolk sac may follow one of three different pathways. The most common condition is the completion of the trilaminar yolk sac, and the subsequent expansion of the exocoelom to produce a completely vascularized splanchnopleuric yolk sac which projects freely into the exocoelom (Figs. 4, 22). This pattern is basically similar to the condition in monotremes, reptiles, and birds, and it occurs in all taxa which develop an epitheliochorial or endotheliochorial placenta, with the exception of the Talpidae and Soricidae (Table III and Fig. 23).

In the insectivoran families Talpidae, Soricidae, and Erinaceidae (collectively, the suborder Erinaceota, sensu Van Valen, 1967) and in sciurid and aplodontid rodents, extraembryonic mesoderm never completely envelops the yolk sac endoderm, and the abembryonic portion of the bilaminar omphalopleure persists throughout gestation, similar to the condition in marsupials (Table III and Figs. 23, 24). This persisting bilaminar omphalopleure probably retains a functional role throughout gestation. In addition to its role in the absorption of uterine secretions, there is experimental evidence that the bilaminar omphalopleure functions in the uptake of immunoglobulins in sciurids (Wild, 1971). A region of bilaminar omphalopleure also persists in the chiropteran families Vespertilionidae and Phyllostomatidae (Mossman, 1937).

The most specialized condition of the yolk sac is found in geomyoid, muroid, and hystricognath rodents, lagomorphs, solenodontid insectivorans, and dasypodid edentatans. In these taxa, a bilaminar omphalopleure is temporarily retained in early stages, then both the trophoblastic and endodermal layers become disrupted and disappear. As a result, the endodermal epithelium of the persisting vascular splanchnopleure of the yolk sac becomes closely apposed to the endometrium and comprises a completely inverted yolk sac placenta (Mossman, 1937). Both experimental and ultrastructural evidence indicate that the endodermal cells of the inverted yolk sac placenta in the guinea pig and rabbit are involved in the uptake of immunoglobulins and other proteins (Brambell, 1970; King and Enders, 1970). Mossman (1937) has presented an excellent discussion of the morphocline polarity leading to the derived condition of a completely inverted yolk sac, based on comparative studies of the yolk sac in rodents.

The intimate apposition of the vascularized trilaminar omphalopleure of the yolk sac to the endometrium to form a temporary choriovitelline placenta occurs in some families of

TABLE III

A comparison of selected primitive and derived character states of conservative fetal membrane features associated with different placental types.

| | Primitive | Intermediate | Derived |
|---|---|---|---|
| 1. Chorioallantoic placenta | | | |
|    Epitheliochorial | x | | |
|    Endotheliochorial | | x | x |
|    Hemochorial | | | x |
| 2. Bilaminar omphalopleure | | | |
|    Temporary | x | | |
|    Permanent | | x | |
|    Lost or inverted | | | x |
| 3. Temporary choriovitelline placenta | | | |
|    Present | x | | |
|    Absent | | | x |
| 4. Amniogenesis | | | |
|    Folding | x | | |
|    Cavitation | | | x |
| 5. Allantoic vesicle | | | |
|    Large | x | | |
|    Moderate-sized | | x | |
|    Reduced or vestigial | | | x |

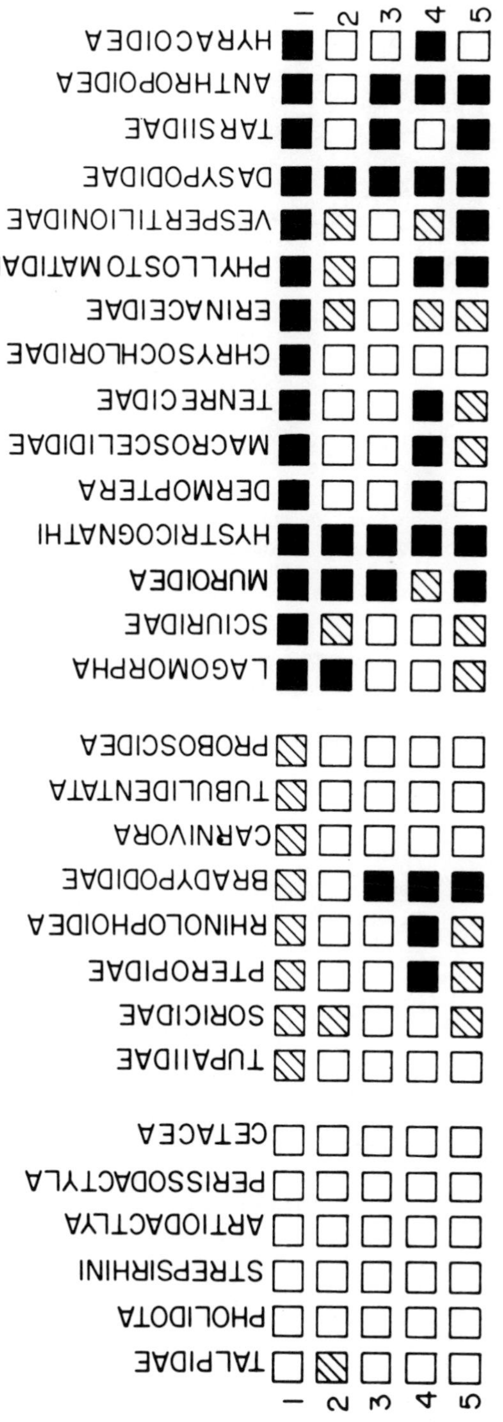

Fig. 23. Comparison of primitive and derived states of selected fetal membrane features associated with epitheliochorial, endotheliochorial, and hemochorial placentae in eutherians. 1 = placental type; 2 = fate of bilaminar omphalopleure; 3 = choriovitelline placenta; 4 = amniogenesis; 5 = allantoic vesicle. The primitive and derived character states of each of these features are listed in Table III.

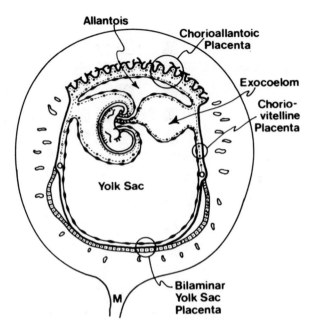

Fig. 24.   Diagram of fetal membrane relationships during mid-
           pregnancy in the European mole, Talpa europaea. In
           later stages, the allantois expands and completely
           disrupts the choriovitelline placenta, whereas the
           abembryonic region of bilaminar yolk sac persists.
           Compare with the definitive relationships in marsupials
           (Figs. 19, 20).

every order of eutherian mammals, with the probably exception of
the Edentata (excluding Pholidota) (Table III and Figs. 3, 23, 24).
In contrast to the condition in marsupials, in which the choriovitelline placenta maintains a permanent relationship during
intrauterine life, the choriovitelline placenta of eutherians is
always a transitory organ.  It is replaced both temporarily and
functionally by the chorioallantoic placenta, usually concomitant
with the expansion of the exocoelom (Figs. 3 and 4).  The absence
of a transitory choriovitelline placenta in edentatans, haplorhine
primates (Fig. 22), and "higher" rodents is correlated with the
precocious differentiation of the exocoelom and body stalk mesoderm, and with the precocious establishment of the chorioallantoic
placenta (Luckett, 1974).  There is no direct correlation between
the fate of the bilaminar omphalopleure and the occurrence of a
choriovitelline placenta (Fig. 23), and these should be considered
as two distinct types of yolk sac placentation.

    Allantoic vesicle. The development of a large allantoic
vesicle (Figs. 3, 4) characterizes all eutherians which possess an
epitheliochorial placenta (Table III and Fig. 23); this primitive

condition is shared with all monotremes, reptiles, and birds. At the other extreme, the allantoic diverticulum is greatly reduced or vestigial in many orders or families which possess a hemochorial placenta. In haplorhine primates (Fig. 22), rodents, and bradypodids, the vestigial nature of the endodermal allantois is correlated with the precocious differentiation of body stalk (=allantoic) mesoderm (Hill, 1932; Luckett, 1974). This relationship emphasizes the important concept that it is the vascularized allantoic mesoderm which is the essential component of the allantois for the functional development of the chorioallantoic placenta. In those taxa with a moderate-sized allantoic vesicle (Lagomorpha, Sciuridae, Macroscelididae, Erinaceidae, and Pteropidae), the expansion of the allantoic vesicle is necessary for the apposition and fusion of the vascular allantoic mesoderm with the chorion to initiate chorioallantoic placentation. This developmental relationship characterizes all eutherians with a large or moderate-sized allantoic vesicle, as well as all monotremes, reptiles, and birds.

Chorioallantoic placenta. Eutherian chorioallantoic placentas are categorized according to their gross shape, location, degree of invasive activity, and finer morphology (Mossman, 1937). The finer morphology is classified according to the number and character of the layers which separate maternal and fetal bloodstreams in the chorioallantoic placenta (Gross, 1909). In general, the fetal component (vascularized chorioallantois) remains intact, whereas differences are due to the reduction in number of the maternal layers (Fig. 25). In the simplest relationship, the epitheliochorial placenta, there is no loss of maternal tissue, and the uterine epithelium is closely apposed to the chorionic epithelium. A loss of maternal epithelium and the underlying connective tissue results in the intimate apposition of maternal capillary endothelium (with its basement membrane) and the chorion to form an endotheliochorial placenta. A hemochorial placenta is established when maternal epithelium, connective tissue, and capillary endothelium are lost from the placental barrier, and maternal blood bathes the chorionic trophoblast. In most, if not all, taxa which develop an endotheliochorial or hemochorial relationship, the chorionic trophoblast becomes differentiated into an outer syncytial layer (syncytiotrophoblast) and an inner, relatively undifferentiated layer, the cytotrophoblast.

The presumed primitive and derived conditions of chorioallantoic placentation have generated more controversy than any other aspect of mammalian placentation. The diffuse epitheliochorial placenta, characteristic of Artiodactyla, Perissodactyla, Pholidota, strepsirhine Primates, and talpid Insectivora, is the simplest condition developmentally and has been considered as the primitive eutherian condition (Turner, 1877; Grosser, 1909; Hill,

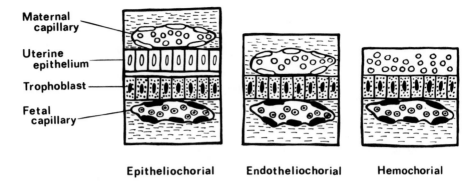

Fig. 25.   Chorioallantoic placental types in eutherian mammals.

1932).   On the other hand, Hubrecht (1908) and Wislocki (1929)
believed that an invasive hemochorial placenta represents the
primitive condition, because this relationship occurs in the most
"archaic" or "primitive" mammals: insectivores, bats, rodents,
hyracoids, edentates, and Tarsius.   They further emphasized that
it would be improbable for "highly specialized" mammals such as
ungulates and strepsirhines to have retained a primitive type of
placentation; instead they suggested that diffuse placentation
must be a secondary simplification from a more primitive invasive
type.   This hypothesis ignores the concept of mosaic evolution,
the occurrence of both primitive and derived characters within
every taxon, and it also fails to consider the nature of the
other fetal membrane characters associated with these two placental types.

Mossman (1937) suggested that a zonary endotheliochorial
placenta, similar to that which occurs in most carnivorans, arose
early during the evolution of viviparous (=therian) mammals and
that this may have been the ancestral condition for both the
diffuse epitheliochorial placenta on the one hand, and the discoidal hemochorial placenta on the other.   Thus, an unusual situation exists in that each of the three principal character states
of a morphocline has been considered as the primitive condition.
However, cladistic analysis of all the mammalian fetal membranes
corroborates the hypothesis that diffuse epitheliochorial placentation represents the primitive eutherian state (Luckett, 1975).

All unrelated eutherian taxa with a diffuse or cotyledonary
epitheliochorial placenta (Artiodactyla, Perissodactyla, Pholidota,
Strepsirhini) exhibit a detailed homology in the morphogenesis of
all their fetal membrane characters (Fig. 23 and Table III).   Such
detailed homologies would be expected in cases of primitive retention of the ancestral condition in distantly related descendant
taxa, whereas they would be highly unlikely if similarities due to

convergence had occurred.  On the other hand, differences in the
character states of at least some of the fetal membranes are
evident in all higher taxa which develop either an endotheliochorial or hemochorial placenta (Fig. 23).  In addition, the
endotheliochorial placentas which develop in Carnivora, Probos-
cidea, Tupaiidae, Soricidae, Bradypodidae, and several families
of Chiroptera are the result of differing patterns of morphogenesis, and the same is true for the development of a hemochorial placenta in unrelated mammalian taxa.  Thus, ontogenetic
evidence also supports the concept of convergent evolution of
hemochorial and endotheliochorial placentation in eutherians.

It has been repeatedly emphasized (Mossman, 1937, 1967;
Luckett, 1974, 1975) that the finer morphology of the eutherian
placental "barrier" is but a single character for phylogenetic
studies, and that it is the entire ontogenetic pattern of all the
fetal membranes and placenta which should be utilized in phylogenetic reconstruction.  Nevertheless, the controversy over the
morphocline polarity of this single feature (the structure of the
placental barrier) has led some paleontologists (Gingerich, 1974;
Simons, in press) to dismiss the use of placentation in assessing
phylogenetic relationships, without evaluating the evidence which
surrounds the controversy.  It should be obvious that pure
phenetic descriptions of characters (whether placental or dental),
in the absence of any consideration of the theoretical basis for
similarities and differences among taxa, are of minimal value in
assessing phylogeny.  Conversely, character analyses of organ
systems which are the product of great genetic complexity, such
as the fetal membranes, provide strong evidence of phylogenetic
relationships, particularly when considered in conjunction with
other data from both "hard" and "soft" anatomy (see Luckett and
Szalay, in press, for a phylogeny of the Primates based on such
a multidisciplinary character analysis).

Morphotype reconstruction of the developmental pattern of
the fetal membranes and placenta in Eutheria. Cladistic analysis
of the entire developmental pattern of the fetal membranes and
placenta in all orders and suborders of Eutheria, as well as a
consideration of the pattern in monotremes, reptiles, and birds,
suggests that the primitive eutherian condition included: (1) paraembryonic and noninvasive attachment of an expanded bilaminar
blastocyst, (2) a large vascular yolk sac during early ontogenetic
stages, (3) development of a temporary choriovitelline placenta,
(4) amniogenesis by folding, (5) a large allantoic vesicle, and
(6) a noninvasive, diffuse epitheliochorial placenta.  Difficulties arise, however, when the morphotype of the eutherian fetal
membranes is compared with that of the marsupials.  Previous
investigators generally have not considered the differences in
fetal membrane morphogenesis between marsupials and eutherians

when speculating about the primitive conditions in Eutheria,
despite the fact that most paleontologists accept a common
(pantothere) origin for the two taxa.

The morphotype of the definitive fetal membranes in marsupials, in particular, the persisting bilaminar omphalopleure
abembryonically and the moderate-sized allantois, is strikingly
similar to the relationships in sciurid rodents (and the rodent-
lagomorph morphotype) and in the morpyotype of the insectivoran
taxon Erinaceota (Erinaceidae, Soricidae, and Talpidae) (Fig. 24).
If the precocious occurrence of parturition in marsupials (and
monotremes) represents the ancestral mammalian and therian condition, then the definitive nature of the fetal membranes in
marsupials probably represents a pedomorphic retention of the
relationships which occur during the early incubation phase of
prototherians (compare Figs. 13, 17 and 18). Such pedomorphic retention in marsupials may be functionally related to the altered
nutritive and respiratory needs of the embryo which result from
the incorporation of the prototherian incubation phase of development into the intrauterine phase. Thus, pedomorphic retention of an extensive bilaminar omphalopleure in marsupials
appears to be correlated with its continued functional activity
as the primary nutritive organ during pregnancy. Retention of a
functional bilaminar omphalopleure is made possible by a concomitant retardation in expansion of the exocoelom and allantois.

Embryonic respiration is accomplished primarily by the
vascular choriovitelline placenta in marsupials, and this relatively inefficient relationship may be one contributing factor
to the failure of extending intrauterine life in marsupials. The
functional differentiation of complex organ systems during the
late fetal period requires a greater oxygen supply than can be
provided by simple apposition of a limited region of choriovitelline membrane. However, the nutritional needs of the embryo
preclude the expansion of the allantois, even in those cases
(Phascolarctidae, Peramelidae) where the allantois exhibits a
limited area of fusion with the chorion.

If the metatherian arrangement of the fetal membranes represents the ancestral therian condition, then the eutherian
pattern (such as occurs in the morphotype of the Erinaceota and
Lagomorpha-Rodentia) could have been derived from the therian
morphotype by the prolongation of intrauterine development, associated with replacement of the choriovitelline placenta by a
"villous" chorioallantoic placenta during ontogeny, and by the
retention of a functional bilaminar omphalopleure. The development of allantoic mesodermal "villi" in the eutherian chorioallantoic placenta is a major advancement over the condition in
peramelids, in that it greatly increases the surface area available for placental transport of oxygen and nutrients. The trans-

fer of both respiratory and nutrient functions to the chorioallantoic placenta is probably related to the reduction in functional importance of the bilaminar omphalopleure in late stages
of erinaceotans and sciurids, and its replacement or loss in most
other eutherians.

The possibility that differing rates of morphogenesis
(heterochrony) and pedomorphosis may have played an important
role in the evolution of mammalian fetal membranes complicates
any attempt at morphotype reconstruction of the ancestral eutherian stock.  Nevertheless, if the morphotype of the fetal membrane
developmental pattern in the Erinaceota and Glires (=Lagomorpha
plus Rodentia) approximates the ancestral therian-eutherian condition, then the morphotype of all other extant eutherian taxa
could have been derived from this pattern by further ontogenetic
elaboration of the fetal membranes.  This would entail further
expansion of the exocoelom and allantoic vesicle, so that the
yolk sac would become completely vascularized and reduced in
later stages.  A comparison of two stages in the ontogeny of the
fetal membranes in strepsirhine Primates illustrates the temporary
development of a pattern (Fig. 3) strikingly similar to the
morphotype of marsupials, erinaceotans, and glires, whereas the
definitive relationships of the yolk sac and allantois in later
stages (Fig. 4) are homologous to the condition in reptiles and
birds.

In summary, the occurrence of precocious hatching or parturition in monotremes and marsupials (and probably in their common
prototherian ancestors), in contrast to a more prolonged period
of prenatal life in reptiles, birds, and eutherians, may have
played a significant role in the evolutionary development of the
fetal membranes in the ancestral therian and eutherian stocks.
Thus, differences in the suggested morphotypes of metatherian and
eutherian fetal membranes are postulated to be the result of
functionally related pedomorphic retentions in marsupials.

DISCUSSION

Origin of the Cleidoic Egg in Amniotes

The evolutionary origin of the amniote fetal membranes was
probably associated with the development of a cleidoic egg in the
ancestral reptilian stock.  As emphasized by Romer (1957), the
deposition of eggs on land rather than in the water doubtlessly
evolved as a protective adaptation for avoiding aquatic predation.
Although extant amphibians are often categorized by their aquatic
mode of reproduction, in fact many species have acquired modifica-

tions in their life history associated with terrestrial reproduction and development' (Noble, 1931; Needham, 1942; Goin, 1960).
These adaptations may include internal fertilization, an increase
in the yolk content of the egg, deposition of eggs in sheltered
places on land, incorporation of the aquatic larval phase into the
pre-hatching stage of development, or the retention of the developing embryos within the parent's body for a variable period.
Although all these specializations do not occur in any single
species, a consideration of the variability in amphibian reproductive habits provides insight into the possible adaptations
which led to the evolution of a cleidoic egg in the ancestral
reptilian stock.

Cladistic analysis of the varying patterns of reproduction
and development in reptiles, birds, and prototherian mammals
suggests that there were at least five prerequisites for the
evolutionary origin of the cleidoic egg:

(1)   Internal fertilization, a necessary prerequisite for
      the laying of a shelled egg;

(2)   Attainment of a megalecithal egg, with concomitant
      development of a trilaminar yolk sac for respiratory
      gas exchange;

(3)   Secretion of protective and supportive secondary egg
      envelopes by the oviduct and "uterus";

(4)   Incorporation of the free-swimming larval stage into
      the pre-hatching phase of development, thus prolonging
      gestation; and

(5)   Deposition of the fertilized egg on land in a sheltered
      spot.

The first three of these specializations have evolved convergently in Chondrichthyes.  In addition, many chondrichthyes
provide further protection for their shelled eggs by retaining
them in utero for varying periods, resulting in the attainment
of an ovoviviparous or viviparous condition (Needham, 1942;
Amoroso, 1952).  Although the megalecithal eggs of elasmobranchs
develop a trilaminar yolk sac which may form a "pseudoplacenta"
(Jollie and Jollie, 1967) that is involved in both fetal nutrition and respiration, the shelled eggs are never deposited on
land, and additional fetal membranes comparable to the amnion,
chorion, and allantois of amniotes do not develop.  These data
support the hypothesis that the evolutionary origin of the amnion,
chorion, and allantois was concomitant with the origin of a
cleidoic egg which was deposited on land and subject to desiccation.

Conversely, there is no evidence that the initial appearance of the amniote fetal membranes was related to evolutionary trends toward viviparity.

Each of the five listed prerequisites for the evolution of a cleidoic egg occurs to a varying degree in some amphibians (Needham, 1942; Goin, 1960), and it is likely that all were present in the amphibian ancestor of the reptiles.

Although it is unclear as to which of the orders of extant Amphibia is cladistically closest to the ancestral reptilian stock, the Apoda is characterized by internal fertilization and the development of a megalecithal egg (Goin, 1960). Furthermore, their pattern of cleavage and gastrulation closely approaches that of extant reptiles (Kerr, 1919). Unfortunately, the developmental and reproductive biology of apodans has received minimal study during the past 50 years, and renewed examination of this poorly known tropical group should provide further data for speculation on the origin of the amniote cleidoic egg.

Ontogeny and Phylogeny of Amniote Fetal Membranes

Yolk sac. The yolk sac is the first of the fetal membranes to differentiate during ontogeny, and it was probably also the first to appear during vertebrate phylogeny. Its origin was concomitant with the development of a megalecithal egg and the subsequent sequestering of yolk extracellularly during cleavage. A yolk sac has evolved convergently in reptiles and in some elasmobranchs, teleosts, and amphibians. The primary role of the yolk sac during ontogeny and phylogeny is nutritive in that it is the initial site of mobilization and intracellular digestion of stored yolk for embryonic nutrition. The yolk sac mesoderm is also the initial site of hematopoiesis and angiogenesis; this is related to its role in providing a pathway for the efficient movement of stored yolk nutrients to the embryo.

Mossman (1967) has emphasized that the yolk sac undergoes considerable morphogenetic change during prenatal life, and all the variations in vertebrate yolk sac morphology can be seen during the ontogeny of the fetal membranes in eutherian mammals such as the strepsirhine Primates (Figs. 3 and 4). The development of unilaminar (ectoderm only) and bilaminar (ectoderm and endoderm) yolk sacs in therian mammales is correlated with the precocious differentiation of extraembryonic ectoderm (= trophoblast) during blastocyst formation; these stages do not occur in nontherian vertebrates. The trilaminar yolk sac (ectoderm, mesoderm, and endoderm) is the definitive yolk sac of megalecithal elasmobranchs (Fig. 1) and teleosts, and it forms a "pseudoplacenta"

in viviparous elasmobranchs.  In contrast, the trilaminar yolk sac
is only a transitory stage in the development of the amniote yolk
sac, at least in the morphotype of reptiles, birds, and prototherian mammals.  In these taxa, expansion of the exocoelom into
the trilaminar yolk sac results in the separation of somatic and
splanchnic mesodermal layers, so that the inner splanchnopleure
(splanchnic mesoderm plus endoderm) becomes the definitive yolk
sac or yolk sac splanchnopleure, whereas the outer somatopleure
(somatic mesoderm plus ectoderm) becomes a part of the chorion.
Presumably the same ontogenetic changes occurred in the phylogeny
of the reptilian yolk sac from an amphibian ancestor with a megalecithal egg and a trilaminar yolk sac.

Expansion of the exocoelom appears to be a causative factor
in the ontogenetic differentiation of the yolk sac splanchnopleure
in amniotes; in contrast, there is only minimal development of an
exocoelom in elasmobranchs and amphibians.  The development of the
exocoelom has received little consideration in phylogenetic
studies of the amniote fetal membranes, despite its apparent
mechanical role in the differentiation of the yolk sac, amnion,
and chorion (see below).  Presumably, its expansion is associated
with accumulation of fluid within it, although the origin of this
fluid during early somite stages is unclear.  Secretions from
the adjacent mesonephric (and pronephric) kidneys are a possible
source of this fluid, particularly during the period prior to the
union of the mesonephric duct and cloaca.

The trilaminar yolk sac (= choriovitelline membrane) in
therian mammals and some viviparous reptiles generally forms a
choriovitelline placenta, and this may be involved in the feto-
maternal exchange of nutrients, inorganic ions, water, respiratory
gases, or metabolites.  The transitory choriovitelline placenta
is disrupted by the subsequent expansion of the exocoelom in
viviparous reptiles and eutherians; this is an essential prerequisite for the later ontogenetic expansion of the allantois
into the exocoelom.  In contrast, marsupials retain their functional choriovitelline placenta pedomorphically, correlated with
a retardation in the expansion of the exocoelom and allantois.

Amnion and chorion. The amnion is the second of the amniote
fetal membranes to appear during ontogeny, and a consideration of
its morphogenesis in reptiles and prototherians suggests that its
initial development during phylogeny was due primarily to the
movement of a solid ectodermal ridge (= ectamnion).  This hypothesis has been supported previously by Mitsukuri (1891), Dalcq
(1937), and Fisk and Tribe (1949).  As postulated by Fisk and
Tribe (1949), a megalecithal egg laid on land would first have
been surrounded by a resistant shell and shell membrane; this
would enable the egg to retain its shape and impede evaporation.

After the egg is laid, there is a tendency for the yolk and blastoderm to move up through the albumen, due to differences in the
specific gravity of these structures.  As a result, the blastoderm
comes into contact with the overlying shell membrane, and thereby
is exposed to the dangers of adhesion and desiccation.

Fisk and Tribe (1949) have suggested that the evolutionary
origin of the amniochorion in reptiles was correlated with pro-
tecting the embryo from adhesion and desiccation by the extension
of a layer of extraembryonic ectoderm (ectamnion) to envelop the
embryo, thus providing an enclosed, fluid-filled space in which
the embryo could develop free from adhesion, desiccation, and
distortion.  In support of this hypothesis, Dalcq (1937) has
emphasized that the extraembryonic ectoderm of megalecithal
amniotes exhibits a pronounced tendency toward peripheral spreading (epiboly) in order to envelop the yolk mass, and he suggested
that the initiation of an ectamniotic fold in reptiles was also
the result of this same morphogenetic tendency.  An additional
factor in the onset of amniogenesis may be the localized lique-
faction of the yolk immediately beneath the blastoderm to form a
subgerminal cavity.  As a result, there is a tendency for the
heavier embryo, particularly its cranial end, to sink into this
partially liquefied yolk (Mitsukuri, 1891), and to incorporate
the mesoderm-free roof of the yolk sac (proamnion) as the cranial
portion of the developing amnion.

It is probable that both ectamniotic and proamniotic components appeared simultaneously during the evolutionary origin of
the amnion, as they do during amniogenesis in extant reptiles.
The definitive somatopleuric amnion and chorion are not formed
until somewhat later in ontogeny, concomitant with the dorsal
expansion of the exocoelom.  The expansion of the exocoelom during
early somite stages and the concomitant differentiation and
separation of the amnion and chorion were apparently essential
prerequisites for the subsequent functional differentiation of
the allantois as a respiratory organ, as suggested by Mitsukuri
(1891).

Amniogenesis in all reptiles is initiated during presomite
stages, whereas in birds it is somewhat more retarded, commencing
during 12-14 somite stages (see Table I).  As a result, the exocoelom plays more of a primary role in the onset of amniogenesis
in birds, and the extent of the ectamnion is greatly reduced.  The
reason for these differences is unclear, but it may be related to
differences in the rate of embryonic development and the length
of time which elapses between oviposition and the initiation of
amniogenesis.  In the chelonians Chelydra serpentina and Chrysemys
picta (Yntema, 1968; Mahmoud et al., 1973) and in the chick
(Hamilton, 1952), amniogenesis begins about 2-3 days after

oviposition.  However, the rate of embryonic development is more
rapid in homeothermal birds than it is in comparable sized
poikilothermal reptiles, so that the avian embryo is more differentiated at the time of onset of amniogenesis.  Therefore, it
may be that there is a critical time period of 2-3 days following
oviposition during which amniogenesis must begin in order to prevent adhesion of the developing embryo to the overlying shell
membrane.  Although the exact time sequence is unknown, it is
evident that amniogenesis is initiated in monotreme embryos also
shortly after the time of laying.  Those oviparous and viviparous
reptiles (Crocodilia, Squamata) which have prolonged the period
of intrauterine development nevertheless have retained a pattern
of amniogenesis comparable to that of the presumed primitive condition in chelonians.

Fisk and Tribe (1949) postulated that the precocious formation of an amniochorion in Chamaeleo chamaeleon from a circular
ectamniotic ridge represents the ancestral reptilian condition,
but this hypothesis is not supported by the character analysis
of the present study.  All other reptiles which have been studied
exhibit a different pattern in which there is a horeshoe-shaped
head fold composed of both ectamnion and proamnion, and this is
apparently true for other species of Chamaeleo. Thus, the unique
method of amniogenesis in C. chamaeleon is more likely to be a
derived character state, the causative factors of this being unclear.

The solid ectamniotic ridge is the precursor of the epithelium of both the amnion and chorion, and the evolutionary origins
of these two fetal membranes are clearly interrelated.  Expansion
of the exocoelom into the ectamnion is a causative factor in the
separation of the amnion and chorion, and the persistence of an
extensive chorioamniotic connection or raphe throughout much of
gestation in reptiles, birds, and prototherian mammals serves as
a remnant of the originally solid ectamnion.  The chorion completely envelops the other fetal membranes and the embryo; therefore, it is the one tissue which comes into contact with the
surrounding environment (shell membrane and shell or uterus).
Thus, the chorion is in a unique position to play an important
functional role in the passage of substances into or out of the
egg in all amniotes, whether oviparous or viviparous.

Allantois.  The allantois is the last of the fetal membranes
to differentiate; its endodermal component is first evident in
most reptiles, birds, prototherians, and metatherians as a ventral
outgrowth of the hindgut during 18-28 somite stages.  In all these
taxa, the further expansion of the allantois into the exocoelom
and its subsequent fusion with the chorion appear to be causally
related to the functional differentiation of the mesonephric

kidney. With further differentiation the allantois becomes continuous with the urogenital sinus and ultimately with the urinary
bladder. Thus, ontogenetic evidence strongly favors the hypothesis that the evolutionary origin of the allantois was causally
related to its function as a urinary bladder for the storage of
mesonephric excretions. In an aquatic environment, toxic excretory products can readily diffuse out of the embryo, whereas this
is not possible in a terrestrial cleidoic egg.

Although the respiratory function of the allantois develops
later during ontogeny, concomitant with fusion of the allantois
and chorion during 38-50 somite stages in reptiles, birds, and
prototherians, it is difficult to separate the excretory and
respiratory functions when considering the evolutionary origin of
the allantois. The somatopleuric chorion is situated in an ideal
location adjacent to the shell membrane for respiratory gas exchange with the external environment; however, its lack of blood
vessels limits its efficiency as a respiratory organ. The trilaminar omphalopleure differentiates earlier in ontogeny and may
function as a respiratory organ (as it does in elasmobranchs).
However, the yolk sac undergoes gradual reduction during prenatal
development, concomitant with the absorption of stored yolk.
Furthermore, the expanding exocoelom separates the vascular yolk
sac splanchnopleure from the chorion, so that there is only a
very limited area of trilaminar omphalopleure available for respiratory exchange. Indeed, the expansion of the exocoelom in most
extant reptiles extends as far as the peripheral margin of the
vascular omphalopleure (marked by the sinus terminalis), so that
there is scarcely any region of trilaminar omphalopleure (Fig. 8).

In contrast to the yolk sac, the allantois expands during
ontogeny, as a result of its excretory function, and its continued
expansion into the exocoelom brings its richly vascularzed
splanchnopleure into intimate contact with the chorion. Following
its initial fusion with the chorion, the allantois continues to
spread and comes to line the inner surface of most of the chorion.
In this way the chorion is vascularized secondarily by the
allantois, and the tissues of the fused chorioallantois become
greatly thinned and function as the principal site of respiratory
gas exchange between the cleidoic egg and the surrounding environment. In viviparous reptiles and mammals, the richly vascularized chorioallantois forms a placental relationship with the
uterus; this facilitates the fetomaternal exchange of water, nutrients, and metabolites, in addition to its continuing role in
gas exchange.

## Oviparity and Viviparity in Vertebrates

Viviparity, the condition in which the newborn is delivered "alive" from the parent's body, has evolved convergently in some Chondrichthyes, Osteichthyes, Amphibia, Reptilia and Mammalia. Excluding the Mammalia, viviparity has evolved repeatedly within vertebrates; this is evident from the fact that most families and genera which contain viviparous species also contain oviparous forms. A number of intermediate states exist between oviparity and viviparity, and these provide valuable clues regarding the developmental and environmental conditions which led to viviparity. Unfortunately, the term viviparity has been employed in varying contexts to describe developmental relationships in vertebrates, and disagreement exists as to whether morphological or functional considerations should be the primary factor in defining these relationships. Since physiological studies of embryo-maternal exchange in vertebrates are relatively scarce, morphological definitions are utilized in the present study.

Bertin (1952) has provided a useful distinction between two different types of oviparity. For the condition in which ripe ova are released into an aquatic medium where they are subsequently fertilized externally, he has proposed the term ovuliparity. This condition characterizes many cyclostomes, osteichythyes, and amphibians and probably represents the primitive vertebrate condition. The term oviparity can then be reserved for cases in which the egg is laid following internal fertilization. It follows that oviparity (sensu stricto) is a derived condition when compared to ovuliparity, and it must have characterized the common ancestral stock of reptiles, birds, and prototherian mammals. Ovoviviparity represents an intermediate character state between oviparity and viviparity, although frequently this condition is not distinguished from viviparity. Morphologically, ovoviviparity may be defined as a condition in which a shelled egg is retained within the confines of the reproductive tract (or other specialized areas) following fertilization. After a variable period of embryonic development, the egg "hatches" in utero, and the young are subsequently born alive (Nelson, 1953). Many so-called "viviparous" elasmobranchs and reptiles are characterized by such a relationship, in that a shell or shell membrane separates fetal and maternal tissues for much or all of intrauterine gestation, and hatching may occur simultaneously with parturition. Clearly, fetomaternal exchange of respiratory gases, water, inorganic ions, and some nutrients can occur in ovoviviparous elasmobranchs and reptiles (Needham, 1942; Panigel, 1956; Hoffman, 1970), although most ovoviviparous reptiles apparently contain enough yolk in their eggs to provide for the major portion of embryonic nutrition (Weekes, 1935; Bauchot, 1965).

Extensive studies by Weekes (1935) on the occurrence and distribution of ovoviviparous and viviparous reptiles in Australia indicate that viviparity has evolved primarily as a protective device against extreme temperature fluctuations in reptiles which live at relatively high altitudes (above 4000 feet). Most genera which contain ovoviviparous or viviparous species at high altitudes also contain oviparous species at lower altitudes. Similar findings have been reported for the distribution of viviparous reptiles in Europe (Bauchot, 1965). In contrast, the evolution of viviparity in elasmobranchs, teleosts, and amphibians appears to have been primarily a protective mechanism against aquatic predation on the eggs or larvae.

Some investigators have used purely functional criteria to distinguish between ovoviviparity and viviparity in elasmobranchs and reptiles. In this case, ovoviviparity is considered to entail intrauterine development in which embryonic nutrition is derived from stored intraovular yolk, in contrast to viviparity in which the embryo obtains its nourishment principally from the uterus. Such a functional distinction is obscured, however, when the various means by which embryonic nutrition is accomplished in mammals are considered. Thus, the monotremes embryo supplements the stored yolk of its egg by the absorption of nutrients from uterine gland secretions during its prolonged period of intrauterine life. In this case, uterine nutrition of the embryo is associated with an oviparous method of development.

In contrast to the condition in most other vertebrates, viviparity in therian mammals may have evolved primarily for nutritional rather than protective purposes. This is suggested by the occurrence of a microlecithal egg in all therians, a condition which necessitates an extraovarian source of nutrition early during the development of the embryo. The reasons for a phyletic reduction in size of the uvum are unknown. In vertebrates with megalecithal eggs, most of the yolk proteins and lipids are synthesized in the liver and transported to the ovary, where they are taken up by the maturing oocyte through the wall of the ovarian follicle and zona pellucida (Bellairs, 1971). Dalcq (1949) suggested that mutations occurred in the ancestral therian stock which inhibited the passage of yolk precursors across the follicular epithelium and zona pellucida, and he further speculated that the formation of a large follicular antrum (a unique shared and derived character of therians) within the ripe ovarian follicle was causally related to the accumulation of fluids which were prevented from entering the maturing oocyte. Whatever the explanation, it is evident that a partial reduction in size of the ovum also characterizes the extant monotremes, when compared to reptiles of similar body size. It is for this reason that oviparous monotremes exhibit considerable intrauterine growth of the egg before laying, a condition unknwon in oviparous or ovoviviparous reptiles and approached only by specialized viviparous reptiles such as Chalcides chalcides.

The survival of extremely altricial young, such as those produced by all extant monotremes and marsupials (and presumably also by their common prototherian ancestor), necessitated a mechanism of maternal care which would provide for the nutritional requirements of the newborn until they could fend for themselves. Thus, the evolutionary origin of mammary glands, a unique shared and derived feature of all extant mammals, was probably an essential prerequisite for the survival of precociously hatched and incompletely developed prototherian fetuses. It is suggested that the evolution of mammary glands was a compensatory feature associated with a reduction in yolk content of the prototherian ovum, and the inability of the developing embryo to absorb enough nutrients during its intrauterine phase to sustain the completion of fetal development during the abbreviated incubation period.

Ovoviviparity appears to be an intermediate stage in the development of viviparity in all vertebrates, and additional intermediate conditions exist between oviparity, ovoviviparity, and viviparity. Oviparity in monotremes is distinctly intermediate between the typical pattern of oviparity seen in chelonians and birds and the condition in ovoviviparous reptiles, in that the intrauterine period of development is prolonged, and both embryonic nutrition and respiration are carried out between the shelled egg and the maternal uterine environment. In a morphological sense, the marsupial embryo can be considered as ovoviviparous, because the embryo completes about 2/3 of its intrauterine development surrounded by the shell membrane and floating freely within the fluids of the uterine lumen. Hatching subsequently occurs in utero, and a brief period of attachment and choriovitelline placentation occurs before birth. This relationship is only slightly advanced beyond the monotremes pattern of oviparity, in that the incubation phase of monotreme development is incorporated into the intrauterine phase of marsupials, with "hatching" occurring during the later portion of the intrauterine period. The striking similarity in developmental state attained by newbord marsupial and newly hatched monotreme fetuses, and the homologous pattern of differentiation of their fetal membranes during the early embryonic period, provide further evidence to support the hypothesis that ovoviviparity in marsupials was derived from an ancestral pattern of oviparity similar to that which occurs in extant monotremes.

The pattern of true viviparity which characterizes eutherians could have been derived phyletically from an ovoviviparous relationship in the ancestral therian stock (similar to the condition in extant marsupials) by: (1) elimination of the shell membrane, thereby initiating implantation at an earlier developmental stage; and (2) prolonging the intrauterine phase of development and placentation by incorporating part of the lactational period into the prenatal period.  The evolution of prolonged intrauterine gestation in eutherians entailed hormonal modifications of the reproductive cycle in order to lengthen the luteal phase of the ovarian cycle, and probably required the production of an immuno-supressive layer on the surface of the trophoblast as part of an essential mechanism for the prevention of an allograft rejection of the placenta by the mother (for further discussion see Amoroso and Perry, 1975).

## Use of Amniote Fetal Membranes for
## Assessing Phylogenetic Relationships

Although comparative anatomists have long recognized that the common possession of an amnion, chorion, and allantois is an important character complex which indicates a monophyletic relationship among reptiles, birds, and mammals, further attempts at utilizing fetal membrane characters in phyletic studies have been limited primarily to eutherian mammals.  The available evidence indicates that there is little, if any, variation in the developmental pattern of the fetal membranes among the various avian orders.  This conservatism may be related in part to the retention of a similar pattern of oviparous development in all birds.  Within the Reptilia, developmental patterns appear to be constant within each other, although data are lacking for a majority of the genera within the highly successful order Squamata.  Nevertheless, the available evidence indicates that the adoption of ovoviviparity and viviparity in numerous genera of Squamata has not had any significant effect on the morphogenesis of their fetal membranes. This provides further support for the hypothesis that the evolution of viviparity is a relatively recent event in extant reptiles, although clearly it has evolved previously in marine reptiles such as the Jurassic-Cretaceous ichthyosaurs.

Phyletic relationships among the different orders of extant reptiles are not clarified by the available data on their fetal membrane development.  The Chelonia and Rhynchocephalis retain a pattern which is considered to be closest to the morphotype of the Reptilia, and both taxa lay eggs which contain embryos in cleavage-early gastrulation stages, similar to the condition in birds (Table I).  Crocodilia and Lacertilia exhibit tendencies for the retention of fertilized eggs for longer periods within the uterus, even in those taxa which are oviparous.  Histological and temporal aspects of fetal membrane development in crocodilians are poorly known, and this limits the use of such data in assessing the phyletic relationships of Crocodilia to other reptiles and to birds. The apparent reduction in extent of the ectamnion, and the concomitant increase in the role played by the exocoelom in amniogenesis of crocodilians, are intermediate character states between conditions in other reptiles and those of birds, as is the relative delay in closure of the amniotic navel. Retention of the developing egg within the uterus during at least the early somite stages of development, however, is a derived character state of crocodilians which is unlikely to have occurred in the ancestral stock of the birds.

Evidence of the fetal membranes for mammalian phylogeny. During the late 19th-early 20th centuries, certain fetal membrane features, particularly the degree of invasiveness and the gross morphology of the chorioallantoic placenta, were utilized in typological classifications of the mammals. It soon became evident that these classifications, based on relatively few characters, were of little value in reflecting phylogenetic relationships, and they were subsequently abandoned. Further consideration of the phyletic significance of mammalian placentation focused on the presumed primitive or specialized nature of epitheliochorial and hemochorial placentation during the first third of the 20th century, although these discussions generally ignored other features of the fetal membranes and placenta.

Mossman's (1937) comprehensive survey of the available data on fetal membrane morphogenesis in eutherians marked a turning point in the study of mammalian placentation. First, his study encompassed all the developmental characters of the fetal membranes, not just the definitive nature of the chorioallantoic placenta. Next, he evaluated the relative evolutionary conservatism of each of these characters, as indicated by the relative constancy of their developmental pattern in the higher taxonomic categories of eutherians. His emphasis on the utilization of these conservative features (the nature of the allantoic vesicle, yolk sac, and orientation of the embryonic disc, in addition to the structure of the chorioallantoic placenta) as indicators of phylogenetic relationships remains a major contribution to the study of mammalian systematics. However, the phylogenetic tree of the Mammalia constructed by Mossman solely on the basis of fetal membrane characters can be criticized because of its failure to distinguish between primitive and derived character states in assessing relationships, and in part because of the author's underestimation of the possibility of convergent evolution in individual fetal membrane characters. This is evident in his suggestion of possible affinities between sloths (family Bradypodidae) and Anthropoida on the one hand, and armadillos (family Dasypodidae) and Rodentia on the other, based on the common possession of a few derived characters which appear to have evolved convergently in these and other taxa.

Data on fetal membrane development generally have not been incorporated into the major syntheses of mammalian taxonomy, due in part to some of the improbable relationships suggested by Mossman's (1937) study, but also related in great measure to the lack of understanding of these morphogenetic data by many systematists. This naivete is exemplified by equating the evidence of "placentation" with the single character of the definitive chorioallantoic placenta, while ignoring all of the other developmental data available for the fetal membranes. Utilizing the cladistic methodology discussed previously, the relative primitive and derived conditions for each of the mammalian fetal membrane characters can be evaluated, and a morphocline polarity can be established for each character. The morphogenesis of the fetal membranes in all amniotes is considered when evaluating the primitive and derived states of each character and in reconstructing the morphotype of each taxon under study. Such analysis permits the construction of cladograms based on the common possession of shared and derived characters in monophyletic sister groups, following Hennig's (1966) scheme of argumentation of phylogenetic systematics.

The fetal membranes exhibit considerable variability in their morphogenesis among higher categories of eutherians, whereas significant differences are minimal below the family level. This combination of evolutionary conservatism plus variability in the fetal membranes, coupled with the extensive genetic complexity involved in their morphogenesis, provides a character complex which can be of considerable value in assessing phyletic relationships among eutherians. Mossman (1967) has emphasized that it is the complete ontogenetic pattern of the fetal membranes which is available for phylogenetic studies, and an ontogenetic approach facilitates the recognition of parallel or convergent evolution of individual characters, such as the convergent evolution of a hemochorial placenta in some primates, insectivorans, bats, and rodents. At the same time, their genetic complexity minimizes the possibility of convergent evolution in the developmental pattern of all the fetal membrane characters.

Cladistic analyses of an individual organ system, such as the fetal membranes, basicranium, or dentition, should not comprise the sole basis of any phylogenetic hypothesis. Instead, multiple character analyses of both extant and fossil taxa, coupled with a consideration of the temporal and paleogeographic distribution of fossils, offer the best method for the construction of phylogenetic hypotheses which approximate the actual phylogeny of a group. Such a multiple character analysis, utilizing fetal membrane, basicranial, and other anatomical features, has been

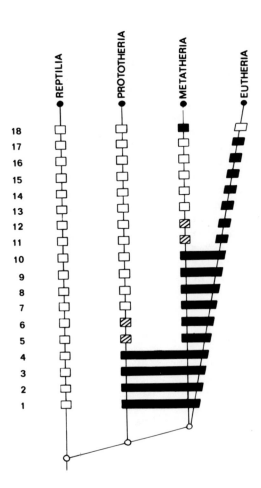

Fig. 26.  Cladogram of selected fetal membrane, developmental, and reproductive characters of reptiles and mammals presented in Table IV.      = primitive;      = derived.

presented elsewhere for the Primates (Luckett and Szalay, in press).

The mammalian orders which exhibit the greatest degree of variability in the morphogenesis of their fetal membranes are the Insectivora, Rodentia, Primates, Chiroptera, and Edentata.  Concomitantly, these taxa provide the greatest opportunity for cladistic analysis of fetal membrane characters and the incorporation of such data into studies of mammalian phylogeny.  The Primates are the only order whose fetal membranes have been evaluated by cladistic methodology to date (Luckett, 1975, 1976), although studies are currently in progress on the Insectivora and

Rodentia.  The following discussion summarizes some of these
data which appear to be useful in reconstructing mammalian
phylogeny.

The evidence of the fetal membranes presented in the present
study, in conjunction with other features of developmental and
reproductive biology, reaffirms the basic dichotomy between
prototherian and therian mammals, and the sister group relationship of the Metatheria and Eutheria (Table IV and Fig. 26).  A
preliminary discussion of some of these data was presented by
Hill (1910), and his astute observations are supported and extended by the present study.  It is essential that cranioskeletal
characters of fossil and extant mammals be evaluated cladistically
in order to corroborate (or refute) this hypothesis.  The fetal
membrane and other soft anatomical data provide further support
for the concept (Hopson and Crompton, 1969) of a monophyletic
origin of the Mammalia.

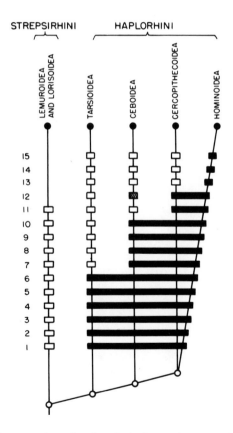

Fig. 27.  Cladogram of primate fetal membrane characters listed
          in Table V. (From Luckett, 1975)

TABLE IV

Primitive and derived character states of selected
urogenital and developmental features of reptiles and mammals

| Primitive | Derived |
|---|---|
| 1. Absence of hair and sweat glands | 1. Presence of hair and sweat glands |
| 2. Absence of mammary glands | 2. Presence of mammary glands |
| 3. No bilaminar blastocyst | 3. Bilaminar blastocyst stage |
| 4. No bulbourethral glands | 4. Bulbourethral glands present |
| 5. Embryonic nutrition intra-ovular | 5. Embryonic nutrition uterine[1] |
| 6. Ureters open into cloaca | 6. Ureters open into bladder[2] |
| 7. Shell present | 7. Shell absent |
| 8. No unilaminar blastocyst | 8. Unilaminar blastocyst stage |
| 9. Ovarian follicles lack antrum | 9. Ovarian follicles with antrum |
| 10. Megalecithal egg with meroblastic cleavage | 10. Microlecithal egg with holoblastic cleavage |
| 11. Oviparous | 11. Viviparous[3] |
| 12. Absence of placenta | 12. Definitive chorioallantoic placenta[4] |
| 13. Absence of polar tropho-blast and inner cell mass | 13. Presence of polar tropho-blast and inner cell mass |
| 14. Vaginae paired | 14. Vagina single |
| 15. Seminal vesicles absent | 15. Seminal vesicles present |
| 16. Ureters pass medial to vaginae | 16. Ureters pass lateral to fused vagina |
| 17. Shell membrane present | 17. Shell membrane absent |
| 18. Extensive chorioallantois | 18. No chorioallantois |

[1] Intermediate character state: Embryonic nutrition both intra-ovular and uterine
[2] Intermediate character state: Ureters open into urogenital sinus
[3] Intermediate character state: Ovoviviparous
[4] Intermediate character state: Definitive choriovitelline placenta

Cladistic analysis of the fetal membranes and placenta supports the hypothesis of any early dichotomy of Eocene-Recent Primates into the suborders Strepsirhini and Haplorhini (Fig. 27 and Table V), and of the sister group relationship of Tarsiiformes and Anthropoidea within the Haplorhini (Luckett, 1975, 1976). Evaluation of a wide range of characters from both extant and fossil primates corroborates these hypotheses, and provides further support for the concept of the monophyletic origin of Anthropoidea from an ancestral tarsiiform stock (Luckett and Szalay, in press). The fetal membrane evidence also indicates that the tree shrews (family Tupaiidae) do not share any special relationship with the Primates, since their placental characters are a combination of primitive eutherian retentions and tupaiid specializations which do not occur in Primates.

Current studies in progress on relationships among the higher categories of rodents, and a morphotype reconstruction of the rodent fetal membranes, support the previous suggestion by Mossman (1937) that the rodents and lagomorphs are closely related, and the fetal membrane evidence is compatible with their sister group relationship as the cohort Glires (See Szalay, this volume for a discussion of the possible Asian ancestry of the Glires). The fetal membranes of the South American caviomorphs are the most specialized of any rodent group, and they comprise a number of unique shared and derived character states which do not occur in other rodents or mammals. A current joint investigation with H.W. Mossman on fetal membrane morphogenesis in the African hystricomorphous families Hystricidae and Bathyergidae demonstrates that these taxa form a sister group with the Caviomorpha, based on the virtual identify of all their shared and derived character states. This concept of a monophyletic clustering of the South American and African hystricognathous rodents as an order Hystricognathi is also supported by numerous similarities in features of their soft and hard anatomy, although most of these characters have not been evaluated cladistically as yet. In contrast, the Afro-Asian hystricomorphous (but not hystricognathous) families Anomaluridae, Pedetidae, and Ctenodactylidae do not share this suite of unique and derived fetal membrane characters (Luckett, 1971), and this supports the concept that these taxa are not closely related phyletically to the Hystricognathi.

The order Insectivora appears to be a paraphyletic (or possibly polyphyletic) assemblage of taxa, based largely on the retention on numerous primitive eutherian features, both in their fetal membrane development as well as other features of soft and hard anatomy. The fetal membrane evidence supports Van Valen's (1967) concept of a suborder Erinaceota which includes the Erinaceidae, Soricidae, Talpidae, and Solenodontidae. All these

TABLE V

Primitive and derived character states
of primate fetal membrane characters

| Primitive | Derived |
| --- | --- |
| 1. Diffuse, epitheliochorial placenta | 1. Discoidal, hemochorial placenta |
| 2. Noninvasive attachment | 2. Invasive attachment |
| 3. No primordial amniotic cavity | 3. Primordial amniotic cavity |
| 4. Choriovitelline placenta | 4. No choriovitelline placenta |
| 5. No body stalk | 5. Body stalk |
| 6. Large, vesicular allantois | 6. Rudimentary allantois |
| 7. Paraembryonic pole attachment | 7. Embryonic pole attachment |
| 8. Amniogenesis by folding | 8. Amniogenesis by cavitation |
| 9. Bicornuate uterus | 9. Simplex uterus |
| 10. Primary yolk sac | 10. Secondary yolk sac |
| 11. No cytotrophoblastic shell | 11. Cytotrophoblastic shell |
| 12. Labyrinthine placental disc | 12. Villous placental disc |
| 13. Uterine symplasma | 13. No uterine symplasma |
| 14. Superficial implantation | 14. Interstitial implantation |
| 15. No decidua capsularis | 15. Decidua capsularis |

taxa retain a bilaminar omphalopleure and exhibit a common anti-mesometrial orientation of the embryonic disc, whereas they exhibit progressive differences in the size of the allantois and their type of definitive chorioallantoic placenta. The association of the families Tenrecidae and Potamogalidae as the superfamily Tenrecoidea is also supported by the fetal membrane evidence, despite differences in the definitive placenta (diffuse and endotheliochorial? in potamogalids and discoidal and hemochorial in tenrecids). It has been recently recognized that both families are characterized by the possession of a unique shared and derived character, a "hematome" in the center of the placenta, although this structure has previously been interpreted as a "placental disc" in potamogalids (Luckett, unpublished). A relationship of the Chrysochloridae to the Tenrecoidea is not ruled out by the evidence of the fetal membranes, but it is unclear as yet whether they share any derived characters. The Macroscelididae and Tupaiidae are quite distinct from each other on the basis of their fetal membranes, and many investigators now prefer to erect separate orders for both taxa. A possible relationship of Macroscelididae to Lagomorpha (or Glires) seems unlikely on the basis of the fetal membrane evidence (but see Szalay, this volume).

In conclusion, the utilization of sound principles of cladistic analysis provides a theoretical framework to support Mossman's (1937) assertion that "evolution of the fetal membranes of mammals is then a fact and, if one could evaluate their characters properly, they might be used to throw light on the phylogenetic relationships of the various groups."

## ACKNOWLEDGEMENTS

The illustrations were prepared by Robert Demarest, Terry Wilcox, and Dr. Nancy Hong.

## REFERENCES

Agassiz, L., 1857, Embryology of the turtle. Contrib. Nat. Hist. USA 2:451-644.

Amoroso, E.C., 1952, Placentation, in: "Marshall's Physiology of Reproduction," (A.S. Parkes, ed.), 3rd ed., Vol. II. pp. 127-311. Longmans, Green and Co., London.

Amoroso, E.C.and Perry, J.S., 1975, The existence during gestation of an immunological buffer zone at the interface between maternal and foetal tissues. Phil. Trans. Roy. Soc., Ser. B. 271:343-361.

Bauchot, R., 1965, La placentation chez les reptiles.  Ann.
     Biol. 4:547-575.
Bellairs, R., 1971, Developmental Processes in Higher Vertebrates.
     Logos Press, London.
Bertin, L., 1952, Oviparite, ovoviviparite, viviparite.  Bull.
     Soc. Zool. Fr. 77:84-88.
Boyd, M.M., 1942, The oviduct, foetal membranes, and placentation
     in Hoplodactylus maculatus Gray. Proc. Zool. Soc. Lond.,
     Ser. A. 112:65-104.
Boyden, E.A., 1924, An experimental study of the development of
     the avian cloaca, with special reference to a mechanical
     factor in the growth of the allantois.  J. Exp. Zool. 40:
     437-472.
Brambell, F.W.R., 1970, The transmission of passive immunity
     from mother to young.  North-Holland, Amsterdam.
Caldwell, W.H., 1887, The embryology of Monotremata and Mar-
     supialia.  Phil. Trans. Roy. Soc., 178:463-486.
Cate-Hoedemaker, N.J. ten., 1933, Beitrage zur Kenntnis der
     Plazentation bei Haien und Reptilien.  Zeit. Zellforsch.
     18:299-345.
Dalcq, A., 1937, Les plans d'ebauches chez les Vertebres et la
     signification morphologique des annexes foetales.  Ann. Soc.
     Roy. Zool. Belg., 68:69-76.
Dalcq, A., 1949, L'apport de l'embryologie causale au probleme
     de l'evolution.  Portug. Acta Biol., Ser. A., 367-400.
Dendy, A., 1899, Outlines of the development of the tuatara,
     Sphenodon. Quart. J. Micr. Sci., 42:1-87.
Dufaure, J.P. and Hubert, J., 1961, Table de developpement du
     lezard vivipare: Lacerta (Zootoca) vivipara Jacquin.  Arch.
     Anat. Micr. Morph. Exp., 50:309-327.
Enders, A.C. and Enders, R.K., 1969, The placenta of the four-
     eyed opossum (Philander opossum). Anat. Rec., 165:431-450.
Farris, J.S., 1966, Estimation of conservatism of characters by
     constancy within biological populations.  Evolution 20:
     587-591.
Fisk, A. and Tribe, M., 1949, The development of the amnion and
     chorion of reptiles.  Proc. Zool. Soc. Lond., 119:83-114.
Flynn, T.T., 1922, The phylogenetic significance of the marsupial
     allanto-placenta.  Proc. Linn. Soc. NSW 47:541-544.
Flynn, T.T., 1923, The yolk-sac and allantoic placenta in
     Perameles. Quart. J. Micr. Sci., 67:123-182.
Flynn, T.T. and Hill, J.P., 1939, The development of the Mono-
     tremata.  Part IV.  Growth of the ovarian ovum, maturation,
     fertilisation, and early cleavage.  Trans. Zool. Soc. Lond.,
     24:445-623.
Flynn, T.T. and Hill, J.P., 1947, The development of the Mono-
     tremata.  Part VI. The later stages of cleavage and the
     formation of the primary germ-layers.  Trans. Zool. Soc.
     Lond., 26:1-151.

Giacomini, E., 1891, Materiaux pour l'etude du developpement du Seps chalcides. Arch. Ital. Biol., 16:332-359.

Gingerich, P.D., 1974, Cranial anatomy and evolution of early Tertiary Plesiadapidae (Mammalia, Primates). Unpublished Ph.D. Thesis, Yale Univ.

Glenister, T.W., 1954, The emperor penguin Aptenodytes forsteri Gray. II. Embryology. Falkland Islands Dependencies Survey, Sci. Repts., 10:1-19.

Goin, C.J., 1960, Amphibians, pioneers of terrestrial breeding habits. Ann. Rep. Smithsonian Inst., 1959:427-445.

Grosser, O., 1909, Vergleichende Anatomie und Entwicklungsgeschichte der Eihaute und der Placenta. Wilhelm Braumuller, Vienna.

Grosser, O. and Tandler, J., 1909, Normentafel zur Entwicklungsgeschichte der Keibitzes (Vanellus cristatus Meyer), in: "Normentafeln zur Entwicklungsgeschichte der Wirbeltiere," (F. Keibel, ed.), Vol. 9. pp. 1-58, G. Fischer, Jena.

Hamilton, H.L., 1952, Lillie's Development of the Chick, 3rd ed. Henry Holt and Co., New York.

Harrison, L. and Weekes, H.C., 1925, On the occurrence of placentation in the scincid lizard Lygosoma entrecasteauxi. Proc. Linn. Soc. NSW 50:470-486.

Hartman, C.G., 1916, Studies in the development of the opossum Didelphys virginiana L. I. History of the early cleavage. II. Formation of the blastocyst. J. Morph., 27:1-83.

Hartman, C.G., 1919, Studies in the development of the opossum Didelphys virginiana L. III. Description of new material on maturation, cleavage and entoderm formation. IV. The bilaminar blastocyst. J. Morph., 32:1-142.

Hartman, C.G., 1928, The breeding season of the opossum (Didelphis virginiana) and the rate of the intra-uterine and postnatal development. J. Morph., 46:143-215.

Haswell, W.A., 1887, Observations on the early stages in the development of the emu (Dromaeus novae-hollandiae). Proc. Linn. Soc. NSW 2:577-600.

Hennig, W., 1966, Phylogenetic Systematics. University of Illinois Press, Urbana.

Hill, J.P., 1895, Preliminary note on the occurrence of a placental connection in Perameles obesula and on the foetal membranes of certain macropods. Proc. Linn. Soc. NSW 10:578-581.

Hill, J.P., 1897, The placentation of Perameles. Quart. J. Micr. Sci., 40:385-446.

Hill, J.P., 1900, On the foetal membranes, placentation and parturition of the native cat (Dasyurus viverrinus). Anat. Anz., 18:364-373.

Hill, J.P., 1910, The early development of Marsupialia, with special reference to the native cat (Dasyurus viverrinus). Quart. J. Micr. Sci., 56:1-134.

Hill, J.P., 1932, The developmental history of the primates.
    Phil. Trans. Roy. Soc., Ser. B., 221:45-178.
Hill, J.P., 1949, The allantoic placenta of Perameles.  Proc.
    Linn. Soc. Lond., 161:3-7.
Hill, J.P. and Martin, C.J., 1894, On a platypus embryo from the
    intra-uterine egg.  Proc. Linn. Soc. NSW 10:43-74.
Hoffman, L.H., 1970, Placentation in the garter snake Thamnophis
    sirtalis.  J. Morph., 131:57-88.
Hopson, J.A. and Crompton, A.W., 1969, Origin of mammals. Evol.
    Biol., 3:15-72.
Hrabowski, H., 1926, Das Dotterorgan der Eidechsen.  Z. Wiss.
    Zool., 128:305-382.
Hubrecht, A.A.W., 1908, Early ontogenetic phenomena in mammals
    and their bearing on our interpretation of the phylogeny
    of the vertebrates.  Quart. J. Micr. Sci., 53:1-181.
Hughes, R.L., 1974, Morphological studies on implantation in
    marsupials.  J. Reprod. Fert., 39:173-186.
Jollie, W.P. and Jollie, L.G., 1967, Electron microscopic
    observations on the yolk sac of the spiny dogfish, Squalus
    acanthias. J. Ultra. Res., 18:102-126.
Kerr, J.G., 1919, Textbook of Embryology. II. Vertebrata.
    Macmillan, London.
King, B.F. and Enders, A.C., 1970, Protein absorption and trans-
    port by the guinea pig visceral yolk sac placenta.  Am. J.
    Anat., 129:261-288.
Krull, J., 1906, Die Entwicklung der Ringelnatter (Tropoidonotus
    natrix Boie) vom ersten Aufstreten des Proamnios bis zum
    Schlusse des Amnios.  Z. Wiss. Zool., 85:107-155.
Lambson, R.O., 1970, An electron microscopic study of the ento-
    dermal cells of the yolk sac of the chick during incubation
    and after hatching.  Am. J. Anat., 129:1-20.
Lemus, D., 1967, Contribucion al estudio de la embriologia de
    reptiles chilenos.  II. Tabla de desarrollo de la lagartija
    vivipara Liolaemus gravenhorti (Reptilia-Squamata-Iguanidae).
    Biologica, 40:39-61.
Lemus, D. and Duvauchelle, R., 1966, Desarollo intrauterino de
    Liolaemus tenuis tenuis (Dumeril y Bibron).  Biologica,
    39:80-98.
Luckett, W.P., 1971, The development of the chorio-allantoic
    placenta of the African scaly-tailed squirrels (family
    Anomaluridae).  Am. J. Anat., 130:159-178.
Luckett, W.P., 1974, Comparative development and evolution of the
    placenta in primates, in: "Reproductive Biology of the
    Primates, Contributions to Primatology," (W.P. Luckett, ed.),
    Vol. 3., pp. 142-234. S. Karger, Basel.
Luckett, W.P., 1975, Ontogeny of the fetal membranes and placenta:
    Their bearing on primate phylogeny, in: "Phylogeny of the
    Primates," (W.P. Luckett and F.S. Szalay, eds.), pp. 157-182.
    Plenum Press, New York.

Luckett, W.P., 1976, Cladistic relationships among primate higher categories: Evidence of the fetal membranes and placenta. Folia Primat., 25:245-276.

Luckett, W.P. and Szalay, F.S., (in press), Clades versus grades in primate phylogeny, in: Proceedings 6th International Congress of Primatology, Cambridge, England. Academic Press, London and New York.

Mahmoud, I.Y., Hess, G.L., and Klucka, J., 1973, Normal embryonic stages of the western painted turtle, Chrysemys picta bellii. J. Morph., 141:269-280.

McCrady, E., Jr., 1938, The embryology of the opossum. Am. Anat. Mem., 16:1-233.

Mitsukuri, K., 1891, On the foetal membranes of Chelonia. J. Coll. Sci., Imp. Univ. Japan, 4:1-53.

Mossman, H.W., 1937, Comparative morphogenesis of the fetal membranes and accessory uterine structures. Contrib. Embryol. Carneg. Inst., 26:129-246.

Mossman, H.W., 1953, The genital system and the fetal membranes as criteria for mammalian phylogeny and taxonomy. J. Mammal., 34:289-298.

Mossman, H.W., 1967, Comparative biology of the placenta and fetal membranes, in: "Fetal Homeostasis," (R.M. Wynn, ed.), Vol. 2, pp. 13-97. New York Academy of Science, New York.

Mossman, H.W., 1971, Orientation and site of attachment of the blastocyst, in: "The Biology of the Blastocyst," (R.J. Blandau, ed.), pp. 49-57. University of Chicago Press, Chicago.

Needham, J., 1942, Biochemistry and Morphogenesis. Cambridge University Press, London.

Nelsen, O.E., Comparative Embryology of the Vertebrates. The Blakiston Company, New York.

Noble, G.K., 1931, The Biology of the Amphibia. McGraw-Hill, New York.

Padykula, H.A. and Taylor, J.M., 1974, Cytological observations on marsupial placentation: The Australian bandicoots (Perameles and Isoodon). Anat. Rec., 178:434.

Padykula, H.A. and Taylor, J.M. (in press), Ultrastructural evidence for loss of the trophoblastic layer in the chorioallantoic placenta of Australian bandicoots (Marsupialia: Peramelidae). Anat. Rec.

Panigel, M., 1951, Rapports anatomo-histologiques etablis au cours de la gestation entre l'oeuf et l'oviducte maternel chez le lezard ovovivipare Zootoca vivipara W. Bull. Soc. Zool. Fr., 76:163-170.

Panigel, M., 1956, Contribution a l'etude de l'ovoviviparite chez les reptiles: Gestation et parturition chez le lezard vivipare Zootoca vivipara. Ann. Sci. Nat., Zool., 18:569-668.

Parameswaran, K.N., 1963, The foetal membranes and placentation of Enhydris dussumieri (Smith). Proc. Ind. Acad. Sci. B. 56: 302-327.

Pasteels, J.J., 1957a, Une table analytique du developpement des reptiles. I. Stades de gastrulation chez les cheloniens et les lacertiliens. Ann. Soc. Roy. Zool. Belg., 87:217-241.

Pasteels, J.J., 1957b, La formation de l'amnios chez les cameleons. Ann. Soc. Roy. Zool. Belg., 87:243-246.

Pasteels, J.J., 1970, Developpement embryonnaire, in: "Traite de Zoologie," (P.P. Grasse, ed.), Vol. 14, Part 3, pp. 893-971. Masson et Cie, Paris.

Pearson, J., 1949, Placentation of the Marsupialia. Proc. Linn. Soc. London, 161:1-3.

Peter, K., 1904, Normentafel eur Entwicklungsgeschichte der Zauneidechse (Lacerta agilis), in: "Normentafeln zur Entwicklungsgeschichte der Wirbeltiere," (F. Keibel, ed.), Vol. 4. G. Fischer, Jena.

Peter, K., 1934, Die erste Entwicklung des Chamaleons (Chamaeleo vulgaris), verglichen mit der Eidechse (Ei, Keimbildung, Furchung, Entodermbildung). Z. Anat. Entwg., 103:147-188.

Peter, K., 1935, Die innere Entwicklung des Chamaleonkeimes nach der Furchung bis zum Durchbruch des Urdarms. Z. Anat. Entwg., 104:1-60.

Reese, A.M., 1908, The development of the American alligator (A. mississippiensis). Smithsonian Misc. Coll., 6:1-66.

Reese, A.M., 1915, The Alligator and its Allies. Putnam, New York.

Renfree, M.B., 1973, The composition of fetal fluids of the marsupial Macropus eugenii. Develop. Biol.,33:63-79.

Romanoff, A.L., 1960, The Avian Embryo. The Macmillan Company, New York.

Romanoff, A.L. and Romanoff, A.J., 1949, The Avian Egg. John Wiley and Sons, New York.

Romer, A.S., 1957, Origin of the amniote egg. Sci. Monthly, 85:57-63.

Romer, A.S., 1967, Major steps in vertebrate evolution. Science, 158:1629-1637.

Schaeffer, B., Hecht, M.K., and Eldredge, N., 1972, Phylogeny and paleontology. Evol. Biol., 6:31-46.

Schauinsland, H., 1899, Beitrage zur Biologie und Entwickelung der Hatteria nebst Bemerkungen uber die Entwickelung der Sauropsiden. Anat. Anz., 15:309-334.

Schauinsland, H., 1903, Beitrage zur Entwickelungsgeschichte und Anatomie der Wirbeltiere. I. Sphenodon, Callorhynchus, Chamaeleo. Zoologica, 16:1-98.

Schauinsland, H., 1906, Die Entwickelung der Eihaute der Reptilien und der Vogel, in: "Handbuch der Vergleichenden und experimentellen Entwickelungslehre der Wirbeltiere," (O. Hertwig, ed.), Vol. 1, Pt. 2, pp. 177-234. G. Fischer, Jena.

Schlafke, S. and Enders, A.C., 1975, Cellular basis of interaction between trophoblast and uterus at implantation. Biol. Reprod., 12:41-65.

Selenka, E., 1887, Studien uber Entwickelungsgeschichte der
    Tiere. IV. Das Opossum (Didelphys virginiana). C.W. Kreidel's
    Verlag. Wiesbaden.
Semon, R., 1894a, Die Embryonalhullen der Monotremen und
    Marsupialier. Denkschr. Med. Naturwiss. Ges. Jena, 5:19-58.
Semon, R., 1894b, Zur Entwickelungsgeschichte der Monotremen.
    Denkschr. Med. Naturwiss. Ges. Jena, 5:61-74.
Sharman, G.B., 1961, The embryonic membranes and placentation in
    five genera of diprotodont marsupials. Proc. Zool. Soc.
    Lond., 137:197-220.
Simons, E.L., (in press), The fossil record of primate phylogeny,
    in: "Molecular Anthropology," (M. Goodman and R.E. Tashian,
    eds.), Plenum Press, New York.
Starch, D., 1959, Ontogenie und Entwicklungsphysiologie der
    Saugetiere. Handb. Zool., 8:1-276.
Tarkowski, A.K. and Wroblewska, J., 1967, Development of blasto-
    meres of mouse eggs isolated at the 4- and 8-celled stage.
    J. Embryol. Exp. Morph., 18:155-180.
Turner, W., 1877, Some general observations on the placenta,
    with special reference to the theory of evolution. J. Anat.
    Physiol., 11:33-53.
Tyndale-Biscoe, H., 1973, Life of Marsupials. Arnold, London.
Tyndale-Biscoe, C.H., Hearn, J.P., and Renfree, M.B., 1974,
    Control of reproduction in macropodid marsupials. J. Endocr.,
    63:589-614.
Van Valen, L., 1967, New Paleocene insectivores and insectivore
    classification. Bull. Am. Mus. Nat. Hist., 135:217-284.
Virchow, H., 1892, Das Dotterorgan der Wirbeltiere (Fortsetzung).
    Arch. Mikr. Anat., 40:39-101.
Voeltzkow, A., 1902, Beitrage zur Entwicklungsgeschichte der
    Reptilien. I. Biologie und Entwicklung der ausseren
    Korperform von Crocodilus madagascariensis Grant. Abh.
    Senckenberg. Naturf. Ges., 26:1-150.
Weekes, H.C., 1927, Placentation and other phenomena in the
    scincid lizard Lygosoma (Hinulia) quoyi. Proc. Linn. Soc.
    NSW 52:499-554.
Weekes, H.C., 1929, On placentation in reptiles I. Proc. Linn.
    Soc. NSW 54:34-60.
Weekes, H.C., 1935, A review of placentation among reptiles,
    with particular regard to the function and evolution of
    the placenta. Proc. Zool. Soc. Lond., Part 2:625-645.
Wild, A.E., 1971, Transmission of proteins from mother to
    conceptus in the grey squirrel (Sciurus carolinensis).
    Immunology, 20:789-797.
Will, L., 1893, Beitrage zur Entwicklungsgeschichte der Reptilien.
    I. Die Anlage der Keimblatter beim Gecko (Platydactylus
    facetanus Schreib.). Zool. Jahrb., 6:1-160.
Wilson, J.T. and Hill, J.P., 1907, Observations on the develop-
    ment of Ornithorhynchus. Phil. Trans. Roy. Soc. Lond., Ser.
    B., 199:31-168.

Wilson, J.T. and Hill, J.P., 1915, The embryonic area and so-
    called "primitive knot" in the early monotreme egg.
    Quart. J. Micr. Sci., 61:15-25.
Wislocki, G.B., 1929, On the placentation of primates, with a
    consideration of the phylogeny of the placenta. Contrib.
    Embryol. Carneg. Inst., 20:51-80.
Witschi, E., 1956, Development of Vertebrates. W.B. Saunders
    Company, Philadelphia.
Yntema, C.L., 1968, A series of stages in the embryonic develop-
    ment of Chelydra serpentina. J. Morph., 125:219-252.
Zehr, D.R., 1962, Stages in the normal development of the common
    garter snake, Thamnophis sirtalis sirtalis. Copeia, 1962:
    322-329.

# SARCOPTERYGII AND THE ORIGIN OF TETRAPODS

Henryk SZARSKI

Dept. of Comparative Anatomy, Jagellonian University

Krupnicza 50, 30-060 Krakow, Poland

## INTRODUCTION

The affinities of land vertebrates have been discussed many times, the history of the issue in itself being an interesting subject. It is not necessary at the moment to go into details; let us only recall that the important paper of Jarvik (1942), as well as the long series of later publications by this author (e.g. 1960, 1975), for a time persuaded the majority of zoologists that the different tetrapod groups arose from crossopterygians in an independent, polyphyletic manner. Jarvik's ideas have been criticised inter alii by Szarski (1962), Parsons and Williams (1963), Remane (1964), Reig (1964), Schmalhausen (1964), and others. They have also been defended by Stensio (1963), Herre (1964), Lehman (1968), Bjerring (1975), Nieuwkoop and Sutasurya (1976), among others.

Recent developments in zoological systematics have somewhat changed the mode of approach to the problem. However, let me state in advance that my conclusions do not differ greatly from my earlier views expressed in a paper written some 15 years ago (Szarski, 1962). I still believe that the ancestors of tetrapods were crossopterygian fishes and that the land vertebrates are a monophyletic group. What has changed then? Two developments seem to me important. Firstly, we have acquired much new information about the structure and the mode of life of fossil and living vertebrates; and secondly, many ideas and terms first proposed by Hennig (1966) have been assimilated into our reasoning. Bearing these two points in mind, I will discuss the following questions: methodology, tetrapod ancestors, and the problem of monophyletism.

## METHODOLOGICAL REMARKS

Although earlier in this volume Hecht and Edwards have discussed the methodology of phylogenetic inference, you may think,
therefore, that it is not necessary to return now to these
problems.  Nevertheless, I feel that I must give you a short
summary of my method of reasoning to explain why I consider some
arguments to be more relevant than others.  I believe that the
basic concept in comparative anatomy and biological systematics is
the notion of homology, defined in the evolutionary manner; viz,
"homologous characters of organisms are those which have been inherited from a common ancestor."  This definition has been accused
of circularity, but the accusation is unfounded (Hull, 1967). The
definition of homology may not be operational, it cannot be
treated as a criterion.  Criteria must be formulated independently
(Remane, 1956; Szarski, 1962; Ghiselin, 1966; Cracraft, 1967;
Campbell and Hodos, 1970; Bock, 1974).  It could, perhaps, be
argued that the concepts ought to be used in accordance with
their original meaning - homology was defined by Owen without any
evolutionary implications.  But the postulate of a permanent sense
of concepts cannot be accepted in the experimental sciences.  Even
in such an exact science as physics the content of concepts undergoes continuous change.  Nobody tries to return to the original
definitions of atoms, electrons, or protons.  New terms in science
emerge when a new detail of reality is perceived.  But it is never
recognised and understood properly until much later.  As knowledge
expands, the contents of the concept must change.  Hull (1967)
quoted Kaplan (1964): "The proper concepts are needed to formulate
a good theory, but we need a good theory to arrive at the proper
concepts...as the knowledge of a particular subject-matter grows,
our conception of the subject-matter changes."

It is true that the original concept of homology had no
evolutionary content.  But at present homology is an evolutionary
term and this cannot and should not be modified.  It is impossible to enumerate the many examples of the evolutionary use of
the concept of homology.  Let me quote only two recent publications.  Alexander (1975) in a textbook defines homologous
maxillae as "...derived from the ancestral maxilla by the
process of evolution."  And, three molecular biologists
(Markert, et al. 1975) discuss at length homologous protein
molecules, obviously assuming that homologous means inherited
from a common ancestor.

A biologist is not interested in the similarities of organisms in the way a mathematician considers the similarities of geometrical figures.  For a biologist a structure may be important as
an expression of function, as the effect of environmental pressure,
as the final result of morphogenetic mechanisms, or, finally, as a
heritage.  If we look on a structure as presumptive evidence of a

common ancestry, its possible functions must also be discussed, since only in this way may we evaluate the probabilities of two conflicting explanations: that the structure arose independently in different forms, or that it has been inherited from a common ancestor.

The concept of homology is, however, sometimes too vague. Attempts have been made to introduce more precise expressions such as "the uniquely evolved character" (Le Quesne, 1974). I think that after the publication of the English translation of Hennig's book (1966) the expressions coined by this author gained the widest acceptance. Let us recall some of Hennig's propositions. Two systematic units are called sister groups if they are more closely related to each other than to any other group. Homologous character states may be either plesiomorph or apomorph. An apomorph character state is present only in two sister groups and is inherited from the nearest common ancestor. A plesiomorph character state may be present in several systematic units and is inherited from distant ancestors. A group is called polyphyletic if its common characters result from convergence. It is paraphyletic if the common characters have only the rank of symplesiomorphies. Only groups distinguished by the possession of synapomorphies are truly monophyletic.

Hennig assumes that when a systematic unit is splitting into two, both resulting groups are dissimilar to the ancestor, although he admits that the speed of change is usually different in different groups and therefore one taxon may be more similar to the ancestor than the other. In Hennig's view it is impossible to classify fossil groups in accordance with his principles. Nevertheless, several of Hennig's followers have attempted to do so. Thus for instance Wahlert (1968) and Nelson (1969) use names for ancestral groups different from those of each of their descendants. There are strong logical arguments for such a stand; e.g., when we write, as we usually do, that the reptiles descended from amphibians, we are unwittingly suggesting that lizards and snakes are descendants of salamanders, or of frogs. Whereas in reality both Recent amphibians and Recent reptiles are very different from the original tetrapods which were the common ancestors of the two groups. One might argue that it is almost as right to say that reptiles descended from amphibians as the opposite - if we consider the Recent animals. The common ancestor of the two groups possessed no derived characters of either group and it could therefore be named at will an amphibian or a reptile. The ancestral tetrapod did not possess embryonic membranes, but neither had it the larval adaptations acquired by amphibians. Although this reasoning is quite sound some consequences of it may be difficult to accept.

According to Hennig (1966), it is necessary to give all groups of similar age a similar rank; e.g., groups which originated in the Cretaceous ought to be called families. Wahlert (1968) and Nelson (1969), who have attempted to introduce Hennig's ideas into the classification of vertebrates, had to coin many new names, such as Rhachipterygii, Archipterygii, and Kinokrania (Wahlert), or "Infraclass Neopterygii," "Division Reptiliomorpha" (Nelson). Here again I must quote Alexander (1975), who writes (p. 19): "A classification is not right or wrong, but it may be good or bad. The best classification is generally the most useful one." Taxonomic terms are the tools of zoological thought, few biological sentences can be formulated without them. It is therefore of the utmost importance that they be precise and unequivocal, and it is also necessary that they are stable and broadly accepted, not multiplied without absolute necessity. The human memory is limited and it cannot master an infinite and ever changing collection of words.

I will use some of Hennig's terms, now generally accepted, without, however, following all the consequences of his philosophy. It is equally justifiable to use the concept of homology not in accordance with Owen's original definition as to employ Hennig's concepts in a manner different from that in which they were originally proposed.

## THE SEARCH FOR A SISTER GROUP TO THE TETRAPODS

If we formulate the problem in this manner by implication we are suggesting that tetrapods are monophyletic. I believe this to be so, but I shall turn to the problem of monophyletism later. The reasons for assuming that tetrapod ancestors were fishes are obvious. It is similarly obvious that Osteichthyes are nearer to tetrapods than Chondrichthyes, and that the tetrapods' ancestors were not among actinopterygians, although the early actinopterygians differed little from early sarcopterygians. It can be said that although actinopterygians and tetrapods do share many homologous characters, these are probably in the majority symplesiomorphies, while a few are possibly convergent. It is impossible to name any character which could be regarded without doubt as synapomorph. On the contrary, several evolutionary tendencies in actinopterygians speak clearly against an affinity with tetrapods. These are, for instance, the specialization of paired fins as stabilizers in swimming, the loss of the oviduct in females, and of the connection between the Wolffian duct and the testis in males, the inability to accumulate urea, and the retention of a low internal osmotic pressure even in the marine environment. These characters can be studied only in Recent forms, but it may be assumed that they arose very early in the actinopterygian line.

We are left accordingly with dipnoans and crossopterygians. Various lineages of these two groups are already separated at their first appearance in the fossil record. Nonetheless, it is evident that they form two natural monophyletic units, though the reasons for uniting them into the Sarcopterygii are less convincing. Thomson and Campbell (1971) have described the presence of several features common to crossopterygians and to the early lungfish Dipnorhynchus. Fox (1965) has stressed the numerous similarities in the structure of the larvae of Dipnoi and Amphibia which are descendants of Crossopterygii. It could be argued, however, that all important characters uniting Dipnoi with Crossopterygii are either symplesiomorphies or are convergent. Let us recall some of them: the ability to accumulate urea (common also to elasmobranchs); the similarity in the structure of paired fins (according to Wahlert 1968 a primitive condition); and the tendency to develop a diphycercal tail fin. Hence it could be argued that the Sarcopterygii are a paraphyletic group, sharing only conservative features, inherited from distant common ancestors. It is nevertheless important to stress the striking contrast between the actinopterygian and the sarcopterygian lines. The use of a common name for dipnoans and crossopterygians therefore seems justified.

Crossopterygians share several features with tetrapods which probably have the rank of true synapomorphies, such as the basic arrangement of the dermal skull bones and the proximal parts of the paired appendages. The unique articulation within the braincase of crossopterygians is absent in tetrapods but traces of it are still present in Ichthyostega (Jarvik, 1952). A striking synapomorphy is the presence of internal nares and of a naso-lacrimal duct in tetrapods and in some crossopterygians. There is also a common tendency for the dentine to become folded in larger forms, and a presumed primitive character is the extension of the notochord well into the base of the braincase. No anatomical details are known which would argue against the affinity between tetrapods and crossopterygians.

It has already been mentioned that even the oldest crossopterygians form several distinct groups. In a recent review Andrews (1973) divides these fishes into five orders: Osteolepidida, Rhizodontida, Onychodontida, Holoptychiida and Coelacanthida. The advocates of tetrapod polyphyletism attempted to demonstrate an affinity between Urodeles and Porolepiformes (= Holoptychiida, Andrews), and the resemblance of the remaining land tetrapods (Eutetrapoda) to Osteolepiformes (= Osteolepidida, Andrews). It would be tempting for a supporter of monophyletism to take a further step and name one crossopterygian order as the common ancestor to all tetrapods, or at least as a sister group to them. I have the feeling, however, that in spite of many

interesting fossils, we know only a few side branches of the main
crossopterygian stock from which land vertebrates arose.  It is
easier to dismiss some crossopterygian groups from an affinity
with land vertebrates.  There are sufficient reasons for reject-
ing Coelacanthidida and Onychodontida for example, as possible
tetrapod ancestors.


## THE ORIGIN OF TERRESTRIAL ADAPTATIONS IN LAND VERTEBRATES

Let us turn now to the origin of features which enable a
tetrapod animal to live in land.  They can be divided into two
main groups.  The first consists of such characters which arose
during the life in water and may be therefore considered as preadaptations to a terrestrial environment.  The second would be
formed by such traits as were developed by animals on land.  Such
a classification is of course artificial, and probably most
anatomical adaptations are based on rudiments present already in
water and subsequently perfected on land.

The lungs were probably present before any other land preadaptations.  Very distant relatives of tetrapods - dipnoans and
actinopterygians - possess organs undoubtedly homologous to the
lungs of land vertebrates.  It is probable that even some extinct
elasmobranchs (Xenacanthids) and placoderms (Bothriolepis) had
homologous air sacs.  Other, analogous organs of aerial respiration arose several times among Teleostei.  A long list of fishes
able to use atmospheric oxygen is contained in the paper by
Gans (1970a), who discussed in detail the consequences for a fish
of the ability to use air for respiration.  Usually the oxygen
deficit, which arises easily in warm and stagnant water, was
quoted as the most important factor which would favour animals
able to use air from above the water surface (e.g., Carter, 1962).
Gans (1970a) drew attention to a different fact.  The quantity of
oxygen dissolved in water is always small, hence the amount of
energy needed to propel the necessary volume of water through the
gills is considerable.  For a fish which usually remains near the
surface it is therefore economical to take advantage of the avail-
ability of air.  Furthermore, the presence of a gas container inside the animal decreases its weight and is useful in many
circumstances.  The advantages of possessing lungs are many, so
that it is easy to understand their early appearance.

It can also be supposed that the main features of such
modifications of the circulatory system as are present in Recent
dipnoans, the division of the heart into two parallel pumping
systems, and the presence of special pulmonary arteries and veins,
were already present in aquatic tetrapod ancestors.

It is not too difficult to imagine the uses to which elongated fins, provided with a complicated skeleton and differentiated muscles, could have been put. They can help the animal in rapid and complicated movements, thus greatly increasing the ability to catch a moving prey. They allow a tortuous path to be followed between obstacles, and are useful when creeping in shallow water. We can again point to numerous teleost fishes which have acquired the ability to travel on land (e.g. Anabas, Clarias, Periophthalmus, etc.). The loss of the basicranial articulation also probably occurred in water and was caused by the elongation of the snout, which decreased the adaptive significance of cranial mobility (Thomson, 1967a).

The origin of internal nares could be a consequence of two different developments. One possibility is the migration of the external, excurrent nostril into the mouth cavity. There are examples of such a development in some teleosts (Echelidae and Ophichthyidae, Atz, 1952) and in elasmobranchs (Wahlert, 1966). Nowadays it is agreed that such was the origin of the internal nares of Dipnoi. Recently Medvedeva (1975) has argued strongly for a similar origin of the crossopterygian internal nares based on embryological observations. But there is also another possibility, namely the formation of a passage between the nasal and buccal cavities, independently of the external nares. We know that such a process occurred at least twice, and probably three times among teleosts. Internal nares are present in Astroscopus (Atz, 1952), in Ichthyoscopus (Mees, 1962), and in Gymnodraco acuticeps (Jakubowski, 1975). Astroscopus and Ichthyoscopus belong to the same family, Uranoscopidae, therefore they could have inherited the internal nares from a common ancestor, although in other investigated genera of Uranoscopidae choanae are absent. But Gymnodraco, which belongs to the Bathydraconidae, is a very distant relative of Uranoscopus and the supposition that its internal nares are homologous to those of Uranoscopus seems untenable.

Jakubowski (1975) failed to find internal nares in two other species of Bathydraconidae, or in seven related genera, and described the probable course of events which led to the formation of internal nares in Gymnodraco acuticeps. This is an abyssal species for which olfactory information is very important. The nasal cavities in Bathydraconidae are large and provided with accessory nasal sacs. In some species the mouth cavity is separated from the accessory nasal sacs by thin, flexible membranes only. Thus pressure changes caused in the mouth cavity by respiratory movements can influence the volume of the nasal sacs and can contribute to water movement through the nasal cavities. The formation of a slit-like opening in the dividing membrane further increased the stream of water through the nose.

The opening functions as an inlet valve allowing water to pass
from nose to mouth but is closed by an increase in pressure in the
mouth cavity (Fig. 1).  Such a one-way valvular action of the
membrane is a necessity, otherwise the internal nares would
interfere with the respiratory stream of water passing through
the buccal cavity into the pharynx.

If the opening between the nose and the mouth cavities arose
twice in teleosts it could also have occurred in crossopterygians,
which probably also relied strongly on olfactory information.  In
a fish partially buried in the bottom mud the internal nares
could have developed later into a route for respiratory water, as
in Astroscopus. But even if the choanae in crossopterygians remained too narrow for the respiratory water current at the time
when they made their first land excursions, they could be transformed easily as a convenient route for less dense air, which
contained a much larger amount of oxygen.

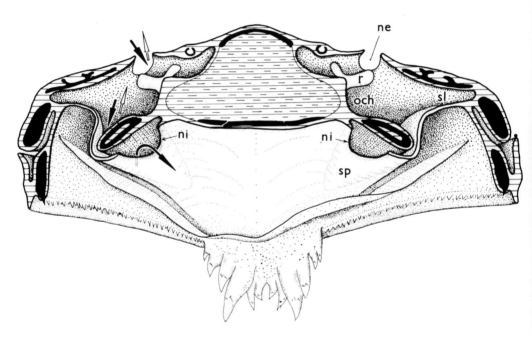

Fig. 1.   Cross-section through the nasal region of Gymnodraco
          acuticeps, after Jakubowski, 1975.  Arrows represent
          water currents.  Abbreviations: ne = external nostril;
          ni = internal nostril; och = olfactory chamber;
          r = olfactory rosette; sl = lacrimal accessory nasal
          sac; sp = palatal accessory nasal sac.

The eyelids, structures indispensable for a land vertebrate, could also have originated in water. Gilbert and Oren (1964) recall that a movable "nictitating membrane" was described by Muller in several selachians as early as 1843. Its function is not known; it may defend the vulnerable cornea against external parasites. The tetrapod eyelids are probably not homologous structures, but the presence of eyelids in one fish group suggests that they could arise in another.

Bergeijk (1966) pointed out that many fishes have airpockets near the membranous labyrinth. Air has a different compressibility than water, therefore the air bubble can act as an amplifier of sound. Hence it can be assumed that such a rudiment of the middle-ear already existed in early crossopterygians. Bergeijk suggests that a cavity described by Jarvik in the skull of Eusthenopteron contained an air bubble.

We have discussed the different features which could have been present in tetrapod ancestors while they were still aquatic, before any selective forces acting on land animals came into action. Further adaptations arose in the new environment. Let us turn first to the factors which could induce a fish to make terrestrial excursions. Some have already been pointed out: a fish living in shallow water (small fishes live in shallows in order to avoid cannibalism) can be trapped in a temporary pool when the only solution is to search for water by crawling on land. It seems that in Paleozoic times one of the safest places was the land, accessible especially during a cool and damp night. In a cool climate basking partially or completely out of water could favour a higher metabolic level. Finally, we must remember that the majority of freshwater habitats are temporary and therefore every freshwater organism must have the ability to emigrate to new surroundings in order to survive. The chance of success lies in exploration which, though dangerous, puts a very high premium on the individuals able to find a new, empty habitat (Lomnicki, 1969). All these factors favoured those animals which tried to leave the water temporarily.

Recent conditions are probably very different; nevertheless it is well to remember that many fishes are at present making excursions onto land. It is well known that the ability to walk on land has made the Asian siluroid immigrant Clarias lazera a pest in some American freshwaters.

Frequent land visits probably influence the structure of paired extremities first. The fins possessed by crossopterygians are very inefficient and easily damaged on land. Thus every tendency leading to a reduction of the perishable marginal membrane and the formation in its place of supporting struts was

strongly favoured by natural selection.  Thomson (1967b) has
suggested that the anterior pair was initially more important
than the posterior, because the pectoral appendages helped to
raise the body above the ground and thus facilitated breathing.

Another early improvement was the ossification of the
vertebral column.  It is most probable that the common ancestor
of crossopterygians and tetrapods had a functional notochord.
The ossification of the centra was shaped by different forces
according to the mode of life of the animal.  Among fishes the
need to improve the resistence of the axial skeleton was caused
by an increase in muscular force mostly correlated with the
general enlargement of the animal.  Among land vertebrates the
most vital task was to carry the heavy trunk above the ground.
As different forces influenced the emerging centra they had a
different structure from the very beginning. Williams (1959)
demonstrated the fallacy of Gadow's attempt to connect in one
scheme the various incomplete centra of early fossil or conservative vertebrates with the anlagen of ossification in Recent
advanced forms.  Gadow's attempt disregarded the possibility of
convergent evolution of the centra and overrated the rule of recapitulation of phylogeny in ontogeny.

Probably even before their first land excursions the tetrapod
ancestors used blood vessels in the skin as an accessory respiratory surface, through which they could get rid of excess $CO_2$.  The
primitive lung was efficient in supplying the blood with oxygen
but it was much less capable as a carbon dioxide eliminating
organ.  In contrast, water loses the ability to supply oxygen
much earlier than the $CO_2$ absorption potential.  Superficial vaso-
dilatation occurs during respiratory distress in Protopterus
(Lenfant and Johansen, 1968; Thomson, 1969).  Gas exchange through
the skin predominates also in some Recent aberrant teleosts
(Jakubowski and Rembiszewski, 1974).  It is often stated that the
scales prevent gas exchange through the skin.  But only scales
covered by enamel, or some enameloid material, are naked, the
others lie under a well-vascularized layer of epidermis and
therefore do not interfere in gaseous exchange.

The early mode of lung ventilation was probably similar to
that reported for Recent dipnoans (Thomson, 1969).  The animal
gulps air into the mouth and then forces it into the lungs by increasing the pressure in the mouth cavity and the pharynx.  Air
movement causes less friction than the movement of water and the
amount of oxygen in a volume of air is much larger than in the
same volume of water.  Thus inspiration through the nasal cavity
could be sufficient and would be more economical than the
laborious mouth opening, which could disappear on land.  The internal naris had, however, to possess a valve, preventing the

back-flow of water so as to allow respiration through the gills and the forcing of air to the lungs. To make the exhalation of air through the nostrils with closed mouth possible a change was needed in the mechanism of the nasal valve. Lung ventilation by a pressure pump was a primitive feature. Gans (1970a, 1970b) gives a convincing argument that it was probably helped very early on by contraction of thoracic muscles which rotated the ribs outwards increasing the thoracic volume and thus lessening the pressure necessary to inflate the lungs. The same author specified the advantages of suction pump breathing. Let us stress here only one: this manner of lung ventilation allows for the elimination of $CO_2$ by the organ, and therefore allows the skin to become a better insulator and isolator of the organism. If the oxygen and the carbon dioxide exchanges are proceeding simultaneously in the same capillary net it is possible to increase the efficiency of both processes by mutual influence. It is also more economical to achieve both aims by driving the blood through a single network of capillaries.

Why then has cutaneous respiration been retained and probably even perfected in Lissamphibia? One supposition is founded on the connection between pressure breathing and vocalization in frogs (Gans, 1970b). Although this mechanism is important, it could explain only the mode of breathing of Anura, whereas all Lissamphibia use the same method. A more general answer lies in the low metabolic rate and the small size of these animals. In such circumstances cutaneous respiration alone may be sufficient, and then it is the cheapest method of gaseous exchange. The lungs are used only on special occasions, playing the role of reserve organs. It was observed recently that amphibians are able to ventilate their lungs even when the mouth is held artificially open (Bentley and Shields, 1973; Lillywhite, 1975). It must be concluded that they possess an imperfect aspiration pump.

We have suggested above that the aquatic tetrapod ancestors already possessed structures allowing at least a partial division of two blood streams in the heart. If, however, we turn to the organization of the circulatory system of Recent vertebrates, we find three different patterns: amphibian, reptilo-avian, and mammalian. They seem to be radically different. The amphibian type may be explained as a secondary adaptation to the extensive use of cutaneous respiration. But the profound differences between avian and mammalian organizations suggest that the common ancestor of these groups still had a very primitive organization both of heart and arterial arches. This observation is an argument in favour of the importance of cutaneous gas exchange in early tetrapods. The large size of Ichthyostega was regarded as an obstacle to the use of skin respiration (Gans, 1970a), but probably Ichthyostega is an exceptionally large descendant of a

much smaller prototetrapod.  Many problems of the origin of reptiles became easier to explain when Carroll (1970) demonstrated
that the first reptiles were small animals, not exceeding 100 mm
in snout-vent length.  The earliest known vertebrate, described
by Bockelie and Fortey (1976) from the early Ordovician was a
tiny animal.  Such small fossils are very rarely found, so we are
led to imagine that the main stem of vertebrates consisted of
large animals - about 1 m long.  This is a result of a biased
sample. If it is accepted, in accordance with Panchen (1972),
that the rhipidistian-tetrapod transition was characterized  by
size reduction followed by a sudden increase, then the hypotheses
on the processes of anatomical reconstruction need to be modified.
Thus the importance of cutaneous gas exchange in early tetrapods
seems greater, while the probability of ossification of vertebral
centra in these animals is smaller, etc.

Another detail of the circulatory system which probably arose
on land was the increase in the number of lymph hearts.  The
pressure of water helps to return the lymph to the veins in
aquatic animals.  For an animal surrounded by air, the outside
pressure gives no help and the numerous lymph hearts are a
necessity.

As late changes which were accomplished by forms that spent
most of their life on land we might consider the definitive change
of the hyomandibular into the stapes and the development of the
tympanum, recently discussed by Shishkin (1973), as well as the
organization of orbital glands, the reduction of lateral line
receptors, etc.  Among the latest terrestrial characters was also
the mobile muscular tongue, which could be of no use for the
collection of food in water.

Finally, several characters of the Recent Lissamphibia were
probably developed after the separation of this line from the
reptilian one.  Here belong the operculum-opercularis muscle
complex, the amphibian skin glands, the many adaptive characters
of larvae, and the mechanism of larval metamorphosis.

## MUTUAL INFLUENCE OF DIFFERENT LAND ADAPTATIONS
## BY POSITIVE FEEDBACK

When the fish which regularly emerged onto land obtained such
modifications of paired appendages that the manner of crawling on
land was improved, it was able to make longer excursions and to
stay longer outside the water.  Thus the pressure of natural
selection for a stronger axial skeleton, for better skin vascularization, for better protection of the eyes against desiccation,
etc., was increased.  A better vascularized skin, or a stronger

vertebral column further augmented the ability to walk on land and hence further increased the selective forces which worked toward the modification of the paired appendages. All the re-constructions which changed the fish into a land animal were interdependent, therefore every progressive improvement in one organ increased the selective pressures on the remaining organs and functions, pressing for further change in the same direction. Thus a multidimensional network of positive feedback circuits was formed and the process of change became self-accelerating. I believe that each important change in the structure of organisms such as the adaptations to flight of birds, pterosaurs, and bats, the jumping ability and the loss of tail in anurans, or the modified structure of the pharyngeal jaw apparatus of cichlid fishes (Liem, 1975), was at least partially driven by a network of positive feedback reactions. I have proposed that such develop-ments should be called chain-evolutionary processes (Szarski, 1967, 1971).

The most important features of chain-evolutionary events are the speed with which the different structural changes occur and the mutual interdependence of the whole process. A very small initial difference may direct the whole chain of events in a new course, as for instance the unknown fact which directed the penguins to use their wings for swimming, and in consequence deprived them of the ability to fly.

## MONOPHYLETISM VERSUS POLYPHYLETISM OF TETRAPODS

The many Recent fishes which are capable of using air for respiration and who are able to walk on land have been mentioned above. It must be inferred that during the millions of years be-tween the Paleozoic and present times numerous other fish species were exploring similar ecological niches and thus were potential ancestors of new lines of land vertebrates. We know, however, that this possibility was never realized. Paleontologists are unanimous in accepting that no tetrapod group is younger than the Devonian. This is a striking fact, which could be explained by two hypotheses. First, when a fish emerges on land, at the present time or when it came on land at any time after the Paleozoic, it must compete with tetrapods which had been living there for a long time and were accordingly better adapted to the terrestrial environment. This does not suggest that there are no empty niches available on continents. It states only that there is no free place on land for an animal built on the fish plan and evolving toward further adaptation to land conditions. Second, Medvedeva (1975) suggested that the various preadaptations to land conditions which were possessed by crossopterygians occurred simultaneously only once. She writes that there is only a

minimal chance that all these preadaptations will appear again in
a single form.

   Probably both assumptions are sound:  most fishes which
occasionally walk on land do not possess all the preadaptations
of crossopterygians, and they find the land niches already filled.
Nevertheless, these considerations do not diminish the possibility
that, more or less simultaneously, two or more crossopterygian
species gave origin to two or more separate lines of tetrapods.
Such a possibility is defended by Jarvik (e.g. 1960), who
especially suggests that the majority of tetrapods are closely
related (Eutetrapoda), and that the urodeles form a separate
branch of land vertebrates.  It is impossible to summarize here
the very extensive discussions.  Most of Jarvik's arguments are
based on differences in the nasal anatomy of crossopterygians,
which in his opinion correspond to differences between Anura and
Urodela (Jarvik, 1942).  Hence, it is well to recall that Jarvik's
conclusions were later opposed by Kulczycki (1960), Thomson (1964),
Jurgens (1971), and Medvedeva (1975), who studied in detail the
nasal anatomy and embryology of lower vertebrates.  Later Jarvik
(e.g. 1975) extended his studies to other organs and came to
similar conclusions, but even now he has not convinced many
zoologists.

   It is not easy to explain in a few sentences how such important differences of interpretation are possible as those existing between the Swedish school and its opponents.  The descriptions of anatomical details are mostly unopposed; both sides
often quote the observations of their antagonists but differ
sharply in the conclusions or inferences drawn from them.  The
approach of polyphyletists is in a way similar to that of
numerical taxonomists.  They describe and compare as many anatomical details as possible, but are reluctant to build hypotheses
on the functions of observed structures and refrain from the
discussion of possible selection forces which could have induced the changes. Against such reluctance one may quote the
following sharp words of Trueb (1973): "... systematics have
utilized characters without fully understanding their variation,
significance, and distribution; without this information, attempts
to designate primitive and derived states border on folly or, at
best, educated guesswork." Numerical taxonomists endeavour to
examine an unbiased sample of characters, whereas polyphyletists
conduct an open search for arguments confirming their assumptions.

   In contrast, the monophyletists focus their attention on a
more restricted number of structures, but try to formulate
hypotheses attempting to evaluate the direction of selective
driving forces.  A large amount of free speculation characterizes
the approach of many monophyletists.  They concentrate on the

search for characters which could be regarded with some degree of certainty as synapomorphies.

Let us enumerate the principal similarities between Urodela and the two other Recent amphibian orders which are regarded as synapomorphies by Szarski (1962), Parsons and Williams (1963), and other authors. Extensive lists of larval similarities and dissimilarities are to be found in Fox (1965). Teeth are divided into two parts by a zone of weakness in which a limited amount of movement is possible in some forms (for exceptions see Estes 1969 ). In Anura and Urodela the fenestra ovalis of the membranous labyrinth is closed by two bones, operculum and plectrum, instead of one as in the remaining tetrapods. The operculum is connected by a special muscle to the shoulder girdle (Monath, 1965). This feature is absent in Apoda, but these animals have no shoulder girdle, and the ontogeny suggests that their middle ear derives from a structure similar to that of the two other orders. All Lissamphibia have very characteristic skin glands and there is an additional sensory receptor in the membranous labyrinth, the papilla amphibiorum. Several peculiarities of karyology are common to all Lissamphibia and separate them from other vertebrates (Morescalchi, 1973, 1975). The retina contains green rods (but Apoda are blind). There are two occipital condyles. The structure of the palate and the skull fenestration in Anura and Urodela are similar, contrasting with reptiles. In the body cavity of Lissamphibia lie the characteristic fat-bodies developed from the germinal fold in the embryo. The nasal cavities have a similar development in Anura and Urodela. Urodela and Apoda possess characteristic polystychous embryonic dentition, which is probably inherited from some early ancestors (Parker and Dunn, 1964). It may, however, be a symplesiomorphy.

The affinities of the Lissamphibia must be judged by weighing the evidence for three possibilities. Characters uniting them were (1) inherited from very distant fish ancestors (symplesiomorphies) or (2) inherited from an ancestor which was a land animal, or (3) result from convergent evolution. Since the second possibility seems the most probable, we regard Apoda, Urodela, and Anura as related. Recently Carroll and Currie (1975) have argued that Urodela and Anura are closely related to each other, being two sister groups, while Apoda are more distant, possibly descending from microsaurs. The most convincing argument presented in support of this hypothesis is the probability that the strong roofing pattern of the skull in Apoda is not a secondary development but a direct inheritance from primitive tetrapods.

The arguments in favour of a common descent of Lissamphibia

and Reptilia from an ancestral tetrapod are less numerous.  The
differences between these two lines are indeed important.  Some
have been specified above.  On the other hand, there are also
several features which are common to all tetrapods but are
absent in crossopterygians and thus could be true synapomorphies,
uniting all land vertebrates.  Most important are the following:
the skeleton of the paired appendages; the structure of the
pelvic girdle; the presence of a sternum; the absence of the
posterior series of skull bones; the presence of the orbital
glands; the nasolacrimal duct; Jacobson's organ; and Bowman's
glands in the olfactory epithelium.  The anurans and most reptiles
possess in common the tympanum and the middle ear.  The absence of
the tympanic cavity in Apoda, Urodela, and some reptiles probably
results from a secondary loss.

    Some of these characters may have originated by convergence.
Jarvik (1960) has stressed that the skeleton of the paired fins of
crossopterygians already had the basic elements of land appendages
and therefore the adaptation to land movement could progress in
several forms in a similar way.  It is also true that the basic
structure of the fore- and hind limb pairs is very similar,
whereas neither the forelimb nor the hind limb is ancestral to the
other.  This could also be regarded as an example of convergent
evolution.

    Let us consider, however, vertebral centra, scales or other
meristic characters.  Neither the vertebral centra nor the scales
are ancestral to each other.  Nevertheless, it would be difficult
to assume that the similarities between single meristic characters
of an organism are consequences of convergence only.  It is more
probable that the genotype of the animal accumulated genes with
the ability to form a fundamental morphogenetic field, which can
be multiplied and put into action in several locations in the
embryo. Thus, for instance, it seems reasonable to assume that the
basic similarities between the paired fins and the second dorsal
and the anal fins of Latimeria result from the application of
similar developmental mechanisms in six different locations.

    However, even if we admit the possibility of a convergent
origin for the skeleton of the appendages, some other similarities
remain rather striking, such as the presence of the sternum, the
structure of the pelvic girdle, and the connections between the
orbit and the nasal cavity.  Finally, the nature of the evolutionary process suggests a monophyletic origin for all tetrapods.

    The advocates of a polyphyletic tetrapod origin are impressed
by the antiquity of differences between Recent taxa.  When new
fossil animals are found they can usually be classified without
difficulty into the already known groups and it is unusual to find

convincing examples of "connecting links," such as <u>Archaeopteryx</u>.
What is true for tetrapods as a whole is also true for smaller
units such as anurans, snakes, or cichlid fishes.  Nevertheless,
the fact that we cannot find the fossils which could be regarded
as the common ancestors of these groups is not a sufficient proof
that these units are polyphyletic assemblages and gained all
their similarities by convergence.  It seems more probable that
these groups are truly monophyletic in the strict sense of the
word; that is each one of them derives from a single ancestral
species.

I assume that every larger animal group has passed through a
period of rapid chain evolution.  Since all the anatomical and
functional changes to which the organisms are submitted during
such a process are interdependent, the final similarity can be
achieved only when the original starting forms are identical and
when the whole process is propelled by identical selective forces.
It is, however, improbable that such an event could be duplicated
in every detail.  Most species have a limited range and, should
a similar evolutionary process run more than once in the same
locality, natural selection would favour divergence and not convergence of forms.  Had the processes started simultaneously on
distant continents they would be under the influence of different
selective forces.  When several lines starting from different
ancestors are shaped even by similar adaptations they retain the
fundamental distinguishing features, as, for instance, the
ant-eaters, the pangolins, and the aardvark.  These mammals are
striking examples of convergence, but despite similarities in
several features, the animals are distinguished by profound
differences and it is therefore assumed by all systematists that
their common ancestor was not a mammal specialized for the eating
of ants and termites and that their similarities are not
synapomorphies.  The differences between the groups of primitive
tetrapods are much less important than those which divide the
several groups of mammals adapted to a diet consisting of ants
and termites.

The self-accelerating nature of chain evolution also favours
the supposition that every divergent group descends from a single
species.  The form which has first acquired an important adaptation to a new mode of life is submitted to a network of mutually
reinforcing pressures, under which it undergoes a rapid reconstruction.  The presence of thos organism prevents similar forms
from entering a similar adaptive niche.  It is therefore unlikely
that two related species could evolve simultaneously in the same
direction.  After achieving an important adaptation to a new mode
of life, the derived form will increase its adaptive ability, will
invade a greater range; it may also diversify into different
niches, giving rise to an evolutionary radiation.  As a result,

the common ancestor of the new group will shortly disappear,
suppressed by the derived species.  One of these will probably
make a further step in the original direction, acquiring a
second important modification.  Then its descendants will in turn
diversify, pressing out of their niches the less perfect species
group derived from the first radiation.  Such events may be
repeated several times (Fig. 2).  In the Recent fauna we mostly
see only the representatives of the latest evolutionary radiations.  Some groups which could be traced to older radiations may
persist longer if they entered special niches, where there is no
competition from the descendants of subsequent radiations.  Thus,
for example, penguins and ostriches are living representatives of
ancient birds; Recent lissamphibians derive from some early
radiation of tetrapods; Apoda and Anura persist because they have
acquired peculiar adaptations.  It is less easy to explain the
persistence of Urodela, but Schmalhausen (1964) suggests that the
fundamental adaptation of these animals lies in their ability to
live in a cold mountain environment, where the competition from
reptiles is small.

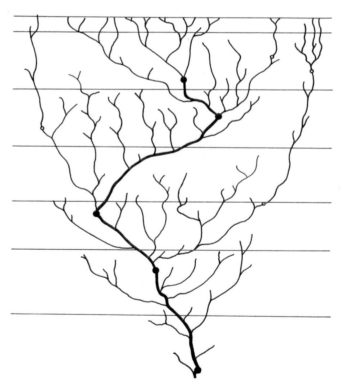

Fig. 2.   Model of a phylogenetic tree.  Black dots represent
          general adaptations, open dots represent special
          adaptations.

## DISCUSSION AND CONCLUSIONS

We can now summarize the probable course of early tetrapod history. The ancestor of the group was some small or medium-sized fish, which - if found as a fossil - would be classified among crossopterygians. It possessed several characters which although acquired in water were also useful on land. Its habit of going ashore released a process of chain evolution under which it became profoundly modified. Its fins were changed into limbs, the ventral part of the shoulder girdle was fortified by a sternum, and the pelvic girdle was greatly enlarged and became attached to one of the vertebrae. The vertebral centra developed ossifications. The posterior series of skull bones was lost. The spiracular gill slit was changed into the middle ear cavity and the hyomandibular into the stapes, while the orbit acquired specialized glands and a connection with the nasal cavity through the nasolacrimal duct. The animal relied on lungs for the supply of oxygen but rid itself of carbon dioxide through the skin.

Such a form gave rise to an evolutionary radiation. Several lines of descendants increased in size, and these left most of the fossils known to present paleontologists. One line of small animals acquired two kinds of poison glands in the skin and pedicellate teeth. They were the ancestors of the Lissamphibia; possibly here belongs Doleserpeton, which, as the only Permian amphibian possesses pedicellate teeth (Bolt, 1969). These animals perfected cutaneous respiration and shortened the period of metamorphosis by tying in this process with thyroid activity. Some descendants of these forms developed various larval adaptations and the operculum-plectrum muscle complex. A few increased in size (e.g. Andrias, Amphiuma).

Another line of small animals lost the larval life period and laid eggs in damp sites on shore. The embryos lacked functional gills and oxygenated the blood in a capillary network which was spread on the yolk surface around the embryo. The early stages were in danger due to the possibility of water loss. Their kidneys soon began to produce uric acid, which was accumulated in the closed cloaca, expanding into the allantoic sac. From the urine the walls of the allantoic sac absorbed water, which could be used several times. Since the expanded and vascularized wall of the allantois also acted as a respiratory surface, the embryo could be sheltered by the amniotic folds of the embryonic membranes closing over its dorsal surface. Later the animals obtained another valuable faculty, the ability to produce a tough egg-shell from glands in the posterior segment of the oviduct. Simultaneously they had perfected suction breathing and cornification of the epidermis. Such animals ought to be regarded as reptiles. They rapidly split into two or possibly more lines, the descendants of which are the Recent turtles, diapsid reptiles with birds, and mammals.

What value has such a hypothetical story? Some biologists will perhaps treat it with contempt, as nothing more than an un-provable speculation. I am less pessimistic. We are witnessing at present a rapid development of biochemical phylogenetics, whose future expansion seems sure. We can hope that many current hypotheses will be confirmed or abolished in the near future. It is well to remember that the results of Salthe and Kaplan (1966), and of Wilson et al. (1974), have confirmed the views of paleontologists and morphologists. Biochemical information taken alone is not the ultimate authority on the affinities of organisms. Haemoglobin occurs in many unrelated taxa, and the peculiar tetro-dotoxin is present in some teleost fishes and in a few newts (Fuhrman, 1967).

Nevertheless, the perspectives opened up by the study of proteins in various organisms are enormous. The comparison of its future results with the reconstruction of the past achieved by paleontologists and morphologists will be very valuable for both sides. It is indeed fortunate that we possess two competing views on the origin of tetrapods. We shall be able to evaluate the merits of the two approaches by comparing their results with the observations of biochemists. This will be a useful test of our methodologies.

## SUMMARY

Tetrapod ancestors, if known, would be classified among crossopterygians. The differences in structure between a fish and a tetrapod have been acquired either as adaptations to a peculiar aquatic environment and later perfected on land, or as adaptations to a new mode of life, which made their appearance only after the animals were terrestrial. The features of tetrapods are examined from this point of view. There are convincing reasons from comparative anatomy which favour the acceptance of Lissamphibia as a natural unit, and less numerous but also strong motives for admitting the monophyletism of all tetrapods. It is demonstrated that there is a high probability that every important systematic group descends from a single species. Such a course of events is a consequence of evolutionary mechanisms. A future comparison of the hypothetical history of early tetrapods built on the evidence of morphology and paleontology with the results obtained by biochemical studies will have very great value for both fields of research.

REFERENCES

Alexander, R. McN., 1975, The Chordates.  Cambridge University
    Press, London.
Andrews, S.M., 1973, Interrelationships of crossopterygians, in
    "Interrelationships of Fishes" (P. H. Greenwood, R. S.
    Miles, and C. Patterson, eds.), Suppl. Zool. J. Linn. Soc.
    53:138-177.
Atz, J.W., 1952, Narial breathing in fishes and the evolution
    of internal nares, Quart. Rev. Biol. 27:366-377.
Bentley, P.J., and Shields, J.W., 1973, Ventilation of toad
    lungs in the absence of the buccopharyngeal pump, Nature,
    Lond. 243:538-539.
Bergeijk, W.A.v., 1966, Evolution of the sense of hearing in
    vertebrates, Amer. Zool. 6:317-377.
Bjerring, H.C., 1975, Contribution a la conaissance de la
    neuro-epiphyse chez les urodeles et leurs ancetres poro-
    lepiformes, avec quelques remarques sur la signification
    evolutive des muscles stries parfois presents dans la re-
    gion neuro-epiphysaire des mammiferes, Coll. int. C.N.R.S.
    218:231-256.
Bock, W.J., 1974, Philosophical foundations of classical
    evolutionary classification, Syst. Zool. 22:375-392.
Bockelie, T., and Fortey, R.A., 1976, An early Ordovician
    vertebrate, Nature, (Lond.) 260:36-38.
Bolt, J.R., 1969, Lissamphibian origins: possible protoliss-
    amphibian from the Lower Permian of Oklahoma, Science
    116:888-891.
Campbell, C.B.G., and Hodos, W., 1970, The concept of homology
    and the evolution of the nervous system, Brain, Behav.,
    Evol. 3:353-367.
Carroll, R.L., 1970, Quantitative aspects of the amphibian-
    reptilian transition, Forma et functio 3:165-178.
Carroll, R.L., and Currie, P.J., 1975, Microsaurs as possible
    apodan ancestors, Zool. J. Linn. Soc. 57:229-247.
Carter, G.C., 1962, Tropical climates and biology, Smithsonian
    Rep. 1961:429-443.
Cracraft, J., 1967, Comments on homology and analogy, Syst.
    Zool. 16:356-359.
Estes, R., 1969, Prosirenidae, a new family of fossil salamanders,
    Nature, (Lond.) 224:87-88.
Fox, H., 1965, Early development of the head and pharynx of
    Neoceratodus with a consideration of its phylogeny, J.
    Zool. 146:470-554.
Fuhrman, F.A., 1967, Tetrodotoxin, Scient. Amer. 217(2):60-71.
Gans, C., 1970a, Strategy and sequence in the evolution of the
    external gas exchangers of ectothermal vertebrates, Forma
    et functio 3:61-104.

Ghiselin, M.T., 1966, An application of the theory of definition
    to systematic principles, Syst. Zool. 15:127-130.
Gilbert, P.W., and Oren, M.E., 1964, The selachian nictitans
    and subocular fold, Copeia 1964:534-535.
Hennig, W., 1966, Phylogenetic systematics, University of
    Illinois Press, Urbana.
Herre, W., 1964, Zum Abstammungsproblem von Amphibien und
    Tylopoden sowie uber Parallelbildungen und zur Polyphy-
    liefrage, Zool. Anz. 173:66-98.
Hull, D.L., 1967, Certainty and circularity in evolutionary
    taxonomy, Evolution 21:174-189.
Jakubowski, M., 1975, Anatomical structure of olfactory organs
    provided with internal nares in the Antarctic fish
    Gymnodraco acuticeps Boul. (Bathydraconidae), Bull. Acad.
    Polon. Sci. Ser. Sci. Biol. 23:115-120.
Jakubowski, M., and Rembiszewski, J.M., 1974, Vascularization
    and size of respiratory surfaces of gills and skin in the
    Antarctic fish Gymnodraco acuticeps Boul. (Bathydraconidae),
    Bull. Acad. Polon. Sci. Ser. Sci. Biol. 22:305-313.
Jarvik, E., 1942, On the structure of the snout of crossoptery-
    gians and lower gnathostomes in general, Zool. Bidrag
    Uppsala 21:235-675.
Jarvik, E., 1952, On the fish-like tail in the ichthyostegid
    stegocephalians, Medd. Gronland 114:1-90.
Jarvik, E., 1960, Theories de l evolution des vertebres, Masson
    et Cie. Paris.
Jarvik, E., 1975, On the Saccus lymphaticus and adjacent
    structures in osteolepiforms, anurans and urodeles, Coll.
    int. C.N.R.S. 218:191-211.
Jurgens, J.D., 1971, The morphology of the nasal region of
    Amphibia and its bearing on the phylogeny of the group,
    Ann. Univ. Stellenbosch, 46, Ser.A., No.2:1-146.
Kaplan, A., 1964, The conduct of inquiry, Chandler Publ. Co.
    San Francisco.
Kulczycki, J., 1960, Porolepis (Crossopterygii) from the lower
    Devonian of the Holy Cross Mountains, Acta Paleont. Polon.
    5:65-104.
Lehman, J.P., 1968, Remarques concernant la phylogenie des
    Amphibiens, in "Current Problems of Lower Vertebrate
    Phylogeny (T. Orvig, ed.), Nobel Symp. 4:307-315.
Lenfant, C., and Johansen, K., 1968, Respiration in the African
    lungfish, Protopterus aethiopicus. I. Respiratory
    properties of blood and normal patterns of breathing and
    gas exchange, J. exp. Biol. 49:437-452.
Le Quesne, W.J., 1974, The uniquely evolved character concept
    and its cladistic application, Syst. Zool. 23:513-517.
Liem, K.F., and Osse, J.W.M., 1975, Biological versatility,
    evolution and food resource exploitation in African cichlid
    fishes, Amer. Zool. 15:427-454.

Lillywhite, H.H., 1975, Physiological correlates of basking in amphibians, Comp. Biochem. Physiol. 52A:323-330.

Lomnicki, A., 1969, Individual differences among adult members of a snail population, Nature, (Lond.) 223:1073-1074.

Markert, C.L., Shaklee, J.B., and Whitt, G.S., 1975, Evolution of a gene, Science 189:102-114.

Medvedeva, J.M., 1975, Olfactory organ in amphibians and its phylogenetic significance, Trudy Zool. Inst. A.N. U.S.S.R. 58:1-174. (In Russian).

Mees, G.E., 1962, Occurrence of internal nares in the genus Ichthyoscopus (Pisces, Uranoscopidae), Copeia 1962:162.

Monath, T., 1965, The opercular apparatus of salamanders, J. Morphol. 116:149-170.

Morescalchi, A., 1973, Amphibia, in "Cytotaxonomy and Vertebrate Evolution (A.B. Chiarelli and E. Capanna, eds.), Academic Press, London and New York, pp. 233-348.

Morescalchi, A., 1975, Chromosome evolution in the caudate amphibia, Evol. Biol. 8:339-387.

Muller, J., 1843, Untersuchungen uber die Eingeweide der Fische. Schluss der vergleichenden Anatomie der Myxinoiden. Abh. K. Akad. Wiss. Berlin 1843:109-170.

Nelson, G.J., 1969, Gill arches and the phylogeny of fishes, with notes on the classification of vertebrates, Bull. Amer. Mus. Nat. Hist. 141:475-552.

Nieuwkoop, P.D., and Sutasurya, L.A., 1976, Embryological evidence for a possible polyphyletic origin of the recent amphibians, J. Embryol. exp. Morphol. 35:159-167.

Panchen, A.L., 1972, The interrelationships of the earliest tetrapods, in "Studies in Vertebrate Evolution (K.A. Joysey, and T.S. Kemp, eds.), Oliver and Boyd, Edinburgh, pp. 65-87.

Parker, H.W., and Dunn, E.R., 1964, Dentitional metamorphosis in the Amphibia, Copeia 1964:75-86.

Parsons, T., and Williams, E.E., 1963, The relationships of modern amphibia: a re-examination, Quart. Rev. Biol. 38:26-53.

Reig, O.A., 1964, El problema del origen monofiletico o polifiletico de los Anfibios, con consideraciones sobre las relaciones entre Anuros, Ameghiniana 3:191-211.

Remane, A., 1956, Die Grundlagen des naturlichen Systems, der vergleichenden Anatomie und der Phylogenetik. Theoretische Morphologie und Systematik. I. Zweite Aufl. Akad. Verlags-gesell. Geest and Portig, Leipzig.

Remane, A., 1964, Das Problem Monophylie-Polyphylie mit beson-derer Berucksichtigung der Phylogenie der Tetrapoden. Zool. Anz. 173:22-49.

Salthe, S.N., and Kaplan, N.O., 1966, Immunology and rates of evolution in the Amphibia in relation to the origin of certain taxa. Evolution 20:603-616.

Schmalhausen, I.I., 1964, Origin of the land vertebrates. "Nauka," Moskva, (In Russian).

Shishkin, M.A., 1973, The morphology of the early amphibians and some problems of lower tetrapod evolution. Trans. Paleont. Inst. A.N. U.S.S.R. 137:1-256, (In Russian).

Stensio, E., 1963, The brain and the cranial nerves in fossil craniate vertebrates. Skr. norske Videns. Akad. Oslo, I. Mat. Naturw. Kl. Ny Ser. V. 13:5-120.

Szarski, H., 1962, The origin of the Amphibia. Quart. Rev. Biol. 37:189-241.

Szarski, H., 1967, Historia zwierzat kregowych, PWN, Warszawa, (In Polish).

Szarski, H., 1971, The importance of deviation amplifying circuits for the understanding of the course of evolution. Acta Biotheor. 20:158-170.

Thomson, K.S., 1964, The comparative anatomy of the snout in the rhipidistian fishes. Bull. Mus. Comp. Zool. Harvard 131:313-357.

Thomson, K.S., 1967a, Mechanisms of intracranial kinetics in fossil rhipidistian fishes (Crossopterygii) and their relatives. Zool. J. Linn. Soc. 46:223-253.

Thomson, K.S., 1967b, Notes on relationships of the rhipidistian fishes and the ancestry of tetrapods. J. Paleont. 41:660-674.

Thomson, K.S., 1969, The biology of the lobe-finned fishes. Biol. Rev. 44:91-154.

Thomson, K.S., and Campbell, K.C.W., 1971, The structure and relationships of the primitive Devonian lungfish Dipnorhynchus sussmilchi (Etheridge). Bull. Peabody Mus. Nat. Hist. 38:1-109.

Trueb, L., 1973, Bones, frogs and evolution, in "Evolutionary Biology of the Anurans" (J.L. Vial, ed.), University of Missouri Press, Columbia, Missouri, pp. 65-123.

Vorobeva, E.J., 1971, The ethmoid region of Panderichthys and some problems of the cranial morphology of crossopterygians, in "Current Problems of Paleontology," Trudy Paleont. Inst. A.N. U.S.S.R. 130:142-159.

Wahlert, G.v., 1966, Atemwege und Schädelbau der Fische. Stutt. Beitrage Naturk. 159:1-40.

Williams, E.E., 1959, Gadow's arcualia and the development of tetrapod vertebrae. Quart. Rev. Biol. 34:1-32.

Wilson, A.C., Sarich, V.M., and Maxson, L.R., 1974, The importance of gene rearrangement in evolution: evidence from studies of chromosomal, protein and anatomical evolution. Proc. Nat. Acad. Sci. U.S.A. 71:3028-3030.

# THE ORIGIN OF THE TETRAPOD LIMB WITHIN THE RHIPIDISTIAN FISHES

Hans-Peter SCHULTZE

Geologisch-Palaontologisches Institut, Universität
Göttingen
Goldschmidtstr. 3 D-3400 Göttingen, W-Germany

In the Devonian, tetrapods (Ichthyostegalia) coexisted with
Dipnoi, Actinistia, and Rhipidistia, the group of fishes which
all recent authorities agree gave rise to the tetrapods.  Any
comparison between Recent and Devonian forms is fraught with uncertainties because of the enormous time lapse between them
(300 Million years).  In his consideration of tetrapods and  fishes
Szarski (this volume) has reaffirmed that Anura and Urodela are
more closely related to each other than to any other tetrapod or
fish.  He also states that the crossopterygians (strictly speaking,
the actinistian _Latimeria_) are the sistergroup of the Recent
tetrapods.  Nevertheless many authors have considered the dipnoans
to be close relatives of certain amphibians.  Two of Szarski's
important preadaptations of fishes for the transition to terrestrial
life - the internal nares (choanae)/nasolacrimal duct and the endoskeletal structure of the paired fins (Fig. 1) - merit further
consideration.

A.  Internal nares (choanae)/nasolacrimal duct:  Olfactory
and respiratory function are separated within gnathostome fish.
Two external nostrils are present (anterior incurrent - posterior
excurrent), situated in the roof of the mouth in Dipnoi, and on
the top of the snout anterior to the orbit in Actinistia.  The
latter position also occurs in porolepiform rhipidistians but in
addition the nasal capsule has an opening into the mouth cavity
(choana).  The choana is also present in the osteolepiform
rhipidistians so that it is a synapomorph character of Porolepiformes, Osteolepiformes and tetrapods.  The presence of choanae
in a few teleosts is a case of parallelism and Szarski has given
some reasons for the interpretation as parallelism.  Osteolepiforms

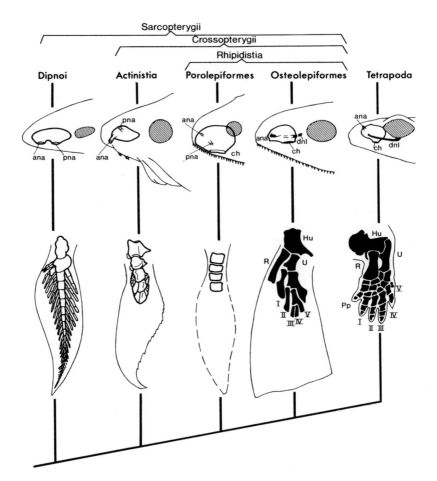

Figure 1.   Interrelationship of "Sarcopterygii" and tetrapods.
A.   Nasal capsule, from left to right:  Neoceratodus (from Günther,
1871, pl. I, pl. XXXIV fig. 3, pl. XXXV fig. 1, and Bing and Burckhardt, 1904, fig. 1), Latimeria (from Millot and Anthony, 1958,
fig. 3B, C, 10 and pl. IV, V), Porolepis (from Jarvik, 1972, fig.
14C, 15A, B, 20A), Eusthenopteron (from Bjerring, 1971, fig. 9A,
B, C), Rana (from Ecker et al., 1904, fig. 139-147, 152, 235).
B.   Endoskeleton of pectoral appendage, from left to right:  Neoceratodus (Günther, 1871, pl. 30, fig. 2); Latimeria (from Millot
and Anthony, 1958, fig. 1); Glypotolepis (from Jarvik 1972, fig.
59C); Eusthenopteron (from Jarvik 1964, fig. 25B, C); Eryops (from
Gregory and Raven, 1941, fig. 24C). - Black synapomorphies of
Osteolepiformes and tetrapods. Abbreviations:  ana  anterior
nostril; ch  choana; dnl  nasolacrimal duct; Hu  Humerus; pna
posterior nostril; Pp  prepollex; R  Radius; U  Ulna; I, II, III,
IV, V, digits I - V.

possess only a single external nostril like tetrapods.  The opening
in the posterior wall of the nasal capsule of the osteolepiform
Eusthenopteron is interpreted by Jarvik (1942) as the opening of
a nasolacrimal duct.  However, there is no functional explanation
for the existence of a nasolacrimal duct in fishes.  A single
external nostril, a choana and a nasolacrimal duct represent an
adaptation series leading towards tetrapods - in Hennig's terminology a synapomorphy of osteolepiforms and tetrapods.

     B.  Endoskeletal structure of paired fins:  Romer (1955) in-
troduced the term "Sarcopterygii" to unite Dipnoi, Actinistia and
Rhipidistia on the basis of the fleshy fin common to all three
groups.  However, the endoskeletal structure of the fleshy fin is
different in these fish.  Dipnoans, actinistians and porolepiform
rhipidistians possess a central axis of skeletal elements, an
archipterygium.  It is extremely difficult to derive the pentadyctal
tetrapod limb from an archipterygium (see Holmgren, 1933, 1949).
Therefore Jarvik (1964) has used the many-rayed pectoral endoskeleton of Sauripterus to derive the pentadactyl limb of Urodela
as opposed to all other tetrapods.  Sauripterus belongs with the
osteolepiform (as used by Jarvik, 1942) rhipidistians (Andrews
and Westoll, 1970b) so that the correspondence in the endoskeleton
between Sauripterus and Urodela (Jarvik, 1964) is a further indication of the monophyletic origin of all tetrapods (Andrews and
Westoll, 1970b).  Besides Sauripterus, it is only other osteolepiforms that have the many-rayed fin skeleton (examples and
functional interpretation in Andrews and Westoll, 1970a, 1970b)
from which a pentadactyl limb could have been derived.

Derived characters in these two important transformation series
to terrestrial life are therefore shared by tetrapods and osteolepiform rhipidistians.  These synapomorphies indicate that the
tetrapods as a whole are more closely related to the osteo-.
lepiforms than to the porolepiforms.

## REFERENCES

Andrews, S.M. and Westoll, T.St., 1970a, The Postcranial Skeleton
    of Eusthenopteron foordi Whiteaves, Trans. R. Soc. Edinburgh
    68:207.
Andrews, S.M. and Westoll, T.St., 1970b, The Postcranial Skeleton
    of Rhipidistian Fishes excluding Eusthenopteron, Trans. R.
    Soc. Edinburgh 68:391.
Bing, R. and Burckhardt, R., 1904, Das Zentralnervensystem von
    Ceratodus Forsteri, Anat. Anz. 25:588.
Bjerring, H.C., 1971, The Nerve Supply to the Second Metamere
    Basicranial Muscle in Osteolepiform Vertebrates, with Some
    Remarks on the Basis Composition of the Endocranium, Acta
    Zoologica 52:189.

Ecker, A., Wiedersheim, R. and Gaupp, E., 1904, Anatomie des
     Frosches auf Grund eigener Untersuchungen durchaus neu
     bearbeitet, 3. Abth., Lehre von den Eingeweiden, dem
     Integument und den Sinnesorganen, 2. Ed., Vieweg and Sohn,
     Braunschweig.
Gregory, W.K. and Raven, H.C., 1941, Studies on the origin and
     early evolution of paired fins and limbs, Ann. New York
     Acad. Sci., 42:273.
Gunther, A., 1871, Description of Ceratodus, a genus of Ganoid
     Fishes, recently discovered in Rivers of Queensland, Australia,
     Philos. Trans. R. Soc. London, 161:511.
Holmgren, N., 1933, On the origin of the tetrapod limb, Acta
     Zoologica, 14:185.
Holmgren, N., 1949, On the tetrapod limb problem - again, Acta
     Zoologica, 30:485.
Jarvik, E., 1942, On the Structure of the Snout of Crossopterygians
     and Lower Gnathostomes in General, Zool. Bidr. Uppsala,
     21:235.
Jarvik, E., 1964, Specializations in early vertebrates, Ann. Soc.
     R. Zool. Belg., 94:11.
Jarvik, E., 1972, Middle and Upper Devonian Porolepiformes from
     East Greenland with special reference to Glyptolepis
     groenlandica n. sp. and a discussion on the structure of
     the head in the Porolepiformes, Medd. Gronland, 187:1.
Millot, J. and Anthony, J., 1958, Anatomie de Latimeria chalumnae,
     Tome 1, Squelette, musches et formations de soutien, Centre
     National de la Recherche Scientifique, Paris.
Romer, A.S., 1955, Herpetichthyes, Amphibioidei, Choanichthyes
     or Sarcopterygii?  Nature, 1976:126.

AN EXCEPTIONAL REPRODUCTIVE STRATEGY IN ANURA: NECTOPHRYNOIDES
OCCIDENTALIS ANGEL (BUFONIDAE), AN EXAMPLE OF ADAPTATION TO
TERRESTRIAL LIFE BY VIVIPARITY

Francoise XAVIER

Laboratoire de Zoologie, Ecole Normale Superieure

46 rue d'Ulm - 75230 Paris, France

## INTRODUCTION

The Nimba ridge is an important area in West Africa for
speciation studies and evolutionary research. It has a large
number of endemics inhabiting it among which is the toad species
Nectophrynoides occidentalis. Discovered by Lamotte in 1942, this
tiny toad (adult length 15-26 mm) is the only entirely viviparous
anuran known. That is to say eggs develop in the oviduct of the
female which gives birth to totally metamorphosed young. Such a
reproductive strategy involves a number of anatomical and physiological adaptations which are quite remarkable in an amphibian.

The Nimba ridge lies on the borders of Guinea, the Ivory
Coast and Liberia 1200-1700 m above sea level. Tropical rain
forest covers liberian part of the mountain. It is only in the
5 km$^2$ of savanna occupying the guinean area that Nectophrynoides
occidentalis occurs. The climate is characterized by alternating
dry and rainy seasons. During the dry season, from November to
March, the toads remain under the soil. From April to October,
during the wet season, they become active and live on the surface,
under the low vegetation (Lamotte, 1959).

## REPRODUCTIVE STRATEGIES

At the end of the rainy season, in October, ovulation and
mating take place. A small number of oocytes are expelled from
each ovary, 2 in the first reproductive cycle rising to about 10
as the females get older. The mature oocyte is yolk-poor and its

545

diameter and weight are only 500-600 um and 200-220 ug respectively.
Fertilization is internal and occurs during mating.  A copulatory
organ is absent but insemination is ensured by mutual apposition
of cloacae.

Embryonic development and metamorphosis take place in the
maternal uterus the gestation period lasting 9 months.  Two
principal phases can be distinguished:  the first one, lasting 6
months, corresponds to the slow development of the embryos during
the underground life of the mother; the second phase is relatively
short, lasting 3 months, and is marked by the rapid growth of the
embryos and their metamorphosis, after emergence of the female in
April (Fig. 1A).

At parturition in June, the new-born size ranges from 7 to 8
mm and its weight from 30 to 60 mg.  Such embryonic growth is
impossible without a maternal supply of nutrients.  As a matter
of fact, exchanges between the embryos and the mother have been
demonstrated experimentally.  Embryos accumulate more than 6% of
the Lysine-$^{14}$C injected into the mother during the first hours
following injection.  Radioactivity is detected at first in proteins of the digestive tract (Xavier, 1971).  However, there is
no placental structure, the embryos are free in the uterine lumen
and they drink a nutritive substance secreted by the uterine
epithelium (Vilter and Lugand, 1959).  Indeed, differentiation of
the gut tube begins early in embryonic development.  The embryos
also show other adaptations to an intrauterine life in the absence
of the usual characteristic organs of aquatic larvae; e.g. external
and internal gills, gill slits, spiracle, horny beak and denticles
(Angel and Lamotte, 1944, Lamotte and Xavier, 1972).

## ANATOMICAL AND PHYSIOLOGICAL ADAPTATIONS TO VIVIPARITY

In order for fertilized eggs to be retained in the female
reproductive tract for their entire embryonic development, modifications must occur.  Changes take place in the nature of the egg,
in the number of eggs produced and in the structure and function
of the genital system.  The successful completion of gestation
with birth of the young at the time of parturition requires, moreover, profound endocrine adjustments.

## The Oviduct

The oviduct is divisible into two principal parts:  tube and
uterus.

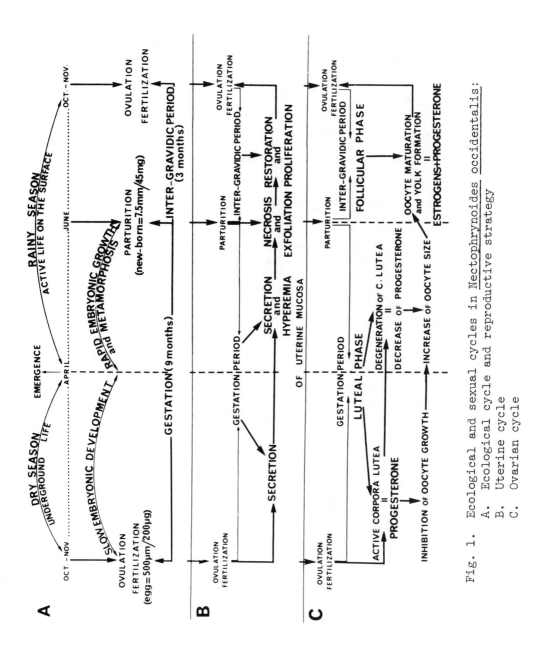

Fig. 1. Ecological and sexual cycles in Nectophrynoides occidentalis:
A. Ecological cycle and reproductive strategy
B. Uterine cycle
C. Ovarian cycle

The tube is never much-coiled unlike most amphibians. Throughout most of the gestation period the tube is inactive although there is hyperactivity of its glandular elements in the days preceding ovulation. As the eggs pass down the tube towards the uterus they receive a mucous coating which is, however, always reduced in N. occidentalis.

The two uteri join distally to form a common uterine segment opening into the cloaca. Unlike the tube, the uterus undergoes continuous changes in order to accommodate and support the embryos. Provision must be made for the nutritional, respiratory and excretory requirements of the retained fetuses, and throughout gestation there must be progressive enlargement of the uterus to permit embryonic growth.

The uterus passes through three essential phases in its adjustments (Fig. 1B): (1) a phase of secretion and hyperemia, which lasts throughout the gestation period; (2) a phase of necrosis and exfoliation, lasting for twelve days following parturition, and corresponding to the elimination of the superficial mucosal layer; (3) a phase of restoration and proliferation, extending from the necrotic phase until the subsequent ovulation, characterized by progressive mucosal regeneration initiated by the deep layer which was not involved in necrosis (Xavier, 1973).

## The Ovary

Amphibian ovaries are generally large and composed of several lobes full of oocytes. Moreover, they only exhibit a follicular phase. The ovary of N. occidentalis, on the other hand is small and contains few oocytes (10-25). Its cycle also has two phases, follicular and luteal (Fig. 1C). The follicular phase, lasting 3 months from July to October, corresponds to oocyte growth and yolk formation in the inter-gravidic period. During this time, follicles secrete both estrogens (theca interna) and progesterone (granulosa layer). Ovulation occurs when the oocyte has reached its mature size (500-600 um in diameter). The collapsed follicle is transformed into a new endocrine structure, the corpus luteum, by proliferation of the granulosa layer. It is the beginning of the luteal phase which lasts throughout gestation. The corpora lutea, relatively stable in size during underground life, decrease progressively after emergence until parturition (Lamotte and Rey, 1954; Lamotte et al., 1964). The corpora lutea contain lipids and 3β-HSDH and can convert pregnenolone to progesterone in vitro throughout gestation. Progesterone produced is maximum at the beginning of gestation but decreases in quantity considerably after emergence (Ozon and Xavier, 1968; Xavier et al., 1970; Xavier and Ozon, 1971).

## Endocrine Adjustments

At first sight the facts set out in the previous section suggest that corpora lutea and, in consequence, progesterone, play a role in the maintenance of pregnancy. However, bilateral ovariectomy, during the first month of gestation, provokes abortion within 2-3 months only in females in their first pregnancy. Therefore the role of corpora lutea cannot be the maintenance of pregnancy, but the secretion of a hormone. The progesterone produced causes differentiation of the uterine pouches and allows the uteri to adapt to embryonic development. Ovarian tissue in a first gestation is 4 times less active in progesterone biosynthesis in vitro than in subsequent pregnancies. With early bilateral ovariectomy in a first gestation there is neither enough time, nor a high enough activity of corpora lutea, to achieve the uterine differentiation necessary to accommodate and support developing embryos. Abortion thus occurs within the next three months (Xavier, 1970a).

In subsequent pregnancies, ovariectomy accelerates embryonic development and advances the time of parturition (April). Progesterone implants in the female, after emergence, slow embryonic development and postpone parturition for 2 to 3 months (Xavier, 1970b). Corpora lutea or progesterone thus inhibit embryonic growth during the period of undergound life and maternal fasting.

If the corpora lutea affect uterine structures on the one hand, and pregnancy evolution and embryonic development on the other, then the pregnant uterus and its contained embryos play a role in the evolution of ovarian structures. Hysterectomy or removal of the embryos in early pregnancy quickly provokes a decrease of the corpora lutea and an atresia of young oocytes. Identical results are obtained after hypophysectomy at the same date (Xavier, 1971). This indicates that pregnant uteri or embryos have an ovarian affect which is mediated through the pituitary gland. Moreover, the presence of growing fetuses induces the gradual enlargement of the uterus. Distension of the uterine lumen is recognized as an important factor conditioning enlargement of this organ. When pregnancy occurs unilaterally in N. occidentalis females, the other uterus which remains empty, does not enlarge to the same degree as the gravid uterus.

Very early on in the pregnancy the embryos can induce the necessary adjustments enabling them to stay in the maternal uterus until the end of their development.

## DISCUSSION AND CONCLUSIONS

Because of the complexity of hormonal correlations during the sexual cycle of this small toad, one is tempted to make comparisons with the phenomena observed in mammals. N. occidentalis possesses eggs which are very poor in yolk, comparable to those of mammals; the steroidogenic activity is likely to be through the hypothalmic-pituitary axis (Zuber-Vogeli and Xavier, 1973). Also, the necrosis followed by exfoliation of the uterine mucosa after parturition in N. occidentalis suggests the casting off of a layer of endometrium as in mammalian menstruation in keeping with the cessation of corpora lutea activity.

It is thus quite remarkable to find the same structures and the same hormones in the Nimba toad and in mammals, along with the same functions, i.e. to contain fetuses in the maternal uterus, to feed them and then to expel them. However, there exist funda- mental differences of which the most important is that there is no placenta in N. occidentalis. Furthermore, if the toad is capable of secreting estrogens and progesterone, as in mammals, it utilizes these hormones in a different way. Contrary to what is observed among mammals, the progesterone acts in conjunction with the estrogens during the follicular phase, whereas it acts alone during the luteal phase. The corpora lutea are mainly responsible for the production of progesterone; this hormone plays a part unknown in mammals and in other non-mammalian viviparous verte- brates, i.e. slowing embryonic development during the maternal fasting period. It is a fact that in all known Elasmobranchii, some Teleostei and Amphibia, and many Reptilia, corpora lutea are formed in both oviparous and viviparous species; they could become functional and be a primary source of estrogens and progesterone. One or both steroids could control the reproductive cycle in ovi- parous species as well as pregnancy in viviparous ones. In a recent review Browning (1973) analyzed in detail the relationships of corpora lutea to gestation in ovoviviparous and viviparous species of non-mammalian vertebrates, and concluded that there cannot be a single evolutionary history of the corpus luteum. It is the uses to which the corpora lutea are put and the controls exerted over them that have evolved and not the structures them- selves.

Viviparity occurs randomly in non-mammalian vertebrate taxa and one may wonder how the Nimba toad, alone among anurans, demon- strates such an extraordinary viviparity with all its complexity, the regularity of its cycle and the succession of its intervening hormones. Living anurans represent a kind of "evolutionary cul- de-sac" and the guinean toad has evolved, independently, a coordin- ated set of structures. The evolution in an isolated anuran species of true viviparity contrasts strikingly with therian mammals in which all members of the subclass have this adaptation presumably derived from a common ancestral stock.  Such an attempt is observed also in urodeles (Salamandra atra), some fishes and Reptilia where placentation may take place.  This proves that the transition from oviparity to viviparity, a very fundamental physiological modification, can occur as a "microevolutionary" step of short duration. It also adds weight to the suggestion that potentially enormous evolutionary steps can occur due simply to the accumulation of changes at the intraspecific level.

Such an adaptation raises a number of questions.  For example, did the toad acquire viviparity at the same time as it was losing its yolk? - or did one of those processes precede the other one and, if so, which one?  Answers to these questions may be found by tracing back the genealogy of the toad from Bufo through the ovoviviparous species of East Africa, such as N. tornieri or N. viviparus from which N. occidentalis quite probably derives.  N. occidentalis would then have acquired viviparity before losing its yolk.  If such is the case, was it viviparous when it colonized the Nimba mountain  or has it become viviparous under the influence of the environment?

The Nimba toad adaptation is obviously linked up with the absence of any areas of water for larval development.  It is a fact that only one other species inhabits the Nimba savanna, Arthroleptis crusculum and this lays large yolk-rich eggs in the ground.  After three weeks young, almost entirely metamorphosed, individuals hatch.  In A. crusculum as in N. occidentalis a small number of oocytes is ovulated compared to the oviparous species. Such a reproductive strategy - a K-strategy - agrees with the remarkable adaptation of the species to an environment without areas of water and with very light predation.  This adaptation has led to the exceptional density of the viviparous species, N. occidentalis.

Finally, the guinean Nectophrynoides shows that some animal groups possess latent characters that are ready to become visible but normally don't manifest themselves until later in the history of the species.  Conversely, other characters are not yet present even potentially.  It is certain that "prophetic" species such as the Nimba toad, possessing anatomical and physiological features ahead of their time, are likely to bring to light valuable speciational and evolutionary information (Wolff, 1971).

## REFERENCES

Angel, F. and Lamotte, M., 1944, Un crapaud vivipare d'Afrique occidentale, Nectophrynoides occidentalis (Angel),  Ann. Sci. Nat. Zool. 6:63-89.
Browning, H.C., 1973, The evolutionary history of the corpus luteum, Bioloty of Reproduction 8:128-157.

Lamotte, M., 1959, Observations ecologiques sur les populations
    naturelles de Nectophrynoides occidentalis (Fam. Bufonides),
    Bull. Biol. Fr. et Belg. 93:355-413.
Lamotte, M. and Rey, P., 1954, Existence de corpora lutea chez un
    Batracien anoure vivipare, Nectophrynoides occidentalis
    Angel; leur evolution morphologique, C.R. Acad. Sc. Paris
    238:393-395.
Lamotte, M., Rey, P. and Vogeli, M., 1964, Recherches sur l'ovaire
    de Nectophrynoides occidentalis, Batracien Anoure vivipare,
    Arch. Anat. micr. Morph. exp. 53:179-224.
Lamotte, M. and Xavier, F., 1972, Recherches sur le developpement
    embryonnaire de Nectophrynoides occidentalis Angel,
    Amphibien Anoure vivipare. I - Les principaux traits
    morphologiques et biometriques du developpement, Ann. Embr.
    Morph. 5:315-340.
Ozon, R. and Xavier, F., 1968, Biosynthese in vitro par l'ovaire
    de l'Anoure vivipare Nectophrynoides occidentalis au cours
    du cycle sexuel, C.R. Acad. Sc. Paris 266:1173-1175.
Vilter, V. and Lugand, A., 1959, Trophisme intra-uterin et
    croissance embryonnaire chez le Nectophrynoides occidentalis
    Ang., Crapaud totalement vivipare du Mont Nimba (Haute
    Guinee), C.R. Soc. Biol. Paris 153:29-32.
Wolff, E., 1971, Les grands problemes poses par un petit crapaud
    vivipare et accoucheur de Guinee, Sciences 74-75:32-38.
Xavier, F., 1970a, Analyse du role des corpora lutea dans le
    maintien de la gestation chez Nectophrynoides occidentalis
    Angel, C.R. Acad. Sc. Paris 270:2018-2020.
Xavier, F., 1970b, Action moderatrice de la progesterone sur la
    croissance des embryons chez Nectophrynoides occidentalis
    Angel, C.R. Acad. Sc. Paris 270:2115-2117.
Xavier, F., 1971, Recherches sur l'endrocrinologie sexuelle de la
    femelle de Nectophrynoides occidentalis Angel (Amphibien
    Anoure vivipare), These de Doctorat es-Sciences C.N.R.S.
    n°A.0.6385, 223p.
Xavier, F., 1973, Le cycle des voies genitales femelles de
    Nectophrynoides occidentalis Angel, Amphibien Anoure vivipare,
    Z. Zellforsch. 140:509-534.
Xavier, F. and Ozon, R., 1971, Recherches sur l'activite endocrine
    de l'ovaire de Nectophrynoides occidentalis Angel (Amphibien
    Anoure vivipare). II- Synthese in vitro des steroides,
    Gen. Comp. Endocrinol. 16:30-40.
Xavier, F., Zuber-Vogeli, M. and Le Quang Trong, Y., 1970,
    Recherches sur l'activite endocrine de l'ovaire de
    Nectophrynoides occidentalis Angel (Amphibien Anoure vivipare).
    I- Etude histochimique, Gen. Comp. Endocrinol. 15:425-431.
Zuber-Vögeli, M. and Xavier, F., 1973, Les modifications cytolo-
    giques de l'hypophyse distale des femelles de Nectophrynoides
occidentalis Angel apres ovariectomie, Gen. Comp. Endocrinol.
20:199-213.

# THE EVOLUTION OF TERRESTRIAL LOCOMOTION

James L. EDWARDS

Dept. of Zoology, Michigan State University

East Lansing, Michigan  48824 USA

## INTRODUCTION

Although the derivation of tetrapods from rhipidistians is now generally accepted (Szarski, 1962 and this volume; Schaeffer, 1965; Schmalhausen, 1968; Jarvik, 1968), the selective pressures which led to terrestrial locomotion are still an area of dispute. Nearly all workers in this field have agreed that the tetrapod limb evolved as a locomotor organ which allowed movement away from the pond or swamp habitat of the rhipidistians and onto the land.  Both biotic and abiotic stimuli have been invoked as the proximate factors pushing the Rhipidistia landward.

Chief among the suggested abiotic factors has been the desiccation of the pond environment.  Romer (1958 and other papers cited therein) pointed to the prevalence of red bed deposits in the Devonian and suggested that they indicated a semi-arid, probably fluctuating, climate.  He then hypothesized that the rhipidistian fin and, later, the tetrapod limb, allowed their possessors to move to a new pond when the one in which they were living dried out.  In a slight variation of this hypothesis, Schmalhausen (1968) suggested that the depletion of oxygen in well-heated ponds with a large quantity of organic material would cause the rhipidistians to leave the pond before the water level had dropped very far.

Biotic factors which have been proposed as stimuli to terrestrialism include predation, overpopulation and the lure of a terrestrial food supply.  Ewer (1955), Goin and Goin (1956) and Warburton and Denman (1961) pointed to the intense competition in

an overpopulated pond. Schmalhausen (1957), Cowles (1958),
Szarski (1962), Cox (1967) and Holman (1969) all stressed the
importance of predation, especially on young or larval stages.
Inger (1957) and Feduccia (1971) pointed out that overpopulation
and predation could combine to produce a potent pressure leading
to overland dispersal in order to find new aquatic environments.
In addition, as noted by Ewer (1955), Gunter (1956), Inger (1957),
Rohdendorf (1970) and Olson (1976), invertebrates, in particular
myriapods, scorpions and other arachnids, may have provided a
terrestrial food supply to lure the rhipidistians away from the
water.

At least two authors have stressed the non-locomotor function
of the rhipidistian fin. Orton (1954) felt that the tetrapod
limb was primarily a digging organ which allowed its owner to
construct a burrow in order to remain in the bottom of a drying
pond, much as the lungfish genera Protopterus and Lepidosiren do
today. However, it should be noted that the paired fins in these
two modern genera are extremely feeble and presumably are used as
sensory receptors (Thomson, 1969); the burrow is constructed entirely by movements of the body and by special secretions of the
skin (Johnels and Svensson, 1954). Neoceratodus, the only genus
of modern lungfish to retain an archipterygial fin, does not construct burrows (Thomson, 1969), but instead uses its fins to provide slow locomotion in water (Dean, 1906). These facts would
thus appear to contradict Orton's hypothesis.

Thomson (1969, 1972) argued that the rhipidistian fin was too
weak to effect locomotion on land, although in the aqueous environment the water would have supported the fish's body sufficiently
to allow the fins to propel the body forward slowly. Thomson also
felt a rhipidistian could always have progressed by lateral undulation on land, much as eels and some catfish do today; therefore,
he suggested that the function of the pectoral fin on land was to
raise the anterior end of the trunk from the ground so that the
weight of the body did not crush the lungs. This hypothesis is
attractive because it also explains why the forelimbs are more
developed than the hindlimbs in Rhipidistia and early Amphibia
such as Ichthyostega (Jarvik, 1960); in later quadrupedal tetrapods the reverse situation, in which the forelimbs are weaker than
the hindlimbs, is found.

It is probable that the evolution of such a complex phenomenon as the invasion of the terrestrial habitat was the result of
several conflicting selective pressures, perhaps including all
those mentioned above. I feel that the elegant scheme of Inger
(1957), which shows how the negative factors of predation and
overpopulation could combine with the positive aspects of dispersal and a terrestrial food supply, best outlines the combination of

selective factors which pushed vertebrates onto the land. At the same time, the arguments of Thomson (1969, 1972) show that although the rhipidistian fin undoubtedly served a locomotor function in the water, its initial function on land may well have been as a prop to aid in respiration.

There has also been some argument about whether the first tetrapodal gait was a walk or a trot. Gray (1968) felt that the walk must have come first, since the walk is a more stable gait, especially at slower speeds. Faber (1956) found that in two species of salamanders, Ambystoma mexicanum and Triturus taeniatus (= T. vulgaris), the first coordinated movements of the limbs involved ipsilateral legs; since ipsilateral movement is not seen in trotting animals, he felt that the walk was the primitive tetrapodal gait. Sukhanov (1974) examined four specimens of one species of Salamander (Salamandra salamandra), observed only walking gaits, and generalized his results in stating that no modern amphibian uses the trot. He further felt that the trot was not evolved until the reptilian grade of organization was achieved.

On the other hand, Howell (1944, p. 224) stated that the trot is the characteristic gait of limbed amphibians and that "the earliest tetrapods, when they emerged from an aquatic habitat, must have progressed much as a salamander does today." Coghill (1929) examined motor behavior in the early embryo of Ambystoma punctatum and reported that the first coordinated movement of the limbs involved contralateral legs moving in the typical sequence of the trot, along with lateral undulations of the trunk that strongly resembled swimming movements in the adult salamander. Coghill therefore generalized that the trot must be the primitive tetrapodal gait, and that this gait was based on swimming movements.

Some insight into the locomotor pattern of the earliest tetrapods may be gained by re-examining the locomotor behavior of their living relatives. Among the modern Amphibia, only the salamanders (Order Caudata) retain what appears to be the primitive mode of locomotion. The saltatory locomotion of frogs is peculiar to the Order Anura, while the caecilians (Order Apoda) lack limbs and girdles entirely.

The exact ancestry of the salamanders is unknown. Jarvik (1942, 1968) has argued that salamanders have a direct derivation from the porolepiform rhipidistians, while all other modern tetrapods are derived from osteolepiform rhipidistians via the labyrinthodonts. Thomson (1968) and Szarski (1962) have found fault with Jarvik's analyses; especially cogent for the present discussion is Thomson's (1968) demonstration that the limbs of

salamanders and all other tetrapods are essentially identical and
provide no evidence for a separate derivation of this group.  Recent evidence (Parsons and Williams, 1963; Estes, 1965; Bolt,
1969) indicates that the salamanders may have arisen from eryopoid
temnospondyls closely resembling the members of the family Dis-
sorophidae.

## SALAMANDER LOCOMOTION

The locomotor behavior of salamanders was examined using
motion pictures (details in Edwards, 1976).  Animals were filmed
from above as they walked across a roughened sheet of Plexiglas
with a centimeter grid affixed to its bottom.  The resulting films
were examined on a Vanguard Motion Analyzer.  A total of 103 individuals representing 48 species of primarily terrestrial salamanders was examined.  Three aspects of salamander locomotor behavior are relevant to the present discussion:  propulsion, gait
analysis and lateral bending of the vertebral column.

### Propulsion

Propulsion in salamanders is achieved by three methods:
girdle rotation, limb retraction and humeral-femoral rotation.

Propulsion by girdle rotation is illustrated in Fig. 1.  The
sequence begins with the vertebral column bent into a lateral arc;
the foreleg on the convex side of the column and the contralateral
hindleg are protracted and in contact with the substrate.  The two
protracted limbs are fixed to the body by isometric contraction of
their protractor and retractor muscles.  The animal then isotonically contracts the axial muscles on the convex side of the
column at the same time that weight is shifted from the retracted
pair of legs.  As the animal rotates around the fixed pair of
limbs, the vertebral column is straightened and then is bent in
the opposite direction from the beginning posture.  Note that in
this form of locomotion, active propulsion is accomplished solely
by the axial musculature.  The extrinsic limb muscles are only
used to isometrically fix the limbs onto the girdles; the protraction and retraction of the limbs are passive results of
girdle rotation.

Active limb retraction is pictured in Fig. 2.  In this type
of locomotion, the axial muscles on both sides of the vertebral
column contract isometrically, thus making the column a rigid rod.
The retractor muscles of the limbs then contract isotonically to
move the animal forward.  This is the general mode of propulsion
in most animals.

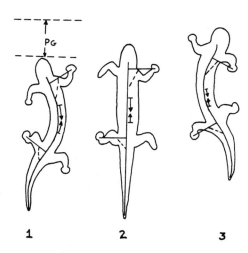

Figure 1.  Propulsion by girdle rotation.  Broken lines represent muscles contracted isometrically, thus fixing the limbs on the girdles.  Solid lines with arrows indicate muscles contracting isotonically.  At time 1, the protracted limbs are fixed and the axial muscles just beginning to contract, leading to retraction of the fixed limbs at times 2 and 3.  Axial muscles also straighten column and then bend it in opposite direction.  $P_G$ = propulsion derived from girdle rotation.  (Modified from Gray, 1968.)

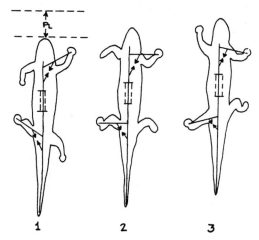

Figure 2.  Propulsion by active limb retraction.  Broken lines represent muscles contracted isometrically, thus making the vertebral column a rigid rod.  Solid lines with arrows represent muscles contracting isotonically.  $P_L$ = propulsion derived from active limb retraction.  (Modified from Gray, 1968.)

The third component of salamander locomotion, humeral-femoral rotation, is also achieved by isotonic contraction of the retractor muscles (see Fig. 3). In the forelimb, a portion of the pectoralis muscle attaches to a ventral process of the humerus called the crista ventralis (Barclay, 1946). In the hindlimb, the retractor muscle attaching to the crista ventralis of the femur is the caudofemoralis muscle. When the limb is in a protracted position, the crista points anteroventrally (see Fig. 3B, 1). Contraction of the appropriate retractor muscles retracts the crista and causes the long bone (humerus or femur) to rotate around its long axis (see Fig. 3B, 2 and 3). However, the next segment of the leg (the radius and ulna or tibia and fibula) is fixed at a right angle to the proximal limb element, so that rotation of the long bone is translated back into antero-posterior movement at the foot (see Fig. 3A). Barclay (1946) calls this a "double crank" mechanism because the reciprocal movement of the crista ventralis is translated first into rotatory motion and then back into reciprocal movement at the foot. Gray (1968, p. 119) has called this system "probably the nearest approach to that of a wheel and axle to be found anywhere in the animal kingdom."

Salamanders use all three types of propulsion in normal locomotion. However, girdle rotation is most effective when the vertebral column is strongly bent, since at this time, the axial muscles on the convex side of the column are stretched. On the other hand, limb retraction is most effective when the column is straight, for at this time, the axial muscles on both sides of the body can most easily fix the column into a straight rod. Humeral-femoral rotation may be considered a special type of limb retraction. Limb rotation is most effective when the column is straight and the limb is at right angles to the body.

My studies show that in a wide range of terrestrial salamanders moving at different speeds and gaits, limb retraction supplies 56-62% of forward propulsion, limb rotation supplies 26-28%, and girdle rotation supplies 10-18%. As an individual salamander increases its speed, the amounts of propulsion from limb retraction and rotation decrease, while the amount of propulsion due to girdle rotation nearly doubles. Even at the slowest speeds, however, the contribution from girdle rotation is substantial (10% of the total).

## Gaits

A gait can be defined as "a regularly repeating sequence and manner of moving the legs in walking or running" (Hildebrand, 1974, p. 510). According to the terminology of Hildebrand (1965, 1966), salamanders use two gaits, the lateral sequence walk and

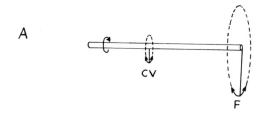

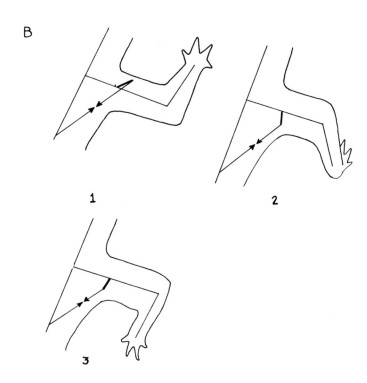

Figure 3.  Propulsion by limb rotation.  (a) "Double crank" mechanism.  Retractor muscles insert on crista ventralis (CV); contraction of these muscles retracts the crista, thereby rotating
the long bone.  Rotation is translated back into retraction at
foot (F).  (Modified from Barclay, 1946.)  (B) Diagrammatic representation of humeral rotation.  When limb is protracted (1), crista
points anteroventrally.  Contraction of retractor muscles (arrows)
first brings crista to a position where it points ventrally (2)
and then posteroventrally (3).  Protractor muscles (not shown)
also insert on crista and reverse the process during protraction
of the limb.

the trot.  Contrary to a previous report by Roos (1964), all walking and trotting salamanders were seen to move with the ventral
surface of the body raised from the substrate.

In the lateral sequence walk (Fig. 4A), the footfall of a
hindfoot is followed by the footfall of the ipsilateral forefoot,
then the contralateral hindfoot, and finally the contralateral
forefoot.  As Gray (1944) noted, this is the most stable of all
tetrapod gaits, for it is the only gait in which the center of
gravity of the animal never leaves the triangle of support provided by the limbs (Fig. 5).  Lateral sequence walks are used by
all tetrapodal vertebrates.  Salamanders use the walk at extremely
slow to medium speeds (0.10-3.16 times the body length per second).

In the trot (Fig. 4b), contralateral feet move together.
Theoretically, this is a less stable gait than the lateral sequence walk, since only two feet are on the ground for a large
percentage of the total stride.  Nevertheless, salamanders use the
trot at very slow as well as fast speeds (0.15-4.78 times the body
length per second).

At faster speeds, salamanders essentially abandon their
limbs, resting the ventral surface of the body on the substrate
and moving by a process I call fast undulatory locomotion.  In
this type of locomotion, the limbs are either folded against the
body or remain extended but only touch the ground occasionally.
The propulsive power in fast lateral undulation is provided by
waves of contraction which pass down alternate sides of the body.
The locomotor sequence is nearly identical to that used by salamanders swimming quickly in water.  Indeed, the animals appear to
be swimming over the substrate, progressing in the manner used by
eels and walking catfish on land.  Fast undulatory locomotion is
an energy consuming, rather non-directional type of movement and
appears to be used in escape and avoidance situations.

As any salamander moves from very slow to very fast speeds,
it follows the sequence:  lateral sequence walk--trot--fast undulatory locomotion.  However, there is a wide range of speeds
over which any individual may use either the walk or the trot.  At
the slowest speeds recorded (less than 0.15 times the body length
per second), all animals used the walk.  At slightly faster speeds,
the gait used appeared to be arbitrary.  At fast speeds (greater
than 3.16 times the body length per second), all salamanders used
either the trot or fast undulatory locomotion, and at speeds
faster than 4.78 times the body length per second, only fast undulatory locomotion was used.  Thus, even though the trot is a
less stable gait than the walk, salamanders are able to use either
gait at slow to moderate speeds.  Lateral bending of the vertebral
column may help reduce the instability of the trot.  A bend to one

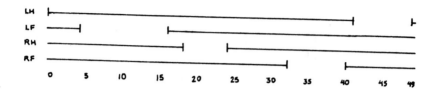

FRAME #: 0    2    4    6    8    10    12    14    16    18    20    22

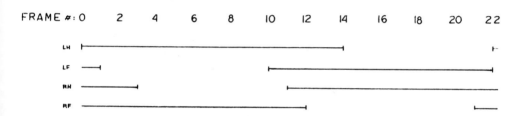

Figure 4.   Footfall diagrams for walking and trotting salamanders.
Horizontal lines represent times when indicated foot was in contact with substrate.   Vertical lines represent frames in which
indicated foot was lifted from or placed on the substrate.   Numbers represent consecutive frames of motion picture film; all
films taken at speed of 24 frames per second.   (a, above) Single
stride of a specimen of <u>Ensatina</u> <u>eschscholtzii</u> using the very
slow walk.   (b, below) Single stride of <u>Pseudotriton</u> <u>ruber</u> using
the moderate walking trot.   Abbreviations:   RF = right forefoot,
RH = right hindfoot, LF = left forefoot, LH = left hindfoot.

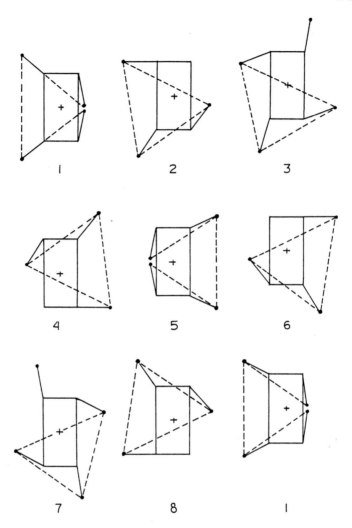

Figure 5.  Stability diagram for the lateral sequence walk.  The
rectangles represent the body of the animal, and the center of
gravity is represented by +.  Dashed lines delineate triangles of
support, formed by the limbs in contact with the substrate.  The
absence of a limb indicates that the limb is being protracted and
the foot is not in contact with the substrate.  Footfall sequence
is right hind (time 1), right front (time 3), left hind (time 5)
and left front (time 7).  Footfalls of forefeet occur when the
center of gravity is at the limits of the triangle of support
formed by the other three limbs (times 3 and 7, respectively); at
all other times, the center of gravity is well within the triangle
of support.  All other footfall sequences result in some periods
during which the center of gravity is outside of the support
triangle.  (Modified from Gray, 1968.)

side in the trunk region can be counterbalanced by a bend to the
other side in the tail, thereby keeping the center of gravity in
the middle of the body.  In fact, the tail is often dragged
passively behind the body in walking animals, while the tail shows
contractions opposite to those of the body in trotting salamanders
(see below).  Therefore, the trot should not be considered a fast
gait only used at high speeds, especially in animals with lateral
bending of the vertebral column.

## Lateral Bending

The isotonic contractions of the axial muscles which lead to
propulsion by girdle rotation (see Propulsion) also produce horizontal movements of the vertebral column which Roos (1964) has
called lateral bending.  The exact nature of this bending has been
a source of dispute.  Slijper (1941) stated that salamanders, like
most fish, use traveling waves, in which points of maximal bending
pass down alternate sides of the body from anterior to posterior.
Hesse (1935), however, maintained that lizards and, by inference,
salamanders employ a standing wave, with points of no lateral
bending (nodes) alternating with areas of maximal bending (internodes).  The nodes, which are also the points of maximal vertebral
rotation, were supposed to occur at the pectoral and pelvic
girdles.

Roos (1964) developed a method for determining the exact
pattern of lateral bending.  He filmed salamanders from above,
and used the insertion of the dorsal fin to indicate the position
of the underlying vertebral column; x-ray cinematography indicated
that this procedure was justified.  In each frame of the film,
Roos determined the curvature of the body by a complicated process
of double graphic differentiation.  The body was divided into
small segments and the bending of each segment was recorded as the
reciprocal of the radius of curvature (in centimeters) of that
segment, with positive values having the center of curvature to
the right of the column and negative values to the left.  Zero
values represented straight portions of the column, and the farther
the value was from zero, the sharper the curve.  The curvature
values were then plotted on a graph in which the abscissa represented time and the ordinate the straightened length of the animal.
Finally, equal values of curvature were connected to give contour
lines which Roos called "isoskolies."

Daan and Belterman (1968) simplified Roos' method by using a
circle template to calibrate the maximal curvature values, while
intermediate values were determined by graphical interpolation.
They also added a new line to the graph, a "shift line," which
connected the points of maximal bending.

Examination of the pattern of isoskolies immediately indicates whether the animal used a traveling or standing wave, or a combination of both. Figure 6a shows the pattern for a traveling wave, in which all isoskolies and shift lines are parallel. In a standing wave pattern (Fig. 6b), the isoskolies form closed concentric ellipses, and the shift lines move in stepwise fashion from one area of maximal bending to another. Horizontal zero isoskolies represent nodes (points of maximal rotation, but no lateral bending), while vertical zero isoskolies represent times at which the vertebral column is straight.

Roos (1964) filmed one specimen of a walking _Triturus vulgaris_ and found that the animal exhibited a nearly perfect standing wave in the anterior trunk and tail regions, while in the posterior trunk region, the pattern was a predominantly standing wave with a slight traveling wave influence. From these meager data, Roos hypothesized that all salamanders use standing waves during terrestrial locomotion.

When I prepared isoskolie diagrams for the salamanders I had filmed, I found that all walking animals used standing waves in the trunk region, while trotting animals used traveling waves. Figure 7 shows the isoskolie diagram for one stride of a specimen of _Aneides lugubris_ using the slow walk. Note that in the trunk region a clear pattern of standing waves is seen, especially in the area near the pectoral girdle. The standing waves change to traveling waves at the sacrum. Figure 8 shows the isoskolie diagram for one stride of _Dicamptodon ensatus_ using the moderate walking trot. Here traveling waves are initiated at the pectoral girdle and pass unimpeded to the tip of the tail. Traveling waves are also used by all salamanders in fast lateral undulation, but no walking salamander was found to use traveling waves, except in the tail. Since the tail is often dragged passively behind a walking animal, it appears that it does not take an active part in propulsion. Therefore the tail can take on any pattern of waves in a walking animal. In trotting animals, however, the tail is used to balance the curvature of the trunk region. Thus traveling waves are found in the tail as well as the trunk of trotting animals, and the centers of curvature of the mid-trunk and tail regions always lie on opposite sides of the animal.

The pattern of lateral bending of trotting animals shows that traveling waves can produce terrestrial locomotion, directly contradicting the speculation of Roos (1964) that terrestrial salamanders use only standing waves. Theoretical considerations indicate why traveling waves are used for the trot and not the walk. Figure 9 shows positive and negative traveling waves alternating down the body of a diagrammatic salamander. At time 1, the pectoral and pelvic girdles are at the shift lines (maximal curvatures)

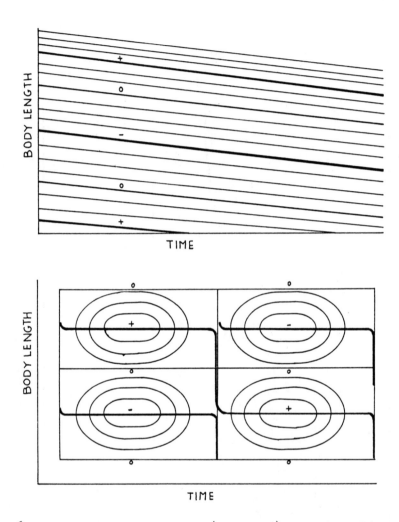

Figure 6.   Isoskolie diagrams for (a, above) pure traveling wave
condition and (b, below) pure standing wave condition.   Heavy
lines are shift lines, which connect points maximal bending.   Positive values are defined as having the radius of curvature to the
right of the animal, negative values with the radius of curvature
to the left.   Zero values represent points of no curvature.

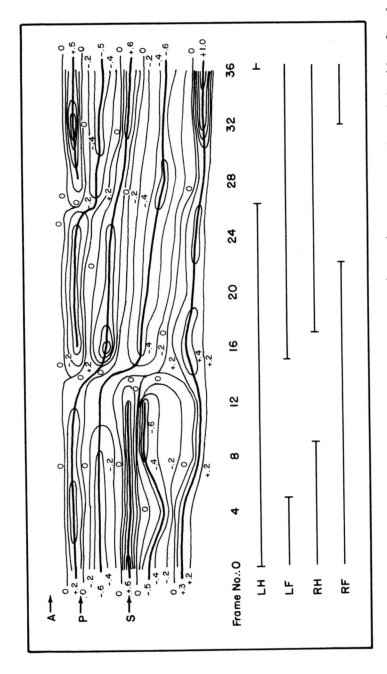

Figure 7.  Isoskolie diagram (above) and footfall diagram (below) for one stride of Aneides lugubris using the slow walk.  Abbreviations:  A = anterior end of animal, P = pectoral girdle, S = sacrum.

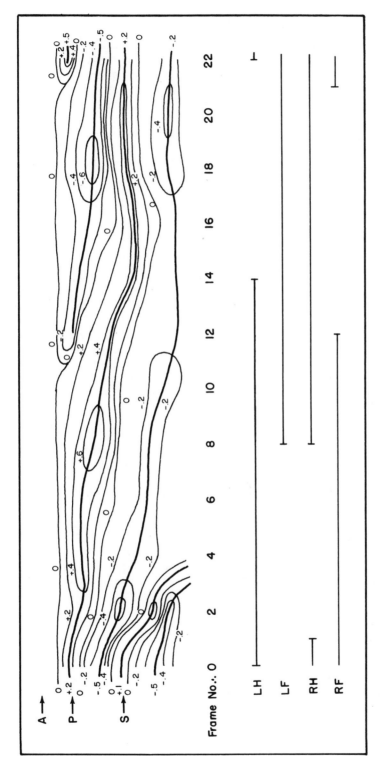

Figure 8. Isoskolie diagram (above) and footfall diagram (below) for one stride of Dicamptodon ensatus using the moderate walking trot. Abbreviations as in Fig. 7.

of positive waves, and the middle of the trunk is at the shift
line of a negative wave.  The short transverse lines represent the
limbs, which are fixed perpendicular to the girdles.  At this
point the animal is ready to protract the left front and right
hind limbs.  At time 2, the girdles are located at nodes, and
protraction is halfway completed.  At time 3, the girdles are now
at the shift lines of negative waves.  At this point, the stride
is half completed, and the right front and left hind limbs are
ready to be protracted.  The positive wave which was at the pectoral girdle in time 1 is now in the middle of the trunk, the
negative wave which was in the middle of the trunk at time 1 is
now at the sacrum, and a new negative wave has been introduced at
the pectoral girdle.  These waves proceed posteriorly until they
are located at the sacrum, the mid-tail region and the mid-trunk
region, respectively, at the end of the stride (time 5), and the
sequence begins again.  Thus, during one stride a single wave of
contraction passes from the pectoral and pelvic girdles, and a
new wave of opposite sign is added at the pectoral girdle midway
through the stride.  Note that contralateral limbs are protracted
together simply by the axial muscle contractions which produce
the traveling waves.  Therefore the animal could theoretically
progress in a trot by relying on the axial musculature alone;
limb muscles would be needed only to fix the limbs onto the
girdles and to lift them off the substrate as they are passively
protracted.  In fact, although trotting salamanders were never
observed to derive more than 20% of their total forward propulsion
from girdle rotation, the pattern of traveling waves needed to
produce a trot solely by girdle rotation is nearly identical to
that actually seen in trotting salamanders (compare Figs. 8 and 9).

Traveling waves alone could not produce a lateral sequence
walk, however.  In the walk, by definition, the contralateral
pairs of limbs must be out of phase by approximately one-quarter
of the stride duration.  It is impossible to accomplish this by
traveling waves alone.


THE LOCOMOTION OF LABYRINTHODONTS

Labyrinthodont locomotion was probably very similar to that
of salamanders.  Muscle reconstructions (Miner, 1925) show that
the axial and limb musculature of eryopoid temnospondyls was
nearly identical to that of salamanders.  Labyrinthodonts undoubtedly propelled themselves forward by limb retraction.  The
presence of cristae ventrales on the humeri and femora of all
known labyrinthodonts (with the possible exception of
Hesperoherpeton (Eaton and Stewart, 1960), as well as the typical
screw-shaped glenoid fossa and S-shaped articular surface of the
head of the humerus in these forms, indicate that labyrinthodonts

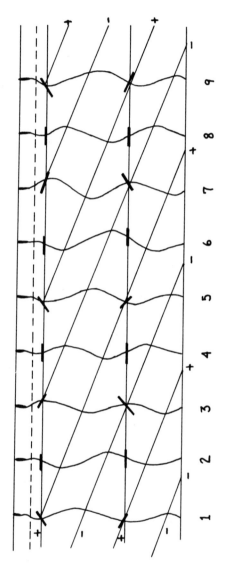

Figure 9. Two strides of a trotting salamander. Vertical curving lines represent the vertebral column of the animal. The four horizontal lines lie at the anterior end of the snout, pectoral girdle, sacrum and posterior end of the tail, while the dashed line marks the occipital-vertebral joint. Short transverse lines represent the limbs. The diagonal lines are the shift lines of traveling waves. Note that a new wave is added to the anterior end of the trunk when the previous wave has reached the middle of the trunk.

also used humeral-femoral rotation. Finally, the sprawling posture seen in all labyrinthodonts implies that they used girdle
rotation. In order for girdle rotation to be effective in propulsion, the girdle must rotate around a long lever. The lever is
provided by sprawling posture, in which the humerus and femur project horizontally from the body. Sprawling posture is found in
most modern amphibians and reptiles. The loss of sprawling posture in such groups as chamaeleons, dinosaurs and mammals is correlated with a change to other methods of propulsion and a loss of
reliance on girdle rotation. Therefore, in the absence of obvious
specializations, such as the rigid column and jumping hindlimbs of
frogs or the constraining shell of turtles, sprawling posture can
be used as evidence that the animal possessing it used propulsion
by girdle rotation. Thus it appears that labyrinthodonts used all
three methods of propulsion found in salamanders today.

The only direct evidence of the gaits used by labyrinthodonts
comes from fossil trackways. Romer (1941) observed that the oldest
known fossil footprints resemble salamander footprints in pointing
forward throughout the stride. Unfortunately, the actual relations between the fossil footprints in a single trackway can be
produced by either a walk or a trot, depending on the trunk
length of the animal which made them. (Peabody, 1959, presented
a good discussion of the effect of trunk length on footprint
patterns.) By analogy with salamanders, it is probable that
labyrinthodonts could use both the walk and the trot; certainly
there is nothing in their known morphology which would prohibit
any labyrinthodont from using both gaits.

## WHAT WAS THE FIRST TERRESTRIAL GAIT?

Thomson (1969) has shown that in open water rhipidistian
fishes doubtless swam using traveling waves which passed down the
body, as the majority of fish do today. It is reasonable to
assume that rhipidistians also used traveling wave locomotion on
land, at least in their initial terrestrial forays. Thomson
(1969, 1972) noted the relatively weak musculature of the paired
fins of rhipidistians, and speculated that such fins would not
have been able to move the animals solely by fin protraction and
retraction in a terrestrial situation. However, as noted above,
a terrestrial gait (the trot) can be produced by traveling waves
even if the limbs act as mere struts (see Fig. 9). If the
traveling waves are so spaced that a new wave is initiated at the
anterior end of the trunk when the immediately preceding wave has
reached the mid-trunk region, then girdle rotation alone will move
the limbs in a trot. The limb acts as a simple strut and the only
limb muscles required are those which fix the limb on the body and
those which lift the limb from the substrate as it is passively

protracted.  The limb muscles of the rhipidistians were probably strong enough to accomplish both of these tasks.  Therefore it is suggested that the first terrestrial gait was a trot.  This hypothesis directly contradicts Sukhanov (1974), who stated that the first terrestrial gait was the walk and that the trot was not evolved until the reptilian grade of organization.  However, the following considerations provide evidence for the priority of the trot:

(1)  All salamanders which I have examined have used traveling wave locomotion, both while swimming and while trotting or using fast undulatory locomotion on land.

(2)  All salamanders have sprawling posture, which has often been considered a less effective propulsive situation than the mammalian condition, in which the limbs are placed vertically under the body.  However, locomotion by traveling waves requires a sprawling posture, and if the limbs are brought under the body, then even a moderate amount of lateral bending leads to instability by removing the body's center of gravity from the triangle of support provided by the limbs.  If the primary tetrapodal gait were the walk, one might expect selection for bringing the limbs under the body, with the suppression of lateral undulation.  If, on the other hand, the primary tetrapodal gait were the traveling wave trot, then there should have been strong selection for retaining sprawling posture.  Thus the hypothesis of the primacy of the trot explains the lateral placement of the glenoid and acetabulum, with the consequent sprawling posture, seen in all limbed fossil amphibians.

(3)  Sukhanov (1974) did not observe the trot in any amphibians.  However, he only examined four specimens of one species of salamander moving at relatively slow speeds. In my own studies, I have found that all terrestrial salamander species examined have used the trot.

(4)  Since rhipidistians had the neural control system to effect traveling wave locomotion in the water, they could have moved on land using the trot without any further neural or morphological specializations.  A walking gait in an organism with sprawling posture is effected by means of standing waves in the vertebral column, and also requires more precise orientation of the limbs.  This more complicated neuromuscular coordination was probably evolved after the invasion of the terrestrial habitat.

Recent work has suggested that vertebrate locomotor patterns, including gaits, are controlled by neural centers located in the spinal cord and brain (see review in Grillner, 1975). Brandle and Szekely (1973) have provided evidence for two kinds of central control of locomotion in salamanders. One mechanism is associated with the production of traveling waves; they call this the "sine wave generator" because, under its influence, the body is thrown into traveling sine waves. This generator appears to be associated with the reticular formation, which extends from the brain stem to the end of the spinal cord. Within the medulla is located, presumably, an oscillator, which sends waves of contraction alternately down each side of the spinal cord.

The second locomotor control mechanism of salamanders is a limb-moving generator, located on each side of the spinal cord in front of each limb. Brandle (1968) and Brandle and Szekely (1973) showed that if a short span of the vertebral column containing only the girdle and limbs was transplanted into the middle of the trunk of Triturus vulgaris, the limbs would not move alternately and in concert with the normal limbs of the walking animal. However, if the two segments rostral to the girdle-containing portion were included in the transplant, the limbs then followed their normal pattern. Therefore, these workers concluded that the two segments anterior to each limb contained the limb-moving generator. The control center for these limb-moving generators appears to be located in the posterior two-thirds of the medulla (Brandle and Szekely, 1973).

Szekely and Czeh (1967) provided some clues as to the nature of the limb-moving control process. They mapped in detail the neurons of the three segments of the spinal cord which give rise to the brachial plexus in Ambystoma mexicanum. They found that the neurons controlling the protractors and flexors of the foot were located in the anterior part of this section, while the neurons controlling the retractors and extensors of the limb and both flexors and extensors of the foot were located in the posterior half of the section. Thus, a simple wave of stimulation passing down the spinal cord would cause the normal movements of the forelimb seen in walking salamanders (Szekely et al., 1969). I suspect that the same arrangement of limb controlling neurons will be found in the segments contributing to the iliac plexus and, thus, controlling the movement of the hind limb.

I hypothesize that these two kinds of control mechanisms, the sine-wave and limb-moving generators, are the producers of traveling and standing waves, respectively, in salamanders. Waves of stimulation produced in the medulla and traveling posteriorly in the reticular system lead to traveling waves in the axial muscles. Waves of stimulation produced by the posterior two-thirds of the medulla and traveling posteriorly in some as yet unidentified neural system lead to standing waves in the axial muscles and also produce the normal pattern of limb movements seen in the lateral sequence walk. At slow speeds, the limb-moving system would predominate, leading to the standing wave pattern seen in salamanders using the walk. At faster speeds the sine-wave generator would predominate, leading automatically to the production of a trot. The animal must have some control over which generator is used at intermediate speeds, however, in order to explain the use of either gait at such speeds.

Presumably the sine-wave generator will be found in all fishes that swim by lateral undulation. This generator would then have been present in rhipidistian fishes and used to produce a trot as they ventured into terrestrial habitats; the limb-moving generator would have been added as the proto-amphibian became better adapted to terrestrial locomotion and added the walk.

## SUMMARY

The selective pressures which led to terrestrialism among vertebrates were probably manifold and conflicting. Negative pressures such as overpopulation and predation in the pond environment may well have combined with such positive pressures as a terrestrial food supply and dispersal. The pectoral fins of rhipidistians, which evolved as an adaptation to slow locomotion in shallow water, were preadapted to aid in respiration on land by holding the anterior end of the body off the substrate. Studies of the locomotor behavior of salamanders, which are the best modern locomotor analog of the first tetrapods, should help reveal further information about the evolution of tetrapody.

In terrestrial locomotion, salamanders use girdle rotation, limb retraction and humeral-femoral rotation as means of propulsion. Limb retraction supplies over half of the total propulsion, and girdle rotation contributes up to one-fifth of the total. Girdle rotation, in which the propulsive power is supplied by the axial muscles, requires a sprawling posture to be effective. Limb rotation is a special kind of limb retraction and is produced by certain retractor muscles which insert onto a process of the humerus or femur called the crista ventralis. Labyrinthodont amphibians probably used all three of these methods of propulsion, since the members of this group had musculature very similar to that of salamanders and also possessed sprawling posture and cristae ventrales.

As salamanders progress from very slow to fast speeds, they switch from the lateral sequence walk to the trot. Walking salamanders use standing waves in the trunk region but may use standing or traveling waves in the tail. Trotting salamanders use traveling waves in both the trunk and tail. Over a wide range of speeds any individual may use either the walk or the trot. Although the gaits used by labyrinthodonts are unknown, there is nothing in their morphology which would prevent them from using either of these salamander gaits.

At very fast speeds salamanders abandon their limbs and move by traveling waves of the trunk which are identical to those used during swimming. This type of movement, called fast undulatory locomotion, is normally used only as an escape behavior.

Rhipidistian fishes could have moved overland by using fast undulation. However, during this type of locomotion the ventral surface of the body rests on the substrate, and in an animal lacking strong ribs the abdominal cavity would be compressed, hindering respiration via the lungs. Theoretical considerations show that even an animal with relatively weak limb musculature can progress in a trot by the use of traveling waves in the axial muscles. All that is required is that the animal fix the limbs on the girdles and space the traveling waves so that a new wave is initiated at the anterior end of the trunk when the immediately preceding wave has reached the mid-trunk region. Therefore it is suggested that the rhipidistians used the traveling wave trot for terrestrial locomotion. Several aspects of salamander locomotion, including sprawling posture and the neural control of locomotor behavior, support this hypothesis.

## ACKNOWLEDGEMENTS

This paper is based on part of a Ph.D. thesis submitted to the University of California, Berkeley. Additional research supported by National Institutes of Health institutional grant 5-S05-FR-06076 and City University of New York Faculty Research Foundation grant 17057.

## REFERENCES

Barclay, O.R., 1946, The mechanics of amphibian locomotion. J. Exper. Biol., 23:177.

Bolt, J.R., 1969, Lissamphibian origins: possible protolissamphibian from the Lower Permian of Oklahoma. Science 166:888.

Brandle, K., 1968, Die Bewegungsweise sechsbeiniger Axolotl mit verlangertem Ruckenmark, Zool. Anz., suppl. 32:448.

Brandle, K., and Szekely, G., 1973, The control of alternating coordination of limb pairs in the newt (<u>Triturus</u> <u>vulgaris</u>), Brain Behav. Evol., 8:366.

Coghill, G.E., 1929, "Anatomy and the problem of behavior,"
    Cambridge University Press.
Cowles, R.B., 1958, Additional notes on the origin of the
    tetrapods. Evol., 12:419.
Cox, C.B., 1967, Cutaneous respiration and the origin of the
    modern Amphibia. Proc. Linn. Soc. Lond., 178:37.
Daan, S. and Belterman, T., 1968, Lateral bending in locomotion
    of some lower tetrapods. Proc. Ned. Akad. Wetten., ser. C,
    71:245.
Dean, B., 1906, Notes on the living specimens of the Australia
    lungfish, Ceratodus forsteri, in the Zoological Society's
    collection, Proc. Zool. Soc. Lond., 1906:168.
Eaton, T.H., Jr. and Stewart, P.L., 1960, A new order of fishlike
    Amphibia from the Pennsylvanian of Kansas. Univ. Kansas
    Publ. Mus. Nat. Hist., 12:217.
Edwards, J.L., 1976, A comparative study of locomotion in
    terrestrial salamanders, unpublished Ph.D. thesis,
    University of California.
Estes, R., 1965, Fossil salamanders and salamander origins.
    Amer. Zool., 5:319.
Ewer, D.W., 1955, Tetrapod limb. Science 122:467.
Faber, J., 1956, The development and coordination of larval
    limb movements in Triturus taeniatus and Ambystoma mexicanum
    (with some notes on adult locomotion in Triturus). Arch.
    Neerl. Zool., 11:498.
Feduccia, J.A., 1971, The origin of terrestrial Amphibia. Tex.
    J. Sci., 22:255.
Goin, C.J. and Goin, O.B., 1956, Further comments on the origin
    of tetrapods. Evol., 10:440.
Gray, J., 1944, Studies in the mechanics of the tetrapod
    skeleton. J. Exper. Biol., 20:88.
Gray, J., 1968, "Animal locomotion," Weidenfeld and Nicolson,
    London.
Grillner, S., 1975, Locomotion in vertebrates: central mechanisms
    and reflex interaction. Physiol. Rev., 55:247.
Gunter, G., 1956, Origin of the tetrapod limb. Science 123:495.
Hesse, R., 1935, Die Tierkorper als selbstandinger Organismus,
    in "Tierbau und Tierleben," (R. Hesse and F. Doflein, eds.),
    Jena.
Hildebrand, M., 1965, Symmetrical gaits of horses. Science
    150:701.
Hildebrand, M., 1966, Analysis of the symmetrical gaits of
    tetrapods. Fol. Biotheor., ser. B, 6:9.
Hildebrand, M., 1974, "Analysis of vertebrate structure," John
    Wiley and Sons, New York.
Holman, J.A., 1969, Predation and the origin of tetrapods.
    Science 164:588.
Howell, A.B., 1944, "Speed in animals. Their specialization for
    running and leaping," reprint by Hafner Publishing Company,
    New York, 1965.

Inger, R.F., 1957, Ecological aspects of the origin of the
    tetrapods. Evol., 11:373.
Jarvik, E., 1942, On the structure of the snout of crossoptery-
    gians and lower gnathostomes in general. Zool. Bidrag
    Uppsala, 21:235.
Jarvik, E., 1960, "Theories de l'evolution des vertebres
    reconsiderees a la lumieres des recentes decouvertes sur
    les vertebres inferieurs," Masson et Companie, Paris.
Jarvik, E., 1968, Aspects of vertebrate phylogeny, in "Current
    problems of lower vertebrate phylogeny," (T. Orvig, ed.),
    pp. 497-527, Almquist and Wiksell, Stockholm.
Johnels, A.G. and Svensson, G.S.O., 1954, On the biology of
    Protopterus annectens (Owen). Ark. Zool., 7:131.
Miner, R.W., 1925, The pectoral limb of Eryops and other
    primitive tetrapods. Bull. Amer. Mus. Nat. Hist., 51:145.
Olson, E.C., 1976, The exploitation of land by early tetrapods.
    Biol. J. Linn. Soc. Lond. (in press).
Orton, G.L., 1954, Original adaptive significance of the
    tetrapod limb. Science 120:1042.
Parsons, T.S. and Williams, E.E., 1963, The relationships of the
    modern Amphibia: a re-examination. Quart. Rev. Biol.,
    30:26.
Peabody, F.E., 1959, Trackways of living and fossil salamanders.
    Univ. Calif. Publ. Zool., 63:1.
Rohdendorf, B.B., 1970, The importance of insects in the evolu-
    tion of land vertebrates. Paleontol. J., 4:5.
Romer, A.S., 1941, The first land animals. Nat. Hist., 48:236.
Romer, A.S., 1958, Tetrapod limbs and early tetrapod life.
    Evol., 12:365.
Roos, P.J., 1964, Lateral bending in newt locomotion. Proc.
    Ned. Akad. Wetten., ser. C, 67:223.
Schaeffer, B., 1965, The rhipidistian-amphibian. Amer. Zool.,
    5:267.
Schmalhausen, I.I., 1957, Biological basis of the origin of the
    tetrapods. Izvest. Acad. Sci. U. R. S. S., 1:3 (in Russian).
Schmalhausen, I.I., 1968, "The origin of terrestrial vertebrates,"
    Academic Press, New York.
Slijper, E.J., 1941, De Voortbewegingsorganen, in "Leerboek der
    vergelijkende ontleekunde van de vertebraten," (E.W. Ihle,
    ed.), pp. 95-222, Utrecht.
Sukhanov, V.B., 1974, "General system of symmetrical locomotion
    of terrestrial vertebrates and some features of movement
    of lower tetrapods," Amerind Publishing Company, New Delhi.
Szarski, H., 1962, The origin of the Amphibia. Quart. Rev. Bio.,
    37:189.
Szekely, G. and Czeh, G., 1967, Localization of motoneurones in
    the limb moving spinal cord segments of Ambystoma. Acta
    Physiol. Acad. Scient. Hung., 32:3.

Szekely, G., Czeh, G., and Voros, G., 1969, The activity pattern of limb muscles in freely moving normal and deafferented newts. Exper. Brain Res., 9:53.

Thomson, K.S., 1968, A critical review of the diphyletic theory of rhipidistian-amphibian relationships, in "Current problems of lower vertebrate phylogeny," (T. Orvig, ed.), pp. 285-305, Almquist and Wiksell, Stockholm.

Thomson, K.S., 1969, The biology of the lobe-finned fishes. Biol. Rev., 44:91.

Thomson, K.S., 1972, New evidence on the evolution of the paired fins of Rhipidistia and the origin of the tetrapod limb, with description of a new genus of Osteolepidae. Postilla Peabody Mus., 157:1.

Warburton, F.E. and Denman, N.S., 1961, Larval competition and the origin of tetrapods. Evol., 15:566.

# THE CONTRIBUTION OF PALEONTOLOGY TO TELEOSTEAN PHYLOGENY

Colin PATTERSON

Dept. of Paleontology, British Museum (Natural History)

London, S.W.7, United Kingdom

## INTRODUCTION

I have been given a title concerning what Medawar (1967:23, 74) called 'a comparatively humble and unexacting kind of science,' research in 'the parish registers of evolution.' To avoid too much parochial detail, I will deal only with the phylogeny of the teleosts as a whole, without going into the different subgroups in any detail, and I shall emphasize generalities, using the teleosts as an example in a discussion of the role of phylogenetic paleontology. I shall try to assess paleontological methods, as practiced by workers on fossil teleosts, with the aim of identifying failures and successes, and of suggesting what contributions to phylogeny can and cannot be expected from paleontologists and their material. With these aims, my approach must be historical, and the main body of the paper is a historical review, with commentary. Since the teleosts are the first group for which a recognizable phylogenetic diagram was proposed (Agassiz, 1844), this review covers a longer period than most reviews of phylogenetic ideas. The review of nineteenth century work is fairly complete, but in the twentieth century, where the literature is more voluminous and well-known, I have cited only selected papers, particularly those containing phylogenetic diagrams, which summarize ideas more concisely than extended quotation.

The fossil record of higher actinopterygians is comparable, in quantity and quality, with that of mammals, as various observers have noted (e.g., Agassiz, 1844: viii; Olson, 1971:519). Regarding quantity, I have not tried to count the named fossil species in teleosts and in mammals, but I am confident that those numbers are of the same order of magnitude (whereas mammals have teeth, teleosts have otoliths). Where quality is concerned, the proportion of fossil teleost species represented by intact skeletons is greater than in mammals. Yet, by almost any standard, there is a gross discrepancy in the influence that fossils have had in phylogenetic work in teleosts and mammals. Some idea of this discrepancy comes from a comparison of the number of extinct higher taxa in Romer's (1966) classification of the two groups: at the family level, 60% of mammalian families are extinct (171 out of 283), whereas only 6% of teleostean families are (23 out of 402); at the ordinal level, 47% of mammalian orders are extinct (15 out of 32), but only 7% of teleostean orders are (2 out of 30). My aim is to explain this discrepancy. I think it has a historical basis, arising from a concomitant time-lag in the analysis of the interrelationships of Recent mammals and teleosts. My principal conclusion is that interpretation of the structure and relationships of fossils is necessarily secondary to those aspects of Recent organisms. Paleontologists do not have unique access to evidence of phylogeny. Rather, the theory of evolution gives them a method of interpreting fossils.

I am grateful to Miss Alison Longbottom for her help with preparing the manuscript and illustrations, and to authors and others who have allowed me to reproduce diagrams.

HISTORICAL REVIEW

The Nineteenth Century: Agassiz to Woodward

The foundation of fish paleontology is Louis Agassiz's Recherches sur les Poissons Fossiles (1833-44). The only text-figure in that work is in the Essai sur la classification des poissons (vol. 1:165-172, 1844). It is reproduced here as Figure 1. Agassiz called this diagram 'genealogy of the class of fishes,' and is explicit about how it was constructed.[1] From the trouble that he took to explain his diagram to the reader, it is clear that Agassiz invented the format, and that this is the first diagram of its type. Today, such diagrams are familiar from many examples (e.g., Figure 2). In Romer's (1966) textbook similar pictures are captioned as 'family trees' (fig. 14), 'phylogenies' (figs. 156, 207, 337), and diagrams of 'evolution and distribution in time' (fig. 26) or 'development' (fig. 67), but never as 'genealogies', as in Agassiz's original. These diagrams are the conventional way of summarizing topics like the title of this paper. But Agassiz's example shows clearly that belief in evolution is not necessary for the production of such diagrams[2]: the information contained in these diagrams is therefore not necessarily concerned with evolution or phylogeny.

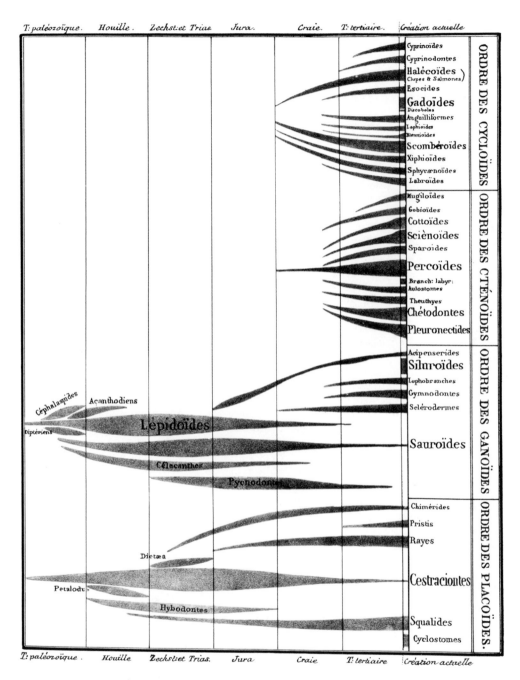

Fig. 1. 'Genéalogie de la classe des poissons.' (From Agassiz, 1844, vol. 1:170.)

If Agassiz did not see the diverging lines in his 'genealogy' as lines of descent, it is reasonable to ask how he did visualize them.  He is explicit about this (1844:170): 'having recognized that the species of each formation are always different from those of other epochs, I have drawn the boundaries of the geological horizons across all the ascending lines of the families to show that the genealogical development of species is repeatedly inter-rupted, and that if, in spite of that, each stock gives us indica-tions of regular progression, this relationship {filiation} is not really the result of a continuous lineage, but of a reiterated manifestation of the order of things "determined in advance".' (my translation).

In his Essay on Classification (1857, 1859; pagination quoted from the later edition) Agassiz amplified his opinion on these matters.  He regarded the supraspecific taxonomic categories not as convenient devices invented by man, but as 'having the same foundation in nature as species' (p. 8).  This foundation is not material - individual organisms alone are material - but ideal or abstract (p. 256), and is expressed in the individual by uniformity or identity of plan and structure.  'Such an agreement in the structure of animals is called their homology, and is more or less close in proportion as the animals in which it is traced are more or less nearly related' (p. 26).  These degrees of relationship are recognized by homologies at more or less fundamental levels, and the more fundamental the homology, the higher the rank of the category represented.  The higher categories are regarded as being successively more long-lived than genera, which in turn are more long-lived than species (pp. 30, 160), and, as represented by fossils, 'all these types bear, as far as the order of their suc-cession is concerned, the closest relation to the relative rank of living animals of the same types' (p. 49).  'It may be said that the earliest fishes are rather the oldest representatives of the type of Vertebrata than of the class of Fishes, and that this class only assumes its proper characters after the introduction of the class of Reptiles {including Amphibia} upon the earth.  Similar relations may be traced between the Reptiles and the classes of Birds and Mammalia, which they precede' (p. 166).

Thus Agassiz's conception corresponds exactly to one illus-trated by Simpson (1961, fig. 19A; cf. Figure 3A).  Of this diagram, Simpson writes 'such a conception is flatly false,' and denotes as the 'correct relationship of temporal-spatial taxa to the phylogeny' the concept illustrated in Figure 3B.  According to Simpson (1961, fig. 19C), diagrams like that in Figure 3A can only be employed without a time-axis, in which case they indicate 'neither phylogeny nor taxa but relationships among the terminal species.'  Simpson did not specify what is meant here by 'relationships', but from later papers (1975, 1976) it seems that he would agree with Mayr's

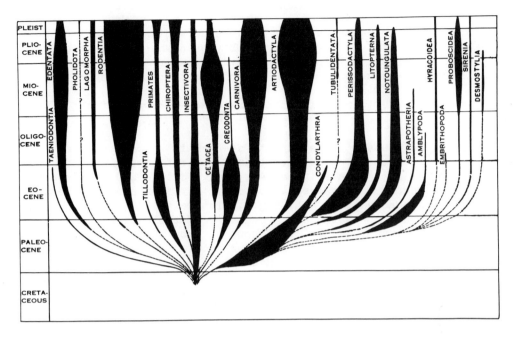

Fig. 2. "The chronologic distribution of the placental mammals."
(From Romer, 1966, fig. 316.)

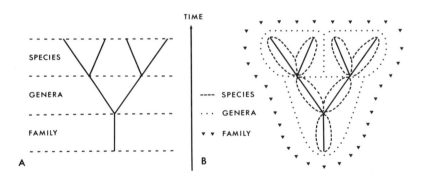

Fig. 3. Diagram contrasting two views of the relationship between
phylogeny and classification. (Modified from Simpson, 1961, fig.
19.)

(1965:79) statement 'when a biologist speaks of phylogenetic relationship, he means relationship in gene content' (cf. Simpson,
1975:7 - 'genetic affinities'). In the same paper, Mayr writes
(1965:94) 'it was legitimate for the pre-Darwinian naturalists to
speak of "affinity" in a vague and ambiguous manner,' but I am not
sure that Agassiz's 'affinity', explained in terms of homologies
at different levels, with embryology the most trustworthy guide to
those levels (1859:127), is any vaguer or more ambiguous than
Mayr's 'phylogenetic relationship' explained in terms of genes in
common.

    I have discussed these aspects of Agassiz's ideas at some
length in order to bring out two points. First, Agassiz's conception of relationship (Figure 3A), though criticized by Simpson
and others, seems fully acceptable today (cf., for example, Read,
1975, fig. 1; MacFadden, 1976, fig.1; Platnick, 1976, fig. 2), and
indeed is argued to be the only concept compatible with monophyletic
taxa (e.g., Lovtrup, 1973:51). I hope to show that the uncritical
assumption that the concept of Fig. 3A should be replaced by that
of Fig. 3B (the latter can be traced back to Darwin's text-figure,
1859:117) is responsible for the unwarranted dominance of paleontology in phylogenetic studies during the century following Darwin.

    Second, although Agassiz was opposed to evolution and ignorant
of phylogeny (a word coined by Haeckel, 1866), his concept of relationship is such that the word 'classification' in his works
(and in those of many other pre-evolutionary taxonomists) can
usually be replaced by 'phylogeny' without altering the sense.
For example, Agassiz's diagram of 'genealogy' (Fig. 1) corresponds
exactly with his classification, at the level specified by the
diagram, in showing 45 families of fishes which fall into four
orders. These orders, Cycloidei, Otenoidei, Ganoidei and Placoidei,
were Agassiz's main contribution to fish classification. He named
and defined the orders on the basis of scale structure, since this
was a feature which could be observed in the fossils: Agassiz's
ordinal scheme can therefore be regarded as a contribution from
paleontology at the highest level of fish taxonomy. Unfortunately,
his scheme was short-lived, for it was immediately replaced by
Muller's (1844, 1845, 1846a,b) 'remarkable system', as Berg (1940:
346) called it.

    The opening sentence in Muller's paper (quotations and pagin-
ation from Griffith's translation) are: 'No branch of natural
history presents a more striking proof of the importance of a
knowledge of extinct fossil genera in the natural classification
{we could write 'phylogeny'} of animals generally, and in particular of living species, than ichthyology. Paleontology has
essentially altered the basis of this part of the system.' However, I read these words as a sop to Agassiz, for Muller's work

destroyed Agassiz's system, and contains criticisms of paleontology which I regard as valid. Muller goes on to discuss the place that paleontology should hold in the classification (again, we could write 'phylogeny') of fishes. He contrasted (p. 502) 'the slight assistance fossils render us' with the information available in Recent fishes, which 'alone are perfectly accessible' (p. 503). Muller was clearly of the opinion that fossils are necessarily subsidiary to Recent animals.

In Muller's paper the Teleostei, Holostei and Chondrostei are named and characterized for the first time. His subclass Teleostei (which is the same as ours) contained all Agassiz's Ctenoidei, most of his Cycloidei and some of his Ganoidei. The Holostei and Chondrostei, as orders, form the restricted Ganoidei, raised to the rank of subclass. Muller's groups were defined on features of the soft anatomy - heart valves, optic chiasma, spiral valve, pseudobranch, swimbladder and its duct, genital ducts, etc. - so that all the characters he used, and the groups he defined, were applicable only in living fishes. Muller concluded (p. 519) 'with existing fish we can decide with absolute certainty, from their anatomy, whether they are ganoids or not. But what characters are to guide us with respect to fossil fish?' He could give no satisfactory answer to that question, but he felt that his reorganization of the system left intact Agassiz's maxim that teleosts (Agassiz's cycloids and ctenoids) do not appear in the fossil record before the Cretaceous. On Agassiz's authority - that they have enamelled scales - Muller treated Jurassic genera like Leptolepis and Thrissops as holostean ganoids.

The relative value of the contribution of Agassiz, based on fossils, and of Muller, based on Recent fishes, may be summed up in two quotations. 'All the subclasses erected by Muller have retained their real significance to the present day' (Berg, 1940: 347); and, of Agassiz's classification, 'as time has gone by it has been found to yield no generalizations of fundamental value' (Jordan, 1905:424).

The aspect of Muller's work that concerns us here is his question of how fossil teleosts and holosteans were to be distinguished. This question occupied paleontologists for the next twenty-five years. The earliest, and most successful, attempt to solve the problem was Heckel's (1850a,b; 1851), based on study of the vertebral column and caudal skeleton. Heckel examined Recent ganoids and teleosts, and recognized that in ganoids the caudal end of the notochord is unossified and unprotected, and that those ganoids which have vertebrae (Polypterus, Lepisosteus, Amia) can add vertebrae to the caudal end of the column throughout life. In teleosts, on the other hand, the caudal end of the notochord is either hidden beneath roof-like bones (Heckel's 'Steguri', uroneurals in modern terminology) or is enclosed in the terminal

vertebra.  Using these criteria, Heckel reviewed the fossils and
proposed that the Jurassic genera Leptolepis, Thrissops, Tharsis
and Anaethalion, which Agassiz had placed in the ganoids and
Muller in the holosteans, with Amia, were true teleosts, and that
teleosts therefore first appear in the Jurassic.

Pictet (1854) took a more equivocal stand than Heckel.  While
accepting Muller's Teleostei, he followed Agassiz's groupings within fossil teleosts, since the characters used by Muller were not
visible in the fossils.  He noted that Muller's transfer of Amia
to the ganoids made it no longer possible to assign fossils to the
ganoids or to the teleosts.  However, he found that there were
only a few genera of doubtful position, and for those he erected
the family Leptolepidae, writing (p. 26) 'are these closer to
Amia or to teleosts?  The question is insoluble in the present
state of science' (my translation).  Commenting on Heckel's
assignment of Leptolepis, Thrissops and Tharsis to the teleosts,
Pictet noted that Megalurus (= Urocles), which Muller had associated with these genera, has the ganoid type of tail, and wrote
(p. 133) 'perhaps there are two types here which should be separated, one remaining in the ganoids and one going into the teleosts.
I would have accepted this modification immediately, if Agassiz
did not positively say that Leptolepis has enamelled scales' (my
translation).  Accepting this last feature as diagnostic of the
ganoids, Pictet placed the Leptolepidae in that group, and included Leptolepis, Thrissops and Tharsis in one leptolepid tribe,
characterized by a stegural tail, and Urocles and Oligopleurus in
another, characterized by a ganoid tail.  Like Muller, Pictet
therefore upheld Agassiz's maxim that teleosts do not appear until
the Cretaceous.

Thiolliere (1854, 1858) studied the Jurassic fishes of Cerin,
France.  At first (1854:5) he questioned the presence of enamel on
the scales of Leptolepis and Thrissops, and the significance of
that character, and regarded these and Urocles, Oligopleurus and
Belonostomus (which has ganoid scales) as teleosts, agreeing with
Heckel that teleosts are found in the Jurassic.  Later (1858),
Thiolliere was prompted by his study of the fossils to reinstate
Cuvier's system, in which the ganoids and teleosts with abdominal
pelvic fins were united in the order 'Malacopterygiens abdominaux.'
Thiolliere's reasoning was that the characters of that group –
absence of fin-spines and abdominal/pelvics – are readily seen in
fossils ('On dira, si l'on veut, que ce caractere est peu philoso-
phique, mais il est sur et simple'), whereas Muller's characters,
drawn from soft anatomy, were not applicable in fossils.

Wagner (1861, 1863), in his monograph on fishes from the
Bavarian Lithographic Stone, repeated the complaint that the
characters used by Muller to differentiate ganoids and teleosts

are inaccessible in fossils, and therefore proposed choosing a skeletal criterion that would apply only to the local fauna he was studying. He found such a character in the constitution of the vertebral column, and assigned to the ganoids all those forms in which the notochord was persistent, and never surrounded by more than a smooth, ring-like centrum. He therefore included in the teleosts all the fossils with fully-developed vertebrae - Leptolepis, Thrissops, Anaethalion, Oligopleurus, Urocles and Macroorhipis (= Ionoscopus).

Huxley (1861) erected the new ganoid suborder Crossopterygidae to include Polypterus (now removed from the Holostei), rhipidistians, coelacanths and some fossil lungfishes. So far as the teleosts are concerned, Huxley attempted to show that they were represented in the Paleozoic by Coccosteus and Pterichthys, which he compared with Recent catfishes (siluroids). Discussing the validity of these comparisons, he wrote (p. 36) 'since we know that a true Ganoid, Amia, completely simulates the outward form of a Clupeoid Teleostean, while retaining all the essentials of its order, - may not Coccosteus be also a true Ganoid which simulates the outward aspect of a Siluroid? To this question it is, perhaps impossible to give any answer, save by asking another, viz.:- Why should not a few Teleosteans have represented their order among the predominant Ganoids of the Devonian epoch, just as a few Ganoids remain among the predominant Teleosteans of the present day?' In other words, Huxley admits that he is lost: the 'essentials' of ganoids are only accessible in Recent fishes, and there is no way of distinguishing fossil ganoids from fossil teleosts.

Lutken (1868, 1869, 1871, 1873) reviewed all earlier work, in neontology and paleontology, on the limits of the ganoids. From this review he concluded (Dallas's translation, 1871:330) 'no one has ever been able to give an exact definition of what is a Ganoid . . . . We must therefore, provisionally at least, limit the name of Ganoids to the indubitable existing types (that is to say, the Lepidostei and Polypteri), and to the fossil types which will naturally group themselves around these.' Lutken decided (p. 336) that Amia 'is a special type, belonging to the true Physostome Teleosteans,' and that 'the Jurassic Teleostei (Leptolepids, Megaluri and Caturi) . . . . are consequently true Physostome Teleosteans . . . . The Leptolepides and Megaluri have the true biconcave vertebrae of the Teleostei, but there is nothing astonishing in the fact that there was among the most ancient Teleostei a type (the Caturi) with a more embryonic spinal column . . . . with regard to the position and rank which the ganoids should occupy in the system, it will be necessary to form with them a suborder of the Physostome Teleostei.' As well as the ganoids, Lutken included the lungfishes in the Teleostei, so that

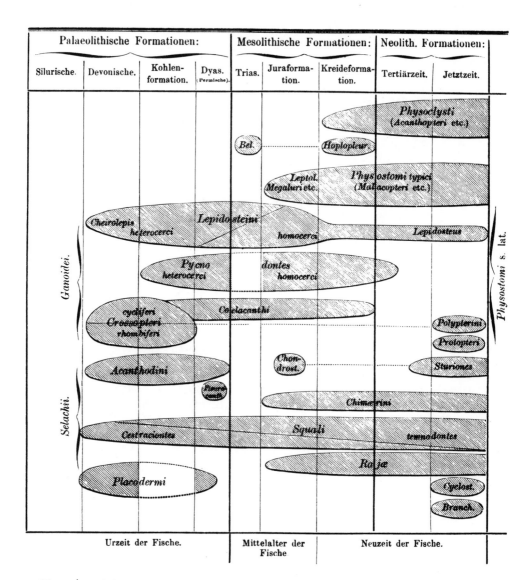

Fig. 4. Diagram of the chronological distribution of the main groups of fishes. (From Lütken, 1873:45.)

his conception of the group was even wider than Owen's (1866)
Teleostomi.  Lutken ends his work with a diagram (reproduced here
as Figure 4) 'of the geological development of the larger groups'
of which he wrote (p. 339) 'when compared with that in the great
work of Prof. Agassiz, will furnish the means of seizing at a
glance the principal progress made in palaeichthyology from 1843
to 1869.'  Comparing Figures 1 and 4, one can only remark that
little progress had been made.

Lutken's work marks the end of an era.  For as well as summarizing the work of the previous 25 years, it is directed towards
the same end as that work - the search for characters that are
applicable in fossils and will delimit major groups of fishes.
That approach had already been in question for ten years when
Lutken's papers were published.

In the first edition of The Origin of Species, Darwin cited
the appearance of the teleosts in the Cretaceous as 'the case
most frequently insisted on by palaeontologists of the apparently
sudden appearance of a whole group of species,' but by the sixth
edition he was able to contrast that opinion, credited to Agassiz,
with the assignment of some Jurassic and Triassic forms to the
Teleostei (cf. Fig. 4), and with Huxley's opinion that the
Devonian placoderms were teleosts (Peckham, 1959, IX, 203-209).
The first attempt to bring evolutionary theory to bear on teleosts
was by Kner (1866).  Having pointed out that all previous attempts
to define the ganoids included words like 'mostly' and 'usually',
Kner concluded that the ganoids were not a natural group, that
recognition of the group had held back the development of a
natural system, that the characters proposed to define the ganoids
were merely primitive, that the fossil ganoids were merely the
ancestors of living ganoids and teleosts, and that the ganoids
'do not represent a single definite order, but rather the whole
amount of development of the Recent Teleostei; they are the expression of the law of progressive evolution in the class of
fishes, whose principal types and great families are already
represented among them by prototypes' (Lutken's translation, 1868).
Kner went into no detail on the supposed relationship of the
various ganoid subgroups to teleosts, but this area was sketched
in during the same year by Haeckel (1866).

Haeckel maintained the ganoids, as a subclass, and divided
them into three legions, Tabuliferi (including sturgeons, placoderms, cephalaspids, etc.), Rhombiferi (including Lepisosteus,
Polypterus, acanthodians, palaeoniscoids, Devonian dipnoans, etc.)
and Cycliferi (including Amia, Urocles, Oligopleurus, coelacanths
and rhipidistians, etc.).  The first of these groups was regarded
as derived from selachians, and as ancestral to the other two
groups.  In turn, the Cycliferi gave rise to the teleosts, through

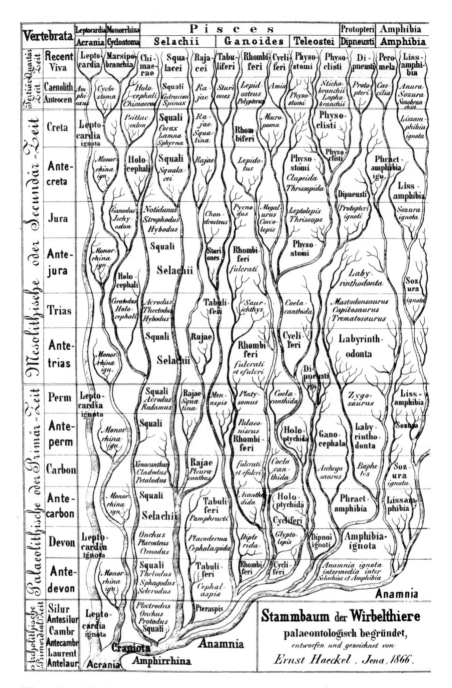

Fig. 5.  Phylogeny of lower vertebrates.  (From Haeckel, 1866, vol. 2, pl. 7.)

a link between the Jurassic megalurids and the leptolepids
(Leptolepis, Thrissops, Tharsis). Haeckel's diagram expressing
these ideas (reproduced here as Figure 5) is now of interest
chiefly for its overtly tree-like form, rather than for its
content. So far as fishes are concerned, Haeckel adapted a somewhat outdated pre-Darwinian taxonomy to the tree format, replacing
'affinity' by 'ancestry'. In fact, the two trees of fishes in
Haeckel's plate 7 show mutually inconsistent branching sequences.
Nevertheless, the form of his diagrams and his introduction of
ancestral groups into classifications were most influential. In
later works (e.g., Figures 6, 7), Haeckel simplified his trees
and complicated his taxonomy, but introduced no new factual justification for his schemes.

     With these diagrams, Haeckel set a pattern that is still
followed, but the fish taxonomy he used was soon superseded. The
first major contribution came from Cope (1871, 1872). On the
basis of his studies of the skeleton of Recent fishes, Cope did
away with the ganoids, distributing the living forms between two
new subclasses, Crossopterygia (Polypterus) and Actinopteri
(Chondrostei, Amia, Lepisosteus), and also dropped the term
Teleostei, distributing the teleosts into two 'tribes', Physostomi
and Physoclisti (names taken from Muller). Amia and Lepisosteus,
each in an order of its own, were included in the Physostomi.
Discussing the phylogeny of the teleosts, Cope regarded the catfishes (his Nematognathi) as (1871:453) 'the nearest ally to the
sturgeons (Chondrostei) among Physostomous fishes, and I imagine
that future discoveries will prove that it has been derived from
that division by descent. In the same way the Isospondylous fishes
are nearest to the Halecomorphi {Amia}, and have probably descended
from some Crossopterygian near the Haplistia, through that order'
(the Haplistia was an 'empty' order named by Cope to contain as
yet undiscovered crossopterygians with lungfish-like dorsal fins).
Here, for the first time, is a clear statement that the teleosts
might be polyphyletic.

     In 1872, Gill published his masterly Arrangement of the
families of fishes (1872b). Like Cope's work, Gill's was based
only on living fishes (mainly on the skeleton), and was evolutionary. Gill reaffirmed the importance of the characters of soft
anatomy that Muller used to separate ganoids and teleosts, and
added two new skeletal characters of ganoids, many separate bones
in the lower jaw and an endoskeletal shoulder-girdle lacking the
three bones found in generalized teleosts. Gill summarized his
conclusions in a diagram (reproduced here as Figure 8) which he
called a 'quasi-genealogical tree.' This diagram is based on a
plan used, and more fully explained, in Gill's classification of
mammals (1872a). It is interesting to compare Gill's 'tree' with
Haeckel's earlier attempt (Fig. 5). Gill's descends on the page,

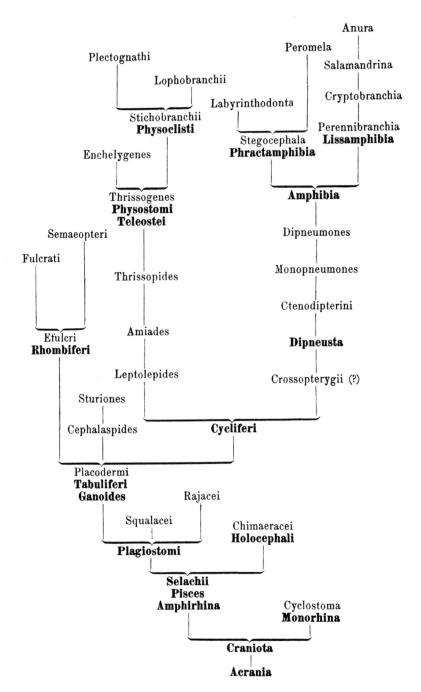

Fig. 6.  Phylogeny of lower vertebrates.  (From Haeckel, 1889:613.)

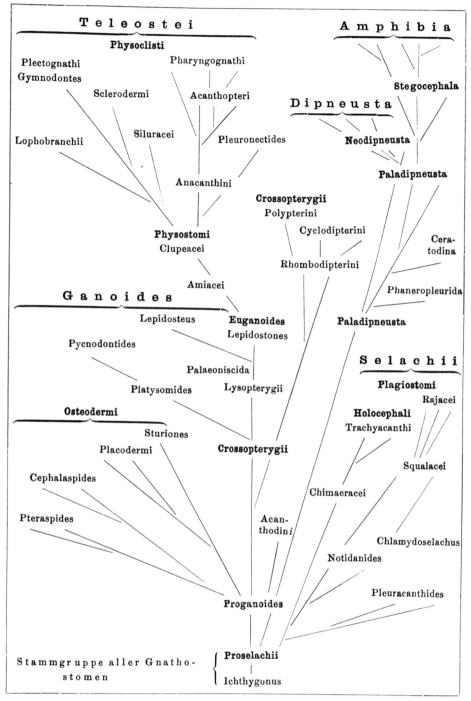

Fig. 7. Phylogeny of fishes. (From Haeckel, 1895:237.)

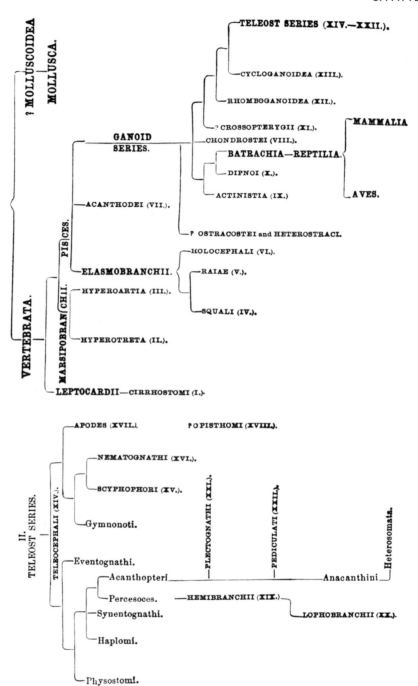

Fig. 8.  "A quasi-genealogical tree' of the vertebrates.  (From Gill, 1872b:xliii.)

like a conventional human genealogy, and has no time-axis.  He
employed the convention that (1872b: xlii) 'in all cases (except
the Vertebrates and Molluscoids), the branch to the left - major
as well as minor - indicates the supposed most generalized type
of the two or more springing or diverging from the same common
stem.'  As Figure 8 shows, Gill, like Haeckel and Cope, regarded
<u>Amia</u> (Cycloganoidei) as the closest living relative of the teleosts.
Unlike cope, Gill retained the Teleostei as a taxon, and he rejected Cope's Actinopteri.

So far as Recent fishes are concerned, Cope's and Gill's
ideas, and their evolutionary approach, set the tone of work for
the remainder of the nineteenth century.  However, one eminent
ichthyologist, Gunther, held out against Darwin's doctrines.  In
1871 he instituted a new classification in which the Teleostei,
as a subclass, were set against all other gnathostome fishes
(subclass Palaeichthyes) and, as Woodward (1891: vii) put it, 'to
emphasize the division all the more clearly, the "Palaeichthyes"
are arranged in ascending series, so far as can be determined,
while the "Teleostei" are treated in precisely the opposite order.'
Gunther maintained this arrangement in his textbook (1880), where
in discussing the ganoids and teleosts he wrote (p. 373) 'there is
no direct genetic relation between these fishes, as some naturalists were inclined to believe.'  This remark led Huxley (1883) to
review once more the characters which Muller had used to distinguish ganoids and teleosts, and having found intermediate or
holostean-like conditions of some characters in various primitive
living teleosts, Huxley felt able to write (p. 139) 'there are no
two large groups of animals for which the evidence of a "direct
genetic connexion" is better than in the case of the Ganoids and
Teleosteans.'  I should emphasize that Huxley was referring here
only to living fishes, and only to those features of soft anatomy
that Muller had used.

So far as fossils are concerned, the major contributions
during this period were by Traquair and Zittel.  Traquair's essay
on the fossil ganoids (1877) was strongly influenced by Huxley's
work.  A quotation illustrates Traquair's approach (p. 7) 'We
cannot doubt that the strict separation of groups, however natural
they may be, must always be more or less arbitrary, and that were
we really acquainted with the entire succession of organic forms
such things as absolutely defined "orders", etc., would be found
to have no existence in nature.'  Zittel, in the fish section of
his <u>Handbuch</u> (1887-8), introduced much new factual material, but
his taxonomy, though avowedly evolutionary, maintained an already
archaic system (see reviews by Woodward (1889) and Cope (1887),
who used his review as an opportunity to reclassify the vertebrates).

We are now approaching the period of the teleost volumes

(1895; 1901) of Woodward's <u>Catalogue of Fossil Fishes</u>, publications which put paleoichthyology into shape to meet the twentieth century, and are better treated as the introduction to that period. Here, I will sum up my impressions of developments in the half century between Agassiz and Woodward.

First, the contributions which, in retrospect, appear most decisive - those of Muller, Heckel, Huxley, Cope and Gill - are entirely (Muller, Gill) or primarily (Heckel, Huxley, Cope) based on anatomical studies of living fishes. Explicitly paleontological contributions, like Agassiz's attempt to classify fishes by their scales, and Huxley's opinion that placoderms are teleosts, were either decisively criticized with evidence drawn from living fishes, as Agassiz was by Muller, or in the main ignored, as was Huxley.

Second, there existed, throughout this period, a rather obvious dichotomy between neo- and paleoichthyology. Paleoich-thyologists were obliged to take note of their neontological colleagues' work, but the reverse was not so. This dichotomy is best exemplified by the <u>Zoological Record</u>; during the first 18 years of its existence (1864-1881), the Pisces section contains no reference whatever to paleontological work. Only in 1882, under Boulenger's Recordership, were fossils admitted. There is, presumably, a link between this neglect of paleontology and the primacy of neontology emphasized in the first point above.

Third, the most profound development during the nineteenth century was certainly general acceptance of the theory of evolu-tion. Only after that acceptance was teleost phylogeny an openly discussed question, but it is the contrast between those evolu-tionary discussions, in the work of Haeckel and Cope, for example, and the work of the pre-Darwinian paleontologists that I find most illuminating. From the publication of Muller's characteriza-tion of the Teleostei (1844) until Lutken's 1869 paper, paleoich-thyologists were struggling with a real problem: by what skeletal characters could fossil teleosts be differentiated from fossil ganoids? In my view, an acceptable preliminary solution to that problem was provided by Heckel in 1850, but through a conjunction of accidents (including the fact that Heckel's papers were not illustrated, and that Pictet's respect for Agassiz led him to associate, in the Leptolepidae, fishes with a holostean and a teleostean type of tail) Heckel's lead was not adequately followed through by pre-Darwinian paleontologists. After Darwin, this problem, still unsolved, simply evaporated. If, as Haeckel, Cope, Gill and Huxley said, the teleosts were linked by descent with the group containing <u>Amia</u>, it was not to be expected that any hard and fast line could be drawn between the two groups. Haeckel's scheme (Fig. 5), for example, is simply Pictet's taxonomy adapted to the

evolutionary doctrine:  where Pictet associated forms with a
holostean tail (Megalurus (=Urocles)) and forms with a teleostean
tail (Leptolepis, Thrissops) in separate tribes of the Leptolepidae,
a family which he could not assign with any confidence to either
the teleosts or the holosteans, Haeckel has turned Pictet's
uncertainty into evidence of descent, and regards the first group
as ancestral to the second, which in turn is ancestral to the
remaining teleosts.  The point I wish to underline here is that
the problem had not changed, nor had the available facts:  all
that had changed was the mental set of the investigators.  This
change is symbolized in Figure 3, where the pre-evolutionary concept - abstract relationships evidenced by homologies - is contrasted with the evolutionary concept - real descent of one group
from another, evidenced by stratigraphic sequence.  As a result of
this change in outlook, what had seemed a crippling practical
difficulty for the paleontologist, inability to assign fossils to
major groups, seems to have been converted into a theoretical
virtue, further evidence of the truth of evolution.

Of these three aspects of nineteenth century teleostean
paleontology - the secondary or subservient role of paleontology,
the dichotomy between neontology and paleontology, and the change
in philosophy leading to a changed view of problems - the third
persisted into the twentieth century, but the first and second
were modified with the publication of Woodward's immensely influential Catalogue.

### The Twentieth Century:  Woodward to the Present

In preparing the four volumes of the Catalogue of Fossil
Fishes, Woodward reviewed all the literature, examined every
specimen in the British Museum, and visited every other museum
or collection of note in the world.  Thus he achieved the same
position of authority as Agassiz 50 years before, that of the man
who has seen everything, but Woodward was working in an explicitly
evolutionary framework.  It was Woodward's task to adapt, in detail,
the descriptive paleontology of his predecessors to the evolutionary
schemes of Recent fishes proposed, principally, by Cope and
Huxley (Woodward 1891: viii).

In his treatment of the fishes that concern us here, Woodward
(1895) followed Cope in adopting the group Actinopterygii, in
placing the Paleozoic actinopterygians with the sturgeons in the
Chondrostei, and in dropping the Teleostei as a separate taxon.
But beyond that, Woodward's interpretation of the Mesozoic actinopterygians (1895: xxi) 'differs so much from any hitherto proposed
that it has not been found possible to arrange a synonymy under
the subordinal and family-headings in the Catalogue. . . . The

existing <u>Acipenser</u>, <u>Polyodon</u>, <u>Lepidosteus</u>, and <u>Amia</u> are now shown
to afford a very inadequate and misleading idea of the actinop-
terygian ganoids, on account of their remarkably specialized
nature.  It is thus no longer scientific to regard the "Acipen-
seroidei" as typical members of the group to which they belong;
they are mere degenerate survivors.  It is equally impossible to
justify the conception of the groups "Lepidosteoidei" and
"Amioidei," most of the extinct fishes which are commonly ascribed
to the former being proved in the Catalogue to be much more
closely related to the latter.'  Here, in categorizing Recent
fishes as atypical and even misleading, Woodward states the claim
of paleontology to lead, rather than follow, in ichthyology.

        The arrangement Woodward adopted was to place the non-
chondrostean Mesozoic actinopterygians in three suborders:
Protospondyli (Semionotidae, Macrosemiidae, Pycnodontidae, Eug-
nathidae, Amiidae, Pachycormidae), Aetheospondyli (Aspidorhynchidae,
Lepisosteidae - 'recognition of this group is a confession of
ignorance.' 1895: xvii), and Isospondyli (Pholidophoridae,
Oligopleuridae, Leptolepidae and (1901) a series of extant
families corresponding to this group as named by Cope and delimited by Gill).  The remaining teleosts were treated in a
conservative way, as six suborders.  The most important innovation
here is inclusion of the Triassic and Jurassic Pholidophoridae in
a teleostean subgroup, the Isospondyli.  The Pholidophoridae, a
family first named by Woodward in 1890, contained fishes which
had previously always been included in the holostean ganoids.  In
the <u>Catalogue</u>, Woodward characterized the Isospondyli principally
by the simple mandible, consisting of only two or three separate
bones, but he had not observed the lower jaw in any pholidophorid
(1895: xix), and included this group in the Isospondyli only because of the 'very close resemblance to <u>Leptolepis</u> in general
aspect.'  The other two Mesozoic groups, the leptolepids and
oligopleurids, correspond to the two leptolepid tribes set up by
Pictet (1854), the first group having a teleostean type of tail
and the second a holostean type:  these are the fishes which had
been shuffled back and forth between the teleosts and the ganoids
by nineteenth century paleontologists.

        The encyclopedic coverage of the <u>Catalogue</u>; the originality
of Woodward's treatment of the Mesozoic actinopterygians in volume
3 (for a contemporary view of this see Traquair, 1896); the detailed integration of fossil and Recent teleosts in volume 4 (see
reviews by Boulenger, 1902, and Newton, 1902); and the hints in
that volume of the antiquity and possible origins of such groups
as the herrings, eels (1901: x, 'It therefore seems probable that
. . . the Apodes . . . are directly derived from some group of
Mesozoic fishes which would be termed "Ganoidei"') and acanthopterygians, were enough to ensure that ichthyologists should begin

to take paleontology seriously.  Woodward's Catalogue had a profound influence, still clearly evident.

Woodward included no evolutionary diagrams in the Catalogue,
or in his other works on fishes.  The next such diagram to be
published (reproduced here as Figure 9) was by Woodward's colleague
Boulenger (1904a,b).  Boulenger retained the Teleostei as a taxon,
unlike Woodward, and wrote (1904a:162) 'The precise definition of
the order Teleostei, as compared with the Holostean Ganoids, is a
matter of some difficulty.  The most important character appears
to be the presence of an ossified supraoccipital bone.  Remnants
of primitive characters, such as ganoid scales, fulcra, rudiments
of a splenial bone, spiral valve to the intestine, multivalvular
bulbus arteriosus, are still found in some lower Teleosteans, but
no longer in that combination which characterizes the preceding
Order.'  Boulenger included fossils in his classification (his was
the first general treatment of teleosts in which this was done),
and followed Woodward's Catalogue in his treatment of them.  The
first division of his Malacopterygii (which is roughly equivalent
to Woodward's Isospondyli) consisted of four Mesozoic families
('connecting forms between Ganoids and Teleosts'), the Pholidophoridae, Archaeomaenidae (a new family erected for a Jurassic
genus which Woodward included in the Pholidophoridae), Oligopleuridae and Leptolepidae:  these fossil groups are differentiated
from other teleosts by the presence of enamelled scales and ring-
like or perforated centra.

Boulenger's diagram (Fig. 9) follows a pattern established by
Gill (Fig. 8) and Haeckel (Fig. 6).  It presents an explicit scheme
of descent of one Recent suborder from another, and whereas
Woodward had suggested a diphyletic origin of teleosts in which
eels arose independently, Boulenger's diagram implies that ostariophysans had an independent origin.  The three-grade pattern shown
in Boulenger's diagram proved a popular one.  Figure 10 shows
Romer's (1945) elaboration of it, and Figure 11 Siegfried's (1954)
conversion of Romer's diagram into a paleontological context, by
placing the central axis of the tree along the floor of the Upper
Cretaceous.

In 1907 Gregory published the views of the New York school on
teleostean phylogeny.  His paper includes (p. 440) an important
exposition of the principles of classification:  'blood relationship' (i.e., recency of common ancestry, which Gregory equates
with 'genetic relationship') is distinguished from 'homological
resemblance' (i.e., phenetic similarity), and Gregory wrote 'in
order to make classification correspond even roughly to degrees of
blood relationship, i.e. to phylogeny, we must assign varying
systematic values to different characters in proportion to their
inferred relative phylogenetic age.'  Gregory also discussed the

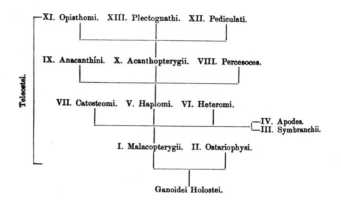

Fig. 9.  Phylogeny of teleostean fishes.  (From Boulenger, 1904a:
162.)

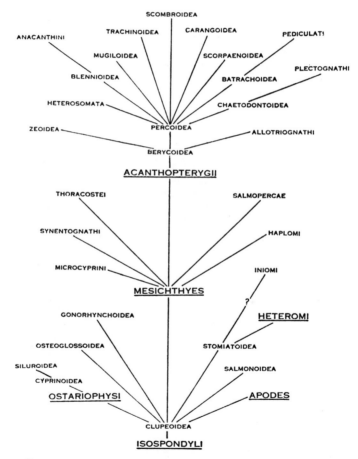

Fig. 10.  "Diagram of the arrangement and relationships of the
main teleost groups.' (From Romer, 1945, fig. 83.)

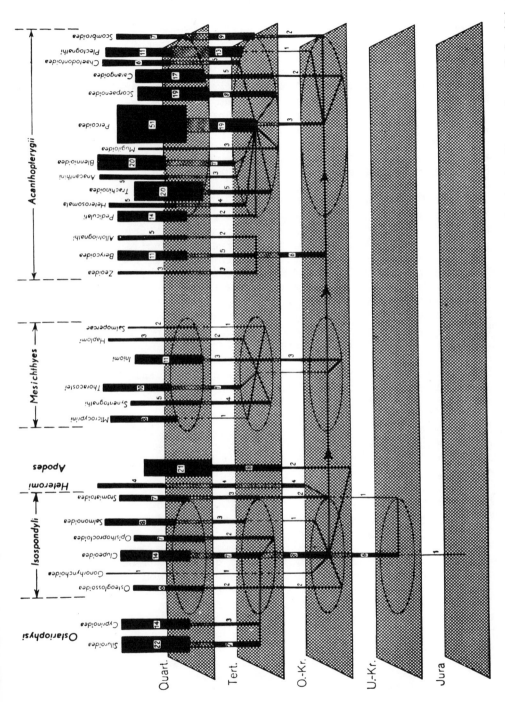

Fig. 11. Scheme of the development of the teleosts. The numerals indicate the number of families in the various suborders. (From Siegfried, 1954, fig. 2.)

problem of ranking:  (p. 443) 'The infraclass is a division recently suggested by Professor Osborn to express the relations of
Marsupials to Placentals. . . The differences between Marsupials
and Placentals do not seem to be more deep-seated than the differences between the Crossopterygii on the one hand and the Actinopteri Cope (all the remaining Ganoids and Teleosts), on the
other.  Hence I regard the Crossopterygii and Actinopteri as subclasses.'  (The Crossopterygii had been ranked as a suborder by
Huxley and Traquair, an order by Gill and Woodward, and a subclass
by Cope.)  On teleost subgroups, Gregory wrote (p. 444) 'such
great groups as the Ostariophysi, the Acanthopterygii proper, the
Haplomi, Isospondyli, etc., which seem in a general way to be of
about the same rank as the orders of Mammalia.'

      Like Boulenger, Gregory maintained the Teleostei as a taxon,
ranked as a cohort.  He followed Woodward and Boulenger in including
the Mesozoic Leptolepidae and Archaeomaenidae in the Isospondyli,
as primitive teleosts, but he transferred the Pholidophoridae to
a new suborder Mesoganoidei included in the order Protospondyli,
while the Oligopleuridae, which Woodward placed with the leptolepids, were placed with _Amia_ in the suborder Halecomorphi of the
Protospondyli.  Gregory summarized his results in a 'diagram of
the phylogenetic relations of the principal families' (pl.29).
That diagram is too complex to reproduce here, but it is similar
to diagrams published by Gregory in 1933 (pl.2) and 1951 (reproduced here as Figure 12).  The caption of the latter, indicating
that virtually the same concepts remained applicable 45 years after
the preparation of the original diagram, gives some idea of the
pace of progress in the understanding of teleostean phylogeny
during the first half of this century.  So far as the fossil forms
are concerned, Gregory's diagrams agree closely with Woodward's
and Boulenger's ideas, as they do in their suggestion that the
ostariophysans and apodans branch off at the very base of the
teleosts.

      Goodrich (1909), in his treatment of actinopterygians in the
Treatise on Zoology, included fossils and relied heavily on
Woodward's Catalogue, like Boulenger and Gregory before him.  He
developed a new classification of teleosts which was dichotomous
where possible.  Goodrich refined this dichotomous scheme in 1930,
and explained his idea of the relationship between phylogeny and
classification (1912: 82) with a paragraph beginning 'The only
"fixed points" in a phylogenetic system of classification are the
points of bifurcation, where one branch diverges from another,'
and with a diagram (1912, fig. 3) resembling Figure 3A.  In 1909,
Goodrich included the Teleostei as an order within the Holostei.
Like Gregory, he excluded the pholidophorids and oligopleurids
from the teleosts, placing both in the Amioidei.  He published a
diagram (1909:370) showing his conception of teleostean phylogeny,

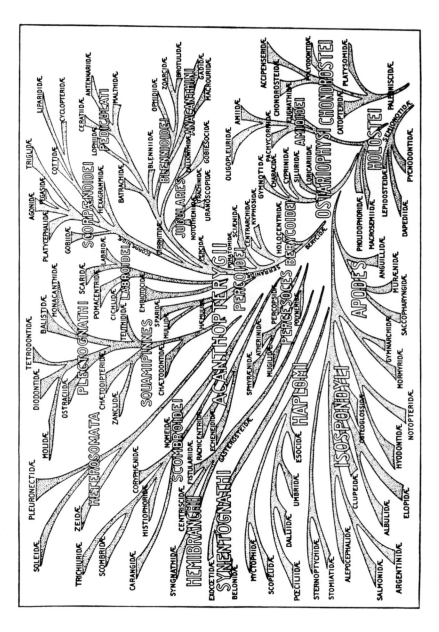

Fig. 12. 'Family tree of the ganoids and teleosts.' (From Gregory, 1951, fig. 9, 1b.) The original caption says that the diagram was drawn in 1926, as a revised version of one published in 1907, and adds 'Subsequent discovery has tended to connect the stem of the higher fishes more closely with those of the amioids, the bony ganoids (Holostei) and the palaeoniscoids.'

in which the Clupeiformes (equivalent to Boulenger's Malacopterygii
and Gregory's Isospondyli), ostariophysans and leptolepids are
successively placed as sister-groups to all other teleosts, and
these groups are ranked accordingly in his classification.  Goodrich did not discuss this phylogeny in any detail, writing (p. 369)
'To classify the Teleostei according to a phylogenetic scheme is
a very difficult matter.  The more highly specialized forms fall
into groups which are fairly well defined, but the position of the
less differentiated families is not yet well determined owing to
lack of palaeontological evidence.'  Goodrich has clearly taken
the hint from Woodward that fossils hold the key in phylogenetic
studies.

      Between 1903 and 1923, Regan published what Greenwood et al.
(1966:345) rightly call 'a long series of brilliant papers' on the
anatomy and relationships of teleostean subgroups.  In those
papers, Regan took care to examine and comment on fossils wherever
possible (e.g., 1911a,b).  He dealt with the phylogeny of the
teleosts as a whole in 1923, and summarized his conclusions in a
diagram, reproduced here as Figure 13.  The format of this diagram
derives from earlier ones by Haeckel (Fig. 7) and Goodrich (1909:
29), and in turn set the style for many later ones (e.g., Gardiner,
1960, fig. 79, reproduced here as Figure 14, whose content may be
compared with that of Regan's diagram).

      In my opinion, Regan was the finest ichthyologist of this
century, and his 1923 paper merits examination in some detail.  The
paper begins with a description of the skeleton of Lepisosteus,
followed by a comparison with Amia, a method which I believe is
the best way of approaching teleostean phylogeny.  In these descriptions and comparisons, as in all his osteological and systematic
work, Regan's style is concise and his observations have stood the
test of time.  In the final part of the paper, Regan discusses the
systematic position of the two living holosteans and their relationship to teleosts:  here, I find the contrast in content
disconcerting.  The discussion (pp. 455-458) relies entirely on
the extinct groups defined by Woodward in the Catalogue and on
their stratigraphic distribution, and is cast in terms of ancestor-
descendent relationships between those groups.  As an example of
the style, an extended discussion of the Semionotidae, as represented by Lepidotes alone, includes guesses at their mode of feeding
and swimming as explanations of structural features, guesses at
the structure of unknown parts by analogy with Amia or Polypterus,
as required, and transformations of structure postulated at will.
I believe that this discussion was the model for much subsequent
work.

      On p. 457 Regan lists characters in which primitive extant
teleosts, such as the elopids, differ from holosteans.  These

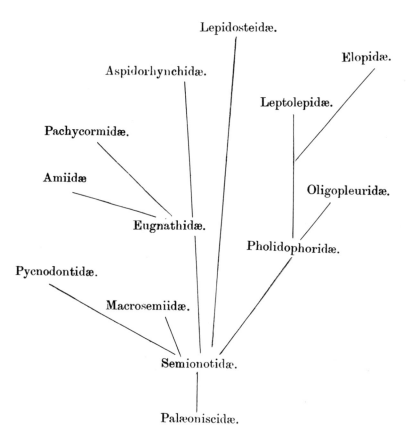

Fig. 13.  Phylogeny of the lower neopterygian fishes.  (From Regan, 1923:460.)

include the structure of the caudal fin and pectoral girdle,
presence of a supraoccipital bone and an unpaired vomer, absence
of surangular, coronoids and prearticular in the lower jaw, and
the loosely attached premaxillae. Regan then writes (p. 458) 'If
only the living forms were considered, it might be held that the
Chondrostei, Holostei, and Teleostei were groups of equivalent
rank. But a study of the fossils shows that this conclusion is
wrong. The mesozoic Pholidophoridae and Oligopleuridae resemble
the Holostei in the structure of the caudal fin (and in the
Pholidophoridae of their scales), but in other characters, such
as the small loose praemaxillaries, the maxillary with convex
oral edge and bearing two supramaxillaries, and the absence of a
"supra-angular", they show clear evidence of relationship to the
Elopidae.' Here Regan cites anatomical characters of the fossils
to support his argument. I criticize them to show the fragile
basis of these apparently authoritative statements. Regan's
Oligopleuridae is the group named by Woodward (1895) for the
genera Oligopleurus, Ionoscopus and Spathiurus. The features of
the oligopleurid skull that Regan cites are taken from Woodward's
Catalogue, in which they are described in Oligopleurus vectensis
Woodward, a species wrongly included in the group and later
(Woodward, 1919) transferred to the genus Pachythrissops, which
Forey (1973a) includes in the Megalopidae. It is therefore not
surprising that the fish appeared to have an elopid-like head.
The caudal skeleton of the oligopleurids is of holostean type, as
Pictet (1854) observed, and the relationships of these fishes are
still unknown (Patterson, 1973:287). Regarding the characters of
the Pholidophoridae cited by Regan, the pholidophorid lower jaw
(Patterson, 1973, fig. 7) and caudal skeleton (Patterson, 1968)
were unknown at that time. Stripped of mistaken and inaccurate
facts, Regan's statement is that oligopleurids have a holostean
caudal fin, and pholidophorids have ganoid scales, small premaxillae and a curved maxilla with two supramaxillae. These facts
(except the last - the supramaxillae of Pholidophorus were first
described by Zittel, 1887: 215) were known to pre-Darwinian
paleontologists, and one may ask by what sleight of mind they
have become 'clear evidence of relationship to the Elopidae.'

Regan goes on to categorize the pholidophorids as descendents
of the semionotids and ancestors of the leptolepids and elopids,
and writes (p. 458) 'the Holostei and Teleostei, therefore, are
one group.' This statement, arrived at through paleontology,
closely resembles Huxley's (1883:139, quoted previously) conclusion
on the same topic, arrived at through study of the soft anatomy.
Regan ends his paper by classifying the groups he has dealt with,
naming a new subclass Heopterygii to include fossil and Recent
holosteans and teleosteans, and a new order Halecostomi to include
the Pholidophoridae and Oligopleuridae.

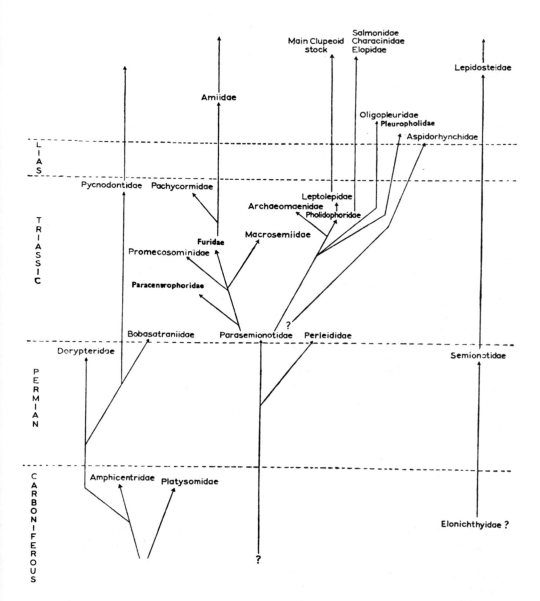

Fig. 14. 'Evolutionary tree of the main holostean groups.' (From Gardiner, 1960, fig. 79.)

I will pass quickly over the forty years between Regan's paper and the middle nineteen-sixties. During that period, three main trends are found in teleostean paleontology. First, anatomical knowledge about certain species of Mesozoic groups like the Pholidophoridae and Leptolepidae was greatly increased (Rayner, 1937, 1941, 1948; Nybelin, 1963, 1966; Griffith and Patterson, 1963; Schultze, 1966, who finally laid to rest the 'ganoid' scales of leptolepids, so frequently cited on Agassiz's authority).

Second, it became generally accepted, following Woodward and Regan, that holosteans and teleosts could not be separated (e.g., Berg, 1940; Lehman, 1966), or could only be separated arbitrarily (e.g., Bertin and Arambourg, 1958; Obruchev, 1964). Thus the holosteans and teleosts came to be regarded as grades rather than taxa, and the word 'holostean' was often printed within inverted commas, when used (e.g., Gardiner, 1960; Gosline, 1971).

Third, in discussions of teleostean phylogeny, the notion of polyphyletic origins from holosteans became generally accepted (see reviews in Gosline, 1965, and Patterson, 1967). The groups involved in these supposed multiple origins included the eels (Woodward, 1942), ostariophysans (Woodward, 1942, from Lycoptera; Gardiner, 1960, from pholidophorids), enchodonts (Woodward, 1942, from eugnathids), acanthopterygians (Woodward, 1942, from semion-otids), clupeids (Arambourg, 1935, 1950, from leptolepids), chirocentrids (Saint-Seine, 1949; Bardack, 1965, from pholidophor-ids), elopids (Saint-Sein, 1949; Gardiner, 1960; Bardack, 1965, from pholidophorids), chanids (Arambourg and Schneegans, 1936) and salmonids (Gardiner, 1960, from pholidophorids). Bardack's (1965) paper is the most thorough discussion of teleostean phylogeny along these lines. It is cast entirely in terms of ancestor-descendent relationships, regards the identification of actual ancestors as the goal of paleontology, and sees the phylogenetic method as the postulation of transformation in stratigraphic series of fossils (P. 15 - 'In theory, phylogeny is deduced by tracing changes in a series of structures from geologically earlier to later members of a family').

During this period, paleontology acquired a monopoly in phylogenetic work on teleosts, as exemplified by Woodward's (1942: 911) statement 'it is evident that no further progress can be made in studying the early evolution of the Teleosteans until large series of well-preserved fossil fishes have been discovered in Lower Cretaceous formations.' New Lower Cretaceous teleost faunas have since been discovered (e.g., Waldman, 1971; Casier and Taverne, 1971), and previously known faunas, such as those from the Santana Formation of Brazil and from Monsech, Lerida, Spain, have been shown to be of Lower Cretaceous age, but these discov-eries have not had the effect that Woodward predicted.

Ideas of teleostean polyphyly, documented by fossils, were generally accepted (indeed, they still have advocates - Blot, 1975; Taverne, 1975a) until Gosline (1965) criticized them by challenging the prevalent notion of teleosts as a grade, and by pointing out that Recent teleosts have certain specializations in common, whose independent acquisition had not been demonstrated in any of the supposed fossil lineages: in other words, his criticism was based on parsimony. Nevertheless Gosline (1965:190) still supposed that fossils held the key to the analysis of teleostean phylogeny (cf. Gosline, 1971:95, writing of 'modern lower teleosts': 'the fossil record is crucial to any attempt to interpret their relationships').

That view is also expressed in the introduction to the next major contribution, Greenwood, Rosen, Weitzman and Myers' paper (1966). These authors excused their concentration on extant fishes by contrasting the situation in teleosts with that in mammals, where fossils hold a more dominant position. They concluded with the statement (p. 347) 'Paleoichthyologists who deal extensively with teleostean fossils are quite aware that the classification of living teleosts must be understood before the fossil record can be properly interpreted.' In my opinion, that comment, rather than imperfections in the fossil record, gives the true reason why the contribution of paleontology to teleostean phylogeny had so far been insubstantial.

Greenwood et al. pointed out that the teleostean classification then in use was largely a nineteenth-century construction, and introduced their own ideas with a diagram, reproduced here as Figure 15. This diagram employs several different conventions. Although its authors were primarily concerned with living teleosts, the diagram includes some fossils and fossil groups in ancestral positions (e.g., Sardinioides, Ctenothrissiformes, pholidophorids), while elsewhere hypothetical fossils are placed as ancestors (the 'presumed Jurassic protoelopoid'). In other places, Recent groups are put in ancestral positions (e.g., Beryciformes, Elopiformes, salmonoids). Polyphyletic origins of one group from another are suggested for the acanthopterygians, paracanthopterygians, and the teleosts as a whole, from pholidophorids. The major contribution made by Greenwood et al. was dismemberment of the old basal group of teleosts, the Isospondyli of Cope, Woodward, Gregory and Regan, the Physostomi of Muller and Gill, the Malacopterygii of Gunther and Boulenger, the Clupeiformes of Goodrich and Berg. Greenwood et al. distributed the members of that group amongst four taxa, shown in Figure 15 as Elopomorpha, Clupeomorpha, Osteoglossomorpha and Division III (the latter group was named Euteleostei by Greenwood et al., 1967). In the original paper, three of these groups were poorly characterized, but the Clupeomorpha was provided with an impressive list of specializations.

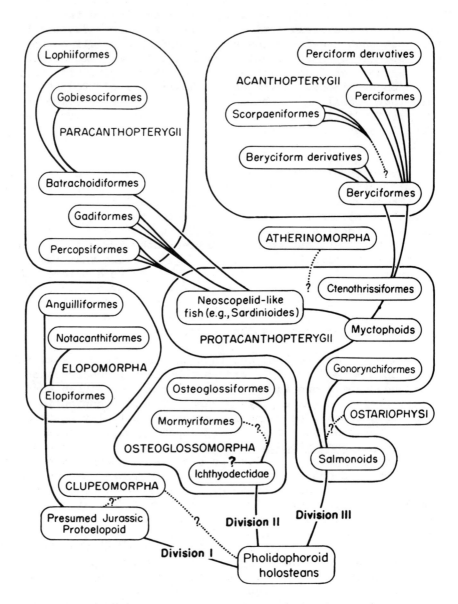

Fig. 15. 'Diagram showing the evolutionary relationships of the principal groups of teleostean fishes.' (From Greenwood et al., 1966, fig. 1.)

Armed with this new information, Greenwood et al. were able to criticize the two supposed independent lineages of clupeomorphs that had been proposed by paleontologists - the ichthyodectid-chirocentrid and clupavid-clupeid lines - pointing out that none of the specializations of Recent clupeomorphs had been demonstrated in the supposed ancestral forms:  in other words, their criticism was based on parsimony.  However, Arambourg, Saint-Seine and Bardack, the paleontologists chiefly responsible for the fossil lineages, are hardly to be blamed, for no ichthyologist had previously shown how herring-like fishes were to be recognized, and the method employed by these paleontologists, the search for stratigraphic sequences of fossils which could be interpreted along ancestor-descendent lines, naturally led to assumptions of extensive parallel evolution.  Greenwood et al.'s characterization of Recent clupeomorphs initiated progress in paleontology, and within a few years several Cretaceous clupeomorphs were described in detail, suggesting the sequence in which the various clupeomorph characters were acquired (Patterson, 1970; Forey, 1973b).

Such immediate progress was not made with the other three groups named by Greenwood et al.  These groups were originally poorly characterized, and only one of them, the Osteoglossomorpha, has since been shown to be definable by several osteological specializations:  as a result osteoglossomorphs have been recognized in the Jurassic (Greenwood, 1970).  Another reason for lack of progress in these groups is suggested by Greenwood et al.'s diagram (Fig. 15), for two of them have Recent subgroups placed in ancestral positions at the base, so that each is in the condition of teleosts as a whole before 1966.  Why basal ancestral groups should be such an effective bar to progress finally became clear in the same year, 1966, with the publication of the English version of Hennig's Phylogenetic Systematics, and of Brundin's exposition of Hennig's methods.

Hennig's major contributions are, in my view, an unambiguous definition of relationship, in terms of recency of common ancestry; an unambiguous method of recognizing relationship, by means of synapomorphies; and an unambiguous method of expressing such relationships diagrammatically, in dichotomous cladograms.  Henng's ideas caught on immediately in ichthyology (see Nelson, 1972, for a review).  One reason for this may be that those ideas arrived at the same time as the paper by Greenwood et al., in which the old basal group of teleosts was broken up for the first time:  a rough analogy would be the situation in mammalogy if the distinction between monotremes, marsupials and placentals had not been recognized until 1966.

Hennig's methods were first applied to the problem of teleostean phylogeny by Nelson (1969a,b:  a cladogram from the second

paper is reproduced as Figure 16). Some of Nelson's comments on
the relevance of paleontology to that problem deserve quoting:
(1969b:22) 'It is a general misconception that the problem of
determining the phyletic relationships among Recent animals can
be directly approached only by the paleontologist. In reality,
the paleontologist, even with a good fossil record, can contribute
very little, if anything, to the solution of this problem. In
fact the paleontologist as a rule cannot do much in determining
the relationships of his fossil species unless there are Recent
relatives whose relationships already are fairly well established.
Certainly, as far as teleosts are concerned . . . . paleontology
has been of little or no direct help in elucidating the phyletic
relationships among any Recent animals. This has been done and
will continue to be done mainly through detailed comparative study
of the diversity of the Recent fauna. The writer does not wish to
underestimate the importance of paleontology, which, after all,
can give us some idea of the absolute ages and past distributions
of the Recent species and taxonomic groups, matters of great
importance to evolutionary studies. In addition, paleontology
often can give us some indication of the direction and magnitude
of phyletic trends. But it is a mistake, all too often committed,
to assume that an earlier known fossil species in structure or
distribution necessarily is more primitive than, or ancestral to,
a later known fossil or Recent species. It is a mistake to believe
even that one fossil species or fossil "group" can be demonstrated
to have been ancestral to another.' Why should acceptance of
Hennig's methods have led Nelson to a view of paleontology (which
I share) so violently opposed to the claims made for it by, for
example, Woodward (1942, quoted previously) and Bardack (1965,
quoted previously)?

The essential point is synapomorphy. If it is acknowledged,
following Hennig, that relationship can only be demonstrated by
synapomorphies, then taxa, groups of related species, can only be
defined by synapomorphies. 'Groups' like the Malacopterygii in
Boulenger's diagram (Fig. 9) and the Elopiformes and salmonoids in
Greenwood et al.'s (Fig. 15), which can be envisaged as ancestral
to other groups, can have no synapomorphies of their own. The
only derived characters common to all the members of such taxa are
the synapomorphies of the larger group comprising the ancestral
taxon and all its descendents - non-ostariophysan teleosts in the
case of Boulenger's Malacopterygii; Elopomorpha in the case of the
Elopiformes of Greenwood et al. The same considerations apply to
extinct ancestral groups, like the Semionotidae and Pholidophoridae
in Regan's diagram (Fig. 13) and the pholidophoroids in that of
Greenwood et al. (Fig. 15). And if the only criterion for assigning
species or groups of species, Recent or Fossil, to higher taxa is
the possession of synapomorphies, how can species be added to
groups like the Malacopterygii, Elopiformes or Pholidophoridae,

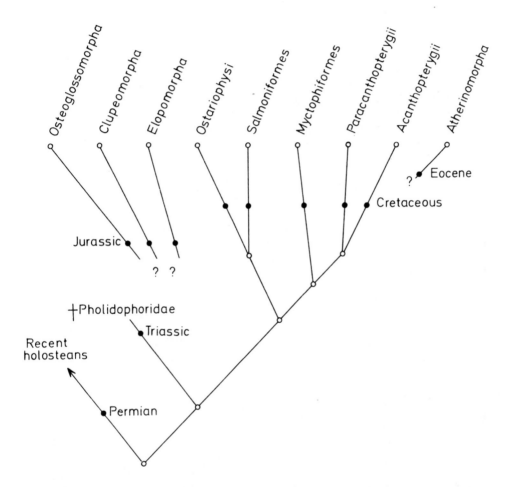

Fig. 16.   'Diagram of the phyletic interrelationships of the major groups of Recent teleosts, with the oldest fossil members indicated.' (From Nelson, 1969, fig. 3.)

which have no synapomorphies?  Only by convention or by arbitrary
decisions.  These ideas, derived from Hennig, make it clear why
Greenwood et al.'s dismemberment of the old basal group of the
teleosts is, in retrospect, so important, and why Nelson viewed
with such disfavor the products of traditional evolutionary
paleontology, ancestral groups and ancestor-descendent relationships, for neither membership in an ancestral group, nor ancestor-
descendent relationship, can be specified by synapomorphy.

Critics of Hennig such as Darlington (1970) and Mayr (1974)
have suggested that the use of synapomorphies as evidence of
relationship is nothing new in taxonomy.  For example, Mayr
(1974:98) writes 'In fact, one can say that most of the better
taxonomists of former eras had applied this principle, as is quite
evident from a study of their classifications.'  If that were so,
the work reviewed here demonstrates that the teleosts have only
attracted second- or third-rate taxonomists, from Agassiz and
Muller onwards.  In fact, it is not the good taxonomists that
exemplify this principle, but the good groups.  Muller's Teleostei,
Cope's Actinopterygii, Sagemehl's Ostariophysi, Regan's Neopterygii
and the Clupeomorpha and Osteoglossomorpha of Greenwood et al. are
groups which have stood the test of time because they have been
found to contain fishes which share synapomorphies, and hence can
be regarded as monophyletic.  Agassiz's Cycloidei, Muller's
Physostomi, Cope's Isospondyli, Woodward's Protospondyli and the
Taeniopaedia of Greenwood, Myers, Rosen and Weitzman (1967) have
not been found to be definable by synapomorphies, and have been
dropped since they yielded no worthwhile generalizations.  In the
same way, the·substance of Regan's claim that the Oligopleuridae
and Pholidophoridae 'show clear evidence of their relationship to
the Elopidae' (1923:458) becomes clear through the concept of
synapomorphy:  amongst the characters of pholidophorids that Regan
cited, two (small, loose premaxillae and two supramaxillae) are
teleostean synapomorphies, but the oligopleurids share no synapomorphies with the teleosts or Elopidae.

The period since 1966 can be viewed, so far as the phylogeny
of the teleosts as a whole is concerned, as one in which the ideas
put forward by Greenwood et al. have been adapted to Hennigian
methods (as exemplified by Fig. 16, which is Fig. 15 translated
into cladogram form) by searching for synapomorphies of the four
groups which they erected, by searching for sister-group relationships between those groups, and, in the paleontological field,
searching for fossil members of those groups.  During that period,
the old dichotomy between neontologists and paleontologists has
waned (e.g., the combined neontological and paleontological studies
by Goody, 1969; Rosen and Patterson, 1969; Greenwood, 1970; Forey,
1973a; Rosen, 1973; Taverne, 1974), but has been replaced by a new
dichotomy between those working in the framework of evolutionary

taxonomy, and the Hennigians or cladists.  Rather than review the
period, I will illustrate that last point in a paleontological context by comparing two recent attempts to solve a major problem in
teleostean phylogeny, the relationships of the Mesozoic leptolepids
and ichthyodectids (Taverne, 1975a; Patterson and Rosen, 1976).

Figure 17 summarizes the solution proposed by Patterson and
Rosen (1976, fig. 54).  The diagram is a cladogram, without a
time-axis, and treats fossil and Recent taxa in the same way.  The
Mesozoic Ichthyodectiformes are shown as a monophyletic group comprising two suborders, Allothrissopoidei (Allothrissops only) and
Ichthyodectoidei: the Ichthyodectiformes are characterized by two
synapomorphies, an endoskeletal ethmo-palatine bone in the floor
of the nasal capsule, and uroneurals which cover the lateral faces
of the last few pre-ural vertebrae.  The Ichthyodectiformes are
placed in a scheme which includes extant teleosts, various
'leptolepids' and two pholidophorids (Pholidophorus bechei and
Pholidolepis dorsetensis), and is based on the most parsimonious
distribution of 48 character-states which are interpreted as
synapomorphies.  In this scheme, the ichthyodectiforms emerge as
the sister-group of all Recent teleosts and one 'leptolepid',
Tharsis dubius.  Two other 'leptolepids', Leptolepis coryphaenoides
and Proleptolepis, are successive sister-groups to that whole
assemblage, and the two pholidophorids are in turn the successive
sister-groups of the entire leptolepid-teleost group.  A 'leptolepid',
Leptolepides sprattiformis, is interpreted as the sister-group to
the extant Clupeocephala (Clupeomorpha and Euteleostei), and other
'leptolepids' (numbered 1,2,6) are incertae sedis at different
levels in the scheme.  So here the 'Leptolepidae' emerge as a
polyphyletic group, since members of the group are related to extant teleosts at four or more different levels, and the ichthyodectiforms are interpreted as a monophyletic group whose relationships
fall in the middle of the leptolepids, as that group is currently
defined.

Figure 18 summarizes Taverne's solution to the same problem,
the relationships of ichthyodectiforms and leptolepids.  As the
format of his diagram indicates, Taverne is working in a different
tradition, that of evolutionary paleontology.  His solution to the
problem is almost the antithesis of that shown in Figure 17.  The
ichthyodectiforms (excluding Allothrissops) are shown as a monophyletic group whose closest relatives are the osteoglossomorphs;
the leptolepids, which Taverne treates as a single genus containing
the subgenera Leptolepis, Tharsis, Ascalabos and Leptolepides, are
indicated as a monophyletic group whose closest relative is
Allothrissops (an ichthyodectiform in Fig. 17); the genera Leptolepis and Allothrissops share a common ancestry in the Lower
Jurassic genus Proleptolepis (Taverne, 1975a: 366; 1975b: 240,243),
and the closest relatives of this extant group are the euteleosteans.

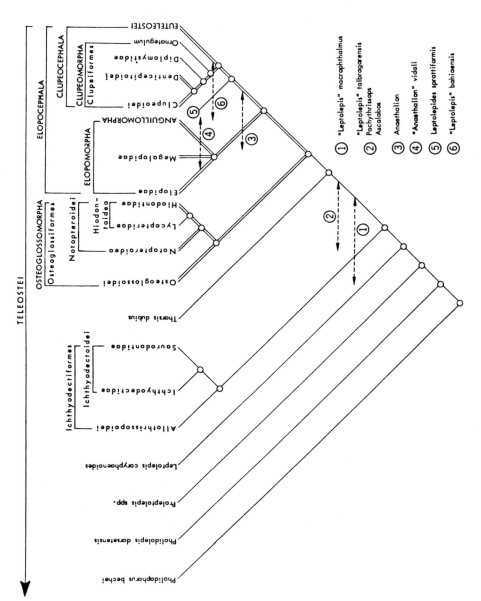

Fig. 17. Cladogram showing a theory of relationships of extant teleosts and certain fossil forms. (From Patterson and Rosen, 1977, fig. 54.)

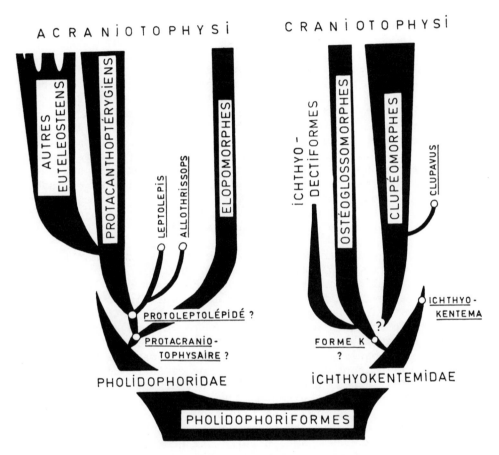

Fig. 18. Diagram showing the relationships and origins of the teleosts, and the systematic position of the genera Leptolepis and Allothrissops. (From Taverne, 1975a, fig. 17.)

Taverne's scheme of the interrelationships of extant teleosts is also different from that shown in Figure 17: following Greenwood (1973), he treats the euteleosts and elopomorphs and the clupeo- morphs and osteoglossomorphs as two sister-group pairs. Taverne regards these two groups of teleosts as independently derived from pholidophoroids, the first group from the pholidophorids, the second from the Ichthyokentemidae (which contains only one species, Ichthyokentema purbeckensis; Griffith and Patterson, 1963).

This is not the place to go into a detailed comparison of these two analyses, but several points deserve mention. First, Taverne and Patterson and Rosen studied much the same fossil material, so that the differences between the schemes in Figures 17 and 18 are due to different methodologies, not to different observations. These differences are not consistent with the opinion expressed by some critics of cladism, that Hennigian and evolutionary approaches generally lead to similar results (e.g., Darlington, 1970:9, 'in most cases the results {of cladistic and evolutionary methods} are different only to the extent that the cladists force them to fit an oversimplified, rigid, unrealistic set of assumptions'; and Mayr, 1974:98, 'There is little argument between cladists and evolutionary taxonomists about the cladogram which results from the cladistic analysis.').

Second, the differences between the proposals summarized in Figures 17 and 18 result from different conceptions of what con- stitutes an explanation in phylogenetic research. To Patterson and Rosen, a phylogenetic hypothesis deals only in monophyletic groups, specifiable by synapomorphies, and explains the distribu- tion of synapomorphies parsimoniously. To Taverne, a phylogenetic hypothesis postulates ancestral conditions, and attempts to identify ancestors as closely as possible, the method that has been applied in teleostean paleontology since the time of Kner and Haeckel. Since ancestral groups are acceptable in that method, that is, groups which are characterized only by primitive features (para- phyletic groups of Hennig), the taxa shown in Figure 18 are not necessarily characterized by synapomorphies (for example, I know of no derived features characterizing the group Acraniotophysi in Fig. 18, nor are there any unique derived features amongst the 65 characters of Leptolepis which Taverne lists; 1975a:337-341). And since the specification of ancestors in that method is by listing primitive characters (e.g., Taverne, 1975a:360-362) and then searching for fossils exhibiting yet more primitive conditions, while the method using in Figure 17 is to specify the derived characters of groups and to search for fossils sharing those synapomorphies, it is natural that Taverne is led to drive his lineages further into the past than are Patterson and Rosen. Thus Taverne postulates (1975a:360) that the common ancestor of his Acraniotophysi (the 'protocraniotophysaire' of Fig. 18) lived at the beginning of the Upper Triassic, so that the extant groups of teleosts must have diverged in the Middle Trias or earlier, while Patterson and Rosen postulate that the extant teleost subgroups (enclosed by double lines in Fig. 17) need not have diverged until late Jurassic times, some 60 million years later.

Finally, it is worth comparing Figure 18 with Agassiz's original diagram (Fig. 1).  The upper part of Figure 18 is drawn using the convention established by Agassiz, that the length and breadth of the lineages corresponds to their time-span and abundance as fossils, but the lower part, the Pholidophoriformes, is clearly drawn by another convention.  For example, the right-hand horn of the pholidophoriforms, representing the Ichthyokentemidae, is hypothetical, since the only known member of that family, Ichthyokentema purbeckensis, is indicated by the open circle at the tip of the horn.  Where Agassiz refrained from linking the converging bases of his lineages (Fig. 1) since he believed that their junction 'may only be sought in the creative intelligence,' the theory of evolution allows a modern author to represent a hypothetical ancestral group as having the same reality as any other major taxon.  Yet the search for ancestors of the teleosts has hardly been successful, for every other major branch point in Figure 18 is also occupied by hypothetical forms or question-marks: these ancestors have no more reality than the abstract synapomorphy-bearers indicated by the open circles in Figure 17.

## SUMMARY

The nineteenth century contributions are summarized under three headings:  the secondary or subservient role of paleontology, a dichotomy between neontology and paleontology, and the changed view of paleontological problems following acceptance of the theory of evolution.  During the first two-thirds of the twentieth century, it is my impression that ideas on teleostean phylogeny were profoundly influenced by Woodward's Catalogue.  For example, all major treatments of Recent teleosts during that period (Boulenger, 1904b; Jordan, 1905; Gregory, 1907, 1933; Goodrich, 1909; Regan, 1929; Berg, 1940; Bertin and Arambourg, 1958) included fossils, whereas none had done so before the Catalogue.  And in the discussions of teleostean phylogeny in those works, and in others, the fossils were given pride of place and treated on the pattern laid down by Woodward.  Yet Woodward's major innovation, inclusion of the Triassic and Jurassic pholidophorids in the teleosts, was not generally accepted.  Instead, the pholidophorids were treated as a holostean (or halecostome) subgroup, and the main paleontological contribution during this period was development of the idea of a polyphyletic origin of teleosts, either from the pholidophorids alone, or from them and other holostean groups.  Woodward, by his own account, associated the pholidophorids with teleosts on very slight evidence, and it is surprising that for many years no attempt was made to investigate the anatomy of members of this supposedly crucial group (Rayner, 1941, was the first such effort). Instead, arguments were largely confined to the question of whether teleosts were descended from pholidophorids, or leptolepids, or oligopleurids, or other groups, and whether the arbitrary line between holosteans and teleosts was crossed once, twice or many times. Discussions of such questions are always inconclusive, and while the pattern of ancestry shown in phylogenetic diagrams became more elaborate (cf. Figs. 9,10,11; Figs. 13,14), the text accompanying those diagrams failed to provide factual backing for them. Conclusions ranged from firm assertions of ancestry (e.g., Rayner, 1941:230 - '_Leptolepis_ was derived from _Pholidophorus_ in the Lias (probably Upper Lias) and gave rise to the basal teleostean stock.') to equally firm contrary assertions (e.g., Nybelin, 1974:195 - 'There are consequently, as far as I know, no group of primitive teleosts which can with certainty be regarded as descendents of the family Leptolepidae s. str. All must belong to other phyletic lineages radiating from unknown or imperfectly known "Pholidophoroidea".'). The only progress evident here is the rejection of one putative ancestor after another. Yet this successive rejection of possible ancestors has not led, as might be expected, to a gradual homing-in on the one true ancestor or sequence of ancestors. Instead, it has always led to postulations of ancestry from still unknown fossils, as in the quotation from Nybelin above, and similar statements about several teleost groups by Taverne (1975a:356,357,358,363). In turn, this leads to the idea that more fossils are needed: no known fossil is the ancestor, but if enough rocks are split the true phylogeny will eventually become clear.

After Woodward had stated the claims of paleontology to lead in phylogenetic work on teleosts, ichthyologists accepted those claims, and sat back to await the answers that fossils alone could produce. But paleontologists failed to deliver the goods - specific ancestors, or any new insight on the interrelationships of extant teleosts: when progress of that sort did come (Greenwood et al., 1966), it was achieved, as in the nineteenth century, by comparative studies of Recent fishes. The reasons for this failure of paleontology were not discussed. It was simply assumed that more fossils were needed before the true phyletic lines became manifest. Hence the attitude embodied in Medawar's words quoted at the beginning of this paper: paleontologists have only to read the patterns of descent revealed by fossils, like the genealogist tracing family history in written records.

During the last ten years, many workers have come to view that attitude as mistaken, and based on a misunderstanding of how phylogeny can be investigated. These new views have come from the development of ideas put forward by Hennig (1966). Various aspects of those ideas are discussed in the remaining sections of this paper.

## DISCUSSION

### Recent Organisms and Fossil Organisms

The point I wish to make here, that the interpretation of fossils is necessarily subsidiary to that of Recent organisms, may seem too obvious to mention. Yet neglect of that truism is responsible for the idea that paleontology has the dominant role in phylogenetic work. In the preceding historical review I have stressed the fact that all the decisive advances in our understanding of teleosts have come from work on Recent fishes, with paleontology following on behind. The primary reason for this is that a fossil is meaningless until it can be interpreted in the light of some Recent model. Darwin (1859:329) made the point 'as Buckland long ago remarked, all fossils can be classed either in still existing groups, or between them.' The success with which a fossil can be interpreted, or made to yield useful information, is directly related to the precision with which it can be classed in still existing groups: once assigned to a group, the fossil can be interpreted by a model drawn from that group. But where the existing group is so extensive or poorly characterized that its morphotype is almost featureless, the fossils will yield virtually no useful information. Examples are provided by fossil animals such as the Silurian Ainiktozoon (Scourfield, 1937; Tarlo, 1960) and the Pennsylvanian Tullimonstrum (Johnson and Richardson, 1969) and 'conodont-bearing animals' (Melton and Scott, 1973). These are all represented by complete, well-preserved fossils, but they cannot yet be assigned to any existing group less extensive than Triploblastica, so that no usable model is available and they remain uninterpretable. A different type of example is found in the Paleozoic mitrate and cornute carpoids (Jefferies, 1975), which Jefferies interprets by a chordate model, and other workers by an echinoderm model, so that there is disagreement on such basic matters as which end is anterior and which side is dorsal.

The basis for assigning a fossil to a Recent group is the discovery in the fossil of one or more of the characters (synapomorphies) of that group. Neglect of that principle - when fossils are interpreted after models drawn from Recent groups with which they share no demonstrable synapomorphies - has been the cause of many unproductive developments in paleontology. Usually, this results from taking a model from a less extensive group than (a subgroup of) the one with which the fossils do share synapomorphies.

Examples include Melton and Scott's (1973) interpretation of their
'conodont-bearing animal' on a protochordate model, with the consequent erection of a new chordate subphylum; Stensio's (1964,
1969) interpretations of heterostracans after a myxinoid model;
and Bardack's (1965) interpretations of various features of
ichthyodectids using Chirocentrus as a model.

Thus fossils are meaningless until they can be interpreted in
the light of Recent models, and the appropriate model can be
selected only by finding synapomorphies which link the fossil with
a Recent group. Fossils can therefore only be properly assigned
to those Recent groups which are characterized by synapomorphies,
that is, to monophyletic (or holophyletic) groups. It follows
that determining the relationships of Recent organisms - recognizing and characterizing the monophyletic groups represented -
must precede allocation of fossils to those groups. In other
words, determining the relationships of fossils is necessarily
secondary to determining the relationships of Recent organisms.
This is the reason why paleontology has to follow, not lead, in
phylogenetic work.

## Recent Groups and Fossil Groups

It is generally taken for granted that when fossil species
are grouped into extinct higher taxa, the latter have the same
status as Recent higher taxa. This assumption is found in all
discussions of teleostean phylogeny, when extinct groups like the
Pholidophoridae, Leptolepidae and so on are treated as having the
same reality as extant groups. Yet the status of the two types
of group, as hypotheses about nature, is not the same.

In neontology, higher taxa are defined by characters which
are assumed to have been present in the common ancestor of the
group (Mayr, 1969:83), that is, by characters which are judged to
be synapomorphies of the group. Membership in the group is open
to test, by checking specimens for the characters in question.
Extinct higher taxa have to be defined in the same way, by the
possession of synapomorphies. But here, membership in the group
is frequently not open to test in the same way, for all the characters in question may not be preserved or accessible in some nominal
members of the group. The usual practice in defining extinct
higher taxa is to base the characterization on one or more comparatively well-known members, and to assume that the relevant
characters were present in other members, represented by fossils
in which not all are preserved or accessible. Any monograph on
an extinct group of fishes (e.g., Nybelin, 1966, 1974; Patterson
and Rosen, 1977) can be cited to illustrate this point, and there
is no need to labor it further.

## Ancestral Groups

The historical survey in the preceding section shows that for the first hundred years of phylogenetic work the accepted scheme of teleostean relationships included a basal group (variously named Physostomi, Isospondyli, Malacopterygii or Clupeiformes) which was thought to have given rise to all or most of the remaining teleosts (Figs. 5-11). This ancestral group was originally imported, without modification, from Muller's pre-evolutionary system. Hennig was the first to point out that such groups, which he called paraphyletic, do not have the same reality as monophyletic groups. In the first place, ancestral groups are not characterized in the same way as monophyletic groups, for they have no synapomorphies: the only synapomorphies characterizing them are those of the entire assemblage comprising the ancestral group and all the descendent groups. The lower limit of the ancestral group is the same as that of the whole assemblage, but the upper limit is arbitrarily drawn by the exclusion of the various derived (descendent) groups, each of which can be characterized by synapomorphies. Thus in Boulenger's scheme (Fig. 9), for example, a malacopterygian is a teleost lacking a Weberian apparatus, or any of the synapomorphies of groups III to XIII: it is recognizable only by lack of synapomorphies. As Hennig points out (1966:146; see also Bonde, 1975:299), the essential difference between a paraphyletic group and a monophyletic group is that the first has no ancestor of its own, no independent history, and no individuality. In other words, since such groups are limited only by convention, they exist not in nature but in the mind of the taxonomist, as abstractions. Yet, if taxonomists can agree that a malacopterygian is a teleost lacking certain synapomorphies, Recent species assigned to the group can be checked for the absence of those characters, and new species can be added to it.

Fossil ancestral groups, like the Pholidophoridae, suffer from all the deficiencies of Recent paraphyletic groups, and have the additional disadvantage, common to all extinct groups, that some nominal members of the group cannot be checked to see whether or not they lack those derived characters whose absence has, by convention, been chosen to define the group (section "Recent Groups and Fossil Groups"). Fossil ancestral groups therefore imply a second level of abstraction, yet they are always discussed as if they have some reality, such as specifiable stratigraphic and geographic ranges.

There seem to be only three ways of dealing with ancestral groups in phylogenetic discussions. The first, and least useful, is to retreat from the problem and use the group names as undefined vernacular terms (e.g., the 'pholidophoroids' of Greenwood

et al, 1966, and Patterson, 1967), which introduces a third level
of abstraction.  The second method is to try to give the groups
greater precision, so that they may serve as better tools in
phylogenetic discussion.  This can only reduce the content of the
group.  For example, Nybelin has written excellent taxonomic re-
visions of the Pholidophoridae (1966) and Leptolepidae (1974), in
which he limits each group to those species which cluster phenetic-
ally about the type-species of each type-genus, and defines each
family sensu stricto.  Those nominal species which he excludes
from the groups, or does not deal with, become incertae sedis.
Yet even in their restricted sense, the groups are defined by
characters drawn only from the better-known members, and including
no synapomorphies.  The third method is to follow this line of
attach further, and demand that all groups be monophyletic in
Hennig's sense.  In this light, ancestral groups melt away, and
their better-known members can be treated individually as taxa
whose relationships may be investigated, while all the remaining
nominal species become incertae sedis:  this is how the Pholidophoridae and Leptolepidae are treated by Patterson and Rosen
(1977) and in section "An Alternative Approach".

## Ancestors and Stratigraphy

As shown in the section "Historical Review", the concepts of
named ancestral groups and identifiable fossil ancestors were
introduced into teleostean paleontology by Kner and Haeckel in
1866.  Those concepts were accepted without discussion or analysis,
and were the basis of all discussions of teleostean phylogeny
during the succeeding hundred years.  Only recently, following
the spread of Hennig's ideas, have these concepts been criticized
(Nelson, 1969a,b, 1970; Schaeffer, Hecht and Eldredge, 1972;
Cracraft, 1974; Eldredge and Tattersall, 1975; Bonde, 1975).
Following this criticism, paleontologists have begun to react,
and explain their rationale (Bretsky, 1975; Campbell, 1975, Lehman,
1975; Harper, 1976).  Harper's paper is the most explicit and
fully-argued of these.  Like the other authors cited, Harper wishes
to retain ancestral (paraphyletic) groups and ancestor-descendent
relationships, and attempts to define 'ancestral taxon' (p. 183).
In order to retain these concepts, he allows 'phylogenetic relationship' to mean several different things (p. 187), regards the rank
of extinct taxa as given a priori (p. 183), and accepts that
stratigraphic succession is no necessary guide to ancestor-descendent relationships, since an ancestral taxon may appear in the
fossil record after its descendent(s) (fig. 5).

Here, I want only to take up that last point.  In the past,
it has been traditional practice amongst phylogenetic paleontologists to demand that putative ancestors should precede their

descendents in the fossil record, and where this condition is not met, to postulate divergence of the taxa concerned from an unknown, earlier ancestor (Figs. 2 and 14 in this paper and figs. 1 and 2 in Harper's demonstrate this principle). Yet Harper allows that imperfections in the fossil record may result in an ancestor appearing later than its descendents. In the same vein, Simpson (1975:6) writes 'ancestral species commonly persist with little or no genetic, hence phylogenetic change, although daughter species have split off from them' and Mayr (1974:110) concurs, with the statement 'The production of a side branch, a new phyletic lineage, does not change the parental species.' The phenomena described here by Mayr and Simpson agree with that envisaged by Harper: occurrence of ancestors later in time than their descendents. The point I wish to emphasize is that if this is acknowledged to be a true description of speciation, then stratigraphic sequence can never be a trustworthy guide to phylogeny, and paleontology must relinquish some of its claims to provide 'both the most direct and most important data bearing on phylogeny' (Simpson, 1975:11). For if ancestral species do not necessarily occur earlier than their descendents, Recent species may be regarded as ancestral to other Recent species, and even to fossil species. Harper's (1976:190) 'Principle of minimal stratigraphic gaps' between ancestral and descendent taxa seems to be ideally met by contemporary taxa, such as the Recent biota.


                          An Alternative Approach

     If fossils cannot provide ancestors, what can they provide? In this section I will outline an alternative approach to teleostean phylogeny, where the contribution of paleontology is not ancestral organisms but ancestral character-states.

     Figure 19 is a cladogram in which the framework of heavy lines expresses the relationships of the major subgroups of extant teleosts (as characterized by Patterson and Rosen, 1977) and their extant sister-group, Amia, representing the Halecomorphi (as characterized by Patterson, 1973). On the basis of a comparison of the skeleton of Amia with those of living teleosts, I compiled a list of skeletal features in which teleosts are more advanced than Amia; that is, a list of skeletal synapomorphies of teleosts.[3] This list included 45 characters, so that in the original cladogram, marked by heavy lines in Figure 19, the point of dichotomy between Osteoglossomorpha and the remaining teleosts was characterized by 45 synapomorphies. A further seven skeletal synapomorphies characterize the other sister-group pairs of Recent teleosts, so that the full list contains 52 characters.

     With this framework established, fossil species and monophyletic

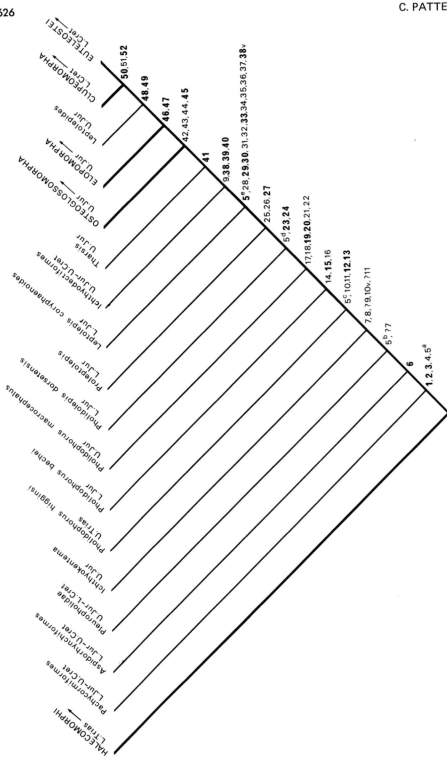

Fig. 19. Cladogram showing a theory of relationships of recent halecostome fishes and certain fossil teleosts. Extant groups are linked by heavy lines and named in capitals. The numbers to the right of the bifurcations refer to synapomorphies listed in the text. For further explanation

groups[4] were added to the cladogram according to the number of teleostean synapomorphies they exhibit. Thus the Pachycormiformes (branching from the dichotomy number 1,2 .. .) share five synapomorphies with Recent teleosts, the Aspidorhynchiformes (dichotomy numbered 6) share those five and one more, and so on up to Tharsis (dichotomy numbered 41), which has all 40 of the synapomorphies listed for taxa branching off below it, and one additional derived character-state. Leptolepides (dichotomy numbered 48,49) has been included to show how this method may be used to intercalate fossil taxa between any two Recent groups (cf. Fig. 17). As a result of this procedure, only four (nos. 42-45) of the 45 synapomorphies which originally characterized Recent teleosts still do so: the remainder are distributed along the line differentiating Recent teleosts from their sister-group, the Halecomorphi. This line is the part of the cladogram corresponding to what Hennig (1966:161) called the period between the age of origin and the age of differentiation of the Recent group, or (1969:32) its stem-group. It is the part of the cladogram (or phylogeny) of any group to which paleontology alone can contribute. But it is important to note that the success with which extinct taxa can be assigned to a stem-group, and ordered there, depends entirely on the success with which three Recent taxa can be characterized, the group under study (here the Teleostei), its sister-group (Halecomorphi), and the higher taxon containing the two (Halecostomi). The synapomorphies of these three groups give the cladogram its outline and credibility, and set limits to the stem-group, and those of the first group are alone available for distribution amongst the stem-group fossils.

Numbers entered in heavy type in Figure 19 indicate that the derived character-state is known in the taxon to the left at that dichotomy, and that the corresponding primitive character-state has been observed in at least one member (specimen) of the fossil taxon to the left at the next lower dichotomy. It does not follow that the primitive character-state has been observed in all the other fossil taxa at lower dichotomies, or that the derived character-state is known in all fossil taxa at higher dichotomies, but it is predicted that these conditions will be met when the character is investigated in those taxa.

Numbers entered in light type in Figure 19 indicate that the derived character-state is known in the taxon to the left at that dichotomy, but that the corresponding primitive character-state has not been observed in the fossil taxon at the next lower dichotomy, because the part in question is not yet accessible in the available specimens. Numbers in light type predict that the derived character-state will be found in all fossil taxa above that dichotomy, but make no precise prediction about conditions in taxa at lower dichotomies. The numbers in light type are

therefore potentially mobile downwards:  future work might show
that several of the derived character-states they represent
occurred in one or more taxa lower down in the cladogram.  If so,
the cladogram and the phylogeny it represents would become un-
parsimonious, and would have to be modified.

Two numbers in Figure 19 are followed by a 'v' (10v, 38v):
this signifies 'variant' and means that the fossil species to the
left at that dichotomy is polymorphic where that character is
concerned, some individuals showing the derived state and others
the primitive state.  In both cases the derived state is found in
all individuals of the fossil taxon at the next higher dichotomy,
and its number is entered there without suffix.  Three numbers are
preceded by a question mark (?7, ?9, ?11).  This means that the
derived character-state has been observed in the fossil taxon at
that dichotomy, but that the observation is open to question because of inadequacies or contradictions in the material.  These
numbers are entered again at higher dichotomies, where the character is unequivocally present in the fossil taxon to the left.

In making this type of analysis, inclusion of fossils leads
one to conclude that some apparent synapomorphies between Recent
groups were arrived at by convergent evolution.  The criterion is
parsimony:  to explain these characters as synapomorphic, rather
than convergent, would demand a less parsimonious distribution of
other characters.  Some examples of convergence between _Amia_ and
teleosts are listed by Patterson (1973:261 - as parallelisms), and
a simple example is shown in Figure 19 by character 5, number of
epurals.  _Amia_ has four epurals, while generalized teleosts have
three, but fossils indicate that these low numbers are due to
independent reduction from some larger number, probably about
nine, as in parasemionotids (Patterson, 1973, fig. 22).  Successive
reductions in the teleost line are indicated in Figure 19 and the
list that follows by 5a-e.

In this list of characters, the derived state is described
first.[5]

1.  Ural neural arches modified as uroneurals, vs. unmodified.
2.  Small, mobile premaxilla, lying lateral to rostral, vs.
fixed premaxilla, lying beneath rostral.
3.  Internal carotid foramen enclosed in parasphenoid, vs. a
notch in margin of parasphenoid.
4.  Pectoral propterygium fused with first pectoral fin-ray,
vs. propterygium free (Jessen, 1972).
5a. Epurals seven, vs. eight or more; 5b, epurals six, vs.
seven; 5c, epurals five vs. six; 5d, epurals four, vs. five (BM
(NH) P.7616); 5e, epurals three, vs. four.

6.  A median tooth-plate covers basibranchials 1-3, vs. smaller asymmetrical or paired tooth-plates (Nelson, 1969a; BM(NH) 37777, P.7626).

7.  Quadratojugal fused with quadrate as a postero-dorsal process, enclosing a groove for the symplectic, vs. quadratojugal free or not recognizable.

8.  A median, unpaired vomer, vs. paired vomers.

9.  A median basihyal tooth-plate, vs. tooth-plate paired or not recognizable (Nelson, 1969a; Patterson, 1975:488).

10.  Foramen of efferent pseudobranchial artery enclosed in parasphenoid, vs. a notch in margin of parasphenoid (BM(NH) P. 3592a).

11.  A median (endoskeletal) supraethmoid ossification present, vs. ethmoid cartilage ossified in one piece, or from paired centers (BM(NH) P.3592a).

12.  Posterior myodome extending into basioccipital, vs. confined to prootics (BM(NH) P.3592a).

13.  Lower jaw without coronoid bones, vs. one or more coronoids present (BM(NH) P.44973-75, P.3592a).

14.  Two supramaxillae, vs. one.

15.  Intercalar and prootic with membrane-bone outgrowths which form a bridge across the subtemporal fossa, vs. no bridge developed.

16.  Long epineurals developed throughout the abdominal region as outgrowths from the neural arches, vs. epineurals absent or confined to the first few neural arches.

17.  A median (endoskeletal) ventral ethmoid ossification present, vs. endoskeletal mesethmoid absent or ossified in one piece.

18.  A median anterior myodome present, vs. anterior myodome paired.

19.  Supraoccipital bone present and extending forwards in roof of otic region, vs. supraoccipital absent or confined to occipital region.

20.  Post-temporal fossa confluent with fossa bridgei, allowing trunk musculature to extend forwards above otic region, vs. fossae separated by a partition, or confluent but fossa bridgei not occupied by muscles.

21.  A urohyal present which is shaped like an inverted 'T' in section, with the plate-like ventral part dermal in origin, vs. urohyal a vertical endoskeletal plate or absent (Patterson, in press).

22.  Pectoral radials four, vs. five or more (Jessen, 1972; BM(NH) P.12070).

23.  Cycloid scales of teleost type, vs. ganoid scales, or cycloid scales of amioid type.

24.  Upper lobe of caudal fin constantly with one unbranched and nine branched principal rays, vs. one and ten, or a larger or variable number.

25.  A vertically-keeled rostrum with paired joint surfaces for the upper jaw, vs. no joint and no such rostrum.

26.  Lower jaw without prearticular bone, vs. prearticular present.

27.  First two hypurals supported by a single centrum, vs. two centra or none.

28.  Auto- and dermopterotic co-ossified, vs. dermopterotic free.

29.  Trigeminofacial chamber with four external openings, vs. three or less.

30.  Spiracular canal ending within postorbital process, vs. canal opens into fossa bridgei.

31.  No endoskeletal parabasal canal, vs. canal present in bone or cartilage of basisphenoid region.

32.  Myodome opens posteriorly, beneath basioccipital, vs. myodome bone-enclosed.

33.  Lower jaw without surangular bone, vs. surangular present.

34.  Upper and lower hypohyals present, vs. hypohyal single.

35.  No clavicle or 'serrated appendage' present, vs. one or more such dermal plates present (Patterson, in press).

36.  Lower lobe of caudal fin constantly with one unbranched and eight branched principal rays, vs. one and nine, or a larger or variable number.

37.  Caudal skeleton containing two ural centre, vs. three or more.

38.  No canals for occipital arteries in basioccipital, vs. canals present.

39.  No spiracular canal or spiracular sense-organ, vs. canal and sense-organ present.

40.  No middle pit-line groove on dermopterotic, vs. groove present.

41.  Epipleural intermuscular bones developed in the middle part of the trunk, vs. no epipleurals.

42.  An endoskeletal basihyal present, vs. absent (Nelson, 1969a).

43.  Three hypobranchials present, vs. four.

44.  Four pharyngobranchials present, vs. three.

45.  Seven hypurals present, vs. eight or more.

46.  Only two uroneurals extend forwards beyond the second ural centrum, vs. three or more.

47.  Epipleural intermuscular bones developed throughout the abdominal and anterior caudal regions, vs. confined to middle part of trunk.

48.  Angular and articular bones co-ossified, vs. bones separate, or fused with each other and with retroarticular in late ontogeny.

49.  Membrane-bone outgrowth developed from anterior face of first uroneural, vs. no such outgrowth.

50.  Retroarticular bone excluded from joint surface of lower jaw, vs. included in joint.

51.  Tooth-plates fused with first three pharyngobranchials and fifth ceratobranchial, vs. tooth-plates independent.

52.  Neural arch on first ural centrum absent or greatly reduced, vs. present and well-developed.

Figure 19 is presented as an abstract cladogram or synapomorphy scheme, and has no time-axis.[6]  It could be converted into the traditional tree by adding a time-axis, and by designating some species as ancestors (Nelson, ms.).  For example, Pholidophorus higginsi, P. bechei, Pholidolepis dorsetensis, Proleptolepis and Leptolepis coryphaenoides are each, in my opinion, still acceptable as ancestors of everything to the right of them in the diagram: each can be viewed as a paleospecies, and none is yet known to have autapomorphic characters which would lead me to reject it from the ancestral line.  Yet if I were to place the five taxa as ancestors in the tree, it could no longer have a time-axis, for P. bechei, Pholidolepis and Proleptolepis are contemporaries, occurring in the same beds at Lyme Regis, and could not be placed on the same horizon yet at successive points on the ancestral line. For this, and other reasons, I prefer to leave the diagram as a cladogram.  Its main function is not to suggest ancestor-descendent sequences, but to give an indication of the sequence in which the various teleost characters were acquired, with an approximate time-scale.  That sequence, in turn, has two main functions.

The first is to serve as a fairly objective framework for speculative discussion about evolutionary trends, adaptive change, functional morphology and so on, the traditional use of a tree. The second, to me the more important, is to offer a research program for paleontology.  The numbers in heavy type in the diagram make predictions about character distribution, which are open to test by new observations.  The numbers in light type direct attention to characters and parts of the skeleton that will repay investigation.  Searching for character-states in the fossils throws up many cases where an apparent one-step difference between the Recent taxa is shown by the fossils to be more complex, involving convergence or a series of intermediate character-states.  Thus further work increases the number of character-states which can be entered in the diagram, and these, in turn, will strengthen or suggest modification of the arrangement.  And while more fossil taxa can be entered in the cladogram, and used to test it, as they are investigated, an analysis of this sort does not, like most discussions in phylogenetic paleontology, lead to the conclusion that more fossils are necessary before answers can be given.

CONCLUSIONS:  THE CONTRIBUTION OF PALEONTOLOGY

The point of view put forward in this paper, that the present is the key to the past, is an ancient one (cf. Ghiselin, 1972:144). I have pressed this view from the cladist standpoint, one which some paleontologists regard with horror - 'a spectre is haunting paleontology, the spectre of cladism' is Campbell's (1975) re-iterated phrase. Yet I do not see the cladist approach as destructive to paleontology. Instead, it seems to me to give direction and purpose to the work.

I believe, as I have tried to show in the section "Historical Review", that phylogenetic paleontology got off on the wrong foot, in teleosts at any rate. Ghiselin (1972:132) wrote 'A phylogenetic theory is more than just systematics of the pre-Darwinian kind with an historical explanation imposed upon it,' but that sentence without the 'more than' seems to me an accurate description of work in the century following Darwin. And during the first two-thirds of the twentieth century, when paleontology was acknowledged as the true guide to teleostean phylogeny, the result was stagnation. The fault lay in neglect of the principle discussed in the section "Recent Organisms and Fossil Organisms", the secondary role of fossils. The fossil record is dumb until it is interpreted by concepts derived from living organisms.

As evolutionary theory was elaborated, it came to be accepted that the fossil record held the key to phylogeny, and could therefore demonstrate how Recent organisms are related. An example is the idea that certain Mesozoic fossils demonstrated polyphyletic origins of the teleosts. Yet that conclusion arose from inversion of an older problem. What was in question was not the status of the teleosts as a monophyletic group, but the relationship of the Mesozoic fossils to that group. It was still the old problem that Muller posed in 1844 - 'what characters are to guide us with respect to fossil fish?'

Hennig's concept of synapomorphy provides a solution to that problem. Fossils may be assigned to a Recent group if they possess synapomorphies of that group. Once this is done, the fossils can contribute to our understanding of that group. I am not advocating the 'notion that paleontology adds nothing but ill-preserved specimens,' criticized by Ghiselin (1972:136). Here, I agree with Simpson (1975:14), who quotes a statement of Schaeffer et al. (1972), 'primitiveness and apparent ancientness are not necessarily correlated,' and replies 'That is true, but they usually are correlated.' That is my experience too. But once we have assigned these ancient and primitive fossils to a group, I suggest that it is their role in character phylogeny that is of value, not their role as possible ancestors. As Schaeffer et al. (1972) emphasize, comparative analysis, not stratigraphy, is the key to character phylogeny.  Fossils often add previously unsuspected primitive character-states to transformation series, and so can be most effective in decisions on the polarity of morphoclines.  Character phylogenies are, in turn, the basis for phylogenies, which are theories of relationship in which the distribution of synapo- morphies is explained parsimoniously.  Fossils can be incorporated in these schemes, and can then indicate the sequence in which characters were acquired, and give an absolute, though approximate, time-scale for those sequences.  And for every group there is a segment of the phylogeny, between the time of origin of the group and its differentiation into extant subgroups, to which fossils alone can contribute.  As shown in the section "An Alternative Approach", the success with which fossils can contribute here is dependent on the success with which Recent groups can be charac- terized, by synapomorphies.

When phylogenies are viewed as schemes of relationship in which the distribution of synapomorphies is explained parsimonious- ly, their empirical basis is clearly and concisely shown, so that they are open to criticism.  Traditional phylogenies, expressed as ancestor-descendent sequences which include some hypothetical ancestors and some named ancestral groups or species, are not open to criticism in the same way.  For neither an ancestral group nor a specific assertion of ancestry can be supported by synapomorphies, but only by convention (conventional limits for a group - the section "Ancestral Groups" - and stratigraphic convention for an ancestor - the section "Ancestors and Stratigraphy").  They are therefore not subject to the parsimony criterion, the only test of a phylogeny that is available to us.  'Parsimony' and 'ancestral group' seem to symbolize two mutually inconsistent approaches to phylogeny.

In conclusion, I believe that the role of paleontology in phylogeny is neatly summed up by Bretsky (1975:113), in these words:  'We cannot expect the "fossil record" to "prove" the truth of some particular inference about the course of phylogeny.  Rather, it is more accurate to view evolutionary theory as a means of organizing and interpreting data from fossils.'

## SYNOPSIS

This paper deals only with the phylogeny of the teleosts as a whole, and does not go into details of subgroups.  The fossil record of higher actinopterygians is comparable to that of mammals, yet the ostensible contribution of paleontology is far smaller in teleosts than in mammals.  From a historical survey it is con- cluded that this apparent failure of paleontology has its roots in

the nineteenth century, when acceptance of evolutionary theory
directed attention away from the search for characters defining
major groups of bony fishes, and towards the search for historical
links between groups.  As a result of this changed approach, it
came to be accepted that paleontology alone could contribute to
teleostean phylogeny, and stagnation ensued.  The basic problem,
which must be tackled before fossils can contribute, concerns the
characterization of major groups of Recent teleosts, and recognition of their interrelationships.  This neglected problem was not
attacked until 1966:  the mammalian analogy would be if the distinction between monotremes, marsupials and placentals was not
recognized until ten years ago.  The traditional method of phylogenetic paleontology, which produces ancestral groups or ancestor-
descendent sequences, is discussed and criticized, and an alternative approach, based on Hennig's concept of synapomorphy and on
parsimony, is proposed and illustrated.  Paleontology must always
be subservient to neontology, and has no fully independent role in
phylogenetic work.

## NOTES

1   'In order to combine the information on the affinities of a
class with that on their succession, we shall need . . . genealogical trees on whose trunks will be written the most ancient
groups, whereas the branches will carry the names of more recent types.  By arranging the proportions of the trunk and
branches and by giving them appropriate sizes, one could even
indicate the exact epoch when each group appeared . . . {and}
the intensity of development of each family in each epoch . . .
Following these principles, I have put together the picture
opposite, which represents the history of the development of
the class of fishes through all the geological formations, and
at the same time expresses the degrees of affinity between the
different families.  At the top of the diagram are written the
names of the four orders that I recognize in this class . . .
Beneath are the names of the families represented in the present
creation . . . The names of families which do not attain the
present creation are written on the trunks that represent them;
those which have no fossil representatives are simply indicated
by strong lines on the horizon marking the present creation.
Finally the convergence of all these vertical lines indicates
the affinity of the families with the main stock {souche
principale} of each order.  However, I have not joined the
side-branches to the main trunks because I am convinced that
they are not descended one from another by direct procreation
or successive transformation, but are materially independent
from one another, although forming integral parts of a systematic whole, whose junction {liaison} may only be sought in the

creative intelligence of its author.' (Agassiz, 1844:170, my
translation.)

2   It could, of course, be argued that the fact of evolution was
    evident to Agassiz, and influenced the form of his diagram, but
    that his conscience held him back from expressing these ideas.
    That interpretation is not consistent with a remark by the aged
    (and still anti-evolutionary) Agassiz: 'at that time I was on
    the verge of anticipating the views of Darwin, but it seemed
    to me that the facts were contrary to the theories of evolu-
    tion' (quoted by Jordan, 1905:381).

3   In compiling this list I was, of course, influenced by know-
    ledge of fossil teleosts and halecomorphs, and some of the
    characters in the list are presented in a form which takes
    account of conditions in fossils, rather than in the 'naive'
    form which comparison with Amia alone might suggest. The
    derived character-states in the list are not all present in
    all Recent teleosts, but are those whose distribution in
    Recent and fossil teleosts is most parsimoniously explained
    by the assumption that they characterized the immediate common
    ancestor of all Recent teleosts.

4   These groups and species are selected from those dealt with by
    Patterson (1973) and Patterson and Rosen (1977), with the
    addition of Pholidophorus higginsi Egerton, from the Rhaetic of
    Gloucestershire, based on unpublished data. Autapomorphies of
    the terminal taxa in Figure 19 are not mentioned here (some of
    the paleospecies have none). The extinct groups are named
    according to the ranks they are conventionally given. A method
    of classifying such a mixture of fossil groups is discussed by
    Patterson and Rosen (1977).

5   To avoid repetition of references, the following papers con-
    tain the evidence for the characters numbered: Patterson,
    1973 (nos. 1,2,5,7,14,16,23,34,37,43,44); Patterson, 1975
    (nos. 2,3,8,10-12,15,17-20,25,28-32,38,39); Patterson and
    Rosen, 1976 (nos. 24,26,27,33,36,40,41,45-52). Other refer-
    ences are given with the characters, and where observations
    of fossils are unpublished, British Museum (Natural History)
    specimen numbers are given.

6   Minimum ages for certain bifurcations can be read off from the
    stratigraphic ranges specified for the taxa: the latest date
    for the dichotomy between halecomorphs and teleosts is basal
    Triassic times, that between Pholidophorus higginsi and higher
    groups is latest Triassic, that between Leptolepis coryphaenoides
    and higher groups is early Jurassic, and extant subgroups of
    teleosts are first known in the late Jurassic.

REFERENCES

Agassiz, J.L.R., 1833-44, "Recherches sur les Poissons Fossiles,"
    5 vols. and supplement, Neuchatel. Dates of publication of
    the parts are given by W.H. Brown, pp. xxv-xxix in Woodward,
    A.S., and Sherborn, C.D., 1890, "Catalogue of British Fossil
    Vertebrata," Dulau, London.

Agassiz, J.L.R., 1857, "Contributions to the Natural History of
    the United States of America. First Monograph," Little,
    Brown & Co., Boston.

Agassiz, J.L.R., 1859, "An Essay on Classification," Longmans,
    London.

Arambourg, C., 1935, Observations sur quelques poissons fossiles
    de l'ordre des Halecostomes et sur l'origine des Clupeides.
    C.R. Acad. Sci., Paris, 200:2110-2112.

Arambourg, C., 1950, Nouvelles observations sur les Halecostomes
    et l'origine des Clupeidae. C.R. Acad. Sci., Paris, 231:
    416-418.

Arambourg, C. and Schneegans, D., 1936, Poissons fossiles du
    Bassin sedimentaire du Gabon. Annls Paleont. 24:139-160.

Bardack, D., 1965, Anatomy and evolution of chirocentrid fishes.
    Paleont. Contr. Univ. Kans., Vertebrata, 10:1-88.

Berg, L.S., 1940, Classification of fishes, both Recent and fossil.
    Trudy zool. Inst. Leningr., 5:85-517.

Bertin, L. and Arambourg, C., 1958, Super-ordre des Teleosteens
    (Teleostei), in "Traite de Zoologie" (P.P. Grasse, ed), 13(3)
    pp. 2204-2500, Masson, Paris.

Blot, J., 1975, A propos des teleosteens primitifs: l'origine des
    Apodes. Colloques int. Cent. natn. Rech. scient., 218:281-
    291.

Bonde, N., 1975, Origin of "higher groups": viewpoints of phylo-
    genetic systematics. Colloques int. Cent. natn. Rech. scient.,
    218:293-324.

Boulenger, G.A., 1902, Fossil fishes in the British Museum.
    Nature, London, 65:388-389.

Boulenger, G.A., 1904a, A synopsis of the suborders and families
    of teleostean fishes. Ann. Mag. nat. Hist., (7) 13:161-190.

Boulenger, G.A., 1904b, Systematic account of Teleostei, in
    "The Cambridge Natural History (S.F. Harmer and A.E. Shipley,
    eds), 7, pp. 541-727, Macmillan, London.

Bretsky, S.S., 1975, Allopatry and ancestors: a response to
    Cracraft. Syst. Zool., 24:113-119.

Brundin, L., 1966, Transantarctic relationships and their signifi-
    cance, as evidenced by chironomid midges. K. Svenska
    VetenskAkad. Handl. (4) 11 (1):1-472.

Campbell, K.S.W., 1975, Cladism and phacopid trilobites.
    Alcheringa, 1:87-96.

Casier, E. and Taverne, L., 1971, Note preliminaire sur le
    materiel paleoichthyologique eocretacique recolte par la

Spanish Gulf Oil Company en Guinee Equatoriale et au Gabon. Rev. Zool. Bot. Afr., 83:16-20.

Cope, E.D., 1871, Contribution to the ichthyology of the Lesser Antilles. Trans. Amer. Phil. Soc., N.S., 14:445-483.

Cope, E.D., 1872, Observations on the systematic relations of the fishes. Proc. Amer. Ass. Adv. Sci., 1871:317-343.

Cope, E.D., 1887, Geology and palaeontology. Amer. Nat., 1887: 1014-1019.

Cracraft, J., 1974, Phylogenetic models and classification. Syst. Zool., 23:71-90.

Darlington, P.J., 1970, A practical criticism of Hennig-Brundin "Phylogenetic Systematics" and Antarctic biogeography. Syst. Zool., 19:1-18.

Darwin, C.R., 1859, "On the Origin of Species by Means of Natural Selection," John Murray, London.

Eldredge, N. and Tattersall, I., 1975, Evolutionary models, phylogenetic reconstruction, and another look at hominid phylogeny. Contrib. Primat., 5:218-242.

Forey, P.L., 1973a, A revision of the elopiform fishes. Bull. Brit. Mus. Nat. Hist. (Geol.) Suppl., 10:1-222.

Forey, P.L., 1973b, A primitive clupeomorph fish from the Middle Cenomanian of Hakel, Lebanon. Canad. J. Earth Sci., 10:1302-1318.

Gardiner, B.G., 1960, A revision of certain actinopterygian and coelacanth fishes, chiefly from the Lower Lias. Bull. Brit. Mus. Nat. Hist. (Geol.), 4:241-384.

Ghiselin, M.T., 1972, Models in phylogeny, in "Models in Paleobiology" (T.J.M. Schopf, ed), pp. 130-145, Freeman, Cooper & Co., San Francisco.

Gill, T., 1872a, On the characteristics of the primary groups of the class of mammals. Proc. Amer. Ass. Adv. Sci., 1871: 284-306.

Gill, T., 1872b, Arrangement of the families of fishes. Smithson. Misc. Collns., 247:1-95.

Goodrich, E.S., 1909, Vertebrata Craniata. First fascicle; cyclostomes and fishes, in "A Treatise on Zoology" (R. Lankester, ed.), 9, A. & C. Black, London.

Goodrich, E.S., 1912, "The Evolution of Living Organisms," T.C. & E.C. Jack, London.

Goodrich, E.S., 1930, "Studies on the Structure and Development of Vertebrates," Macmillan, London.

Goody, P.C., 1969, The relationships of certain Upper Cretaceous teleosts with special reference to the myctophoids. Bull. Brit. Mus. Nat. Hist. (Geol.) Suppl., 7:1-255.

Gosline, W.A., 1965, Teleostean phylogeny. Copeia, 1965:186-194.

Gosline, W.A., 1971, "Functional Morphology and Classification of Teleostean Fishes," University of Hawaii Press, Honolulu.

Greenwood, P.H., 1970, On the genus Lycoptera and its relationship with the family Hiodontidae (Pisces, Osteoglossomorpha),

Bull. Brit. Mus. Nat. Hist. (Zool.), 19:257-285.

Greenwood, P.H., 1973, Interrelationships of osteoglossomorphs.
Zool. J. Linn. Soc. London 53, Suppl., 1:307-332.

Greenwood, P.H., Myers, G.S., Rosen, D.E. and Weitzman, S.H.,
1967, Named main divisions of teleostean fishes. Proc.
Biol. Soc. Washington, 80:227-228.

Greenwood, P.H., Rosen, D.E., Weitzman, S.H. and Myers, G.S.,
1967, Phyletic studies of teleostean fishes, with a provisional classification of living forms. Bull. Amer. Mus.
Nat. Hist., 131:339-456.

Gregory, W.K., 1907, The orders of teleostomous fishes. Ann.
New York Acad. Sci., 47:437-508.

Gregory, W.K., 1933, Fish skulls: a study of the evolution of
natural mechanisms. Trans. Amer. Phil. Soc., 23:75-481.

Gregory, W.K., 1951, "Evolution Emerging," Macmillan, New York.

Griffith, J. and Patterson, C., 1963, The structure and relationships of the Jurassic fish Ichthyokentema purbeckensis. Bull.
Brit. Mus. Nat. Hist. (Geol.), 8:1-43.

Gunther, A.C.L.G., 1871, Description of Ceratodus, a genus of
ganoid fishes, recently discovered in rivers of Queensland,
Australia. Phil. Trans. Roy. Soc. London, 161:511-571.

Gunther, A.C.L.G., 1880, "An Introduction to the Study of
Fishes," A. & C. Black, Edinburgh.

Haeckel, E., 1866, "Generelle Morphologie der Organismen,"
G. Reimer, Berlin.

Haeckel, E., 1889, "Natürliche Schöpfungs-Geschichte," 8th ed.,
G. Reimer, Berlin.

Haeckel, E., 1895, "Systematische Phylogenie der Wirbelthiere
(Vertebrata)," G. Reimer, Berlin.

Harper, C.W., 1976, Phylogenetic inference in paleontology. J.
Paleont., 50:180-193.

Heckel, J.J., 1850a, Ueber das Wirbelsäulen-Ende bei Ganoiden und
Teleostiern. Sber. Akad. Wiss. Wien, 5:143-148.

Heckel, J.J., 1850b, Ueber die Wirbelsäule fossiler Ganoiden.
Sber. Akad. Wiss. Wien, 5:358-368.

Heckel, J.J., 1851, Über die Ordnung der Chondrostei und die
Gattungen Amia, Cyclurus, Notaeus. Sber. Akad. Wiss. Wien,
6:219-224.

Hennig, W., 1966, "Phylogenetic Systematics," University of
Illinois Press, Urbana.

Hennig, W., 1969, "Die Stammgeschichte der Insekten," W. Kramer,
Frankfurt.

Huxley, T.H., 1861, Preliminary essay upon the systematic
arrangement of the fishes of the Devonian epoch. Mem. Geol.
Surv. U.K., Decade, 10:1-40.

Huxley, T.H., 1883, Contributions to morphology. Ichthyopsida.
No.2. On the oviducts of Osmerus. Proc. Zool. Soc. London,
1883:132-139.

Jefferies, R.P.S., 1975, Fossil evidence concerning the origin of the chordates.  Symp. Zool. Soc. London, 36:253-318.

Jessen, H., 1972, Schultergürtel und Pectoralflosse bei Actino-
    pterygiern.  Fossils and Strata, 1:1-101.

Johnson, R.G. and Richardson, E.S., 1969, The morphology and
    affinities of Tullimonstrum.  Fieldiana, Geol., 12:119-149.

Jordan, D.S., 1905, "A Guide to the Study of Fishes," H. Holt &
    Co., New York.

Kner, R., 1866, Betrachtungen über die Ganoiden, als natürliche
    Ordnung.  Sber. Akad. Wiss. Wien, 54:519-536.

Lehman, J.P., 1966, Actinopterygii, in "Traite de Paleontologie"
    (J. Piveteau, ed.), 4 (3) pp. 1-242, Masson, Paris.

Lehman, J.P., 1975, Quelques reflexions sur la phylogenie des
    vertebres inferieurs.  Colloques int. Cent. natn. Rech.
    scient., 218:257-264.

Lovtrup, S., 1973, Classification, convention and logic.
    Zoologica Scripta, 2:49-61.

Lutken, C., 1868, Professor Kner's classification of the ganoids.
    Geol. Mag., London, 5:429-432.

Lutken, C., 1869, Om Ganoidernes Begraensning og Inddeling.
    Vidensk. Meddr dansk. naturh. Foren., 1868:1-82.

Lutken, C., 1871, On the limits and classification of the
    ganoids, Ann. Mag. Nat. Hist. (4) 7: 329-339.  (A translation
    by W.S. Dallas, of a summary of Lutken, 1869.)

Lutken, C., 1873, Ueber die Begrenzung der Ganoiden, Palaeonto-
    graphica 22: 1-54.  (A translation, by R. von Willemoes-Suhm,
    of an updated version of Lutken, 1869.)

MacFadden, B.J., 1976, Cladistic analysis of primitive equids,
    with notes on other perissodactyls, Syst. Zool. 25:1-14.

Mayr, E., 1965, Numerical phenetics and taxonomic theory.  Syst.
    Zool., 14:73-97.

Mayr, E., 1969, "Principles of Systematic Zoology," McGraw-Hill,
    New York.

Mayr, E., 1974, Cladistic analysis or cladistic classification?,
    Z. Zool. Syst. EvolForsch., 12:94-128.

Medawar, P.B., 1967, "The Art of the Soluble," Methuen, London.

Melton, W. and Scott, H.W., 1973, Conodont-bearing animals from
    the Bear Gulch Limestone, Montana, Spec. Pap. Geol. Soc.
    Amer., 141:31-65.

Muller, J., 1844, Ueber den Bau und die Grenzen der Ganoiden und
    uber das naturliche System der Fische, Ber. Akad. Wiss.
    Berlin, 1844:416-422.

Muller, J., 1845, Ueber den Bau und die Grenzen der Ganoiden und
    über das natürliche System der Fische, Arch.Naturgesch., 11:
    91-141.

Muller, J., 1846a, Fernere Bemerkungen über den Bau der Ganoiden,
    Ber. Akad. Wiss. Berlin, 1846:67-85 (English translation by
    J.W. Griffith, 1846, in Scient. Mem., 4:543-558.)

Muller, J., 1846b, Ueber den Bau und die Grenzen der Ganoiden,
      und über das natürliche System der Fische, Phys. Math. Abh.
      K. Akad. Wiss. Berlin, 1846:117-216. (English translation
      by J.W. Griffith, 1846, in Scient. Mem., 4:499-542.)
Nelson, G.J., 1969a, Gill arches and the phylogeny of fishes, with
      notes on the classification of vertebrates, Bull. Amer. Mus.
      Nat. Hist., 141:475-552.
Nelson, G.J., 1969b, Origin and diversification of teleostean
      fishes, Ann. New York Acad. Sci., 167:18-30.
Nelson, G.J., 1970, Outline of a theory of comparative biology,
      Syst. Zool., 19:373-384.
Nelson, G.J., 1972, Comments on Hennig's "Phylogenetic Systematics"
      and its influence on ichthyology, Syst. Zool., 21:364-374.
Nelson, G.J., ms., "A draft of a chapter entitled Classification,"
      1976.
Newton, E.T., 1902, Review of "Catalogue of the fossil fishes in
      the British Museum (Natural History). Part IV, Geol. Mag.,
      London, (4) 9:133-138.
Nybelin, O., 1963, Zur Morphologie und Terminologie des Schwanz-
      skelettes der Actinopterygier, Ark. Zool., (2) 15:485-516.
Nybelin, O., 1966, On certain Triassic and Liassic representatives
      of the family Pholidophoridae s. str., Bull. Brit. Mus. Nat.
      Hist. (Geol.), 11:351-432.
Nybelin, O., 1974, A revision of the leptolepid fishes, Acta. R.
      Soc. scient. litt. gothoburg. (Zool.), 9:1-202.
Obruchev, D.V., 1964, "Osnovy Paleontologii, 11," Akad. Nauk
      SSSR, Moskva. (In Russian.)
Olson, E.C., 1971, "Vertebrate Paleozoology," John Wiley, New York.
Owen, R., "The Anatomy of Vertebrates. Vol. 1. Fishes and
      Reptiles." Longmans, Green & Co., London.
Patterson, C., 1967, Are the teleosts a polyphyletic group?,
      Colloques int. Cent. natn. Rech. scient., 163:93-109.
Patterson, C., 1968, The caudal skeleton in Lower Liassic
      pholidophorid fishes, Bull. Brit. Mus. Nat. Hist. (Geol.),
      16:201-239.
Patterson, C., 1970, A clupeomorph fish from the Gault (Lower
      Cretaceous), Zool. J. Linn. Soc. London, 49:161-182.
Patterson, C., 1973, Interrelationships of holosteans. Zool. J.
      Linn. Soc. London, 53, Suppl., 1:233-305.
Patterson, C., 1975, The braincase of pholidophorid and leptolepid
      fishes, with a review of the actinopterygian braincase, Phil.
      Trans. Roy. Soc. London (Biol.), 269:275-579.
Patterson, C., in press, Cartilage bones, dermal bones and mem-
      brane bones, or the exoskeleton versus the endoskeleton,
      Zool. J. Linn. Soc. London 59, Suppl., 1: in press.
Patterson, C. and Rosen, D.E., 1977, Review of ichthyodectiform
      and other Mesozoic teleost fishes and the theory and practice
      of classifying fossils, Bull. Amer. Mus. Nat. Hist., 158:
      81-172.

Peckham, M., 1959, "The Origin of Species by Charles Darwin. A Variorum Text," University of Pennsylvania Press, Philadelphia.

Pictet, F.J., 1854, "Traite de Paleontologie. Tome 2", 2nd ed., Bailliere, Paris.

Platnick, N.I., 1976, Drifting spiders or continents?: vicariance biogeography of the spider subfamily Laroniinae (Araneae: Gnaphosidae), Syst. Zool., 25:101-109.

Rayner, D.H., 1937, On Leptolepis bronni Agassiz, Ann. Mag. Nat. Hist. (10), 19:46-74.

Rayner, D.H., 1941, The structure and evolution of the holostean fishes, Biol. Rev., 16:218-237.

Rayner, D.H., 1948, The structure of certain Jurassic holostean fishes with special reference to their neurocrania, Phil. Trans. Roy. Soc. London (Biol.), 233:287-345.

Read, D.W., 1975, Primate phylogeny, neutral mutations and "molecular clocks," Syst. Zool., 24:209-221.

Regan, C.T., 1911a, The anatomy and classification of the teleostean fishes of the orders Berycomorphi and Xenoberyces, Ann. Mag. Nat. Hist., (8)7:1-9.

Regan, C.T., 1911b, The anatomy and classification of the teleostean fishes of the order Iniomi, Ann. Mag. Nat. Hist. (8)7:120-133.

Regan, C.T., 1923, The skeleton of Lepidosteus, with remarks on the origin and evolution of the lower neopterygian fishes, Proc. Zool. Soc. London, 1923:445-461.

Regan, C.T., 1929, Fishes, in "Encyclopaedia Britannica," 14th ed., vol. 9, pp. 305-329, London and New York.

Romer, A.S., 1945, "Vertebrate Paleontology," 2nd ed., University of Chicago Press, Chicago.

Romer, A.S., 1966, "Vertebrate Paleontology," 3rd ed., University of Chicago Press, Chicago.

Rosen, D.E., 1973, Interrelationships of higher euteleostean fishes, Zool. J. Linn. Soc. London 53, Suppl., 1:397-513.

Rosen, D.E. and Patterson, C., 1969, The structure and relationships of the paracanthopterygian fishes, Bull. Amer. Mus. Nat. Hist., 141:357-474.

Saint-Seine, P. de, 1949, Les poissons des calcaires lithographiques de Cerin (Ain), Nouv. Archs Mus. Hist. nat. Lyon, 2:1-357.

Schaeffer, B., Hecht, M.K. and Eldredge, N., 1972, Phylogeny and paleontology, Evolut. Biol., 6:31-46.

Schultze, H.P., 1966, Morphologische und histologische Untersuchungen an Schuppen mesozoischer Actinopterygier (Ubergang von Ganoid- zu Rundschuppen), Neues Jb. Geol. Palaont. Abh., 126:232-314.

Scourfield, D.J., 1937, An anomalous fossil organism, possibly a new type of chordate, from the Upper Silurian of Lesmahagow, Lanarkshire, Ainiktozoon loganense, gen. et sp. nov., Proc. Roy. Soc. London (Biol), 121:533-547.

Siegfried, P., 1954, Die Fisch-Fauna des Westfalischen Ober-
    Senons, Palaeontographica (A), 106:1-36.
Simpson, G.G., 1961, "Principles of Animal Taxonomy," Columbia
    University Press, New York.
Simpson, G.G., 1975, Recent advances in methods of phylogenetic
    inference,  Contrib. Primat., 5:3-19.
Simpson, G.G., 1976, The compleat palaeontologist?, Ann. Rev.
    Earth Planet. Sci., 4:1-13.
Stensio, E.A., 1964, Les cyclostomes fossiles ou ostracodermes,
    in "Traite de Paleontologie" (J. Piveteau, ed.), 4 (1),
    pp. 96-382, Masson, Paris.
Stensio, E.A., 1968, The cyclostomes with special reference to
    the diphyletic origin of the Petromyzontida and Myxinoidea,
    Nobel Symposium, 4:13-70.
Tarlo, L.B., 1960, The invertebrate origins of the vertebrates,
    Int. Geol. Congr., 21(22):113-123.
Taverne, L.P., 1974, L'osteologie d'Elops Linne, C., 1766 (Pisces
    Elopiformes) et son interet phylogenetique.  Mem. Acad. r.
    Belg. Cl. Sci. 8° 41 (2):1-96.
Taverne, L.P., 1975a, Considerations sur la position systematique
    des genres fossiles Leptolepis et Allothrissops au sein des
    teleosteens primitifs et sur l'origine et le polyphyletisms
    des poissons teleosteens, Bull. Acad. r. Belg. Cl. Sci.,
    61:336-371.
Taverne, L.P., 1975b, Sur Leptolepis (Ascalabos) voithi (von
    Munster, G., 1839), teleosteen fossile du Jurassique
    superieur de l'Europe et ses affinites systematiques, Biol.
    Jaarb., 43:233-245.
Thiolliere, V.J., 1854, "Descriptions des poissons fossiles
    provenant des gisements dans le Bugey.  Livre 1," Lyon.
Thiolliere, V.J., 1858, Note sur les poissons fossiles du Bugey,
    et sur l'application de la methode de Cuvier a leur
    classement, Bull. Soc. geol. Fr., (2)15:782-793.
Traquair, R.H., 1877, The ganoid fishes of the British
    Carboniferous formations.  Part I.  Palaeoniscidae,
    Palaeontogr. Soc. (Monogr.) 1877:1-60.
Traquair, R.H., 1896, Review of "Catalogue of the fossil fishes
    in the British Museum (Natural History).  Part III,"
    Geol. Mag., London, (4)3:124-127.
Wagner, A., 1861, Monographie der fossilen Fische aus den
    lithographischen Schiefern Bayern's.  Erste Abtheilung:
    Placoiden und Pyknodonten, Abh. bayer. Akad. Wiss., 9:279-352.
Wagner, A., 1863, Monographie der fossilen Fische . . . Zweite
    Abtheilung, Abh. bayer. Akad. Wiss., 9:613-748.
Waldman, M., 1971, Fish from the Lower Cretaceous of Victoria,
    Australia, with comments on the palaeo-environment.  Spec.
    Pap. Palaeontology, 9:1-124.
Woodward, A.S., 1889, Professor Dr. von Zittel on Palichthyology,
    Geol. Mag., London, (3)6:125-130, 177-181, 227-232.

Woodward, A.S., 1890, The fossil fishes of the Hawkesbury Series
    at Gosford, Mem. geol. Surv. N.S.W. (Palaeont. Ser.), 4:
    1-56.
Woodward, A.S., 1891, "Catalogue of the fossil fishes in the
    British Museum (Natural History).  Part II, "British Museum
    (Nat. Hist.), London.
Woodward, A.S., 1895, "Catalogue of the fossil fishes in the
    British Museum (Natural History).  Part III, "British
    Museum (Nat. Hist.), London.
Woodward, A.S., 1901, "Catalogue of the fossil fishes in the
    British Museum (Natural History).  Part IV, "British Museum
    (Nat. Hist.), London.
Woodward, A.S., 1919, The fossil fishes of the English Wealden
    and Purbeck formations, Palaeontogr. Soc. (Monogr.), 1917:
    105-148.
Woodward, A.S., 1942, The beginning of the teleostean fishes,
    Ann. Mag. Nat. Hist., (11)9:902-912.
Zittel, K.A. von, 1887-88, "Handbuch der Palaeontologie.
    Palaeozoologie.  Band III," Lief I (1887), Lief II (1888),
    R. Oldenbourg, Munchen.

# PHYLOGENETIC RECONSTRUCTION: THEORY, METHODOLOGY, AND APPLICATION TO CHORDATE EVOLUTION

Wolfgang Friedrich GUTMANN

Senckenberg Museum

6000 Frankfurt am Main, West Germany

## INTRODUCTION

Evolution is understood to be the process of transformation of organic systems driven by spontaneous mutations and controlled by selection. In most cases selection is considered to be a force exerted by the environment. Although theoretically sound concepts of selection have been put forward by Bock (1959), Hecht (1965), Bock and von Wahlert (1965), Cizek and Hodanova (1971), and Eigen (1971), adaptation and selection are still looked upon predominantly from the environmental point of view.

Adaptational interpretations of small phylogenetic changes can legitimately be based on environmental or behavioral relations alone, since the changes in the evolving biological system are only small. Those phylogenetic reconstructions intended to elucidate the evolution of major taxa or 'bauplan' relations, very often reflect this one-sided view by neglecting the canalizing influence of the evolving organism itself and of the selective pressures effected by the organism and its functional components. This aspect, called 'the internal consistency of the evolving system' by Bock and v. Wahlert (1965), is of the utmost importance in any reasonable attempt to reconstruct the evolution of 'bauplans' which characterize major systematic groups. When the emphasis is placed on the canalizing influence of the evolving biological system itself, it is not suggested that environmental relations are of minor importance or without relevance for major phylogenetic transformations. Just the reverse is true. Phylogenetic reconstructions must be based on adaptational interpretations referring to selective pressures effected by the environment and the organism itself (Gutmann and Peters, 1973a).

This understanding of the evolutionary theory must be used
for the following: (1) to develop a methodology for formulating
phylogenetic theories; (2) to elaborate reasonable criteria for
characterizing phylogenetic theories; and (3) to discriminate between true phylogenetic theories and insufficient or unwarranted
ideas of phylogeny.


## METHODOLOGICAL CONSIDERATIONS

Phylogenetic theories can only be formulated by reconstructing
sequences of adaptational changes and tracing the diverging lines
leading to existing organisms and their characters. Only those
features and characters whose functions are known, and for which
the value of the adaptational changes can be assessed, can be
utilized in phylogenetic reconstruction. A phylogenetic theory
can neither be constructed by comparison of morphological configurations nor by character analyses, although both methodological
approaches may be used in preliminary stages of the phylogenetic
study. This means that morphological comparison and character
analyses are only useful for ordering the material.

Phylogenetic theory is never contained in a dendogram but
only in the phylogenetic model based on a reconstruction of the
adaptational process. The phylogenetic dendrogram can only give
an abbreviation of the relationships contained in the model on
which it is based. A dendogram remains an empty scheme if the
phylogenetic theory on which it is based is not clearly outlined.
As adaptational changes are only tacitly assumed in most phylogenetic hypotheses, a revision of the majority of theories is
urgently needed. Only those theories which are explicitly formulated can be adequately criticized, tested, and improved.

The adaptational changes explained in every phylogenetic model
are determined by the mutual interaction of the components in the
organism and the organism's relations to the environment. So both
aspects of selection and adaptation, the environmental and the
functional relations within the organism, have to be inherent in
the phylogenetic reconstruction although the relative contribution
may change and can be concentrated on one aspect or another. It
is, in principle, possible to reconstruct some changes in internal
functional relations without referring to any specific environmental
relation. The internal relations, the functional consistency, the
feasibility of the organism's construction, and the relations of
the organism with the environment, act together in the selective
strategy. It is this complex of changing structural and functional
relations that must be reconstructed in phylogenetic theories.

The possibility of attaching priority of any kind to the

environmental relations is clearly incompatible with a strict
understanding of the evolutionary theory. All adaptational changes
are related to environmental conditions. The properties of the
organism as a functioning unit composed of many organs and structures without direct environmental relevance, limit the environmental conditions in which these organisms can live. This point
can be emphasized by pointing to organisms that have evolved by
becoming adapted to a certain environment. Evolutionary changes
take place in their bodies that allow them to live in a new ecological situation differing, possibly profoundly, from the one in
which the organisms had evolved. Neither the prevailing environmental conditions nor the constructional relations are of greater
importance. Changes in the construction allow new environmental
relations to be sought by the organisms. Thus improvement (better
adaptation) of the organism in one environment may produce the
prerequisites for it to live in another.

If the organism is understood to be a functional system which
consists  of internal and external relations, evolution consists
of gradual changes in some or all of the relations. These comprise  structural, environmental, and behavioral properties, all
interdependent and imposing limitations on each other. As a
consequence there is no criterion that could be deduced from the
evolutionary theory allowing a definition of adaptation in relation
to one aspect, either the environmental or the internal and structural, alone. The phylogenetic process that is reconstructed cannot in any instance or any phase of its progression be understood
separately from the environment in which the organisms live and
evolve. Nor can the canalizing and controlling influence of the
functional requirements of the organism itself be dispensed with.
If the process of phylogenetic change is driven on by mutations
in a gene pool under pressure from competition, selection for
efficiency and economy of the organisms is the inevitable result.
Selection can therefore be understood as "the evaluation of energy
required for structures in relation to their function" (Bock and
v. Wahlert, 1965; Cizek and Hodanova, 1971; Eigen, 1971). The
outcome of the competition between organisms requiring different
amounts of energy to live and reproduce is adaptation.

Selection is therefore a process with many facets. It means
that those changes not fitting into any relation of the organism
are eliminated because the organisms bearing these characters
cannot survive or are impaired in their functional performance.
Some genotypic changes may even be incompatible with normal
chromosome activity in reduplication, transcription, and cell
division. Thus elimination by selection can already be effected
at this functional level of the organism (Whyte, 1965, 1967).
Other stumbling blocks for mutational changes lie in the ontogenetical, physiological, behavioral, and ecological requirements

(Gutmann and Peters, 1973a). Only those changes generated by
mutations that fit into all functional relations of the extremely
complex functional network of the organism are operated on by the
positive mechanism of selection and can potentially be beneficial.
The organisms not eliminated because of functional shortcomings in
their basic structures participate in the competition for the
chance to transmit their genetic information to the offspring.
This determines the selective value of mutational changes. According to this definition of non-random differential reproduction of
genotypes (Bock and v. Wahlert, 1965:291), selection is brought
about by the action and functioning of all structures of the
organism. From this definition it can clearly be deduced that any
structural changes caused by mutations that improve the efficiency
of the organism, or eliminate unnecessary structures, must be
advantageous. Major adaptational changes are always concerned with
the energy consuming biological apparatus. This may become adapted
to a certain environment. However, the adaptational transformations acquired in one environment may allow the organism to change
its environmental relations and continue a different adaptational
process in the new habitat.

The multitude of requirements that every mutational change
must fulfill favors a gradual process of evolution. Every major
mutational change which would result in a profound change in body
construction must necessarily and inevitably be detrimental and
selectively negative. Thus phylogenetic gradualism is based on
the fact that complex living systems will only tolerate minor
phylogenetic changes. Those transformations fitting into the
complex relations of the organism's functional components, and
into the environmental, ecological and behavioral relations are
evaluated with reference to the energy already required for the
formation and running of the structures which the organism
possesses. Every improvement of a functional component in relation
to the energy invested in it must constitute an adaptive advantage.
This means that all structures without functional relevance or with
diminishing functional importance must gradually be eliminated in
the adaptational process. Reduction of structures means economiza-
tion of the construction and is part of the process of adaptation.
All organs of growing importance must be enlarged or improved by
changes that make them more efficient. So adaptation must comprise
an increase in complexity of some structures and simplification of
others. Both processes may occur together as the complication of
some structures can run parallel to the simplification and reduction of others. Thus during the course of evolution the energy
used to build up complex organs is redistributed by diverting it
from those organs that are undergoing reduction of simplification.
The interrelationships of the structures in a complex organism do
not allow any functional system to attain its optimal form. The
compromises between the functional requirements of different, but

interdependent, structures, is of course also enforced by selection.

The methodological consequences of these theoretical considerations must be applied with some severity in phylogenetic theories.
Only changes that are supposed to be effected by mutations can be
accepted as relevant.  Numerous gradual changes may accumulate in
time and lead to profound transformations of body construction.
The phylogenetic theory must explain the canalizing effects exerted
by the construction itself and by the environmental relations as
well. As changes in the complex relations during evolution cannot
be reconstructed in their totality, a model must be designed by
singling out some relations that can be traced through their
gradual adaptational changes.  The models must be constructed in
the same way as every scientific theory and tested against observable relations and structures in living or fossil organisms
(Bock, 1973).  One can start with morphoclines which might be
representative of stages in a phylogenetic lineage.  The polarity
of morphoclines can then only be decided on the basis of adaptational interpretations.  These have to be reconstructed on the
basis of insight into the working of the structural components.
Those transformations that increase efficiency may be assessed as
advantageous.  A decision on the direction of the adaptational
process is reached only in this way.

The rise of efficiency reconstructed in a phylogenetic model
can have a direct relationship to the environment when organs are
concerned which perform a biological role.  In other cases the
efficiency of internal organs may only have an indirect influence
on the performance in the environment.  Internal organs may improve
the efficiency of the whole system by equipping it with more
effective mechanisms or by eliminating structures of low efficiency.
Therefore different types of phylogenetic theories may be the
outcome of phylogenetic reconstructions.  Most phylogenetic changes
are characterized by changes in both the environmental relations
and the organism.  However, environmental changes without any
transformation of the organism itself may occur.  This is the case
when new environments are entered by organisms that evolved in a
different ecological situation.  Another case would be an organism
not undergoing a change in its environmental relations at all, but
becoming better adapted by an increase in the efficiency of the
machinery of the organism.  As a consequence of this there might
be evolution by transformation by the organismic system alone
(Gutmann and Peters, 1973b).

If this concept of adaptation is applied to phylogenetic reconstructions, morphological sequences of organs or organisms
(morphoclines) put together arbitrarily or by reference to similarity alone can no longer be considered as phylogenetic theories.
Only those gradual models that describe and explain adaptational

changes and demonstrate the feasibility of the intermediate forms
attain the rank of phylogenetic theories. They are not permitted
to contain tacit interpretations of the adaptational change, but
are required to give explicit accounts of the adaptational changes.
This can only be achieved by the reconstruction of the phylogenetic
process (Peters and Gutmann, 1971, 1973; Franzen et al., 1976).

To achieve this goal of phylogenetic reconstruction no strict
methodology can be given. Phylogenetic theories must be tentatively
formulated and tested against observations, while the explanatory
powers of the theory are tested in a strict way (Bock, 1973;
Popper, 1969, 1972; Feyerabend, 1972; Stegmuller, 1969; Medawar,
1972). The evolutionary theory itself and the structural and
functional properties of organisms are the basis of a phylogenetic
strategy that can be applied as a kind of methodology. All phylogenetic theories must be presented in the form of continuous
adaptational models. Morphological comparison and character
analyses continue to be essential and indispensable methodological
instruments that are used in the preliminary stages of attempts to
solve phylogenetic problems. They do not, however, decide on the
phylogenetic theory and the directive adaptational process as
nearly unanimously as is maintained by phylogeneticists. The
establishment of phylogenetic relationships and the reconstruction
of the phylogenetic changes must be identical: the former can
only be achieved by the latter.

## The Reconstruction of the Ancestors of Chordates

An example of the reconstruction of a phylogenetic process
will now be given in the form of a model for some of the transformations that led to the chordate body construction. It is only
one part of a broad theory based on reconstructions and explanations of the 'bauplan' relations of most higher metazoans (Gutmann,
1972, 1974; Bonik et al., 1976a, 1976b; Vogel et al., 1976).

The mechanical systems reconstructed in the model are built
up in a somewhat deductive way by using functional components
observed in the locomotory system of existing metazoans: muscles,
collagenous fibers, and a filling of fluid or parenchymal cells.
The connective tissue fibers form a continuous mechanical system,
the elements of which are inextensible. Changes in shape during
movement are allowed by fiber rearrangements. Because of the
properties of muscles (in particular the fact that they require
to act in antagonistic couplets or groups) certain mechanical requirements must be catered for. An arrangement is required in
which a mutual mechanical relation between muscles of different
orientation is provided. The answer is incorporated in the series
of fluid-filled cavities which make up the hydrostatic skeleton.

The muscles generate hydrostatic pressure in the coelomic cavities. As the volume of the fluid contained in the coelom is constant, contraction of one part of the musculature must be coupled with extension in another part. The former shape of the body is regained by a reversal of the muscle action. The three elements (muscle, collagen fibers, and fluid-filling) and the requirements for their proper functioning exert a restrictive influence by excluding many morphological arrangements and leaving only a narrow path for phylogenetic change.

A schematized, mechanically-closed concept is constructed, according to some aspects of the methodology of Dullemeijer (1974). The assembly of functional components composing the model is designed in accordance with natural laws. The sequence of model stages represents the phylogenetic process. In every stage of the phylogenetic sequence the evolving construction is conceived as a system of mutually dependent components organized in a way that is determined by both functional and environmental requirements. The functional relations of the organism can only be studied in abstractions of real organisms and by selecting certain parts of a complicated system. All stages are shown to be functionally feasible and constitute a continuous process of adaptational change.

The construction of rather simplistic and not too detailed models is at variance with the methods proposed by Dullemeijer (1974). The relations and functional elements used in it are nevertheless very strictly defined and characterized by law-like properties or even laws of physics (biomechanics and hydrostatics). The shortcomings of the extremely simplistic, initial model are compensated for by later tests against additional observations (Bock, 1973). A model for the evolution of the locomotor system such as the one presented for the chordates, must allow integration of the other internal organs. That it does indeed allow integration of the gut, the kidneys, and the circulatory system, and can thereby stand the tests, is demonstrated elsewhere (Gutmann, 1969a, 1973, 1975b, 1975c, 1976). The chordate motor system consists basically of a longitudinal axial component (notochord or vertebral column) and compact, longitudinal muscles organized into metameric units (myotomes) separated by connective tissue sheets. The muscle fibers of the myotomes run from one connective tissue septum to the next. The axial component keeps the body constant in length so that the only deformations of the body the longitudinal muscles can effect are bending movements; any kind of peristalsis is impossible. The primary function of the axial component is therefore not its ability to receive mechanical force produced by the muscles as there is scarcely any insertion of muscles on it. It cannot be bent under the direct mechanical influence of muscles, only within the whole complex of

muscles and collagenous fibers in which it guarantees constancy
of body length.  This situation reminds us of features in the
hydrostatic skeleton of worm-like organisms in which the muscles
are organized into a motor unit by collagenous fibers.  The integrated system of muscles and connective tissue fibers normally
enclose fillings of fluid or parenchymal cells (Chapman, 1958;
Clark, 1964; Gutmann, 1966, 1972).

The resemblance of the motor systems in lower chordates to
the hydrostatic skeleton suggests that the forerunners of chordates
were worm-like animals with a hydrostatic skeleton and perhaps
with metameric, fluid-filled compartments.  That this is a compulsory postulate allowing no mechanically feasible alternative can
be shown if we perform a theoretical experiment and imagine that
the axial component of the body is removed from a chordate.  The
normal function of the body muscles would then be impeded.  The
body would shorten but, without an antagonistic system being able
to restore the initial situation, locomotion would be impossible.
Antagonistic muscles showing a circular orientation, together with
radially-disposed muscles in connective tissue partitions crossing
the body transversely in a dense sequence, are required, if this
organism without an axial component should perform undulations or
other peristaltic motor actions.  In this case the coelomic cavity,
divided into metameric compartments, functions as the fluid-filling
of the hydrostatic skeleton.

The Phylogenetic Model for the<br>
Evolution of the Chordate 'Bauplan'

In primitive stages of metazoan evolution the whole shape of
the body and its locomotory capabilities must have been controlled
by longitudinal, transverse and dorsoventral muscles.  Such an
organism had to use its whole body for locomotion, since muscular
force generated in one part of the body could not be transmitted
to other parts.  This constructional and functional limitation
governs the organization of all lower metazoans including primitive
coelomates.  The body of these animals can only have been consoli-
dated by a lattice of muscles containing a jelly-like connective
tissue.  The only configuration the newly developing coelomic
cavities could attain in a gradual way in such a mechanical situation, was a metameric one.  Narrow spaces left between a dense
bracing of muscle sheets controlling the body shape became the
fluid-filled cavities, thus constituting an essential structure in
the developing hydrostatic skeleton of the coelomates.  It was the
muscular lattice that enforced the metameric organization of the
hydrostatic fluid cavities and the corresponding arrangement of
muscles.  A metameric arrangement of the coelomic cavities also
allowed the muscular system to retain its orientation in the main

directions of space.  Deriving the metamerism of the coelom from
a monomeric or oligomeric forerunner would require saltatory
changes.  Hence a theory starting with a metameric coelom has the
advantage of gradualism.

From the beginning of coelomate evolution movement by lateral
undulations of the body must be assumed (Bonik et al., 1976a, 1976b).
When lateral bending movements are performed in a slender animal,
the sagittal plane of the body is only bent but not deformed in
other directions.  By contrast the lateral regions of the body are
involved in much greater deformations.  If therefore the coelomates
evolved from organisms in which the body was supported by jelly-
like connective tissue, one can suppose that the fluid-filled
cavities of the coelom had to form in the lateral parts where
deformations were strongest.  In the sagittal plane muscle and
connective tissue would have persisted.  The arrangement of a
sagittal mesentery and lateral coelomic cavities conforms to these
functional requirements and can be observed in all coelomates including chordates (Fig. 1A).

The lateral bending movements necessary for propulsion were
brought about by longitudinal muscles.  The remaining muscles are
required to control the shape of the body preventing those energy
consuming deformations not contributing to the force of propulsion.
Thus a great amount of energy is used by muscles simply for the
control of body form.  Any transformation of the muscular system
allowing the proportions of these muscles to be diminished and/or
replaced by structures using less energy must have provided a
strong selective advantage in the course of evolution.  One can
imagine different ways of decreasing the amount of muscle necessary
for the control of the body shape.  However, chordate evolution
must have been canalized in one narrow phylogenetic path that led
from the primitive metameric hydrostatic skeleton to a construction
with a notochord and myotomes.

Initially a small amount of muscle derived from the phylogenetically ancestral muscular lattice in the mesentery dorsal to
the gut may have taken over the function of controlling the length
of the body.  This thick, rod-like structure which evolved was the
notochord (Fig. 1B).  Thus constancy of body length was guaranteed
mechanically in a much more economical way by using less muscle
than in a typical hydrostatic skeleton.  Enormous amounts of circular and transverse muscle could be reduced making economies in
the mechanical system.  The notochord further developed as a hydrostatic organ within the hydrostatic skeleton.  It consisted of a
fluid-filled rod surrounded by a sheath of connective tissue
supplied with muscles controlling its shape (as in Branchiostoma).
Swelling of the hydrostatic notochord was prevented by the muscles
and the connective tissue sheath, which thereby must have prevented

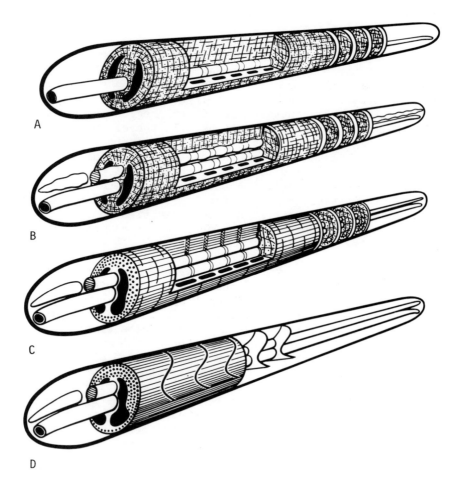

Fig. 1A-D.  Models representing stages in the evolution of the
chordate construction:  A: A worm-like coelomate construction with
metameric coelomic cavities and muscles running in all directions
of space controlling body shape.  B: The notochord develops as a
hydrostatic organ in the dorsal mesentery restricting body movements
to lateral undulations.  C: The notochord assumes the function of
keeping the body constant in length.  The system can now become
more economical by reducing muscles that control body shape and more
efficient by increasing longitudinal muscles which produce the sinuous locomotory movements.  D: A primitive chordate construction
with thick metameric longitudinal muscles (myotomes) and connective
tissue sheets (myocommata), suspending the notochord in the tube-
like body.

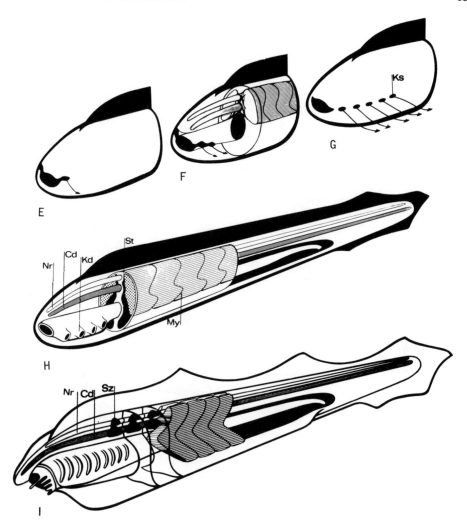

Fig. 1E-I. Models representing stages in the evolution of the
chordate construction: E, F, G: The evolution of the branchial
basket. Lateral slits from the angle of the mouth become divided
into gill clefts by transverse bars improving the continuous water
current utilized in filter-feeding. The gill slits can only have
evolved after the hydrostatic skeleton had been reduced and after
the notochord and myotomes were fully functional. H: A primitive
chordate state with a branchial basket (Kd), a notochord (Cd) functioning as a hydrostatic organ, and metameric myotomes (My). The
nerve cord (Nr) became concentrated on the dorsal side where the
body muscles have been formed into the myotomes. I: The 'acrania-
stage' of chordate evolution. The hydrostatic notochord is retained
and vestiges of the metameric coelomic cavities persist in the form
of sclerocoels (Sz) near the axis.

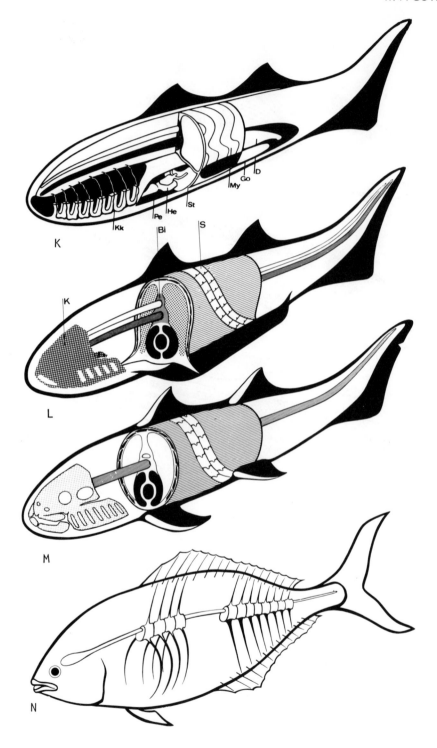

shortening of the whole body. The action of the longitudinal, somitic muscles is antagonized in a dynamic way by notochordal muscles as well (not as in the higher vertebrates by a static vertebral column).

Other transformations in the newly created situation contributed to an enormous rise in locomotory efficiency. They were caused by an increase of longitudinal muscles effecting more powerful bending strokes during swimming. The efficiency of body movements could rise to a level unattained by a hydrostatic skeleton (Figs. 1C and 1D). These transformations were made possible by the evolution of a narrow, 'economical' axial component which could antagonize a considerable amount of longitudinal muscle organized into metameric units (myotomes). In a true hydrostatic skeleton such a disproportionate enlargement of longitudinal muscles would not have been possible, because changes of one muscular element would have required the balanced increase of other muscles too.

The notochord could only evolve in a hydrostatic system with densely arranged, transverse partitions. These were an appropriate means of suspending the evolving notochord in the center of the fluid-filled, tube-like organism. The transverse partitions prevented bending of the notochord within the tube-like body (Fig. 2). Uncontrolled bending deformations of the notochord would have caused changes of length of the body rendering the notochord useless. The notochord could not have functioned without structures providing for its mechanical integration into the body of a hydro-

Fig. 1K-N. Models representing stages in the evolution of the chordate construction: K: A chordate close to the origin of the vertebrates. The coelom has lost its hydrostatic function and is transformed in accordance with the requirements of the internal organs, viz. freedom of movement. The heart (He) is situated in the pericardial cavity (Pe), the general body cavity contains the gonads (Go), and allows the gut (D) to perform peristaltic movements. L: The 'agnathan-stage' of fish evolution. The head (K) is formed, metameric cavities around the axis are filled by connective tissue (Bi), rhomboid scales (S) evolved as stabilizing structures. M: Primitive gnathostome fishes of all phylogenetic lines are equipped with a scale covering (S) of the body. N: The teleostean stage in which the rhomboid scales are transformed into round scales that have no mechanical influence on the body musculature. The teleosteans are representatives of several lines of fishes that lost their rhomboid scale covering, when the bony internal skeleton of the vertebral column was developed.

static skeleton.  It could not even have evolved if a metameric
division of the coelom by transverse partitions had not already
existed before its phylogenetic formation was initiated.  The conclusion that a metameric hydrostatic skeleton must have preceded
the construction of the chordates can be supported by very strict
biomechanical arguments.

When the longitudinal myotomal muscles of chordates increased,
the transverse partitions which had lost their muscles persisted
as connective tissue sheets dividing the longitudinal muscles into
metameric units, the somites.  These sheets, or myocommata, are
attached to the notochord forming the mechanical connection with
the somitic musculature.  The notochord continues to be mechanically
integrated and suspended in the body by transversely organized,
connective tissue structures.  However, as long as the notochord
retained muscles it still remained a rather uneconomical structure
(as in the Acrania).  Efficiency was enhanced by the reduction of
notochordal muscles; thus the sheath of connective tissue fibers
alone prevents swelling of the axial component.  The filling of the
notochordal sheath continues to form a hydrostatic structure.  This
is only gradually reduced when the vertebral column is formed by
skeletal structures becoming attached to the notochord.

Fig. 2.  A:  The structural situation in a metameric coelomate in
which the notochord developed and required transverse connective
tissue structures to suspend the evolving notochord in the tube-
like, fluid-filled body.  B:  In a worm-like organism lacking transverse partitions the notochord would have been able to bend within
the body and its functional advantage could not have been exploited.

Another organ characteristic of chordates deserves being integrated into the model for the evolving body architecture. The branchial basket (Figs. 1E, 1F and 1G) consists of perforations of the body wall in the foregut region allowing water which entered the mouth to leave through the gill clefts. These organs cannot have been feasible in a hydrostatic system with fluid-filled cavities contained in a turgescent body wall. The body wall cannot simultaneously have fulfilled the functions of a muscular coat for a hydrostatic skeleton and of a frame for the gill clefts. The gill clefts could only develop as perforations of the body wall after the notochord and the myotomes had replaced the phylogenetically more primitive metameric hydroskeleton in evolution.

After the coelomic cavities had lost their functional role as a hydrostatic organ they are either reduced or modified. They are reduced and ultimately lost dorsally where the myotomes were situationed (Fig. 1I). Ventrally the metameric divisions were obliterated and a continuous coelomic cavity developed around the gut (Fig. 1K). The coelomic cavity must always have allowed free peristaltic movement of the gut and of other internal organs. This function was retained after the hydrostatic function was eliminated.

The lower vertebrates, especially living agnathans, are representatives of a later stage of chordate phylogeny. Though the notochord is no longer supplied with muscles, it remains a hydrostatic organ in some fishes. When in gnathostome fishes vertebrae are formed the notochord is restricted to the intervertebral section where it forms fluid- or tissue-filled, hydrostatic cushions between the cartilaginous or osseous vertebral elements. The vast majority of myotomic muscle fibers are attached to the myocommata in fishes. It is only later in the course of vertebrate phylogeny, in the reptile, avian, and mammalian motor systems, in which osseous skeletal elements have developed around the notochord and in the body wall, that the body muscles can attach to the skeletal elements. They are then able to convey their mechanical force directly to the skeleton. The typical musculo-skeletal system of the 'higher' vertebrates must therefore be understood as the late outcome of a long phylogenetic history starting with a hydrostatic skeleton in which only after the completion of the notochord and the myotomes, have stiffening skeletal elements evolved. These gradually restricted the freedom of movement of the primitive hydroskeletal system, till the motor performance is confined to the relation between muscles and stiff skeletal elements (Gutmann, 1975c).

The body construction of the forerunners of the chordates cannot have been composed of functional components different from those used in the model. All conclusions drawn are based on the restrictive influence of the functional components occurring in

the bodies of all metazoans.  Tracing the phylogenetic history
back to such primitive organisms as the metameric coelomates and
even to their forerunners is the only way to fulfill the requirements of the evolutionary theory.  Direct morphological studies of
lower chordates and fishes, living or fossil, can only reveal the
functional structures such as body metamerism, muscle arrangement
and the hydrostatic action of the notochord, but cannot contribute
to a phylogenetic explanation.  This depends on the functional and
structural conditions of the ancestors that enforced these morphological patterns.  Therefore, reconstructions of the ancestral
organisms, as exhibited in the example, are the only way to apply
the evolutionary theory.  Only phylogenetic reconstructions showing the limitations and the functional requirements for later
phylogenetic stages can explain the morphological structure
observed in recent and fossil animals.

After the basic chordate construction had evolved, different
lines of adaptations gave rise to further modifications leading
to other extant animals (Figs. 1L, 1M and 1N), the different taxa
of vertebrates, the Acrania, and the hemichordates (Gutmann, 1969b,
1972, 1975a).

The craniate chordates must have evolved from transparent,
pelagic organisms, swimming by means of sinuous movements.  The
body musculature gained considerably in efficiency when the coelomic
cavities in the somites, still present in the acraniate stage of
phylogeny, became filled with connective tissue (Fig. 1L).  This
tissue provides the material necessary for cartilage and bone
formation around the axial component.  The myomeres also adopted
a more complicated shape with much more folding of the myocommata.
This yielded the advantage of a tighter 'bandaging' of the whole
motor apparatus by the collagenous fibers.  The efficiency of
propulsion was now increased to such a degree that the whole body
need no longer be involved in locomotion.  The rostral part could
be transformed into an immobile, basically non-locomotory, body
region.  Stiffness of the head region was secured by cartilage
developing around the tip of the notochord to form a mechanically
solid framework for the paired sense organs and brain which
developed as important phylogenetic innovations.  For a full account
of the evolution of the vertebrate head see Gutmann (1969a).

Another functional system, the scales, may be mentioned here,
as these are the only structures preserved in many fossil fishes.
Most fossil agnathans and early gnathostomes are covered by bony
shield-like plates in the head and by rhomboidal scales in the
trunk and tail.  They were often interpreted as protective devices
or as excretory products deposited in the skin.  However, the
scales inevitably exert a mechanical influence on locomotory movements (Fig. 3).  They are arranged in oblique rows overlapping from

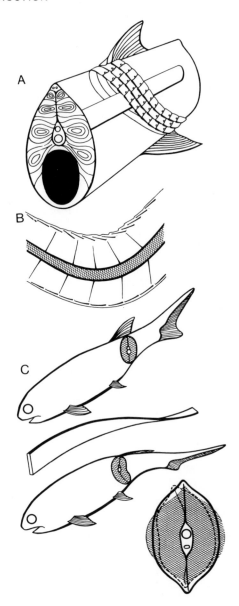

Fig. 3.  Diagrams showing the biomechanical influence of the rhomboid scales on the body muscles of lower fish-like vertebrates.
The functioning of the scales depends on the existence of an axial
component (notochord) providing constancy of length and a dense
packing of longitudinal myotomal muscles.  A: The scales form oblique rows; within each row overlapping of the scales is prevented.
B: When the body bends the rows of scales overlap.  C: Deformations
of the body cross-section are prevented or suppressed by the scales.

front to rear.  Thus they leave freedom of movement for the lateral
undulation required in swimming.  The axial component within the
body keeps it constant in length but cannot provide for stability
of shape.  The oblique rows of scales provide the necessary additional restriction against distortion of the oval cross-section of
the body without impeding lateral bending.  As the rows of either
side come into contact in the mid-dorsal and mid-ventral lines they
form a series of complete encircling bands that fit into each
other.  Since the cross-section of the bands of rhomboidal scales
is oval they prevent turning of one band relative to another.
Torsion around the longitudinal axis is thus excluded.  This
interpretation receives some support if we look at those fossil
fishes that must have had a body which was circular in cross-section.
The scales on these fishes could not have ensured stability of body
shape and are partly or completely reduced.  Rhomboidal scales are
most conspicuous in those fishes with a deep body or a laterally
flattened cross-section.  A second stabilizing influence is
effected in combination with the underlying muscles and the tight
connective tissue structures attaching the scales to the body
musculature.  Oblique bending and uncontrolled changes of body
shape are therefore restricted by the mechanical strength of the
scales.  Supporting evidence for this mechanical interpretation is
provided by different evolutionary lines of fishes exhibiting a
reduction of the rhomboidal scales.  As the internal stabilizing
components (bony vertebral column, vertebral processes and ribs)
are increasing in size and complexity the external stabilizing
covering is reduced.  The stabilizing effect of the scales provided for so little distortion in the vicinity of the notochord
that the mechanical requirements for the transformation of connective tissue into bone were fulfilled (Pauwels, 1965).  Skeletal
structures could only evolve in regions of the body in which deformations were already suppressed.  Chondrosteans, fishes without
skeletal elements around the body axis, have only a notochord but
a complete scale covering; holosteans show a considerable advancement in the evolution of the vertebral column within a body that
is still covered by rhomboidal scales; teleosteans lack rhomboidal
scales but the bony axial elements have long processes projecting
into the muscles.  Similar phylogenetic changes must have occurred
in crossopterygians and dipnoans.  This interpretation of primary
scale function fits into the adaptational process described by the
model.  In a hydrostatic skeleton low efficiency is caused by lack
of mechanical stability.  The evolution of the notochord and the
elaboration of the myotomes provided for growing stabilization.
The stabilizing influence of the rhomboidal scales must have
caused an additional rise in the efficiency in lower vertebrates.

The process of the evolution of the vertebrate 'bauplan'
cannot be studied directly in any aspect since it is fully developed
when the fossil record sets in.  Nevertheless the reconstruction

presented here relies on functional needs and structural limitations conforming to basic laws of physics and the properties of the functional elements used to reconstruct the intermediate stages. The bodily organization is translated from morphology into functional and structural diagrams. The model cannot of course give a realistic picture of the extinct animals: it does not even strive for this aim. It provides what may be called blue prints for the structural relations and the functional changes in the evolving systems.

The question could be raised as to whether the amount of speculation involved in phylogenetic reconstructions is too great, and if so should science be restricted only to observable facts? However, in no branch of science can hypotheses be formulated without implying events or relations which can never be directly seen or controlled. Phylogeny itself is no different; it is based on events that cannot be directly observed but can only be reconstructed. To attain the rank of scientific theories phylogenetic models must be given in an explicit form and based on unequivocal explanations. As with every other scientific theory a phylogenetic model must be open to criticism and should be tested against observable phenomena. These requirements are evidently fulfilled by the model outlined here, as many structural relations of the lower chordates and the fish-like vertebrates are knit into the adaptational changes reconstructed.

## The Methodological Significance of the Example

The evolution of major taxa means in reality the appearance of changes in gross morphological relations in biological machinery which is, in most cases, characteristic for a great number of species. Evolution of this kind of course occurs at the species level - microevolution - and consists of numerous gradual changes. The essential aspects in phylogenetic reconstructions of the evolution of major taxa are therefore neither the systematic names given to different animals nor the character states through which they have to pass but changes in the construction of the animals. These must be described in biomechanical, physiological, behavioral, and ecological terms. The features and the characters which result from transformations during the evolutionary process are often used to define major taxa. Their delimitation and naming is of no phylogenetic relevance: it is only the adaptational explanation of the continuous changes that counts. Even the patterns of the dendrograms do not convey any phylogenetic information of their own (Peters et al., 1974). All essential arguments for the changes and the branchings of the phylogenetic process are given in the written and depicted model. The disconnected character states into which a model can be divided represent intermediate constructions

through which the phylogenetic process must have passed.  Discriminations of plesiomorphic and apomorphic character states are
achieved only as a side effect within the context of the reconstructions.  The decision on what is an ancestral and what a
derived character is arrived at by adaptational interpretations.

In most papers on the problem of homology (Siewing, 1967,
1972; Remane, 1956; Osche, 1966; or in the theory of Hennig, 1950,
commented upon by Schlee, 1971), any hints on how to decide between plesiomorphic and apomorphic character states are lacking.
The reasoning showing the necessity of an adaptational interpretation, demonstrated by Peters and Gutmann (1971), was not accepted
by defenders of Hennig, but was used by Mayr (1974) against some
postulates of Hennig.

The claim that phylogenetic theories have to be formulated as
adaptational explanations ('Lesrichtungs-Kriterium' - Peters and
Gutmann, 1971) of morphoclines or phylogenetic models, resembles
a suggestion made by Bock (1959) that a conditional or qualifying
phrase is to be stipulated when structures are to be homologized.
The homology of the fore leg of a horse and the wing of a bat can
only be maintained on the condition that they are both anterior
extremities of tetrapods.  This implies their derivation by adaptational changes from a common ancestral structure, which can only
be supported by an explicit model in the sense of the 'Lesrichtungs-
Kriterium'.

Neither comparison alone nor character analyses can by themselves disclose phylogenetic relationships.  Only the reconstruction of the process of phylogeny as a series of adaptational
transformations of the organism in the course of time can solve
phylogenetic problems.  The task of the phylogeneticist is not to
erect systematic entities or to design dendrograms but to reconstruct the continuous process explaining it in adaptational terms
and thereby basing the decisive arguments on requirements of the
evolutionary theory.  No kind of systematic organization of living
beings can disclose the phylogenetic changes, because the static
order achieved in systematics cannot contain any information on
the sequence of changes leading to the features characterizing the
different taxa.  The phylogeneticist may leave the important work
of classifying and ordering organisms to taxonomists, who will
attempt to constitute a systematic order by using phylogenetic
relations in addition to the other criteria that cannot be provided by phylogeny.

CONCLUSION

The evolution of the chordate construction is an example of

a phylogenetic process leading to a major taxon.  It is based upon
adaptational changes continuously controlled by environmental requirements and the selective pressures exerted by the evolving
apparatus and its functional components as well.  The theoretical
points to be emphasized are that neither selection from the
environment nor the selection effected by the evolving construction itself can be dispensed with.  Minor changes at the species
level may be adequately described and explained as adaptations to
environmental conditions alone.  However, in every major change of
an organism's construction the consistency of the functional requirements of the internal organs and of the construction itself
are involved.

The methodological aspect of the reasoning contained in this
paper consists in outline of a new approach to phylogenetic problems,
especially those in which the evolution of major taxa or 'bauplan'
constructions is involved.  The constructionist methodology is
designed as an attempt to apply evolutionary theory with some
figorousness to the evolution of morphological structures.  The
methodological and theoretical postulates of constructive phylogenetics were developed in cooperation between several authors
(Franzen, Grasshoff, Gutmann, Mollenhauer, and Peters), based
mainly on the theoretical work of Bock (1959, 1965, 1969), Bock
and v. Wahlert (1965), and Hecht (1965), remodeled according to
the ideas of Eigen (1971) and Cizek and Hodanova (1971).  The
latest developments in the field of the philosophy of science contributed by showing the necessity of starting with tentative
attempts to formulate a phylogenetic theory, that has to be
corroborated in the sequential advance of observation and theor-
izing (Popper, 1969, 1972; Feyerabend, 1972, Medawar, 1972;
Stegmuller, 1969; Gutmann et al., 1975).  The postulates of the
constructivist method of phylogeny have already been extensively
applied in different fields of phylogenetic research:  Bonik et al.,
(1976); Franzen (1972, 1977); Gutmann (1971, 1976); Heine (1976);
Lauterbach (1973, 1974); Maier (1976); Peters and Gutmann (1976a,
1976b); Regenfuss (1973); Richter (1975); and Vogel et al., (1976).

As an extension of the theoretical analyses on which the
methodology for reconstructing the process of phylogeny is based
it might perhaps be possible to present an improved formulation of
the concept of living systems.  On the basis of an adequate idea
of the organism as a living system phylogenetic transformation
might under certain conditions be shown to occur by necessity.
This would mean that a discrete evolutionary theory might not be
required as evolution could be a characteristic attribute of the
physical properties of living organisms (Mollenhauer, 1976).

Finally a new concept of phylogeny, requiring phylogenetic
theories to be organized in accordance with evolutionary theory,

should consist of models for the continuous phylogenetic processes that are explained as adaptational changes leading to existing organisms. These models should incorporate the structure of fossil and recent organisms.

## SUMMARY

Phylogenetic relationships of major taxa must be based solely on adaptational interpretations of transformations of the bio-mechanical, physiological, and ecological relations of organisms. Reconstructions of the phylogenetic transformations leading to major 'bauplan' relations are envisaged as continuous models describing and explaining the process of adaptation by reference to selective pressures exerted by the environment and by the internal functional relations of the organism itself. The model-like reconstructions cannot be based on the comparison of morphological configurations or on character analyses alone because these give no indication of polarity. Only those morphoclines based on adaptational transformations can be considered. The evolving organism is, in every stage of the phylogenetic transformation, conceived of as a network of functional components. A model for the evolution of the chordate 'bauplan' is given which shows a derivation from ancestors with a metameric, hydrostatic skeleton that swam by lateral undulations. In the transitional stage a notochord developed to ensure constancy of body length, so that circular and transverse muscles became reduced as adaptations towards economy while longitudinal muscles increased, raising locomotor efficiency. The metameric pattern shown by the myotomes is derived from the transverse connective tissue partitions of the forerunners with a hydrostatic skeleton. After the loss of transverse muscles attached to the partitions these structures suspended the notochord in the body, at the same time dividing the longitudinal muscles into myotomes.

The rhomboidal scales of lower vertebrates are interpreted as structures improving locomotor efficiency by preventing distortion of the body. The evolution of an internal, bony skeleton provides internal stabilization and was followed by the reduction of the rhomboidal scales. The typical motor system of higher vertebrates, characterized by muscles, transmitting their force to the skeletal elements, must be viewed as the final stages of a phylogenetic process starting with a metameric hydrostatic skeleton.

## REFERENCES

Bock, W.J., 1959, Preadaptation and multiple evolutionary pathways. Evolution, 13:194-211.

Bock, W.J., 1965, The role of adaptive mechanisms in the origin
    of higher levels of organization. Syst. Zool., 14:272-287.
Bock, W.J., 1969, Discussion: The concept of homology. Ann.
    N.Y. Acad. Sci., 167:71-73.
Bock, W.J., 1973, Philosophical foundations of classical evolu-
    tionary classification. Syst. Zool., 22:375-392.
Bock, W.J. and Wahlert, G.v., 1965, Adaptation and the Form-Function
    Complex. Evolution, 19:269-299.
Bonik, K., Grasshoff, M. and Gutmann, W.F., 1976a, Die Evolution
    der Tierkonstruktionen I u. II, Natur u. Museum, 1o6:129-143.
Bonik, K., Grasshoff, M. and Gutmann, W.F., 1976b, Die Evolution
    der Tierkonstruktionen III, Natur u. Museum, 1o6:178-188.
Chapman, G., 1958, The hydrostatic skeleton in the invertebrates.
    Biol. Rev. Cambridge Phil. Soc., 33:338-364.
Cizek, F. and Hodanova, D., 1971, Evolution als Selbstregulation.
    VEB G. Fischer-Verlag, Jena.
Clark, R.B., 1964, Dynamics in metazoan evolution. The origin of
    the coelom and segments. Clarendon Press, Oxford.
Dullemeijer, P., 1974, Concepts and approaches in animal
    morphology. Assen, Van Gorcum.
Eigen, M., 1971, Selforganization of matter and the evolution of
    biological macromolecules. Naturwissenschaften, 58:465-522.
Feyerabend, P.K., 1972, Uber die Interpretation wissenschaftlicher
    Theorien, in "Theorie und Realitat" (Albert, H., ed.),
    pp. 59-66, Mohr, J.C.B., Paul Siebeck, Tubingen.
Franzen, J.L., 1972, Wie kam es zum aufrechten Gang des Menschen?,
    Natur u. Museum, 1o2:161-172.
Franzen, J.L., 1977, Die Primaten als stammesgeschichtliche Basis
    des Menschen, in "Enzyklopaedie des Menschen", Kindler-Verlag,
    Munchen.
Gutmann, W.F., 1966, Zu Bau und Leistung von Tierkonstruktionen
    4-6, Abh. senckenb. naturf. Ges., 51o:1-1o6.
Gutmann, W.F., 1969a, Die Entstehung des Vertebraten-Kopfes, ein
    phylogenetisches Modell, Senckenbergiana biol., 5o:433-471.
Gutmann, W.F., 1969b, Acranier und Hemichordaten, ein Seitenast
    der Chordaten, Zool. Anz. 182:1-26.
Gutmann, W.F., 1971a, Was ist urtümlich an Branchiostoma?, Natur
    u. Museum 1o1:34o-356.
Gutmann, W.F., 1971b, Der biomechanische Gehalt der Wurmtheorie,
    Z. wiss. Zool. 182:229-262.
Gutmann, W.F., 1972, Die Hydroskelett-Theorie, Aufsätze u. Reden
    senckenb. naturf. Ges. 21:1-91.
Gutmann, W.F., 1973, Der Konstruktionsplan der Cranioten: ein
    phylogenetisches Modell und seine methodisch-theoretischen
    Konstituentien, Cour. Forsch.-Inst. Senckenberg (CFS) 3:1-36.
Gutmann, W.F., 1974, Die Evolution der Mollusken-Konstruktion:
    ein phylogenetisches Modell, Aufsätze u. Reden senckenb.
    naturf. Ges. 25:1-84.
Gutmann, W.F., 1975a, Das Tunicaten-Modell, Zool. Beitr. N.F.
    21:279-3o3.

Gutmann, W.F., 1975b, Das Schuppenhemd der niederen Wirbeltiere
    und seine mechanische Bedeutung, Natur u. Museum lo5:169-185.
Gutmann, W.F., 1975c, Konstruktive Vorbedingung und Konsequenz in
    der phylogenetischen Entwicklung des Körperstammes der
    Cranioten, in "Ontogenetische und konstruktive Gesichtspunkte
    bei phylogenetischen Rekonstruktionen," Aufsätze u. Reden
    senckenb. naturf. Ges. 27:4o-56.
Gutmann, W.F., 1976, Aspekte einer konstruktivistischen Phyloge-
    netik: Postulate und ein Exempel, in "Evoluierende Systeme
    I u. II," Aufsätze u. Reden senckenb. naturf. Ges. 28:165-183.
Gutmann, W.F., Mollenhauer, D. and Peters, D.S., 1975, Wie
    entstehen wissenschaftliche Einsichten?  Die hypothetiko-
    deduktive Methode der Wissenschaft speziell in der Erforschung
    der Phylogenetik, Natur u. Museum lo5:335-374.
Gutmann, W.F. and Peters, D.S., 1973a, Konstruktion und Selektion:
    Argumente gegen einen morphologisch verkürzten Selektionismus,
    Acta biotheoretica 22:151-18o.
Gutmann, W.F. and Peters, D.S., 1973b, Das Grundprinzip des
    wissenschaftlichen Procedere und die Widerlegung der phylogene-
    tisch verbramten Morphologie, in "Phylogenetische Rekonstruk-
    tionen - - Theorie und Praxis," Aufsätze u. Reden senckenb.
    naturf. Ges. 24:7-25.
Hecht, M.K., 1965, The role of natural selection and evolutionary
    rates.  Syst. Zool., 14:3o1-317.
Heine, H., 1976, Organphylogenese am Beispiel des Wirbeltier-
    herzens, in "Evoluierende Systeme I und II," Aufsätze u.
    Reden senckenb. naturf. Ges. 28:69-78.
Hennig, W., 1950, Grundzüge einer Theorie der phylogenetischen
    Systematik, Berlin.
Lauterbach, K.-E., 1973, Schlüsselereignisse in der Evolution der
    Stammgruppe der Euarthropoda, Zool. Beitr. N.F. 19:251-299.
Lauterbach, K.-E., 1974, Über die Herkunft des Carapax der
    Crustaceen, Zool. Beitr. N.F. 2o:273-327.
Maier, W., in press, Die Evolution der bilophodonten Molaren der
    Cercopithecoidea, Morphologie 67.
Mayr, E., 1974, Cladistic analysis or cladistic classification,
    Z. zool. Syst. Evolutionsforsch 12:95-128.
Medawar, P.B., 1972, Hypothesenbildung und produktive Phantasie,
    in "Die Kunst des Lösbaren" (Medawar, P.B., ed.), VR Kleine
    Vandenhoeck-Reihe 365:115-139.
Mollenhauer, D., 1976, Systemtheorie und botanische Systematik
    Drei Betrachtungen, in "Evoluierende Systeme I u. II,"
    Aufsätze u. Reden senckenb. naturf. Ges. 28:32-68.
Osche, G., 1966, Grundzüge der allgemeinen Phylogenetik, in
    "Handbuch der Biologie," (L. v. Bertalanffy, ed.) 3:817-9o6.
Osche, G., 1971, Mechanismen der Evolution und die Mannigfal-
    tigkeit der Organismen, Studium Generale 24:191-2o1.
Pauwels, F., 1965, Gesammelte Abhandlungen zur funktionellen

Anatomie des Bewegungsapparates, Springer Verlag, Berlin, Heidelberg, New York.

Peters, D.S., 1976, Evolutionstheorie und Systematik, J. Orn. 117, 3:329-344.

Peters, D.S. and Gutmann, W.F., 1971, Über die Lesrichtung von Merkmals- und Konstruktions-Reihen, Z. zool. Syst. u. Evolutionsforsch. 9:237-263.

Peters, D.S. and Gutmann, W.F., 1973, Modelvorstellungen als Hauptelement phylogenetischer Methodik, in "Phylogenetische Rekonstruktionen - - Theorie und Praxis," Aufsätze u. Reden senckenb. naturf. Ges. 24:26-38.

Peters, D.S. and Gutmann, W.F., 1976, Die Stellung des "Urvogels" Archaeopteryx im Ableitungsmodell der Vogel, Natur. u. Museum 1o6:265-275.

Popper, K., 1969, Logik der Forschung, 3rd. ed., Mohr, J.C.B., Paul Siebeck, Tübingen.

Popper, K., 1972, Objective Knowledge. An Evolutionary Approach, Clarendon Press, Oxford.

Regenfuss, H., 1973, Beinreduktion und Verlagerung des Kopulations- apparates in der Milbenfamilie Podapolipidae, ein Beispiel für verhaltensgesteuerte Evolution morphologischer Strukturen, Z. zool. System. Evolutionsf. 11:173-195.

Remane, A., 1956, Die Grundlagen des natürlichen Systems der vergleichenden Anatomie und Phylogenetik, Geest and Portig KG, Leipzig.

Richter, G., 1973, Zur Stammesgeschichte pelagischer Gastropoden, Natur u. Museum 1o3:265-275.

Schlee, D., 1971, Die Rekonstruktion der Phylogenese mit Hennig's Prinzip, Aufsätze u. Reden senckenb. naturf. Ges. 2o:1-62.

Siewing, R., 1967, Diskussionsbeitrag zur Phylogenie der Coelomaten, Zool. Anz. 179:132-176.

Siewing, R., 1972, Zur Deszendenz der Chordaten. Erwiderung und Versuch einer Geschichte der Archicoelomaten, Z. zool. Syst. Evolutionsforsch. 1o:267-291.

Stegmuller, W., 1969, Probleme und Resultate der Wissenschaftstheorie und Analytischen Philosophie, 1 Wissenschaftliche Erklärung und Begründung, Studienausgabe Teil 2, Springer Verlag, Berlin, Heidelberg, New York.

Vogel, K., Zorn, H. and Gutmann, W.F., 1976, Die Evolution der Tentakulaten, Natur u. Museum 1o6:316-317.

Whyte, L.L., 1965, Internal factors in evolution, G. Braziller, New York.

Whyte, L.L., 1967, Directive agencies in organic evolution, J. Theoret. Biol. 17:312-314.

# INSULARITY AND ITS EFFECT ON MAMMAL EVOLUTION

P. Y. SONDAAR

Geological Institute

Oude Gracht 320, Utrecht, Netherlands

## INTRODUCTION

When the car ferry "Rethymnon" sails from Piraeus towards Crete one can already feel the strong endemic atmosphere of the island. Rethymnon itself is a beautiful historical town on Crete in an area containing many Pleistocene fossil mammal localities which have yielded endemic deer, elephants and murids. A striking thing on board the ferry are notices written in Japanese which suggest that the ship was probably not launched under the name Rethymnon and served in her earlier days on the Japanese islands. For a paleontologist this is a remarkable coincidence since the Pleistocene of Japan has also yielded un-balanced endemic faunas with a very uniform composition of elephant and deer like the fauna of Crete.

In order to understand the effect of insularity on mammal evolution some basic information about islands is necessary. Zoogeographically, they are often divided into oceanic and continental (Darlington, 1957). The former are assumed to have arisen from beneath the sea, the latter originated through sub-sidence of the isthmus of a peninsula. Due to the usually complex geological history of islands, it is not possible in paleozoogeographical studies to draw a sharp boundary between the two types, and it will be necessary to add more specific informa-tion about each island.

The size of an island is another important aspect and even continents are sometimes zoogeographically considered as oceanic islands; for example South America and Australia. In these cases the term island continent is used.  Both island continents possess rain forests, deserts, tropics, and snow covered mountains.  The wide climatic range proved to be ideal for evolutionary diversification which resulted in the production of balanced mammal faunas with many ecological variants having convergent functional and anatomical features in varying degrees.

Madagascar with its area of 592,000 km$^2$ is the third largest island in the world but may be considered the largest oceanic island (Mahe, 1972; Simpson, 1940; Millot, 1952), as New Guinea and Borneo were connected till the Late Pleistocene with Australia and S. E. Asia, respectively.  Though Madagascar has not, and had not in the past, the climatic variety of the continents there are considerable differences from the tropical North to the temperate South.  At present, the vegetation runs from tropical rain forest to near desert scrub and cool upland forest.  Mammalian diversification is restricted to a few groups only which suggests colonization by a small number of ancestral types and adaptive radiation of these groups (lemurs, tenrecs, soricids, viverrids, hippos, and some rodents) within the habitats of this large island.  Many endemics were exterminated by man who has also introduced new mammals to the islands.  Although the mammal fauna of Madagascar before the arrival of man was impoverished and unbalanced compared to the continents, the groups present showed great diversification (tenrecs, 13 genera - 29 species; lemurs, 8 genera - 14 living species; Bigalke, 1972).  The mammalian fauna of Madagascar illustrates the result in evolutionary terms of several million years of isolation but we cannot follow this step by step since the fossil record is too poor and the diversification too complex.

The mammal faunas of South America, Australia and Madagascar are often used to demonstrate the effect of insularity on mammal evolution.  Far less is known of the mammalian faunas of smaller "oceanic" islands , for example the Mediterranean islands, Ryukyu islands (Japan), Caribbean islands and some islands of the Indonesian archipelago.  This might be due to the fact that these endemic faunas are extinct, being mostly exterminated by man after his arrival, and so are only known by fossils.  These mammal faunas were much more impoverished and unbalanced than that on Madagascar.  The climate is, and must have been, quite uniform for the same island.  In some cases the mainland ancestor of the endemic island form and its approximate time of arrival is known.

In this paper the effect of insularity on mammal evolution will primarily be approached by considering the endemic fossil mammals of these smaller "oceanic" islands.  To offset the disadvantage of only having fossils to show the effect of isolation, we can point out here that it is in many cases possible:

(a)  to compare the endemic fossils with their direct
     mainland ancestors and explain the evolutionary
     meaning of the changes the animals underwent as the
     environments on the islands are known;

(b)  to compare the endemic mammals of different islands
     and see if the evolutionary effect follows parallel
     patterns in similar environments;

(c)  to speculate about the rate of evolution on islands
     as in some cases the approximate time of arrival of
     the ancestors is known;

(d)  to study the relation between ecology and adaptation
     more readily because the effect of insularity on
     mammalian evolution on smaller islands is less
     complex because of the quite uniform environment
     and the very small number of genera.


## THE ISLANDS AND HOW THEY WERE REACHED

From the Pleistocene a host of unbalanced endemic island
faunas are known in which large mammals such as elephant, deer
and hippopotamus are often present.  Islands in this category
can be found in the Mediterranean, the Japanese archipelago,
the Phillipines, and the Indonesian archipelago.  They have not
changed very much in physical geography since the Pleistocene and
must have been mountainous during that time.  An extensive
literature on the fossil mammals from these islands has accumulated.  However, most of these papers deal mainly with the
morphology of the fossil mammals, although often a discussion
considers how the mammals reached the islands.  This implies
some speculations relative to the paleogeography.  The presence
of large mammals, such as elephant and deer, is often used as an
argument in favor of landbridges (Vaufrey, 1929; Audley-Charles
and Hooijer, 1973; Kuss, 1973).  Against a too general use of
the landbridge idea we can suggest that it does not provide us
with an explanation to account for the composition of the faunas
with elephant and deer.  Furthermore, the idea of landbridges is
often difficult to reconcile with what is known about the
island's geology.  If we accept the landbridge conception we are
implying that big tectonic changes have occurred since the
Pleistocene on all islands with an endemic Pleistocene elephant -
deer fauna.  If we consider the Ryukyu islands (south of Japan)
and Cyprus this must have been of a catastrophic nature,
unsupported by any geological evidence.

If we do not accept the landbridge concept then the mammals must have crossed the sea by natural rafts, swimming or drifting by the so-called sweepstakes routes (Simpson, 1940). The possibility that smaller mammals might cross considerable distances on natural rafts is generally accepted. The Galapagos endemic rodents are a good example of this: their ancestor crossed 500 miles of ocean (Carlquist, 1965). However, some authors consider seas even when they are only 20 miles wide to be insurmountable barriers to large mammals such as elephants (Hooijer, 1972).

If we take into account that those mammals which are found on the islands, such as elephant, deer, and hippopotamus, belong to the group of the large mammals with good swimming capabilities, only overseas dispersal can explain the uniform composition of the endemic island faunas in which elephants and deer are often present. Elephants, for example, love bathing and are not afraid of swimming (Carrington, 1962). It is reported that they went island hopping for two hundred miles in the Bay of Bengal, sometimes across more than a mile of open ocean. The extraordinary swimming capacity of deer is reported in many handbooks on mammals, and of the hippopotamus it is recorded that this aquatic mammal crossed from the mainland of Africa to Zanzibar (Joleaud, 1920), and even Madagascar was reached in the Pleistocene (Mahe, 1972). Only when the crossing is an exception can an endemic form evolve, and so the overseas distances must not be too short as they are, for example, between the islands in the Bay of Bengal where the individuals remain in contact with the mainland population (Fig. 1). The above also implies that the landbridge concept is very improbable from a zoogeographical point of view, as one would expect different faunal composition. The dispersal clearly followed a sweepstakes route which Simpson (1965) defines as one which is highly improbable for most animals. Although a great barrier exists it might occasionally be crossed by some species (Fig. 2). A factor influencing the migration route might have been the lowering of the Pleistocene sea-level in which case the distance overseas would have been smaller. Also, the behavioral activity of the colonizer may have played a role; for example, the nomadic way of life of elephants; overpopulation in herds of hippos and elephants, and the consequent search for new territories, (Sondaar and Boekschoten, 1967).

The presence of an "elephant-deer" fauna in which both groups are not necessarily present most probably implies an overseas dispersal. When there has been a land connection the composition of an island fauna is different. As an example we can take Sardinia with an endemic Pleistocene fauna but its direct ancestor of Pliocene age is also found on the island. This

Fig. 1. An artist's view of island hopping. Some mammals can pass easily over a sea barrier which for others is an insurmountable barrier. Endemic forms will not evolve because the crossings are in both directions.

Fig. 2. An artist's view of the sweepstakes route. For some mammals the barrier can occasionally be crossed by swimming or drifting. This must be a very rare extent if an endemic form is to evolve.

Pliocene fauna was not so impoverished as the Pleistocene and resembled more the mainland fauna of the same age (Pecorini, et al. 1973). A land connection with Europe must have been present during the Upper Miocene/Lower Pliocene. The faunal composition is clearly distinguishable from the elephant-deer faunas from Mediterranean islands such as Cyprus, Crete and Sicily/Malta. The faunal succession in Sardinia is not totally clear because of the lack of sufficient fossil evidence. From older to younger deposits we can note a decrease in the number of families present.

The paleozoogeographic history of islands can also be very complex. Crete, for example, must have been connected with the mainland during the Late Miocene, as balanced mainland faunas are known from that age. No mammals are known from the Pliocene, probably most of the island vanished beneath the sea as some geological evidence indicates (Meulenkamp, 1971). In the Pleistocene we see endemic elephant, deer and hippo, but no other large mammals are found till the arrival of Neolithic man. For Japan we can draw a similar picture: a mainland fauna in the Miocene; mainly marine deposits in the Pliocene; mainly an elephant-deer fauna in the Pleistocene; and then a mainland fauna in the Late Pleistocene. Probably during the last glacial period the sea-level was lowered to such an extent that a sea barrier did not exist (Sondaar and Boekschoten, in preparation). Japanese fossil deer and elephants are not always endemic, some species are also found on the mainland of China. Perhaps the barrier was not always sufficient to maintain the endemic character of those mammals who were good swimmers. New invasions take place and the island populations interbreed with the mainland stock. The barrier was, however, great enough to bar other mammals, such as large carnivores and perissodactyls.

On Cyprus no terrestrial fossil mammals are found before the Pleistocene and it may be considered as a good example of a true oceanic island. The endemic fauna consists basically of a very small hippo and a dwarf elephant. The island of Kythera has also yielded some Pleistocene elephant and deer remains though they are not endemic. Here also the barrier was not great so the island and mainland populations remained in contact (Sondaar, 1971).

Finally, we may sum-up by stating that the faunal history of the islands from which endemic fossil mammals are known is often complex. If we find an endemic elephant-deer fauna we may assume that these mammals came by sweepstakes routes. If the barrier was too small no endemic forms would have developed. We may also note that large carnivores are missing in the elephant-deer faunas except for a find of Ursus on Malta. This genus is known for its swimming abilities and also was not solely a predator. Table I lists the main islands and their faunal characteristics.

TABLE I
"ELEPHANT-DEER" FAUNAS AND THEIR OCCURENCE

| ISLAND | AGE OF FAUNA | LARGE MAMMALS PRESENT | ENDEMIC OR MAINLAND SPECIES | LITERATURE |
|---|---|---|---|---|
| Cyprus | Pleistocene | elephant hippopotamus | endemic | Bate, 1906 Boekschoten & Sondaar, 1972 |
| Tilos | Pleistocene | elephant cervid | endemic | Symeonides, 1972 |
| Rhodes | Late Pleistocene | elephant | endemic | Symeonides & Marinos, 1973 |
| Karpathos | Pleistocene | cervid | endemic | Kuss, 1967 |
| Kasos | Pleistocene | cervid | endemic | Kuss, 1969 |
| Crete | Pleistocene | elephant hippopotamus cervid | endemic | Bate, 1905 Kuss, 1970 Sondaar & Boekschoten |
| Delos | Pleistocene | elephant | endemic | Melentis, 1961 |
| Naxos | Pleistocene | elephant | endemic | Mitzopoulos, 1961 |
| Kythera | Pleistocene | cervid/elephant | mainland | Kuss, 1967 |
| Kefalinia | Pleistocene | hippopotamus ruminant | mainland | Psarianos, 1953 |
| Sardinia | Miocene | primate/suid/ cervid/bovid/ | endemic | Dehaut, 1920 |
| Sardinia | Pleistocene | elephant/canid elephant/cervid | endemic | |
| Sicily | Pleistocene | hippopotamus | endemic | Malatesta, 1970 Vaufrey, 1929 Ambrosetti, 1968 |
| Malta | Pleistocene | hippopotamus elephant/ursid cervid | endemic | Falconer, 1862 Accordi, 1955 |
| Mallorca | Pleistocene | bovid | endemic | Bate, 1914 |
| Menorca | Pleistocene | bovid | endemic | Andrews, 1915 |
| Gargano | Miocene | ruminant | endemic | Freudenthal, 1971 |

| | | | | |
|---|---|---|---|---|
| Corfu | Pleistocene | deer | endemic? | not published |
| Philippines (Mindanao and Luzon) | Pleistocene | elephant | endemic | Nauman, 1890<br>v. Koenigswald, 1956 |
| Celebes | Pleistocene | elephant/suid. | endemic | Hoijer, 1975 |
| Flores | Pleistocene | elephant | endemic | Hoijer, 1972 |
| Timor | Pleistocene | elephant | endemic | Sartono, 1969 |
| Sta. Rosa (California) | Pleistocene | elephant | endemic | Stock & Furlong, 1928 |
| Ishigaki | Pleistocene | cervid | endemic | Otsuka & Hasegawa, 1973 |
| Miyako | Pleistocene | cervid/elephant | endemic | Tokunaga, 1936, 1941<br>Hasegawa, Otsuka & Nohaza, 1973 |
| Ie | Pleistocene | cervid | endemic | Tokunaga & Takai, 1939 |
| Okinawa | Pleistocene | cervid/elephant | endemic | Tokunaga & Takai, 1939<br>Nohaza & Hasegawa, 1973 |
| Japan | Early Pleistocene till Late Pleistocene | cervid/elephant | mainly endemic | Takai, 1936 |
| | Late Pleistocene | balanced mainland fauna | mainland | Otsuka, 1968, 1972 |

PATTERNS IN EVOLUTIONARY CHANGES AND THEIR POSSIBLE MEANING

## Size Change

The aberrant size of the fossil island mammals (pony-sized elephants, pig-sized hippos and rabbit-sized mice) has caught the attention of researchers since their discovery more than a hundred years ago. A variety of theories have been proposed to explain this size difference. Dwarfism was usually regarded as a result of a degenerate or pathological condition (Leonardi, 1954; Kuss, 1965). The founder population could never have been very big and so genetic degeneration could be considered plausible, the more so since, in the absence of large predators, the "pathological" individuals were not selected and could have inter-bred. Therefore, fossil island mammals are often considered as a kind of paleontological curiosity instead of an example of an evolutionary process which made the animal better adapted for island life. If the changes, however, were caused by degeneration or some pathological condition one would expect a random effect. Instead, we see that a number of characters change in the same way on different islands with similar environments. If we compare the island dwarfs and giants with the mainland ancestors we note that the change in size did not produce scale models. This is not unusual since with a change in the surface/volume ratio (Gould, 1966) the same function could have been achieved by a different shape. A good example of this is the allometrical change in the skull of Elephas falconeri (Fig. 3), a dwarf elephant, from Sicily better known from the studies of Ambrosetti (1968). This elephant was about four times smaller than its ancestor, E. namadicus, from the mainland. The enormous cranial muscles of elephants are attached to an outgrowth of the surface of the cranium formed by pneumatic bone tissue. Most of the skull consists of this tissue which was developed in elephants during their evolutionary trend toward gigantism. The reduction in size in the island elephant made pneumatic bone tissue superfluous and the general shape of the skull changed drastically (Sondaar and Boekschoten, 1967). Accordi and Palombo (1971) go further into changes of the brain size. They find that it was relatively bigger in the dwarf which agrees with the pattern that small animals have relatively larger brains (Gould, 1971). Another character which accompanies decrease in size is that the cheek teeth have fewer plates and proportionally thicker enamel than in E. namadicus. The low number of plates could also be explained by supposing a more primitive ancestor, but Maglio (1973) states that this character is a direct response to the small size of the molars. The lower number of plates is caused by the fact that the skull and jaw could not accommodate molars with the same number of plates as the ancestor unless the

enamel was reduced below the critical minimum thickness required
for efficient mastication.

In the limbs a further character is observable which might
be correlated with size change.  In large mammals such as
elephants, and also in Hippopotamus, the femoral shaft is twisted
through about 50° from proximal articulatory surface to distal
articulatory surface presumably enabling the feet to be placed
more under the body, a necessary factor in supporting the
enormous weight of these large mammals.  For E. falconeri,
Ambrosetti (1968) gives a torsion value of 30° and the values for
the dwarf elephant from Tilos and the dwarf hippo from Cyprus are
similar.  The lesser degree of torsion in dwarfs might be explained by the fact that here the movement was more along the
body which was more effective in locomotion in which a weight
supporting function was less emphasized.

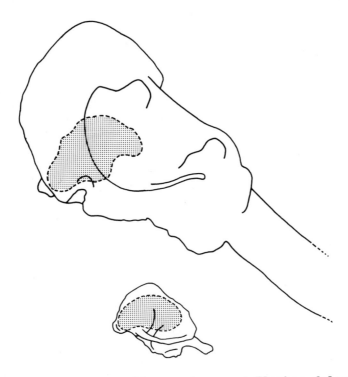

Fig. 3.  Skull of Elephas antiquus above and Elephas falconeri
below.  After Accordi and Palombo, 1971.  Both skulls are drawn
to the same scale and show allometrical changes.  In the dwarf
form E. falconeri from Sicily the braincase is relatively bigger,
while the pneumatic bone tissue for the attachment of the muscles
is mostly lost.

    The deer from the Mediterranean islands are often considered
to be descendants of the giant deer <u>Megaloceros</u> from the mainland.  This would mean that the endemic island form is also a
dwarf.  Recent studies cast doubt on this, as the deer from Crete,
for example, is already so distinct from all mainland deer that
it is not possible to designate a particular genus of the mainland as the ancestor, particularly as there is considerable
variation in the Pleistocene island deer.  The most common type,
however, is a short-legged, small, heavily-built form resembling
the dwarf elephant and hippo in these aspects.  We can thus note
a relative shortening of the leg in deer, elephant and hippo
compared with the mainland ancestors.  This indicates that the
dwarfism cannot be considered as being due to ontogenetic scaling
or paedomorphic changes.  The relative shortening of the legs is
not directly connected with dwarfism but must have a special
adaptational meaning to be considered in the next section.  Small
size can also be observed in the endemic Pleistocene ground
sloths from the Caribbean (Matthew and Paulo Couto, 1959;
Hooijer, 1962).  In explaining the aberrant sizes of island
mammals compared with the mainland, Thaler (1973) mentions the
study of Valverde (1964) who suggested that there is a relation
between predator size and prey size.  It is clear that size plays
an important role in protective adaptation.  Small mammals can
hide easily; bigger ones, like deer and boar, can run fast and
also defend themselves; elephants are so big that they have
practically no enemies at all.  Medium-sized prey are scarce and
if they do occur they have special adaptations; e.g. spines in
the hedgehog, running in rabbits and hares.  Predation will
exert a selective pressure maintaining the size differences or
special adaptations.  In the absence of carnivores this selective
pressure on the size of large mammals is missing.  Large size
therefore lost its advantage and smaller size becomes more
advantageous since it allows: (1) greater mobility; (2) decrease
in the quantity of food necessary; (3) reduction in territorial
area; and (4) increased heat exchange with the surrounding air
or water due to a proportionally larger body surface.  The increased size of rodents might also have some correlation with the
lack of carnivores, as on the "empty" island there are many unoccupied ecological niches and the selective pressure for small
size is missing.  Large size in rodents is therefore not only
advantageous for occupying a new ecological niche but also for
avoiding predators.  Island rodents will still be the targets for
mainland birds of prey which are present in large numbers on the
islands.  A characteristic of fossil island avifaunas is the
great number of birds of prey.  The owl is a common fossil on
islands and larger size might be explained as advantageous in
order to escape owl predation, but most probably the answer will
be more complex.  Ballmann (1973) described giant owls from
Gargano and deduced from the osteology of the foot that they

spent much of their time on the ground, an adaptation commonly found in birds on islands where carnivores are missing. The largest owl from Gargano, Tyto gigantea, is found in the bio-stratigraphically higher levels and was contemporaneous with the largest murid. This might indicate that there was a direct relation between the change in size of prey and predator, which was, oddly enough, probably provoked by the lack of carnivores. This favored an increase in size of the smaller sized, and opened the possibility for the birds to move without danger on the ground.

Another example of gigantism on islands is the largest insectivore yet known, Deinogalerix koenigswaldi, from Gargano. Freudenthal (1972) supposes that this animal, which exceeded the size of a fox, was not insectivorous but preyed on rodents although its skeleton does not suggest a fast moving mammal. Inasmuch as the ancestor of this insectivore is unknown it is difficult to speculate about the changes the animal underwent through its isolation. There are also mammals which did not change very much on the islands; for example, most of the insectivores, lagomorphs and some of the rodents which were contemporaneous with the giants. They probably occupied a similar ecological niche on the islands as on the mainland. The natural enemies of the insectivores were the same - birds of prey.

### Low Gear Locomotion

A common character of the endemic island ruminants (Fig. 4) and also in a lesser degree of the dwarf elephants and hippos (Fig. 5) is shortening of the distal part of the leg. This shortening is accompanied by the fusion of some of the bones in the foot. In ruminants fusion of the metatarsus with the naviculo-cuboid restricts movement between the metatarsus and tarsus (Leinders and Sondaar, 1974). Together with the short metapodials and phalanges this gives the foot a solid construction advantageous for low speed locomotion in a varied mountainous environment. If we want to evaluate the meaning of this "low gear" locomotion we must compare endemic island ruminants with mainland ruminants in which there is some rotation possible between the tarsus and metatarsus resulting in the ability to manoeuvre at full speed in order to escape predators (Leinders and Sondaar, 1974; Barone, 1968). The long distal parts of the legs also augment the animal's speed when escaping from predators. On the islands we see ruminants with short metapodials and phalanges which means that the animal must have lost its speed. On the other hand, the center of gravity was lowered and the foot was less vulnerable because of the sturdiness produced by the fusion of tarsus and metatarsus. This construction must have been advantageous in the mountainous environment of the islands.

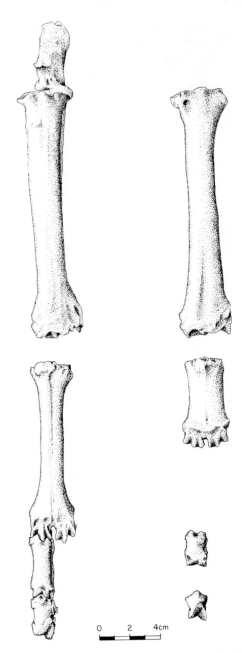

Fig. 4.  Radius, metacarpal and phalanges from left, a Recent goat
(left), and Myotragus balearicus, an endemic extinct bovid from
Mallorca (right).  A striking feature is the relative shortening
of the metacarpal and phalanges in Myotragus, if we take in account
that this part of the leg of the goat is also short compared to
other bovids.

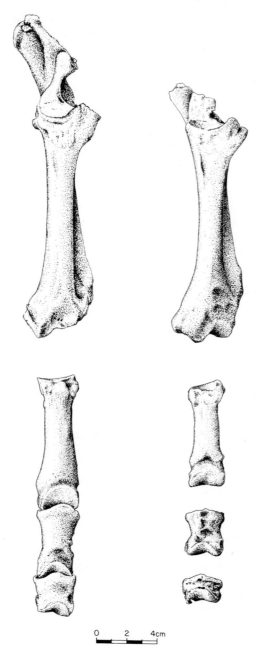

Fig. 5. Radius, metacarpal and phalanges of <u>Choeropsis</u> <u>liberiensis</u> the Recent small hippo from West Africa (left) and <u>Phanourios</u> <u>minor</u>, the dwarf hippo from the Pleistocene of Cyprus (right). Note the relative shortening of the metacarpals and phalanges in <u>Phanourios</u> <u>minor</u>.

Speed and zigzag movement lost its significance on the islands
because big carnivores were lacking and the selective pressure
favored low gear locomotion.

In dwarf elephants and hippos the modifications facilitating
low gear locomotion are less obvious.  One would expect more
slender limb-bones in the dwarfed forms but the converse is true.
Ambrosetti (1968) states that Elephas falconeri had very short
limbs in relation to body length when compared with living forms.
The same author notes a firm fusion between radius and ulna, and
between tibia and fibula which resulted in a reduction in the
degree of lateral mobility as compared to the mainland elephants.
In the dwarf hippos we can observe similar characters.  The metapodials and phalanges are shortened in relation to the mainland
forms.  This is abundantly clear in the Cyprus dwarf in which
some tarsal fusion was also found (Leinders and Sondaar, 1974).
Changes in the dwarf elephant and hippo suggest adaptations
parallel to those of ruminants, namely, a foot construction which
was sturdy and solid and adapted for walking over uneven terrain
to reach the more rugged grazing areas.

From Water to Land

On islands mammalian genera are found which are semi-aquatic
on the mainland; e.g. Hippopotamus and some rare finds of Lutrines.
The islands do not provide all the requirements necessary for
Hippopotamus amphibius or its Pleistocene ancestor which was
quite similar.  Lakes are scarce, large rivers are absent and much
of the islands consists of barren limestone rocks often mountainous.
On Crete endemic hippo fossils are found in coastal caves and
beach deposits but also in the Katharo, a Pleistocene mountain
basin situated more than 1000 meters above sea level.  This means
that the endemic hippo must have been capable of climbing mountain sides, not feasible for the Pleistocene H. amphibius, its
ancestor.  The endemic island hippos were smaller than H.
amphibius suggesting greater mobility but also their locomotory
apparatus was more adapted for walking.  In some functional
aspects the foot resembles that of Choeropsis liberiensis (the
smaller hippo from Liberia which is less aquatic in habit than
H. amphibius) but in general morphology it is more similar to
H. amphibius; e.g. the small amount of reduction of the laterals;
the alternation of carpals on the metacarpals; the form of the
astragalus, etc.  In other words, the locomotor apparatus evolved
from an H. amphibius like foot. Lateral movement of the limbs was
restricted while anterior-posterior movement was better developed.
The short phalanges are a striking character of the island
hippos.  If we look at the foot of Hippopotamus amphibius we can
observe a large digital cushion which is tender and precludes

rambling through mountainous regions. The short phalanges in the island hippos indicate that some of this foot pad was lost and from an examination of the articulatory surfaces of the phalangeal joints we could conclude that the phalanges made a bigger angle with the ground than they do in H. amphibius. This latter fact adds weight to our suggestion that a smaller foot pad was present reducing substrate friction. The whole aspect of the foot is sturdy, which parallels the ruminants in some respects.

Hippopotamus amphibius is a good swimmer. Indeed historical observations are known of Hippopotamus swimming in the Mediter-ranean near the river Nile. Since elephant and deer are common one can ask why this semi-aquatic mammal is not more frequently found on islands. Perhaps the answer lies in the fact that the distribution of Hippopotamus on the mainland is more bound to big rivers and lakes and the natural meadows near it. Once on the island Hippopotamus possibly could not find a directly suitable biotope; e.g. rivers, to shelter during the day to avoid the heat. In some cases they had to cover greater distances to reach the islands. That they could still adapt to a totally different environment is clearly demonstrated by the dwarf hippo which was present in huge quantities during the Pleistocene on Cyprus.

Another example, though less striking, of a semi-aquatic mammal which became more adapted to life on land is the otter. We have fossil material from Malta, Corsica and Gargano while from Crete an articulated skeleton has been recovered giving us the opportunity to make speculations on the mode of life of this otter. Superficially, it resembles Lutra, a possible ancestor, but the otter from Crete was more adapted to land than Lutra lutra. Symeonides and Sondaar (1975) suppose that the endemic otter from Crete probably became adapted to a somewhat different ecological niche, as carnivores were absent on the island. It might be that the otter changed to a somewhat different diet, viz. more rodents, which were available in large quantities on Pleistocene Crete. This seems plausible as recent otters do not refuse rodents either (Harris, 1968). Another possibility is that the Cretan otter might be related to the Recent genus, Lutrogale, (van Bree, personal communication) now living mainly in India, Indonesia and an isolated occurrence in the Middle East. If this is the case the changes are not so obvious as Lutrogale is also more adapted to land life than Lutra lutra. Further studies are necessary in order to show whether the resemblence between the Cretan otter and Lutrogale is a case of parallel evolution or of relationship.

## Increase in Hypsodonty

In several endemic island genera among elephants, ruminants and rodents we can see an increase in hypsodonty as compared with mainland ancestors. The most striking example is <u>Myotragus</u>, a bovid from the Baleares which besides very hypsodont teeth also possessed continuously growing incisors like rodents. We can also observe an increase of hypsodonty in the deer from Crete. Freudenthal (1971) notes that the murids from Gargano show an extreme increase in size and hypsodonty while Hooijer (1973) observes an increase in hyposodonty in the <u>Stegodon</u> from Timor compared with its possible ancestor from Java. He concludes from this that it could live longer than its ancestor from Java. However, in general, hypsodonty in island mammals cannot be connected with longevity, but must be considered as an adaptation to the mastication of harder food such as grasses. Increased hypsodonty in island mammals might therefore be better explained as an adaptation to the changed environment with more abrasive food.

## Some Further Changes

Hypsodonty does not always develop as one of the evolutionary effects of insularity; donty also occurs as in the teeth of the dwarf hippo from Cyprus. The unworn molar of <u>Phanourios</u> already has a lophodont aspect, but the oblique ridges from the main cusps are hardly developed and do not reach the grinding surface. The transverse valley becomes still clearer when the teeth grind down to an extent not found elsewhere in the family <u>Hippopotamidae</u>, although older members of this family also show a somewhat lophodont aspect. This wear pattern suggests a jaw action with a predominantly transverse movement (Fig. 6).

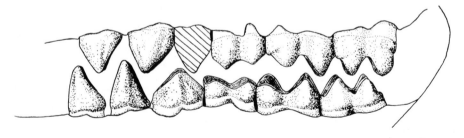

Fig. 6. Reconstructed occlusion of upper- and lower jaws of <u>Phanourios</u>, the dwarf hippo from the Pleistocene of Cyprus. The jaw movement was predominantly transverse together with the usual vertical component. In all specimens the upper fourth premolar is lacking. An imaginary fourth premolar is indicated by the cross-hatching and this shows how its presence would have impeded this jaw movement.

Another observation concerning this dwarf hippo from Cyprus is that the upper fourth premolar is absent, presumably correlated with the changed jaw movement. In $\underline{H}$. $\underline{amphibius}$ the conical $P^4$ occludes with the flat $M_1$. The wear facets on $P^4$ are perpendicular to the axis. The premolars of $\underline{Phanourios}$ show wear facets oblique to the axis of the premolars. The occlusion is different and shows that when $\underline{Phanourios}$ closed its jaws the premolars alternated. Were $P^4$ still present then it would contact $M_1$ (as in $\underline{H}$. $\underline{amphibius}$). If we suppose that an extreme transverse jaw movement was essential for the mode of feeding in $\underline{Phanourios}$, the presence of a $P^4$ would have obstructed this mastication. The absence of $P^4$ then seems to have been an advantage for the function of this type of mastication. Lophodonty is normally found in mammals with a diet of twigs and leaves and this suggests that the hippo from Cyprus had a mode of life somewhat like a leaf-eating pig (Boekschoten and Sondaar, 1972).

In island deer simplification of the antler occurs in some cases resulting in a degree of resemblance between antlers from different islands. Kuss (1973) observed that the morphology of antlers from Crete and Japan is alike but a close relationship between Japanese and Cretan island deer is out of the question. It might be a case of parallel evolution producing an antler useful in display for intraspecific recognition and aggression in island deer. Other explanations which have been proposed include the suggestion that the simple antler is a primitive character. Kuss (1973) uses this as an argument to relate the island deer from Crete with Tertiary deer from the mainland. However, in morphology the antlers do not resemble the Tertiary forms at all, while geological and paleozoogeographical evidence make this supposition improbable.

## Unusual Wear Markings on Fossils

Bones from Crete, mainly of deer, are often found having been, in some way, fashioned into the shape of forks. Kuss (1969) describes such bones in detail and interprets them as artifacts of an osteokeratic culture of paleolithic man from Crete. Similar fossils are found on the Ryukyu islands (Japan) and have been interpreted likewise as the proof of the Pleistocene presence of man (Tokunaga, 1940). What use the bones may have been put to is difficult to explain and not a single other sign is found of the presence of man on the islands.

Sutcliffe (1973) shows, in a well documented study, that these bones have been chewed by deer and refers to an extensive literature on Recent ruminant chewing. It is known that deer chew bones in order to obtain an additional phosphorus supply.

Sutcliffe figures a recent reindeer metatarsal which has been
chewed by a Norwegian reindeer. This chewed bone strikingly
resembles the fossil "forks" from Crete and Japan. It is hardly
surprising that chewed bones are fairly common on islands such as
Crete and probably also on the Ryukyu islands as the fauna consisted mainly of deer. The bones are often found in caves which
suggests that deer probably used caves also as shelters.

Another strange wear marking is found on the tusks of endemic
elephants from Sicily (Ambrosetti, 1968); Crete (Kuss, 1965) and
Japan. A deep, V-shaped, groove is visible near the apex of some
tusks. Specimens have been found which showed that the groove
could cause a break at that point of the tusk. Kuss thinks that
this feature might be explained by a grazing habit of the island
elephant in which the tusks were used to extract grass sods. The
similar V-shaped grooves on the tusks of endemic elephants from
different islands can be explained as parallel behavior in
grazing habits.


Concluding Remarks

From the preceding remarks on the evolutionary changes undergone by island mammals, it appears that they should be considered
together as adaptations to an island environment lacking in large
carnivores. Patterns of adaptations occur in parallel on different islands and these are not restricted in geological age or
geography.


THE DEGREE OF CHANGES AS REFLECTED IN THE SYSTEMATICS

The changes that occurred in mammals due to isolation on
islands varied greatly in extent. Some changed only slightly,
while the changes in others are so big that it is difficult to
judge to which family they may belong, e.g. the artiodactyls from
the Miocene of Gargano.

In order for the degree of changes and the classification of
the island mammals to be considered we need to look at several
characters:

1.  characters which are inherited from their mainland
    ancestors; e.g. the elevated eyes of the dwarf hippos
    of the Mediterranean islands which became more
    terrestrial in mode of life; the slight reduction of
    the lateral metapodials in the island hippos as in
    H. amphibius. These characters are important in
    tracing the ancestor.

2. Characters specific to the endemics which can be
   explained as adaptations to the changed environment.
   These characters can be found in different groups;
   e.g. low gear locomotion and small size. If these
   changes are found in related groups on different
   islands the forms may resemble each other to a high
   degree due to parallelism not necessarily indicating
   a direct relation. This situation creates taxonomic
   problems, especially since in some cases the ancestor
   will be the same and the endemic island mammals will
   be strikingly alike in those morphological aspects
   which are available to the paleontologist. This
   problem can best be illustrated by the Mediterranean
   dwarf elephants which are usually considered to be
   descended from "E. antiquus" = E. namadicus, (Maglio,
   1973). Vaufrey (1929) has placed all Mediterranean
   dwarf species, when equal in size, in the same species.
   Thus dwarf elephants from Cyprus, Crete and Sicily are
   considered as conspecific. The same method is also
   often used for the Pleistocene Mediterranean island
   deer and hippos. This classification more or less
   implies the concept of interbreeding populations and
   of the existence of landbridges. As stated before, it
   is obvious that the ancestral colonizers came by
   sweepstakes routes into the islands, thus the genetic
   input of the ancestors was different for each separate
   case. Moreover, the time of arrival was not necessarily
   the same.

   The resemblances between island forms are the result of
   parallel developments in several aspects of which the
   clearest is the reduction in size. These forms constitute separate branches and should not be considered
   conspecific outside an island or island group. Undue
   extension of the species concept to unite dwarf forms
   that are certainly not directly related (as proven by
   paleogeography alone) under one name should be avoided.
   The small goundsloths from the Caribbean are another
   clear case of how mammals from different islands are
   alike. Hooijer (1962:59) remarked "Considering the
   great distance between Curacao and the three Greater
   Antillean islands it is somewhat surprising that the
   end product in Curacao should be so similar." It is
   suggested that the ancestral form must have been the
   same and most probably the resemblance is the result of
   parallel evolution.

3. A common characteristic in island mammals is a change in
   size; these changes are sometimes allometrical.

4.  The changes resulting from an adaptation to the dif-
    ferent island environment can give rise to characters
    which are not found in the direct ancestor but occur in
    more distant relatives.  The changes in endemic island
    mammals can be so big that the original ancestral
    species, genera or even families may not be clear.  Or
    it may be that the island form resembles in some re-
    spects a form which is certainly not its ancestor;
    e.g. Kuss (1973) has cited the resemblance between the
    antlers of Cervus japonicus and the Cretan deer. Another
    case is the dwarf hippo from Cyprus which shows in its
    lophodont dentition more similarity with Choeropsis
    liberiensis, the small Recent hippo from Liberia, and the
    small fossil hippo, H. aethiopicus, from Pliocene/
    Pleistocene deposits of Lake Rudolf (Coryndon and
    Coppens, 1975).  Other morphological characters show
    that the former cannot be its ancestor (Boekschoten and
    Sondaar, 1972) while on paleozoogeographical and
    ecological grounds the latter is highly improbable.  H.
    aethiopicus is thought to have had habits like Choeropsis
    liberiensis, solitary and less aquatic than H. amphibius
    (Coryndon and Coppens, 1975) which made it a worse over-
    seas traveller than H. amphibius of which it is known
    that it can reach Zanzibar and once reached Madagascar.
    The way of life, amphibious, living in a herd, and
    sometimes suffering from overpopulation, points to H.
    amphibius as the ancestor of Cyprus hippo, especially as
    there is ample fossil material of H. amphibius from
    around the Mediterranean while H. aethiopicus is so far
    only known from East Africa.  On morphological grounds,
    H. amphibius could very well have been the ancestor.
    The lophodont teeth of the Cyprus hippo can be explained
    by parallel adaptations in chewing, in which the island
    form resembles more its "grandfather" than its "father"
    as lophodonty is a more common character in the strati-
    graphically older members of the Hippopotamidae while in
    other characters the island hippo is advanced.  In the
    island elephants we see that there are fewer ridges in
    the teeth which is an archaic character in elephants,
    but is also to be explained as an adaptation (Maglio,
    1973).

5.  In some groups there is great size variation.  This
    variation far exceeds the species variation when com-
    pared with the mainland but it is often not possible to
    delineate separate groups.  This is especially clear in
    Myotragus, a Pleistocene bovid from the Baleares (Fig. 7).
    This was the only ungulate on the island and a number of
    different ecological niches must have been available to

it.  It is not clear whether we can explain this variation as an example of an extremely variable species or
whether we are simply unable to recognize different
forms in the material.  Research under way by Dr. de Vos
on the fossil deer from Crete suggest the latter possibility as he has recognized several antler types.

On the mainland the niches are occupied by species belonging to different groups and therefore are not difficult to distinguish.  On the contrary, island forms
are restricted to only a limited number of genera and
probably are descended from the same mainland ancestor
making it difficult to separate forms.  (The various
species of finches on the Galapagos islands would be
hardly recognizable if found only as fossil remains
since in fossil birds the beak is normally not preserved.)
In some cases, however, the differences in one genus
are clearly seen; e.g. the fossil owls from Gargano.
Ballmann (1973) using size distinguishes three owl
species in the highest levels of the Miocene on Gargano.
Size is an important feature in the adaptation of owls
to prey on the varied endemic rodents of Gargano.  The
largest owl is only found in the highest level together
with the largest murid.  In the case of the Gargano owls
a character in which the forms differed had an adaptational meaning, but in other island fossils it is not
always possible to find such features.

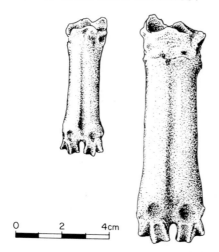

0   2   4cm

Fig. 7.  Size variation in Myotragus balearicus metatarsals from
Muleta cave (Mallorca).  The bones are from the same
level.  All intermediate sizes are also found and no
grouping of bones is possible into different size
ranges.

The classical parameters used to distinguish the mainland species will often not be particularly useful in the classification of island forms. The ecological niches are occupied by closely related forms and it will be necessary to look for morphological features with an adaptational meaning.


## SELECTION MECHANISM, RATE OF EVOLUTION AND EXTINCTION

The changes undergone by the island elephants, hippos and ruminants were not bound to geography and time but were independent products of evolution occurring in similar environments and resulting in parallelism on different islands. If we want to explain this phenomenon and to find a mechanism for these patterns of adaptation, then it should first be stressed that successful colonization is infrequent and that the arriving population will be numerically small. This original population would have contained only a small fraction of the genetic variation of the parent species. From such a population the island forms must have evolved and the "founder principle" of Mayr (1942) may very well have applied.

Due to the limitations inherent in the fossil record we cannot follow the changes step by step or see how great the changes were until the invader found a suitable niche and survived the "genetic revolution." As the islands were probably empty for the greater part there would have been niches available and there is abundant evidence in the literature of successful founder populations which were very small; even the establishment of new colonies by single founders is quite likely (Baker and Stebbins, 1965). The offspring can be very successful and there need not necessarily be harmful effects due to inbreeding (Mayr, 1975). In the study of Dobzhansky and Pavlovsky (1957) several parallel lines of Drosophila populations were exposed to the same selective pressure. One of the results was the smaller the starting population the greater degree of indeterminacy.

If we study the fossil mammal material from different islands we note that there exists a great degree of parallelism in morphology. The changes can be explained as an adaptation to the island environment. As the genetical composition of the founder population was different on the different islands, we would expect genetic drift to have caused random fluctuations. But, the evolution is clearly directed so that there must have been a strong selective pressure exerted on the island mammal populations. It is possible to find evidence for mass starvation such as bone beds. Thus, the selective pressure could have been overpopulation followed by food shortage. At one Pleistocene locality on Crete most of the deer bones exhibited osteoporosis. (Osteoporosis is

a rarefying condition of bones which lose much of their mineral
matter and become fragile and often deformed.) The bones are
thin, the nutritive canals are very wide and the articulatory
surfaces are sharp-edged (Fig. 8). Evidently there was a failure
in bone formation while bone loss was proceeding normally.
Clearly the sick animals were not selected by predation due to
the lack of big carnivores. A similar defect, though less pronounced, is seen in sheep suffering from chronic malnutrition
(probably associated with a deficiency of vitamin D). With the
bone bed evidence suggesting frequently occurring mass starvation,
it may point to a food shortage which occurred repeatedly and
sometimes over a long period. An explanation might therefore be
that on the islands the population was not kept in balance by
carnivore predation. Overpopulation could have occurred in turn
causing overgrazing and damage to the vegetation, resulting in a
destruction of otherwise suitable biotopes. That such overpopulation can occur is clearly demonstrated by many cases; e.g.
the Kaibab reservation in Arizona where, in less than twenty years
after the kill of all predators in 1906, the reservation was overpopulated and mass starvation occurred (Basler, 1972). Another
case is the arrival of the elk on Isle Royale in northern Lake
Superior, North America (Grimzich, 1970) where the population
grew quickly in number and damaged the island vegetation resulting
in mass starvation. After the arrival of the wolf the population
was stabilized at one-fifth of the previous peak value.

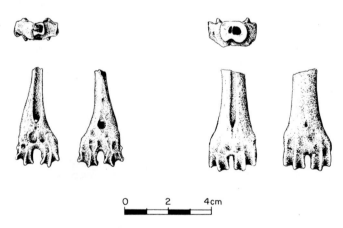

Fig. 8. Distal metatarsal fragments of Cervus cretensis (section
        through diaphysis above; anterior and posterior views
        below) from the Pleistocene of Rethymnon in Crete. The
        figures at the left are drawn from an individual suffer-
        ing from a severe osteroporosis. The figures to the
        right are from a normal individual.

On the Pleistocene islands overpopulation could have been the main cause which exerted selective pressure and gave advantage to those individuals that needed less food or could reach different food resources. That overpopulation not only occurred in the founder population is shown in the bone bed in which we found fossils of already adapted endemic mammals, and the deer fossils showing osteoporosis are already of the endemic type. This means that overpopulation must have occurred several times and so the population density fluctuated considerably through time. This would have favored genetic turnover (Haldane, 1956) and might have influenced the rate of evolution. In small populations a gene that elsewhere is rare may occur in a higher frequency, which might be an important factor during rapid evolution in small populations. No clear fossil evidence is found of intermediate forms between the mainland and island type and probably the main change took place in a relatively short time. Once the island form had come into existence the evolutionary changes were less drastic and probably involved rather small modifications of the primary adaptation. In Fig. 9 this is shown in a scheme following the arrival of the mainland colonizer when we see a rapid rate of adaptation and a high degree of plasticity. After the major change had taken place we can note that the rate of change decreased considerably. The first step is a change in evolutionary direction as from water to land, low gear locomotion, etc. The next is a further continuation of the existing direction of change; e.g. the gradual increase of hypsodonty in the murids of Gargano (Freudenthal, 1971). Of course, several variations on the scheme are possible, such as those given by Sondaar and Boekschoten (1967), which are dependent upon the degree to which the island environment differed from the mainland. For most insectivores this is little, the predator is the same (bird of prey) and so is the food. Thus, the changes in these mammals are small. In elephants there is perhaps some indication that the primary change did not proceed as fast as in other mammals, such as hippo and deer, because on Sicily probable transitional forms are found. Perhaps the changes proceeded more slowly since the rate of reproduction is very low in this group.

The reasons for extinction may be multiple. Important causes are submergence of the island, or a part of it, under the sea and also the establishment of a landbridge with the mainland. The island forms would have been easy prey for mainland predators. There are clear examples of both as the island of Gargano was mostly submerged in the Pliocene (Freudenthal, 1971), while the mainland connection of Sardinia at the end of the Miocene must have finished off the endemic island fauna. In the Late Pleistocene of Japan a balanced mainland fauna is found; this points to a land connection in this period. The migration of the mainland mammals to the former islands caused the extinction of

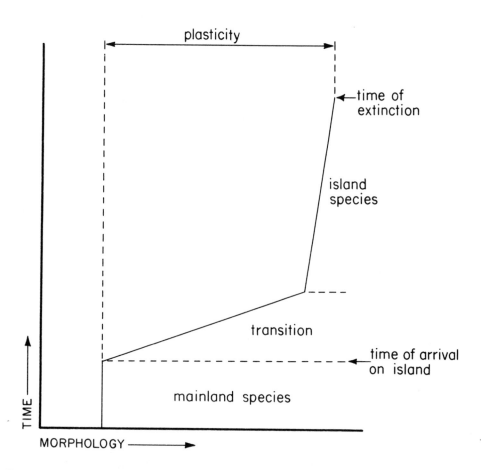

Fig. 9.  Schematic diagram depicting the stages in the evolution of island mammals (after Sondaar and Boekschoten, 1967).

the endemics. In the Pleistocene of the Mediterranean man, and
other organisms introduced by him, may be considered the main
cause of extinction of these island mammals (Fig. 10).


## DISCUSSION AND CONCLUSIONS

The effect of isolation on mammal evolution is considered on
small to medium-sized Pleistocene "oceanic" islands. A number of
species from several islands are studied. As the invading species
and the environment were mostly similar, it is possible to construct a model for the evolutionary effect. It is shown that the
same large mammals dominate the fauna: elephant, deer and hippo
in the Pleistocene of the Mediterranean; deer and elephant on
some islands from Japan and S.E. Asia. The most striking character
of the faunas is the lack of large carnivores. The uniform
composition of the fauna can be explained in that only those
large mammals with good swimming abilities could reach the
islands. This is the case with elephant, deer and Hippopotamus.
Smaller mammals could reach the islands by natural rafts; e.g.
the rare cases of small carnivores, such as the viverrids in
Madagascar (Carlquist, 1965), or by swimming like the otter
(Symeonides and Sondaar, 1975).

Endemic forms are only found on islands when the distance to
the mainland was not too small and the arrival was an exception
via a sweepstakes route. There is a degree of similarity between the endemic fossil forms from the different islands in such
things as small size, low gear locomotion and land adaptation of
semi-aquatic mammals like the hippo. This is explained by the
parallel evolutionary effects of isolation. The endemic island
species must have evolved from a small founder population with
little genetic variation which, however, was different on each
island. The similarities in morphology cannot therefore be explained by a direct relationship nor as being due to an identical
genetic composition but must be considered to be parallel adaptations to similar environments which differed essentially from the
mainland in the lack of large carnivores.

No clear intermediates are found between the island and
mainland forms although in some cases the fossil record is good.
It is supposed that the primary change which was one of different
evolutionary direction must have proceeded quickly. This transformation like the change from water to land in the hippo was
followed by a trend which might be explained as a perfection of
the first change like increase of hypsodonty in the murids of
Gargano. Some mammals from the mainland with a high degree of
specialization in behavior and adaptation proved still to possess
a great plasticity and could change drastically. The best

Fig. 10. An artist's view of the effect of a landbridge on an island fauna.

example of this is the dwarf hippo from Cyprus with a lophodont
dentition and a locomotion adapted for climbing and walking in
which it differs completely from its semi-aquatic mainland
ancestor, H. amphibius.

In island mammals we can note some characters which are more
common in their more distant phylogenetic relatives (grandfather)
than in their direct ancestors (father) which suggests an inversion of evolution. This might better be interpreted thus: characters which resemble those of their "grandfather" are newly
attained and are adaptations to a different environment which
resembled in some aspects that of its "grandfather" rather than
its "father."

Deleterious characters which occur only rarely in the mainland population might be advantageous in the adaptation to an
island environment and might then become a stable character in the
island form; e.g. the lack of the upper fourth premolar in the
dwarf Cyprus hippo. In general, the lack of a tooth in the upper
jaw will allow excessive outgrowth of the opposing tooth in the
lower jaw, and mastication may be obstructed. Many cases of this
defect are known in recent and fossil mammals (Hussain and
Sondaar, 1967). In the dwarf hippo from Cyprus, however, the
converse is true. The lack of the upper fourth premolar was
necessary for the changed chewing habit (Fig. 6) and had an
adaptational function (Boekschoten and Sondaar, 1972).

Certain changes are sometimes the consequence of other
changes. Reduction in size caused the loss of pneumatic bone
tissue in the skull of dwarf elephants as it lost its function
(Fig. 3). The pneumatic bone tissue was also not present in the
small ancestor of the elephants. Also the low lamellar formula of
dwarf elephants, more common in geologically older giants from
the mainland, might have had a functional significance. The
changes cannot be explained as paedomorphic nor by resemblance to
an early ontogenetic stage because the total shape is different.
In early ontogenetic stages in ungulates the distal parts of the
limbs are full-grown first while in the island forms this part is
mainly shortened for "low gear locomotion."

Changes such as small size and fusion of footbones cannot be
explained as degenerative processes since one should then expect
more random changes. In trying to explain the selection mechanism
behind the evolution of island mammals it is suggested that
malnutrition and periodical overpopulation probably caused by the
lack of large predators led to overgrazing with the consequent
destruction of vegetation. After this the population density
probably decreased drastically and under those circumstances
small size with low gear locomotion might have been advantageous

in the search for food, and selection may have favored those
characters.  It is also supposed that the populations fluctuated
greatly as the fauna was unbalanced and this might have influenced
the rate of evolution.

These changes and even more strikingly the land adaptations
of a semi-aquatic animal like the hippo on the islands point to a
change in evolutionary direction starting from a small population
in a new environment.  This proceeded relatively quickly and was
followed by a trend which might be considered as a further more
gradual perfection of the first change.  The high degree of
parallelism between forms from different islands emphasizes that
there are only a few definite lines of evolutionary change
possible in adaptation to a certain environment. The model outlined for the evolution of island mammals might also be valid for
the mainland.  On islands, the "founder population" encountered a
new environment and sometimes succeeded in adapting to it.  On the
mainland the environment can change through many factors like
climatological changes, change of paleogeography, new arrivals,
etc., and consequently an existing, flourishing population might
be exiled to an environment to which it was not primarily adapted.
The population density would decrease drastically to a small
number to which the "founder principle" may apply.  The new
environment would exert a strong selective pressure and it is not
unlikely that, if the population survived, evolution would proceed quickly and might have a different direction as an adaptation
to the new environment.  We could take this comparison with the
evolution of island mammals even further.  The same change in
environment would be evident over a greater area on the mainland
and would reduce local populations of the same form.  Isolated
local populations might consequently evolve separately and produce
a parallel outcome in form and/or function comparable with the
evolutionary effect on islands.  This might be a reason why
"after major evolutionary change" we see a great radiation in the
group with the new character which could be caused by separate
parallel evolution in the same taxon.  In other words, a newly
obtained character which made the group better adapted to a new
environment need not be restricted to one evolutionary event, but
might have happened more than once in time and space.

If we take the evolution of the horse as an example we see a
big radiation after this family entered the grass biotope in which
there are several parallel developments like Hipparion and
Neohipparion.  In Parahippus from Florida the carpal bones are
built on a different plan which parallels in function later horses.
Leinders and Sondaar (1974) suggest that this branch might have
evolved separately.  This implies a polyphyletic origin and it is
questionable whether we should use the generic name of Parahippus
for these horses.  Parallel evolutionary developments are probably

more common than one would expect as they are difficult to detect
in fossils where only hard parts can be studied. Gould (1971)
reached similar conclusions studying the Pleistocene land snails
from the Bermudas and emphasized the taxonomic problems of convergence which might confound the efforts of the taxonomist working on organisms which leave only relatively simple hard parts as
fossil remains.

## SUMMARY

The effect of insularity on mammal evolution is approached by
considering endemic fossil mammals of smaller "oceanic" islands.
It is shown that in general the same mammalian types will arrive
on the islands by sweepstakes dispersal. Among the larger mammals
this includes the elephant, hippopotamus and deer; all mammals
known to have good swimming capabilities. The evolutionary changes
these mammals underwent follow parallel patterns on different
islands restricted neither to geological age nor geography and can
be explained as an adaptational process to an island environment
lacking in large carnivores. The changes vary to a great degree:
size reduction, low gear locomotion, increase of hypsodonty, etc.,
and the adaptational significance of these changes is discussed.
It is suggested that the evolution of island mammals was clearly
directed involving a strong selective pressure which was similar
on different islands. There is fossil evidence indicating mass
starvation and malnutrition which could point to overpopulation
through lack of predators and the consequent overgrazing and
destruction of otherwise suitable biotopes. After this the
population density declined drastically which might have been an
important factor promoting the rapid evolution of island mammals.
An evolutionary model for island mammals is proposed and it is
suggested that this model might also be valid for the mainland.
The model starts from the principle that after a change in the
environment population density decreases which influences the
rate of evolution. On the mainland small local populations might
consequently evolve separately and produce a parallel outcome in
form and/or function and the character which made the group better
adapted to a new environment need not be restricted to one evolutionary event in time and geography.

## ACKNOWLEDGEMENTS

The writer is indebted to Prof. N. Symeonides, Drs. G.J.
Boekschoten, D. Mayhew and J. de Vos for reading the manuscript
and participating in many inspiring discussions on island mammals
in general. Further I wish to thank Dr. H. Otsuka for information
about the Japanese fauna. Figures 1-5 and 7-10 were drawn by
Mr. W. Smit, and fig. 6 by J. Reumer for which I am grateful.

## REFERENCES

Audley-Charles, M.G. and Hooijer, D.A., 1973, Relation of
     Pleistocene migrations of pygmy stegodonts to island arc
     tectonics in eastern Indonesia. Nature, 241:197-198.
Accordi, B., 1955, Hippopotamus pentlandi v. Meyer del
     Pleistocene della Sicilia. Palaentographia Italica, 50:1-52.
Accordi, F.S. and Palombo, M.R., 1971, Morfologia endocranica
     degli elefanti nani pleistocenici di Spinagallo (Siracusa)
     e comparazione con l'endocranio di Eldphas antiquus. Rend.
     Acc. Naz. Lincei, Roma, 51(1-2):111-124.
Ambrosetti, P., 1968, The Pleistocene dwarf elephants of
     Spinagallo (Siracusa, South eastern Sicily). Geol. Rom.,
     Roma, 7:277-398.
Andrews, C.W., 1915, A mounted skeleton of Myotragus balearicus.
     Zeol. Magazine, London, 6(2):337-339.
Baker, H.G. and Stebbins, G.L., eds., 1965, The genetics of
     colonizing species. Academic Press, New York.
Ballmann, P., 1973, Fossile Vögel aus dem Neogen der Halbinsel
     Gargano (Italien). Scripta geologica, Leiden, 17:75.
Barone, R., 1968, Anatomie comparée des mammifères domestiques.
     Laboratoire d'Anatomie école nationale veterinaire, Lyon.
     2 tomes.
Basler, E., 1972, Strategie des Fortschritts. Umweltbelastung,
     Lebensraumverknappung und Zukunftserforschung. Huber & Co.,
     Frauenfeld, p. 140, 15 afb.
Bate, D.M.A., 1905, Four and a half month in Crete in search of
     Pleistocene mammalian remains. Geol. Magazine, London,
     5(2):193-202.
Bate, D.M.A., 1906, The pigmy Hippopotamus of Cyprus. Geol.
     Magazine, London, 5(3):241-245.
Bigalke, R.C., 1972, Evolution, mammals and southern continents.
     Eds., Allen Keast, Frank C. Efk, Bentley Glass. State
     University of New York Press, Albany.
Boekschoten, G.J. and Sondaar, P.Y., 1972, On the fossil mammalia
     of Cyprus. Proc. Koninkl. Ned. akademie van wetensch, Am-
     sterdam, series B, 75(4):306-338.
Carlquist, S., 1965, Island life, a natural history of the islands
     of the world. Natural History Press, Garden City, New York,
     451 p.
Carrington, R., 1962, Elephants. Penguin Books, Harmondsworth,
     Middlesex, 285 p.
Coryndon, S.C., and Coppens, Y., 1975, Une espèce nouvelle
     d'Hippopotame nain du Plio-pléistocène du bassin du lac
     Rodolphe (Ethiopie, Kenya). C.R. Acad. Sc., Parris, 280
     (série D):1777-1780.
Darlington, P.J., 1957, Zoogeography: The geographical distribu-
     tion of animals. Harvard University.

Dehaut, E.G., 1920, Contribution a l'étude de la vie vertébrée
    insulaire dans la région méditerranéenne occidentale et
    particulièrement en Sardaigne et en Corse. Le Chevalier,
    ed., Parigi.
Dobzhansky, T. and Pavlovsky, O.A., 1957, Indeterminate outcome
    of certain experiments on Drosophila populations. Evolu-
    tion, 7:198-210.
Falconer, H., 1862, On the fossil remains of Eléphas melitensis,
    an extinct pigmy species of Elephant; and of other mammalia,
    etc., from the ossiferous caves of Malta. Palaeontological
    Memoirs and Notes, London, 2:292-308.
Freudenthal, M., 1971, Neogene vertebrates from the Gargano
    Peninsula, Italy. Scripta Geologica, Leiden, 3:1-10.
Freudenthal, M., 1972, Deinogalerix koenigswaldi nov. gen.,
    nov. spec., a giant insectivore from the Neogens of Italy.
    Scripta Geologica, Leiden, 14:1-10.
Gould, S.J., 1966, Allometry and size in ontogeny and phylogeny.
    Biol. Reviews, Cambridge Phil. Soc., 41:587-640.
Gould, S.J., 1971, Geometric similarity in allometric growth:  A
    contribution to the problem of scaling in the evolution of
    size. Am. naturalist, Chicago, 105(no. 942):113-136.
Gould, S.J., 1971, Precise but fortuitous convergence in Pleisto-
    cene land snails from Bermuda. Journal of Paleontology, 45
    (no. 3):409-418.
Grzimek, B., 1970, Grzimeks Tierleben, Enzyklopädie des Tierreiches,
    Kindler Verlag AG, Zürich.
Haldane, J.B.S., 1956, The relation between density regulation
    and natural selection. Proc. Roy. Soc., London (B), 145:
    306-308.
Harris, C.J., 1968, Otters, a study of the recent Lutrinae, 397 pg.
Hasegawa, Y., Otsuka, H. and Nohara, T., 1973, Fossil vertebrates
    from the Miyako Islands. Studies of the palaeovertebrates
    faunas of Ryukya Islands, Japan, Part I. Mem. Natur. Sci.
    Mus., Tokyo, 6:39-52.
Hooijer, D.A., 1962, A fossil ground sloth from Curacao, Netherland
    Antilles. Proc. Koninkl. Nederl. akademie van wetensch.,
    Amsterdam, series B, 65(no. 1):46-60.
Hooijer, D.A., 1972, Stegodon trigonocephalus florensis Hooijer
    and Stegodon timorensis Sartono from the Pleistocene of Flores
    and Timor. Proc. Koninkl. Nederl. akademi van wetensch.,
    Amsterdam, series B, 75:12-33.
Hooijer, D.A., 1973, Reuzenschildpadden en dwergolifanten. Muse-
    ologia, Amsterdam, no. 1, 9-14.
Hooijer, D.A., 1975, Quaternary mammals west and east of Wallace's
    Line. Netherlands Journal of Zoology, 25(1):46-56.
Hussain, S.T. and Sondaar, P.Y., 1968, Some anomalous features in
    Euroasiatic Hipparion dentition. Proc. Koninkl. Nederl.
    akademie van wetensch., Amsterdam, series B, 71(no. 2):137-143.
Joleaud, L., 1920, Contribution à l'étude des Hippopotames fossiles.
    Bull. Soc. Géol. France, series 4, 20:13-26.

Koenigswald, G.H.R. Von, 1956, Fossil mammals from the Philippines. Proc. Fourth Far-Eastern Prehistory Congress, Quezon City, 1:339-362.

Kuss, S.E., 1965, Eine pleistozäne Säugetierfauna der Insel Kreta. Ber. Naturf. Ges., Freiburg i. Br., 55:271-348.

Kuss, S.E., 1967, Pleistozäne Säugetierfunde auf den ostmediterranen Inseln Kythera und Karpathos. Ber. Naturf. Ges., Freiburg i. Br., 57:207-216.

Kuss, S.E., 1969, Die paläolithische osteokeratische "Kultur" der Insel Kreta/Griechenland. Ber. Naturf. Ges., Freiburg i. Br., 59:137-168.

Kuss, S.E., 1969, Die erste pleistozäne Säugetierfauna der Insel Kasos (Griechenland). Ber. Naturf. Ges., Freiburg i. Br., 59:169-177.

Kuss, S.E., 1970, Abfolge und Alter der pleistozänen Säugetierfaunen der Insel Kreta. Ber. Naturf. Ges., Freiburg i. Br., 60:35-83.

Kuss, S.E., 1973, Die pleistozänen Säugetierfaunen der ostmediterranen Inseln. Ihr Alter und ihre Herkunft. Ber. Naturf. Ges., Freiburg i. Br., 63:49-71.

Leinders, J.J.M. and Sondaar, P.Y., 1974, On functional fusions in footbones of Ungulates. Z. f. Säugetierkunde, Hamburg, 39 (Heft 2):109-115.

Leonardi, P., 1954, Les Mammifères nains du Pleistocène méditeranéen. Annal. Paleontol., Paris, 40:189-201.

Maglio, V.J., 1973, Origin and evolution of the Elephantidae. Transactions Am. Phil. Soc., Philadelphia, new series, 63 (part 3):149.

Mahe, 1972, The Malagasy subfossils. Biogeography and ecology in Madagascar eds. R. Battistini and G. Richard-Vindard. Junk, The Hague, 339-366.

Malatesta, A., 1970, Cynotherium sardous Studiati, an extinct canid from the Pleistocene of Sardinia. Mem. Ist. It. Paleont. Umana., Roma, I:1-72.

Matthew, W.D. and de Paula Couto, C., 1959, The Cuban edentates. Bull. Amer. Mus. Nat. Hist., 117:1-56.

Mayr, E., 1942, Systematics and the origin of species. Columbia University Press, New York.

Melentis, J.K., 1961, Studien uber fosille Vertebraten Griechenlands. Die Dentition der pleistozanen Proboscidier der Beckens von Megalopolis im Peleponnes. Ann. Geol. Pays. Hellen., 12: 153-262, 20 figs., 17 pls., 19 tables.

Meulenkamp, J.E., 1971, The Neogen in the Southern Aegean Area. In A. Strid. Evolution in the Aegean, Opera Botanica, 30: 5-12.

Millot, J., 1952, La faune Malgache et le mythe Gondwanien. Mém. Inst. Sci. Madagascar, Sér. A, 7:1-36.

Mitzopoulos, M.K., Über einen pleistozänen Zwergelefanten von der Insel Naxos (Kykladen). Prakticka Akad. Athenon, 36:332-340, 1 fig., 2 pls.

Naumann, E., 1890, Stegodon mindanensis, eine neue Art von Ueber-
    gangs-Mastodonten.  Zeitschr. Deut. Geol. Ges., 42:166-169.
Otsuka, H., 1972, Elaphurus shikamai Otsuka (Pleistocene cervid)
    from the Akashi Formation of the Osaka Group, Japan, with
    special reference to the genus Elaphurus.  Bull. Nat. Sc.
    Museum, Tokyo, 15(no. 1):197-210.
Psarianos, P., 1953, Ueber das Vorkommen von Hippopotamus auf
    Kaphallinia (Griechenland).  Praktika Akad. Athinon, 28:
    408-412.
Recorini G., Rage, J.C. and Thaler, L., 1974, La formation con-
    tinentale de Capo Mannu, sa faune de vertébrés pliocènes et
    la question Messinien Sardaigne.  305-321.  Rendiconti del
    Seminaria della Facolta di Scienze del l'Universita di
    Cagliari (suppl. vol. XIII, 1973).
Sartono, S., 1969, Stegodon timorensis:  A pygmy species from
    Timor (Indonesia).  Proc. Kon. Ned. akademie v. vetensch.,
    Amsterdam, series B, 72:192-202.
Simpson, G.G., 1940, Mammals and land bridges.  Jour. Washington
    Acad. Sci., 30:137-163.
Simpson, G.G., 1965, The geography of evolution.  Chilton Books,
    Philadelphia and New York.
Sondaar, P.Y., 1971, Paleozoogeography of the Pleistocene mammals
    from the Aegean.  Opera Botanica, no. 30.  In A. Strind.
    Evolution in the Aegean, 65-70.
Sondaar, P.Y. and Boekschoten, G.J., 1967, Quaternary mammals in
    the South Aegean Island arc, I and II.  Proc. Kon. Ned. akad.
    v. wetensch., Amsterdam, series B, 70:556-576.
Stock, C. and Furlong, E.L., 1928, The Pleistocene elephants of
    Santa Rosa Island, California.  Science, 68:140-141.
Sutcliffe, A.J., 1973, Similarity of bones and antlers gnawed by
    deer to human artefacts.  Nature, 246(no. 5433):428-430.
Symeonides, N., 1972, Die Entdeckung von Zwergelefanten in der
    Höhle, "Charkadio" auf der Insel Tilos.  Annales Géol. Pays
    Helléniques, 24:445-461.
Symeonides and Marinos, G., 1973, Erstmalige Funde von Zwergelefanten
    auf der Insel Rhodos.  Anz Österr. Akad. Wiss. Mathem.  Nat.
    Kl, Ig 1973, Wien.
Symeonides, N., and Sondaar, P.Y., 1975, A new otter from the
    Pleistocene of Crete.  Annales géol. des pays Helléniques,
    Athènes, 27:11-24.
Takai, F., 1939, On some Cenozoic mammals from Japan.  Jour. Geol.
    Soc. Japan, 45(no. 541):745-763.
Thaler, L., 1973, Nanisme et gigantisme insulaires.  La Recherche,
    4(no. 37):741-750.
Tokunaga, S., 1936, Bone artifacts used by ancient man in the
    Riukiu Islands.  Proc. Imp. Acad., Tokyo, 12(no. 1):353-354.
Tokunaga, S., 1941, A fossil elephant tooth discovered in Miyako-
    jima, an island of the Ryukyu Archipelago, Japan.  Proc.
    Imp. Acad., Tokyo, 16(no. 3):122-124.

Valverde, J.A., 1964, Remarques sur la structure et l'évolution
    des communautés de vertébrés terrestres.  La terre et la vie,
    Paris, 111:121.
Vaufrey, R., 1929, La question des isthmes méditerranéens
    pleistocènes.  Rev. Géogr. Phys. Géol. Dynam. Paris.

# EVOLUTION OF THE CARNIVOROUS ADAPTIVE ZONE IN SOUTH AMERICA

Larry G. MARSHALL

Department of Geology, H.S. Colton Research Center
Museum of Northern Arizona
Flagstaff, Arizona   86001 U.S.A.

## INTRODUCTION

The role assumed by an animal group in a community is best understood when the structure of the community is viewed in historical perspective (MacArthur, 1972; Pianka, 1974).  In many environments it is not physiological and behavioral limitations which guide a group's evolutionary strategy, but the opportunity to exploit available adaptive zones, which because of the nature of the fauna, were vacated (Hecht, 1975:248).  It is often assumed or implied that a particular systematic group can, in its adaptive radiation, fill all the available niches within a particular adaptive zone (Hecht, 1975:247).  The type of analysis of a "taxocene" (Whittaker, 1972:218) which promotes these views has obvious shortcomings.  It is now evident that mammals, for example, were unable to equally exploit and partition available adaptive zones on every continent.  The failure of mammals to do this to a complete degree in Australia (see Hecht, 1975:247) and in South America (see below) has some historical basis in the development of the stratification of the vertebrate fauna.

The term _adaptive zone_ is used as defined by Van Valen (1971: 421)..."the _niche_ of any taxon, especially a supraspecific one." The successive replacement through time of different animals or groups of animals stemming from different lineages but occupying the same adaptive zone has been called "Eco-replacement" or "Evolutionary Relay."

Van Valen (1971:421) has distinguished adaptive zones from the ways of life of taxa.  Adaptive zones are part of the environment

and exist independently of a taxon to exploit them. The way of
life of a taxon must be fitted to some adaptive zone, but may
occupy only a small part of a larger, relatively undivided zone.
We cannot infer the boundaries of an adaptive zone from the boundaries of an animal's way of life.

Here I am particularly concerned with terrestrial animals
which ate other animals, either by active predation or by scavenging. For purposes of discussion I have subdivided the terrestrial
carnivore adaptive zone into four categories: that for 1) a small
to medium sized omnivore or carnivore, 2) a large omnivore,
3) a large carnivore, and 4) a saber-tooth carnivore (Fig. 1).

## THE CARNIVOROUS ADAPTIVE ZONE IN SOUTH AMERICA

It is well established that South America was an island continent during most, if not all, of the Tertiary Epoch (65 through
2 mybp) (Marshall, in press). As a consequence of this isolation
marsupials evolved to fill the ecological roles of terrestrial
mammalian carnivores in the Cenozoic of South America (and
Australia), an adaptive zone then filled on northern continents
and Africa by placental carnivores (first certain creodonts {to
include Hyaenodontidae and Oxyaenidae} and members of other orders,
and later members of the Carnivora {to include miacids}). These
faunal differences are clearly related to the long isolation of the
South American continent and to the initial presence of marsupials
and absence of placental carnivores.

The large predaceous South American marsupials are known from
beds of Riochican (late Paleocene) through Chapadmalalan (latest
Pliocene) age. In beds of pre-Huayquerian age marsupials were the
only medium to large mammalian carnivores on the South American
continent. Some dog-like borhyaenids (family Borhyaenidae) were
of small to medium size, had omnivorous to carnivorous diets and
were semiarboreal. Others were large terrestrial omnivores-
carnivores or strictly carnivores. All were short legged and none
display marked cursorial adaptations. One specialized group, the
Pliocene saber-tooth thylacosmilids (Family Thylacosmilidae), converged upon placental machairodont saber-tooth cats.

Extinction of the Borhyaenidae was not restricted to a short
interval of geologic time (Marshall, in press). There is compelling
evidence documenting a protracted decline of the group beginning
in the Deseadan (early Oligocene), with accelerated extinctions at
the end of the Santacrucian (early Miocene) and Montehermosan
(late Pliocene). Most of these events are correlated with the
appearance and/or diversification of other animal groups.

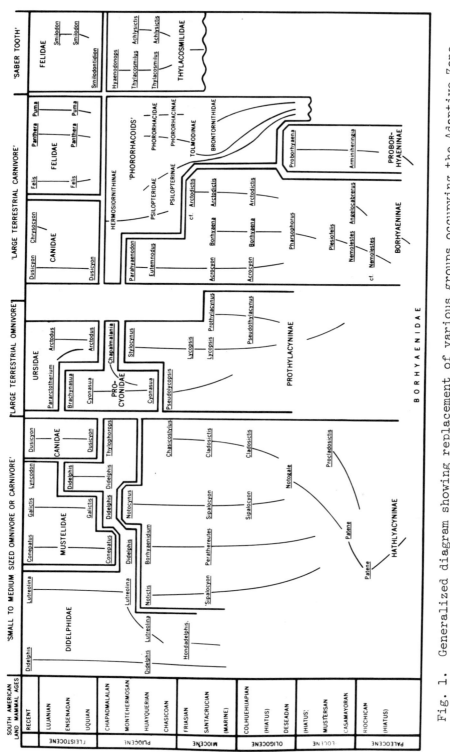

Fig. 1. Generalized diagram showing replacement of various groups occupying the Adaptive Zone for 1) a small to medium sized omnivore or carnivore, 2) a large terrestrial omnivore, 3) a large terrestrial carnivore, and 4) a saber-tooth. For explanation see text. Stratigraphic range of non-marsupial mammalian genera follows Pascual et al. (1966, chart opposite p. 202).

The Proborhyaeninae reached and maintained a large size in the early Tertiary.  The last member of this group, Proborhyaena gigantea, is the largest species of borhyaenid known with adults having skulls up to two feet in length.  Contemporaneous members of the Borhyaeninae were also large, but dwarfed in comparison to individuals of Proborhyaena. With the disappearance of Proborhyaena at the end of the Deseadan, the adaptive zone for a large terrestrial mammalian carnivore in South America became vacated.

Arctodictis, a form derived from a smaller Deseadan member of the genus Pharsophorus, appears in beds of Colhuehuapian (late Oligocene) age.  Individuals of Arctodictis are the largest known members of the subfamily Borhyaeninae and appear immediately following the disappearance of Proborhyaena. Adults of Arctodictis were smaller than those of Proborhyaena, yet were the giants of their time and are the largest post-Deseadan borhyaenids known. Arctodictis converged upon Proborhyaena in its large size, in development of large massive canines, and in fusion of the mandibular symphysis in adults.  Proborhyaena appears to have occupied an adaptive zone in the Deseadan which was later partially filled by Arctodictis in the Colhuehuapian and Santacrucian. Whether Arctodictis evolved in response to the disappearance of Proborhyaena, or if it actively replaced Proborhyaena is not known.

A major change occurred in the large native herbivore fauna between the Deseadan and Colhuehuapian, and this change may be partially (or wholly) responsible for the disappearance of the Proborhyaeninae.  At the close of the Deseadan the condylarths declined sharply in diversity, and the Pyrotheria and Isotemnidae became extinct.  Many groups make their first appearance in the Deseadan, including mesotheriids, homalodotheriids, leontiniids and toxodontids among the Notoungulata, and mylodontids among the Edentata.  All of these groups radiated by the Colhuehuapian and megatheriids first appear in beds of that age (Patterson and Pascual, 1972:283).

A marked feature of borhyaenid evolution is the disappearance of large forms in the later Tertiary (post-Santacrucian).  This is best exemplified in the subfamily Borhyaeninae, where three taxa are known from the Santacrucian (Acrocyon, Borhyaena, and Arctodictis), but only one line represented by Parahyaenodon (Montehermosan) continued into the Pliocene, and another (cf. Arctodictis sp.) survived into the Friasian (Marshall, 1976b).

The other occupants of the terrestrial carnivore adaptive zone were members of a group of large cursorial ground birds, the phororhacoids, distant relatives of the living South American Cariama (Order Gruiformes).  Three families, assorted into medium, large and gigantic size groups by mammalian standards, are recognized.  They are known to have ranged from Deseadan through

Montehermosan. The Brontornithidae are as yet unknown subsequent
to the Santacrucian. The other two families (Psilopteridae and
Phororhacidae, both with two subfamilies) appear in the Deseadan
and are well represented in the Pliocene (Patterson and
Kraglievich, 1960). Although not yet recorded in the Pleistocene
of South America, there is proof that they survived into that
epoch in North America as remains of one of them has been described from the early Pleistocene of Florida (Brodkorb, 1963).
Two of the families, Phororhacidae and Psilopteridae, were swift
and rather lightly built; and the third, Brontornithidae, included
ponderous forms with large massive beaks. The former two families
were evidently the dominant cursorial carnivores of their time
(Patterson and Pascual, 1972:262).

The phororhacoids may have actively contributed to the disappearance of such large borhyaenids as Proborhyaena, and checked
size increase in Arctodictis, preventing its descendents from
attaining a size comparable to its Deseadan counterpart, Proborhyaena. These points are clearly speculative, although it does
appear that these birds either actively or passively replaced large
borhyaenids in the later Tertiary, at least on the savanna grasslands of Argentina.

Differences between the La Venta fauna (Friasian) of Colombia
and the Santa Cruz faunas (Santacrucian) of Argentina lend support
to this possibility. The faunas from these areas suggest marked
differences in habitat, with the Santa Cruz being a savanna grassland and the La Venta being more densely forested. cf. Arctodictis
sp. is known from the La Venta fauna, but large phororhacoids are
not (Marshall, 1976b). This large borhyaenid survived into a
later fauna in more forested areas in the absence of phororhacoids.

By the close of the Friasian the large herbivorous lineages
represented by the Condylartha, Astrapotheria and Leontiniidae,
became extinct. This coincided with the last appearance of the
giant borhyaenid cf. Arctodictis sp. and the semiterrestrial deep-
skulled crocodyle Sebecus (Langston, 1965:141). Sebecus, like the
large extinct varanid Megalania in the Pleistocene of Australia
(see Hecht, 1975), may have occupied the highest trophic level in
the food pyramid in the La Venta fauna.

Only large forms of the Santacrucian Prothylacyninae with
omnivorous habits (e.g. Stylocynus) continued into later faunas.
In contrast, Prothylacynus patagonicus, a form having a number of
specializations suggestive of a more carnivorous diet, is unknown
after the Santacrucian.

All species of Prothylacyninae and Borhyaeninae for which
post-cranial skeletons are known were completely terrestrial. This

fact, coupled with dental specializations favoring a meat eating diet (all Borhyaeninae, some Prothylacyninae), suggests potential competition with phororhacoids.

Practically all Santacrucian lineages of Hathlyacyninae can be traced into later faunas. These animals were of small to medium size and were adapted for generalized to carnivorous feeding habits. They had semiarboreal habits as evidenced by structure of the pes and manus, suggesting little if any potential competition with the large cursorial phororhacoids.

The decline to extinction of the more omnivorous members of the Prothylacyninae, namely Pseudolycopsis and Stylocynus, appears correlated with arrival into South America of the family Procyonidae, Order Carnivora. Procyonids came from North America as waif-immigrants sometime in the early to mid-Pliocene, and first appear in beds of Huayquerian age in Argentina. The earliest known form was Cyonasua, a medium sized omnivore (Patterson and Pascual, 1972). The last appearance of the borhyaenid Pseudolycopsis in beds of Chasicoan age and the first appearance of Cyonasua in subsequent faunas appears significant.

Stylocynus was large and certainly the most bear-like of all known Borhyaenidae. This animal appears in beds of Montehermosan age of Argentina and probably represents the end of a lineage extending back to a Lycopsis-like form in the Santacrucian.

In the Chapadmalalan appeared a large specialized procyonid, Chapalmalania, an enormous omnivore, markedly bear-like in many features. In fact, it was originally described as a bear, although new material enabled Kraglievich and Olizabal (1959) to make a correct familial assignment. In their respective faunas Stylocynus and Chapalmalania were the large mammalian omnivores and converged broadly in feeding specializations. No other animal in either fauna approached the size or feeding adaptations of these taxa.

Factors influencing the decline of the Hathlyacyninae are complex, although a major feature appears to involve an ecological interplay with large didelphids. For example, small to medium sized borhyaenids (Hathlyacyninae) are unknown in the La Venta fauna (Friasian) of Colombia. Instead, there occurs a very large didelphid, Hondadelphys, which is similar in size and shares dental specializations with Santacrucian Hathlyacyninae, such as Sipalocyon and Cladosictis. Hondadelphys clearly converged upon small to medium sized borhyaenids in both dental and basicranial features (Marshall, 1976b). Hathlyacynines were certainly present during Friasian times as many Santacrucian lineages can be traced into post-Friasian faunas. The adaptive zone occupied by small borhyaenids in the Santacrucian of Argentina was thus filled by large didelphids in the Friasian of Colombia. This evidence suggests possible competitive exclusion between these groups, and may explain the absence of large Didelphinae in the taxonomically rich Santa Cruz faunas of Argentina.

Two lines of Santacrucian Hathlyacyninae survived into the Huayquerian, Notictis and Borhyaenidium, and another, Notocynus, into the Montehermosan. A number of large and specialized Didelphinae appear in beds of Huayquerian age in Argentina, and increase in diversity through the Chapadmalalan.

Notictis is similar in size and dental structure to the didelphid Lutreolina. Both occur in beds of Huayquerian age, although in different faunas; their distribution is apparently allopatric. Lutreolina is known from deposits in Catamarca Province, while Notictis is known only from near Parana, Entre Rios Province. Notictis is unknown from later faunas, while Lutreolina persists and is recorded from the Montehermosan and Chapadmalalan, and is a common element in the Recent temperate-grassland fauna of Argentina, Paraguay, Uruguay, and southern Brazil.

The genus Didelphis first appears in beds of Huayquerian age in the Province of Catamarca, Argentina. The one recognized species, D. pattersoni (see Simpson, 1974:5), occurs together with a species of Borhyaenidium (Notocynus of Riggs and Patterson, 1939:149). These species differ in size and basic feeding adaptations (Didelphis is clearly more omnivorous) such that their joint occurrence is not unexpected.

A number of species of Didelphis were described from the Monte Hermoso Formation (Montehermosan), Buenos Aires Province, Argentina. Only two of these species appear valid: D. biforata (including Paradidelphys parvula), and D. inexpectata (including all others - for synonymies see Simpson, 1972:5). Didelphis biforata has dental specializations more indicative of a meat eating diet than does D. inexpectata, and it broadly resembles Borhyaenidium in size and dental features. A correlation clearly exists between the last appearance of Borhyaenidium and the subsequent appearance of D. biforata.

The borhyaenid Notocynus is larger than Notictis, Borhyaenidium, and contemporaneous didelphids. In the Chapadmalal Formation, between Mar del Plata and Miramar, a multitude of large and specialized didelphids have been found, including Lutreolina crassicaudata, Didelphis brachyodonta, the very large D. reigi, and the enormous, coyote-sized, Thylophorops chapalmalensis which is the largest known species of late Cenozoic Didelphidae (Simpson, 1972). Both D. reigi and T. chapalmalensis rival Notocynus in size and dental

specializations.  The last appearance of <u>Notocynus</u> immediately
precedes the first appearance of these very large and specialized
didelphids.

The repetition in last appearance of small borhyaenids followed by first appearance of similarly adapted didelphids, suggests
ecological replacement of one group by another.  Extinction of
small to medium sized borhyaenids may have resulted from active
competition with large Didelphinae or, alternatively, the smaller
borhyaenids, for some as yet unknown reason, became extinct and
were replaced by large Didelphinae.  The replacement of one group
by the other may have, thus, been either active or passive.  Whatever the cause, the change was affected over a considerable period
of time.  It began in the Huayquerian and was completed by the
Chapadmalalan, the marked change occurring between the Montehermosan and Chapadmalalan.

Large Chapadmalalan didelphines are basically similar in
dental structure, and differences in size represent important
criteria in distinguishing the taxa.  Their diversity in size but
conservatism in structure suggest a rapid and almost explosive
radiation at or shortly before this time.  A radiation possibly
stemming from one or two ancestral forms in the Chasicoan or
Huayquerian and triggered by the vacated or opening adaptive zone
for small to medium sized terrestrial or semi-arboreal mammalian
carnivore, and linked with the mid- to late Pliocene decline to
extinction of the borhyaenid subfamily Hathlyacyninae.  The first
appearance of cricetid rodents in South America at this time
(first known forms appear in beds of Montehermosan age in Argentina) may also have had an influence on the carnivorous groups
which came to exploit this new food source.

During the early and middle Pliocene, faunal interchange between North and South America occurred by waif-dispersal over the
then narrow marine sea-way (Bolivar geosyncline) extending across
the northwestern corner of Colombia.  A limited interchange during
this period is evidenced by the presence of procyonids in beds of
Huayquerian age in South America (see above), and megalonychid
ground sloths in beds of early Hemphillian age in North America
(Hirschfeld and Webb, 1968).

Sometime during the late(?) Pliocene, the Panamanian land
bridge came into existence, forming a direct dry land connection
between North and South America (Patterson and Pascual, 1972).
Across this portal an extensive interchange of faunas occurred
between the two continents.  One immigrant group to South America
was the placental Order Carnivora, including the families Ursidae,
Canidae, Felidae and Mustelidae.  Representatives of the first
three families are first recorded in the South American fossil

record in beds of Uquian (early Pleistocene) age (Patterson and Pascual, 1972), while a single species of Mustelidae (<u>Conepatus altiramis</u>) is first recorded from the Chapadmalalan (Reig, 1952).

It must be realized that the dog-like family Borhyaenidae was apparently extinct before these carnivoran groups first appear in the South American fossil record. There is thus no compelling evidence indicating that the appearance of these immigrant forms of the Carnivora were in any way responsible for the decline to extinction of borhyaenid marsupials.

A possible explanation for these pre-land bridge faunal changes appears to be linked with concomitant climatic changes in South America. During the Pliocene the Argentinian sedimentation center shifted from Patagonia to the pampas and northwestern regions (Patterson and Pascual, 1972:251). The sediments changed from predominantly pyroclastic (tuffs and bentonitic clays) that characterize pre-Chasicoan units, to predominantly clastic (silts, sands and loess) which predominate during the Pliocene and Pleistocene of the pampean region (Pascual, 1965; Pascual and Odreman Rivas, 1971:399). This change of sediment type coincided with a late Miocene (post-Friasian) phase of Andean uplift that was to result in elevation of the Main Cordillera (Herrero-Ducloux, 1963). Elevation of the Cordillera acted as a barrier to moisture-laden Pacific winds (Patterson and Pascual, 1972:251). The southern South American habitat changed from primarily savanna-woodland (which predominated during the early to middle Tertiary - Eocene through Miocene) to drier forests and pampas, ranging from forests in the northern parts of the continent to grasslands in the south (Pascual and Odreman Rivas, 1971:399). Pampas environments similar to those prevailing today probably came into prominence at about this time. Many subtropical savanna-woodland forms retreated northward, and new opportunities arose for those mammals able to adapt to a plains environment (Patterson and Pascual, 1972:251).

A remarkable feature of the South American Tertiary fossil record is the variety of groups composing the "large herbivore" categories (see Patterson and Pascual, 1972:283, Table 6). In addition to seven ungulate orders, five families of edentates and two of rodents have contributed to the total. The total number of families (12) is a little higher in the Deseadan than later because of the presence of a few archaic groups that made their last recorded appearance in that age. Thereafter, the number settled around 10 until the end of the Tertiary. Although the number of families remained relatively constant, the groups contributing to the total changed (Patterson and Pascual, 1972:283). The food source for the large terrestrial carnivores was thus in a continued state of fluctuation, with marked changes occurring in the herbivore fauna at the end of the Deseadan and end of the Friasian as noted

above.  Seven large ungulate groups continued well into or to the
end of the Pleistocene.  Patterson and Pascual (1972:284) stress
this point because there appears to be an impression that a notable
reduction of large South American herbivores followed upon or
shortly after establishment of the land connection with North
America.  By the time the Panamanian land bridge was established
over half of the large herbivores were non-ungulates, and of those
the large edentates survived but not the gigantic rodents, although
the families to which they belong did.  Much of the large ungulate
extinction took place during the later Tertiary and had, as far as
the record goes, nothing to do with the faunal interchange
(Patterson and Pascual, 1972:284).

    The picture regarding the marsupial saber-tooth family
Thylacosmilidae is quite different.  Placental saber-tooths first
appear in the South American fossil record in beds of Uquian age
(along with herbivore prey of North American origin; e.g. two
horses, a tapir, a llama and possibly a mastodon, see Pascual
et al. 1966, chart opposite p. 202) and continue through beds of
Lujanian (latest Pleistocene) age.  The last appearance of marsupial saber-tooths in beds of Chapadmalalan age (represented by
Hyaenodonops) and the first appearance of placental saber-tooths
in immediately overlying beds, suggests ecological replacement of
the endemic thylacosmilids by immigrant machairodonts (Marshall,
1976a).

    The arrival of the four families of placental carnivores
(Mustelidae, Canidae, Felidae and Ursidae) appears to have had a
profound effect on other members of the established South American
fauna.  Decline of the phororhacoids may be linked to competition
with large Felidae and certain Canidae.  The disappearance of the
large procyonid Chapalmalania appears linked with the appearance
of the Ursidae, as it may have failed in competition with the
immigrant Arctodus (Patterson and Pascual, 1972:292), an animal of
comparable size.  The decline in diversity of large and small
specialized didelphids is correlated with the appearance and diversification of mustelids, and smaller canids and felids.


                               SUMMARY

    During the Cenozoic in South America the ecological roles of
terrestrial carnivores (i.e. the carnivorous adaptive zone) were
shared at various times, but not equally, by reptiles, birds and
mammals.  In the early Tertiary (pre-Oligocene) these roles were
occupied by dog-like marsupials of the family Borhyaenidae, and by
the semiterrestrial deep skulled crocodyle Sebecus. During the
middle Tertiary (Oligocene through Miocene) large predaceous
phororhacoid ground birds became the dominant terrestrial carnivores.

By the end of the Miocene these birds completely replaced the large
terrestrial carnivorous borhyaenids of the subfamily Borhyaeninae,
at least on the savanna grasslands of Argentina.  Large terrestrial
omnivorous borhyaenids of the subfamily Prothylacyninae were replaced in the middle Pliocene by immigrant procyonids.  Small
generalized borhyaenids of the subfamily Hathlyacyninae were replaced in the mid- to late Pliocene by similarly adapted opossums
of the family Didelphidae.  The ecological roles filled by didelphids, procyonids and phororhacoids in the later Pliocene became
filled by immigrant members of the order Carnivora in the early
Pleistocene.  Coincident with the changes in the groups assuming
the roles of terrestrial carnivores were changes in the groups
filling roles of large terrestrial herbivores.  A correlation
further exists between these faunal changes and concurrent climatic
changes.  The latter resulted in the appearance of new dominant
environments and afforded new opportunities for those animals able
to adapt to them.

Whether this "relay" of various carnivorous groups through
time was due to active competition between the successive groups
filling these roles, or whether this relay was the result of the
disappearance of one carnivorous group (possibly linked with concurrent environmental changes) with subsequent passive replacement
by another group which came to fill a similar role in a later
fauna, or if the faunal changes were the result of a combination
of these possibilities or others, is unknown.


## ACKNOWLEDGMENTS

A generous research grant (no. 1329) from the National
Geographic Society, Washington, D.C., and supplemental financial
support from both the Department and Museum of Paleontology,
University of California, Berkeley, made this study possible.
Appreciation for this support is greatfully acknowledged.


## REFERENCES

Brodkorb, P., 1963, A giant flightless bird from the Pleistocene
     of Florida.  Auk., 8:111-115.
Hecht, M.K., 1975, The morphology and relationships of the largest
     known terrestrial lizard, Megalania prisca Owen, from the
     Pleistocene of Australia.  Proc. Roy. Soc. Vict., 87:239-249.
Herrero-Ducloux, A., 1963, The Andes of western Argentina.  Am.
     Assoc. Pet. Geol., Mem., 2:16-28.
Hirschfeld, S.E. and Webb, S.D., 1968, Plio-Pleistocene megalony-
     chid sloths of North America.  Bull. Florida State Mus.,
     Biol. Sci., 12(5):213-296.

Kraglievich, J.L. and Olizabal, A.G. de, 1959, Los procionidos
    extinguidos del genero Chapalmalania Amegh.  Rev. Mus. Arg.
    Cienc. Natur., 6:1-59.
Langston, W. Jr., 1965, Fossil crocodilians from Colombia and the
    Cenozoic history of the Crocodilia in South America.  Univ.
    Calif. Publ. Geol. Sci., 52:1-157.
MacArthur, R.H., 1972, Geographical Ecology:  Patterns in the
    distribution of species.  Harper and Row, New York, 269 pp.
Marshall, L.G., 1976a, Evolution of the Thylacosmilidae, extinct
    saber-tooth marsupials of South America.  PaleoBios., no.
    22:1-30.
Marshall, L.G., 1976b, New didelphine marsupials from the La Venta
    Fauna (Miocene) of Colombia, South America.  Journ. Paleont.,
    50(3):402-418.
Marshall, L.G., in press, Evolution of the Borhyaenidae, extinct
    South American Predaceous Marsupials.  Univ. Calif. Publ.
    Geol. Sci.
Pascual, R., 1965, Las Toxodontidae (Toxodonta, Notoungulata) de
    la formacion Arroyo Chasico (Plioceno inferior) de la
    Provincia de Buenos Aires.  Caracteristicas geologicas.
    Ameghiniana.  4:101-129.
Pascual, R., et al., 1966, Paleontografia Bonaerense.  Fasciculo
    IV:  Vertebrata.  Com. Invest. Cient. Prov. Bs. Aires. La
    Plata.
Pascual, R. and Odreman Rivas, O.E., 1971, Evolucion de las
    comunidades de los vertebrados del Terciario argentino.
    Los aspectos paleozoogeograficos y paleoclimaticos
    relacionados.  Ameghiniana.  8(3-4):372-412.
Patterson, B. and Kraglievich, J.L., 1960, Sistematic y nomenclatura
    de las aves fororracoideas del Plioceno argentino.  Pub. Mus.
    Munic. Cienc. Natur., Mar del Plata.  1:1-49.
Patterson, B. and Pascual, R., 1972, The fossil mammal fauna of
    South America, in "Evolution, Mammals, and Southern
    Continents (A. Keast, F.C. Erk and B. Glass, eds.), pp. 247-
    309.
Pianka, E.R., 1974, Evolutionary Ecology.  Harper and Row, New
    York, 356 pp.
Reig, O.A., 1952, Sobre la presencia de mustelidos mefitinos en la
    Formacion de Chapadmalal.  Rev. Mus. Munic. Cienc. Natur.
    Trad. Mar del Plata.  1(1):45-51.
Riggs, E.S. and Patterson, B., 1939, Stratigraphy of the late
    Miocene and Pliocene deposits of the Province of Catamarca
    (Argentina) with notes on the fauna.  Physis., 14:143-162.
Simpson, G.G., 1972, Didelphidae from the Chapadmalal Formation
    in the Museo Municipal de Ciencias Naturales of Mar del
    Plata.  Publ. Mus. Munic. de Cienc. Natur. Mar del Plata.
    2(1):1-39.
Simpson, G.G., 1974, Notes on Didelphidae (Mammalia, Marsupialia)
    from the Huayquerian (Pliocene) of Argentina.  Amer. Mus.
    Novit., 2559:1-15.

Van Valen, L., 1971, Adaptive zones and the orders of mammals.
    Evolution, 25:420-428.
Whittaker, R.H., 1972, Evolution and measurement of species
    diversity.  Taxon., 21(2-3):213-251.

GEOGRAPHICAL AND ECOLOGICAL DISTRIBUTION

OF THE EARLIEST TETRAPODS

A.L. PANCHEN

Department of Zoology, The University

Newcastle upon Tyne NE1 7RU  England

## INTRODUCTION

Acceptance of the theory of Continental Drift was generally welcomed by vertebrate paleontologists who had concerned themselves with the problems of paleogeography (e.g. McKenna, 1973). This is perhaps particularly the case with those concerned with the fauna of early tetrapods which first appears in the fossil record in the late Paleozoic.  In the last century Cope (1868) noted similarities in both Upper Carboniferous and Early Permian tetrapod faunas between Europe and North America.

Perhaps the first attempt to explain these resemblances in terms of continental drift was that of Nopcsa in 1934.  Subsequently both Westoll (1944) and Romer (1945) made detailed comparisons between continental 'Permo-Carboniferous' vertebrate faunas.  Both concluded that direct terrestrial connections were present between the two continents, but while Westoll was unequivocal in suggesting drift as the explanation, Romer, more cautiously, clung to the idea of intercontinental land bridges as a possible alternative explanation.

It is now generally accepted, however, that in the late Paleozoic the Atlantic Ocean did not exist and North-Western Europe and North America were united in a single land mass.  During the Carboniferous period itself closure of the reconstructed Mid-European and Eurasian Seas apparently formed the great continent of Laurasia.  This consisted of the whole of North America united with most of Europe and Asia (Johnson, 1973).  Most authorities also agree that by the early Permian Laurasia was coupled with the

southern continent of Gondwanaland to form the super-continent of
Pangea (Smith et al., 1973) which did not fragment until well into
the Mesozoic.

Another aspect of the theory of continental drift and its
development into plate tectonics is the phenomenon of "polar
wandering", interpreted as a relative movement through time of the
continental plates relative to the magnetic poles (Creer, 1970).
With this concept came the possibility of plotting a paleoequator
in relation to the reconstructed position of the continents at any
period during geological time and study of the contemporary flora
and fauna in relation to that equator.

An early attempt to do this concerned a major group of
Paleozoic tetrapods.  Irving and Brown (1964) rather ambitiously
attempted to show not only that the Paleozoic labyrinthodont
Amphibia had a largely paleotropical distribution, but also that
the diversity of genera and abundance of individuals increased
significantly towards the equator.  This effect was said to be
much less marked in the Triassic.  The data are inadequate for
anything but the first conclusion and the authors were strongly
criticized by Stehli (1966; rejoinder: Irving and Brown, 1966).
There is, however, little reason to doubt the equatorial distribution of most early tetrapods including the Labyrinthodontia.

The labyrinthodonts include the majority of fossil Amphibia
from the late Paleozoic (Devonian {?}, Carboniferous, Permian) and
Triassic and share a number of features with the crossopterygian
fishes from which they sprang.  The distribution of the latter was
plotted by Thomson (1969) in relation to Devonian and Carboniferous
paleoequators, using paleolatitude data from Irving (1964).
Thomson notes the tropical distribution of the group but his map
shows only some major Laurasian localities.  He does, however,
tabulate the majority of "sarcopterygian" (crossopterygian +
dipnoan) genera, with their localities and horizons, including
Paleozoic forms from Gondwanaland.

Thomson's map is based on a simple Mercator projection but a
more realistic idea of distribution during the period of the
existence of Pangea is given by a plot of the reconstructed "pre-
drift" position of the continents.  This also allows the extrapolation of reconstructed paleoequators from North America to
Europe or vice-versa when the data are less numerous and less
reliable on one side or the other, and also gives a truer picture
of the distribution of fossil sites in terms of their original
geography.

A landmark in such reconstructions was that of Bullard et al.
(1965) of the continents now surrounding the Atlantic.  The

'pre-drift' position of these was plotted using the 500 fathom (ca. 900 m) isobath to represent the edge of the continental shelf and thus of the continental plate in each case. This gave a very good fit of all the peri-Atlantic continental outlines and was consistent with other data.

## TETRAPOD DISTRIBUTION IN THE DEVONIAN AND CARBONIFEROUS

In 1970 I attempted to plot the distributions of one important group of labyrinthodont amphibians using the reconstruction of Bullard et al. (1965) of the relative positions of Europe and North America. The positions of the Carboniferous and Permian equators were computed by Dr. D.H. Tarling of the School of Physics, University of Newcastle upon Tyne, from North American paleomagnetic data then available and extrapolated across to Europe.

The group of labyrinthodonts were the Anthracosauria (s.s.: Anthracosauria, Embolomeri sensu Romer, 1947, 1966; Panchen, 1975). The Anthracosauria in the broad sense, together with the reptile-like Seymouriamorpha, constitute the Batrachosauria, one of the two major groups of labyrinthodonts (Panchen, 1975). The other is the Temnospondyli, which includes the vast majority of known forms. Separated from the Batrachosauria and Temnospondyli are the primitive and aberrant Ichthyostegalia, known only from an Upper Devonian or lowermost Carboniferous locality in East Greenland (Jarvik, 1952).

Figure 1 represents a revision of this original distribution map for the anthracosaurs. The paleogeographical data have not been revised but since it was first drawn members of a primitive non-embolomerous group of anthracosaurs, the Herpetospondyli, have been described from the Edinburgh district in Britain and from Greer, West Virginia in the U.S.A. (Hotton, 1970; Romer, 1970; Panchen, 1975). These roughly contemporary forms are of Upper Visean, Lower Carboniferous age, while all other certainly described anthracosaurs are Upper Carboniferous. They are plotted in Figure 1. A third group of anthracosaurs are the Gephyroste-goidea (Caroll, 1970), small terrestrial forms lacking the specializations of the Seymouriamorpha. Until recently gephyrostegids were known only from sites in Czechoslovakia and Ohio, U.S.A., from which embolomeres were also known, but recently a very early (Namurian B, early Upper Carboniferous = basal Pennsylvanian) gephyrostegid has been described from the Ruhr coalfield in Germany (Boy and Bandel, 1973). This is also plotted in Figure 1. The ichthyostegid site is included for reference .

The European Upper Carboniferous is equivalent to the whole

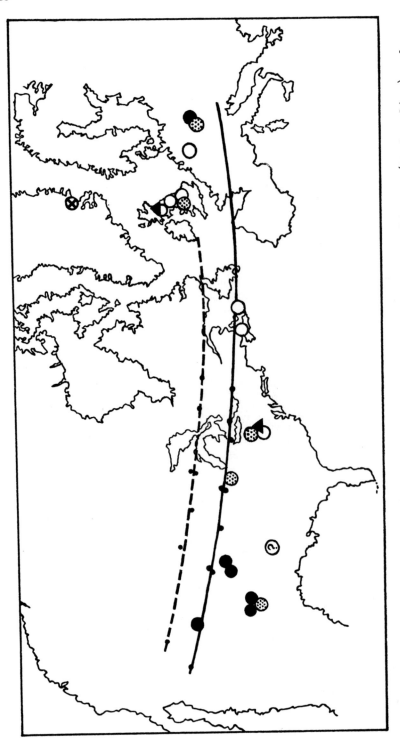

Fig. 1.  Distribution of anthracosaur Amphibia in relation to Carboniferous (broken line) and Permian (solid line) equators and "pre-drift" position of Europe and North America. Lower Carboniferous, triangles ; Namurian-Westphalian C, open circles; Westphalian D - Stephanian, stippled circles; Permian, solid circles. X = ichthyostegid locality.  (Modified after Panchen, 1970.)

of the North American Pennsylvanian and also the topmost Mississippian (correlation chart in Panchen, 1970). It is divided into the Namurian, the Westphalian (subdivided A to D with Westphalian A the earliest) and the Stephanian, which immediately preceded the Permian period. It will be seen from Figure 1 that from their first occurrence in the Lower Carboniferous, through to the end of Westphalian C times, known anthracosaurs were confined to an area bounded in the east by the Ruhr and in the west at the level of eastern Ohio, U.S.A. During the succeeding Westphalian D, Stephanian and Lower Permian, they appear to have extended their range westward to Texas and New Mexico and eastward to the area near Prague, Czechoslovakia. Anthracosaurs disappear from the fossil record during the Lower Permian and are known from nowhere else in the world.

One anomaly existed in this tentative picture of anthracosaur radiation and is marked by a query on the map: current theory was that a single embolomere specimen, Eobaphetes kansensis, was derived from the earliest Pennsylvanian, possible Namurian of Arkansas, rather than from an unknown horizon in Kansas as originally reported. However, this proves not to be the case. Spore data and coal analysis show Eobaphetes to have come from a much higher horizon, possibly Stephanian, whatever its locality (Panchen, 1977).

Thus the tentative picture of anthracosaur origin and distribution from within an area composed of north-western Europe, Greenland and eastern North America seemed not unreasonable. It was also consistent with the lack of any fossil tetrapods outside the area of anthracosaur distribution until well into the Permian. Thus in a review of Carboniferous tetrapods (Panchen, 1973) maps were drawn on the same basis for the remaining known Carboniferous tetrapods; one for the labyrinthodonts as a whole, one for the remaining Amphibia, known as the Lepospondyli, and one for the Reptilia. These three, with the addition of the gephyrostegid site in the Ruhr, have been combined as Figure 2. The overall area of distribution is identical with that of the Carboniferous anthracosaurs, with most of the localities having anthracosaurs represented.

This is not altogether surprising as the map to some extent merely represents the position of suitable collecting areas for Carboniferous vertebrates, strongly correlated, in the Upper Carboniferous at least, with areas of coal-mining and related industries. However, the total lack of any record of Carboniferous or pre-Carboniferous tetrapods from elsewhere reinforced the picture of the earliest tetrapods radiating from a centre near the longitude of the East Greenland ichthyostegid locality and maintaining a closely equatorial distribution. This was further corroborated by the closeness of the East Greenland site to the

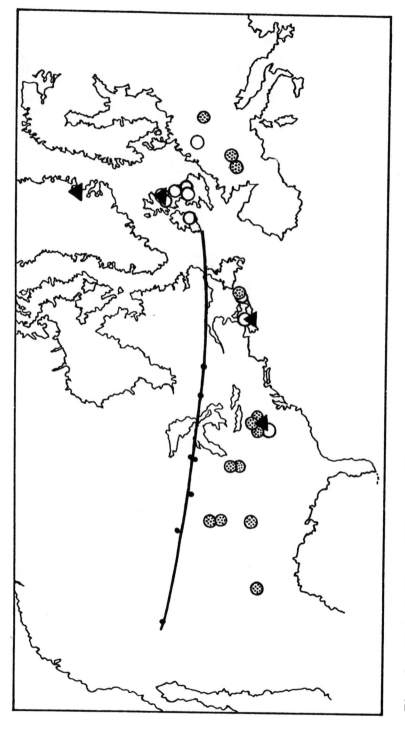

Fig. 2. Distribution of Devonian(?) and Carboniferous tetrapods in relation to Carboniferous equator and "pre-drift" position of Europe and North America. Devonian(?) and Lower Carboniferous, triangles, other conventions as in Fig. 1. (Composite after Panchen, 1973.)

Upper Devonian - Lower Carboniferous equator (Smith et al., 1973). It therefore came as a not inconsiderable shock to receive an account of evidence of tetrapods from the Upper Devonian of Australia!

Warren and Wakefield (1972) report and illustrate what appear to be tetrapod trackways from the Devonian of eastern Victoria, Australia. The horizon is apparently of Upper Devonian, Frasnian age and thus predates the ichthyostegids, whether the latter are regarded as uppermost Devonian (Jarvik, 1950) or lowermost Carboniferous (Westoll, 1940). Both identification and horizon are probably correct and thus the whole picture of the geography of tetrapod origin is overturned.

A little more may be said, however. It has already been noted that crossopterygian fish are known from the Devonian of 'Gondwanaland' including Australia (Thomson, 1969) with the important addition of superb material from the Frasnian Gogo formation in Western Australia (Brunton, Miles and Rolfe, 1969). It is thus not impossible that the origin of tetrapods occurred in Gondwanaland at some time during the earlier Devonian. In the provisional paleogeographic maps presented by Smith et al. (1973) the latitude of Australia is shown as tropical, lying between 0° and 30° South in both the Lower Devonian and Lower Carboniferous. Furthermore in the Lower Carboniferous (Smith et al., Map 6) the area of known Carboniferous tetrapods lies entirely within the same band of paleolatitudes. Thus on this provisional reconstruction any Australian tetrapod fauna in Devonian and early Carboniferous would have had a similar equatorial distribution to the Laurasian one. On present evidence it would appear rash, however, to say whether low latitude migration of tetrapods between the two areas was possible.

There is evidence to show that during the Carboniferous Australia must have moved rather rapidly to a much higher latitude so that by the Permian, as reconstructed in map 5 of Smith et al. (1973)., it lies between latitudes 35° and ca. 63° with Victoria higher than 60°. This movement represents a southward migration of the whole of Gondwanaland and was undoubtedly correlated with the Permo-Carboniferous glaciation of that continent (Crowell and Frakes, 1972). Following earlier glacier formation that continental ice sheets spread across Australia late in the Carboniferous (Stephanian) to reach their maximum extent in the Lower Permian (Crowell and Frakes, 1971).

A recent review by Crowell and Frakes (1975) reconstructs the south polar wandering curve in relation to Gondwanaland in the Carboniferous and Permian and shows that in the early Permian South East Australia moved to within very few degrees of the pole.

Thus it is probable that whatever the area of origin of the tetrapods all but northern Gondwanaland had become uninhabitable to them by the Lower Permian. They do not appear in the Australian fossil record until late in the Upper Permian (Cosgriff, 1969). A similar phenomenon occurs with the distribution of fossil plants (Chaloner and Lacey, 1973). In the early Carboniferous there was a cosmopolitan flora (the "Lepidodenropsis flora") occurring on all continents including Australasia, with a distinct flora only in Siberia. By the retreat of the Gondwana ice sheet the well-known Glossopteris flora had differentiated in Gondwanaland and remained endemic to that continent until the beginning of the Triassic when it became extinct.

## PALEOECOLOGY OF UPPER CARBONIFEROUS AND LOWER PERMIAN TETRAPODS

In reconstructing the anatomy of fossil animals it is usually possible to propose some hypotheses about their functional morphology, behaviour and ecology. Further information on ecology and biogeography is usually available from the nature of the sediments in which the fossil is found and from the associated fauna and flora. Fruitful comparisons may also be made using the contrasting features of the fauna of similar and contemporary sites in different geographical areas.

In the case of the early tetrapods comparisons between the faunas of Europe and North America would be of particular interest but these only become possible at a fairly late stage in the Carboniferous. The described faunas in the Lower Carboniferous and the Namurian are very limited. In the early Westphalian there are two notable sites with a rich fauna, one at Jarrow, County Kilkenny in Ireland (Westphalian A) and the other in North America at Joggins, Nova Scotia (Westphalian B). They are, unfortunately not comparable: the Jarrow site represents a small ox-bow lake (Rayner, 1971), while the Joggins fauna is an assemblage of relatively terrestrial forms trapped in hollow tree stumps (Carroll, 1967). It is not until the late Westphalian that one has tetrapod sites on both continents with faunas which are approximately contemporary, ecologically similar and sufficiently rich for worthwhile comparison.

Even in the late Westphalian (Westphalian D), however, only one site on each side of the Atlantic is sufficiently rich for significant comparison, but a further site on each side, though poorer, reinforces the conclusions drawn from the principal ones. When one moves into the Stephanian and early Permian the picture changes radically. A large number of sites is known on each side of the Atlantic and the overall fauna is strikingly diverse. The

Upper Carboniferous Stephanian fauna does not differ significantly from the Early Permian ones. Such a comparison between European and North American faunas was therefore made initially by Dr. Milner and the author in the Stephanian to Kungurian in Europe and the corresponding Missourian to Leonardian in North America (Francis and Woodland, 1964; Smith, 1964). This also eliminated problems of the Carboniferous - Permian boundary on both continents. We then returned to the Westphalian sites to corroborate our ideas (Milner and Panchen, 1973).

The Stephano-Permian localities are shown in Figure 3 on the same 'pre-drift' projection as before. In comparing the tetrapod fauna the family rank seemed to us the most realistic one to use, as more work is urgently needed to compare the genera of Paleozoic tetrapods between Europe and North America, but some sort of concensus can be reached on family attribution after recent revision by several authorities.

41 families of tetrapods (labyrinthodont, lepospondyl and reptile) have been recognized in the Stephanian and Lower Permian of North America and 20 in Europe, and at first glance (allowing for sampling error particularly of rare forms) it appears that the European fauna is a smaller sample of that which existed in America. It is possible, however, to classify the fossils into apparently mainly terrestrial forms and aquatic ones. The criteria used are skull shape and dentition, as a clue to feeding habits, presence or absence of lateral line grooves in the skull (aquatic and terrestrial respectively), presence, at least at some stage, or absence of external gills and structure and proportions of the post-cranial skeleton, particularly the limbs. On these criteria all the reptiles and most of the lepospondyl microsaurs were considered terrestrial together with some of the labyrinthodonts; the remaining lepospondyls and labyrinthodonts were considered aquatic. The overall results of the comparison are set out in Table I.

Even allowing for differences due to geographical distance, which would enhance the effect, it seemed probable that some major ecological barrier existed to the dispersal of aquatic tetrapods between Europe and North America which did not affect terrestrial forms. This hypothesis was corroborated by the fact that two of the six aquatic families occurring in common were relicts of the Westphalian fauna, not found in the Permian, and more importantly by the phenomenon of ecological replacement. Thus, particularly among the aquatic labyrinthodonts, similar ecological niches in Europe and North America appear to have been occupied by representatives of different endemic families. For example, the niche for a long-snouted piscivore, filled by the gharials today, was filled by Archegosaurus in Europe and Chenoprosopus in America, that for an alligator-like form by Actinodon and related genera in Europe

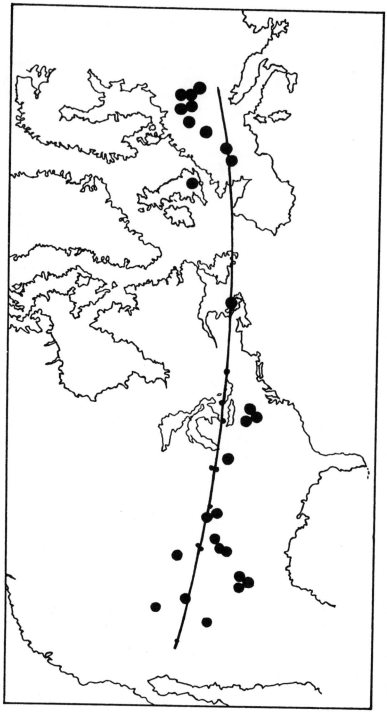

Fig. 3. Localities of principal Stephanian and Lower Permian tetrapod localities in relation to Permian equator and "pre-drift" Europe and North America. (Modified after Milner and Panchen, 1973.)

TABLE I.   Comparison of tetrapod faunas from the Stephanian and
           Lower Permian of Europe and North America at the family
           level.

|                 | TERRESTRIAL | AQUATIC | TOTALS |
|-----------------|:-----------:|:-------:|:------:|
| IN COMMON       | 9           | 6       | 15     |
| EUROPE ONLY     | 1           | 4       | 5      |
| N. AMERICA ONLY | 19          | 7       | 26     |
| TOTALS          | 29          | 17      | 46     |

and by <u>Saurerpeton</u> and <u>Trimerorhachis</u> and members of their respective families in America, and that for a long-bodied short-limbed
microsaur by <u>Microbrachis</u> in Europe and the distantly related
molgophids and lysorophids in North America.

Turning to the Westphalian D sites noted above, the same distribution phenomenon appeared to prevail.  The principal European
site was that at Nyrany, near Prague, supported by data from Tremosna in the same area, while the American one is at Linton, Ohio,
with supporting data from Mazon Creek, Illinois.  Linton represents
a small pond (Baird, 1962) with Nyrany apparently similar.  The
others are apparently less certain, but no family certainly occurs
at either of the subsidiary sites without occurring at the corresponding major one.  Their geographical position is shown in Fig. 4.

Of the terrestrial tetrapods, five families are common to
Europe and North America.  Of the two families apparently endemic
to Europe one is a microsaur family (the Gymnarthridae) common in
the American Lower Permian, the other is an amphibian or reptilian
family of disputed status represented by only two specimens.  Only
one family (the reptilian Ophiacodontidae) is exclusive to North
America and is represented by a single specimen.  Thus there is no
detectable significant difference in the tetrapod fauna.

There are seven common aquatic families, plus three confined
to Europe and six to North America.  Of the seven families held in
common, six have histories extending back at least to Westphalian
A times, as all occur at Jarrow, and one of these, the labyrinthodont Loxommatidae, is known from the Lower Carboniferous.  Thus
their relict status may reduce their significance.

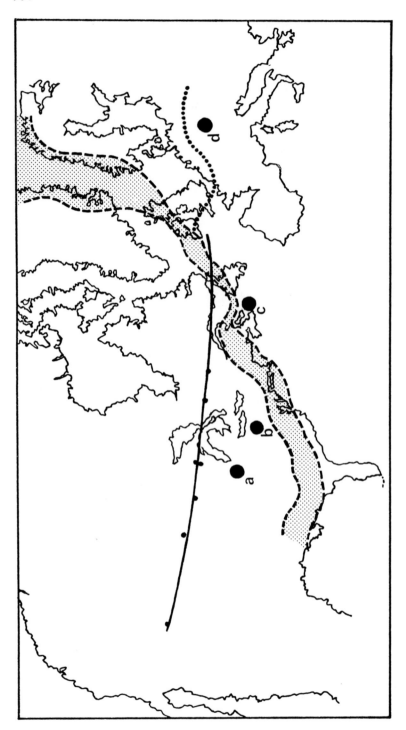

Fig. 4.  Principal Westphalian D localities in "pre-drift" Europe and North America in relation to Carboniferous equator.  Axis of Appalachian/Caledonian range after Eardley (1951) and Dewey (1969), broken lines and stipple; Hercynian Front suture after Johnson (1973), dotted line.
(a), Mazon Creek; (b), Linton; (c), Florence; (d), Nyrany and Tremosna. (Modified after Milner and Panchen, 1973.)

It seems therefore that the hypothesis of a barrier to aquatic tetrapods is supported by the evidence from Westphalian D, and the obvious candidates are extensive deserts or a mountain chain. There is no evidence of the former, but there is a well-established case of the latter which separates all the North American from all the European sites considered above. This is the Appalachian/Caledonian mountain system (Dewey, 1969), which ran as a single chain from Scandinavia, through Scotland and eastern Canada to join the course of the present Appalachians in North America (Figure 4).

One series of localities was of particular interest to our hypothesis of the Appalachian/Caledonian barrier between American and European faunas. Nova Scotia, now part of Eastern Canada, and thus North America, lay on the European side of the barrier. Tetrapod localities are known in Nova Scotia from the earliest Carboniferous (Tournasian) to the Upper Westphalian (Figure 2). In addition there is a series of tetrapod localities on Prince Edward Island which may be Stephanian or Permian (Figure 3). However, the fauna of the latter appeared neutral to our hypothesis (Milner and Panchen, 1973).

Of more interest was the fauna of Florence, Cape Breton Island, Nova Scotia, which is of Westphalian D age and thus comparable in time to those considered above (Figure 4). Most of the tetrapods are reptiles, but one labyrinthodont, Cochleosaurus, represented an aquatic form only certainly known from Nyrany and Tremosna. Thus on this rather slender evidence the Florence fauna seemed of European type, corroborating our hypothesis: however, it now seems possible that Cochleosaurus also occurs at Linton (Dr. A.R. Milner, personal communication) and the Florence fauna would also then be neutral.

In our original review we considered only the Appalachian/ Caledonian barrier, but there seems little doubt that physical features in Europe must have had an important effect on the dispersal of tetrapods in the Carboniferous. The late Paleozoic Hercynian orogeny produced a mountain belt separating Northern from Central and Southern Europe. This is interpreted by Johnson (1973) as the result of the closure of the Mid-European Sea (or more correctly ocean) referred to above. Thus a marine barrier was replaced by a partial upland one during the Carboniferous.

It is certainly significant therefore that the only tetrapod locality in continental Europe before Westphalian D time is that in the Ruhr (Figure 2) which probably lay to the north of the Hercynian Front suture which marks the boundary of the northern European plate (Figure 4). Future work on the distribution of Paleozoic tetrapods in Europe will certainly have to take account of the geological events associated with the Hercynian orogeny.

## REFERENCES

Baird, D., 1962, A haplolepid fish fauna in the early Pennsylvanian of Nova Scotia, Palaeontology 5:22-29.

Boy, J.A. and Bandel, K., 1973, Bruktererpeton fiebigi n. gen. n. sp. (Amphibia: Gephyrostegida) der erste Tetrapode aus dem rheinischwestfalischen Karbon (Namur B; W-Deutschland), Palaeontographica (A) 145:39-77.

Brunton, C.H.C., Miles, R.S. and Rolfe, W.D.I., 1969, Gogo expedition 1967, Proc. geol. Soc., Lond. 1969:79-83.

Bullard, E.C., Everett, J.E. and Smith, A.G., 1965, A symposium on continental drift. IV The fit of the continents around the Atlantic, Phil. Trans. R. Soc. (A), 258:41-51.

Carroll, R.L., 1967, Labyrinthodonts from the Joggins Formation, J. Paleont. 41:111-142.

Carroll, R.L., 1970, The ancestry of reptiles, Phil. Trans. R. Soc. (B) 257:267-308.

Chaloner, W.G. and Lacey, W.S., 1973, The distribution of Late Palaeozoic floras, in "Organisms and Continents through time" (N.F. Hughes, ed.), Spec. Pap. Palaeont. 12:271.

Cope, E.D., 1868, Synopsis of the extinct Batrachia of North America, Proc. Acad. nat. Sci. Philad. 1868:208-221.

Cosgriff, J.W., 1969, Blinasaurus, a brachyopid genus from Western Australia and New South Wales, J.R. Soc. W. Australia 52:65-88.

Creer, K.M., 1970, A review of palaeomagnetism, Earth-Sci. Rev. 6:369-466.

Crowell, J.C. and Frakes, L.A., 1971, Late Paleozoic glaciation: part IV, Australia, Bull. geol. Soc. Am. 82:2515-2540.

Crowell, J.C. and Frakes, L.A., 1972, Ancient Gondwana glaciations, Proc. 2nd Gondwana Symp., S. Afr.:571-574.

Crowell, J.C. and Frakes, L.A., 1975, The late Palaeozoic glaciation, in "Gondwana Geology, papers presented at the Third Gondwana Symposium, Canberra, Australia, 1973" (K.S.W. Cambell, ed.), pp. 313-331, Australian National University Press.

Dewey, J.F., 1969, Evolution of the Appalachian/Caledonian orogen, Nature, Lond. 222:124-129.

Eardley, A.J., 1951, "Structural geology of North America", Harper, New York.

Francis, E.H. and Woodland, A.W., 1964, The Carboniferous Period, in "The Phanerozoic Time-scale" (W.B. Harland et al. eds.), pp. 221-232. Geol. Soc., London.

Hotton, N., 1970, Mauchchunkia bassa gen. et sp. nov. an anthracosaur (Amphibia, Labyrinthodontia) from the Upper Mississippian, Kirtlandia No. 12:1-38.

Irving, E., 1964, "Palaeomagnetism," Wiley, New York.

Irving, E. and Brown, D.A., 1964, Abundance and diversity of the labyrinthodonts as a function of paleolatitude, Am. J. Sci. 262:689-708.

Irving, E. and Brown, D.A., 1966, Reply to Stehli's discussion of labyrinthodont abundance and diversity, Am. J. Sci. 264:488-496.

Jarvik, E., 1950, Note on middle Devonian crossopterygians from the eastern part of Gauss Halvo, East Greenland, Meddr Gronland 149(6):1-20.

Jarvik, E., 1952, On the fish-like tail in the ichthyostegid stegocephalians, Meddr Gronland 114(12):1-90.

Johnson, G.A.L., 1973, Closing of the Carboniferous Sea in Western Europe, in "Implications of Continental Drift to the Earth Sciences," (D.H. Tarling and S.K. Runcorn, eds.), pp. 843-850, Academic Press, London.

McKenna, M.C., 1973, Sweepstakes, filters, corridors, Noah's arks and beached viking funeral ships in palaeogeography, in "Implications of Continental Drift to the Earth Sciences," (D.H. Tarling and S.K. Runcorn, eds.), pp. 295-308, Academic Press, London.

Milner, A.R. and Panchen, A.L., 1973, Geographical variation in the tetrapod faunas of the Upper Carboniferous and Lower Permian, in "Implications of Continental Drift to the Earth Sciences" (D.H. Tarling and S.K. Runcorn, eds.), pp. 353-368, Academic Press, London.

Nopcsa, F., 1934, The influence of geological and chronological factors on the distribution of non-marine reptiles and Stegocephalia, Q. Jl geol. Soc. Lond. 90:76-140.

Panchen, A.L., 1970, "Teil 5a Anthracosauria. Handbuch der Palaoherpetologie," Fischer, Stuttgart.

Panchen, A.L., 1973, Carboniferous tetrapods, in "Atlas of Palaeobiogeography" (A. Hallam, ed.), Chapter 13, pp. 117-125, Elsevier, Amsterdam.

Panchen, A.L., 1975, A new genus and species of anthracosaur amphibian from the Lower Carboniferous of Scotland and the status of Pholidogaster pisciformis Huxley, Phil. Trans. R. Soc (B) 269:581-640.

Panchen, A.L., 1977, On Anthracosaurus russelli Huxley (Amphibia: Labyrinthodontia) and the family Anthracosauridae, Phil. Trans. R. Soc. (B), in press.

Rayner, D.H., 1971, Data on the environment and preservation of late Palaeozoic tetrapods, Proc. Yorks. geol. Soc. 38:437-495.

Romer, A.S., 1945, The late Carboniferous vertebrate fauna of Kounova (Bohemia) compared with that of the Texas Red Beds, Am. J. Sci. 243:417-442.

Romer, A.S., 1947, Review of the Labyrinthodontia, Bull. Mus. Comp. Zool. Harv. 99:1-368.

Romer, A.S., 1966, "Vertebrate paleontology," 3rd Edn. Univ. Press, Chicago.

Romer, A.S., 1970, A new anthracosaurian labyrinthodont, Proterogyrinus scheelei, from the Lower Carboniferous, Kirtlandia No.10:1-16.

Smith, A.G., Briden, J.C. and Drewry, G.E., 1973, Phanerozoic
    World Maps, in "Organisms and Continents through time,"
    (N.F. Hughes, ed.), Spec. Pap. Palaeont. 12:1-42.
Smith, D.B., 1964, The Permian Period, in "The Phanerozoic
    Time-scale," (W.B. Harland et al., eds.), pp. 211-220,
    Geol. Soc., London.
Stehli, F.G., 1966, Discussion - labyrinthodont abundance and
    diversity, Am. J. Sci. 264:481-487.
Thomson, K.S., 1969, The biology of the lobe-finned fishes, Biol.
    Rev. 44:91-154.
Warren, J.W. and Wakefield, N.A., 1972, Trackways of tetrapod
    vertebrates from the Upper Devonian of Victoria, Australia,
    Nature, Lond. 238:469-470.
Westoll, T.S., 1940, (In discussion on the boundary between the
    Old Red Sandstone and the Carboniferous), Advanc. Sci., Lond.
    1:258.
Westoll, T.S., 1944, The Haplolepidae, a new family of late
    Carboniferous bony fishes, Bull. Am. Mus. nat. Hist. 83:
    1-121.

# PART 3

# Phylogeny and Classification of
# Vertebrate Taxa

# CLADISTIC CLASSIFICATION AS APPLIED TO VERTEBRATES

Niels BONDE

Institut f. Historisk Geologi & Palaeontologi
University of Copenhagen
Østervoldgade 10, DK-1350 København K, Denmark

## INTRODUCTION

It is the object of this paper to point out some consequences and problems arising from the use of cladistic classifications. However, we must initially clarify what is implied in the terms 'phylogenetic systematics' and 'cladistics'. Phylogenetic systematics (exemplified by Hennig, 1966) encompasses the whole theory of classification in accordance with phylogeny (biological evolutionary history), while cladistics (which alludes to Rensch's kladogenesis (1947, 1959) and to Huxley's term clade (e.g. 1959) for a strictly monophyletic group) could perhaps with some advantage be restricted to the practical methodological approach, as developed by Hennig (e.g. 1966), Brundin (e.g. 1968), Nelson (e.g. 1971), Miles (1973), Patterson (1973), Rosen (1973), and others. Cladistics, however, has been used as a designation for the theory of phylogenetic systematics (e.g. by Mayr, 1974); although more often the word cladism has been so used (e.g. Mayr, 1969; Darlington, 1970; Nelson, 1971). Instead of having two more or less synonymous concepts it might be worthwhile restricting them to one as suggested above. The merit of this action is to recognize that workers other than the ones employing cladistic methods are convinced that they produce phylogenetic classifications (e.g. Simpson, 1959: 300). Thus phylogenetic systematics appears to be a goal that many post-Darwinian taxonomists have striven for (Nelson, 1971: 376; 1973: 372), while cladistics is a practical approach to that end. The practitioners of such an approach may well be termed cladists.

Classification acquired an explanatory function when the typological 'Linnean' concepts were ousted by evolutionary 'Darwinian'

concepts.  Classifications were more 'natural' which tended to mean
more in agreement with phylogeny (Nelson 1973:  370).  A great
improvement occurred with the integration of systematics, genetics
and paleontology into a 'neodarwinist' or 'synthetic' approach in
the 1930's and '40's.  The ideas on phylogeny and classification
developed by the synthetic school (which have come to be closely
associated with the names of Simpson (1943, 1945) and Mayr (1942)
were seriously challenged in the '60's with the wider circulation
of Hennig's cladistic viewpoint.  At the same time radical criticism came from another front, namely the numerical taxonomists or
pheneticists (e.g. Sokal and Sneath, 1963).

Concerning the views of Simpson and Mayr, it should be remembered that both have primarily been concerned with studying
'higher' vertebrates - Mayr a neo-ornighologist and Simpson a
paleomammalogist - fields which perhaps easily predispose their
students to attribute especially great importance to the very
advanced, expressed in classification in a way that may seem distinctly anthropocentric (see Mayr's dislike (1969) of classifying
Pan and Homo in the same family, and (1969:  234) - "the evolutionary distinctness of Man surely justifies recognition of a separate family"; and Simpson (1959b:  270) - "this representation
(with hominids a family) of the history is admittedly biased by
human interests in Homo").

Mayr's and Simpson's philosophies on biosystematics, however,
are not identical.  Mayr (e.g. 1969, 1974) using a concept of genetic similarity is not far removed from pheneticists; while Simpson
(e.g. 1961, 1976) is somewhat closer to cladists in attitude.  This
was well pointed out by Nelson (1972) and Ruse (1973).  The philosophies concerning classification, of the pheneticists, of Mayr,
and of Simpson, in decreasing degree, comprise obvious pre-evolutionary (pre-Darwinian) elements which may well be termed typological (as by Brundin(1966, 1972) and Kiriakoff (1959)).  Typological here should indicate that classification is entirely or
partly based upon degree of morphological similarity and difference
which implies conforming to a certain essential type of evolutionary level or grade, and not as in cladism (Nelson, 1971) entirely
upon genealogy (or phylogenetic relationships).

Whether the appearance of phylogenetic systematics or numerical taxonomy will some day in retrospect be called a revolution
(in the sense of Kuhn, 1970) is doubtful, although Heywood (1964)
apparently believes numerical taxonomy to be such a potential revolution, called the 'realist' phase in taxonomy.  The present situation perhaps can be described as a pre-paradigmatic period of
evolutionary biosystematics with three competing paradigms:  (1)
'phenetic' (see Farris, in this volume); (2) 'synthetic' which by
some is termed 'evolutionary' (see Bock, in this volume); and (3)
'cladistic' or phylogenetic systematic as set out here.

In an earlier paper (Bonde, 1975: 312) I mentioned as my conviction that in twenty years or so systematics and phylogenetic studies will be dominated by the cladistic paradigm; phenetics will be used for special purposes; while the synthetic approach will mostly be of historical interest.

## SOME REMARKS ON BIOSYSTEMATICS AS A SCIENCE

Biosystematics is the ordering and naming of living beings. To justify its status as (natural) science one of the main demands is objectivity. This normally means that phenomena, relations and objects of the external world which it deals with should be considered 'real', i.e. existing independently of man. Another demand is that the observations made within an objective field of science should be repeatable independently by other workers. This naturally leads to considerations of how to communicate observations and experiences unequivocally. All arguments, hypotheses and theories must be consistent, and in this context it appears to me that the word 'consistency' has been used in an unsatisfactory way by Simpson (1961: 107) for a much more vague concept of correlation between classification and interrelationships of organisms, which could just as well have been termed 'some correlation'. It has important consequences for Simpson's choice (1961: 124) of a definition of monophyly (see below). Certainly Simpson's use does not imply a one-to-one correspondence (isomorphism) between classification and phylogeny, which I would rather call consistency.

Scientific theories should so far as possible approach axiomatic systems. For such deductive systems Gödel in 1931 (see Nagel and Newman, 1959) showed that they cannot at the same time be consistent and complete ('complete' means that one can deduce from the axioms all true statements that can be expressed in the system). A choice must be made between consistency and completeness, and almost always consistency will be preferred. This means that if one succeeds in constructing a consistent axiomatic biosystematic theory (and I believe phylogenetic systematics is at least semi-axiomatic; Farris et al., 1970; Cracraft, 1974; Løvtrup, 1973, 1975 (on aspects of cladistics; Gregg, 1954 (on taxonomy); and Woodger, 1952 (on many aspects of biology)), one must expect that there are some relevant statements which remain indeterminate (neither true nor false). They can only be determined (the paradoxes resolved) by further expanding or generalizing the theory and this is an important feature in the development of any science.

Coupled with axiomatic systems, deductions and consistency, is another demand, not of a strictly logical nature, but rather a necessary scientific principle, namely that statements should be derived in a parsimonious way, i.e. as simply as possible. If certain deductive systems can be applied to describe nature (if they

have models in the real world), the statements - axioms and theories - acquire the character of assumptions and hypotheses, which can
be tested against nature.  An inevitable demand is that hypotheses
be derived in a parsimonious way, not because this is a better
guarantee of true statements about nature (which may not be parsimonious at all), but because, if not upheld as a principle, an infinite number of equally valid hypotheses concerning the same
phenomenon could be justified.

Parsimony is sometimes called a 'test' of hypotheses; e.g. by
Bock (1973) referring to the 'consistency test' of Wilson (1965).
Perhaps one could test hypotheses for internal (logical) consistency, however, I believe that this is a slightly unfortunate use of
the word test.  It should rather be reserved to mean the confrontation of hypotheses with nature (the phenomena of the real world),
i.e. an empirical test by experiments and/or observations.  Wilson's
(1965) consistency test of phylogenies is certainly nothing more
than the application of 'Occam's razor' (Bonde, 1975:  298; 1975a),
which demands that in an hypothesis about evolutionary history there
should be as few reversals and convergences as possible.

Testing hypotheses ia an important scientific endeavor and
can stem from different philosophical viewpoints.  A fairly naive,
but commonly held, view is that from nature one can obtain data
which are true and objective and constitute direct evidence.  A
sufficient quantity of such data more or less automatically reveals to us which true relations must exist in nature.  Further
confrontations between ideas and other data or observations can
simply prove or disprove the ideas definitely.  Such viewpoints
unfortunately appear implicit in some of the writings about evolutionary phenomena such as phylogeny (e.g. Simpson (1971:  357) -
"complete phylogeny" obtainable - "objectively from a simple array
of the data").  They are the basis for many claims that paleontology
is the only direct evidence of phylogeny and such commonly held
beliefs that one can study actual evolutionary histories from
fossil evidence (Gilmour, 1940:  471), or that paleontology 'proves'
the evolutionary theory.

Modern natural sciences were almost exclusively developed
within the tradition of analytical philosophy.  Popper (e.g. 1973)
has developed a philosophy of 'objective knowledge' concerned with
empirical science which he tries to differentiate from non-empir-
ical sciences in which statements cannot be seriously tested by
observations.  Popper claims to have shown that induction is logically unjustified as a method for use in empirical science.  So
he uses deductions from hypotheses which are testable, while the
other schools mentioned above use empirical verification (or confirmation) to indicate the scientific value of an hypothesis.  Only
the refutation of an hypothesis tells anything definite about the
hypothesis.

It is important to note Popper's claim that there are no such phenomena as data that are independent of previous hypotheses about the world. Therefore the scientific procedure starts with ideas or hypotheses, however vague, from which others follow, perhaps not as strictly logical implications, but as what I would like to call 'bio-logical' implications, or biologically relevant deductions. These statements, and others derived from them in a parsimonious way, form hypotheses that can be tested by confrontation with observations from nature, which, like experiments, can in principle be repeated by different persons. If observations agree with the hypothesis the latter is said to be confirmed or corroborated and it should be tested again. If the hypothesis and observations/experiments do not agree, the former or the latter or both are wrong; the observations should be repeated and/or the hypothesis should be modified. If subsequently further observations are still found to disagree with original hypothesis, it is said to be falsified and it must be rejected or altered. This hypothetico-deductive procedure can be performed in a very precise way within the framwork of phylogenetic systematics as exemplified by Miles, 1975 (see below). However, as mentioned by Farris (1973: 398) and Bonde (1975a: 563), the ideal of absolute falsification (disproof) is not possible in phylogenetic studies. Therefore it might be more correct to talk about refutation. In much evolutionary biology, statements, arguments and conclusions are often coined in such a way as to make them practically untestable. Sometimes it appears as if an author tries to protect his own hypotheses from testing (see Miles (1975: 137) on Jarvik's and Säve-Söderbergh's immunizing strategies).

It should be mentioned here that Ghiselin (1969) interprets Darwin's work in the light of an hypothetico-deductive method, and that Popper (1976) looks upon Darwin's evolutionism as a nontestable metaphysical framework of explanation rather than as a proper theory. Therefore it is of some importance that phylogenetic systematics can fulfill most ideal demands of the hypothetico-deductive methodology in a more satisfactory way than other biosystematic approaches (Wiley, 1975; Miles, 1973, 1975; Bonde, 1975a).

In concluding this section it can be pointed out that an ideal for many hypotheses of phylogeny and classification has often been the capacity to 'predict'. Mayr (1969, 1974) claims that this is true of evolutionary classification; however, I think the predictiveness of these kinds of hypotheses can only be very general. Predictions such as 'this scheme of relationships will hold good also in the light of future finds or tests' are all that are possible. The ability to predict features of organisms; e.g. of relatives and ancestors of certain groups, as pheneticists also would like their classifications to allow, appears to involve a rather atypical and narrow concept of prediction (cf. Farris et al. (1970a) discussing such statistical predictivity). However, distribution patterns of unknown relatives can be predicted, like Nelson's

(1969a) prediction of a primitive relative of _Hiodon_ in northern
Asia and Greenwood's (1970) confirmation showing that the Chinese
Jurassic teleost _Lycoptera_ is such a relative.

## AXIOMATIC STRUCTURE OF CLADISTIC THEORIES

The theory of phylogenetic systematics was not explicitly
structured as an axiomatic theory by Hennig (1950, 1966).  With
the purpose of making a quantitative analysis of phylogenetic
relationships, Farris _et al_. (1970) have tried to axiomatize
methodological principles derived from Hennig (1966).  One axiom
corresponds to Hennig's 'auxiliary principle' (1966:  121), apomorphous features must a priori be supposed to indicate kinship
between the species sharing them (merely a variant of Occam's
razor, the principle of parsimony, see below and fig. 1); a second
concerns the order of change of characters; the third defines a
criterion for strict monophyly of groups; and the forth concerns
quantification.  This last axiom also has to do with the hypothetico-deductive method and parsimony, namely the more synapomorphies
supporting a group's monophyly the better corroborated is the
hypothesis (equivalent to stating that the hypothesized synapomorphies lead to the most parsimonious distribution of character
states (features), i.e. the one demanding the least number of
changes of features during the postulated phylogeny).

Farris _et al_. (1970) so far succeeded in producing quantifi-
cations permissible under Hennig's methods, and developed from the
method of most parsimonious trees by Camin and Sokal (1965), and
from Wilson's (1965) ideal demands for unique and unreversed character states (cf. Farris, this vol.).

Cracraft (1974) has also developed some axioms or basic conditions for phylogenetic models involving statements about:  (1)
nature of kinship relationships of taxonomic units; (2) origin and
diversification of such units; and (3) clustering of such units
into lineages.  He then contrasted the more specific statements
under these three headings in a 'phylogenetic systematic' model
(e.g. like that of Hennig (1966) or Farris _et al_. (1970)), and
an 'evolutionary systematic' model (like that of Bock, this vol.).
Concerning (1), Cracraft's analysis (1974) emphasizes the differences in ancestor concepts in the two models.  In the cladistic
model ancestors are abstractions which cannot be recognized or
identified as fossil (or recent) specimens (but see the modification in this paper).  In the evolutionary model ancestral species
and especially ancestral groups are identified (and classified),
not only among fossils but also among recent organisms (e.g. when
discussing para- and sympatric speciation models among insects
and plants; e.g. Gottlieb, 1976).

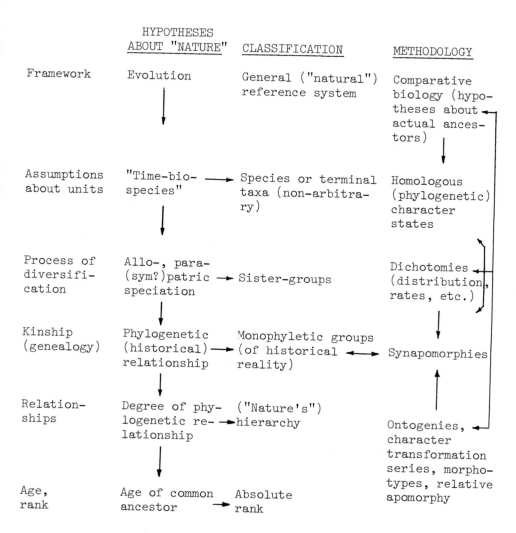

Fig. 1. Internal relations in Hennig's theory of phylogenetic systematics. Fat arrows mean "bio-logical implications", thin arrows should be read "leading to", arrows with cross bar mean "by convention converted to". (Modified after Bonde, 1975 and 1976).

While cladists only talk about speciation at branching points
of the phylogenetic tree (Cracraft, 1974, (2) above), in evolutionary systematics it is also accepted that one species in an undivided lineage evolves into another species (phyletic speciation).
Though Cracraft claims that there is no specific concept about the
nature of lineage branching, I believe it is fair to say that in
such an evolutionary model the ancestor almost always survives the
branching point, while giving off a divergent descendent species
as a side branch; e.g. Simpson (1975:  6), Throckmorton (1965).

Concerning clustering into groups in Cracraft's (1974) analysis, cladists relate species based upon shared derived features -
synapomorphies (see below and Hecht and Edwards, and Patterson,
both in this vol.), while he states that in evolutionary systematics 'weighted' similarity is used.  In the latter case there are
two components to be expressed, 'vertical' and 'horizontal' relationships (Simpson (1961) equivalent to Mayr's (1968) genetic and
genealogical relationships), only the latter being equivalent to
the cladistic relationship.  Stratigraphic position is given high
weight by Simpson in the evaluation of vertical relationships, as
is general in paleontology.  In fact it is often not much more than
the stratigraphic argument combined with an estimate of overall
similarity that is involved in many paleontological evolutionary
lineages.  The stratigraphic argument is almost always used to
indicate direct ancestor-descendent relationship, i.e. an important
indication of primitiveness (relative plesiomorphy) of the features
of the oldest fossil.  But, as pointed out by Schaeffer et al.
(1972), Nelson (1973:  368; 1973a:  89), Cracraft (1974:  77) and
Bonde (1975:  302), such an argument is irrelevant.

In short, one of the most important implications of analyses
such as the one by Cracraft (1974) is that whatever information
one may think has been put into a classification like a Linnean
Hierarchy, comprising names in a certain sequence and subordination (relative rank), nothing can be retrieved from this system
except hierarchical structures.  Degree of similarity or divergence
(or genetic similarity of Mayr) or other phenetic similarity
measures are in general not structured hierarchically but, as admitted by all taxonomists, show a reticular structure.  This cannot be stored in a hierarchy without distortion (i.e. loss of information), and certainly nobody has shown that it can be retrieved from such a hierarchy (unless already in the input only
hierarchically structured similarities had been used).

Phylogenetic relationships, as abundantly shown by Hennig
(1966), Nelson (1974), Griffiths (1974) and others, are structured
as a hierarchy, therefore if information about phylogenetic relationships are used as input in a Linnean hierarchy using some well-
defined conventions (e.g. about ranks), then exactly that same
pattern of relationships can be retrieved again.  If speciation was

never sympatric and organisms did not move much, this would be
true for distributional data also (see Hennig, 1966; Brundin, 1966;
Croizat et al., 1974; Nelson, 1974; and Rosen, 1976; and see below
on speciation and biogeography). Croizat's far-reaching biogeographical studies have shown that, to an astonishing degree such
hierarchical patterns can still be traced in the distribution of
recent organisms.

At this juncture the very interesting and slightly provocative
contributions on the axiomatization of cladistic classifications
by Løvtrup (1973, 1975) should be mentioned. He attempts to produce a theory for a strictly dichotomous hierarchy by starting from
the 'beginner' of the hierarchy (the root of the phylogenetic tree)
and proceeding towards the smallest groups, his "terminal taxa"
(instead of the more usual method of beginning at the latter level,
the specific (or sub-specific) taxa, the observable unity, and
grouping these into larger and larger taxa). Løvtrup increases
the rank from the beginner to the terminal taxa which has the consequence, as his ranks are determined by number of dichotomies
since the beginner, that the terminal taxa need not have the
same rank. Being a species thus has nothing to do with having a
specific rank in the system. This clash with the Linnean categories
will perhaps be the most difficult consequence for most systematists
to accept.

Løvtrup (1975: 500) begins with two initial methodological
premises, the second being - "Barring convergence, the varying
degree of similarity obtaining with respect to the properties
possessed by various animals are expressions of phylogenetic kinship". But how can one exclude convergence from the concept of
degree of similarity? This premise also does not exclude symplesiomorphies (shared primitive features) as indicators of phylogenetic
kinship, and, unless it in fact acts as a definition of phylogenetic kinship, this concept is never properly defined. Phylogenetic
classification is, however, defined (Løvtrup, 1975: 502, D5) by
using the undefined phylogenetic kinship, which is a relation later
characterized in a theorem (T38: 510) corresponding to Hennig's
(1966) definition of degree of phylogenetic relationship. This if
phylogenetic kinship happens to be based on shared primitive characters, this could contradict the theorem.

Løvtrup's first axiom (1975: 501, A1) states that - "classification involves the distinction and naming of classes of animals,
defined by certain properties possessed by all members of a set".
His first definition (D1) states that a - "taxon is defined by a
set of properties distinguishing a particular class of animals.
All individuals past, present and future, possessing the whole set
of properties are ... members of the taxon". On the whole this
appears to be a fairly classical (Linnean) and typological introduction to a theory of classification, and, from the initial premise,

apparently leads to a quantitative approach in agreement with Sokal
and Sneath (1963) that phenetic groups are usually monophyletic
(which they often are not in the strict cladistic sense, see below).
Løvtrup's definition (Dl), as well as his second axiom, appear not
to take into account that taxonomic characters can change. This
is admitted by Løvtrup, but I fail to see how the second axiom can
then be accepted as axiomatic. I also fail to see why there should
be identity between the number of dichotomies and the number of
changes of characters (steps) as implied in a theorem (1975:  507).

Løvtrup's (1975:  501) first theorem - "only extant animals
can be classified", I strongly disagree with (see Nelson, 1972,
1974; Rosen, 1973; Patterson, 1973; Bonde, 1976; and especially
Patterson and Rosen, 1977). Further (1975:  502) the statement
that fossils have - "no impact on classification" (cf. Hennig,
1966, 1969; Griffith, 1974; on the connection between ages and
ranks of taxa), needs some comment especially since the whole
purpose of the present paper is cladistic classification of recent
and fossil organisms.

## ASSUMPTIONS IN THE THEORY OF PHYLOGENETIC SYSTEMATICS

It was suggested above that phylogenetic systematics could be
considered at least a semi-axiomatic theory, based upon a few assumptions or general statements as follows:  (1) Biological diversity is the result of evolution; (2) Individual organisms have
characteristics (or features) which may be transferred (inherited)
from one generation to the next, either changed or unchanged; (3)
Objective units of biological diversity exist above the level of
the individual; (4) Processes exist due to which some such units
are divided, creating increased diversity; (5) Some such units may
become extinct, decreasing biological diversity; (6) Life as known
today had a unique origin. The last two assumptions may seem more
or less superfluous, but (5) helps explain why diversification (4)
can still go on; while (6) is necessary to prevent the argument
that all species (or most of them) with or without change of
features have existed since organic life originated. (6) implies -
as should be obvious later on - that all organisms are (phylogenetically) related.

### On Changes

The concept of biological evolution basically means no more
than (historical) changes from one generation of organisms to the
next (ancestor to descendent). Such changes are mediated through
information passed from ancestor to descendent and through interaction with the environment and occur at two levels:  genotype
(mutations) and phenotype. The reconstruction and systematization
of this evolutionary history (or phylogeny) is part of the task of

phylogenetic systematics, namely that part which has direct bearing upon classification in a general reference system (or 'natural'
system).  The diversity exists, i.e. similar (or identical) versus
different features, at both the geno- and phenotypic level.  All
are results of changes during phylogeny, and the pattern of similarities and differences must be evaluated by the same sort of analysis.  Whether it is justified to weight one set of data - say
the genotypic features - as more important than other features
must be a matter of dispute so long as the exact interdependence
between genotypic and phenotypic features is known in only a minute fraction of the cases that have to be analyzed.

The entire distribution of features still has to be explained
as the outcome of one unique evolutionary history in the simplest
way possible, which probably means involving as few changes and
as small changes as possible.  This can almost be done without
knowledge of the underlying causal relations and specific processes
by an analysis leading to a 'minimum-step model' of the phylogeny.
This line of reasoning must be closely similar to that of Fitch
and Margoliash (1967), Fitch (1976, and this vol.), Camin and
Sokal (1965), and Farris et al. (1970).  Specifically it is worth
stressing that any prior understanding of the functional relationships or adaptive advantages of features is completely unnecessary
(or rather irrelevant; cf. Farris, 1969) for the reconstruction of
a phylogeny.  But after the reconstruction it is obviously interesting if a functional interpretation can also be added (e.g.
Schaeffer and Rosen (1961) on fish feeding; von Wahlert (1968) on
gnathostome adaptations; Simpson (1951a) and Sondaar (1968) on
horses).  However, to make the reconstruction of a phylogeny a
priori dependent upon a knowledge of functional anatomy, physiology,
ethology, or other sorts of 'adaptiogenesis', is turning the case
upside down.  Anyway it is unrealistic or even impossible for most
organisms, which cannot in any way mean that attempts to reconstruct
their phylogeny should not be made.

## On Features

The term feature has repeatedly been used above for something
that can be similar or dissimilar; something pertaining to both
geno- and phenotype; and something that can change (synonyms include:  character, characteristic, attribute, character state).
It simply means any descriptive statement that can be compared
with other such statements to show whether they are identical,
similar or different (in the organisms or groups of organisms concerned).  As such, features are purely conceptual; they are arbitrarily chosen and delimited; they may overlap or be exclusive of
each other; they may be interdependent or independent; they may be
simple and indivisible or complicated and divisible with ease or
difficulty into simpler (sub-) statements; and very often they are
generalizations (e.g. mean values, ranges).

Obviously features are symbols supposedly corresponding to some real phenomena, but the degree to which phenomena of the genotype and the phenotype are interwoven and dependent on each other is unknown in the majority of cases. Specifically such factors as the pleiotropic effects of a gene; the degree to which one phenotypic feature is the expression of many genes or one gene, or whether it could result from some other combination of genes or different single genes; and the degree to which changes in genes are dependent on each other, are known in so few cases that they can hardly count for much in bio-classification. Probably such details for most organisms will remain obscure for the foreseeable future.

What we employ as features depends on the sort of analysis we want to do, and it is difficult a priori to exclude some features or weight them very low in classification, because all have to be explained as the results of one evolutionary history. Hecht and Edwards (1976, and this vol.) have provided some 'rules of thumb' for weighting features which are not unreasonable. Such weighting provides a scale on which features can be measured, so that for example, complex features count much more than simple ones. But how much more is a matter of dispute; presumably the weight must be inversely related to an intuitively measured probability of a convergence in the feature.

A very different sort of weighting concept is the one suggested by Farris (1969) in which the features are evaluated for cladistic reliability, i.e. degree of correspondence between feature and ('true') phylogeny. As this phylogeny (a tree) obviously can only be justified from an analysis of the features, it is accordingly called 'true cladistic relationship', this computer technique should be viewed as one which assures discovery of a hierarchic structure (a tree) that is indicated by correlation (hierarchic correlation) of the largest number of features. These features are then said to have 'cladistic reliability', while features which show much convergence, parallelism and reversal on that tree are unreliable, they show little hierarchic correlation.

In Farris' analysis (1969) he also refers to properties of 'unit characters', meaning the discrete steps (or changes) in a specific transformation series of features. However, often the concept of a unit character is used as an ideal by numerical taxonomists (e.g. Sokal and Sneath, 1963) to indicate a feature which cannot be further subdivided; this appears rather meaningless in a biosystematic context. Logically morphological features, for example, could be subdivided into statements about atomic and subatomic structures, and it might be difficult to decide where to end this regression.

## On Definitions

Feature being defined as a theoretical term above, is in accordance with Ghiselin's (1966, 1966a) attitude towards definitions which are simply words to replace and (hopefully by being more immediately understandable or less ambiguous) to explain other words. The very common expression definition of a particular species (or group of species) by enumeration of some of its features is an unfortunate use of the theoretical term definition (e.g. by Løvtrup, 1973, 1975). Such definitions are simply descriptions or diagnoses. The theoretical or abstract definition of feature is of the 'non-operational' type, which means that it does not refer to any criterion by which a feature can be recognized in practice. Such non-operational definitions should always be preferred, because, contrary to operational definitions, they do not impose unnecessary restrictions on the concepts being defined.

## On Species

The 'biological species' (Mayr, 1963, 1969: 25) is defined as - "a reproductive unit of individuals, which is reproductively isolated from other such units". Gene flow between semi-isolated populations is supposed to keep these together in a cohesive species and to have a homeostatic effect upon the combined gene pool. Therefore a species is presumed to possess a characteristic gene pool with a somewhat limited variation. The importance of gene flow has, however, been questioned by Ehrlich and Raven (1963) and Endler (1973); and some pheneticists doubt that the biological species is such a fundamental evolutionary unit (e.g. Sokal, 1974; Sokal and Crovello, 1970). Hennig (1966: 47) accepted the biological species concept for recent bisexual organisms, but adds that distribution in space should also be taken into account in defining the species as a - "complex of vicarying communities of reproduction".

The species is considered to be a real phenomenon (and therefore also a historical unit) by most biologists; but quite often, especially among philosophers, it has been thought of as a class concept with individuals, as members of species, being the only real things. The reality, and therefore the logical individuality, of species is also emphasized by Ghiselin (1974) who stresses the logical consequence that individual species cannot then be defined (e.g. by possession of some properties), but only discovered, described and given a proper name. An important aspect of such an approach is that - "similarity of the individuals of a species through which they form a logical class, is an indicator, but no criterion of their belonging to the same species" (Löther, 1972: 296); any typological or phenetic 'operational'definition can thus be rejected.

The temporal aspect is left out of the biological species concept much to the dissatisfaction of paleontologists.  Simpson (1961)
therefore defined the 'evolutionary species' as a - "lineage (an
ancestral-descendent sequence of populations) evolving separately
from others and with its own unitary evolutionary rôle and tendencies".  Although this is intuitively appealing, the phrase 'evolutionary rôle and tendencies' is not immediately understandable and
pheneticists (e.g. Sokal and Crovello, 1970) have criticized the
concept heavily.  Furthermore, it does not give any indication
about the delimitation of the time-extended species.  According
to Simpson (1961) it has to be arbitrarily delimited by considering the amount of change through time.  Many paleontologists and
neontologists believe that this arbitrariness is necessary to
divide the evolutionary continuum into 'species' (see Sylvester-
Bradley (1956); George (1971:  227) and Mayr (1963, 1969:  35) -
"The vertical delimitation of species in the time dimension should
in theory be impossible").  Westoll (1956:  60) suggested, however,
that the branching points of the evolutionary continuum were ideal
limits of a time-extended species to be called - "a chronospecies
throughout, no matter what change might occur within its range".
Simpson (1961:  161) objected to this method of delimitation as -
"undesirable and impractical".  Nevertheless he (1961:  168) remarks (about a stem lineage giving rise to two lineages) that -
"in principle, the best solution is available when the lineages
can be divided into three species ... separated by the point of
branching".

Hennig (1966), and all cladists, simply assume that species
are delimited at the speciation events.  The species concept resulting from this is the only logical extension in time of the
concept of the integrated gene pool.  At speciation this gene-
pool is disintegrated and two (or more) new sister species originate, while the original species becomes extinct.  In this way the
species becomes a time-extended (historical) unit with, in principle, non-arbitrary limits, and with the biological species (bio-
species) as a time transect.  This is the fundamental unit of the
phylogenetic system and (as the term 'phyletic species' has been
used for another concept) it might well be called a 'time-bio-
species'.

If phylogeny is visualized as a phylogenetic tree every internode (or segment) represents one species irrespective of the changes
taking place between the nodes.  In this manner degree of similarity within a species is irrelevant.  Obviously it is a definition
which is barely 'operational', and this is exactly the point, we
are now free to choose between any practical criterion that might
suit our purposes to indicate when a real species is met with in
nature.  In fact the biological species concept is not quite as
ideal being to some degree an operational definition taking into
account only one criterion of a species, namely that of being

reproductively isolated.  Problems of asexual species therefore
cannot be solved under the bio-species concept, neither does
Simpson's (1961) evolutionary species concept allow this (although
Simpson and also Maslin (1968) think so) because it refers to populations which probably have to be defined in terms of interbreeding (e.g. Strickberger 1968:  693).

An important aspect of any species definition whether in
neontology or paleontology is that any statement that particular
individuals (or fragmentary specimens) belong to a certain species
is an hypothesis (not a fact).  More particularly it is an hypothesis about a phylogenetic relationship.  The uncertainty about
specific status remaining in paleontology becuase of fragmentary
information is probably the greatest stumbling block to pro-posing
phylogenetic ideas based upon fossils.  Even in principle no-one
ever knows for sure whether two fossil specimens are representatives
of the same species.  In most cases, however, the position is not
very much better with extant organisms.

## PROCESS OF DIVERSIFICATION

Diversity among organisms can originate by different means:
(1) by multiplication of species through the disintegration of
former species (speciation proper by branching - cladogenesis);
(2) by divergence of species without branching (change within
single lineages - anagenesis (phyletic evolution)); (3) by increased variability within single lineages.

In reconstructing phylogeny cladistically the first step is
to try to determine the speciations (branching points or 'nodes'
of the phylogenetic tree), because not until this is done are the
different lineages well defined.  Actually the lineages can best
be seen as a consequence of the analysis of phylogenetic relationships.  So that only the proper speciations will interest us here
because they can be precisely reflected in classifications, while
what happens between the nodes, adaptiogenesis, is left for other
types of studies as it cannot be unequivocally expressed in hierarchical classifications (see Hennig, 1975:  245; Nelson, 1974a;
Farris, this vol.; and Cracraft, 1974).

The speciation process (as visualized by Mayr, 1963, 1969,
1970) involves allopatric speciation in which some physical barrier splits the original unified species.  After such (macro or
micro) geographic isolation, reproductive isolating mechanisms may
occur.  If later on these isolates happen to meet and cannot successfully interbreed, the speciation process has been completed
and - at least in animals - is generally supposed to be irreversible.

It has often been questioned whether allopatric speciation is

really such a general feature of animal evolution and whether
sympatric speciation is occurring.  Some geneticists have claimed
that there exists a type of speciation process between allo- and
sympatric speciation, the so-called parapatric (stasipatric) speciation, inferred from the distribution patterns of closely related species with close or adjacent borders, but without mixing
or perhaps with narrow hybrid zones containing less viable hybrids
(e.g. While, 1968, 1973, 1973a, 1974; and Key, 1968, 1974, on
grasshoppers; Nevo et al., 1974, on fossorial rodents; and perhaps
Fryer and Iles, 1972, for some cichlids).

I assumed (1975:  295) that for animals there was no reason
to take any speciation model other than the allopatric one into
account.  However, Bush (1975) has recently suggested that sympatric speciation is quite likely for vast numbers of animal species, notably all parasites.  Sympatric speciation has even been
suggested for fishes (Frost, 1965) by the development of temporal
isolation of breeding periods (allochronic speciation).

How can we evaluate the consequences for phylogenetic systematics?  First of all the possibility that some species could spe-
ciate sympatrically does not mean that many of them necessarily
arose that way.  Secondly, since phylogenetic analyses are retro-
spective, it does not matter whether species originated by allo-
or parapatric speciation, the resulting patterns of both phylogenetic relationships and distribution will be of the same kind in
both cases.  Concerning the pattern developing from a strictly
sympatric speciation process, the phylogenetic relationships will
probably still be analyzed in the usual way in as much as the two
or more descendent species (including what, in a non-cladistic
terminology, would be called the persisting ancestral species or
original species) must be considered each others' phylogenetically
closest relatives - sister-species.  The only real difficulty might
be if a species periodically (or constantly) gave rise to identical
sympatric species (e.g. the effect of identical mutations); but
presumably it is extremely difficult to find criteria indicating
such occurrences.  Perhaps most sympatric speciations could even
be considered 'microgeographical' (e.g. changing host), thereby
indicating the essential similarity of the developing patterns of
phylogenetic relationships.  But even in this way biogeographical
analyses will probably be disturbed by such occurrences which must
still within these working methods be called sympatric.  This
phenomenon, sympatry, cannot then always be used as a primary
criterion of dispersal in a 'Croizatian-type' analysis (Croizat
et al., 1974; Nelson, 1973c, 1975, 1976; Rosen, 1976).  Croizat
et al., (1974:  278) believe that most cases of sympatric speciation will probably show a vicariance pattern of distribution anyway, so perhaps one can avoid some of the pitfalls, but, I would
estimate, not all.

I therefore conclude that some of the chorological (distributional) relationships elucidated by Croizat's method of generalized tracks may have to be re-analyzed if sympatric speciation turns out to be a likely interpretation for the origin of many animals. However, the analyses of phylogenetic relationships - contrary to my earlier impression - will be very little, if at all, affected by cases of origin by sympatric speciations.

## MONOPHYLY AND RELATED TERMS

A monophyletic group (in the strict sense of Hennig (1966: 73); equivalent to Huxley's (1957, 1958; clade) consists in the recent biota of all and only the descendents of one ancestral species (a group of unique descent). To extend it into the time dimension to suit paleontology it should be phrased: A monophyletic group includes (only) a species and all its descendents. Consequently the species in a monophyletic group are more closely (phylogenetically) related to each other than to any species outside the group. In that sense it is a 'natural' group, probably of a kind that most biologists intuitively can appreciate as a basis for a classification. Such a strictly monophyletic group I shall call a 'clade'.

The kinship relationship within a clade is currently called genealogical relationship (e.g. Mayr, 1974). Genealogy is a reticulate pattern and, therefore, a slightly unfortunate analogy. Nevertheless this designation serves to keep it distinct from what the 'evolutionists' (and some pheneticists) would call genetical relationship (e.g. Mayr, 1969, 1974) which means genes in common or degree of similarity of genotype, and which therefore is a purely phenetic (typological) relationship. However, Hennig (1966) employs genetical relationship as a designation for both ontogenetic relationships (between his semaphoronts - extremely short lifestages of an individual), and tokogenetic relationships (depending on interbreeding and therefore constituting the genealogy).

The special kinship relationships or phylogenetic relationships between the species of a clade means that it is a real phenomenon of nature or a natural entity (Hennig, 1966: 77-83; Griffiths, 1974; Woodger, 1952: 21; Brundin, 1972: 108), despite remarks by Beckner (1959: 68) and others to the contrary. It is as real as a species even in the sense of being one gene pool. A species, naturally, is also a clade (the smallest), but above the species level the original ancestral gene pool is disintegrated and distributed among the descendents. Logically a clade, just like a species, can be treated as an individual (cf. Ghiselin, 1966a, 1974; Hull, 1974), as it is a real object deserving a proper name. Exactly as in the case of a species, such a whole or individual (the particular group) cannot be defined in terms of the common properties of its parts; being real it can be discovered and described but not defined. Quite a different (practical) question is how clades can be recognized, and this is exactly what Hennig's cladistic methodology is all about.

A competing definition of monophyly is that by Simpson (1961: 124) - almost identical with Beckner (1959) and Gilmour (1940: 469) - "monophyly is the derivation of a taxon through one or more lineages (temporal successions of ancestral-descendent populations) from one immediate ancestral taxon of the same or lower rank". This makes monophyly of a group dependent upon a predetermined rank for both the group itself and for its ancestral group; further it assumes that ancestral groups can be identified. Simpson called his concept 'minimal monophyly', and this concept was generally accepted by the synthetic school (e.g. Mayr, 1969).

With proper adjustments of rank any group can be made monophyletic under this definition (e.g. a 'superclass' Aves + Mammalia, as it is derived from Reptilia generally called a 'class'). A serious drawback is that Simpson never provided precise rules for ranking; actually he stated (Simpson, 1961: 145) that - "classification is concerned first with the delimitation and then with the ranking of taxa". So we move in a complete circle - we must know the groups before we can determine the rank! A basic fallacy of evolutionary taxonomy and traditional paleontological (and some phenetic) arguments is that some supraspecific taxa ('groups') and their categorical rank are accepted a priori without the slightest evidence for either rank designation or group delimitation. 'Evolutionists' in this way simply begin by assuming what they supposedly hope to 'prove' (or indicate) in their phylogenetic analyses and classifications, namely that particular named groups (or supraspecific taxa) are real and have a particular rank. Acceptance of such groups seems to be based on convention and tradition in order to keep classifications stable. These preconceived 'groups' are always established with reference to degree of similarity and as such are an expression of a pure phenetic approach at some 'basic' level. The basic level generally singled out, especially by paleontologists, is generic. Probably this is a pre-evolutionary leftover legitimated by the Linnean binomial nomenclature imparting an 'essentialistic' and phenetic impression. It is why cladists, with some justification, call 'evolutionary' taxonomy a typological approach with pre-evolutionary elements.

A recent example of such a paleontological analysis is that of Harper (1976) in which problems are encountered in attempting to explain what an ancestral supraspecific taxon means. If the relevant questions are asked, namely 'what is his concept of a genus?' and 'is it a monophyletic group?' and 'how were the original fossils grouped into the initial 'genera' (or other higher

taxa)?', then many of Harper's arguments (the non-cladistic ones)
are shown to be merely pheneticism and can for that reason alone
be dismissed as pre-evolutionary or a philosophically defect (see
Hull, 1970). Harper's (1976: 184) arguments are based upon the
fallacy of conceiving taxa as classes defined intentionally, instead of as real phenomena, existing entities which cannot be defined , but only discovered, described and related (cf. Griffiths,
1974; Ghiselin, 1974).

Simpson (1961: 113) has argued that such a concept of minimal monophyly can be employed in classifications which are said to
be 'consistent' with phylogeny - "a consistent evolutionary classification is one whose implications, drawn according to stated
criteria of such classifications, do not contradict the classifier's
views as to the phylogeny of the group". However, this prescription
would seem to allow almost any relation between phylogeny and a consistent classification depending on the stated criteria. No precise
criteria for this consistency with phylogeny were presented by
Simpson (1961: 114-46). Non-arbitrariness of a taxon is an ideal,
which evidently has to do with degree of resemblance, i.e. a phenetic clustering. Age of a group is dismissed as having little
relevance for ranking and even the possibility of using age in a
consistent way to indicate rank is denied. The only advice given
is well summarized (Simpson 1961: 119) - "it is, of course, a
question more of taste or skill in the art of classification than
of theory in the science of taxonomy"!

An advantage of applying the cladist's strict monophyly is
that para- and polyphyletic groups (see below) can be clearly distinguished and kept apart from clades, which is not possible with
Simpson's definition of monophyly. Another advantage is that inter-
group relationships between clades are of the same kind as interspecific relationships, namely degrees of phylogenetic relationship,
and all species of a clade have the same degree of phylogenetic
relationship to all other groups. If groups are not clades, which
counts for many of Simpson's minimally monophyletic groups, then
one cannot talk about their phylogenetic relationships as groups,
because their constituent species do not possess the same degree
of phylogenetic relationship to all other groups.

I have earlier (Bonde, 1975: 294) considered Simpson's definition of monophyly - "the single most unfortunate choice of the
'evolutionists'". Thus the development of Simpson's concept of
monophyly will be followed below because of its importance in
phylogenetic theories.

Many 'evolutionary' paleontologists; e.g. van Valen (1960),
Reed (1960), or Simpson (1960), have suggested more or less arbitrary choices of features (conventions) to distinguish Mammalia
from something else (Reptilia). The boundary being fixed somewhere

between the original speciation leading to the mammals (and sauropsids) and the last (or immediate) common ancestor of all recent
mammals, or even (Simpson, 1959a; Romer, 1966) at some arbitrary
points after the latter ancestor had split into subgroups. These
suggestions are all based typologically upon some 'essential' features characterizing 'true' mammals. Simpson (1959a) even went
as far as suggesting that the class Mammalia (a grade and not a
clade) should be accepted although being a polyphyletic group
(but a non-monophyly of recent mammals was not at all properly
demonstrated, as noted by Hennig (1966: 214)). Such delimitation
was accepted by Romer (1966, 1968).

In the same year Simpson (1959b: 267) did in fact define
monophyletic in the strict way (as cladists do) - "clades are
monophyletic higher categories" and also - "crossing of a grade
line may be monophyletic (grade and clade coincide)", as opposed
to - "polyphyletic (they do not coincide)". Nevertheless he still
considered mammals as being a grade - "not a natural group as regards their origin", and he wanted to retain this 'not natural'
taxon in what he would still call a truly natural or phylogenetic
classification. In the following year, Simpson (1960: 389), possibly influenced by Beckner (1959), changed his terminology radically to make 'monophyly' the proper designation by definition for
the concept he the year before called 'polyphyletic'. Further he
dismisses completely the sort of relationship that he had called
monophyletic before (which was the cladistic concept; see Simpson,
1961: 123).

That the claimed polyphyly was not properly demonstrated by
Simpson (1959a), but was mainly due to a confused terminology, was
perhaps later hinted at by Mayr (1969: 75) - "some authors have
referred to 'polyphyly' of a taxon when only a polyphyly of a diagnostic character of the taxon was involved, the taxon itself being
monophyletic". This describes precisely what had happened:
Simpson used a postulated 'polyphyly' (or multiple origin, as I
prefer to call it) of the mammalian jaw articulation to indicate
that the taxon Mammalia was polyphyletic. However, this does not
necessarily follow, because it does not disprove that all (recent)
mammals are most closely related to each other, one of the conditions
that Mayr (1974: 104) now considers a criterion of monophyly.

Such confusion between character-phylogeny and group (or taxon)
phylogeny is often met with (e.g. in Jarvik's 'diphyly' of Tetra-
poda (1968)). This stresses the need for a non-ambiguous terminology, which should restrict words (and their associated concepts)
like mono-, para-, di-, and polyphyletic to designations for the
relationships of groups (or taxa) and not for the origin of single
characters. Characters should be described by quite different des-
gnations such as 'homologous' (which can be used to indicate strictly monophyletic groups) or 'homoplastic' (that is non-homologous;

according to Simpson (1961:  78) including parallelism, convergence, analogy, mimicry, and chance similarity, and therefore, indicating non-monophyletic groups).

The reason why I have considered Simpson's monophyly at some length is not only because of the enormous influence of this concept on taxonomic methods (monophyletic - or at least non-polyphyletic - taxa have become the ideal for most classifiers, regardless of how the concept may be defined); but also because Simpson's definition is apparently easily accepted by biologists, although it makes little sense without unequivocal rules (better than artistic feelings and traditions) for determining rank of taxa.

'Evolutionists' like Mayr (1974) have criticized Hennig for giving new meaning to old well established concepts such as monophyly; but, as Nelson (1974b:  454) has argued, it may well be that Simpson (1961), and before him Gilmour (1940) and Beckner (1959), gave the concept a new meaning.  To be exact Simpson (1961:  124) called the concept 'minimal monophyly', a concept defended by Hull (1970), Ashlock (1971), Mayr (1969, 1974), and Jardine and Sibson (1971), as the correct definition of monophyly.  Using this concept Mayr's (1965, 1969, 1974) essentially phenetic attitude towards relationships appears phyletic since almost all groups of some similarity can be postulated to form a 'monophyletic' group. Mayr (1974) has confused the concept monophyletic even more by stating two conditions necessary for the recognition of monophyletic groups; - "(1) the component species, owing to their characteristics, are believed to be each other's nearest relatives; and (2) they are all inferred to have descended from the same common ancestor".  Despite Mayr's claim of unambiguity, point (2) has no use - even assuming that the common ancestor is a species, not a group - as it counts for any grouping of organisms; and point (1) depends upon what is meant by 'nearest relatives ... owing to their characteristics'.  It is however, clear that Mayr has only purely phenetic groups of similarity in mind.  Paraphyletic groups are therefore as good as any strictly monophyletic group in Mayr's system, and the distinction he claims is of no relevance.  The obvious advantage (from an analytical viewpoint) in being able to distinguish three different groupings (mono-, para- and polyphyletic) instead of only two is lost.

Other authors have characterized monophyletic groups in the vaguest possible terms; e.g. Hecht (1976:  337) - "all members of that group share a common ancestor".  As mentioned above this covers any group of organisms (as we assume that life originated only once on Earth).  The concept is not made more useful by the statement that - "fundamentally such common ancestor should be a single biological species", especially when it is added that - "this is an unrealistic goal" since the - "level of resolution required is too low in fossil material", and - "monophyly is generally an

applicable interpretation in fossil vertebrates only at generic
level or above".  Such again implies that paraphyletic ancestral
groups are included in the concept, as is evident also from a
statement (Hecht, 1976:  346) about - "the fact that the equid
lineage is derived from this genus" (Hyracotherium).  For classificatory purposes Hecht (1976:  338) apparently uses phenetic taxa
- "species or other high category recognized by the morphological
similarities of the contained members".

     Hull (1964, 1970, 1974) has analyzed Simpson's (1961) concepts
of 'monophyly' and 'consistency', concepts that Hull considered
characteristic of what he called phylogenetic taxonomy.  Hull
(1964:  7) has no doubt as to what monophyly originally meant -
"a taxon was considered monophyletic if and only if it was derived
from a single species" (my emphasis).  He considered that should
this definition be taken literally - "even contemporary classifications would have to be so reorganized that they would cease to
perform the primary task of any classification, that of organization and simplification".  Therefore - "abandon ... strict monophyly or else phylogenetic taxonomy cease to be useful as a classification" (both statements I believe have been refuted by cladistics).
Hull (1964:  7, 10) suggested that Simpson's redefinition provided
the necessary modification making sure that - "polyphyly at higher
levels never ... conflicts with the Linnean hierarchy", and - "any
empirical situation ... can be accommodated by a redistribution of
neighboring taxa".  Hull points to three phases in the progressive
weakening of the relation between classification and phylogeny,
from classification (1) expressing phylogeny, to (2) based upon
phylogeny, to (3) being consistent with phylogeny.  And he claims
- "any further weakening of the relation ... would mean that nothing at all would be implied by any ... classification".

     Nevertheless, Hull (1964:  8) finds good reasons for the
changes that Simpson had made as - "this is the result of the form
of the Linnean hierarchy and not ... of purposes of phylogenetic
taxonomy or phylogeny itself".  This is inspired by Simpson (1959:
301; 1961, 1964) also quoting Darwin (1872:  484), stating that
phylogeny cannot be represented on a piece of paper.  However, this
is certainly wrong, as shown by all cladistic analyses (Hennig,
1966; Nelson, 1973; Rosen, 1973; Patterson, 1973; Løvtrup, 1975;
Cracraft, 1974; Farris et al., 1970; Farris, in press; Bonde, 1975;
etc., also below).

## RELATIONSHIPS

     Relationship, as pointed out by Hennig (e.g. 1975) has two
distinct meanings:  form relationship (similarity as generally
used) and blood relationship (genealogical).  The latter is characteristic of living beings, while both animate and inanimate

phenomena can show  form relationship.  When the word 'related' is
used by biologists it is often impossible to decide which of the
two meanings is being used, and in 'evolutionary' taxonomy both
meanings are deliberately used at the same time creating confusion.

The unique kinship relationships of organisms are a consequence
of biological evolution and therefore used in phylogenetic systematics.  The genealogical relationships of a species are broken up
by the speciation process into segments, called sister-species
above, and when these disintegrate a hierarchy is formed.  Now the
phylogenetic relationship is defined as a relationship between
such segments (or species, see above) so that sister-species are
considered each others closest phylogenetic relatives.  Further,
degree of phylogenetic relationship is defined so that two species
A and B are more closely (phylogenetically) related to each other
that to a third species C, when A and B have a common ancestor (or
stem-species) which is not ancestral to C, i.e. A and B share a
more recent ancestor than either of them do with C.  This means
that degree of phylogenetic relationship is measured by recency of
common ancestry (Hennig, 1965:  98, 1966:  72).  This is probably
what Darwin meant by 'propinquity of descent' (see for example
Simpson's (1959) very phylogenetic interpretation of Darwin).

Consequently, in a monophyletic group all members are more
closely related to each other than to any species outside the group.
Phylogenetic systematics aims at a hierarchic classification with
units which are 'natural' groups of relationship.  Monophyletic
groups (clades) can now be understood as such natural groups with
a unique history of their own, and they are therefore used as units
in the cladistic classifications, which can thus be called natural
classifications.

The whole theory of phylogenetic systematics could be looked
upon as a semantic analysis of the word 'related' with a precise
and unique biological meaning, and the consequences of this analysis for bio-classifications (Bonde, 1975:  312).

PARAPHYLETIC AND POLYPHYLETIC GROUPS

The antonym of monophyletic is polyphyletic which designates
that a 'group' originated through several distantly related lineages by convergence or parallel evolution.  Such groups have
generally not been considered 'natural' groups and are not normally
accepted in classifications (but see Simpson, 1959a, and Mayr, 1974).
Polyphyletic was given a precise formal definition by Nelson (1971a,
1973) in relation to a phylogenetic kinship diagram (see fig. 3)
and equals a (strict) monophyletic group (his 'complete sister-
group system') exclusive of two or more monophyletic subgroups (or

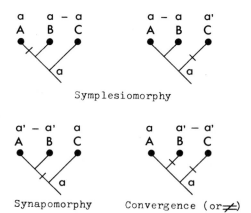

Symplesiomorphy

Synapomorphy          Convergence (or≠)

Fig. 2.  Relationships as indicated by "similarities" divided into
symplesiomorphy, synapomorphy, and convergence (or parallelsim).
A, B, C are species (or monophyletic groups); a is plesiomorph,
a' the corresponding apomorph feature; cross bar indicates origin
of apomorph feature.  Synapomorphy is the most parsimonious solution indicating close phylogenetic relationship as symplesiomorphy
is ambiguous.

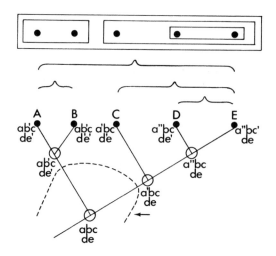

Fig. 3.  Assumed "true" phylogeny of 5 species (or monophyletic
groups) A-E.  Plesiomorph features are a-e, corresponding apomorph
ones a'-e'; a, a' a" form a trend, which indicates that D+E is
monophyletic and C its sister-group.  A+B is monophyletic (b'
synamorph).  A+B+C+D (c symplesiomorph) is paraphyletic.  B+C is
polyphyletic (d' convergent).  Arrow points to area in which a
species with only plesiomorph features could be linked to the
dendogram.  Two corresponding hierarchies shown.  (From Bonde,
1975).

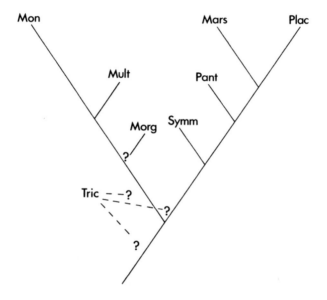

Fig. 4. Phylogeny of mammals based mainly on Kermack & Kielan-Jaworowska (1971) and simplified considerations of tooth and jaw structures. Hypothesis in slight disagreement with McKenna (1975) concerning early possible Prototheria. See the corresponding cladistic classification in the text.

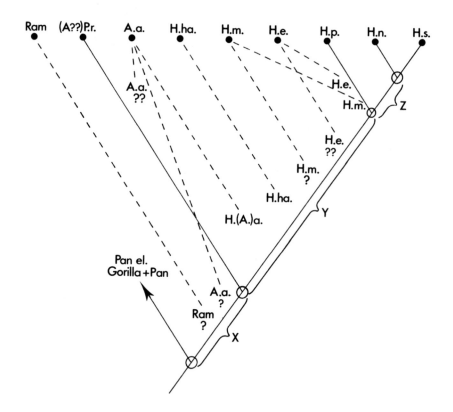

Fig. 5. Models of the phylogeny of Man. Solid lines indicate the phylogenetic relationships of species which are almost certainly monophyletic. Relationships of more problematic fossil species (or parts of species) are indicated by the broken lines. Names of "species" which cannot be excluded as possible ancestors of other species are also shown near their "ancestral position" on the lines. More ? indicates less likely interpretations. x, y, z are ancestral species. A = Australopithecus, a = africanus, e = erectus, H = Homo, ha = habilis, m = modjokertensis, n = neanderthalensis, P = Paranthropus, p = pekinensis, r = robustus, Ram = "ramapithecines", s = sapiens. (Simplified from Bonde, 1976, fig. 5).

species), which combined do not form one monophyletic group. Therefore polyphyletic groups contain two or more subgroups which are more closely related to groups not included.

A third category of taxonomic groups - paraphyletic - falls conceptually in between mono- and polyphyletic groups.  In Nelson's (1971b, 1973) precise definitions a paraphyletic group means a monophyletic group exclusive of one of its monophyletic subgroups (or one species), again it is in relation to a phylogenetic kinship diagram (see fig. 3).  As for polyphyletic groups some (at least one) of the contained groups in a paraphyletic group have closer relatives outside the group than inside it.

Hennig (1966, 1975) attaches a somewhat different meaning to poly- and paraphyletic groups, relating the concepts to the way in which the groups are kept together by features shared by their constituent species.  In his conception paraphyletic groups are indicated by symplesiomorphic features:  polyphyletic groups are based upon convergence.  In terms of structure of phylogenetic relationships the two kinds of groups may be identical as shown by Hennig (1975, fig. 1).  This is an unfortunate distinction, because it naturally means that species of a particular 'group' can share symplesiomorphies in one feature, but be convergent in another feature.  Should it then be called para- or polyphyletic or both? An example would be the recent Pongidae (excluding Hylobatidae) which, if one of my preferred hypotheses (Bonde, 1976 and below, fig. 5) is correct and Pan (or Pan + Gorilla) is the sistergroup of Homo, then the Pongidae in its classical conception (Pan + Gorilla + Pongo) is paraphyletic because of all the plesiomorphic (non-human) features, but polyphyletic because of the large shovel-shaped front teeth, which most likely must be considered convergent.

For this reason I prefer Nelson's formal redefinition of paraphyletic, which makes paraphyletic groups a category conceptually midway between mono- and polyphyletic groups.  Hennig's definition (which has priority) tends to transfer the concept from being a designation of a group to being a characteristic of the features of a group if inconsistencies are to be avoided.  Why not instead simply call the group 'symplesiomorphic'?

An interesting class of paraphyletic groups comprises most of the extant so-called 'stem-groups' (e.g. Pongidae, Reptilia, Lacertilia, Sarcopterygii, Pisces, Protochordata, and perhaps Agnatha). A few, however, may be polyphyletic in Nelson's sense (e.g. Insectivora); while almost none of them is in Hennig's sense (but see Pongidae as explained above).

'Ancestral groups' with fossil forms taken into account are all paraphyletic in Hennig's sense.  This includes all the groups mentioned above and also such as Prosimii (lemuroids) and

Tarsioidea, both of which groups can be called 'ancestral' and paraphyletic only because of inclusion of some dubious fossils (Prosimii are probably, and Tarsioidea are certainly, good monophyletic groups as represented by living forms) and entirely fossil groups like 'Eohippus' and some other horse ancestor, Symmetrodonta, Therapsida, Captorhinomorpha, Rhipidistia, Palaeonisciformes, Leptolepis, and Calcichordata. Quite a few of the extinct ones on close scrutiny may well be polyphyletic in Nelson's sense, and this also counts for Reptilia (including Therapsida and Pelycosauria) which equals Amniota excluding Aves and Mammalia. Nevertheless a rather peculiar case can be made out for supposed true ancestral species like some Leptolepis spp. (see Patterson, in this vol.), Archaeopteryx, lithographica and Homo (Australopithecus) africanus; in a way they could be called paraphyletic in Hennig's sense (see further below). Tuomikoski (1967) in fact uses such terminology.

Non-monophyletic groups are 'non-historical' groups (they have no history which is only theirs) which cannot become extinct in the same precise sense as monophyletic groups, and which cannot be used in a biogeographic analysis (Hennig, 1971; Rosen, 1976). Polyphyletic groups are generally dismissed, and perhaps an extreme example of paraphyletic groups can best illustrate why they are not very informative groupings: viz. as distinct from any single living or extinct species all remaining organisms form one huge paraphyletic group! Certainly this must make the claims of 'evolutionists' (e.g. Ashlock, 1971; Mayr, 1974) that paraphyly should be considered some sort of monophyly (and strict monophyly then called holophyly) unattractive to biosystematists, especially when both Ashlock and Mayr (and even Tuomikoski (1967) a cladist) want to use such groups in classification (criticized by Nelson (1973b) and Hennig (1975): see also Bonde (1975: 299)).

A classification of the above mentioned extreme kind has been suggested by Huxley (1958, 1959) contrasting man as Psychozoa with a grade containing all remaining animals, but this was meant for a secondary grade system as a supplement to, and distinct from, the normal zoological classification. Evolutionists in general (Simpson, 1961: 129; Mayr, 1974) fail to see the relevance of such a proposal (also suggested by Griffiths (1972) to take care of divergence) as they believe that both clades and grades can be consistently incorporated into the same hierarchical classification. Mayr's (1974) attitude of allowing the inclusion of paraphyletic groups has also been criticized by Nelson (1974) with specific reference to the classification of Hominoids.


## PLESIOMORPHY AND APOMORPHY

In Hennig's (1950, 1966) terminology, plesiomorphy and apomorphy, as designations for features, are terms free of prejudice created to substitute for what is generally called 'primitive'/ 'original' and 'advanced'/'derived' respectively. Hennig's terms are relative concepts with a precise meaning when applied to discussions of character transformation or change of features. In a character transformation series an earlier state of a feature is relatively plesiomorph compared to any later change of the feature, which then is termed relatively apomorph. The features of the common ancestor of a group of species by definition are plesiomorph for that group.

They are concepts of great analytical value because they help distinguish between several concepts included in the broad, general term 'similarity' as used in comparative biology. Fig. 2 illustrates this central point in the theory of phylogenetic systematics. Similarity of features can be of different origin: either a plesiomorph feature is preserved unchanged in two or more descendents of an ancestral species - symplesiomorphy of the feature; or a unique change of a feature can be transferred to two or more descendents - synapomorphy; finally a particular change in a feature might originate twice or more independently - convergence.

A central argument in phylogenetic systematics is, therefore, that while animals obviously must be considered related because of features they share (similarity), phylogenetic relationship is only indicated by shared derived features (synapomorphies), not by shared primitive features (symplesiomorphies) and not by convergence. The last point has always been realized (convergence is defined in this way), but symplesiomorphies are very often erroneously considered indications of phylogenetic relationship. The rather vague term 'affinity' seems to cover any sort of undifferentiated similarity, and therefore can form the basis for a typological approach enumerating similarities, including phenetic procedures which measure some overall similarity. Pre-evolutionary affinities are probably of this kind as well.

The argument for only using synapomorphies as indicators of relationship is based simply on the parsimony principle. Fig. 2 shows that symplesiomorphy may or may not give the correct estimate. Consideration of the number of changes necessary to produce the distribution patterns of features called synapomorphy and convergence shows that synapomorphy (only one change) is a simpler assumption than convergence (at least two changes from a to a' or one change a to a' plus one reversal a' to a).

Hennig (1965: 104; 1966: 121), therefore, stresses as a fundamental principle that shared apomorph features must a priori be considered synapomorphies and as such indicators of phylogenetic relationships (monophyletic groups). Only when there is some strong, independent evidence that this is not the case, is it acceptable to interpret shared derived features as convergence. If

this principle is not upheld analyses of phylogenetic relationships
are impossible.

As a methodological principle, therefore, models of phylogenetic relationships must contain a minimum of convergence (or
parallelism) and a minimum of reversals from apomorph to plesiomorph
states (cf. Wilson's 'consistency test' (1965)). The problems of
how to recognize apomorph and plesiomorph features are reviewed by
Hennig (1966), Nelson (1970, 1971, 1972, 1973), Schaeffer et al.,
(1972), Hecht (1976), and Hecht and Edwards (1976, and this vol.).


ANCESTORS

The following statement by Nelson (1970: 377) - "ancestral
species, by design and by necessity, are purely hypothetical con-
structs. The extent to which they can be reliably constructed is
the same as that to which the history of life can be reliably por-
trayed ...", and other similar statements by cladists, stress the
important point that, from a methodological viewpoint in phylogenetic analyses, ancestors are only hypothetical or models.

Nelson (1970) introduced the very useful concept of a 'morphotype' (modified from Zangerl, 1948) for the hypothetical reconstructed ancestor of a group (the model of the ancestor). This
morphotype can be reconstructed in as great detail as our hypothesis
of apomorph and plesiomorph features will permit by projecting the
plesiomorph features into it. It follows, naturally, that fossils
are less well suited than recent organisms for morphotype reconstruction since our knowledge of fossil features is much less detailed. Fossils are, therefore, not a priori the most important
evidence for estimates of ancestral structures, the opposite view
to that normally expressed by 'evolutionary'systematists, and also
by some phenoticists (Sokal and Sneath, 1963).

Cladists claim, furthermore, that even if one finds that fossil specimens exactly conform with the detailed, reconstructed ancestors, there is no way of showing that such fossils are representatives of the actual ancestral species - several other models
are just as likely. This is because such an ancestor only has (by
definition) plesiomorph (primitive) features of the group it is
supposed to be ancestral to. For this reason its phylogenetic relationships are not precisely indicated, and it could just as easily be the closest relative to one of the basal subgroups in the
group under consideration (see fig. 3). In a classification the
position assigned to such 'ancestral' fossils may be erroneous,
which incerta sedis appended to its name can indicate (see below).
This difference in attitude towards 'ancestors' is of paramount
importance since it has led to such fundamental differences between
cladists and 'evolutionary' systematists concerning evidence from

fossils, and, therefore, also concerning their classification. Cladists assert that the futile 'search for the ancestor' under- taken by many paleontologists should be replaced by another stra- tegy, the 'search for the sister'group' (Hennig, 1966:  139; Patterson, 1968:  95; Brundin, 1968:  474; Nelson, 1972a; Rosen, 1973:  500; Cracraft, 1974; Schaeffer et al., 1972; Bonde, 1975: 301).

Despite the intimation of Miles (1973:  64) that the search for ancestors is a "supposed problem of logic", cladists in gen- eral, including myself (Bonde, 1975), have claim that fossil an- cestors cannot be identified.  However, I shall revise my position on these matters in the following assuming (with Popper, 1973, 1976) that the best hypotheses are the most testable ones, and, further- more, that they must be as simple (parsimonious) as possible.

Fig. 3 shows that an 'ancestor' with features a b c d e can- not be placed in the dendrogram with reference to the identical last common ancestor of the whole group.  Several positions are possible for it if we visualize the diagram as a phylogenetic tree, i.e. if we extend each species (circle) to include the internode (line) below it.  The ancestor could be placed:

1)  as the stem-species itself
2)  as a side branch of the stem-species
3)  in the internode (in the phylogenetic tree representing the disintegrating species at the speciation process)
4)  on the left main branch before the change of b to b'
5)  on the right main branch before the change of a to a'
6-7)  as side branches from left and right main branches.

In the same way the upper right hand 'ancestor (a" b c d e) can be placed in as many positions before the origin of e' in D and c' in E, but after the 'intermediate' ancestor (a' b c d e).

If the two first mentioned 'ancestors' (i.e. the morphotypes) were convincingly reconstructed in an analysis involving only the groups (or species) A, B, D, and E (C was unknown), and then a fossil with features a' b c d e was found (assuming that a to a' to a" was a well-argued character transformation series), then this fossil will be somewhat less uncertain in position in the tree as only two positions are possible:

I)  below a" b c d e directly ancestral to D and E
II)  as an extinct side branch between a" b c d e and a b c d e. (We assume our fossil is older than D and E).

These two models of the phylogenetic relationship of the fossil a' b c d e seem, from the standpoint of the analyses, to be equally valid hypotheses.  We may then ask if one of the models

can be said to be more testable than the other.  Model (I) indicated that the fossil was directly ancestral to D and E; model (II)
that it represented the sister-group of D and E.  If more features
should be known of the involved fossil and recent species, then
these features are potential tests of such phylogenetic hypotheses.
However, it can be seen that only model (I) is potentially refut-
able, by any feature of the fossil that could not be accepted as
an ancestral feature of D and E.  Hennig (1969:  34) rightly remarked that it is impossible to prove but possible to refute, that
certain fossil species are ancestral to other species (Popper
(1973, 1976), naturally, would only comment that this is true for
all scientific hypotheses).  Model (II) could never be refuted; so
(I), implying that the direct ancestor was found, is the boldest
hypothesis (Popper).  Furthermore, model (I) is more parsimonious
than (II), because the latter implies one extra speciation process
during the phylogeny.

     Using the above models we seem to have justification for the
use of the concept of a 'direct fossil ancestor'.  However, we
are still left with the problem of how to classify this 'true ancestor'.  It was exactly this theoretical possibility of finding
'true' ancestors that was the main reason why Hennig (1966:  72)
had reservations concerning mixed classifications, believing that
fossils cannot be incorporated in the phylogenetic system of recent organisms.  The same applies to Brundin (1966); Crowson (1970),
who strongly advocates separate classifications for each time horizon; and Løvtrup (1973, 1975) who does not allow fossils into his
classifications.  This is obviously unacceptable to paleontologists,
and would seem to have been a major reason for the unpopularity of
phylogenetic systematics among their ranks.  When an ancestral
species is defined by reference to a kinship diagram the problem
with (fossil) ancestors is fortunately not insurmountable, and
Nelson (1972, 1974) has provided some classificatory conventions
that can cope with it (see below).

     While the above arguments to some degree can validate the
'true' ancestral species for the paleontologist, the popular concept of an ancestral group is difficult to justify.  As pointed
out above, an ancestral group is generally paraphyletic (sometimes polyphyletic), since at least one of its included species
is more closely related to groups not included.  The 'evolutionary'
taxonomists' acceptance of such incomplete groups in their classifications is justified from their viewpoint because the concept
of monophyly being used by them covers both paraphyly and strict
monophyly, and to some degree even polyphyly (see above).  An
evolutionist (Ashlock, 1971) has in fact proposed that paraphyly
should be looked upon as a sort of monophyly, but cladists (e.g.
Hennig, 1975) strongly disagree (cf. Nelson, 1973).

     Hennig (1969) discussed ancestors in some detail and called

fossil species with a phylogenetic relationship as in the above
'ancestor' models (I or II) a stem-group (Stammgruppe) as a non-
genealogical compromise. He said that species occupying positions
1 and 2 (of the 7 possible positions outline above) were forming
two "unechte Stammgruppen" (false stem-groups of A - E and D - E
respectively)


## PHYLOGENETIC TREES

A 'Stammbaum' ('stem-tree'), evolutionary or phylogenetic
tree, is a branching pattern illustrating the course of the evolutionary history of a group of organisms as it is supposed to have
occurred; it is one kind of model of a phylogeny. The 'internodes'
or segments of the tree symbolize the species (following Hennig,
1966) - here called 'time-bio-species' (see above) - delimited by
speciation processes, symbolized by the 'nodes' or bifurcations.
A phylogenetic tree is not identical with a dendrogram or a cladogram, both of which symbolize some specified relationship between
their end points; e.g. terminal taxa, species, sub- or supraspecific groups, communities, biota, distribution, particular features,
etc.

The first pictorial phylogenetic trees of Haeckel (1866) implied strict monophyly of groups, i.e. clades (but the species
concept was not well defined and well understood at that time),
and for that reason transfers some precise information. The only
figure in Darwin (1872) is a 'diagram' interpreted in the text as
a phylogenetic tree, though it looks much like a schematic cladogram. Time is indicated in terms of number of generations.

Here we are concerned only with cladograms symbolizing degree
of phylogenetic relationship between species and clades. The
cladogram (see figs. 2 and 3) differs from a phylogenetic tree
(cf. Hennig, 1966, figs. 14 and 15) in that the nodes designate
the species and the end points either species or clades, while the
interconnecting segments (lines or arrows) symbolize the (phylogenetic) relationships, that some species or group descended from
some other species. Time need not be specified in the cladogram,
but its direction is implicit in it if the tree has a 'root' (a
beginner), such a cladogram is equivalent to a hierarchy. There
are other, more unusual cladograms or trees like the "Wagner-trees'
used by Farris et al., (1970); here a root need not be specified,
therefore time is not implicit and a unique hierarchy is not indicated.

It is a straightforward matter to convert a 'rooted' cladogram into a phylogenetic tree by extending all the species (or
terminal clades) 'back' towards the beginner, but stopping short
of the next node (species); time then is explicit. Fig. 4 here

can be interpreted as either a phylogenetic tree or a cladogram, while fig. 5 is merely a cladogram, representing relationships, but if 'true ancestors' are included in the species X, Y and Z on the lines it is easily visualized as a phylogenetic tree. Hennig's 'Stammbäume' (1969) are in the main slightly modified cladograms, while those in his introduction are fairly typical cladograms.

If the cladogram symbolizes relationships different from the phylogenetic ones this conversion is not generally possible. However, phenomena such as the pathway of actual genetic material (cf. Sneath, 1975), and, as a good approximation, changes in certain macromolecular structures (cf. Fitch, 1976, and this vol.; Goodman and More, 1974) can be represented by cladograms which can be converted into phylogenetic trees. Likewise, distribution can be expressed in a phylogenetic tree (cf. Simpson's (1951a) diagram of horse evolution) since distribution and cladograms can easily be connected (Brundin, 1966; Nelson, 1972, 1974; Croizat et al., 1974; Rosen, 1976).

If the evolutionary history of species is reticulate, i.e. if a new species originates by hybridization between species (such occurs at least among plants (Grant, 1971)); the phylogenetic tree must have anastomosing branches (cf. Sneath, 1975). The corresponding cladograms then are also networks (and do not correspond to a strictly 'Linnean' hierarchy (cf. Nelson, 1974)).

Phylogenetic trees, when correctly formulated, convey very precise items of information. However, they are often misused and misunderstood, not only because an arbitrary species category may be used, but mainly perhaps due to mixing para- and polyphyletic groups with the clades. Much popular literature contains misuses of phylogenetic trees; e.g. Romer (1962); Davis et al., (1976) with a fig. (7-1) showing dinosaurs as a sistergroup of amniotes. But more specialized texts contain identical errors; e.g. Romer (1966, fig. 207) on archosaur phylogeny, indicates that birds, crocodiles and Saurischia form a monophyletic group - no evidence exists for this. Other examples include Mayr (1969), containing several figs. (4-4, 4-5, 4-7) from which similar conclusions may be drawn. Fig. 4-7 shows the problems of having the paraphyletic class 'Reptilia' in the diagram, in which it is impossible to follow the history of any reptilian group but the crocodiles. The 'family tree' of all vertebrate 'classes' (Romer, 1966, fig. 14) is completely misleading with up to 50% of the included groups being paraphyletic.

Sometimes divergence of morphology or other features can be shown by a phylogenetic tree (e.g. Simpson, 1961, fig. 12; Mayr, 1969, fig. 10-13; and Hennig, 1966, fig. 15) and also relative abundance (e.g. species numbers) can easily be shown, as in numerous 'romerograms' in Romer (1966).

Phylogenetic trees, properly designed to represent the supposed pathways of evolutionary history along the branches, can, like cladograms, immediately indicate a consistent classification with relative ages and ranks.

## REMARKS ON SOME METHODOLOGICAL PROBLEMS

Methods for the reconstruction of phylogenies are discussed by Hecht and Edwards (this vol.), and at length by Hennig (1950, 1957, 1965, 1966, 1969), Schlee (1971), Brundin (1966, 1968), Nelson (1970, 1971, 1973), Schaeffer et al., (1972), Cracraft (1972, 1974), and were reviewed by Bonde (1975). Only a few remarks are necessary here on points which have been widely misunderstood.

### Dichotomies

The dichotomies branching pattern of cladograms (and phylogenetic trees) is criticized for being unrealistic or 'simplistic' or simply an erroneous representation of natural events. Actually it has very little to do with what is going on in nature; the dichotomous branching pattern representing the speciation processes is only a methodological principle (Hennig, 1966: 210; 1975; Brundin, 1968: 480; Nelson, 1971; Bonde, 1975: 302). But this principle is unavoidable and necessary as it is part of a scientific strategy which aims at the maximum possible information about phylogenetic relationships. Resolving such interrelationships into dichotomies (corresponding to the strategy of 'search for the sister-group') gives maximum information, and all other patterns (trichotomies, etc.) are not worth analyzing since they cannot be distinguished from lack of knowledge of possible dichotomous splits.

To illustrate this point we can refer to the diagram of relationships between three groups (fig. 2). If resolved into dichotomies, as shown, five monophyletic groups can be distinguished at three levels. The alternative a trichotomy, has less information content only distinguishing four groups at two levels, or it can be considered to represent three equally likely models, each with five groups but with uncertainty about one group. Furthermore, it is easy to construct a true dichotomous phylogenetic pattern with a character distribution which cannot resolve this pattern (Hennig, 1966, fig. 63; Bonde, 1975, fig. 6). Such a distribution of features, therefore, cannot be taken as evidence that a 'radiation' (trichotomy) with multiple simultaneous splits has occurred.

These considerations are at an epistemological level which are independent of what may happen at the ontological level in nature in terms of speciation processes. This is completely misunderstood by Darlington (1970), Lehman (1975) and Mayr (1974).


## Stratigraphy

The stratigraphic occurrence of a fossil can only indicate a minimum age of the clade to which it belongs. Which clade a fossil belongs in can only be indicated by shared apomorph features, not by similarity or even identity of features which are relatively plesiomorph even though these might be all that is known from a particular fossil. Critical indication that a clade existed at a particular time in earth's history also indicates that its sister-group existed, whether fossils belonging to the latter group are known or not.

Age of a clade may mean: (1) the age of origin of the clade, i.e. when it split from its sister-group; or (2) the age of differentiation of the clade, i.e. origin of its basal subgroups. There may be a considerable time span between these two events; e.g. the recent mammals differentiated during the Triassic, while the group (in the sense of the Theropsida) arose in the mid Carboniferous or earlier. During this time many groups arose by numerous splits at speciation processes, which Patterson and Rosen (1977) call 'plesions' and which Hennig (1969) included in his paraphyletic 'Stammgruppen'.

Age overlap or synchronism of two (or more) closely related clades naturally means that the two cannot be ancestral to each other. This is obvious, but worth pointing out, because in some cases at the specific level, when comparisons are made between very close relatives it may be difficult to find convincing apomorphies to characterize each of two groups. The time overlap may be the only clear indication that the two are sister-groups, not ancestor and descendent. An example is the time overlap between 'Peking-man' and a more 'modern' type (_Homo_ _heidelbergensis_ in Europe), this makes it impossible that the former (which is probably conspecific with typical 'erectus-forms' of the so-called 'pithecanthropine group', as the typical forms have peculiar, straight, thick brow ridges as possible shared specializations) is the actual ancestor of modern man, a view not subscribed to by many authorities, because it is not generally agreed that any erectus feature is too specialized to be ancestral to structures of modern man (cf. Bonde, 1976).

It becomes clear then, that age (or stratigraphy) is extremely important for the reconstruction of phylogenies, but, as indicated by Nelson (1973), Schaeffer et al., (1972), Cracraft (1974), and Bonde (1975: 302), greater age must not be misused to indicate the primitive (relatively plesiomorph) end of a morphocline. Such a correlation is correct in most cases, but certainly not all,

and for that reason cannot be used as an important argument.  If
the argument is, nontheless, used, then what should be the conclusion of a comparative analysis is simply assumed a priori (notwithstanding Hennig's remarks (1966:  55)).


## Morphoclines

Morphoclines or character transformation series and the problems of determination of their 'polarity', i.e. which state is
most primitive, are essential for the cladistic procedure and are
reviewed by Hecht and Edwards (this vol.), Nelson (1970, 1973),
and Schaeffer et al., (1972).  We have already pointed out here
that stratigraphy is an irrelevant argument for primitiveness
(relative plesiomorphy), and that the primitive state need not be
at either end of the morphocline; e.g. the morphocline of tetrapod
digits runs from 8 (some ichthyosaurs) to 1 (horses) or even 0
(snakes), but the plesiomorph condition was 5 (or perhaps 6 or 7
on anatomical grounds).

Morphotype reconstruction, i.e. estimating as many features
as possible for the hypothetical last common ancestor of a group,
depends upon the analyses of morphoclines, and then of the distribution of apomorph features.  Therefore, if the hypothesis of
the phylogenetic relationships of a group is changed, then its
morphotype will change - in this sense an ancestor is only hypothetical.

For the evaluation of morphocline polarity, i.e. the direction
of evolutionary trends, cladists base their judgement primarily on
the construction of a model which gives the most parsimonious distribution of features.  It is evident, however, that many 'evolutionists' and especially functional anatomists like Bock, Dullemeijer and Barel, and Gutmann (all in this vol.) believe the parsimony principle to be irrelevant in these evaluations.  They
stresss the importance of understanding the postulated directions
of changes in terms of functional adaptation and selective value
for the organisms.  Such functional, adaptational statements which
claim to support a certain direction of evolution appear to be of
two kinds.  Firstly, statements expressed in terms of mechanics,
energetics, and the like, that could be viewed as features whose
distribution could be analyzed under a parsimony criterion.  Secondly, statements belonging to a type of science other than natural
science, namely the science of interpretation (hermeneutics - Radnitsky, 1968, 1969; Habermas, 1968).  This concerns statements
about another sort of 'understanding'; e.g. the often quite convincing feeling of what must be of advantage to an organism of a
particular design.  Essentially the observer tries to understand
the object from the 'inside', often in terms of needs.  The
scientists attempts to break down the boundary between observer

and object; e.g. by trying to imagine himself in the situation of
the object.  The success of this method must depend on both observer and observed having had similar experiences.  Probably such
concepts as objectivity, proof, parsimony, tests in a strict sense,
and repeatability, are hardly relevant, though perhaps predictions
can be attempted.

The common experience as land-living tetrapods makes Szarski's
(this vol.) explanation of the advantages of lungs and limb modifications in the earliest tetrapods, easily understandable and convincing.  On the other hand, it is much easier to disagree completely with Gutmann's (this vol.) interpretation of evolutionary
trends in early chordates simply because we have no significant
experiences in common with such water-dwelling animals.  Also,
since most of Gutmann's mechanical 'constructions' have models in
the living world, neither is it easy to be convinced about the
adaptational advantages of some stages compared with others - all
seem happily to survive.  Thus Gutmann's functional arguments also
perhaps fail to be convincing in this case.  I am merely suggesting
that some functional statements are intuitively appealing for
reasons which may have little to do with testable, empirical
science.

## Tests

The problems associated with testing phylogenetic hypotheses
are reviewed by Nelson (1970), Wiley (1975), and are most elegant-
ly expressed by Miles (1973, 1975).  Within the framework of
Popper's (1973, 1976) hypothetico-deductive procedure (see above),
Miles has developed a method of using the distribution of apomorph
features to test phylogenetic models (or hypotheses).  The best
corroborated hypothesis is the one that is refuted by as few
shared apomorph features as possible.  Or alternatively the hypothesis having the lowest number of apomorphies to contradict it
(i.e. to be considered convergent).  In a more conventional, non-
Popperian philosophy the corresponding statement would read:  the
best hypothesis is indicated by the largest number of synapomorphies.  This simply is the hypothesis involving a minimum number
of changes of features during the evolutionary history.  These
statements form the basis for the remarks made by Hennig (1965:
110; 1966) about the numerical nature of the cladistic methods,
and why Farris et al., (1970) have attempted to quantify these
methods (see above).

## Synapomorphy and Homology

As both Nelson (1970) and Wiley (1975) point out the tests
for phylogenetic hypotheses, i.e. for ideas about synapomorphies,

are the same as those for hypotheses of homologies (phylogenetic homologies referring to common ancestors as opposed to purely topographic homology defended by Jardine, 1969). Therefore the concepts of synapomorphy and (phylogenetic) homology are almost identical (see Patterson, this vol.). In the German literature; e.g. Remane (1952, 1956), Rensch (1947), and von Wahlert (1968, 1972), synapomorphies are, in fact, often referred to as "Spezial-homologie'. Hennig's reason (1966: 95) for coining the term synapomorphy is that a feature may be the loss of a structure, and it seems artificial to talk about the homology of a non-exis-tent structure.

In fact, if homology statements were always followed by the appropriate conditional phrases as recommended by Bock (1959, 1969, 1973), then they would form a hierarchy of nested statements which would be equivalent to the hierarchy of synapomorphies (character-izing the morphotypes of more and more inclusive monophyletic groups). This substantiates the belief that the concepts of syna-pomorphy and homology represent the same relation. (Actually a term seems to be missing from Hennig's vocabulary, namely one for the sharing of an apomorph feature - whether convergent of homolo-gous. The term 'synapomorphy' might in fact be used for this more inclusive concept, while the older term homology could be reserved for Hennig's 'synapomorphy'. But perhaps this would create chaos for the moment. Several more terms for the different possibilities of evolving similarities and differences have been proposed by Løvtrup, 1975).

## Ontogeny and Parsimony

Nelson (1973b) has formulated the two most important criteria for evaluating relative plesiomorphy of a feature: (1) the 'in-direct' method, which involves comparison with groups other than those immediately under consideration (the 'out-group- comparison, see Hecht and Edwards, this vol.); and (2) the 'direct' method of studying ontogeny. Nelson (1973b) also demonstrated why it is generally preferable to accept an early ontogenetic stage as 'prim-itive' because this minimizes the number of ontogenetic changes that will have to be accepted during phylogeny. This clearly shows that Haeckel's 'biogenetic law' (1866) is simply a parsimony principle minimizing the number of ontogenetic 'reversals' (paedo-morphosis, neoteny) accepted under a phylogenetic hypothesis, and is unavoidable as an a priori principle. (Much lengthy discussion in Remane (1952, 1956) for example, thus becomes rather irrelevant). Likewise 'Dollo's law' minimizes the number of reversals in char-acter transformation series or evolutionary trends to be accepted during a phylogeny (see Farris, 1977, and Wilson's (1965) principle of the 'unreversed' features, and the methodologies of Farris et al. (1970)). Therefore both 'laws' can be seen to be parsimony principles which lead to minimum-step phylogenetic models.

## LINEAGES AND PALEONTOLOGICAL SPECIES

Paleontologists frequently talk about 'lineages' of fossils implying that they can follow fossil populations through time. The 'evolutionary species' concept of Simpson (1951, 1961: 153) involves lineages defined as 'ancestral-descendent sequences of populations'. Simpson (1943) had earlier called such a sequence a 'chronocline', a concept later altered (Simpson, 1961; see Schaeffer et al., (1972) on this concept). The word lineage is rarely defined but is generally also applied to lineages of species (as in Simpson, 1961, fig. 12) and I have used it above to refer to successive 'time-bio-species'. Lines from level I to X in Darwin's (1872) figure on p. 117, also consist of species as he repeatedly indicated in his explanation. I find it impossible to understand 'lineages' composed of supraspecific groups; e.g. genera in succession each with several species, as, for instance, Throckmorton's (1965) Drosophila to Idiomya and other genera demands. Likewise, the paleontological example of 'Eohippus' through Merychippus to Equus or Hipparion. Such lineages cannot describe something that happened in nature, and certainly the meaningfulness does not increase because the taxonomic level is raised. One cannot derive a natural group from another unless the latter is a species. Cracraft (1974) did not define his term but apparently used 'lineage' as synonymous with taxonomic group (of species). Perhaps such a usage is defendable with reference to phylogenetic trees in which the terminal branches (lineages) may be groups (preferably clades).

Many evolutionists (including the majority of invertebrate paleontologists, especially micropaleontologists (e.g. Scott, 1976) see also a quotation from Matthew in Schaeffer et al. (1972: 39) and several statements by Romer (1949)) firmly believe that they can 'dig lineages right out of the rocks'. Perhaps they can - but how can that be shown by anything other than a cladistic analysis? It is by following lineages that one reads the phylogeny, but this can generally only be done after a cladistic analysis in which the evolution that has occurred within the lineages is implied simply by the differences in features between nodes or between nodes and terminal taxa. The nodes represent morphotypes reconstructed in as great detail as possible. Normally nothing can be known of the sequence of events within each internode. New finds of fossil or recent organisms can give extra nodes, but only if a series of actual ancestors indicated as outlined above are found is it possible to claim that a posteriori one is dealing with the 'lineage' itself. This is what Gingerich (1976), working on Lower Tertiary mammals, tried to do, applying 'species' arbitrarily delimited in time and using stratigraphic arguments. I have attempted (Bonde, 1976) to

use the 'time-bio-species' (therefore using features and age differently, and, I believe, more critically) within a strictly cladistic analysis in order to supply a framework for following lineages of fossil hominids.  I wanted to indicate for which of the
fossils there is good evidence of their phylogenetic relationships
and for which it is poor or lacking or misinterpreted (different
models of these relationships as cladograms/phylogenetic trees
as shown in fig. 5).

Only the 'time-bio-species' can be used in such hypotheses
because it is delimited by speciation processes and therefore is
the only unit composed of parts all of which may be said to have
exactly the same phylogenetic relationship to all other fossil or
recent species (or parts of these), including for that matter ancestor-descendent relationship (which might be measured in terms
of number of speciation processes lying between two species).  It
is the only concept with non-arbitrary delimitation of species
which can be used as a paleontological species concept (cf. Westoll,
1956).  The evolutionary species of Simpson (1961:  153) is arbitrarily delimited in time and corresponds to anything from a fragment of the 'time-bio-species' to several 'time-bio-species', depending on the changes in its 'evolutionary rôle and tendencies'.
If its rôle was considered a unique phylogenetic position and its
tendencies considered the tendencies towards coherence (because of
the homeostatic effect of gene-flow between populations said to
share a gene-pool; see Eldredge and Gould (1972:  114) and Hennig's
(1966:  52) characterization of species as a state of equilibrium)
then obviously Hennig's and Simpson's species definitions mean the
same.

Eldredge and Gold (1972) and Hennig (1966) disagree on an important aspect of this 'equilibrium state' of the species.  While
the former believe that the state of homeostatic equilibrium is the
normal state, Hennig suggests that such a condition is 'rarely
found in nature'.  Perhaps this view would alter our conception
of a phylogenetic tree into one in which the segments representing
'true' species extended in time may be very short, while the speciation periods, of uncertainty as to specific status, were much
longer.  A phylogenetic tree like this and such a species concept
with its dynamic implications is diametrically opposed to the pre-
evolutionary and static view of the species.

The dynamic conception comprises both the biological species
concept and its extension in time, the 'time-bio-species'.  It
emphasizes the gene-pool and stresses the importance of the local
population as a unit (Ehrlich and Raven (1969), and Endler (1973)),
lessening the emphasis on the gene-flow.  In this way it abolishes
the contradiction between the opposing views of Eldredge and Gould
(1972) and Hennig (1966).  If true, this dynamic species conception
makes identifications both in neontology and paleontology in

principle quite problematic. In a cladistic analysis it is, furthermore, a methodological implication that in an extended zone surrounding each speciation process it may be impossible to indicate
the exact phylogenetic relationship of a fossil. A test on the relative length of the phase of being a species and being in the process of speciation is probably difficult to establish. Perhaps the
only empirical finding bearing upon the problem is the relative number of living 'unproblematic' species versus the cases in which the
specific status of vicarying populations is uncertain. The paleontological species which at any time level will be indicated only by
morphology and at most a fragmentary knowledge of distribution and
relative isolation of populations (and often some uncertainty of synchronism), will probably have very little bearing on this test problem.


## CLASSIFICATIONS

The problems concerning the conversion of phylogenetic information into classifications must now be considered. Biological classification has, since Linnaeus, been a hierarchical structure and
the practical problems encountered are: (1) which groups can be
used as taxa; (2) what rank should the taxa be given; (3) what names
should be applied to the taxa. With reference to the first question,
all taxa in cladistic classifications are clades or strictly monophyletic groups (whether some paraphyletic groups should be included
as temporary 'waste-baskets' is a matter of dispute). The second
question is not so simply answered. Cladists have suggested some
different conventions for assigning rank to taxa in a precise way.
Hennig (1966: 154) argues that sister-groups must have the same absolute rank, and that the only non-arbitrary way to assign this
rank is to take the absolute age of origin of the group (and its
sister-group) - the greater the age, the higher the rank.

This principle is accepted by practically all cladists. However, Løvtrup (1973, 1975) uses quite a different ranking system assigning the lowest rank to the 'beginner' (the apex of the hierarchy
or the foot of the phylogenetic tree). Rank is elevated by one at
each dichotomous speciation process (or perhaps even any time there
is a change in a 'taxonomic character'). Thus Løvtrup has completely
abandoned the Linnean categories to the degree that his terminal taxa (corresponding to species or subspecies) are not of the same rank
(as there may have been many more changes in one lineage than in
another).

Farris (1976) proposes a system in which, as usual, more inclusive taxa have higher (Linnean) rank, but where sister-taxa seldom
have the same absolute rank. In extant taxa, rank is determined by
age of differentiation (instead of origin): in an extinct group,
rank is determined by the 'life span' from differentiation to extinctio

This is quite different from Hennig's suggestions, and has been specifically designed to avoid the many redundant categories of monotypic classifications and to avoid giving a very high rank to small fossil groups that arose very early. Paleontologists apparently find this latter effect of Hennig's (1966) system difficult to accept. For instance, nearly all trilobite species would have to be placed directly into 'classes' with no categories in between (as theses categories did not yet 'exist' in the Lower Paleozoic, see Hennig, 1966: 186).

An example of the difference between the proposals of Hennig (1966) and Farris (1976) is provided by the basal classification of Chordata. Hennig (1966: 155) recognizes two classes: Acrania with a single family; an Vertebrata with two subclasses each with numerous families. Between Acrania and the family level the so-called 'obligatory' Linnean categories are often intercalated. This gives no extra information and is avoided by Farris (1976) who under (phylum?) Chordata would simply include family Branchiostomidae as the sister-group of the class Vertebrata. The convention adopted is that, within their common inclusive group, a sister-group of lower rank is always listed before its sister-group of higher rank.

Hennig (1966) and Farris (1976) thus both treat categories as age-classes, but use different conventions for converting absolute age into categorical rank both methods being logically consistent. It should be recalled, nevertheless, that fossils only provide minimum ages of group origin as well as differentiation.

Nelson (1971: 234) has formulated the most general rule for transforming hypotheses of phylogenetic relationships into a classification, namely - "related species may be classified together". This rule covers the methods of both Hennig and Farris and does not put restriction on rank. Other measures of rank are possible in principle; Griffiths (1974) treats the problem of ranking according to morphological divergence, stressing that the groups to be used must be the same as in phylogenetic systematics, namely clades, while their rank might be measured differently. Hennig (1966: 155) had already emphasized that this must be done within the limits of phylogenetic systematics; e.g. a group cannot be of lower rank than one of its phylogenetic subgroups, i.e. Aves cannot be a class if Archosauria is of lower rank. Both Hennig and Griffiths state however, that a precise measurement of divergence has not yet been found, although Griffiths does mention some future comparisons of DNA and RNA in large numbers of organisms as a possibility for such measurement.

Many 'classical' cladistic systematists have stressed that is is not possible to classify both fossil and recent organisms into one system. Hennig (1966: 72) expressed reservations because

of problems with possible true ancestors.  Hennig (1969), later
discussed the inclusion of fossils in classifications of recent
organisms and to that end created his 'Stammgruppen' (stem-groups)
comprising fossils related to a recent taxon more closely than its
recent sister-group is.  This means that such fossils split off
between the origin and differentiation of a recent group.  Examples
would be; Archaeopteryx, Therapsida plus Pelycosauria (? plus some
Captorhinomorpha), Thecodontia, Anthracosauria (? plus some Captorhinomorpha), and a 'group' of all fossil genera of Equidae.
This last case in particular shows that the stem-group is a very
peculiar concept in phylogenetic systematics; it is paraphyletic
and, therefore, a typological group.  Accordingly Hennig (1969)
did not give it the same formal status as a recent (monophyletic)
group, but suggested that 'stem-groups' should be placed in intermediate categories between the categories of the recent groups,
without disturbing the latter.  Paraphyletic groups cannot be used
in the same way as monophyletic groups in other types of research;
e.g. they are completely misleading in biogeography (cf. Hennig,
1971).  As non-natural in an otherwise natural system they could
be put in quotation marks or marked in some other way; e.g. Patterson (1973:  298) who also distinguished between such grade groups
and other fossil groups either polyphyletic or of 'unknown status'.

   It should be clear that one type of paraphyletic fossil group
is very common, the paraphyletic genus; e.g. probably most of the
fossil horse genera Hyracotherium, Orohippus, Epihippus, and Meso-
hippus (despite MacFaddens cladistic analysis (1976), which does
not indicate that these genera are anything but grade groups since
species were not discussed at all and no 'genus' was shown to have
synapomorphies not found in the succeeding more advanced 'genus' -
some like Merychippus are probably polyphyletic; cf. Thenius (1969)).
This may well be due to the nomenclatorial rules which demand that
any species when described has to be assigned to a genus.  The
easiest thing to do then is to put it into a 'genus' from which it
does not differ much in morphology.  Such a genus is often completely typological.  Certainly the obligatory 'genus' is the
Linnean category that creates most inconvenience.  In paleontological discussions it often gets a wholly 'essentialistic' character;
e.g. Romer (1966) and Cooper (1970).

   A single name for the species and then names for the higher
monophyletic categories would seem to be consistent with the philosophy of phylogenetic systematics.  Hennig (1969) has himself
dropped the Linnean categorical names and employed a numbering
system in his phylogeny of insects.  This might appear to be a
minor point of practical convention, but as I shall show it certainly can create problems, and so can the Linnean type concept
with which it is connected.

   Returning to the inclusion of fossils in classifications

Crowson (1970) has suggested that different classifications for
each time level is the best solution, but in such a system the
rank of the group will change from one time level to the next.
Patterson and Rosen (1977) show that this is troublesome and creates
problems, although it does have the advantage that 'groups' which,
in classifications from rather late time levels may be paraphyletic
or incerta sedis, are monophyletic and therefore classifiable close
to the time level at which they originated.  Furthermore, fossils
are removed from the classification of recent organisms (but what
about subfossils and recently extinct species such as the dodo or
great auk?).  Problems indicated by Patterson and Rosen (1977) are:
(1) a large number of categorical levels will be necessary; and
(2) some 'genera' cross several time levels.  They therefore suggest
instead that fossils should be classified with recent organisms
after the models provided by Nelson (1972, 1974).  He has shown
how an hypothesis of a phylogeny including fossils can be converted
into an hierarchical classification which precisely reflects the
phylogeny by using the practical conventions of subordination and
sequencing.  Nelson (1974) discusses alternative systems:  (A) in
which ancestor-descendent relationships are recognized, i.e. ancestors are grouped together; and (B) in which only descendents
are grouped, as in Hennig's system.  System (A), with ancestors
grouped (sometimes together with descendents), to be efficient and
generally applicable uses both of the only two resources for expression in the hierarchy, namely 'subordination' (meaning that
groups are included in or subordinated to other groups) and sequencing (meaning that groups at the same categorical level are
written after each other, sometimes in specified order).  Uncertainty of relationship in the system can be expressed by using an
incerta sedis category.  In system (B), in which descendents are
grouped, only subordination is needed to make the classification
efficient and generally applicable.  Therefore sequencing can be
used for other purposes.

     Nelson (1974), therefore, suggested, and Patterson and Rosen
(1977) agree, that sequencing of recent clades be used to express
uncertainty of relationships of some groups at a certain level in
the hierarchy.  In this way trichotomy, etc., in the phylogenetic
tree or in a cladogram, signifying uncertainty of precise relationships, can be reflected in an exact way in the classification.  No
specified order is necessary for the sequencing as the uncertainty
always involves more than one group.

     We are still left in system (B) with one aspect of sequencing
(in specified order) and with the category incerta sedis available
to be used for other purposes.  Nelson (1974) suggested (and Patterson and Rosen, 1977, have expanded upon in great detail) that
these two aspects can be used for fossil (extinct) groups of known
and uncertain phylogenetic relationships respectively.  To distin-
quish sequenced fossil groups from recent groups the fossils are

given a prefix; e.g. a cross (+).

Fossil groups of uncertain relationships correspond to the problem with 'true ancestors' or other fossil subgroups within a recent group when the fossils do not have any synapomorphies with recent subgroups. Within an extinct monophyletic group subordination and sequencing could be used as for recent groups.

If fossils and recent organisms are classified together in this way in a system where the 'primary' relation, subordination, is reserved for recent groups and sequencing used as suggested above, then all fossils can be removed from the system without any effect upon the system of recent groups. Fossils in this way do not create a proliferation of categorical ranks, they are simply ranked at the same level as their closest recent relatives.

Obviously the above conventions do not tell anything about absolute rank, but, as mentioned above, this can be decided by another convention using the age of a monophyletic group; either the age of origin (like Hennig, 1966); or the life-span from group differentiation to extinction (or present day) like Farris (1976).

### CLASSIFICATION IN PRACTICE

Nelson (1969) in his classification of vertebrates used some of the above mentioned considerations when including a few fossil groups. Later (Nelson, 1972a), he classified in some detail fossil and recent esocoid fishes down to the species level (a group of 21 spp.) with methods as outlined above. In Greenwood et al. (1973) several cladistic phylogenies were presented: Greenwood, Forey, Miles, Schaeffer, Rosen, and Patterson; but only the last two used some of the conventions mentioned above; e.g. an incerta sedis category for fossils included in a recent group at as low a categorical level as possible when considering synapomorphies. Also the sequencing of recent groups was used by Rosen and that of fossil groups by Patterson with a 'trend' from advanced to primitive groups of different status.

However, it should not be forgotten that some years ago, Goodrich (1909) produced a classification of all fishes as a detailed dichotomous, key-like arrangement (the list of contents); with fossil and recent groups classified in the same way; with incerta sedis categories used for both sorts of groups; and with dichotomous subordination used for both recent and fossil groups, but sequencing used when there was doubt about the phylogenetic relationships of some groups. The remarks made by Goodrich (1909: 397) leave little doubt that he wanted 'phylogenetic value' to be expressed by the groups.

Winterbottom (1974) has produced a classification of recent and fossil 'families' (a rather arbitrary, traditional category) of tetraodontiform teleosts using both subordination and sequencing, but not in precisely the same way as Nelson, and also without naming or indicating all recent clades.

Other vertebrate groups have not received quite the amount of attention afforded to fish. Nevertheless, Gaffney (1976) has produced cladistic classifications of chelonians including fossils, and Cracraft (1974) has discussed some aspects of subordination and 'phyletic sequencing' in ratite birds. I have (Bonde, 1976) produced a detailed classification based upon the most likely phylogeny of the Hominoidea. A rather similar, but less detailed, phylogeny based on cladistic evidence can be found in Eldredge and Tattersal (1975); and a strict cladistic classification of the Catarrhini was produced after a predominantly cladistic analysis by Delson and Andrews (1975: 440).

Before I deal with my own, slightly modified, hominoid classification below, let us look at a more simple example. The five groups (or species) in fig. 3 can obviously be classified in a hierarchy as symbolized in two different ways above the cladogram. The classification can be written:

Group A - E

    Subgroup A - B
        Division A . . . . . . . . . . . . . . . . . . . A
        Division B . . . . . . . . . . . . . . . . . . B

    Subgroup C - E
        Division C . . . . . . . . . . . . . . . . . . C
        Division D - E
            Subdivision D . . . . . . . . . . . . . . D
            Subdivision E . . . . . . . . . . . . . . E

This exactly reflects the phylogenetic relationships of A, B, C, D, and E. Hennig's (1966) convention that sister-groups are of the same rank has been employed throughout, although absolute ranks have not been indicated.

Fig. 4 is a phylogenetic tree or a cladogram representing the relationships of the major groups of Mammalia (in the strict sense, i.e. related to the diversification of recent mammals) in the Mesozoic and Cenozoic. Based upon the recent groups a classification could read:

        Class Mammalia
            Subclass Prototheria
                Infraclass Monotremata

```
            Subclass Theria
                    Infraclass Marsupialia
                    Infraclass Placentalia
```

- or without deciding upon absolute rank:

```
            Group Mammalia
                    Subgroup Monotremata
                    Subgroup Theria
                            Division Marsupialia
                            Division Placentalia
```

If we want to include the fossils we might give a particular rank and name to every split indicated, as does MacKenna (1975). (+ designates a fossil taxon in the following):

```
            Group Mammalia
                    Incerta sedis + Triconodonta
                    Subgroup        Prototheria
                            Division  + Morganucodontia (?)
                            Division    Unnamed
                                    Subdivision + Multituberculata
                                    Subdivision   Monotremata
                    Subgroup        Theria
                            Division  + Symmetrodonta
                            Division    Trechnotheria
                                    Subdivision + Pantotheria
                                    Subdivision   Tribosphenida
                                            Section   Marsupialia
                                            Section   Placentalia
```

The new names and some of the extra categories demanded by this classification could be avoided by using Nelson's (1972, 1974) conventions:

```
            Group Mammalia
                    Incerta sedis + Triconodonta
                    Subgroup        Prototheria
                            Division  + Morganucodontia (?)
                            Division  + Multituberculata
                            Division    Monotremata
                    Subgroup        Theria
                            Division  + Symmetrodonta
                            Division  + Pantotheria
                            Division    Tribosphenida
                                    Subdivision  Marsupialia
                                    Subdivision  Placentalia
```

The convention for sequencing the fossil groups states that the first listed group is the sister-group (often relatively

<u>plesiomorphic) of the following groups combined</u>.  If one did not have clearly marked specializations to show that a group like the Symmetrodonta, for example, is a monophyletic group, it could be placed in inverted commas, the same might apply to the Triconodonta.

Such a sequenced fossil group under the above convention is called a 'plesion' by Patterson and Rosen (1977) regardless of its status as a monophyletic or paraphyletic group.  A plesion thus corresponds to Hennig's 'Stammgruppe' or subdivisions of such a group and may contain only one species; e.g. <u>Archaeopteryx litho-graphica</u> or <u>Paranthropus robustus</u> (being probably paraphyletic and monophyletic respectively).

In the above examples Triconodonta might illustrate the pro-blem of a possible 'true' ancestor, since one of the triconodont species may be such a true ancestor, the group as conceived here then might well be even polyphyletic.

The ancestors are much more of a problem in the classification of fossil man (see fig. 5).  The mixture of a phylogenetic tree and cladogram shown represents several models of the phylogeny of man, a different one for each alternative line indicated, and the less likely models with most question marks.  A detailed classifi-cation with subordination only of groups considered phylogenetical-ly well-known (solid lines) could be written as follows:

```
Hominoids
     Group Pongo pygmaeus (e.g. Pongidae)
     Group Gorilla and Pan and Homo (e.g. Hominidae)
          Subgroup Paninae
               Division (?) Gorilla gorilla
               Division      Pan troglodytes and paniscus
          Subgroup Homininae
               Division      Paranthropus robustus
               Division      Homo
                    Subdivision + Homo (Pithecanthropus) pek-
                                                       inensis
                    Subdivision   Homo (Homo)
                         Section + Homo (Homo) neanderthalensis
                         Section   Homo (Homo) sapiens
```

If we want to include other fossils in the classification one like this could follow (and there is still ample scope for dispute over the absolute ranks since there are many Linnean cat-geories around the genus which are certainly not too well suited for a cladistic classification)

```
     Pan (or Pan and Gorilla)
     Homo
          + Homo ("Ramapithecus") ?
```

+ <u>Homo</u> (<u>Paranthropus</u>) <u>robustus</u>
+ <u>Homo</u> "(<u>Australopithecus</u>) <u>africanus</u>"
+ <u>Homo</u> (unnamed) "<u>habilis</u>"
+ <u>Homo</u> (<u>Pithecanthropus</u>) <u>erectus</u>
+ <u>Homo</u> (<u>Homo</u>) <u>neanderthalensis</u>
  <u>Homo</u> (<u>Homo</u>) <u>sapiens</u>

This indicates all the fossils as 'plesions' and has poly- or paraphyletic groups (uncertain species) in inverted commas.

There has still been no convention indicated to distinguish between a very uncertain group like <u>Australopithecus</u> <u>africanus</u> and the less uncertain <u>Homo</u> <u>habilis</u> (which is quite likely our direct ancestor), and the 'direct ancestors' X, Y, and Z, have not been named (and synonymized). We might consider the most likely model of the phylogeny in fig. 5 (without question marks but for 'Ramapithecines') and try to incorporate 'true' ancestors in a classification with a minimum of sequencing, and ranking from Hominidae:

Family Hominidae
    Subfamily Paninae
    Subfamily Homininae
        Incerta sedis + <u>Ramapithecus</u> (and 'ramapithecines') ?      (X)
        Tribe            + Paranthropina: <u>Paranthropus</u> <u>robustus</u>
                                          (syn. <u>boisei</u>)

        Tribe            Hominini
        Incerta sedis + <u>Australopithecus</u> <u>africanus</u>                      (Y)
                                          (syn. <u>habilis</u>)
        Genus            + <u>Pithecanthropus</u>: <u>Pithecanthropus</u> <u>erectus</u>
                              (syn. <u>pekinensis</u>, <u>modjokertensis</u>, etc.)

        Genus            <u>Homo</u>
        Incerta sedis + <u>Homo</u> <u>heidelbergensis</u>
                                          (syn. <u>steinheimensis</u>)       (Z)
        Species          + <u>Homo</u> <u>neanderthalensis</u>
        Species          <u>Homo</u> <u>sapiens</u>

The subfamily Homininae might also have been ranked in this way:

Incerta sedis (?) + <u>Ramapithecus</u>                                      (X)
Genus             + <u>Paranthropus</u>: <u>Paranthropus</u> <u>robustus</u>
                                          (syn. <u>boisei</u>)

Genus            <u>Homo</u>
    Incerta sedis + <u>Homo</u> (<u>Australopithecus</u>) <u>africanus</u>                (Y)
                                          (syn. <u>habilis</u>)
    Subgenus     + <u>Homo</u> (<u>Pithecanthropus</u>): <u>Homo</u> (<u>Pithecanthropus</u>)
                      <u>erectus</u> (syn. <u>pekinensis</u>, etc.)
    Subgenus     <u>Homo</u> (<u>Homo</u>)

```
Incerta sedis + Homo (Homo) heidelbergensis
                      (syn. steinheimensis, etc.)           (Z)
Species        + Homo (Homo) neanderthalensis
Species          Homo (Homo) sapiens
```

These absolute ranks would still be far too high compared with
many other animal groups but one might, as shown above, simply decide that every species closer to Homo than to any other 'genus'
should be called Homo.   In fact all these 'homininae' are plesions
to a species Homo sapiens.   If we use 'all possible' categories
for every clearly indicated split we might get the following (considering that neanderthalensis is only subspecifically distinct
from sapiens):

```
Family Hominidae
   Subfamily Hylobatinae
   Subfamily Homininae
      Tribe Pongini
      Tribe Hominini
         Genus Pan
            Subgenus Pan (Pan)
               Species Pan (Pan) troglodytes
               Species Pan (Pan) paniscus
            Subgenus Pan (Gorilla): Pan (Gorilla) gorilla
         Genus Homo
            Incerta sedis + Homo ("Ramapithecus") several spp.
            Subgenus      + Homo (Paranthropus): Homo (Paranthro-
                               pus) robustus
            Subgenus        Homo (Homo)
               Incerta sedis + Homo (Homo) africanus
               Species       + Homo (Homo) erectus
               Species         Homo (Homo) sapiens
                  Subspecies  + Homo (Homo) sapiens heidelbergensis
                  Subspecies  + Homo (Homo) sapiens neanderthal-
                                                     ensis
                  Subspecies    Homo (Homo) sapiens sapiens
```

In this last classification, however, it is not entirely clear
what the sequence of the three subspecies of Homo sapiens signifies,
because phylogenetic relationship is not defined for subspecies,
it is a relation between species and higher categories.   It is also
not obvious what the crosses mean in front of heidelbergensis
and neanderthalensis:   in what sense are they extinct under this
hypothesis, especially in what way can the supposed 'ancestral'
subspecies (? one population) heidelbergensis be said to have died
out?   So at this point the analysis loses precision.   It also
suggests that there is a problem in calling 'true' ancestral species extinct (since they can be looked upon as paraphyletic species, this turns out to be a general problem with paraphyletic
groups).   A solution to this problem might be the placing of the

cross of the 'true' ancestor in parentheses; in this way it could
be unambiguously indicated which of the plesions (species) could
be directly ancestral to the following sequenced groups.

We might apply this convention to the models of human phylogeny (fig. 5), fully realizing all the uncertainties concerning
'australopithecines' and early 'pithecanthropines', and modify the
extremely sequenced classification, again accepting neanderthalensis as a good species.  The distinction between the very uncertain
species (or groups) and the less uncertain 'true' ancestor is now
possible:

Pan (or Pan and Gorilla)                           2 (or 3) species
Homo
      + Homo "(Ramapithecus)" and other 'ramapithecines' ?
      Incerta sedis + "Homo (Australopithecus) africanus"
      + Homo (Paranthropus) robustus
      (+) Homo (unnamed) habilis
      Incerta sedis + "Homo (unnamed) modjokertensis"
      + Homo (Pithecanthropus) erectus
      (+) Homo (Homo) heidelbergensis
      + Homo (Homo) neanderthalensis
      Homo (Homo) sapiens

The subgeneric names are, in fact, unnecessary here.  However,
if ramapithecines one day should turn out to be a monophyletic
group with several species some use of subgeneric names could be
justified.  Likewise, if Paranthropus consists of two species or
if africanus and robustus are sister-species.  The category incerta sedis, as usual, signifies uncertainty in relation to the
immediately succeeding speciation, and the minimum number of speciations indicated can now be read from the classification.  The
monophyletic groups (species) indicate one speciation each (minus
one for each terminal pair of species); the non-monophyletic groups
in quotation marks indicate one or more speciations apart from
those incerta sedis which may not indicate any.  The 'true' ancestors (+) are not evidence of speciations.  The incerta sedis
'species' are not necessarily evidence of a separate 'time-bio-
species'; e.g. africanus might be the same species as robustus or
habilis.

The above examples show that it is possible for a given phylogeny, even with some uncertainty concerning the relationships of
certain groups, to produce cladistic classifications which precisely reflect the phylogeny.  Which of the many possible classifications
to prefer is a question of choice of conventions for the transformation of phylogeny to classification.  As soon as the convention
is known the two are isomorphic.  I, therefore conclude with Dendy
(1924: 241) - "that the taxonomic tree and the phylogenetic tree

are, after all, one and the same thing, for we should arrange all organisms strictly in accordance with the course of their evolution", and Darwin (1872) - "our classifications will, so far as they can be so made, become genealogies".

## FINAL REMARKS

I have emphasized above, that contrary to my earlier beliefs, the details of the processes of speciation are not important for the phylogenetic systematic theory because the patterns resulting from allo-, para- and sympatric speciations can be analyzed in the same way in terms of degrees of phylogenetic relationship (in terms of analysis of distribution patterns it does make a difference, however). I would hope that such a statement might reduce the importance of population genetic thinking, which for many biologists is the central discipline in relation to evolutionary theory (e.g. Ruse, 1973; Hull, 1974; and any genetics text). The development of the species concept is one area in which population genetics has been of paramount importance. However, if Mayr's (1969) 'biological species concept' (leading to the 'time-bio-species' used here) and the pheneticists' 'local population' concept (e.g. Ehrlich and Raven, 1963; Sokal and Crovello, 1970; Endler, 1973; and Sokal, 1974) can both be viewed as different phases of the same dynamic species evolving through time, then again population genetics details may lose significance from the viewpoint of phylogenetic analyses.

There is also a conflict between viewpoints derived from infraspecific and supraspecific research. The former, as exemplified by population genetics, is concerned with the analyses of processes; while the latter is concerned with pattern analyses, studying the end products of these processes. The methods used for the two kinds of research are quite different, analytic as opposed to systematic and correspondencies between the two levels of study are not always clear. The phylogenetic systematic philosophy with its cladistic methods is such a systematic approach (cf. Griffiths, 1974), which can proceed quite comfortably without a too detailed knowledge of the basal processes behind the recognized patterns. The unifying principle of the two types of research is the framework within which theories in both areas have to be explained, namely the idea of biological evolution. It is also true to say that Hennig's approach to systematics in some important respects, particularly concerning the use of 'similarities', is the only one known which is without pre-Darwinian (or pre-evolutionary) attitudes. However, Hennig's theory has adopted the only pre-evolutionary taxonomic ideas worth preserving: (1) that there is a basic unit of diversity greater than the individual called a species (which in Hennig's theory is related to other such units in a way specific for organisms although the exact practical distinction between species is a

matter of dispute); and (2) that a hierarchical system is convenient for classification (this happens to be the type of system that
Linnaeus had chosen). With the acceptance of the evolutionary
theory, such a hierarchic classification can be seen to be the only
one which can exactly indicate the evolutionary history (or phylogeny) by reflecting the so-called phylogenetic relationships.

## CONCLUSION

Biosystematics was for some years dominated by the proponents
of the 'synthetic' or 'evolutionary' approach (associated closely
with the names of Mayr and Simpson). The taxonomic approach of
these workers is, however, very weak and imprecise in its formulations, and in fact no strict rules or conventions for converting
a phylogeny into one (or several) classifications are provided.
Thus classifications which should be used as general reference
systems arising from this approach are difficult to compare and
evaluate. An even more serious fault is that the accepted classifications act back on the reconstruction of the phylogenies by
providing the groups which are compared in such reconstructions.
Cladistics is an attempt to remedy this defect in the exploration
of the evolutionary history (phylogeny) of life by providing more
precise terms, definitions and methods.

I have attempted to show that cladistic or phylogenetic classifications are possible, and that phylogenetic relationships can
be retrieved unequivocally from such classifications, provided
some simple conventions are agreed upon. Several workable conventions have been put forward by, among others, Nelson (1974),
Hennig (1969), Farris (1976), Patterson and Rosen (1977), and
Crowson (1970).

Reasons have been given why, compared with other techniques,
cladistic methods are preferable: they stand out as logical consequences of the theory of biological evolution and a semantic analysis of the concept of biological relationships. Cladistic methods preserve such desirable features as (1) leading to a hierarchical classification; (2) having species (here the 'time-bio-species')
as the basic units of classification; and (3) forming a single,
combined system for recent and fossil organisms, (though based
primarily upon a framework of interrelationships between recent
clades).

So the Linnean type of hierarchy (cf. Buch and Hull, 1966) is
not, in principle, restricted in the ways supposed by Simpson (1961)
and Hull (1964); it can easily 'express' phylogenetic relationships.
Løvtrup's (1973) criticisms of Linnean classification is directed
specifically against the Linnean system as understood by the 'evolutionists' (e.g. Mayr, 1969; Simpson, 1961). There is no consistent

way in which the classificatory Linnean hierarchy can contain both the phenetic and phylogenetic information, only the latter can be stored and retrieved again without distortion, and even the distortion of phenetic data is minimized in a cladistic (phylogenetic) classification.

Cladistic classification can even include the paleontologist's 'true' ancestral species. Statements about fossils as 'true' ancestors are shown under certain circumstances after a cladistic analysis to be at least as reasonable as alternative statements in terms of sister-group relationships.

The term 'monophyly' is of paramount importance for classifications since monophyletic taxa appear to be an ideal for most if not all sorts of biosystematics. The apparent change in the meaning of the concept from a phylogenetic one (strict monophyly) to an 'evolutionary' one (minimal monophyly) removed all phylogenetic significance from phylogenies and the classifications derived from them. Strictly monophyletic groups (clades) can be comprehended as real existing entities (wholes), which therefore can be the units of a 'natural' system (Löther, 1972; Griffiths, 1974; Hennig, 1966).

Reluctance to accept the natural phylogenetic system as a general reference system for evolutionary biology is an odd feature of the recent debate. An analogy would be chemists denying the importance of the periodic system of elements. Like this system, the phylogenetic one is there, ready to be discovered in nature, and not to be invented.

## ACKNOWLEDGEMENTS

I am very grateful to the editors for their great patience, and especially to Dr. Goody who corrected and edited the manuscript. I am also indebted to Mrs. A. Dawes, E. Møller-Hansen, and A. Panning for typing and to Mrs. H. Egelund for the drawings.

I profited greatly from discussions at the Advanced Study Institute with Drs. S.J. Farris, C. Patterson, W. Fitch, J. Edwards, M.K. Hecht, S. Løvtrup, and W. Gutmann, and Drs. P. Forey and G. Young during a subsequent stay in London, which was financially supported by my Institute. Discussions with Prof. K.G. Wingstrand, Copenhagen, and Mr. K. Thomsen, Aarhus, have contributed towards my viewpoints on cladistics. This certainly does not imply that the persons mentioned agree with my interpretations.

Finally I would like to take this opportunity to pay tribute to the late Prof. Willi Hennig (whom I never had the good fortune to meet) for the great inspiration he has provided in the field of biosystematics. Prof. Hennig died unexpectedly in November 1976.

REFERENCES

Ashlock, P.D., 1971, Monophyly and associated terms.  Syst. Zool.,
    20: 63-69.
Ayala, F. (ed.), 1976, Molecular evolution . (Sinauer) Mass.
Beckner, M., 1959, The biological way of thought.  Columbia Univer-
    sity Press, New York.  200 pp.
Bock, W.J., 1959, Preadaptation and multiple evolutionary pathways.
    Evolution, 13: 194-211.
Bock, W.J., 1969, Discussion.  The concept of Homology.  Ann. N.Y.
    Acad. Sci., 167: 71-73.
Bock, W.J., 1973, Philosophical foundations of classical evolution-
    ary classification.  Syst. Zool., 22: 375-392.
Bock, W.J., 1977, Foundations and methods of evolutionary classi-
    fication.  NATOASI, 'Major Patterns in Vertebrate Evolution'.
    (This vol.).
Bonde, N., 1975, Origin of "higher groups":  viewpoints of phylo-
    genetic systematics.  Coll. Int. C.N.R.S., no. 218: 293-324.
Bonde, N., 1975a, Review of "Interrelationships of fishes" by
    Greenwood, P.H., et al., 1973.  Syst. Zool., 23: 562-569.
Bonde, N., 1976, Nyt om menneskets udviklingshistorie (News of
    Man's evolutionary history).  Geol. Soc., Denmark, Arsskr. f.
    1975: 19-34.
Brundin, L., 1966, Transantarctic relationships and their signifi-
    cance, as evidenced by chironomid midges.  K. Svenska Vetensk.
    Akad. Handl. (4), 11: 1-472.
Brundin, L., 1968, Application of phylogenetic principles in syste-
    matics and evolutionary theory.  Pp. 473-495 in:  Ørvig, T.
    (ed.).
Brundin, L., 1972, Evolution, causal biology, and classification.
    Zool. Scripta, 1: 107-120.
Brundin, L., 1972a, Phylogenetics and biogeography.  A reply to
    Darlington's "practical criticism" of Hennig - Brundin.  Syst.
    Zool., 21 :  69-79.
Buck, R.C. & Hull, D.L., 1966, The logical structure of the Linnean
    hierarchy.  Syst. Zool., 15: 97-111.
Bush, G.L., 1975, Modes of animal speciation.  Ann. Rev. Ecol.
    System., 6: 339-364.
Camin, J.H. & Sokal, R.R., 1965, A method for deducing branching
    sequences in phylogeny.  Evolution, 19: 311-326.
Cooper, G.A., 1970, Generic characters of brachiopods.  Proc. N.
    Amer. Paleont. Conv., C: 194-263.
Cracraft, J., 1972, The relationships of the higher taxa of birds:
    Problems in phylogenetic reasoning.  The Condor, 74: 579-592.
Cracraft, J., 1974, Phylogenetic models and classification.  Syst.
    Zool., 23: 71-90.
Croizat, L., Nelson, G. & Rosen, D.E., 1974, Centers of origin and
    related concepts.  Syst. Zool., 23: 265-287.
Crowson, R.A., 1970, Classification and biology, 350p. London.

Darlington, P., 1970, A practical criticism of Hennig-Brundin
    "Phylogenetic Systematics" and Antarctic biogeography.  Syst.
    Zoo., 19: 1-18.
Darwin, C.R., 1872, The origin of species.  6th ed., 592 p. (1956,
    reprint, Oxford University Press).  London.
Davis, S.N., Reitan, P.H. & Pestrong, R., 1976, Geology:  Our phy-
    sical environment.  (McGraw Hill) London, N.Y. 470 pp.
Delson, E. & Andrews, P., 1975, Evolution and interrelationships
    of the catarrhine primates.  Pp. 405-446 in:  Luckett, W.P. &
    Szalay, F.S. (eds.), Phylogeny of the Primates.  Plenum, N.Y.
Dendy, A., 1923 (1924), Outlines of evolutionary biology (3rd ed.).
    Constable & Co.  London. 481 pp.
Ehrlich, P.R. & Raven, P.H., 1963, Differentiation of populations.
    Science, 165: 1228-1232.
Eldredge, N. & Gould, S.J., 1972, Punctuated equilibria:  An alter-
    native to phyletic gradualism, 82-115.  In Schopf, T.J.M. (ed):
    Models in paleontology, San Francisco.
Eldredge, N. & Tattersal, I., 1975, Evolutionary models, phylogen-
    etic reconstruction, and another look at hominid phylogeny.
    Contrib. Primatol., 5: 218-242.
Endler, J.A., 1973, Gene flow and population differentiation.
    Science, 179: 243-250.
Farris, J.S., 1969, A successive approximation to character weight-
    ing.  Syst. Zool., 18: 374-385.
Farris, J.S., 1973, A probablistic model for infering evolutionary
    trees.  Syst. Zool., 22: 250-256.
Farris, J.S., 1976, Phylogenetic classification of fossils with
    Recent species.  Syst. Zool., 25: 271-282.
Farris, J.S., 1977, On the phenetic approach to vertebrate classi-
    fication.  NATO-ASI, Major patterns in vertebrate evolution,
    (this vol.).
Farris, J.S., Kluge, A.G. & Eckardt, M.N., 1970, A numerical ap-
    proach to phylogenetic systematics.  Syst. Zool. 19:  172-189.
Farris, J.S., Kluge, A.G. & Eckardt, M.N., 1970a, On predictivity
    and eficiency.  Syst. Zool., 19:  363-372.
Fitch, W., 1976, Molecular evolutionary clocks.  Pp. 160-178 in
    Ayala, F. (ed.).
Fitch, W., 1977, The phyletic interpretation of macromolecular se-
    quence interpretation.  NATO-ASI.  1976, 'Major Patterns in
    Vertebrate Evolution', (this vol.).
Fitch, W.M. & Margoliash, E., 1967, Construction of phylogenetic
    trees.  Science, 155: 276-284.
Forey, P.L., 1973, Relationships of elopomorphs, Pp. 351-368 in
    Greenwood, P.H., et al., (eds.).
Frost, W.E., 1965, Breeding habits of Windermere charr, Salvelinus
    willughbii (Günther), and their bearing on speciation of these
    fish.  Proc. R. Soc. Ser. B. Biol. Sci., 163: 232-284.
Fryer, G. & Iles, T.D., 1972, The Cichlid Fishes of the Great Lakes
    of Africa.  Their Biology and Evolution.  TFH, Hong Kong.
    641 pp.

Gaffney, E.S., 1976, Phylogeny and classification of the higher
    categories of turtles.  Bull. Amer. Mus. Nat. Hist. 155 (5):
    387-436.
George, T.N., 1971, Systematics in palaeontology.  J. Geol. Soc.,
    Lond., 127 (3): 198-245.
Ghiselin, N.T., 1966, An application of the theory of definitions
    to systematic principles.  Syst. Zool., 15: 127-130.
Ghiselin, M.T., 1966a, On psychologism in the logic of taxanomic
    controversies.  Syst. Zool., 15: 206-215.
Ghiselin, M.T., 1974, A radical solution to the species problem.
    Syst. Zool., 23: 536-544.
Gilmour, J.S.L., 1940, Taxonomy and philosophy.  Pp. 461-475 in
    Huxley, J. (ed.).
Gingerich. P.D., 1976, Paleontology and phylogeny:  Patterns of
    evolution at the species level in Early Tertiary mammals.
    Amer. J. Sci., 276: 1-28.
Goodman, M. & Moore, G.W., 1974, Phylogeny of hemoglobin.  Syst.
    Zool., 22 (4): 508-532.
Goodrich, E.S., 1909, Vertebrata Craniata, first fasc., Cyclostomes
    and Fishes.  In Lankaster, R. (ed.).
Gottlieb, L.B., 1976, Biochemical consequences of speciation in
    plants.  Pp. 123-140 in Ayala, F. (ed.).
Grant, V. 1971, Plant speciation, 435 pp. (Columbia University
    Press), New York.
Greenwood, P.H., 1970, On the genus Lycoptera and its relationships
    with the family Hiodontidae (Pisces, Osteoglossomorpha).  Brit.
    Bull. Brit. Mus. Nat. Hist. (Zool.), 19: 257-285.
Greenwood, P.H., 1973, Interrelationships of osteoglossomorpha.
    Pp. 307-331 in Greenwood, P.H. et al., (eds.).
Greenwood, P.H., Miles, R.S. & Patterson, C. (eds.), 1973, Inter-
    relationships of fishes.  536 pp. (Academic Press), London.
Gregg, J.R., 1954, The language of taxonomy.  An application of
    symbolic logic to the study of classificatory systems.  70 pp.
    (Columbia University Press), New York.
Griffiths, G.C.D., 1972, The phylogenetic classification of Diptera,
    Cyclorrhapha, with special reference to the male postabdomen.
    Ser. Entomol., 8.  Junk, The Hague.
Griffiths, G.C.D., 1974, On the foundations of biological system-
    atics.  Acta Biotheoretica, 23: 85-131.
Gödel, K., 1931, Ueber formal unentscheidbare Sätze der Principia
    Mathematica und verwandter Systeme I. Monotshft. Math. Phys.,
    38: 173-198.
Gutman, W.F., (this volume), Phylogenetic reconstruction - theory,
    methodology, and application to chordak evolution.  NATO-ASI,
    1976, 'Major Patterns in Vertebrate Evolution'.  (This vol.).
Habermas, J., 1968, Technik und Wissenschaft als "Ideologie".
    Suhrkamp Frankfurt a.M.  (Norwegian ed., 1969:  Vitenskap som
    ideologi. Gyldendal Oslo. 109 pp.).
Haeckel, E., 1866, Generelle Morphologi der Organismen, II. Berlin.
    462 pp.

Harper, C.W., 1976, Phylogenetic inference in Paleontology.  J. Paleont., 50: 180-193.

Hecht, M.K., 1976, Phylogenetic inference and methodology as applied to the vertebrate record.  Evolutionary Biol., 9: 335-363.

Hecht, M.K. & Edwards, J.L., 1977, The methodology of phylogenetic inference above the species level.  NATO-ASI. 1976.  'Major Patterns in Vertebrate Evolution'.  (This vol.).

Hecht, M.K. & Edwards, J.L., 1976, The determination of parallel or monophyletic relationships, the proteid salamanders:  A test case.  Amer. Nat., 110: 653-677.

Hennig, W., 1950, Grundsüge einer Theorie der phylogenetischen Systematik. Berlin.  370 pp.

Hennig, W., 1957, Systematik und Phylogenese.  Ber. Hundertj. deutsche entomol. Ges. (1956), 50-71.

Hennig, W., 1965, Phylogenetic systematics.  Ann. Rev. Entomol., 10: 97-116.

Hennig, W., 1966, Phylogenetic systematics, 263 p.  Urbana, Ill.

Hennig, W., 1969, Die Stammesgeschichte der Insekten. Senckenberg Naturf. Ges. Frankfurt a.M.  436 pp.

Hennig, W., 1971, Zur Situation der biologischen Systematik, 7-15. In Siewing, R. (ed.):  Methoden der Phylogenetik, Erlanger Forschungen B, Naturwiss., 4, 88 p.

Hennig, W., 1975, "Cladistic Analysis or Cladistic Classification?": A reply to Ernst Mayr.  Syst. Zool., 24: 244-256.

Heywood, V.H., 1964, I. General Principles.  In "Phenetic and phylogenetic classification".  Syst. Assoc. Publ. No.6: 1-4.

Hull, D.L., 1964, Consistency and monophyly.  Syst. Zool., 13: 1-11.

Hull, D.L., 1970, Contemporary systematic philosophies.  Ann. Rev. Ecol. Syst., 1: 19-54.

Hull, D.L., 1974, Philosophy of biological science.  (Prentice Hall:  Englewood Cliffs).

Huxley, J.S. (ed.), 1940, The new systematics.  Oxford University Press, 583 pp.

Huxley, J.S., 1957, The three types of evolutionary process. Nature, 180: 454-455.

Huxley, J.S., 1958, Evolutionary processes and taxonomy with special reference to grades.  Pp. 21-39 in Hedberg, O. (ed.):  Systematics of todya.  Uppsala. Univ. Arsskr., 1958, 6.

Huxley, J.S., 1959, Clades and grades.  Publ. Syst. Assoc., 3: 21-22.

Jardine, N., 1969, The observational and theoretical components of homology:  a study based on the morphylogy of the dermal skull-roofs of rhipidistian fishes.  Biol. J. Linn. Soc., 1: 327-361.

Jardine, N. & Sibson, R., 1971, Methematical taxonomy. London.

Jarvik, E., 1968, Aspects of vertebrate phylogeny.  Pp. 498-527. In Ørvig, T. (ed.).

Kermack, K.A. & Kielan-Jaworowska, Z., 1971, Therian and non-therian mammals. Pp. 103-115 in Kermack, D.M. & Kermack, K.A. (eds.): Early Mammals. Suppl. No. 1, Zool. J. Linn. Soc., vol. 50.

Key, K.H.L., 1968, The concept of stasipatric speciation. Syst. Zool., 17: 14-22.

Key, K.H.L., 1974, Speciation in the Australian moraline grasshoppers - taxonomy and ecology. Pp. 43-56 in White, M.J.O.). (ed.).

Kiriakoff, S.G., 1959, Phylogenetic systematics versus typology. Syst. Zool., 8: 117-118.

Kuhn, T.S., 1970, The structure of scientific revolutions (2nd ed., University Chicato Press), Chicago. 210 pp.

Lankester, E.R. (ed.), 1909, A treatise on Zoology. Adam & Black, London.

Lehmann, J.P., 1975, Quelques réflexions sur la phylogenie des vertébrés inférieurs. Coll. Int. C.N.R.S., no. 218: 257-264.

Löther, R., 1972 (1974), Die Beherrschung der Mannigfaltigkeit. Fischer Verlag Jena. 285 pp. Summary transl. 1974, Syst. Zool., 23: 291-296.

Løvtrup, S., 1973, Classification, convention and logic. Zool. Scripta, 2: 49-61.

Løvtrup, S., 1975, On phylogenetic classification. Acta Zool. Cracov. 20: 499-523.

Luckett, W.P. & Szalay, F.S. (eds.), 1975, Phylogeny of the primates. Plenum, New York.

MacFadden, B.J., 1976, Cladistic analysis of primitive equids with notes on other perissodactyls. Syst. Zool., 25: 1-14.

Maslin, T.P., 1968, Taxonomic problems in parthenogenetic vertebrates. Syst. Zool., 17: 219-231.

Mayr, E., 1942, Systematics and the origin of species. Columbia University press, New York. 334 pp.

Mayr, E. 1963, Animal species and evolution. Harvard University Press. Cambridge. 797 pp.

Mayr, E., 1965, Classification and phylogeny. Amer. Zool., 5: 165-174.

Mayr, E., 1968, Theory of biological classification. Nature, 220: 545-548.

Mayr, E., 1969, Principles of systematic zoology. McGraw-Hill, New York. 428 p.

Mayr, E., 1974, Cladistic analysis or cladistic classification? Z. f. Zool. Syst. Evolutionsf. 12: 94-128.

McKenna, M.C., 1975, Toward a phylogenetic classification of the Mammalia. Pp. 21-46, in Luckett, W.P. & Szalay, F.S. (eds.).

Miles, R.S., 1973, Relationships of acanthodians, 63-103 in Greenwood, P.H. et al., (ed.).

Miles, R.S., 1975, The relationships of the Dipnoi. Coll. Int. C.N.R.S., no. 218: 133-148, Paris.

Nagel, E. & Newman, J.R., 1959 (1964), Gödel's Proof. Routledge & Kegan, Paul. London. 118 pp.

Nelson, G.J., 1969, Gill arches and the phylogeny of fishes, with notes on the classification of vertebrates. Bull. Amer. Mus. Nat. Hist., 141: 475-552.

Nelson, G.J., 1969a, The problem of historical biogeography. Syst. Zool., 18: 243-246.

Nelson, G.J., 1970, Outline of a theory of comparative biology. Syst. Zool. 19: 373-384.

Nelson, G.J., 1971a, 'Cladism' as a philosophy of classification. Syst. Zool., 20: 373-376.

Nelson, G.J., 1971b, Paraphyly and polyphyly: Redefinitions. Syst. Zool., 20: 471-472.

Nelson, G.J., 1972, Phylogenetic relationship and classification. Syst. Zool., 21: 227-231.

Nelson, G.J., 1972a, Cephalic sensory canals, pitlines, and the classification of esocoid fishes, with notes on galaxiids and other teleosts.. Amer. Mus. Novitates, 2492: 1-49.

Nelson, G.J., 1973, Comments on Hennig's 'Phylogenetic systematics' and its influence on ichthyology. Syst. Zool., 21: 364-374.

Nelson, G.J., 1973a, The higher-level phylogeny of vertebrates. Syst. Zool., 22 (1): 87-91.

Nelson, G.J., 1973b, 'Monophyly again?' - A reply to P.D. Ashlock. Syst. Zool., 22: 310-312.

Nelson, G.J., 1973c, Comments on Leon Croizat's biogeography. Syst. Zool., 22: 312-320.

Nelson, G.J., 1974a, Classification as an expression of phylogenetic relationships. Syst. Zool., 22: 344-359.

Nelson, G.J., 1974b, Darwin-Hennig classification: A reply to Ernst Mayr. Syst. Zool., 23: 452-458.

Nelson, G.J., 1975, Historical biogeography: An alternative formalization. Syst. Zool., 23: 555-558.

Nelson, G.J., 1976, Reviews: Biogeography, the vicariance paradigm, and continental drift. Syst. Zool., 24: 490-504.

Nevo, E., Kim, Y.J., Shaw, C.R. & Thaeler, C.S., Jr., 1974, Genetic variation, selection and speciation in Thomomys talpoides, pocket gophers. Evolution, 28: 1-23.

Ørvig, T., (ed.), 1969, Current problems of lower vertebrate phylogeny. Nobel Symposium 4, 539 pp., Stockholm.

Patterson, C., 1968, The caudal skeleton in Mesozoic acanthopterygian fishes. Bull. Brit. Mus. Nat. Hist. (Geol.), 17: 47-102.

Patterson, C., 1973, Interrelationships of holosteans. Pp. 233-305 in Greenwood, P.H., et al., (ed.).

Patterson, C., 1977, The contribution of paleontology to teleostean phylogeny. NATO-ASI, 'Major Patterns in Vertebrate Evolution', (this vol.).

Patterson, C. & Rosen, D.E., 1977, Review of ichthyodectiform and other Mesozoic teleost fishes and the theory and practice of classifying fossils. Bull. Amer. Mus. Nat. Hist. (In press).

Popper, K.R., 1973 (1974), Objective knowledge. An evolutionary approach (2nd ed.). (Clarendon Press, Oxford). 380 pp.

Popper, K., 1976, Unended quest. An intellectual autobiography.
    Fontana Collins Glasgow. 255 pp. Rev. ed. of 'Autobiography
    of Karl Popper', 1974.
Radnitsky, G., 1968 (1973), Contemporary schools of metascience.
    Göteborg, New York. 2 vols. (3rd ed., 1973, 446 pp., Chicago).
Radnitsky, G., 1969, Ways of looking at science: on a synoptic
    study of contemporary schools of 'metascience'. Scientia,
    104: 49-57.
Reed, C.A., 1960, Polyphyletic or monophyletic ancestry of mam-
    mals, or: What is a class? Evolution, 14: 314-322.
Remane, A., 1952, Die Grundlagen des natürlischen Systems der ver-
    gleichende Anatomie und der Phylogenetik. Geest & Portig
    Leipzig.
Remane, A., (1956, 1971, Die Grundlagen des natürlischen Systems,
    der vergleichende Anatomie und der Phylogenetil. (2. Aufl.)
    Verl. O. Koeltz Taunus. 364 pp. (Reprint of 2nd ed. 1956,
    Geest & Portig Leipzig).
Rensch, B., 1947, Neuere Probleme der Abstammungslehre F. Enke
    Stuttgart, 407 pp. (1973, 3rd ed., 468 pp).
Rensch, R., 1959, Evolution above the species level. Columbia
    University Press, New York. 419 pp. (Transl. of 'Neuere
    Probleme der Abstammungslehre'. 2nd ed., 1954).
Romer, A.S., 1949, Time series and trends in animal evolution,
    103-120. In Jepsen, G.L. et al., (eds.): Genetics, paleon-
    tology and evolution. Princeton.
Romer, A.S., 1962, The vertebrate body. (3rd ed.). Saunders,
    London.
Romer, A.S., 1966, Vertebrate paleontology. 3rd ed., 468 p.,
    Chicago.
Rosen, D.E., 1973, Interrelationships of higher euteleostan fishes.
    Pp. 397-513 in Greenwood, P.H. et al., (ed.).
Rosen, D.E., 1976, A vicariance model of Caribbean biogeography.
    Syst. Zool., 24: 431-464.
Ruse, M., 1973, The philosophy of biology. (Hutchinson University
    Library, London). 231 pp.
Schaeffer, B., Hecht, M.K. & Eldredge, N., 1972, Phylogeny and
    paleontology. Evol. Biol., 6: 31-46.
Schaeffer, B. & Rosen, D., 1961, Major adaptive levels in the evol-
    ution of actinopterygian feeding mechanisms. Amer. Zoologist,
    1: 187-204.
Schlee, D., 1971, Die Rekonstruktion der Phylogenese mit Hennig's
    Prinzip. Aufsätze Reden Senckenberg. Naturf. Ges., 20, 62 p.
Schopf, T.J.M. (ed.), 1972, Models in paleontology, Freeman,
    San Francisco
Scott, G.H., 1976, Foraminiferal biostratigraphy and evolutionary
    models. Syst. Zool., 25 (1): 78-80.
Simpson, G.G., 1943, Criteria for genera, species and subspecies
    in zoology and paleozoology. Ann. N.Y. Acad. Sco., 44 (2):
    145-178.

Simpson, G.G., 1945, The principles of classification and a classification of mammals. Bull. Amer. Mus. Nat. Hist., 85: 1-350.

Simpson, G.G., 1951, The species concept. Evolution, 5: 285-298.

Simpson, G.G., 1951a, Horses. Oxford University Press, N.Y., 247 p.

Simpson, G.G., 1959, Anatomy and morphology: Classification and
evolution: 1859 and 1959. Proc. Amer. Phil. Soc., 103 (2):
286-306.

Simpson, G.G., 1959a, Mesozoic mammals and the polyphyletic origin
of mammals. Evolution, 13: 405-414.

Simpson, G.G., 1959b, The nature and origin of supraspecific taxa.
Cold Spring Harbour Symp. Quant. Biol., 24: 255-271.

Simpson, G.G., 1960, Diagnosis of the classes Reptilia and Mammalia.
Evolution, 14: 388-392.

Simpson, G.G., 1961, Principles of animal taxonomy. 245 p., New
York & London.

Simpson, G.G., 1964, The meaning of a taxonomic statement, 1-31.
In Washburn S.L. (ed.): Classification and human evolution.
N.Y. Wenner-Gren Found. Viking Fund Publ. Anthropol., 37 (1963).

Simpson, G.G., 1971, Status and problems of vertebrate phylogeny.
(In Alvaredo, R., Gadea, E. & de Haro, A. (eds.): Simposio
Internacional de Zoofilogenia). Acta Salmantic., Cienc., 36:
353-368.

Simpson, G.G., 1975, Recent advances in methods of phylogenetic
inference. Pp. 3-19 in Luckett, W.P. & Szalay, F.S. (eds.).

Simpson, G.G., 1976, The compleat paleontologist? Ann. Rev. Earth
Planet. Sci., 1976: 1-13

Sneath, P.H.A., 1975, Cladistic representation of reticulate evolution. Syst. Zool., 24: 360-368.

Sokal, R.R., 1974, The species problem reconsidered. Syst. Zool.,
22: 360-374.

Sokal, R.R. & Crovello, T.J., 1970, The biological species concept:
A critical evaluation. Amer. Naturalist, 104: 127-153.

Sokal, R.R. & Sneath, P.H.A., 1963, Principles of numerical taxonomy. 359 p. Freeman, San Francisco.

Sondaar, P.Y., 1968, The osteology of the manus of fossil and recent
Equidae with special reference to phylogeny and function.
Verh. Konik. Nederl. Akad. Wetensch., Afd. Natuurk., I, 25,
No. 1: 76 pp (p pls, 25 figs.).

Strickberger, M.W., 1968, Genetics. 868 p., New York.

Sylvester-Bradley, P.C. (ed.), 1956, The species concept in palaeontology. Syst. Assoc. Publ. 2.

Szarski, H., 1977, Sarcopterygii and the origin of tetrapods.
NATO-ASI, 'Major Patterns in Vertebrate Evolution', (this vol.).

Thenius, E., 1969, Stammesgeschichte der Säugetiere (einschliesslich
der Hominiden). Handbuch der Zoologie, Band 8, Lief. 47/48,
722 pp.

Throckmorton, L.H., 1965, Similarity versus relationship in Drosophila. Syst. Zool., 14: 221-236.

Tuomikoski, R., 1967, Notes on some principles of phylogenetic
systematics. Ann. Ent. Fenn., 33 (3): 137-147.

Valen, L.v., 1960, Therapsids as mammals.  Evolution, <u>14</u>: 304-313.

Wahlert, G.v., 1968, <u>Latimeria</u> und die Geschichte der Wirbeltiere. 125 p. Stuttgart.

Wahlert, G.v., 1972, The definite approach:  Phylogeny an ecological process, 20 p. (preprint).  XVII.  Congr. Int. Zool., Theme 2. (M.S.).

Westoll, T.S., 1956, The nature of fossil species, p. 53-62.  I Sylvester-Bradley, P.C. (ed.).

White, M.J.D., 1968, Models of speciation.  Science, <u>159</u>: 1065-1070.

White, M.J.D., 1973, Animal Cytology and Evolution.  (3rd ed.), Cambridge University Press, London.

White, M.J.D., 1973a, Chromosonal rearrangements in mammalian population polymorphism and speciation.  Pp. 95-128 in Chiarelli, A.B. & Capanna, E. (eds.):  Cytotaxonomy and vertebrate evolution.  Academic Press, London.

White, N.J.D., (ed.), 1974, Genetic mechanisms of speciation in Insects.  Australia & N. Zealand Book Co.  Sydney. 170 pp.

Wiley, E.O., 1975, Karl R. Popper, systematics, and classification: A reply to Walter Bock and other evolutionary taxonomists. Syst. Zool., <u>24</u>: 233-243.

Wilson, E.O., 1965, A consistency test of phylogenies based on contemporaneous species.  Syst. Zool., <u>14</u>: 214-220.

Winterbotton, R., 1974, The familial phylogeny of the Tetraodontiformes (Acanthopterygii:  Pisces) as evidenced by their comparative myology.  Smithsonian Contr. Zool., <u>155</u>: 1-201, 185 figs.

Woodger, J.H., 1952, From biology to mathematics.  Brit. J. Phil. Sic., <u>3</u>:  1-21.

Zangerl, R., 1948, The methods of comparative anatomy and its contribution to the study of evolution.  Evolution, <u>2</u>: 351-374.

PHYLOGENETICS:  SOME COMMENTS ON CLADISTIC THEORY AND METHOD

Søren LØVTRUP

Department of Zoophysiology, University of Umea

S-901  87   Umea, Sweden

Phylogenetics occupies a central position between traditional - or Linnean - systematics and the neo-Darwinian theory of population genetics.

In Linnean systematics (living) organisms are classified for the purpose of identification and information retrieval.  Phylogeneticists also classify animals and plants, but with the specific aim of bringing out genealogical relationships.  Phylogenetic classifications therefore constitute theories dealing with the <u>history</u> of this evolutionary process.

The theory of evolution based on population genetics concerns certain aspects of the <u>mechanism</u> of evolution.  This problem can be studied only through observations made on living organisms. The validity of the results obtained for explaining phylogenetic evolution rests entirely on the premise that it is permissible to extrapolate to past evolutionary events.

Phylogenetics is thus a biological discipline with a program (and a methodology) distinct from those of the other two disciplines. However, for various reasons some of the concepts in systematics and population genetics have been introduced into phylogenetics where, in fact, they do not properly belong.  In my opinion they have done much harm by obscuring several important issues.  One of the main purposes of the following discussion is to provide examples demonstrating this point.  The present paper is a condensed version of ideas presented elsewhere (Løvtrup, 1973; 1974; 1975; 1977).

THEORY

## The Basic Premises

The conviction that it is possible to classify living organisms in a way revealing their phylogenetic kinship is based on two premises: (1) All organisms, living and extinct, have a common phylogenetic origin. (2) Barring convergence, the varying degrees of similarity of the properties possessed by various organisms are expressions of phylogenetic kinship.

The first premise implies that all living organisms have originated in the course of a unique evolutionary process.

The second premise, which presumes the validity of the first, indicates the method whereby phylogenetic relationships may be established, i.e. by comparative studies of the various properties in which similarities are displayed. It should be noticed that the premise does not imply that 'overall similarity' should be used for this purpose. We shall return to this point, and also to the reservation about convergence, in the following text.

## Monophyletic Taxa

It was suggested in premise (2) that the properties possessed by living organisms are used for their classification. The properties thus used may be called 'taxonomic characters.' Traditionally, Linnaeus and his successors used morphological features for this purpose. But this is a matter of convention and convenience, in fact all properties possessed by an organism are potential taxonomic characters. In Linnean systematics a limited number of taxonomic characters may often suffice, but in phylogenetics the demands on classification are higher, and therefore the repertoire employed is often much larger. It may therefore be expected that all new information acquired about plants and animals will be used by phylogeneticists, at least for testing their classifications.

In the classification of fossils we must rely on properties exhibited by hard tissues. Even in Linnean systematics it would not be possible to work only with such features, and in phylogenetic systematics it is impossible. Therefore the possibilities for correct phylogenetic classification of fossil forms are strictly limited; the best we can hope for is to establish their approximate location in the hierarchy of living organisms.

Starting with Linnaeus, classification is performed by the

erection of taxa.  A taxon is a concept, defined by a set of properties distinguishing a particular class of organisms.  All individuals, past, present or future, possessing the whole set of properties are usually said to be members of the taxon.

The various properties found in living organisms differ with respect to the number of individuals in which they are found.  Some are present in all organisms, some only in subsets, some only in subsets of the subsets, etc.  This fact implies that it is possible to erect taxa whose members are members of different subordinate taxa.  This was used by Linneaus to erect systematic hierarchies, constructed by placing at the apex the taxon comprising the largest number of organisms and at the base those taxa whose members have no properties in common other than those defining higher levels in the hierarchy.

The principles outlined here are also adopted in phylo-genetics, with the important reservation that the classification should reveal phylogenetic kinships.  If it is granted that living organisms have originated through one unique evolutionary process, it thus follows that the goal of phylogenetics is to establish the unique hierarchy which faithfully reflects the history of this process.

The constraint thus placed on the endeavours of the phylo-geneticist implies that the taxa in his system must be monophyletic, i.e. their members shall comprise all those individuals which are descendents of the first organisms endowed with all the taxonomic characters of the taxon.  The erection of monophyletic taxa is thus the primary concern of the phylogeneticist.  Fortunately, he does not have to start from scratch.  There are a number of Linnean taxa; e.g. Echinoidea, Nematoda, Oligochaeta, Gastropoda, Myxinoidea, Dipnoi, Salientia, Mammalia and many others, both higher and lower ones, which are so clearly monophyletic that they do not pose any problems.  It is only when the relationships be-tween these taxa have to be established that the difficulties be-gin.

## Cladograms

A phylogenetic hierarchy may be represented diagrammatically in a cladogram.  The shape of the latter must of course in some way reflect the way new taxa arise from old ones.  To account for the relationships between those taxa which are descendents of the same taxon Hennig (1966:139) has introduced the concept 'sister groups,' defined as follows: 'Species groups that arose from stem species of a monophyletic group by one and the same splitting process may be called "sister groups".'  Experience has shown that sister groups usually come in pairs.

I shall later argue that Hennig's terminology is wrong, or at least confusing. Replacing 'sister groups' by 'twin taxa', I submit that the following definition is preferable: A pair of taxa that arise through a process of dichotomy from the taxon in which they are included may be called twin taxa.

A cladogram thus generally is a dichotomous dendrogram. It is natural to accept, as proposed by Hennig, that each bifurcation involves the introduction of a new taxonomic rank. This convention implies a radical break with Linnean systematics in which ranking corresponds to set categorical levels. Of these there are some twenty at most, and this number is far too small for phylogenetic classification.

It is an empirically observed fact that the frequency of dichotomy varies in different parts of a cladogram. This has an important consequence for the numbering of the taxonomic levels. Hennig has suggested, quite rightly, that 'sister groups' are given equal rank, and further claimed that 'every monophyletic group, together with its sister group (or groups), forms - and forms only with them - a monophyletic group of higher rank' (1966: 139). This proposition implies, in conformance with a convention in Linnean systematics, that the numbering begins with the terminal taxa at the base of the hierarchy (the species) and proceeds towards the apex. But if the premise about the frequency of dichotomy is correct, then Hennig's suggestion is impracticable. It is easy to see that if the number of dichotomies is different in a pair of twin taxa, and we begin the numbering with the terminal taxa, then the twin taxa cannot have the same rank.

The only way to reach a numbering which is consistent, and gives the same rank to twin taxa, is to begin at the apex of the hierarchy and go downwards. This means that the most comprehensive taxa have the lowest numerical rank. Since it may feel awkward to call kingdoms, phyla, etc., 'lower taxa,' and genera and species 'higher taxa,' I have proposed using the names 'superior' and 'inferior' to remedy this terminological dilemma. It should be noticed that the numbering suggested has the practical advantage that it properly reflects the course of evolution which, clearly, has proceeded from the apex to the base of the hierarchy. It has, however, the consequence that the terminal taxa will be of different rank.

An example of a dichotomous cladogram is shown in Figure 1. It is seen that following a generally accepted practice I have turned the dendogram upside down, implying that movement from the bottom to the top of the figure represents the course of evolution. The 'ordinate' thus represents time, but it must be emphasized that separate scales would be necessary for most or all of the taxa.

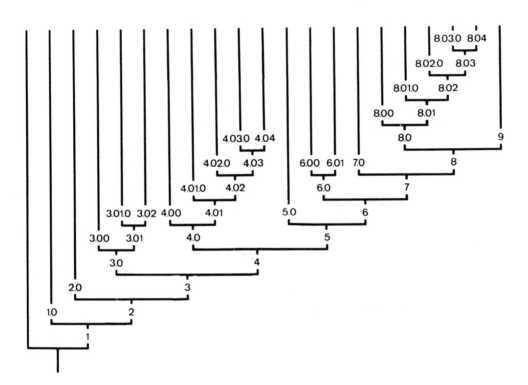

Figure 1. A cladogram. The example shown represents the phylogenetic classification of Vertebrata, as proposed in Løvtrup
(1977). The taxon furthest to the left represents the invertebrate twin taxon to Vertebrata. The remaining taxa, in alphabeti-
cal order, are: Actinistia, Amiidae, Aves, Caudata, Chelonia,
Chondrostei, Cladistia, Crocodylia, Dipnoi, Gymnophiona, Holoce-
phali, Hyperoartii, Hyperotreta, Lepisosteidae, Mammalia, Rhynco-
cephalia, Salientia, Selachii, Squamata and Teleostei. The various
taxa have been numbered according to a digital code.

In the cladogram the horizontal lines represent the processes of splitting (and isolation) through which new taxa arise, and the vertical lines represent the various taxa, and the temporal aspect of their history. Some taxa are distinguished by a number of taxonomic characters. If it is assumed that these have been acquired independently at separate occasions, then a number of dichotomies corresponding to the number of characters must have occurred in the course of time represented by a vertical line. Since those organisms no longer exist which did not acquire the taxonomic characters in question we may conclude that to each independent taxonomic character corresponds an extinct 'twin taxon', representing the branch which did not acquire the property in question.

Phylogenetic cladograms clearly are very different from Linnean hierarchies. Yet, it should be recalled that all monophyletic Linnean taxa will be found in the phylogenetic classification, but that their relative ranks will often be subject to radical changes.

Taxonomic Age

The above interpretation of the cladograms implies that the defining characters of a taxon change with time, that, in fact, the definition of the taxon as we know it today was established only when it was split into two new taxa. Therefore, if by the age of a taxon we mean the time it has existed as a fully defined taxon, then we must use this time of splitting. It is true that they began their history at the same time, but since the frequency of dichotomy varies, it is not at all likely that they underwent divergence at the same time.

Hennig (1966) has suggested that the absolute rank of a taxon bears relation to its age. According to the ranking procedure proposed here, this principle should imply that the age of a taxon is inversely proportional to its numerical rank. Although this unquestionably holds within certain bounds, it cannot be a general rule. Even if we should adopt the rule that the age of a taxon dates from the time of its inception, i.e. from the time of the dichotomy at which its history began, this generalization would not be true. This follows from the premise that the frequency of dichotomy varies within the hierarchy. Therefore reference to fossils cannot help us at all in the determination of the rank of the taxon. This characteristic, being a function of its phylogenetic relationship, is determined by means of the distribution of the various properties possessed by the organisms, without the slightest reference to fossils. However, fossils may give an approximate idea about the age of the various superior taxa. This information may be used to adjust the length of the vertical lines

in the cladograms such that the time ordinate linear.  But I would
like to suggest that if fossil data are used in this way, then they
should be accepted even when they do not fit preconceived notions.

If thus we cannot compare the age of taxa in different parts
of the cladogram, we can still make one interesting statement
about taxonomic age.  If we define a phylogenetic line (or lineage)
as a series of taxa, $T_1$, $T_2$, $T_3$ ... $T_n$, each of which is included
in the preceding one, then it holds that $T_1$ is older than $T_2$, $T_2$
is older than $T_3$, etc.

## Dichotomy

The number of branches at each level in the phylogenetic
hierarchy is, of course, determined by fortuitous, evolutionary
events.  We cannot a priori exclude that trichotomies, etc., occur.
And therefore it is not possible to establish dichotomy as a
'necessary and unavoidable methodological principle' (Bonde, 1975,
p. 302).  I also think it is a mistake to invoke the principle of
parsimony as an argument in favour of dichotomy.  In fact, I think
the statement:  'irrespective of how speciations may happen in
"Nature", we can analyze them only as dichotomous splittings'
(Bonde, 1975, p. 303) should be replaced by:  the shape of our
cladograms is determined by the data available.  If they indicate
the occurrence of plurichotomies we must accept them, at least for
the time being.  But if we suspect that all branchings in the cladograms should be dichotomies, a reasonable assumption, then we must
go on looking for data which permit further resolution in our cladograms.

For those phylogeneticists who think in neo-Darwinian, or for
that matter, in Linnean terms, the taxa arising from a dichotomy
are species.  The difficulties caused by this conventional thinking
is illustrated by the following quotation:  'That the two sister-
species are new and the ancestral species becomes extinct at the
speciation is most practical for a consistent terminology' (Bonde,
1975, p. 295).

There is a grain of truth in this contention, but this can
never be borne out when discussed in terms of 'species'.  Suppose
the bifurcation involves part of the original species acquiring a
new character x, while the remainder is unchanged.  These two
groups being the sisters, it follows that the unchanged one should
at the same time be a new and an extinct 'species'.  This interpretation seems to defy all common sense.

Stated in terms of the present theory the ancestral taxon is
an incipient taxon, $T_j$.  When some of its members acquire the

property x, the taxon $T_j$ becomes fully defined, and we get two new incipient taxa $T_{j+1}$, one of which is distinguished by x, the other by $\bar{x}$, i.e. absence of x. So the unchanged taxon is indeed new, but at a higher numerical rank. And therefore we should not talk of 'speciation' in phylogenetic contexts, but only about taxonomic 'divergence', 'splitting', 'dichotomy', etc.

## The Terminal Taxa

Linnean systematics is based on some simple rules of convention. Among these is that the number of categories is determined in advance. To be sure, the categorical levels in the original hierarchy have turned out to be too few, and accessory ones have been introduced. But even this has been done on a purely conventional basis.

Phylogenetic classification cannot be based on any a priori conventions, for the shape of the unique phylogenetic hierarchy is determined by the historical course of evolution. And therefore the phylogeneticist has no other choice than to construct his cladograms according to the dictates of Nature.

The Linnean convention also concerns the terminal taxa. In the original Linnean hierarchy these are species, involving that every distinguishable group of animals and plants constitutes a species. So far, nobody has ever succeeded in establishing absolutely objective rules for the delimitation of this category, and yet, paradoxically enough, most biologists claim that the species assumes a unique position among the Linnean categories. The aim of Linnaeus was evidently to ensure that all terminal taxa have the same rank. However, his successors have in many instances distinguished a number of separate groups within a species. For the purpose of classifying these organisms various subordinate categories, like subspecies, races, etc., have been introduced, thus upsetting, it seems, one of the fundamental Linnean principles of systematics. This example shows that in spite of its few constraints, numerous difficulties are encountered when the Linnean system is confronted with the realities of Nature.

This very fact suggests that even greater problems will obtain if the species concept is used outside its proper territory. Those evolutionists who concern themselves with the mechanism of evolution base their argument on the theory of population genetics. Under these conditions it is natural to deal with gene pools, selection coefficients, mutation frequencies, fitness, etc. But is it necessary to deal with species? In fact, owing to the shortcomings of the Linnean species, these evolutionists have introduced the 'biological species', supposed to be the 'real' unit of evolution. This concept is defined as groups of actually or potentially

interbreeding natural populations, which are reproductively isolated from other such groups (Mayr, 1964, p. 120). But this definition shows the weakness of the concept, for as shown in Figure
2, potential interbreeding often reaches far above the limit of
the species, and actual gene flow most often ceases far below this
level; on rare occasions it may be above. Even if we disregard
that an essential part of the organic world--those organisms which
do not interbreed to reproduce--has been excluded in order to fit
the synthetic theory, it is evident that the attempt to accommodate
a Linnean category in a theory of evolution has led to confusion.
Would it not be simpler to deal merely with interbreeding populations?

The situation is even more clear-cut as far as the phylogeneticists are concerned. Their task is to establish relationships
between monophyletic taxa and, as we have seen, in their cladograms
we shall find most, probably, of the Linnean taxa. However, the
Linnean categories lose all meaning and must be abolished. Why
then preserve the lowest category, particularly if we know that the
important evolutionary events usually take place at levels inferior to that of the species?

In fact, I believe that once we understand the implications
of the cladograms it will be clear that the notion about the unique
status of the species is due to a misunderstanding. For one thing
it should be noted that as we go down the cladogram, and thus back
in time, we shall find only taxa of lower numerical rank, corresponding to higher category Linnean taxa, and no species. It will
be argued that in the past these taxa were in fact diverged to the
extent that a contemporaneous Linnean systematist would be able to
subdivide the members of a class, as recognized by us, into orders,
families, genera and species. This argument fails for three reasons.
First because even if the class had not undergone any divergence,
it would still be necessary to name all the categories mentioned.
Thus, through convention the class would also be species. Second,
if divergence had occurred such that it was a 'true' species which
gave rise to the inferior taxa in the class, then all the properties possessed by these organisms, distinguishing all the taxonomic
levels to which they belonged, including the species characters,
would today be class characters, provided they had been preserved.
Third, whether or not species have existed in the past, we shall
never be able to test at this point. And I believe that it is un-
sound in any empirical science to operate with unfalsifiable propositions.

It seems to me that this confusion might be avoided if the
phylogeneticist made clear what is, and what is not, his task.
His concern is, as stated above, to establish a theory about the
historical course of evolution, and this he may do by the constru-
tion of cladograms. The phenomenon which he primarily studies is

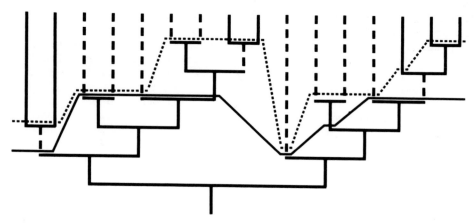

Figure 2. A hypothetical cladogram, resolved so as to demonstrate
subspecific taxa. The twelve Linnean species are represented by
vertical dashed lines. The limit of <u>potential</u> reproduction--or
interfertility--is represented by a fully drawn line. This is
thus the limit of irreversibility in evolution. The line of
<u>actual</u> reproduction is represented by a dashed line. It is thus
below the level of this line that evolution occurs. The area between these two lines represents the region of the possible, but
not very likely, reversibility of evolution.

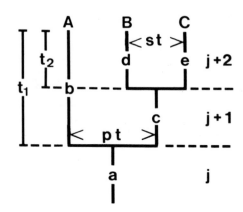

Figure 3. A basic classification. The terminal taxa are A, B
and C. The lines a - e represent the five taxa in the classification, of these (a) has the rank j, (b) and (c) the rank j + 1
and (d) and (e) the rank j + 2. (b) and (c) are the primary twins
(pt), (d) and (e) the secondary twins (st). The times from the
splittings to the present day, $t_1$ and $t_2$, are indicated at the
left side.

taxonomic divergence, i.e. the splitting of taxa.  He does not deal
with the mechanism of evolution, and therefore it is not necessary
to work with concepts like 'species' and 'speciation', as little
as with those belonging to the population-genetics repertoire,
gene pool, gene frequencies, etc.

In the cladogram the limit of potential reproduction--or
interfertility--in sexually reproducing organisms may be represented by a wavering line (Figure 2).  This line represents the
limit of irreversibility of evolution.  The unique feature about
the species is that this line never passes below the species level.
In those cases where a species is not subdivided into inferior
taxa, the species is unique in one phylogenetic line, insofar as
it is a terminal, not yet fully defined, taxon.  But if the species
is divided into subspecies, the latter assume this unique position,
while the species becomes a taxon among other phylogenetic taxa.

The line of interfertility has been moving, and is still moving today, towards the base of the phylogenetic hierarchy, as the
consequence of mutations interfering with reproduction.  And therefore we may conclude that in the course of phylogenetic evolution,
interfertility has prevailed at all taxonomic levels.

METHOD

### The Basic Classification

Discussion of the methodology of phylogenetic classification
is most easily surveyed when based on the simplest possible classification, one involving three terminal taxa.  This model, which
may be called a basic classification, involves five taxa altogether,
representing three different numerical ranks (Figure 3).  The classification comprises two sets of twin taxa, of the ranks j+1 and
j+2.  The two sets may be called the primary and the secondary twin
taxa, respectively.  The particular value of this model is that the
phylogenetic classification of any number of taxa can be carried
out as a succession of basic classifications.

It is seen that a prerequisite for phylogenetics is that, barring trichotomy, the phylogenetic kinship between any three monophyletic taxa can be established in a basic classification.  When
this has been done, we will know that the secondary twins have ancestors in common that were not ancestors to the third taxa.  The
task of phylogenetics thus involves ascertaining the secondary twins
among the three possible pairs in a basic classification.

As already discussed, phylogenetic relationships must be decided on the basis of characters distinguishing the monophyletic

taxa to be classified.  But not all characters are of equal value
for this purpose.  To clarify this I shall introduce some concepts,
based on Hennig's terminology, although slightly modified.  The
characters defining the taxon $T_j$ and all the superior taxa (of
lower numerical rank) in the phylogenetic line to which it belongs,
are called plesiotypic.  Barring losses, these should be found in
all three terminal taxa.  These are therefore of no value for the
establishment of a basic classification.  All the characters which
have been acquired in each of the terminal taxa, are called teleotypic.  These are likewise of no classificatory value.  Finally,
those characters defining the primary twin taxon containing the
secondary twins are called apotypic.  They are found only in two
taxa, and if some of them can be recognized, we also know which of
the three pairs of taxa is the secondary twins, for only these
should have characters in common.

## Apotypy and Non-Apotypy

It would thus appear that a very simple approach to the establishment of a basic classification would be to look for apotypy,
i.e. for characters common to two of the three taxa.  Yet, when
this is done it will generally be observed that all three pairs
of taxa have some characters in common.  The phenomenon that two
taxa, one primary twin and one secondary twin, have characters in
common may be called non-apotypy.  Table 1 surveys the various ways
in which two characters may end up by being common to two of the
taxa in a basic classification.  Altogether 10 different ways are
imaginable; four involving two mutations (convergence) and six involving one mutation.  Among the first, two (3 and 9) will unite
the pair of secondary twins.  These are indistinguishable from true
apotypy.  The other two, (4 and 10), are to be rated as non-apotypic.

Among the one-step changes four (1, 5, 6 and 7) are to be rated
as apotypic.  Two cases (5 and 6) are true apotypy, i.e. they represent changes in the primary taxon comprising the secondary twins;
the remaining one (1 and 7) are the outcome of gains or losses in
the other primary taxon.  The last two cases (2 and 8) constitute
non-apotypy.

An analysis of character distribution thus does not necessarily give a clear-cut answer.  In order to proceed it is, in fact,
necessary to adopt one of the following premises:  (1) it is possible to decide which characters are apotypic and which are not;
or (2) the numerical distribution of characters can settle the
issue.

## The Qualitative Approach

When analyzing the first case, it should be recalled that all characters that are indisputably plesiotypic, i.e. those present in all three taxa, have been ignored. Yet, some plesiotypic characters still appear in our lists in two cases (1 and 2), and in two other cases (7 and 8) the absence of a teleotypic innovation probably means the presence of a plesiotypic one. It is seen that the difficulties encountered in phylogenetic classification arise from the two cases of non-convergent non-apotypy (2 and 8), where one of the secondary twins, the 'derived' one, loses a plesiotypic or gains a teleotypic character, thereby 'uniting' the other secondary with the primary twin.

If, under these circumstances, it is found that one taxon has many characters in common with both of the others, then it may not be difficult to decide which of the two sets of characters consists of plesiotypic characters, therefore uniting one of the secondary twins with the primary twin. It is important, however, to stress that we must have two sets of characters in order to carry through this procedure. If only one set is found, they must be accepted even if most or all are suspected to be plesiotypic.

I believe that this method, when applied with judgement, is of value in certain situations. A typical case is represented by a basic classification involving two taxa of related, morphologically primitive animals and an advanced taxon to which the others are also related, for instance, the two orders of Cyclostomata and the Gnathostomata. There is no doubt whatsoever that the characters which primarily led to the erection of the class Cyclostomata are plesiotypic ones, in the form of absence of Gnathostome characters of various kinds, notably morphological ones. What is required to test the validity of this traditional classification is to search for properties, by necessity non-morphological ones, which unite one of the orders with Gnathostomata. As shown elsewhere (Løvtrup, 1977) it seems possible to show that in fact Hyperoartii and Gnathostomata are secondary twins in this classification.

I do not think, however, that this method can always be applied, in particular this holds in basic classifications involving superior taxa; e.g. phyla and classes. In a phylogenetic classification these must also be classified, and very often the distinctions of the taxa are so different that it is very difficult to find characters in common, at least on the morphological plane. In this case, I believe there is no possibility of arriving at an objective classification other than by a numerical approach.

## The Quantitative Approach

If it is assumed that characters arise with a mean frequency $F$, then the number of taxonomic characters defining a certain taxon should be $Ft$, where $t$ is the time during which the taxon has been under creation. Since this period is generally unknown, I shall here let $t$ represent time from the separation of the taxon from its twin to the present day. This means that when we are dealing with superior taxa, the taxonomic characters should not be those defining only these taxa but, rather, those representing a phylogenetic line, leading from the bifurcation in the dendrogram to some terminal taxon (species) which is included in each of the taxa under investigation.

If these stipulations are accepted, and the ages of the various taxa in the basic classification are assumed to be $t_1$ for the terminal primary twin taxon, $t_2$ for the secondary twin taxa, and thus $t_1 - t_2$ for the non-terminal primary twin taxon (figure 3), then the characters in common for the non-convergent cases in Table 1 can be calculated, as shown in the last column of the table.

The frequency of convergence may be estimated in the following way. Let the total number of potential mutations in the phylogenetic lines originating in a set of twin taxa be $N$. Let further the number of mutations which actually have occurred be $n$ $(= Ft)$. The number of identical mutations, i.e. the frequency of convergence, will have a hypergeometrical distribution with the mean value $n^2/N$. Thus, as listed in Table 1, the frequency of apotypic convergence in a basic classification will be $F^2 t_2^2/N$., and of non-apotypic convergence $F^2 t_1 t_2/N$.

Our calculations thus show that non-apotypic and apotypic convergence are of the same order. Furthermore, presuming $N \gg Ft$, convergence must be a relatively rare phenomenon.

Hennig has stated '"that phylogenetic systematics would lose all the ground on which it stands" if the presence of apomorphous characters in different species were considered first of all as convergencies' (1966, p. 121). And with this one cannot but agree, for this assumption would involve a serious violation of the principle of parsimony. In any case it is certain that we shall never be able to distinguish apotypic convergence from apotypy, at least if we are really dealing with the same character in the former case. But as the preceding calculations show, the phylogeneticists need not worry too much about convergence. For one thing, apotypic and non-apotypic convergence tend to cancel each other out, and secondly, both of them will be much less frequent than the corresponding non-convergent cases.

From the values listed in Table 1 we may calculate that the

Table 1. The different ways in which a taxonomic character may end up being common to two of three taxa in a basic classification.

| | A | B | C | Numbers of taxonomic characters AB AC | BC |
|---|---|---|---|---|---|
| (1) Plesio-apotypy | $\langle\bar{p}\rangle \leftrightarrow$ | $p \leftrightarrow$ | $p$ | | $\underline{F}\,\underline{t}_1$ |
| (2) Plesio-non-apotypy | $p \leftrightarrow$ | $p \leftrightarrow$ | $\langle\bar{p}\rangle$ | $\underline{F}\,\underline{t}_2$ | |
| (3) Convergent plesio-apotypy | $p$ | $\langle\bar{p}\rangle$ | $\langle\bar{p}\rangle$ | | $\underline{F}^2\underline{t}_2^2/\underline{N}$ |
| (4) Convergent plesio-non-apotypy | $\langle\bar{p}\rangle$ | $\langle\bar{p}\rangle \leftrightarrow$ | $p$ | $\underline{F}^2\underline{t}_1\underline{t}_2/\underline{N}$ | |
| (5) Apotypy | $p \leftrightarrow$ | $p \leftrightarrow$ | $\langle\bar{p}\rangle$ | | $\underline{F}(\underline{t}_1 - \underline{t}_2)$ |
| (6) Apotypy | $\bar{a} \leftrightarrow$ | $a \leftrightarrow$ | $a$ | | $\underline{F}(\underline{t}_1 - \underline{t}_2)$ |
| (7) Teleo-apotypy | $\bar{t} \leftrightarrow$ | $\bar{t} \leftrightarrow$ | $t$ | | $\underline{F}\,\underline{t}_1$ |
| (8) Teleo-non-apotypy | $\bar{t} \leftrightarrow$ | $\bar{t} \leftrightarrow$ | $t$ | $\underline{F}\,\underline{t}_2$ | |
| (9) Convergent teleo-apotypy | $\bar{t}$ | $t$ | $t$ | | $\underline{F}^2\underline{t}_2^2/\underline{N}$ |
| (10) Convergent teleo-non-apotypy | $t$ | $t$ | $\bar{t}$ | $\underline{F}^2\underline{t}_1\underline{t}_2/\underline{N}$ | |

A is the primary, B and C the secondary, twin taxa. An arrow between two characters indicates non-convergence, a bar above a letter means absence of the characters in question. Uncircled this absence implies 'non-gain', circled 'loss'. The small letters refer to the origin of the characters: a, apotypic; p, plesiotypic; t, teleotypic. To designate the various phenomena some names have been proposed, the derivation of which seems to be straightforward. For the calculations of the numbers of taxonomic characters, the reader may refer to the text.

difference between the number of non-convergent characters uniting
the secondary twins and those common to any of the other pairs is:

$$\underline{D} = 2\underline{Ft}_1 + 2\underline{F}(\underline{t}_1 - \underline{t}_2) - 2\underline{Ft}_2 = 4\underline{F}(\underline{t}_1 - \underline{t}_2)$$

The requirement for the dendrogram being a basic classification
rather than a trichotomy is that $\underline{t}_1 - \underline{t}_2 > 1/4\ \underline{F}$, the time necessary
for a character to arise in either of the primary twin taxa. The
requirement for the dendrogram being a basic classification rather
than a dichotomy is that $\underline{t}_2 > 1/4\ \underline{F}$, the time required for a character to arise in either of the secondary twin taxa.

The quotient between the two sets of characters is:

$$Q = \frac{\underline{Ft}_1 + \underline{F}(\underline{t}_1 - \underline{t}_2)}{\underline{Ft}_2} = 2\underline{t}_1/\underline{t}_2 - 1$$

For $\underline{t}_2 \to \underline{t}_1$, we have $Q \to 1$, so that the numbers of apotypic
and non-apotypic characters are the same, as would be expected.

On the basis of these very crude assumptions it thus seems
possible to establish a quantitative method of phylogenetic classification which involves that among the three pairs of taxa in
a basic classification, the secondary twins have the maximum number of characters in common. This result could also be stated in
the dictum that non-apotypy should be minimized.

The approach outlined here, to decide the classification on
the basis of the numerical distribution of characters, is, I believe, similar to a suggestion made by Hennig (1966, p. 121):
'The more characters certainly interpretable as apomorphous (not
characters in general) that there are present in a number of
species, the better founded is the assumption that these species
form a monophyletic group'. However, I cannot agree with this
quotation for, in fact, if only one character is certainly interpretable as apomorphous (apotypic), then we do not have to assume,
then we know, that we are dealing with a monophyletic taxon. Rather,
it is in all those cases where we are not absolutely sure about the
nature of the characters that we must adopt the numerical approach
implied by the above quotation.

Sneath and Sokal (1973) stress that 'invariant' characters
should be excluded. If this means that, as here, all surely plesiotypic characters are ignored, then the present method also coin-
cides with the methodological principle of numerical taxonomy.

A number of objections have been raised against the numerical
or quantitative approach, but I am not so sure that the confidence

with which Hennigian cladists perform their classifications is always justified.  Phylogenetic classification is difficult because
evolution has been so frequently determined by chance, and therefore
the only safe way to proceed is to use all methods and all characters available.

## CONCLUSIONS

The present paper is a survey of ideas presented elsewhere.
These are based largely on the cladism of Hennig, but deviate from
this theory on a number of points.  Among these may be mentioned:

(1) A consistent numbering of taxonomic rank can be obtained only
if it begins at the apex of the hierarchy.  If this procedure is
adopted, twin taxa (sister groups) will have the same rank.  However, since the frequency of dichotomy varies in the different
parts of the cladogram, the terminal taxa (species, subspecies,
etc.) cannot have the same rank.

(2) The age of the taxon cannot be reckoned from the time it parted
from its twin taxon, but from the time it itself gave rise to two
new taxa.  Owing to the variation of frequency of dichotomy we
therefore cannot state anything about the relative ages of twin
taxa.

(3) For the same reason it does not hold that the absolute rank
of a taxon bears relation to its age.

(4) The species is a terminal taxon in the Linnean system, but not
necessarily in phylogenetic cladograms.  Therefore there is no
justification for assuming a unique position for this category,
rather, a species is a taxon among other taxa.  When it is a terminal taxon, then it represents the contemporary level of evolution,
but this position is often assumed by subspecific taxa.  It is concluded that the use of the Linnean species in phylogenetic and evolutionary contexts can only lead to confusion.

(5) It is shown that all phylogenetic classifications can be reduced to basic classifications involving three taxa.  The decision
between the three alternative classifications may be reached by
looking for apotypic characters uniting the pair of taxa having the
most recent common ancestors, called the secondary twins.  For
reasons outlined it will generally be found that all three pairs
of taxa have characters in common.

(6) In order to establish the basic classification it is therefore
necessary to adopt either one of two premises:  (a) it is possible
to decide which are, and which are not, apotypic characters; or

(b) this insight is not available.  In the latter case a numerical approach must be used, involving that the pair of taxa having most characters in common are the secondary twins.  This method is based on the principle of parsimony, and seems to coincide with the methodology employed in numerical taxonomy.

## REFERENCES

Bonde, N., 1975, Origin of "Higher groups":  Viewpoints of phylo-
     genetic systematics.  Coll. internat. C.N.R.S.  218:  293-
     324.
Hennig, W., 1966, Phylogenetic Systematics.  University of Illi-
     nois Press, Urbana.
Løvtrup, S., 1973, Classification, convention and logics.  Zool.
     Scr.  2:  49-61.
Løvtrup, S., 1974, Epigenetics.  Wiley, London.
Løvtrup, S., 1975, On phylogenetic classification.  Acta Zool.
     Cracov.  20:  499-523.
Løvtrup, S., 1977, The Phylogeny of Vertebrata.  Wiley, London
     (in press).
Mayr, E., 1964, Systematics and the Origin of Species.  Dover,
     New York.
Sneath, P.H.A. and Sokal, R.R., 1973, Numerical Taxonomy.  Freeman,
     San Francisco.

ON THE PHENETIC APPROACH TO VERTEBRATE CLASSIFICATION

James S. FARRIS

Dept. of Ecology and Evolution, State Univ. of New York

Stony Brook, L.I., New York  11794 U.S.A.

## INTRODUCTION

I consider the general subject of phenetic classification to possess two major subdivisions.  The first is the matter of definition:  what is meant by phenetic classification?  The second is the matter of motivation:  on what grounds do pheneticists advocate their particular methods for constructing classifications? The question of motivation can be looked at in two ways.  First, what principles are involked by pheneticists in selecting the methods which they advocate; and second, what drawbacks do pheneticists ascribe to the methods of classification proposed by other schools of taxonomy?  The definition of phenetic taxonomy is necessarily purely a matter of convention, and I shall therefore consider it only in enough detail to avoid ambiguity.  The motivations of phenetic taxonomy are of much greater importance, for they touch on the long-standing debate among taxonomists of the phenetic, phylogenetic, and evolutionary schools concerning the proper basis upon which to select classificatory methods. This debate has been perpetuated at least in part by the tendency of some reviewers (for example, Mayr, 1974; Sokal, 1975) to criticize the principles of other schools of taxonomy on a superficial, terminological level.  I shall devote most of my discussion to attempts to elucidate what appear to me to be the most fundamental principles of phenetic taxonomy and to obviate the purely terminological aspects of the debate through an evaluation of both phenetic and non-phenetic taxonomic methods on the basis of these principles.

## THE MEANING OF PHENETIC CLASSIFICATION

No authoritative definition of "phenetic classification" appears to be available.  The term "phenetic" itself has often been defined implicitly through definitions of such quantities as "phenetic relationship."  Sneath and Sokal (1973, p. 29) define "phenetic relationship" as "similarity (resemblance) based on a set of phenotypic characteristics of the objects of organisms under study."  The definition of Sneath and Sokal would seem to suggest that any arrangement or classification of organisms based on their observable features is a "phenetic" classification.  Indeed, when Sneath and Sokal state, for example, that "the sources of data for cladistic {phylogenetic} inferences are invariably phenetic," (Sneath and Sokal, 1973, p. 319) it is hard to interpret "phenetic" as meaning anything more specific than simply "observable."  Such a broad conception of "phenetic" is quite useless for purposes of the present discussion.  Other schools of taxonomy certainly use methods which are based on observable features of organisms.  In order to discuss the differences between "phenetic" and non-"phenetic" taxonomies, it is necessary to use a definition of "phenetic classification" which admits of the existence of empirically meaningful, non-"phenetic" classifications.

Much more restrictive meanings of "phenetic classification" have apparently been intended by most reviewers.  Cain and Harrison (1960, p. 3) define "phenetic relationship" as "arrangement by overall similarity, based on all available characters without any weighting ..."  Mayr (1974, p. 95) characterizes "phenetics" as systematics in which organisms are classified on the basis of overall similarity.  Jardine and Sibson (1971, p. 136) describe "pheneticism" as taxonomy whose sole aim is to "produce classifications which reflect as accurately as possible the relative similarities or dissimilarities of populations..."  Some ambiguity, however, remains:  what precisely is meant by "arrangement" or "grouping" according to overall similarity, or by "reflecting relative similarities?" The answer to this question seems not to have been incorporated in any formal definition of phenetic taxonomy, possibly because authors have felt the matter too obvious to require explicit clarification.  I shall try to answer it indirectly by abstracting the common properties of methods which pheneticists advocate for the construction of phenetic classifications.

There are two minor pitfalls in this process of abstraction. Pheneticists (for example, Sokal and Sneath, 1973) sometimes discuss quantitative methods of phylogenetic analysis.  These methods might well be used to construct classifications, but pheneticists do not advocate them for this purpose, and so

(although they are based on observable features of organisms) I
do not regard them as phenetic methods.  Pheneticists (Sneath
and Sokal, 1973; Blackith and Reyment, 1971; Clifford and
Stephenson, 1975) frequently also employ and recommend techniques
of multivariate statistical analysis such as ordination, seria-
tion, and various techniques of graphical analysis in order to
describe what Sneath and Sokal (1973) term "taxonomic structure."
Whatever is meant by "taxonomic structure," techniques such as
these cannot be regarded as methods of phenetic classification,
simply because they do not produce classes - that is, groups of
taxa.  Some phenetic methods - emphasized particularly by
Jardine and Sibson (1971) and also discussed by Sneath and Sokal
(1973) - do produce groups of taxa, which however are not
hierarchically arranged, and so cannot be regarded as comprising
a classification in the sense in which "classification" is
usually meant in biological systematics.  The methods used by
pheneticists which do produce hierarchic arrangements of taxa
interpretable as classifications, and which are advocated by
pheneticists for that purpose, form a relatively small and homogeneous group of techniques.  Almost all of the hierarchic
classificatory techniques commonly used by pheneticists consist
of what Rohlf (1970) has termed a hierarchic clustering strategy
performed on a matrix of overall similarity.  For brevity, I shall
refer to such techniques as phenetic similarity clustering. This
usage being understood, I shall consider phenetic classification
to mean "classification performed by phenetic similarity
clustering."

## PHENETIC SIMILARITY CLUSTERING

As not all participants may be familiar with techniques of
phenetic similarity clustering, I shall discuss them briefly.  A
phenetic classification is produced in two major steps.  The first
step converts descriptive information on the studied taxa into a
measure of overall similarity between pairs of taxa.  The input
data for this step is considered to comprise a rectangular table

| | | Taxa | | |
|---|---|---|---|---|
| Variables | 1 | 2 | ... | t |
| 1 | $x_{11}$ | $x_{12}$ | ... | $x_{1t}$ |
| 2 | $x_{21}$ | $x_{22}$ | ... | $x_{2t}$ |
| . | | . | . | . |
| n | $x_{n1}$ | $x_{n2}$ | ... | $x_{nt}$ |

in which each of t terminal taxa - corresponding to a column of
the table - is described by a value of each of n variables -
corresponding to the rows of the table.  The values of the variables are frequently termed states.  The entry $x_{ij}$ of this table
of data represents the state shown for the ith variable by the
jth taxon.  The result of this step is a square table in which
both rows and columns correspond to terminal taxa.

Taxa

| Taxa | 1 | 2 ... | t |
|---|---|---|---|
| 1 | $s_{11}$ | $s_{12}$ ... | $s_{1t}$ |
| 2 | $s_{21}$ | $s_{22}$ ... | $s_{2t}$ |
| . | . | . | . |
| t | $s_{t1}$ | $s_{t2}$ ... | $s_{tt}$ |

The entry $s_{jk}$ of this table - or matrix - is called the <u>coefficient</u>
<u>of overall similarity</u> between the jth and kth taxa.  Each overall
similarity value $s_{jk}$ is computed from the jth and kth columns of
the original table of data.  The coefficient s is intended to
measure the average degree to which $x_{ij}$ resembles - or is close
in value to - the corresponding $x_{ik}$.  A variety of methods have
been suggested by pheneticists for computing the similarity values
$s_{jk}$ from the original table of data.  A typical method applicable
in those cases where the original data table entries are numerical
values is given by formula (1).

$$s_{jk} = 1 - \frac{\sum_{i=1}^{n} \frac{|x_{ij} - x_{ik}|}{R_i}}{n} \qquad (1)$$

$$s_{jk} = 1 - \sqrt{\frac{\sum_{i=1}^{n} \left(\frac{x_{ij} - x_{ik}}{R_i}\right)^2}{n}} \qquad (1a)$$

Here $R_i$ represents the range of variation of the ith variable - the
numerical difference between the largest and smallest data table
entries for that variable.  The purpose of the divisions by $R_i$ and

by n in the computation of $s_{jk}$ is to insure that $s_{jk}$ lies between 0 and 1. Division by $R_i$ also serves to render the coefficient of overall similarity independent of the original scales of measurement of the data variables. A great many other techniques for computing average or overall similarity between pairs of taxa have been advocated by various pheneticists (reviewed by Sneath and Sokal, 1973; Clifford and Stephenson, 1975), but no consensus appears to have been reached as to which method is most preferable, and I present formula (1) only as an example. One alternative method is described by formula (1a).

The second step in the production of a phenetic classification consists of cluster analysis: the application of a hierarchic clustering strategy to a matrix of similarity coefficients in order to produce a hierarchic classification of the studied taxa. Hierarchic clustering strategies generally proceed as follows. A pair j, k of distinct taxa is identified which are mutually most similar. That is, there is no taxon m distinct from both j and k for which either $s_{mj}$ or $s_{mk}$ is greater than $s_{jk}$. Taxa j and k are then united to form a new taxon denoted as, say, p. The rows and columns of the similarity matrix corresponding to taxa j and k are then deleted from the matrix. A new row and column corresponding to the new taxon p is computed and added to the similarity matrix, which then has one fewer rows and columns than previously. The process of uniting pairs of taxa is then repeated until only one row and column remain in the matrix. The order in which the original terminal taxa are united into successively larger taxa determines a complete hierarchic classification of the original taxa.

Almost all phenetic similarity clustering methods can be implemented through simple repetitive procedures of this sort. In fact, most such methods are customarily performed in just this way, and differ only in the technique used to compute the similarity coefficients for a newly-formed taxon p united from pre-existing taxa j and k. Several methods for this computation are in fairly wide-spread use. Perhaps the most common utilizes formula (2)

$$s_{pm} = \frac{N_j s_{jm} + N_k s_{km}}{N_j + N_k} \qquad (2)$$

$$s'_{pm} = \frac{s_{jm} + s_{km}}{2} \tag{2a}$$

$$s''_{pm} = \max \{s_{jm}, s_{km}\} \tag{2b}$$

$$s'''_{pm} = \min \{s_{jm}, s_{km}\} \tag{2c}$$

to compute for each other taxon m the similarity between m and taxon p from the similarities between m and taxa j and k. Here $N_j$ denotes the number of original taxa of the data contained in the jth taxon, so that $N_p = N_j + N_k$. (It may be that the jth taxon is itself one of the original terminal taxa of the data in which case $N_j = 1$.) Formula (2) is used in the "unweighted pair-group analysis" (UPGMA) of Sneath and Sokal (1973). Several other hierarchic clustering strategies (reviewed by Sneath and Sokal, 1973; Clifford and Stephenson, 1975) use methods closely analogous to formula (2), differing only in the way in which the similarities $s_{jm}$ and $s_{km}$ are averaged. Some of the more common methods are shown in formulae (2a) through (2c).

## PRINCIPLES OF PHENETICS

The most obvious question raised by the construction of phenetic classifications through phenetic similarity clustering is: why is it done in just that way? Why is one formula for computing overall similarity to be preferred over another, and why is one hierarchic clustering strategy considered superior to others? Why, for that matter, is a classification to be constructed by any sort of phenetic similarity clustering, when some totally different means might be employed? But these questions are only superficial, for any answer to them necessarily presupposes that more fundamental issues have already been settled. Before the relative merits of any methods of classification may be judged, one must first specify what properties are desirable in a general biological classification. Only once this has been done, can alternative classificatory methods be compared as to their efficacy in achieving the specified goal. The significant question to be considered then is that of what basic principles of classification are considered to motivate the choice of phenetic classificatory techniques. Many pheneticists seem to have avoided this important issue. Jardine and Sibson (1971) argue that quantitative classificatory procedures should be selected on the basis of computational or formal mathematical convenience. The mathematical criteria used by those authors are developed within a framework which presupposes clustering by overall similarity, and so cannot be directly applied to the broader question of desirability of phenetic similarity clustering. Indeed, Jardine and Sibson (1971, p. x) contend that no extrinsic criteria exist for judging methods of biological classification. Clifford and Stephenson (1975) go a step further, eschewing the use of any criteria for judging classificatory methods and claiming that methods of classification should be chosen according to the intuitive appeal of their results. The best that one could say of such an attitude is that it is of no interest for the present discussion, for it would remove from the area of rational discourse all issues pertaining to comparison of alternative approaches to biological classification. Sneath and Sokal (1973) confess uncertainty as to which methods of phenetic classification are to be most preferred; to their credit they do suggest criteria through which this issue might eventually be resolved. These criteria, it would seem, constitute the basic principles of phenetic classification, to the extent that those principles have ever been formulated.

The first criterion discussed by Sneath and Sokal (1973) is that a general biological classification should be "natural" in the sense in which that term was used by Gilmour (1961). Gilmour conceived of a classification as being "natural" to the extent to which the classification serves as many separate purposes as possible. Gilmour's formulation of his concept has frequently been misunderstood. It is often commented, for example, that a "purpose" could be anything a taxonomist says it is, and from this it might be concluded that Gilmour's "naturalness" corresponds to no objectively definable property of a classification. But of course Gilmour had nothing so trivial in mind, and while he might have been more explicit, the proper interpretation of his concept of naturalness is quite evident from his papers. Gilmour and Walters (1963, p. 12) give as an example of a classification serving a single purpose the division of plants into "woody" and "herbaceous." Thus it is seen that what is meant by a classification's "serving a purpose" is just that a class - or group - of the classification describes the distribution among organisms of some feature. Accordingly, the most natural classification in Gilmour's sense is that classification whose constituent groups describe the distributions among organisms of as many features as possible. Natural classifications in this sense may be readily recognized in practice because the characters of organisms often fall into tightly-correlated groups, so that a single taxon may serve to describe the distributions of several correlated features. Similar points have been made by Sneath (1961), who emphasized

that we consider the taxon Mammalia to be a natural group in
Gilmour's sense because it can be characterized by a large suite
of well-correlated features.  Of course, as was recognized by
both Gilmour and Sneath, correlations between features are seldom
perfect, so that it is often necessary to recognize a taxon which
describes to a very good approximation the distributions among
organisms of a large set of features, whereas the distribution of
any one feature in the set may be not quite perfectly reflected
by the membership of the taxon.  This corresponds to no more than
the common observation that while the presence of hair is usually
taken to distinguish mammals, a hairless but otherwise obviously
mammalian species would indeed be classified in the Mammalia, and
that the presence of hair would continue to be regarded as
characterizing mammals generally.

It must be emphasized that while the correspondence between
the membership of a Gilmour-natural taxon and the distribution of
a feature considered described by that taxon is allowed to be not
quite perfect, the taxon cannot very well be said to describe the
distribution of the feature unless the correspondence is kept as
close as possible.  This implies that the set of correlated features upon which a natural taxon is based must not only be
shared by the members of the taxon, but must also be largely restricted to the taxon.  Sneath (1961, p. 122) points out that the
hypothetical taxon consisting exactly of horses, mice, and
rabbits is not a natural taxon, even though its members share
many features.  The features common to horses, mice, and rabbits
are common also to other mammals, and so that suite of features
determines the natural taxon Mammalia.  A taxon consisting just of
horses, mice, and rabbits could be a natural taxon only if there
were some set of well-correlated features which distinguished just
this group of organisms.

The second criterion suggested by Sneath and Sokal (1973) is
a more mathematical one. It consists of measuring the degree of
correspondence of the relative similarities among taxa as implied
by a classification to the actual overall similarities calculated
from the observed features of the taxa.  Ideally the similarities
implied by the classification should be as well correlated with
the observed similarities as possible, a concept reminiscent of
Jardine and Sibson's characterization of pheneticism, cited
earlier.  The way in which the implied similarities between taxa
are computed from a classification constructed by phenetic similarity clustering is readily understood in terms of my previous
description of hierarchic clustering strategies.  Each new higher
taxon p is formed through uniting two pre-existing taxa, between
which a similarity value $c_p$ obtains.  If taxon p is formed by
uniting taxa j and k, then $c_p = s_{jk}$.  Thus with every higher taxon
p formed during the process of hierarchic clustering, there is

naturally associated a corresponding similarity level $c_p$.  The
classification-implied similarity between any two of the original
data taxa j, k is then taken to be the similarity level associated
with the smallest higher taxon that includes both j and k. De-
noting as $c_{jk}$ the classification-implied similarity between
original taxa j and k, and supposing that higher taxon p is the
smallest taxon containing original taxa j and k, $c_{jk} = c_p$.  The
similarity value $c_{jk}$ is usually called the <u>cophenetic</u> similarity
between taxa j and k.  The degree of correspondence between the
cophenetic similarities c and the original similarities s is
assessed through computing the linear product-moment correlation
coefficient between c and s over all pairs of distinct data taxa.
The resulting correlation coefficient is called the <u>cophenetic
correlation coefficient</u>. The cophenetic correlation coefficient
has been utilized in a large number of phenetic studies for the
purpose of evaluating the relative merits of alternative methods
of phenetic similarity clustering (reviewed by Sneath and Sokal,
1973).

     Pheneticists seem not to have studied whether the cophenetic
correlation coefficient measures the desirability of a classification in a way compatible with Gilmour's concept of naturalness.
If it did not, of course, there would be a basic inconsistency in
the phenetic theory of taxonomy.  I shall not pursue this issue
here and shall assume the two phenetic criteria to be largely
equivalent.


NATURAL CLASSIFICATIONS

     Given the criterion of Gilmour-naturalness as a means for
evaluating classifications, it is natural to ask by what sort of
methods classifications optimal under this criterion might be
constructed.  Since pheneticists of the Sokal-Sneath school
(cf. Sneath, 1961; Sokal and Sneath, 1963) have long advocated
both naturalness of classifications and clustering by overall
similarity, one might be tempted to suppose that the connection
between the criterion and the method is already well-established,
or that the principle of clustering by overall similarity had been
deduced from the principle of Gilmour-naturalness.  Curiously,
such is not the case.  The method of phenetic similarity clustering arose originally through slight modification of grouping
techniques that had long been in use in such areas as psycho-
metrics and phytosociology.  The applicability of clustering by
overall similarity to the problem of constructing Gilmour-natural
classifications appears to have been accepted on a largely intuitive basis, and it seems that no logical derivation of a
phenetic similarity clustering method from the premise of Gilmour-
naturalness has ever been developed.  It would then appear that

one might well ask just what connection exists between clustering
by overall similarity and Gilmour-natural classifications.

To begin studying this question, consider the hypothetical
data of Table 1 with 8 taxa and 14 variables. Application of the
coefficient of overall similarity of formula (1) yields a
similarity matrix from which application of the UPGMA clustering
method produces the classification represented as a tree-diagram
(phenogram) in Figure 1. The matrix of cophenetic similarity
values corresponding to the classification of Figure 1 turns out
to be identical to the original matrix of overall similarities,
so that the cophenetic correlation coefficient attains its maximum
possible value of 1. Judging by the cophenetic correlation coefficient, the classification of Figure 1 would appear to be an
excellent phenetic classification. The same classification would
seem also to be a good natural classification for the data of
Table 1. It is seen that each of the features corresponding to
state 1 of one of the variables has its distribution among the
terminal taxa perfectly described by one of the taxa of the
classification of Figure 1. For example, the feature represented
by state 1 of variable 9 corresponds exactly in its distribution
to the higher taxon A of Figure 1 which is united from terminal
taxa 1 and 2.

The distributions of the features corresponding to the 0
states of most of the variables of Table 1 are not well described
by any of the taxa of the classification of Figure 1. This
might seem to indicate that the classification is less than
ideally natural, but it actually means that some refinement of
the concept of naturalness is necessary. The two features represented by the two alternative states of any of the variables of
Table 1 are mutually exclusive and exhaustive - that is, just one
occurs in each of the taxa of the data. For any pair of exhaustive,
mutually exclusive features, it is possible exactly to represent
the distributions of both through two taxa in the same hierarchic
classification, only when those taxa represent the highest-level
subdivision of the entire group classified. For the classification of Figure 1, this occurs for variables 13 and 14 of Table 1,
each state of each of which is characteristic of one of the two
major groups of the classification. But the other variables of
Table 1 are not perfectly correlated with variables 13 and 14, and
so it is impossible in the same classification for both states of
any of these other variables each to characterize a taxon. But
this limitation does not lead to any reduction of our ability to
describe the distributions of the values of the data variables in
terms of the taxa of the classification. The distribution among
the taxa of Table 1 of state 0 of variable 9 can be quite concisely
described in terms of the classification of Figure 1 by noting that
that feature characterizes the complement of taxon A - that is,

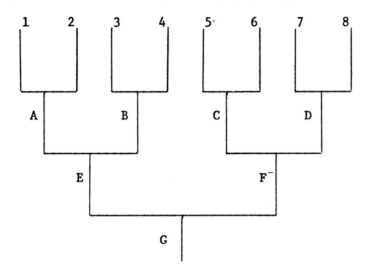

Fig. 1.   Classification of eight hypothetical terminal taxa 1–8.
Letters A–G designate groups.

the set of all terminal taxa not included in A.  With these points
in mind we see that the classification of Figure 1 is as natural
a hierarchic representation of the data of Table 1 as it is
possible to construct, and that in fact the data of Table 1 may
be perfectly described in terms of the taxa of this classification.
Gilmour's criterion of naturalness therefore agrees with the cophenetic correlation coefficient in judging the classification of
Figure 1 an ideal representation of the data of Table 1.

    Given that phenetic similarity clustering can produce natural
classifications, we should next like to know whether it generally
does so.  As a first step in answering this question, consider a
new hypothetical data set produced from the data of Table 1 by
replicating each of the variables of that table the number of
times specified by the column labeled "Factor."  The new data set
then has 5 variables identical to variable 1 of Table 1 and a
total of 34 variables.  Application to this new data set of the
overall similarity coefficient of formula (1) produces a new
matrix of overall similarities.  This matrix has a complex system
of ties at the highest similarity level (0.82), as a result of
which a hierarchic clustering strategy is unable to resolve any
isolated groups.  Application of phenetic similarity clustering
therefore leads to the "classification" represented in Figure 2.
The cophenetic similarity values computed from the classification
of Figure 2 are all identical in the off-diagonal elements, so
that by a limiting argument the cophenetic correlation coefficient

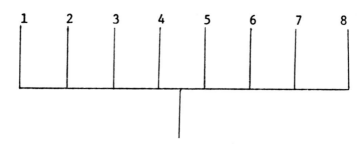

Fig. 2.   Classification of eight hypothetical terminal taxa 1-8.


is 0, assuring us that this is indeed a poor phenetic classification.   The classification also appears unpromising according to
Gilmour's criterion of naturalness.   The single higher taxon of
Figure 2 is united from all the terminal taxa of Table 1, and
plainly does not usefully describe the distribution of any of the
features of the original data.   Each of the eight terminal taxa
1 through 8 aptly describes the distribution of one of the types
of features - corresponding to state 1 of one of the variables 1
through 8 of Table 1 - distinguishing that taxon.   But this component of the naturalness of the classification of Figure 2 is
trivial, for those same terminal taxa - with the same powers of
description - would be present in any classification of those
terminal taxa.   Thus the classification of Figure 2 is seen to
possess as little Gilmour-naturalness as is possible for any
classification of these terminal taxa.

Of course we wish to know whether modification of the original data of Table 1 through replication of some of its variables
has brought out a deficiency of phenetic similarity clustering in
particular, or whether the replication has indeed made it more
difficult to recognize a natural classification for these data.
To resolve this issue we need only note that replication introduces no new types of distribution of features into the data,
but only alters the relative frequency with which the various
types of distributions of features are represented.   Accordingly,
the classification of Figure 1 - which has already been judged
to be a perfectly natural classification of the original data of
Table 1 - remains a perfectly natural classification for the data
even after replication, since the distribution of the values of
any of the variables of the data may be exactly described in terms
of one of the taxa of the classification.

It is thus seen that it is possible for phenetic similarity
clustering to produce a poor classification as judged by phenetic
criteria, even in cases where a natural classification can readily
be recognized.   We then wish to know which aspects of the method

of clustering by overall similarity give rise to this deficiency.
It would appear that the weakness lies in the concept of overall
similarity itself. Recalling formula (1), we see that a match –
or a certain degree of similarity – between any pair of taxa in a
certain variable contributes the same amount to the measure of
overall similarity between that pair of taxa, regardless of the
state – or value – in which the match occurs. Attempting to construct natural classifications through clustering by overall
similarity therefore corresponds simply to utilizing the principle
that the members of a natural group are expected to share many
features. But as I have already emphasized, the features upon
which a natural group is based must not only be shared by the
members of the group, but also must distinguish the group. As we
have already seen, it is generally impossible – even in a perfectly natural classification – for every value of every variable
directly to characterize one of the taxa of the classification,
since some values may characterize the complements of taxa. Thus
in considering on an equal basis the contributions to similarity
between a pair of taxa from all states of all variables, a coefficient of overall similarity includes in its assessment of the
total similarity between the two taxa much information that is
irrelevant to the issue of whether those two taxa should be
united to form a natural group. Two taxa united on the basis of
overall similarity in the course of phenetic similarity clustering
might indeed share many features distinguishing a natural group,
and thus be properly united. But given only that two taxa share
many features overall, it is also quite possible that the features
common to these two taxa do not comprise the set of features distinguishing a natural group, and therefore do not constitute a
proper basis for uniting those two taxa.

If the inefficiency of clustering by overall similarity for
recognizing natural classifications is in fact due to the inclusion in coefficients of overall similarity of components of
similarity corresponding to features not characterizing natural
groups, it should be possible to achieve a more effective technique
by utilizing a similarity measure which excludes these irrelevant
similarity components. This might be accomplished by identifying
for each variable of the study an uninformative state – a value
corresponding to a feature taken not to characterize groups of
the natural classification. The list of these uninformative
states – one for each variable – could be thought of as the
description of an artificial terminal taxon (it might of course
also be an actual terminal taxon). I shall call a taxon whose
description is determined in this way a reference point. That
component of similarity between two terminal taxa relevant to
uniting those taxa into a natural group could then be isolated by
subtracting from the overall similarity between the two taxa the
similarity shared by the two with the reference point. A measure

of similarity computed in this way has been used by Farris, Kluge, and Eckardt (1970). Their coefficient, denoted here a, can be expressed in terms of the coefficient of overall similarity s of formula (1), as shown in formula (3).

$$a_{jk} = \tfrac{1}{2} (1 + s_{jk} - s_{jr} - s_{kr})$$

(3)

Here r denotes the reference point. The constant term 1 and the constant factor $\tfrac{1}{2}$ serve to restrict a to vary between 0 and 1. It is seen from formula (3) that $a_{jk}$ differs from $s_{jk}$ of formula (1) just in excluding those similarity components due to the shared possession by taxa j and k of uninformative states, as only these uninformative states figure in the coefficients $s_{jr}$ and $s_{kr}$. The coefficient a of formula (3) is not a coefficient of overall similarity as that term is generally understood, since similarities in some features are not counted; a could perhaps be called a coefficient of special similarity.

Returning for an example to the hypothetical data of Table 1, we note that the groups of the natural classification of Figure 2 are characterized by the 1 states of the data variables. Hence we should be able to construct the classification of Figure 1 utilizing the special similarity coefficient a with a reference point described by all 0 states. The special similarity value $a_{jk}$ is in this case just the fraction of data variables in which taxa j and k share a 1 state (this was proved by Farris, Kluge, and Eckardt, 1970). Application of unweighted pair-group analysis to the matrix of special similarities, produced either from the original data of Table 1 or from Table 1 modified through replication of variables as in the preceding example, produces the classification of Figure 1 as desired, the cophenetic correlation coefficient in both cases being unity. Clustering by special similarity thus seems to avoid the drawbacks of clustering by overall similarity in this case.

The example arrived at through replications of the data variables of Table 1 is of course artificial. To assess more realistically the relative effectiveness of clustering by special and overall similarity, I have applied both techniques to each of 50 sets of data on real organisms. In each case, similarity matrices were computed using formulae (1) and (3) and were analyzed using unweighted pair-group analysis. The phenetic optimality of each classification constructed was evaluated using the square of the cophenetic correlation coefficient. No attempt was made at optimal selection of reference points, the reference point used in each special similarity analysis being arbitrarily

TABLE I
Hypothetical Data Matrix

| Variables | Taxa | | | | | | | | Factors |
|---|---|---|---|---|---|---|---|---|---|
| | 1 | 2 | 3 | 4 | 5 | 6 | 7 | 8 | |
| 1 | 1 | 0 | 0 | 0 | 0 | 0 | 0 | 0 | 5 |
| 2 | 0 | 1 | 0 | 0 | 0 | 0 | 0 | 0 | 1 |
| 3 | 0 | 0 | 1 | 0 | 0 | 0 | 0 | 0 | 5 |
| 4 | 0 | 0 | 0 | 1 | 0 | 0 | 0 | 0 | 1 |
| 5 | 0 | 0 | 0 | 0 | 1 | 0 | 0 | 0 | 5 |
| 6 | 0 | 0 | 0 | 0 | 0 | 1 | 0 | 0 | 1 |
| 7 | 0 | 0 | 0 | 0 | 0 | 0 | 1 | 0 | 5 |
| 8 | 0 | 0 | 0 | 0 | 0 | 0 | 0 | 1 | 1 |
| 9 | 1 | 1 | 0 | 0 | 0 | 0 | 0 | 0 | 3 |
| 10 | 0 | 0 | 1 | 1 | 0 | 0 | 0 | 0 | 1 |
| 11 | 0 | 0 | 0 | 0 | 1 | 1 | 0 | 0 | 3 |
| 12 | 0 | 0 | 0 | 0 | 0 | 0 | 1 | 1 | 1 |
| 13 | 1 | 1 | 1 | 1 | 0 | 0 | 0 | 0 | 1 |
| 14 | 0 | 0 | 0 | 0 | 1 | 1 | 1 | 1 | 1 |

selected as one of the terminal taxa of the data.  In every case
the classification constructed by clustering according to special
similarity was found to be superior to that produced through
clustering by overall similarity.  The mean value of the squared
cophenetic correlation coefficient for classifications produced
by clustering according to overall similarity was .63, and the
corresponding mean for analyses based on special similarity was
considerably higher at 0.89.

It would thus seem that clustering by special similarity is
generally superior to clustering by overall similarity, as judged
by the phenetic criteria of Gilmour-naturalness and the cophenetic
correlation coefficient.  I do not mean to suggest, however, that
this method is free of defects, or that it is the most effective
method for recognizing natural classifications.  It is beyond the
scope of this presentation to develop a general theory of natural
classification (for a fuller treatment, see Farris, in press).  I
have here considered this area in only enough detail to demonstrate one point which will be useful in evaluating pheneticists'
criticisms of other schools of taxonomy.  That point is that
clustering by overall similarity does not seem to be an optimal
classificatory method according to the principles of phenetics.

## PHENETICISTS' CRITICISMS OF OTHER SCHOOLS OF TAXONOMY

So far I have considered phenetic taxonomy essentially
in isolation and have concentrated on describing its fundamental
principles from an internal viewpoint.  But phenetic taxonomy does
not exist in isolation, and many features of pheneticists' overall philosophy of classification can be understood only as
criticisms of other schools of systematics, that is as arguments
concerning respects in which phenetics is supposed to be superior
to other approaches to classification.  Not all of these arguments
are equally cogent; some are little more than expressions of
personal interests and motivations.  But others touch on significant questions concerning the choice of a basis for classification.
I shall summarize the main types of pheneticists' arguments against
other approaches to taxonomy and attempt to isolate the most
meaningful issues.

Pheneticists differ from most other taxonomists primarily in
holding that phylogenetic analysis should play no role in the construction of general biological classifications, and most of their
criticisms of other schools of taxonomy are concerned with the
undesirable consequences of admitting phylogenetic "speculation"
into the classificatory process.  The drawbacks attributed by
pheneticists to phylogenetic methods as contrasted with phenetic
methods concern three main subjects:  the unreliability of

phylogenetic analysis, the instability of phylogenetic classifications, and the loss of phenetic information imposed by phylogenetic classification.

For purposes of the present discussion, it is unnecessary to
inquire into the reasons underlying pheneticists' belief that
phylogenetic relationships cannot be reliably inferred. We will
be interested only in the logic of criticisms based on that belief. Such a criticism has been voiced by Sneath and Sokal
(1973, p. 53), who state that, "... the difficulty with the use
of the phylogenetic approach in systematics is that we cannot make
use of phylogeny for classification, since in the vast majority of
cases phylogenies are unknown and possibly unknowable." By contrast, Sneath and Sokal (1973, p. 56) maintain that, "Because
phenetic classifications require only description, they are possible for all groups..." These statements misrepresent the difference between phenetic and phylogenetic approaches in two
respects. First, phenetic classifications do not require only
description. A phenetic classification can be obtained only by
combining descriptions of organisms with a phenetic classificatory
method. That method is selected on the basis of a theoretical
framework, and the framework can hardly be said to be "observed"
in the same sense as features of organisms are observed. Second,
phylogenetic classifications are not based on "unknown phylogenies."
They are based on inferred systems of kinship relationship - that
is phylogenetic hypotheses. Those hypotheses in turn are constructed by methods of phylogenetic analysis which depend ultimately upon empirical observations. Thus phenetic and phylogenetic classification are just alike in that in each case data
is analyzed by a particular method in order to construct a classification. Now it is plainly true that a phylogeneticist following a single consistent methodology and provided with certain data
would construct one particular phylogenetic classification -
whether the kinship relationships corresponding to that classification were true or not. Therefore, the supposed unreliability of
phylogenetic inferences, and the unknowability of phylogenies are
seen to be irrelevant to the process of phylogenetic classification, when that process is thought of from a phenetic viewpoint -
that is simply as a method for analyzing data. Accordingly, the
claim that phenetic classification is superior to phylogenetic
classification in that the former is empirical whereas the latter
is not, cannot be accepted as well-founded.

Sneath and Sokal (1973, p. 56) state that phenetic classification is to be preferred because, "...a stable and consistent
system of classification and nomenclature must be developed and
maintained." They content on the other hand that by insisting on
the application of phylogenetic principles to classification,
"...conventional systematists perpetuate a system which is

inherently unstable and hypothetical by the very nature of its
operations." (Sneath and Sokal, 1973, p. 56). The issue raised
by this criticism is not immediately clear. One might suppose
that Sneath and Sokal intend that phylogenetic classification is
unstable because it possesses a hypothetical element in the sense
that it corresponds to a phylogenetic hypothesis. But this
property could distinguish phylogenetic classification from
phenetic classification, only if phenetic classification were
regarded as lacking a hypothetical element. This would not seem
to be the view of Sneath and Sokal, for they discuss such topics
as "problems of estimating phenetic relationships" (Sneath and
Sokal, 1973, p. 31), and "corroboration of a classification by
biochemical methods." (P. 297). It seems relevant to inquire in
just what sense the existence of a hypothetical aspect to a
classification can be supposed to induce instability of the
classification. A phylogeneticist might modify a phylogenetic
hypothesis, and consequently the corresponding phylogenetic
classification, if on obtaining new observations he discovered a
new phylogenetic hypothesis to be better supported by available
information than the old hypothesis. From a methodological viewpoint this simply means that different phylogenetic classifications might be produced from different sets of data concerning the
same organisms. But phenetic classification - whether it is regarded as hypothetical or not - certainly also has the property
that different classifications of the same set of organisms may
be constructed from different sets of data on those organisms. It
would appear that the possible interpretation of a phylogenetic
classification as a hypothesis of phylogenetic kinship has nothing
intrinsically to do with the matter of the relative stability of
phenetic and phylogenetic classifications. The only meaningful
question raised by Sneath and Sokal's criticism would seem to be
whether the use of phenetic or of phylogenetic methods of classification leads to a more stable system. This matter has recently
been studied empirically by Mickevich (in prep.), who computed
both phenetic and phylogenetic classifications for each of
several different sets of characters for each of a large number of
groups of organisms. She found that phenetic classifications of
a single group of organisms based on several different sets of
characters invariably differed much more drastically from one
another than did phylogenetic classifications prepared from the
same sets of data. From these findings it would appear that
phylogenetic analysis actually leads to greater stability of
classification than does the application of phenetic techniques.
Applying the premise of Sneath and Sokal, we could then conclude
that on this basis phylogenetic classification is to be preferred
over phenetic classification in selecting a general reference
system for organisms.

Perhaps the most basic phenetic criticism of phylogenetic approaches to classification is that suggested by Jardine and Sibson (1971, p. 136) when they state, "The application of the strict monophyly criterion to both living and fossil populations appears to be an unsatisfactory method of classification, because it results in the allocation of fossil populations to taxa without regard to their relative dissimilarities to other living and fossil populations." Sneath and Sokal (1973, p. 56) comment, "In most cases where primitive patristic elements {symplesiomorphies(!)} constitute an appreciable portion of the phenetic similarity, the cladistic and phenetic classifications will largely coincide. However, in the cases where this is not so, the general superiority of a phenetic over a cladistic classification becomes even more evident." Sneath and Sokal go on to consider hypothetical examples from which they conclude that phenetic groups have higher "information content" and "predictivity" than phylogenetic groups. The sense of these statements is simply that phenetic classification is to be preferred over phylogenetic classification because phylogenetic classification is supposed not to satisfy the aims of phenetic taxonomy. (The properties of high information content and predictivity are associated by Sneath and Sokal with Gilmour-naturalness.) These sentiments of pheneticists have probably arisen in reaction to the idea, often expressed by phylogeneticists, that the sole aim of phylogenetic classification is to represent as nearly as possible the branching sequences of genealogical relationship. This point is mentioned both by Jardine and Sibson (1971, p. 135) and by Sneath and Sokal (1973, p. 310). A pheneticist might argue that since a phylogenetic classification consists of nothing more than a collection of hypothesized monophyletic groups, it is devoid of any "phenetic" information on the features or the similarities of the classified organisms. But such an argument cannot bear close inspection. No classification, even a phenetic one, is by itself any more than a collection of groups of organisms. No grouping of organisms, however arrived at, can directly express anything more about the organisms than the membership of the groups. Classifications may possess "information content" only in the indirect sense that a classification may be used as a reference system. If each taxon of a classification refers to a diagnosis, the diagnoses may be used to convey information about the organisms themselves. Diagnoses, however, can be provided for the taxa of any classification, regardless of how the classification was arrived at, simply by listing the features common to the members of each taxon. Thus it is misleading to single out phylogenetic classification as having low "information content" solely on the grounds that its advocates express interest in recognizing monophyletic groups. The real issue, as before, is more a matter of methodology than philosophy. Given that both phenetic and phylogenetic classifications may be caused to possess "information content" in just the

same way through constructing diagnoses, we wish to know whether
the phenetic or the phylogenetic method of constructing a classification leads to a more "informative" system in the sense in which
pheneticists understand that term, that is as judged by the
phenetic criterion of Gilmour's concept of naturalness.  Although
I cannot provide a definitive answer to this question, some of my
earlier points seem relevant to establishing a tentative one.  I
have already noted that phenetic methods do not seem generally to
produce optimal classifications as judged by the criterion of
Gilmour-naturalness, and in particular that the technique of
clustering by special similarity seems superior for this purpose
to the common phenetic technique of clustering by overall similarity.  Further, clustering by special similarity appears to be a
mathematical analog of the classificatory method used by phylogeneticists.  The distinctive feature of the phylogenetic method
is that taxa are grouped according to apomorphic similarity (cf.
Hennig, 1966).  If in the technique of clustering by special
similarity we take the reference point to be characterized entirely by plesiomorphic features, then the similarity components
counted by the special similarity coefficient of formula (3) are
exactly the synapomorphies of the two taxa compared.  This was in
fact the original application of the similarity coefficient of
formula (3) by Farris, Kluge, and Eckardt (1970).  From these
considerations it would seem that phylogenetic methods may actually be superior to phenetic methods for the purpose of satisfying
the phenetic criterion of Gilmour-naturalness of classifications.


### PHENETIC, EVOLUTIONARY, AND PHYLOGENETIC CLASSIFICATION

Having considered the principles of phenetic taxonomy, I
should like finally to show how these principles may be applied
with the aim of resolving the debate among the phenetic, evolutionary, and phylogenetic schools of systematics.  This debate
has been in progress for some time (cf. Sokal and Sneath, 1963;
Mayr, 1965; Sneath and Sokal, 1973; Mayr, 1974; Sokal, 1975;
Hennig, 1975); it is in fact of much greater antiquity than is
perhaps generally recognized, as the disputes among the modern
schools correspond in many respects to controversies over the
proper nature of systematics that have been debated since well
before the time of Darwin (reviewed by Sneath and Sokal, 1973;
Hennig, 1966).  While many systematists believe that this controversy is now close to being settled, it appears to me that it
is being settled - if at all - through consensus of opinion.  I
consider such a situation unsatisfactory, both because a consensus is likely to prove at best ephemeral and at worst illusory,
and because I would hold that scientific issues should be settled
by reason, rather than by majority vote.

In order rationally to analyze the debate among the schools of systematists, it is necessary to recognize that the perpetuation of that debate is due largely to the tendency of many of the proponents of the various schools to concentrate on issues not directly related to the question of how to classify organisms. When Mayr (1969, p. 75) criticizes Hennig's (1966) definition of monophyletic on the grounds that it is "completely contradicted by common sense," the effect can only be to obscure the real issue of whether a classification is to include only groups that are inferred to be monophyletic in the sense in which Hennig uses that term. When Sokal (1975) contends that a phylogenetic classification would be of little interest even if it were available, the effect can only be to confuse debate by confounding the merits of phylogenetic classification with those of motivation of phylogeneticists and the interpretations that phylogeneticists place on their results. It would seem in fact that much of the disagreement among the schools of systematics can be traced to attempts by their proponents to resolve the controversy by arguing the merits of motivations and interpretations. In the logical development of any one school, of course, its methods must be deduced from its basic principles or motivations, so that the principles appear to be of primary importance and the methods, secondary. It is probably for this reason that the advocates of the various schools have emphasized abstract principles in their discussions. But the basic principles of a school of systematics are frequently grounded ultimately in premises that are essentially judgments of interest, and disagreements over such judgments may often be difficult to resolve on an objective level. Hence in comparing different schools of systematics, it is probably more productive to consider first of all the properties of their methods, since these may be established objectively, and to resort to consideration of motivations only when this is made necessary by an irreconcilable difference in method.

The approach of comparing schools of systematics through their methods rather than their motivations seems to be of particular benefit in comparing the phenetic and evolutionary schools. These schools certainly differ drastically in their basic principles as those are usually stated by their proponents (Sneath and Sokal, 1973; Mayr, 1974), but the differences in their methods appear relatively slight. Mayr (1974, p. 95) states that evolutionary systematists agree with pheneticists in "grouping by a largely phenetic approach," and differ from them primarily in the weighting of characters. Sneath and Sokal (1973, p. 26) hold that equal weighting of characters is not essential to a phenetic approach, and Sokal (1975) criticizes Mayr (1974) mostly on the grounds of a subjective element in Mayr's weighting of characters. But the methods which Mayr advocates for weighting

characters are not actually subjective - at worst they might be
termed complex.  Mayr (1969) has discussed several empirically-
based criteria for character weighting.  Jardine and Sibson
(1971, p. 138) noted that despite differences in motivation, most
of Mayr's criteria are "in fact the same as those which arise
naturally in the course of computing phenetic dissimilarity..."
Thus - aside from the trivial matter of degree of automation -
there does not seem to be any pronounced methodological differences
between the phenetic and evolutionary schools of taxonomy.  Once
this is realized the apparent philosophical differences between
the two schools are seen to be less significant than they might
otherwise seem.  Philosophically, evolutionary taxonomists differ
from pheneticists primarily in the importance they would ascribe
to evolutionary interpretations of their classifications.  The
theoretical importance attached by Mayr to certain types of
character weighting which pheneticists might well use anyway can
reasonably be seen simply as a means of facilitating those interpretations.  Given the agreement on methods of classification,
even evolutionary interpretations of classifications would seem
unobjectionable - or even desirable - from a pheneticist's viewpoint.  Sneath and Sokal (1973, p. 429), for example, cite with
approval the view of Silvestri and Hill (1964) that the formulation of new scientific hypotheses may be an important application of phenetics.  The differences between the phenetic and
evolutionary schools then seem to be more a matter of degree than
of kind, and the two might very well be regarded as variant
theoretical justifications of a single approach to classification.

    The differences between the evolutionary/phenetic school and
the phylogenetic school cannot be explained away quite so. easily.
Phylogenetic methods of classification do differ from phenetic
and evolutionary ones, and what is worse, often produce different
classifications.  An example is shown in Figures 3 and 4.  Figure
3 depicts the phylogenetic relationships of several species of
ape, as proposed by Simpson (1963).  Figure 4 shows in the form
of a tree-diagram the evolutionary classification of these species
as recommended by Simpson (1945).  Taking Simpson's reconstruction
to be correct, the tree-diagram of the phylogenetic classification
of these species would have the same form as Figure 3.  Since the
two classifications cannot be reconciled, it is necessary to consider questions of principle or motivation in order to choose
between them.  I would suggest, however, that it is fruitless to
consider these matters at the level of generality that has
characterized much past debate.  It would serve little purpose,
for example, for a phylogeneticist to argue that the classification of Figure 3 is to be preferred over that of Figure 4 because
it more accurately represents kinship relationships, not because
this is not true, but because an evolutionary taxonomist might
well reply that he considers it more important that a classification represent something other than kinship relationships.  We

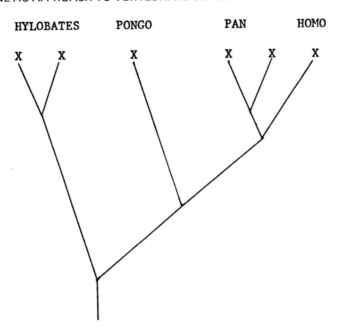

Figure 3.   Phylogenetic relationships of several apes, following
Simpson (1963).

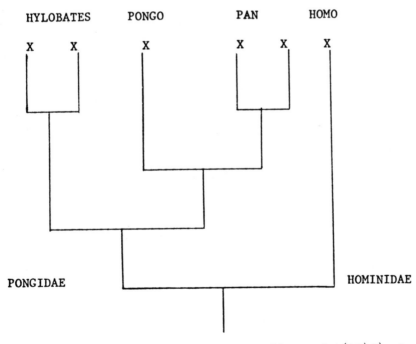

Figure 4.   Tree-diagram representation of Simpson's (1945) classification of apes.

would thus be left to choose between two different judgments of
interest or value, and such a dispute could never be settled on
a purely logical basis.  To avoid this trap, we turn perforce to
consideration of motivation at a more concrete level.  We note
that in this case the two different classifications are based
ultimately on the same data, and that therefore they must in some
sense represent different aspects of those data.  We might inquire then, just what properties of the data evolutionary taxonomists seek to emphasize in selecting the classification of
Figure 4, and why they feel that these lead to an improvement in
classification.  Evolutionary taxonomists' arguments for the
superiority of their techniques over those of phyleticists
largely parallel those of pheneticists in that greater information content of classifications is claimed.  Thus Mayr (1974,
pp. 107-108) holds, "...cladists seem to think that they have to
make a choice, in the delimitation of taxa, between basing them
either on branching points or on degrees of evolutionary divergence.  They fail to appreciate the added amount of information
{obtained?} by utilizing both sources of evidence."  On the
example at hand Mayr (1969, p. 234) states, "The study of chromosomes, of numerous biochemical characteristics, and of their
parasites, has revealed so much similarity between Man and the
African apes (Pan) that some authors have suggested placing them
in a single family.  Yet Man has entered so unique and strikingly
distinct an adaptive zone that Huxley even suggested recognizing
for him a separate kingdom (Psychozoa).  This would seem to go
too far, but the evolutionary distinctness of Man surely justifies
recognition of a separate family."  As it is of course organisms
rather than adaptive zones that are directly observed, the empirical basis for the separation of man from apes would seem to
comprise the distinctiveness of man.  Simpson (1971, p. 370)
makes this more explicit:  "It is abundantly established that
anatomically, behaviorally, and in other ways controlled or influenced by total genetic makeup, Homo is very much more distant
from either 'Pan' or {Pongo} ...than they are from each other...
The distinction is real, and it still justifies the classical
separation of Pongidae and Hominidae in classification."  Thus it
seems that the evolutionists' motivation for assigning man to a
separate family is to indicate through the familial categorical
rank the degree of difference between Homo and its relatives.
This decision on ranking is in fact solely responsible for the
differences between the evolutionary and phylogenetic classifications.  In the evolutionary classification the group consisting
only of Homo is assigned the same categorical rank - family - as
the group consisting of all other hominoids.  Since in the
Linnean system the family Hominidae could not be subordinated to
the family Pongidae, the categorical rank of the Hominidae effectively forces deletion from the classification of the group united
from Homo and Pan and the group united from that group and Pongo.

The difficulty with attempting to use categorical rank to convey information about degree of distinctiveness of taxa is that in order for it to work the categorical level of separation between the two taxa obviously must generally be interpretable in terms of degree of difference. But Homo is separated from Hylobates at the same categorical level as that at which Homo is separated from Pan, conveying the "information" that Homo is equally distinct from Pan and Hylobates. Homo, however, is admitted to be much more similar to Pan than it is to Hylobates. Basing categorical rank on degree of difference is thus seen to distort or conceal the very kind of information it was intended to express. A similar situation would arise in any case in which rates of evolutionary divergence differ sharply between phylogenetically closely-related groups. The aim of expressing degree of difference, therefore, does not seem to be an acceptable argument for preferring evolutionary over phylogenetic classification. This conclusion is not obtained from the premise that there is anything undesirable about information on degree of difference in itself. Rather the conclusion follows from the fact that expressing degree of difference through categorical level of separation is a very inefficient means of conveying that information. Indeed, Mayr (1969, p. 231) seems to have realized as much. He stated, "In most cases, however, the categorical separation of an aberrant species (as a separate genus) or an aberrant genus (as a separate family) leads to a fractioning of the system and lowers its information-retrieval capacity."

Another way to look at the evolutionary classification of Figure 4 is that Pongo, Pan and Hylobates are united at the same level at which they are separate from Homo because they are more similar to each other than any of them is to Homo - that is, they are grouped by similarity. The evolutionary classification is then seen to be closely analogous to one produced by phenetic similarity clustering. The difficulty encountered by evolutionists in seeking to express relative divergence by grouping by similarity is likewise very similar to one I discussed earlier, when I pointed out that phenetic clustering by overall similarity is inferior to clustering by special similarity as judged by the cophenetic correlation coefficient. Correspondingly, grouping by synapomorphy better reflects similarities than grouping by shared characters in general. This corresponds to a fact of which phylogeneticists have long been aware. A system of synapomorphies is naturally suited to a hierarchic representation, whereas a suite of general similarities is not.

Mayr (1974, p. 96) discusses another sense in which he holds evolutionary classifications to express more information than phylogenetic classifications. He cites a concept of "best" classifications due to Mill (1874), and holds that phylogenetic

classifications do not satisfy Mill's criterion.  Mill was an
early proponent of the view of classification that I have already
discussed under the name of Gilmour-naturalness (he is cited by
Gilmour and Walters, 1963), and it is unnecessary to consider this
material again.  As I have already demonstrated a close analogy
between phylogenetic methods and recognizing Gilmour-natural
classifications, I can only conclude that Mayr's claim is incorrect.  In the particular case of the alternative classifications of apes of Figures 3 and 4, we see that there is little
reason to consider the evolutionary classification of Figure 4
more natural in Gilmour's sense than the phylogenetic classification of Figure 3.  Elevating the group consisting only of Homo to
high categorical rank can in no way improve the ability of that
group to reflect the distinctive features of man.  Further, dis-
membering the group of Figure 3 that is united from Homo and Pan
completely destroys the ability of the classification to reflect
the distributions of the large number of features - which Mayr
admits to exist - which distinguish Pan and Homo from other
primates.  It would seem, therefore, that the phylogenetic classification of Figure 3 is considerably more natural in Gilmour's
sense than the evolutionary classification of Figure 4.

## CONCLUSIONS

As a result of my study of the principles of phenetic taxonomy,
I should draw three principle conclusions.  First, the dispute
among phenetic, evolutionary, and phylogenetic systematists has
been unnecessarily prolonged by the tendency of all concerned to
concentrate on essentially unresolvable arguments on matters of
abstract principle and motivation, whereas much more decisive results can be obtained quite easily through considering the properties of methods of classification.  Second, there does not
appear to be any justification for phenetic taxonomy as it is
currently practiced.  Pheneticists espouse the goals of a stable
reference system and a classification highly natural in Gilmour's
sense, but these aims can be more effectively attained through
phylogenetic - rather than phenetic - methods of classification.
Third, the justification of evolutionary taxonomy appears equally
dubious.  Evolutionary taxonomy is methodologically closely
analogous to phenetics, and evolutionary taxonomists have criticized phylogenetic systematics on much the same grounds as have
pheneticists - that their classifications possess greater "information content" than the phylogenetic system.  Their claims are
false for the same reason that the claims of the pheneticists are
false.  Phylogenetic classifications appear to be more natural in
Gilmour's sense than either phenetic or evolutionary groupings.

On the basis of criteria that have been proposed for selecting

methods of classification, one can only conclude that the phylogenetic system is the only defensible choice for a general reference system for biology.

REFERENCES

Blackith, R.E. and Reyment, R.A., 1971, Multivariate morpho-
    metrics. Academic Press, New York.
Cain, A.J. and Harrison, G.A., 1960, Phyletic weighting. Proc.
    Zool. Soc. Lond. 135:1-31.
Clifford, H.T. and Stephenson, W., 1975, An introduction to
    numerical classification. Academic Press, New York.
Farris, J.S., Kluge, A.G. and Eckardt, M.J., 1970, A numerical
    approach to phylogenetic systematics. Syst. Zool. 19:
    172-189
Farris, J.S., 1977, Gilmour-natural classifications. Syst. Zool.
    In Press.
Gilmour, J.S.L., 1961, Taxonomy, in "Contemporary botanical
    thought," (A.M. MacLeod and L.S. Cobley, eds.), Quadrangle
    Books, Chicago.
Gilmour, J.S.L. and Walters, S.M., 1963, Philosophy and classfi-
    cation, in "Vistas in botany," (W.B. Turril, ed.), Vol. 4.
    Pergamon Press, London.
Hennig, W., 1966, Phylogenetic systematics. Univ. Illinois
    Press, Urbana.
Hennig, W., 1975, "Cladistic analysis or cladistic classifica-
    tion?": a reply to Ernst Mayr. Syst. Zool. 24:244-256.
Jardine, N. and Sibson, R., 1971, Mathematical taxonomy. Wiley,
    London.
Mayr, E., 1965, Numerical phenetics and taxonomic theory. Syst.
    Zool. 14:73-97.
Mayr, E., 1969, Principles of systematic zoology. McGraw-Hill,
    New York.
Mayr, E., 1974, Cladistic analysis or cladistic classification?
    Z.f. zool. Systematik u. Evolutionsforsch. 12:94-128.
Mickevich, M.F., 1977, Taxonomic congruence. Syst. Zool. In Press.
Mill, J.S., 1874, A system of logic, ratiocinative and inductive,
    being a connected view of the principles of evidence and
    the methods of scientific investigation, 8th ed. Longsman,
    Green, & Co., London.
Rohlf, F.J., 1970, Adaptive hierarchic clustering schemes. Syst.
    Zool. 19:58-82.
Silvestri, L.G. and Hill, L.R., 1964, Some problems of the taxo-
    metric approach, in "Phenetic and phylogenetic classification"
    (V.H. Heywood and J. McNeill, eds.), Syst. Ass. Pub. 6.
Simpson, G.G., 1945, The principles of classification and a
    classification of mammals. Bull. Amer. Mus. Nat. Hist.
    85:i-xvi, 1-350.

Simpson, G.G., 1963, The meaning of taxonomic statements, in
    "Classification and human evolution," (S.L. Washburn, ed.),
    Viking Fund Publ. in Anthropology. 37:1-31.
Simpson, G.G., 1971, Rémarks on immunology and catarrhine classi-
    fication.  Syst. Zool. 20:369-370.
Sneath, P.H.A., 1961, Recent developments in theoretical and
    quantitative taxonomy.  Syst. Zool. 10:118-139.
Sneath, P.H.A. and Sokal, R.R., 1973, Numerical taxonomy. W.H.
    Freeman & Co., San Francisco.
Sokal, R.R., 1975, Mayr on cladism - and his critics. Syst. Zool.
    24:257-262.
Sokal, R.R. and Sneath, P.H.A., 1963, Principles of numerical
    taxonomy.  W.H. Freeman & Co., San Francisco.

# FOUNDATIONS AND METHODS OF EVOLUTIONARY CLASSIFICATION

Walter J. BOCK

Dept. of Biological Sciences, Columbia University

New York, New York  10027 U.S.A.

## INTRODUCTION

Biological classification has always been concerned with expressing relationships between organisms, living and fossil, in a system that provided a foundation for further study and generalization.  The foundations and mode of expression of the resulting classification have not always been the same.  The Linnean hierarchy provided a formal scheme of ranking groups within groups, but the exact rules of the hierarchy depended on the accepted theory underlying biological classification.  With the acceptance of organic evolution after 1859, the theory underlying classification modified from that of ideal typology to organic evolution.  Yet the exact nature of classificatory schemes varied with the state of knowledge of evolutionary mechanisms and with the acceptance of diverse combinations of these mechanisms as crucial for biological classification.  Thus all of the major modern approaches to biological classification are regarded as evolutionary, and the designation of one as classical evolutionary classification, or more simply as evolutionary classification, does not imply that the other approaches of phenetics or cladistics (phylogenetic classification) are not based upon organic evolution.

Evolutionary classification - an eclectic approach - is based upon all aspects of evolutionary theory and reflects the entire evolutionary history of organisms, not just the history of branching points in the phylogeny of organisms.  As such, a clear distinction must be made between the formal scheme of classification and the phylogenetic diagram.  These are not equivalent in

evolutionary classification.  A one-to-one relationship does not
exist between the formal classification and the phylogenetic
diagram; each contains information not possessed by the other.
Thus, the phylogeny of organisms cannot be read directly from an
evolutionary classification.  Both modes of expression are required to provide the maximum summary of information about the
evolutionary history of organisms.  Omission of either at the
completion of a systematic study means that that analysis is still
incomplete.

This paper is an extension of my earlier paper on
"Philosophical foundations of classical evolutionary classification" (Bock, 1974).  I wish, as in my earlier study, to express
my acknowledgements to the contributions by Ernst Mayr and
George Gaylord Simpson to the foundations of evolutionary classification.  These workers have contributed more to our understanding of biological classification than any other living
systematist.  In the present analysis, I remain firm in my conviction that the approach of evolutionary classification provides the best system of classification for the needs of all
biologists working in areas from molecular biology to sociobiology.  However, I have changed or gone beyond many of the
ideas expressed in my earlier paper and must warn the reader that
the statements and conclusions in my earlier paper must be compared carefully with those in the present one.

Although the theoretical and philosophical discussions of
classification are most fascinating to me and many other systematists, and much attention has been given to these esoteric topics,
I believe that the most important point of focus is the analysis
of methods, especially that portion which may be called
"character analysis."  I believe that too little attention has
been given to the details of character analysis in the systematic
literature over the past three decades, as well as in this
Institute.  Further, I believe that much of the misunderstandings
and differences of opinion between the several modern approaches
to classification will decrease with a clear exposition of
methods at the level of character analysis.


THEORETICAL FOUNDATIONS OF BIOLOGICAL CLASSIFICATION


Underlying Theory for Biological Classification

Biological classification deals with living organisms and
their attributes; it serves to compress the diversity of characteristics exhibited by these organisms into a comprehensible
scheme.  These organisms and their attributes are the result of

their past evolutionary history and hence any pattern or order shown by the organisms is dependent upon the order of events in their evolutionary history and on the action of the diverse mechanisms of organic evolution. Thus the underlying theory for biological classification is that of organic evolution and in particular that of the synthetic theory developed over the past forty years. No theory of biological classification exists in itself; nor can classification be developed in the absence of any theory.

It is, of course, possible to base classification upon some other biological theory and this has been done in the past. Prior to 1859, most classification was done under the theory of ideal typology, and in those times no theory of classification existed independently of ideal typology. Some biologists, phenetists and more recently cladists, have expressed the belief that biological classification is simply another form of a general approach to classification that can be used for all objects, living and nonanimate, natural and man-made, or that classification is simply a reflection of patterns of concurrences independent of evolutionary or any other theory. These essentialist ideas are a form of ideal typology, and classifications based upon them would be difficult, if not impossible, to combine with evolutionary theory.

Classification, being based upon organic evolution, must be based upon all aspects of evolutionary theory. Moreover, the validity of classification will depend upon the validity of evolution as a scientific theory. It is necessary to examine the demarkation of evolutionary theory as a scientific theory and to consider the empirical tests of the validity of evolutionary change in organisms and of the particular mechanisms of change. The ideas of demarkation advocated by Karl Popper (1968a, b) are most useful in this connection.

## Evolution As A Scientific Theory

Much discussion exists in the literature of the philosophy of science on the question of whether the theory of organic evolution differs from other scientific theories such as the theory of gravitation. Much of this discussion reflects a real lack of understanding of biology and evolution, and is not very relevant to the question. Without going into detail, I would like to state that I believe that organic evolution is a scientific theory identical to those used in physics and other sciences, e.g., gravitation, with the only possible difference being the greater complexity of evolutionary theory. A special social and religious problem does exist in the demarkation of evolution

as a scientific theory, but here again the criteria for demarking
evolution as a scientific theory does not differ from others.  It
is necessary, as advocated by Popper, to delimit evolution as a
scientific theory without recourse to the distinction between
meaningful and meaningless sentences.

The demarkation of evolutionary theory as a scientific one
is not a trivial point because the whole approach to evolutionary
classification as advocated in this paper falls as a scientific
method if evolutionary theory can be shown to lie outside of the
limits of science.

Without providing any details, I would like to state the
following:

a)  That evolutionary theory is scientific as it is falsi-
    fiable by empirical observations - the criterion of
    Popper (see Bock, 1974:381-383).

b)  That evolutionary theory is not a tautology.

c)  That evolutionary theory has been thoroughly tested
    against a large number of empirical observations, that
    these tests have been severe ones, and that these tests
    have dealt with all essential parts of evolutionary
    theory; hence evolutionary theory possesses a high
    degree of corroboration in the sense of Popper.

d)  That a thorough, severe testing of evolutionary theory
    does not mean that considerable disagreements do not
    exist on the mechanisms of evolutionary change, but
    these disagreements do not affect the tests of the
    theory or organic evolution.

#### Application of Popper's Philosophical
#### Concepts to Biological Classification

In my earlier paper (Bock, 1974), I advocated strongly the
application of the ideas on philosophy of science advanced by
Karl Popper (1968a, b, 1972) to work in biological classification.
While still accepting his criterion for the demarkation of
science and still finding many of his ideas very useful in the
study of evolution and classification, it is clear that a strict
following of Popperian philosophy would result in serious problems
in these studies.  Moreover, it is clear that many workers, e.g.,
phylogenetic taxonomists, who advocate Popper's ideas very
strongly, simply do not use them in their actual work.  These
discrepancies between philosophies and methods advocated and the

actual practices has caused much confusion in biological classification during the past several years.

Popper argues that science is strictly deductive and that
induction does not exist as a method in science.  His position has
considerable merit, although other philosophers argue that
Popper's system simply readmits induction through "the back door,"
Popper's position may serve as the ideal approach to science that
should be strived for.  Such an ideal is valuable as it can
serve as a standard against which the methods used in a particular
study can be measured.  However, the complete rejection of induction as a method in science does not appear justified when
considering practical approaches and especially when considering
the problems of formulating narrative explanations which constitute a major part of explanation in evolutionary biology and
classification.  A classification or a phylogenetic diagram, or
the analysis of the evolutionary history of any group are all
narrative explanations, attempting to understand the past history
of the group of organisms.  Formulation of these explanations,
which are testable scientific statements, are far more complex
than the general covering law in science (which is of primary
concern to Popper), and are best done by some means of inductive
generalization, or of retroduction (Hanson, 1958).  However, much
more attention must be given to the testing of these narrative
explanations and to the independence of the hypothetical statements to be tested and the empirical observations used to test
them.

Popper advocates falsification as the means of demarking
science from nonscience and as the means of testing all scientific
hypotheses.  Again, the second stems clearly from the first.
Little disagreement exists between philosophers of science that
testability against empirical observations is the hallmark of
science; this idea did not originate with Popper.  More argument
exists on whether this testing must always be falsification
(negative testing) or whether it could also be verification
(positive testing).  Both may be acceptable depending upon the
particular type of scientific statement being tested.  Popper is
concerned mainly with the testing of covering laws, and here
falsification may be the only acceptable approach.  Testing
appears to be quite different for narrative explanations of the
sort that permeates all aspects of classificatory and phylogenetic
analysis.  Herein, much of the usual testing, e.g., concordance of
homologues, appears to be verification in the usual sense of the
word although the tests can always be stated in the opposite way
to make them appear to be falsifications.  Nevertheless, falsification can and should be used in tests of narrative explanations
and it appears that the most severe tests in classification and
phylogeny are cladistic tests involving falsification (Bock,
1974:383; 390-391).

## Goal of Science

The goal of empirical science is to explain phenomena, but
this is the goal of all types of human inquiry.  Science explains
in a very definite way and that is by the formation of hypotheses -
theories, covering laws, mechanisms, narrative explanations -
which are actually or potentially testable against empirical
observations.  According to Popper's demarkation, testability in
science is by falsification.  It may not be possible to test a
hypothesis at the time it is proposed because of technological
problems, but the scientist must, at the very least, indicate how
the hypothesis could be tested.  Any hypothesis is simply not
scientific if it cannot be tested because the proposer did not
provide a method whereby it may be validly tested against empirical
evidence or because the hypothesis has been put forth in such a
way that it cannot be tested.

Thus a major part of scientific methodology in any area of
science is to formulate procedures of testing hypothetical
scientific statements.  This is the area of concern of philosophers
of science; they are interested in ways in which hypotheses are
tested validly against empirical observations.  Little interest
exists in how one reaches or formulates the hypothesis to be
tested (but see Hanson, 1958), except for practical considerations
which are primary in narrative explanations.  One simply cannot
test all possible phylogenetic hypotheses about 100 taxa, nor
does one test classificatory or phylogenetic hypotheses that are
formulated randomly.  Philosophers of science are almost completely uninterested in how hypotheses are formulated because of
their overriding concern with general theories and covering laws.
Most reject inductive approaches, either absolutely or except as
a practical technique.  Philosophers generally agree, and
correctly, that no generalizations have been reached on how
scientists formulate hypotheses and that no overall guide-lines
or methods can be offered on that part of science that can be
labelled "theory generation."  This is clearly the subjective or
"arty" area of science, including narrative explanations.

## Tests

It is the testing of hypotheses and explanations which is of
the essence in science.  And it is this area - methods of valid
tests using empirical observations - that has not been of sufficient concern to systematists including many of the contributions
to this Institute.

Properties of tests. Tests and the empirical observations used in the tests constitute the objective aspect of objective science. The basic property of these tests is that the empirical observations are public in that they can be made by any person having the ability and training to do so. Observations that can be made by only certain people, perhaps because of certain insights, are private observations and would be subjective and invalid as tests for scientific hypotheses. The hypotheses, themselves, are not objective and should have other properties such as boldness - the explanation of as many phenomena, especially unrelated or unknown ones, as possible.

The tests must be valid in that they must serve to test properly the particular hypotheses. Validity of tests depends upon their bridges to the hypothesis and to the theory underlying the hypothesis. Determination of the validity of tests is not simple, nor can absolute rules be provided.

The tests should be severe in that they should provide the strongest possible testing of the hypothesis. Scientists should try always to demonstrate as hard as possible that a hypothesis is incorrect. The strength of a hypothesis increases with the number of times workers have failed to disprove it with severe tests.

The tests need not provide the correct answers every time. Tests may be perfectly valid ones and still fail us much of the time. Indeed, in evolutionary biology, classification and phylogeny, most of our valid tests have low resolving power to distinguish between correct and incorrect hypotheses. The tests fail us much of the time and we simply do not know when they do; they fail to reject incorrect hypotheses at an alarmingly high rate. The basis of the low resolving ability of tests lies in the complexity of the narrative explanations to be tested and in the complexity in the bridges between the theory, the explanations and the tests.

Most systematists have worried too much about the low resolving ability of tests and have confused validity of tests with the accuracy of tests (ones with high resolving powers that provide correct answers most of the time). One consequence has been the rejection of completely valid tests such as the use of the stratigraphic position of a fossil to test the polarity of a transformation series. Systematists should not worry too much about errors; they will be made and hopefully even corrected. What is of concern is to indicate the tests used for any particular hypothesis and to show why those tests are valid ones.

<u>Testable statements versus empirical observations</u>. Testable
Statements:  Testable statements are all sorts of scientific
statements that attempt to explain phenomena and include general
theories, covering laws, mechanisms, hypotheses, narrative explanations and the like.  Examples of general theories would be
evolutionary theory while covering laws or mechanisms would be
those of speciation, phyletic evolution, adaptation, formation of
genetical variation and natural selection.  Hypotheses and narrative explanations are of many types.  One category includes
classificatory hypotheses about groups and may be the recognition
of taxa of different ranks and the arrangement of these taxa into
a Linnean hierarchy.  Another includes phylogenetic hypotheses
about groups and may be the recognition of sister-group relationships or the recognition of ancestral-descendent relationships
(e.g., <u>Archaeopteryx</u> is the ancestor of all other known birds)
and the arrangement of these groups into a phylogenetic diagram.
A third category are hypotheses about characters and includes
statements about homologues, transformation series, plesiomorphs
versus apomorphs, and synapomorphs versus convergent features.

Procedure of testing:  Tests are made by formulating predictions from the testable statements and then comparing the
predictions against empirical observations.  The procedure may
involve a long chain of argument and may include secondary and
further hypotheses.  For example, hypotheses about groups are
"tested" against hypotheses about characters (secondary hypotheses)
and these may be tested against still further hypotheses
(evolutionary) before the final comparison against empirical
observations.  What is important is that at the end of the chain
of argument and secondary hypotheses are empirical observations.

Empirical observations are things that all of us can observe.
They must be objective and hence exclude all private observations.
They must be properties of objects in nature which include properties (e.g., form-function) of features, the position of an
organism in space (=location on the surface of the earth) and the
position of an organism in time (=the present for Recent organisms
or the stratigraphic position of a fossil which can give its
absolute or relative age).

Empirical observations are, of course, theory-laden.  But
in order to serve as a valid test of a hypothesis, the observation
must be <u>independent</u> of the hypothesis under test.  Otherwise, the
test would be simply and directly circular.  Thus if a particular
homologue is to be a valid test of a classificatory hypothesis
about groups (taxa), then the test of that homologue must be
against observations that do not depend at all on those taxa or
any other taxa for that matter.  The independence of empirical
observations from the hypothesis under test is a critical point

because many commonly used methods in classification and phylogeny
do not exhibit this independence.  All tests which fall under the
general heading of commonality or of the distribution of a
character over taxa, or out-group comparisons, or the use of a
"higher phylogeny" are dependent upon empirical comparisons that
are not independent of the hypothesis under test and hence do not
constitute valid empirical tests of the hypothesis.

Parsimony:  Parsimony has been cited frequently in the
systematic literature as a criterion, or even as the criterion,
for testing hypotheses or testing the internal logic of theories;
this use is widespread in phylogenetic systematics as well as in
many of the contributions to this Institute.  Unfortunately this
use of parsimony is wrong.  Parsimony is not a test of the
validity of hypothetical statements or of the internal consistency
of theories.  Internal consistency is one of the basic properties
that theories must have before they are seriously considered for
testing.  The validity of hypothetical statements, as well as
any undetected inconsistencies, are determined by testing against
empirical observations.

Parsimony is one of the criteria by which one of two or more
still valid theories is chosen as the one that will be accepted
and used until further tests invalidate the excessive theories.
But it must be noted that the several theories must have been
tested against observations and found to be still valid; they
are not tested with the criterion of parsimony.  Moreover, other
criteria exist that may be of even greater value than parsimony
in deciding which of several competing theories to use.  Popper,
for example, argues that boldness of a theory is most important.
One should search, he argues for theories that attempt to explain
as many and as wide a range of phenomena as possible.  Such bold
theories contain a maximum of empirical content and are potentially
the most testable.  Such theories are almost certainly not the
most parsimonious.

Parsimony has been used by most systematists, including by
most contributors to this Institute, as a method by which inductive
generalizations are reached from their data.  Thus the use of
parsimony to generate phylogenetic diagrams with the minimum
number of independent origins and parallelisms of characters or the
minimum number of mutations when using biochemical data (proteins).
This use of parsimony may be quite reasonable within an inductive
approach to the formulation of hypotheses, especially when they
are part of a narrative explanation.  But they do not serve as a
means of testing the validity of the hypothesis.  The presentation
of Dr. Joysey on the phylogeny of mammalian myoglobins in this
Institute is an elegant demonstration of the erroneous conclusions
that can be reached through the misuse of parsimony in classifi-

cation and phylogeny; I concur completely with his argument and conclusions.

## WHAT IS A CLASSIFICATION?

### Definition

A classification is a system of arranging the objects studied in some area of science according to theory applicable to that science and for some heuristic purpose. Generally the heuristic purpose is to provide the basis for the formulation of further generalizations - further testable statements - in that science. Part of this notion of the basis of further generalizations is the ordering and summary of observations into a convenient system and the development of information storage and retrieval systems. The classification must be based upon some theory. If an incorrect theory or a less useful or less general theory is chosen as the basis of classification, then the system of classification would be wrong or less useful. The history of classification in any science is a history of correcting and improving the system of classification as the underlying theories are improved. At any point in time, the particular classification in use may be a hodge-podge of patches stemming from different stages of theory development.

### Research Strategy

The classification itself is not a theory, nor is there a theory of classification. The several approaches to biological classification, e.g., cladistics, phenetics and evolutionary classification, represent different research strategies. Only one theory exists in my opinion - that of organic evolution - on which biological classification can be based today. Other theories would result in classifications of considerably less usefulness. The particular research strategy is based upon the theory of underlying biological classification. Each strategy deals with how to group organisms according to certain ideas, e.g., on similarity only, on genealogy only, or a combination of similarity and genealogy.

### Resulting Classification

The classification is the result of using a particular research strategy. It is a testable scientific statement, not an empirical observation. The method of presentation must conform to

accepted conventions, e.g., the Linnean hierarchy, otherwise its
heuristic value will be lessened.  As a testable scientific statement, classifications are not objective.  Rather, they are subjective statements that are tested against objective observations.
As a testable scientific statement - a hypothesis - the formulation of a classification may well be linked to an art because one
must deal with a large number of factors, interrelations between
these factors and a complex theory in order to propose a classification that is sufficiently worth the investment of the time and
effort required to test it thoroughly.  Not all classifications
and not all phylogenies are sufficiently well thought out to
justify their testing.  The test of a classification or of a
phylogeny is a long, arduous task if done properly.

Classifications and phylogenies are formulated by a prior
analysis of characters and are generally in the form of inductive
hypotheses of narrative explanations.  Often they are reached by
very rapid evaluations of possible homologues using rapid and
often superficial analyses of similarities, and frequently by
hurried determinations of possible synapomorphs without considering all of the necessary steps.  I have no objections to the use
of induction to formulate testable classificatory and phylogenetic
hypotheses and to the use of hurried, superficial analysis of
characters in the process.  But, it should be pointed out that it
is not possible to formulate classifications inductively and
claim, at the same time, to follow Popper's approach to the
philosophy of science.  Moreover, the shallow analysis of
characters that may be permissible in the inductive formulation
of hypotheses is not sufficient for the subsequent testing of
these hypotheses.  Unfortunately, the level of character analysis
used in the testing of classifications and phylogenies is often
at the level that is barely acceptable in the initial inductive
phases of study.

## PURPOSES AND NATURE OF CLASSIFICATION

### Classifications As Man-Made Devices

Being man-made devices, biological classifications must be
primarily heuristic and must be heuristic for all biologists.
Judgment of the usefulness of classifications is not in terms of
the systematists who construct them, but with respect to the
other biologists who are dependent upon classifications as the
foundation for their comparative studies and the formation of
their generalizations.  Thus systematists provide an extremely
essential service for 95+% of all biologists; the practicality of
this service cannot be overrated.  Although this service to other

biologists is not the only task of systematists, it is so
essential that should it not be provided or should it not be
provided well, we systematists may soon be out of business.

Being useful devices, classification must be carefully
balanced between several conflicting desirable attributes.  It
should reflect as precisely as possible the evolutionary history
of organisms, assuming that this is the foundation of biological
classification.  But attempts for extreme precision for the sake
of a particular approach or construction of the classification
beyond the resolving powers of the character analysis decreases
the value of the classification.  Increase in the number of
categorial levels and with it increase in the number of names of
supraspecific taxa will, of course, increase the amount of information that can be carried in the system.  Such an increase
is not warranted if it goes well beyond the resolving powers of
the character analysis.  Nor would such an increase of categories
and taxa names be useful if each step carried with it little
information compared with the increase in names.  A properly
constructed phylogenetic classification requires one less higher
taxa names than the number of species being classified.  Thus,
although only 20 categorial ranks are needed to classify just
over one million organisms, a total of over one million higher
taxa (above the species level) are required in the classification;
thus the number of names required to classify the species is
doubled.  This problem was not considered by McKenna (1975) in
a recent attempt to provide a phylogenetic classification of
mammals which was not a proper phylogenetic one and did not go
below suborders.  In an attempt to provide a phylogenetic classification of birds (again, the classification was not a proper
phylogenetic system), Wolters (1975) was unable to find enough
generic-level names in the literature to provide names for all of
the subgenera recognized; he left these taxa unnamed.  Thus, even
for a highly oversplit group such as birds, not enough names are
available to permit the naming of the taxa required by an incomplete phylogenetic classification.  It appears very doubtful
that the increased expression permitted by a phylogenetic classification will compensate for the confusion and problems that will
accompany the proliferation of categorial levels and taxa names.

Thus, to have maximum usefulness, a classification must be
balanced between the precision of expression, the resolution of
the character analysis and the required number of categorial
levels and taxa, along with their names.  Another consideration,
but more difficult to judge, is the stability of the classification with increasing knowledge.  It is desirable for classifications to reflect increased levels of knowledge, but it is best if
this reflection can be accomplished with minimum alterations in
the formal system of classification.

## Ways to Classify

No single absolute way exists to classify living organisms. The decision on the formal system depends upon several factors above the decision on the basic theory and the approach within this theory. It depends upon the basic purpose of the classification, whether a single, multipurpose classification is desired or many classifications, each best suited for a particular purpose. And it depends upon the objects classified. A system considering only Recent organisms will be quite different from one classifying only fossils, and each of these will differ from one treating fossils and Recent organisms together. And classifications will differ, even accepting a tightly constructed set of rules, for groups with a large fossil record compared with those with a poor or nonexistent fossil record. If the record is especially good at the time of origin and initial radiation of the group, the classification will differ again. Thus for organisms with a poor fossil record, the taxa may be largely "vertical" ones, but for groups, such as mammals, with a rich fossil record, some of the taxa may be horizontal ones. In such cases, primitive features (symplesiomorphs) may be more useful in establishing the classification than synapomorphs in spite of the general greater desirability of the latter type of character.

## Natural Classification

Definition of natural. Most biologists accept the idea that classification should be natural, but most workers are vague on the meaning of natural. The best meaning and use of "natural" is "in agreement with underlying theory." Natural has always had this meaning of being in agreement with theory, and was so used most consistently by typologists prior to the acceptance of evolutionary theory. Thus the meaning of natural has not changed for natural biological classification, only the theory underlying the classification has altered.

This meaning of natural is not the same as that expressed by Gilmore (1940) and discussed by Farris in his presentation. Gilmore's concept of natural does not appear to be associated with an underlying theory. And if it is theory-related, then it is related to ideal typology, not to organic evolution; hence I would reject his notion of natural classification because I reject the underlying theory.

Natural as applied to classifications is not equal to objective because classifications, being hypotheses, are not objective.

Natural classification:  If classification is based on evolutionary theory, a natural classification must be in agreement with evolutionary theory and with the whole of evolutionary theory, including all laws, mechanisms of change and subfactors thereof.  By the whole of evolutionary theory, I mean all factors and mechanisms.  In particular, I mean all studies of function, biological role, behavior and environmental factors required to understand the evolutionary mechanisms.  And I mean all mechanisms including that of phyletic evolution, especially all factors of the formation of genetical mechanisms and of natural selections underlying evolution in single phyletic lineages.  Speciation is not the only evolutionary mechanism of importance to classification contrary to statements of many phylogenetic systematists.

Evolutionary classification:  Of the several major approaches to biological classification, the most natural is evolutionary classification because this approach is based on all parts of evolutionary theory.  Any classification based upon only part of evolutionary theory would automatically be less natural than one based on the whole.  Phenetic classifications are based only on a measure of similarity which is an index to the amount of phyletic evolution.  Cladistic classifications are based only on phylogenetic branchings which are equated with speciation, and many cladists deny the existence of phyletic evolution.  I would like to emphasize that I define phyletic evolution as a change in a single phyletic lineage, and that a phyletic lineage is the passage of a species through time by successive reproductions.  Phyletic evolution is not the same as phyletic gradualism as no rate of change is implied in phyletic evolution.

## Classification in Other Sciences

The problems, demands and arguments relating to biological classification, especially those factors that fall under the general heading of "artiness in classification," are common to systems of classification used in all fields of science.  Biological classification is not unique with respect to these problems.  If one examines the periodic table in chemistry, this system of classification does not express all ideas in chemistry and clearly cannot be used as the basis of all explanation in chemistry, even limiting oneself to elements.  The arrangement of types of stars based upon their mass to show the time course of change in stars only expresses some, not all, properties of stars.

The common feature of classification is that they appear to be almost essential for all fields of science.  It is virtually impossible to avoid schemes of classification in a science if workers hope to reach general causal principles of explanation.

The lack of a classification for the objects investigated in a particular science is usually a sign that that field is still in an early stage of development and that fundamental theories are lacking.

### Diverse Purposes of Biological Classification

Merely to assert that classifications are heuristic is insufficient; one must show exactly how they are useful. Hennig states (1966:9) that classifications serve as general reference systems in biology which for most cladists means an index to the phylogenetic relationships. Other ideas have been an index to maximum concurrence of characters as expressed by pheneticists or some idea of information storage and retrieval. Considerable confusion exists on the last point because many workers believe that the information is actually stored in the classification. This is not true, nor is the classification, itself, an information storage-retrieval system. Rather, the classification serves as the foundation on which an efficient information storage and retrieval system can be constructed, be it a book, the arrangement of specimens in a museum, a computer system or whatever. Warburton (1967) in a discussion of the purposes of classification suggests that the most important is "to construct classes about which we can make inductive generalizations." (p. 243). A similar, but distinctly different expression was proposed by me (Bock, 1974:379-380) and by Mayr (1975:95-96) under the heading of the basis of comparative study in biology. The best expression is that the most important purpose of biological classification is to serve as a hypothesis generation mechanism with stress placed on deductive hypotheses rather than inductive ones.

Under the heading of classification serving as a hypothesis generation mechanism are a number of interrelated activities. These include the use of classification as the basis for the construction of causal covering laws and of narrative explanations in biology. It can serve for the construction of causal variables used in nonhistorical comparisons as done by workers such as Oxnard, Gould and Raup. Classifications serve as a means of collecting and presenting factual data in an efficient way, thus eliminating the need of writing a book the size of "Gray's Anatomy" for every species of vertebrates. And in like fashion, it can serve for the construction of information storage and retrieval systems. All of these may simply be different ways of expressing the basic hypothesis generation attribution of classification of generalizing beyond the known facts, be they hypotheses about unknown features in other organisms (extrapolation of conclusions) or the extension of generalizations to nonobvious properties of organisms. In any case, it could be argued that

classifications to be useful should permit maximum boldness in
generating hypotheses.

## Single Versus Multiple Classifications

The remaining question about the nature of biological
classification is whether a single scheme of classification
should be advocated (irrespective of the approach) or whether
multiple systems of classifications should be advocated, each of
which is best suited for particular biological studies.  Both
positions have been argued, the second advocated especially
strongly by some proponents of phenetic classification.

I would argue for a single system based on theory and using
a clearly stated approach (research strategy).  It is clear that
a single system of classification will not provide the best
possible classification for each biological study, but neither
does the periodic table provide the best classification for all
studies in chemistry.  If we wish to develop biology as a unified
science, if we believe that organic evolution is the major
unifying theory in biology, and if we wish to be able to relate
conclusions reached in different areas of biology, then it is
essential to have and use a single system of classification.

If different schemes of classification are used in different
areas of biology, then the conclusions reached in these several
fields are not compatible, and they cannot be compared.  In-
compatibility will be greater if the several classifications are
constructed according to different approaches.  The fact that the
results reached in different fields are highly compatible suggests
that the classifications and the research strategies used to
reach them are quite similar regardless of theoretical discussions
to the contrary.

It must be realized that we are far from agreeing on a
single actual classification of all organisms, even for most
small groups of vertebrates.  But this lack of agreement on actual
classifications does not eliminate the possibility of accepting,
in principle, the idea that only one system of classification is
possible, and further accepting the idea that one definite
approach (research strategy) can be postulated.

## THE BEST CLASSIFICATION - HOW JUDGED

### Introduction

If the position is accepted that only one system of classification can be used in biology and hence that it must be based
on a particular approach, then we are faced with the very difficult task of choosing the best classification.  Because most,
but not all biologists, today accept organic evolution as the
theory underlying classification, the choice of the best classification is really that of the best research strategy.  The many
different approaches may be grouped together under three headings
of phenetics, cladistics and evolutionary.  Some problems arise
because of hybrid approaches of phenetic cladistics and other
problems arise in the delimitation of evolutionary classification
from those on either side.  I will use the notion that a system
concerned only with similarity - a reflection of phyletic evolution - is phenetic and one concerned only with phylogenetic
branching points - a reflection of speciation events - is
cladistic.  Evolutionary classification is based upon a combination of these two mechanisms of evolutionary change although a
fixed ratio cannot be given.  It seems clear, however, that
cladistic and phenetic systems must be pure.  The statement that
a system is mainly cladistic and only slightly phenetic means
that is is an evolutionary approach.  Perhaps a major objection
to evolutionary classification is that it spans such a great
range between phenetics and cladistics and that the contribution
of phyletic evolution and of speciation varies greatly between
the classificatory schemes of different systems.  The consequence
is indeed a sloppy system of evolutionary classification, but
this is only a reflection of the sloppy material that must be
classified.

These three approaches to classification are really not
three different paradigms in the usual sense of that term.  These
are simply different research strategies based upon the same
paradigm - that of evolutionary theory.  Different paradigms in
the sense of Kuhn are not understandable to workers comprehending
one or another of the several paradigms.  Yet, it seems clear,
at least to me, exactly what is meant by systematists advocating
different approaches to classification.

### Criteria for Judging

A large literature exists in the philosophy of science on
criteria with which different theories and/or research strategies
can be compared and the best chosen.  These debates are exceedingly

complex, and although they center around heuristic notions, it is clear that simple measures of evaluation do not exist.  Several important factors emerge.

The most important criterion may be boldness (sensu Popper) which describes the ability of a theory to explain hitherto un-explained phenomena.  This may be the best criteria and may be directly correlated with other criteria as would be argued by Popper.

A second criterion is content which describes the number of phenomena and relationships that are actually covered by the theory.

The third criterion is testability which measures the number of possible tests that may be used to examine the validity of the theory.  It is clear that testability is directly related to content and also to boldness.

A final measure is a vague pragmatic one which would be how many useful hypotheses may be formulated from the classification. This may be the same or closely related to the measure of boldness.

## Evolutionary Classification

With these criteria in mind, it is clear that evolutionary classification is the best approach because it is the most natural.  It is the approach that is dependent upon the whole of the synthetic theory of evolution and hence would be the boldest, has the greatest content and is the most testable.  Other approaches are dependent upon only part of the synthetic and hence would be less natural.

It would be possible to make some direct comparisons of the classifications resulting from different approaches, especially using the criterion of number of useful hypotheses.  Nelson (1969:534) proposed a phylogenetic classification of the major groups of vertebrates.  This may be compared with a standard evolutionary classification such as that advocated by A.S. Romer in his texts (e.g., "The vertebrate body").  Nelson divides the gnathostomes into two groups, the Elasmobranchiomorphi and the Teleostomi.  The Teleostomi are divided into two groups the Actinopterygii and the Sarcopterygii.  The latter contains the group usually called the sarcopterygian fishes and that called the tetrapods.  One subgroup of Nelson's Sarcopterygii, the Choanata, contains the Rhipidistia and the tetrapods (his Batrachomorpha and Reptiliomorpha).  The Sauropsida contains three groups, the Chelonia, the Lepidosauria and the Archosauria with

the latter including the Crocodilia and the Aves.  Although
Nelson's classification contains at one rank or another, the
groups of Romer's classification, a quick comparison of the two
classifications shows that the classical system advocated by
Romer is far superior to that proposed by Nelson in the number of
possible generalizations and hypotheses that may be generated from
each.  Indeed some of the groups in Nelson's system are almost
devoid of useful generalizations.

## BASES OF EVOLUTIONARY CLASSIFICATION

### Nature of Evolutionary Classification

Evolutionary classification is a system of taxa arranged in
a Linnean hierarchy.  The taxa must be monophyletic in the sense
of Simpson and their rank reflects the attempt to maximize
simultaneously the two semi-independent variables of amount of
phyletic change as reflected in the degree of similarity and the
phylogenetic sequence of events as reflected in the pattern of
phylogenetic branching and ultimately in the pattern of speciations.  No one-to-one relationship or correlation exists between
the evolutionary classification and the phylogenetic diagram of
the groups contained in the classification.

Phyletic evolution is the change in individual phyletic
lineages and reflects the degree of adaptive evolutionary change.
It is based upon mechanisms associated with the formation of
genetical variation and those associated with selection.  Because
selection arises from the physical environment and the biotic
environment, the strength of directional selection is dependent
to a large extent upon the numbers and type of species present
(= the biotic environment).  Thus phyletic evolution is dependent
strongly upon speciation.

The phylogenetic sequence of events is based upon branching
points and ultimately on speciation.  It must be emphasized that
speciation cannot occur without phyletic evolution.  Of importance
is that the magnitudes of splittings must be kept in perspective.
The nature of the phylogenetic branchings that may be resolved in
an analysis of higher-ranked taxa in studies of either fossils or
of Recent organisms are far above the level of species splitting.
To claim that these phylogenetic branchings are speciation events
and that the segments between them are species is invalid because of greatly different magnitudes involved and because of the
difficulties of resolving most major adaptive changes down to the
level of speciations.

Lastly, different types of comparison must be recognized and included in the analysis. These were discussions in my earlier contribution to this Institute and will not be repeated here.

## Phenotypic Similarity

Measurement of the amount of phyletic evolution is difficult because most workers feel that this evolutionary change should be in terms of genetical modification. I am not at all sure that this requirement is necessary because selection acts on the phenotype and the proper measure of adaptive evolution, for example, is modification of the phenotype. Nevertheless, phenotypic similarity can be used as a good index to genotypic similarity and the amount of phenotypic change can be used to judge the amount of genetical change. The correlation between the genotype and the phenotype is not exact, but it is reasonably good, and it is basically the only index available for broad comparisons. The closeness of this correlation will vary according to the gene-phenotype link during development and as a result of physiological (somatic) adaptation in the adult. But the influence of these factors is presumed to be small compared to the overall degree of similarity of features in organisms placed in the same groups and to the differences seen in organisms placed in different groups.

## Rates of Evolutionary Change

These rates can be measured in several ways as discussed by Hecht in this Institute, and should include both the rate of phyletic evolution and the rate of phyletic branching or speciation. The rate of phyletic evolution, which results in the degree of difference (inverse of similarity) varies mainly with the strength of directional selection. The strength of directional selection depends more on the biotic environment, reflected in interactions between species (Bock, 1972). Thus phyletic evolution is not independent of phyletic branching. The rate of phyletic evolution varies between and within lineages and it cannot be ignored in any comparative study or classification.

The rate of branching is ultimately the rate of speciation and is dependent upon the opportunities for speciation. This varies with the pattern of geographic-ecological barriers and with the abilities of organisms to disperse and cross these barriers. Speciation can occur only with phyletic evolution, hence the rate of speciation depends on the rate of phyletic evolution. Because much of the directional selection for phyletic evolution arises from species interactions, the degree of this

selection and hence the rate of phyletic evolution goes up increasingly rapidly as the number of species increases.  And this
will in turn increase the rate of speciation.

The argument of many phylogenetic systematists that phyletic
evolution does not exist cannot be taken seriously.  Almost all
the work of animal and plant breeders as well as the experimental
studies of population geneticists document phyletic evolution.
Longer range periods of phyletic evolution has been shown in
observations of microfossils in deep-sea cores (Kellogg, 1975)
and in mammalian fossils on islands as discussed by Sondaar on
some of the Mediterranean Islands in this Institute.

## Meaning of Semi-Independent Variables

In my earlier paper (Bock, 1974:377-378), I characterized
(a) the amount of phyletic evolution and (b) the sequency of
branching as semi-independent variables and claimed that the goal
of evolutionary classification was to recognize groups that
maximize both variables simultaneously.  The meaning of semi-
independent was apparently obscure.  It is clear that the evolutionary mechanisms of phyletic evolution and of speciation are
related in a causal sense.  Phyletic evolution is needed for
speciation, although phyletic evolution can proceed without
speciation.  However, the rates of both types of evolutionary
mechanisms are related as discussed above.

The meaning of semi-independent seemed to be clear and was
used in the sense of being partly correlated, but that the degree
of correlation varied greatly between groups.  If these variables
were absolutely correlated, they would be redundant and either
one or the other could be used as the sole basis of classification.
The history of classification shows that complete correlation between these variables does not exist.

If these variables were completely independent of each other,
then no correlation could exist between them.  In this case, we
would have to use one or the other as the basis of classification
as it would make no sense to use both.  Again, the history of
classification shows that complete independence between these
variables does not exist.

A degree of commonality must exist between the rate of
speciation and the rate of phyletic evolution as a prerequisite
for their comparison and correlation.  This commonality stems
from the just mentioned interactions between these mechanisms of
evolutionary change.  Little direct observation exists on the
degree and variation of the correlation.  Indirect observations

suggest that the correlation varies considerably which is the major basis for the difficulties experienced in actual studies of biological classification.

## METHODS OF ANALYSIS

### Introduction

The methods of classical evolutionary classification must allow the formulation and testing of hypotheses about the existence of taxa (=groups of organisms), their classification (=formal arrangement in a Linnean hierarchy according to the rules of evolutionary classification) and their phylogeny (=formal arrangement in a phylogenetic diagram). Because the approach of evolutionary classification does not insist on a one-to-one correspondence of the classification and the phylogeny, separate sets of hypotheses about groups are needed to cover both aspects of relationships. And it is necessary at the completion of an evolutionary classificatory analysis to present the conclusions in a formal classification and in a phylogeny. Many workers omit the phylogeny; this omission causes problems for others who may need the exact phylogeny for their studies and yet cannot obtain this information from the classification.

The formulation and testing of these classificatory and phylogenetic hypotheses must be from "scratch" - the complete lack of knowledge of any existing hypotheses - if one wishes to develop a complete set of methods. The only information available to the systematist at the outset of her work is (a) a knowledge of the organisms and their attributes; (b) a knowledge of the position of the organisms in space and time; and (c) a knowledge of evolutionary theory. In reality organisms means individuals as it is necessary to group the individuals into species; however, I will start with the assumption that the species are known because the thrust of my analysis is at the supraspecific levels.

The same methods must be used for Recent and for fossil organisms. Fossils are not logically or otherwise secondary to Recent organisms, and hence they must be studied and classified in the same way. To be sure, differences in the details of study exist on practical grounds between fossils and Recent organisms because of the available information. But, vast differences exist in the details of study between various Recent organisms for the same reasons. Certainly many fossil organisms have been studied in more detail than Recent organisms; one could easily list scores of fossils that are better known than 90% of the Recent species of animals.

As mentioned above, in order to provide a general statement, the methods of study must not be dependent on the long history of study on classification and phylogeny of organisms. This is especially critical for the formulation of hypotheses about groups, but also for some of the testing procedures in the character analysis. On practical grounds, however, one cannot ignore past studies and usually begins an investigation with hypotheses based upon previous classification and phylogenies. This is a perfectly valid approach, but it must be emphasized that hypotheses stemming from previous studies do not have any special status. The status of a hypothesis is dependent almost strictly upon the tests to which it has been subjected and hence upon its degree of corroboration.

## Definition and Recognition

A clear distinction must be made between the definition of words and the recognition of objects that correspond to these words. The word being defined stands for a concept which is set forth in the definition. Thus the biological or nondimensional species stands for the concept of the species in evolutionary theory and is defined in terms of a reproductive community of actually or potentially interbreeding individuals which is reproductively isolated from other such reproductive communities (Mayr, 1963). One can discuss the species concept further and demonstrate, for example, that its validity or meaning decreases as one progresses further away from a single interbreeding population both in space and in time. Actual species, taxa that we call species, are only rarey recognized on the basis of their reproductive behavior but are almost always recognized on the basis of morphological attributes in that a sharp morphological gap is usually found between members of different species.

Words are defined with the definition outlining the concept for which the word stands. It is possible to consider the definition of a word in terms of defining criteria (Bock, 1974: 383-385), although I realize that objections can be raised against the term "defining criteria." Definitions must relate to theory and must not be circular -- that is, the defining criteria must not be circular with respect to the word being defined and they must relate to the theory under which the concept falls. Definitions are, in themselves, arbitrary as it is generally possible to subdivide a theory in different ways, but they must be useful in that some methods of subdividing a theory may be overly cumbersome. Although definitions are arbitrary, one cannot define words arbitrarily; words must be treated and defined within the historical development and tradition of a science if one wishes to maintain maximum communication. Lastly, it must be emphasized

that definitions are not hypotheses and hence they are not tested
against empirical evidence.

Several different types of definitions are possible.  I will
use definitions that may be termed theoretical in which the words
are defined within the context of a particular theory and the
definition does not necessarily provide the means of recognizing
objects corresponding to the defined word.  The definitions of
most (almost all) words used in classificatory and phylogenetic
studies does not provide any clue as to how to recognize objects
corresponding to these words.  Operational definitions have been
used by some systematists, mainly those advocating phenetic
approaches.  In operational definitions, the definition must include the exact set of instructions needed to recognize the
objects corresponding to the defined word.  Strong philosophical
objections have been raised against operational definitions by
Hemple and others, but I do not wish to concern myself with that
discussion.  Rather, I wish to argue that operational definitions
are not the only acceptable definitions in science, that opera-
tionalism is not equal to useful and that words that elude an
operational definition may generally be defined according to the
above theoretical approach.  In general, I believe that operational
definitions are not useful in classification and phylogeny in
which a major portion of explanation is narrative.

Given a definition of a word, objects corresponding to that
word must be recognized.  These objects are generally recognized
using criteria of recognition which must relate to the theory
under which the word is defined and which must relate to the
criteria of definition.  Almost always the criteria of recognition
are not the same as the criteria of definition.  Species are defined in terms of reproductive isolation between groups of interbreeding individuals.  They are recognized in terms of morphological gaps between continuously varying morphological features.
The criteria of recognition cannot be circular in themselves.

Thus it is essential to avoid circularity within the realm of
defining criteria and within the realm of recognizing criteria.
Homology is defined in terms of phylogeny and phylogeny is defined
in terms of evolution.  But homologous features are recognized on
the basis of similarities and phylogenies are recognized, in part,
on the basis of homologous features, and the evolutionary history
of a group is shown, in part, on the basis of phylogeny.  No
circularity exists within the definitions or within the recogni-
tions, and the recognizing criteria, the defining criteria and
the underlying theory - that or organic evolution - are all
interconnected by bridging links.

It must always be remembered that the objects that are

recognized, be it a particular taxon, a phylogeny, or a type of feature (homologues or synapomorphs, for example) represent scientific hypotheses and hence they must be tested against empirical observations. The criteria of recognition are basically the rules by which valid tests of these objects against empirical observations are undertaken. In operational definitions, the defining and the recognizing criteria are interwoven. The reason for the difficulties of operational definitions in classification and phylogeny lies in the low resolving power of the recognizing criteria.

## Hypotheses About Groups

Types of Hypotheses. Two distinct sets of hypotheses about groups exist in classificatory and phylogenetic investigations. These hypotheses and the associated character analyses (=hypotheses about characters) are closely interrelated, but they are not identical. The failure to distinguish clearly between these sets of hypotheses which may be termed classificatory hypotheses and phylogenetic hypotheses has resulted in endless confusion. In phylogenetic systematics, only one set of hypotheses about groups exists as it is assumed in that approach that the classificatory hypotheses are identical with the phylogenetic hypotheses. This is a perfectly valid assumption under this approach to biological classification. But it is absolutely wrong to conclude, as phylogenetic systematists have, that this is the only valid assumption. Separation of classificatory and phylogenetic hypotheses is a basic assumption under the evolutionary approach to classification and it would be equally wrong to conclude that this is the only valid assumption. The decision of which of the assumptions is accepted depends strictly upon which approach to biological classification is accepted, and that decision is made on totally different grounds as discussed above under "Best classifications - how judged."

I would argue that a phylogenetic analysis is not only possible, but it is an essential part of any study of evolutionary classification. Thus it is necessary to formulate phylogenetic hypotheses in addition to classificatory hypotheses. Thus no disagreement exists between evolutionary systematists and phylogenetic systematists on the necessity of phylogenetic hypotheses about groups. But just to state a belief that a phylogenetic analysis is an essential and integral part of any classificatory study does not make it true. Many (all?) phenetic systematists reject the notion of valid phylogenetic hypotheses about groups (and characters) and argue that only classificatory hypotheses exist. Colless (1967) discussed this under the notion of "The phylogenetic fallacy" and argued that phylogenetic hypotheses are empty because they contain no information not completely included in (phenetic) classificatory hypotheses. Colless' argument has been answered (e.g., Bock, 1969b), but not as well and as thoroughly as it should be because the issue he raises is a very important one in classification.

Classificatory hypotheses:  Classificatory hypotheses about groups are those hypotheses about the evolutionary relationships of groups as expressed within the framework of a formal system of classification according to the rules and conventions accepted for this system under the particular approach to biological classification.  Thus for evolutionary classification, the formal classification would be a Linnean hierarchy and the rules and conventions for recognizing the ranks of groups would be those that maximize simultaneously the degree of genetic similarity and the sequence of phylogenetic events.

Included in classificatory hypotheses are:

(1)  The recognition, delimitation and ranking of individual taxa;

(2)  The nature of the taxa.  Each must be monophyletic in the sense that all members have evolved from a common ancestor - the definition of monophyly of Simpson, not of Hennig; and,

(3)  The hierarchy of taxa which must be arranged in a strictly inclusive, non-overlapping hierarchy in which each group at a particular categorial rank contains one or more groups of the next lower rank.  This is a system that is usually called a Linnean hierarchy.

Phylogenetic hypotheses:  Phylogenetic hypotheses about groups are those hypotheses about the phylogenetic branching re-lationships of groups as expressed within the framework of a formal phylogenetic diagram according to the rules and conventions accepted for these diagrams.  Thus for evolutionary classification, the formal phylogenetic diagram should be a dendrogram that is as close as possible to a cladogram in the sense of Hennig.  The problem in phylogenetic diagrams arise from those branching points that cannot be resolved into dichotomies with strong evidence.

Included in the phylogenetic hypotheses are:

(1)  The recognition and delimitation of the branching points in the phylogeny.  It is not necessary to assume that these branching points equate to speciations and it is wrong to regard the segments of phyletic lineages

between branching points as species;

(2)  The nature of the phylogenetic units which are closed
descendent groups consisting of a species, at the base
of the group, and all of its descendents.  These
phylogenetic units have generally been called taxa
but this is a term that has been taken over from
classificatory hypotheses and has been incorporated
into phylogenetic hypotheses with much resulting confusion.  I propose that a closed descendent group be
called a phylon (based on phylo - in the same way that
taxon is based on taxo -) with the clear recognition
that classificatory hypotheses dealing with the arrangement of taxa in a classification differ sharply from
phylogenetic hypotheses dealing with the arrangement of
phyla in a phylogenetic diagram.  It is possible under
some approaches to classification (e.g., cladistics)
that all phyla are equivalent to taxa and under other
approaches to classification (e.g., evolutionary) that
only some of the phyla are equivalent to taxa.  A
phylon has particular properties among which is that it
is holophyletic using Ashlock's (1971) term (=mono-
phyletic in the sense of Hennig and other phylogenetic
systematists);

(3)  The phylogenetic hypotheses may be about sister-group
relationships or they may be about ancestral-descendent
relationships.  Many cladists argue that ancestral-
descendent relationships do not constitute valid
scientific hypotheses (=testable) or that it is not
possible to postulate ancestors.  Such arguments are
simply at odds with the philosophy of science advocated
by Popper which is accepted by these cladists.  An
ancestral-descendent hypothesis is exceedingly easy to
falsify - indeed it has such a high degree of testability
that it constitutes a most desirable hypothesis for
Popper.  Statements that some organism is the ancestor
of a group, e.g., Archaeopteryx is the ancestor of all
birds, is again a perfectly testable hypothesis and a
most desirable one in Popperean philosophy.  And it is
a valid hypothesis - one that corresponds to truth until it has been falsified.  In this case the hypothesis
that Archaeopteryx is the ancestral bird has been tested
numerous times (see Ostrom, 1976 and cited references)
without being falsified; hence it would be a well
corroborated hypothesis.

A further word must be given on the term monophyly.  This term
was originally coined and defined to express a certain attribute of

taxa which was ascertained after the taxon had been recognized and
delimited.  It indicated those taxa whose members evolved from the
same common ancestor.  In his definition, Simpson included a statement about the nature of the common ancestor.  But this problem
could, and perhaps would have been better to, have been dealt with
in a separate definition.  Clearly monophyly is a concept that
applies to taxa under the general heading of classificatory
hypotheses.  Hennig (1966 and elsewhere) changed the entire application and hence meaning of monophyly in using it to express a
certain attribute of an ancestor which was recognized first after
which the closed descendent group evolving from the ancestral
species was considered.  Thus monophyly was used as a concept
applying to ancestors under the general heading of phylogenetic
hypotheses.  Such a radical change in the use of a word is really
not justified in any human endeavor if one hopes to maintain
maximum communication.  For this reason, I argue strongly against
the redeployment of the term monophyly by the phylogenetic
systematists and urge that Ashlock's "holophyly" be used to
describe the desired property of closed descendent groups.

   Categories and groups:  After a number of clear expositions
on the distinction between category and group, many systematists
still confuse these terms.  Categories, such as species, genus,
family and order, are words and hence are defined.  These are
ranks in the Linnean hierarchy used in evolutionary classification.
Good, clear definitions exist for all categories, although only
the species definition can be affixed to a definite biological
phenomena.  Taxa are groups of organisms within the scope of
classificatory hypotheses and hence are real objects in nature
which are recognized, delimited and described, but never defined.
Although taxa may be regarded as real objects in nature their
recognition represents hypotheses (narrative explanations) and
hence must be tested against empirical observations.  Each taxon
corresponds to a particular categorial level in the Linnean
hierarchy but no absolute rule exists as to how to decide on the
level of a taxon.

   Phyla are also groups of organisms but within the scope of
phylogenetic hypotheses and hence are real objects in nature which
are recognized, delimited and described, but never defined.
(Unfortunately, the plural of phylon is the same as the plural of
phylum - a taxon - but I doubt that this will ever cause any
confusion.)  Although phyla may be regarded as real objects in
nature, their recognition represents hypotheses and hence must
be tested against empirical observations.  It is clear that phyla
exist at different ranks in the phylogenetic diagram, and that
they can be arranged in a hierarchy.  This is indeed what has been
done in phylogenetic systematics in which the phyla have been
forced into a Linnean hierarchy with much resulting confusion.

Tests of group hypotheses:  With the distinction made between classificatory and phylogenetic hypotheses about groups, it is clear that the tests of each type of hypothesis may overlap considerably, but they are not the same.  In either case, the hypotheses about groups are tested by comparing the predictions arising from these group hypotheses against particular hypotheses about characters possessed by members of the groups.  For example, members of a monophyletic taxon should possess a certain type of characteristic that is designated as homologous.  Members of a holophyletic phylon should possess a certain type of characteristic that is designated as autapomorphous.

These hypotheses about characters must then be tested further - either directly against empirical observations or against still further hypotheses in a chain and eventually against empirical observations.  <u>Herein lies the real core of methods in classification and phylogeny - that is character analysis</u>.  The formulation of classificatory and phylogenetic hypotheses about groups is not the important step, nor is the designation of particular features as homologues or as synapomorphs.  What is important is how these hypotheses are tested against empirical observations because it is this part of the analysis that makes the study of classification and phylogeny an objective science.  Unfortunately, the methods of character analysis is perhaps the least discussed aspect of classification and phylogeny in the theoretical literature.  And it is the single facet that has been least discussed in this Institute with the exception of a few papers such as those by Szalay, Luckett and Joysey.

A special comment must be made about the method of phylogenetic analysis.  By phylogenetic analysis, I mean the character analysis used to test phylogenetic hypotheses about groups.  Only one valid method of phylogenetic analysis exists regardless of the approach to classification used and the resulting classification.  Thus I reject the notion that separate and distinct methods of phylogenetic analysis exist for the approaches of evolutionary classification and of phylogenetic classification.  Indeed how can different valid methods of phylogenetic analysis exist if we accept the idea that only a single phylogeny of organisms exists?  The method of phylogenetic analysis presented below is, I believe, in full agreement with that advocated by Hennig (1965, 1966).  Seeming differences are, I believe, more apparent than real and result largely from Hennig's cryptic style and expression.  My disagreement with phylogenetic systematists is not with the idea of a phylogenetic analysis, but with the details of the method of phylogenetic analysis they advocate.  Almost all of the methods advocated by phylogenetic systematists working with vertebrates are either "commonality" observations or use parsimony as the criterion, both of which are not valid

empirical tests of hypothetical statements (see above under
Procedure of testing and Parsimony).

Formulation of hypotheses about groups: The first step in
any classificatory or phylogenetic study is to formulate the
appropriate hypotheses about groups. This can be done in a
number of ways. It is possible to generate these hypotheses
randomly, given the species under consideration, and to test all
of the possible hypotheses. Such an approach is foolish although
it is suggested by some cladists who present cladograms of all
possible phylogenies of a small number of groups (usually four
being the maximum). Or one can start practically and use the
classifications and phylogenies that exist as the hypotheses to
test, but this is almost the same as random choice and tells us
nothing of the general methodology for reaching useful hypotheses
to test.

The most reasonable approach is some inductive method, or
perhaps some retroductive method (as advocated by some philosophers
such as N.R. Hanson), as is actually done by almost all systematists. This is clearly contrary to the ideas of those philosophers, such as Popper and his followers, who argue that inductive
methods have absolutely no place in science. The best inductive
methods for recognizing taxa, for arranging taxa into classification, and even for recognizing and arranging phyla into phylogenetic diagrams are phenetic ones based upon overall comparisons
of similarities (Bock, 1974:383). These methods have been in use,
perhaps in a crude form, long before Darwin introduced evolution
into the thinking of taxonomists, and have been greatly refined
by the methods advocated by numerical taxonomists. These inductive methods are based upon the recognition of homologous
features and the arrangement of species into taxa and a classification based upon the shared possession of homologues. It is clear
that phylogenetic systematists, even those following Popper, use
inductive methods to advocate hypotheses about groups.

Hypotheses about characters. Hypotheses about characters constitute the character analysis part of study and is the real heart
of what is done in classification and phylogeny; perhaps stars
should be placed around the heading. Hypotheses about characters
are used as tests of classificatory and phylogenetic hypotheses
about groups and hence should be divided into these two classes of
hypotheses on the level of character analysis. In the following
discussion, I can treat each step only briefly, but will endeavor
to indicate for each step which type of tests of hypotheses against
empirical observations are valid and should be used. I have not
included all possible tests of classificatory hypotheses about
groups. It suffices to say that I believe that the most severe
tests of classificatory hypotheses about groups are phylogenetic

(=cladistic) tests (Bock, 1974:383;390-391) but these are not
simply the determination of synapomorphs.  It appears that other
tests can be developed such as the use of paradaptations (Bock,
1967, 1969a).  Such methods must be explored and their discussion
is left to future papers.

Homology:  Homology is the central concept in character
analysis and is basic to all approaches to classification.  It is
used in the inductive formulation of hypotheses about groups.  It
serves as an important test of classificatory hypotheses and as
the start of the phylogenetic analysis.  Individual homologues
must be recognized, they cannot be read automatically out of
nature.  The following discussion of homology is based on my
earlier treatments of this topic (Bock, 1963, 1969a, 1969c, 1974).

The generally accepted definition of homology and the one I
advocate may be stated as follows:  Features (or conditions of a
feature) in two or more organisms are homologous if they stem
phylogenetically from the same feature (or the same condition of
the feature) in the immediate common ancestor of these organisms.
Thus homology is defined in terms of phylogeny and phylogeny can
be defined in terms of evolution (Bock and von Wahlert, 1963); it
is a noncircular and nonoperational definition.  This definition
eliminates the need to distinguish between characters and
character states.

Homology is not defined in terms of similarity - the word
does not appear anywhere in the definition.  Hence homologous
should not be used as an adjective simply to designate similar
biological attributes in two or more species.  Biologists wishing
to designate similar features in different organisms should use
the term "similar."

Homology is a relative concept, hence it is always necessary
to state the nature of the relationship when talking about particular homologous features.  This statement is the conditional
phrase and it describes the presumed nature of the common ancestor
from which the homologues stemmed phylogenetically.  Any statement about homologues lacking the conditional phrase is incomplete
and meaningless.  For example, it makes no sense to say that the
femur of the gorilla is homologous to the femur of the chimpanzee.
Rather, one must say that the wing of birds and the wing of bats
are homologous as the forelimb of tetrapods (describing the
structure of this feature, of course) or that avian wings and
chiropteran wings are not homologous as aerodynamic planes.

Hierarchies of homologues are established by establishing
hierarchies of conditional phrases, and testing again and again
whether the features are still homologous.  These hierarchies

indicate what workers usually mean by expressions that features
are more homologous or have a higher degree of homology.  The
hierarchies of conditional phrases are usually set up in very
definite sequences which are completely analogous to the later
step described as transformation series.  Yet, most workers were
unaware that they were establishing such hierarchies of conditional phrases, not to mention failing to be concerned with the testing of these hierarchies which constitute scientific hypotheses.
The nature of these hierarchies also shows that homologies were
established for horizontal comparisons and for vertical comparisons (ancestral-descendent relationships).

The test of homology - the recognizing criterion - is by
means of similarities of all kinds between the presumed homologous
features.  Shared similarity is the <u>only</u> <u>valid</u> <u>empirical</u> <u>test</u> <u>of</u>
<u>homology</u> and it is restricted strictly to a comparison of the
homologous features under consideration.  The basis of this test
stems from evolutionary theory as I have shown earlier (Bock,
1974:387).

This test is not a good one in that it has low resolving
power, but it is the only valid test.  Two types of errors result,
the most serious being the recognition of nonhomologous features
as homologues; and the number of such effors is alarmingly high
especially in morphologically uniform groups (Bock, 1963).

Uses of homologous features:  A divergence of methods exists
after the initial determination of homologues, regardless of how
careful the characters have been compared and the hypotheses about
homologues tested.

The homologous features, which are arranged in hierarchies
dependent on the hierarchies of the conditional phrases, can be
used to erect classificatory hypotheses about taxa in an inductive
way.  These classifications may be regarded by some workers to be
phenetic classifications, and indeed, systematics within a phenetic
approach to classification works in just this fashion.  But note!
A phylogenetic factor has already crept into the analysis in the
hierarchal nature of the conditional phrases; this cannot be
avoided if one approaches the statement and testing of homologies
properly.  Pheneticists actually include hierarchies of conditional
phrases tacitly, but do not realize it.  Hence, I would question
whether it is possible, in principle, to construct purely
phenetic classifications.

The homologues are used to formulate phylogenetic hypotheses
about groups inductively, and also serve in the first step of the
phylogenetic analysis of characters.  Because similarity is the
only available valid test of hypotheses of homology, the use of

similarity by phylogenetic systematists is unclear to me as I have
stated previously (Bock, 1974:389).  Expressions that similarity
does not appear in the definitions of phylogenetic relations are
clear and correct, but most workers go further and appear to
exclude similarity from the phylogenetic analysis or are unclear
where similarity comes into the method.

Compare the statement by Bonde (1975:295) that "This
definition of phylogenetic relationship is completely devoid of
notions of similarity." with his other statement (p. 297) that
"If evaluations of phylogenetic relationships between species
and groups has to be based on similarities and differences of
character states - and what else could it be based on? -..."  I
interpret the latter statement to mean that phylogenetic relationships are based upon similarities between features and find it
difficult to see the difference between evolutionary and phylogenetic approaches claimed to exist by cladists.  And indeed most
cladists, including Hennig, are vague on the role of the concept
of homology in phylogenetic analysis or deny that homology has
any role.

The vagueness of the use of homology by cladists stems in
part from statements such as "the concept of resemblance" and the
division of the simple concept of resemblance into various categories, e.g., symplesiomorphy, synapomorphy and convergence
(Hennig, 1965:102).  The term "concept of resemblance" has no
meaning to me, nor can I see how this concept can be broken up
into categories.  These ideas do have some meaning if the term
"concept of resemblance" is a misnomer for the "concept of
homology" but in that case the defining and recognizing criteria
have been confused.

Because similarity is the only valid test of homologies, because the hierarchies of conditional phrases, which must be included, are directly analogous to the idea of transformation series
(morphoclines) and because the determination of homologies is the
first step in any phylogenetic analysis, I cannot see how a phylogenetic classification can be formulated and tested without a
major input of similarity into the final product.  To say that
phylogenetic relationships are based on synapomorphs and to claim
that these synapomorphs are not determined mainly by similarities
is simply misleading.  If the synapomorphs are determined mainly
by similarities, then too are the phylogenetic relationships between groups.

Hence, at least for me, it is difficult to see sharp distinctions between the three main approaches to classification.  Whatever distinctions exist in the theoretical expression of the
approaches appear to break down and merge in the methods considering only the central concept of homology which must be used in

any approach to biological classification.

The most important use of homologues is to test classificatory hypotheses about taxa and the arrangement of taxa into a
Linnean hierarchy.  This may be done by the study of other features and the discovery of new homologues or by a more exact
analysis of existing homologies.  Great care must be exercised
if homologies are used both for the inductive formulation of the
classificatory hypotheses about groups and for their subsequent
test.  (Such a statement will make the hair of most philosophers
of science stand on end even if he is bald!)  Yet such a dual
role of hypotheses about characters, especially homology, can
scarcely be avoided in classification and phylogeny which are
narrative explanations.  And it is for this reason the methods
used to test hypotheses about characters must be strictly independent of the hypotheses about groups, e.g., methods of
commonality must be ruled invalid.  Tests of classificatory
hypotheses about groups using homologies are not very severe ones,
but can be strengthened by amassing large numbers of homologies.
Even then the hypotheses about the groups are severely tested
only if the hypotheses about the homologues are severely tested.
One will have more confidence in the classification if it has
been tested against complex homologies taken from different
systems of characters or from different parts of the organism.

Phylogenetic analysis.  Under the term of phylogenetic analysis, I mean those methods by which phylogenetic hypotheses about
features are formulated and tested against empirical hypotheses.
I am here especially concerned with how these methods can be used
to test phylogenetic hypotheses about groups.  But the same method
of phylogenetic analysis may be used to test various aspects of
classificatory hypotheses about groups, such as the monophyly of
taxa, either directly or via special tests such as the use of
paradaptations (Bock, 1974:383).  I would like to state again that
it is my belief that the same method of phylogenetic analysis must
be used in evolutionary classification and in phylogenetic classification; the assertion that different phylogenetic methods exist
for these two approaches to classification is spurious.  Further,
I believe that the method to be described below is in close agreement with that of Hennig (1965, 1966) although it is unclear to me
whether and how he uses the concept of homology and how similarity
as a test for hypotheses is used in his method.  My method is not
in agreement with that used by most phylogenetic systematists the
details of which will be discussed elsewhere.

Phylogenetic analysis must begin with phylogenetic hypotheses
about groups; these hypotheses may be generated in any way such as
using existing phylogenies in the literature or by formulating
them by inductive methods using homologues as discussed above.

These hypotheses recognize phyla or holophyletic groups (=closed descendent groups or monophyletic groups in the sense of Hennig) and recognize sister-group relationships between these holophyletic groups. The results of phylogenetic hypotheses about groups can be shown in a phyletic diagram which should ideally be a dichotomous branching diagram. How far the actual diagram is from a strict dichotomous one is a reflection of our ability to analyze the known features of the members of the group. Sister-groups have the same rank and it is erroneous to refer to one of the two sister-groups as the plesiomorphous group and the other as the apomorphous group.

Homology: The first step in phylogenetic analysis of characters is recognition of homologues which are arranged in hierarchies according to the nature of their conditional phrases. The empirical test of hypotheses about homologous features is similarities of all sorts between the homologues as discussed above. This step is usually done at the onset of any study and homologous features may have been used to reach inductively the phylogenetic hypotheses about groups; hence, workers forget that it as done and argue that determination of homologues is not the initial step in phylogenetic analysis. But it must be there otherwise how is it known which features to arrange in transformation sequences.

Those features concluded to be homologues (these include true homologues and nonhomologues that have been incorrectly recognized as homologues) are retained for further analysis and the features concluded to be nonhomologues are thrown away.

Transformation series: The second step is to arrange the homologues into transformation series or morphoclines. This may have been done already with the establishment of the hierarchies of conditional phrases of the homologous features. Transformation series reflect the pattern of change in the features according to our knowledge of the morphology, function, behavior, adaptive significance, embryology, ecological interaction and other attributes of the features. Thus the arrangement of features into a transformation series depends upon our judgment of how the feature could change during evolution. Transformation series are not arranged by chance or by the caprice of the investigator. And the arrangement of features into transformation series is clearly a hypothesis about ancestral-descendent relationship about features which is a short step away from ancestral-descendent hypotheses about groups.

It is clear from his discussion that Hennig (1966:88-89) ascertains the homology of features and arranges these homologues in transformation series as the first step in his analysis.

Transformation series represent scientific hypotheses and must be tested. Observations of similarities do not represent valid tests of transformation series. Rather the tests are via secondary hypotheses about how the particular feature could evolve and via still further hypotheses about the mechanisms of evolutionary change, notably of phyletic evolution. The empirical observations that test transformation series are thus those observations used to test the hypotheses about the evolution of the particular feature and the hypotheses of basic evolutionary mechanisms. Such tests are difficult and valid empirical observations are not easy to ascertain. Central to these tests is study of the function and adaptive significance of the features.

Polarity of transformation series: The polarity of a transformation series is a phylogenetic polarity and is expressed in terms of plesiomorphic - apomorphic relationships of features in the series. Plesiomorphy and apomorphy of features can be ascertained only with respect to a stated transformation series as these concepts are relative to one another only within this context (Hennig, 1966:89). It makes no more sense to designate a feature as an apomorph than to say it is a homologue. Thus the apomorphy-plesiomorphy nature of features can be determined only after transformation series are established.

Because plesiomorphy and apomorphy are terms bearing a relationship to one another, they can be defined together and with respect to one another. Within a given transformation series, an apomorphous feature is phylogenetically derived with respect to a plesiomorphous feature. Starting at the plesiomorphous end of a transformation series, each step is more apomorphic than the last. An apomorphous feature is presumed to have evolved from the previous plesiomorphous feature in the series. Plesiomorphy is synonymous with primitive but not with generalized (a rather vague evolutionary term). Apomorphy is synonymous with advanced and derived, but not with specialized (a rather vague evolutionary term). Plesiomorphy and apomorphy are words that have strict phylogenetic meaning and I would urge their use which is unambiguous. The substitution of generalized and specialized for plesiomorphy and apomorphy by some cladists is a retrograde step.

The testing of the polarity of transformation series is difficult and I do not accept most of the methods commonly used because they involve commonality, or the observations are not independent of the hypothesis under test, or parsimony is cited as the criterion. The methods cited by Maslin (1952) in his analysis of morphoclines fall under the first two categories mentioned. Tests of the phylogenetic polarity of transformation series that appear valid to me fall into two categories.

The first is the use of out understanding of how the features could evolve and hence the tests would be against secondary hypotheses about evolutionary change of the features and eventually about the mechanisms of evolutionary change (phyletic evolution, mainly) and finally against the empirical observations used to test these hypotheses.  As tests for transformation series, these depend upon knowing as much about the function, adaptive significance and other aspects of the biology of the features as possible. Many of these tests depend upon a comprehension of the probability of evolutionary change in one direction as opposed to the probability of change in the reverse direction.  Such notions were expressed in my examples demonstrating the validity of phylogenetic analyses (Bock, 1969c).  For example, the probability of a limbless tetrapod having a limbed descendent is so small that it would invalidate the polarity of a transformation series leading from the absence of limbs to fully developed ones.  Likewise, evolution from an akinetic vertebrate skull to a kinetic skull is so highly improbable that it would invalidate any morphocline expressing this direction of change.  I must emphasize that such tests can be done only after the function and adaptive significance of the features have been studied.

The second valid test of morphocline polarity is the stratigraphic position of fossils possessing plesiomorphous or apomorphous features.  A fossil possessing the plesiomorphic condition is found early in the stratigraphic record (earlier in time) than a fossil or Recent organism possessing the apomorphic condition provides a valid empirical test of the phylogenetic polarity of the morphocline.  This method has been rejected by most cladists based largely on the argument presented by Schaeffer, Hecht, and Eldredge (1972).  Unfortunately the argument presented in that paper confused inductive formation of hypotheses and empirical tests of hypotheses, and rejected the use of the stratigraphic position of fossils in the mistaken belief that this information was used in the inductive formulation of hypotheses.  Further, they confused the nature of valid tests with those of high resolving power.  As all other tests of classificatory and phylogenetic hypotheses, the stratigraphic position of fossils may have low resolving power.  As a general rule the resolving power of this test is dependent upon the age span between the stratigraphic position of the organism showing the plesiomorphic condition and that of the organism showing the apomorphic condition with respect to the total age of the group.

Once the polarity of the transformation series is established, one is interested only in the apomorphs because plesiomorphs possessed in common by different groups (=symplesiomorphs) do not provide a valid test of phylogenetic hypotheses.  It should be noted that symplesiomorphs are usually good homologues and that

they may be valuable as tests for classificatory hypotheses about
groups.  Further, it should be noted that the determination of
homologues does not automatically provide synapomorphs or
symplesiomorphs for that matter.  In the introduction of his
system, Hennig quite correctly (although tacitly) argued that it
served to further analyze the concept of homology.

Synapomorphy and automorphy:  Recognition of apomorphs is
not the last step in the phylogenetic analysis because common
possession of apomorphs does not provide a sufficient test of
phylogenetic hypotheses about groups.  It is necessary to analyze
apomorphs held in common and to distinguish between synapomorphs
(which includes autapomorphs) and convergent features as stated
clearly by Hennig (1965:102, and Fig. 1; 1966).  Curiously most
phylogenetic systematists omit this step using the erroneous
definition of synapomorphy as shared possession of apomorphous
features and claiming that the analysis of convergence can be
done only after the phylogenetic analysis is completed.

Apomorphs possessed in common by two groups are of two types:

(a)   Synapomorphs are homologous apomorphs and may be defined
      as those apomorphs in two groups which stem phylo-
      genetically from the same feature in the immediate
      common ancestor of these two groups.  Thus the apomorph
      is present in the ancestral species and in all descendent
      groups, but sometimes in a modified condition.  It is
      clear that the term synapomorph is the same as homologous
      apomorphs and I would urge this association because it
      clearly links the concept of homology with that of
      apomorphy.

      Autapomorphs are the same as synapomorphs but at one
      higher stage in the phylogeny.  The existence of a holo-
      phyletic phylon is shown by the possession of at least
      one autapomorph by members of the group.  Grouping of
      two holophyletic phyla is done by the shared possession
      of at least one synapomorph by members of the two
      groups.  This synapomorph becomes the autapomorph of
      the single phylon containing these two sister groups.

(b)   Convergent features are nonhomologous apomorphs and may
      be defined as apomorphs in two groups which does not
      stem phylogenetically from the same feature in the
      immediate common ancestor of these two groups.

The relationship of these definitions to that of homology is
obvious, and the necessity to include the notion of the conditional phrase in discussing synapomorphs and convergent features

is clear. Thus in the definition of convergent features, the feature in the eventual common ancestor of two groups sharing a
nonhomologous apomorph would have a structure different from the
apomorph in the two groups.

Because the first step in the phylogenetic analysis was the
recognition of homologues, after which the nonhomologous features
were discarded, it is possible to assume that most shared apomorphs
will be homologous apomorphs or synapomorphs. Hence there is a
firm basis for this assumption which is quite different from saying simply "It must be recognized as a principle of inquiry for
the practice of systematics that agreement in characters must be
interpreted as synapomorphy as long as there are no grounds for
suspecting its origin to be symplesiomorphy or convergence."
(Hennig, 1965:104). In the absence of ascertaining homology, one
can only ask why and why cannot characters equally well be
assumed to be convergent?

Although homologues have been determined and nonhomologues
have been discarded, many convergent features (nonhomologous
apomorphs) may still be included in the analysis because of the
low resolving powers of the available tests of homologous features. Thus the last step in the phylogenetic analysis is to
identify and separate convergent features from synapomorphs which
must be done prior to the completion of the phylogenetic analysis.
The most correct method would be to examine each set of shared
apomorphs separately and reach a decision, with proper tests
against empirical evidence, as to whether it is a synapomorph or
a convergent feature. But, one is at the end of a long arduous
analysis and having tested each of the previous stages properly,
it is permissible to cut corners a bit at this stage. The usual
approach is to examine only those shared apomorphs that are in
conflict, assuming that where no conflicts exist one was correct
in the original decisions about the homologues. Such an approach
may be a bit too simplistic because in most actual studies, one
must deal with a complex pattern of partially overlapping conflicts
of shared apomorphs which necessitates treatment of each individually.

Given two sets of shared apomorphs which are in conflict, the
judgment of the synapomorph and the empirical test of this hypothesis of synapomorphy is always based on evolutionary theory and
used as the empirical test those observations that serve to test
the pertinent aspects of evolutionary theory. In general the
method is based upon the different probabilities of certain evolutionary changes occurring independently two or more times. And
the tests will depend upon studies of ontogenetical development of
features, population genetics, functional studies and environmental
interactions. Thus in the case of conflict between two sets of

shared apomorphs, one will be judged to be the synapomorph if:

(a)  It is a complex feature rather than a simple one be-
     cause the probability of a complex feature with many
     interconnected parts evolving independently two or
     more times is less than that of a simple feature with
     few parts.

(b)  It is a new feature, be it simple or complex rather
     than the reduction or loss of a feature because the
     probability of a new feature evolving independently
     two or more times is less than that of the reduction
     or loss of a feature.

(c)  It is a feature that serves in a wide range of bio-
     logical roles rather than having a single biological
     role because the probability of a feature serving a
     wide range of roles evolving independently two or more
     times is less than one that has only a single bio-
     logical role and tight correlation with a single or
     narrowly delimited environmental factor.

(d)  It is a feature whose evolution depended upon the inter-
     workings of a complex pattern of structural and func-
     tional relationships rather than a feature with few
     such structural interrelationships for the same reasons.

(e)  It is a feature which has an ontogenetical development
     that depended upon a complex pattern of embryological
     modifications and interconnections with other develop -
     ing features rather than one that has a simple ontogeny
     for the same probability reasons.

All of these arguments are dependent on relative probabilities of evolutionary change and hence they are not absolute tests.
The empirical observations used in these tests cover a wide range
from embryology, genetics to functional morphology.  Least important is a comparison of the morphology of the features among
groups.  And it should be noted that these tests are quite
similar to the ideas of weighing characters presented in this
Institute by Edwards and Hecht and have been used by many taxonomists for decades.

Tests of hypotheses about groups:  Having gone through the
procedure as outlined above and having made decisions about autapomorphs and synapomorphs - actually having ascertained a hierarchy
of autapomorphs which differs little, if at all, from the hierarchies of homologues based upon conditional phrases - one can use
these characters to test the initial phylogenetic hypotheses about

groups.  Each phylon must possess at least one unique autapomorph.
and each sister group relationship between two phyla must possess
at least one unique synapomorph which becomes the autapomorph of
the phylon containing the two sister groups.

The single important attribute of the procedure of phylogenetic analysis of characters described above is that each
stage I endeavored to show the valid tests of hypotheses.  In
some cases, the tests involved complex arguments and chains of
secondary hypotheses necessitating considerable study beyond a
comparison of the morphological form of the features.  But the
subject matter and the desired explanations of classificatory and
phylogenetic studies are very complex; the expectation of simple
procedures and solutions is unrealistic.

## CONCLUSIONS

A.    The theory on which biological classification is based is
that of organic evolution.  Classifications and phylogenies are
narrative explanations of the past evolutionary history of
organisms; as hypotheses they must be tested against empirical
observations.  No theory of classification exists distinct from
evolutionary theory.  And the several approaches to classification
represent different research strategies.  A particular classification is the result of using one of these research strategies.

B.    Classifications are scientific hypotheses and as such are
not objective; they should have other properties such as boldness
or maximum explanatory value.  Tests of the classification must
be against empirical observations and hence are objective.  Each
test of any hypotheses within classification and phylogeny must
be valid for that hypothesis.  Tests depending upon commonality or
upon distribution among taxa including outgroup comparisons and
those using parsimony as the criteria are not empirical observations and hence are not valid tests.

C.    A classification is a system of arranging objects studied in
some field of science according to theory applicable to that
science and for some heuristic purpose.  The most important
heuristic function of classification is as a mechanism to generate
the maximum number of useful generalizations - testable statements
about unknown properties of the objects studied in that science.
As a useful system, a classification must balance complexity
against utilization.  A system that is too complex, contains too
many units and names, and makes no distinction between the contents of the units will be less useful even if it is more precise.
If biology is to be a unified science with a single system of
explanation, then only one system of classification is possible,
and criteria must be established on which to judge this best system.

D.  Evolutionary classification would be judged to be the best
system to biological classification because it is the most
natural.  By most natural is meant that evolutionary classification is based upon all aspects of evolutionary theory (phyletic
evolution and speciation); hence it is the most testable system
and it permits formulation of the largest number of useful biological generalizations and hypotheses.  Unfortunately this
approach results in a sloppier system than the neat, but partly
forced, system resulting from using the approach of phylogenetic
classification.  But the nature of the universe of biological
objects to be classified is complex and defies simple arrangements.  A sloppy classification that is a realistic reflection of
the evolutionary history of organisms is the better choice over
a neat classification that is an overly precise reflection of
only the branching pattern of their phylogeny.

E.  Evolutionary classification is not a one-to-one expression of
the branching phylogeny; hence, it is necessary to present a
phylogenetic diagram in addition to the formal classification.

F.  Two types of hypotheses about groups must be formulated in
any systematic study.  The first are classificatory hypotheses
about taxa (the formal classification) and the second are phylogenetic hypotheses about phyla (singular=phylon - the phylogenetic diagram).  The unit of phylogenetic diagrams is the phylor
- a closed descendent group - which should be distinguished from
the taxon which is a unit of the classification.  Both types of
hypotheses about groups must be tested against empirical observation - the objective step in systematic studies.

G.  The real work in systematics is testing hypotheses about
characters - this is the area of character analysis.  Central to
all character analysis is the concept of homology.  Those features
in different organisms that stem phylogenetically from the same
feature in their immediate common ancestor are homologous.  Similarities of all sorts between the features hypothesized to be
homologous serve as the only valid test of this hypothesis.  Statements about homologoues must include a conditional phrase describing the homology; hierarchies of homologies are obtained by
establishing hierarchies of conditional phrases.  The homologues
can be used to test classificatory hypotheses about taxa.

H.  Phylogenetic analysis must start with the hypothesis and
testing of homologies.  Next the homologies are arranged in transformation series which reflect the probable sequence of evolutionary change.  The plesiomorphic-apomorphic polarity of the transformation is then determined in which an apomorphous feature is
phylogenetically derived (=evolved) from a plesiomorphous feature
in the same transformation sequence.  Tests of the hypotheses of

transformation series and of the polarity of these series are based upon secondary hypotheses of possible evolutionary change of the features and finally upon the empirical observations testing these evolutionary hypotheses. The final step is to separate shared apomorphous features into two categories - synapomorphs (=homologous apomorphs) and convergent features (nonhomologous apomorphs). The distinction of synapomorphs from convergent features is done usually only when there are sets of shared apomorphs in conflict because of the initial determination of homologues and throwing away the nonhomologues. The tests of hypotheses of synapomorphy is usually via secondary evolutionary hypotheses and on the probability of the independent origin and evolution of one set of apomorphs against the other.

I. The distinction between the several major approaches to classification as expressed in theoretical statements appear to break down when considering the details of procedures. This is especially true for the concept of homology which is basic for all approaches to classification. The requirement of always in- cluding a conditional phrase and the result of having a hierarchy of conditional phrases (=phylogenetic statement) and the use of similarity as the only valid test of homology breaks down the separations between evolutionary and phenetics and cladistics to either side.

J. The most essential part of systematics - classification and phylogeny - is a clear exposition of the procedure by which hypothetical statements are tested and of the empirical observa- tions that serve as valid tests. In actual studies, the impor- tant step is not the designation of certain features as homologues or as synapomorphs, but the reasoning and the demonstration of the actual empirical observations that tested the hypotheses about these features. Unless this is done, biological classification cannot claim to be a science.

ACKNOWLEDGEMENTS

The ideas expressed in this paper were developed over many years with much input from numerous friends. I would like to express my appreciation to all and to thank especially Ernst Mayr, F.E. Warburton, Ernest Nagel, Gerd von Wahlert, and Arthur Caplan for their help in many ways, but mainly to have kept the analysis on the straight and narrow path. The theoretical ideas about classification and phylogeny were developed during the course of a number of systematic studies on birds done with the assistance of financial support from the National Science Founda- tion. This study was done with the support of a research grant BMS-73-06818 from the N.S.F.

## REFERENCES

Ashlock, P., 1971, Monophyly and associated terms.  Syst. Zool.,
    20:63-69.
Bock, W.J., 1963, Evolution and phylogeny in morphologically
    uniform groups.  Amer. Nat., 97:265-285.
Bock, W.J., 1967, The use of adaptive characters in avian
    classification.  Proc. XIV Internat. Ornith. Cong., pp. 61-74.
Bock, W.J., 1969a, Comparative morphology in systematics, in
    "Systematic Biology," Nat. Acad. Sciences, pp. 411-488.
Bock, W.J., 1969b, Nonvalidity of the "phylogenetic fallacy."
    Syst. Zool., 18:111-115.
Bock, W.J., 1969c, The concept of homology.  Ann. N. Y. Acad.
    Sci., 167:71-73.
Bock, W.J., 1972, Species interaction and macroevolution, in
    "Evolutionary Biology," (Dobzhansky, Hecht and Steere, eds.),
    Vol. 5, Appleton-Century-Crofts, New York, pp. 1-24.
Bock, W.J., 1974, Philosophical foundations of classical
    evolutionary classification.  Syst. Zool., 11:375-392.
Bock, W.J. and von Wahlert, G., 1963, Two evolutionary theories -
    a discussion.  British Journ. for the Phil. of Sci., 14:140-
    146.
Bonde, N., 1975, Origin of "higher groups": viewpoints of
    phylogenetic systematists.  Colloque internal. C.N.R.S.,
    218:293-324.
Colless, D.H., 1967, The phylogenetic fallacy.  Syst. Zool.
    16:289-295.
Gilmore, J.S.L., 1940, Taxonomy and philosophy, in "The new
    systematics," (J. Huxley, ed.), Clarendon Press, Oxford,
    pp. 461-474.
Hanson, N.R., 1958, "Patterns of discovery," Cambridge Univ.
    Press, Cambridge, ix + 241 pp.
Hennig, W., 1965, Phylogenetic systematics.  Ann. Rev. Entomol.,
    10:97-116.
Hennig, W., 1966, "Phylogenetic systematics."  Univ. Illinios
    Press, Urbana, 263 pp.
Hennig, W., 1969, "Die Stammesgeschichte der Insekten."
    Senckenberg-Buch 49, Frankfurt/M.
Kellogg, D.E., 1975, Character displacement in the radiolarian
    genus, Eucyrtidium. Evolution, 29:736-749.
McKenna, M.C., 1975, Toward a phylogenetic classification of
    the Mammalia, in "Phylogeny of the Primates," (Luckett and
    Szalay, eds.), Plenum Press.
Maslin, T.P., 1952, Morphological criteria of phyletic relation-
    ships.  Syst. Zool., 1:49-70.
Mayr, E., 1963, "Animal species and evolution."  Harvard Univ.
    Press, Cambridge, Mass. 797 pp.
Mayr, E., 1975, Cladistic analysis or cladistic classification?
    Z. f. zool. Syst. u. Evolutionsforsch., 12:94-128.

Nelson, G.T., 1969, Gill arches and the phylogeny of fishes, with notes on the classification of vertebrates.  Bull. Amer. Mus. Nat. Hist., 141:475-552.

Ostrom, John H., 1976, Archaeopteryx and the origin of birds. Biol. J. Linn. Soc., 8:91-182.

Popper, K., 1968a, "The logic of scientific discovery." Harper Torchbooks, New York and Evanston, 2nd Edition, 480 pp.

Popper, K., 1968b, "Conjectures and refutations:  The growth of scientific knowledge."  Harper Torchbooks, New York and Evanston, xiii + 417 pp.

Popper, K., 1972, "Objective knowledge.  An evolutionary approach." Oxford Univ. Press, Oxford, x + 380 pp.

Schaeffer, B., Hecht, M. and Eldredge, N., 1972, Phylogeny and paleontology, in "Evolutionary Biology," (Dobzhansky, Hecht, and Steere, eds.), vol. 6:31-46.

Warburton, F.E., 1967, The purposes of classification.  Syst. Zool., 16:241-245.

Wolters, H.E., 1975, "Die Vogelarten der Erde." Verlag Paul Parey, Hamburg.